KEY CONCEPTS IN GEOMORPHOLOGY

KEY CONCEPTS IN GEOMORPHOLOGY

Paul R. Bierman
University of Vermont

David R. Montgomery
University of Washington

H. Bäsemann/Picture alliance/Newscom

W. H. Freeman and Company Publishers

A Macmillan Higher Education Company

Associate Publisher: Jessica Fiorillo

Senior Acquisitions Editor: Bill Minick

Associate Editor: Heidi Bamatter

Senior Marketing Manager: Alicia Brady

Marketing Assistant: Alissa Nigro

Senior Project Editor: Vivien Weiss

Cover and Interior Designer: Victoria Tomaselli

Photo Editors: Ted Szczepanski, Hilary Newman

Illustrations Coordinator: Janice Donnola

Illustrations: Precision Graphics, Luke Reusser

VP, Director of Production: Ellen Cash

Composition: Aptara

Printing and Binding: RR Donnelley

Library of Congress Control Number: 2013934736

ISBN-13: 978-1-4292-3860-1
ISBN-10: 1-4292-3860-7

Copyright © 2014 by W. H. Freeman and Company Publishers
All rights reserved.
Printed in the United States of America
Third printing

W. H. Freeman and Company Publishers
41 Madison Avenue
New York, NY 10010
www.whfreeman.com

Contents

Foreword xi

PART I
Geomorphology and Its Tools 3

1 Earth's Dynamic Surface 5

Introduction 5

Geosphere 6
   Isostasy 6
   Tectonics 11
   Lithology and Structure 13

Hydrosphere 14
   Climate and Climate Zones 14
   Hydrologic Cycle 16

Biosphere 16
   Geographical Distribution of Ecosystems 18
   Humans 20

Landscapes 20
   Process and Form 22
   Spatial Scales 22
   Temporal Scales 23

Unifying Concepts 24
   Conservation of Mass 24
   Conservation of Energy 24
   Material Routing 24
   Force Balances and Thresholds 25
   Equilibrium and Steady State 27
   Recurrence Intervals and Magnitude-Frequency Relationships 28

Applications 28
Selected References and Further Reading 30
Digging Deeper: Why Is Earth Habitable? 31
Worked Problem 34
Knowledge Assessment 34

A Brief History of Geomorphology 36

2 Geomorphologist's Tool Kit 43

Introduction 43

Characterizing Earth's Surface 44
   Field Surveys 44
   Active Remote Sensing 46
   Passive Remote Sensing 47
   Digital Topographic and Landscape Analysis 47

Relative Dating Methods 48
   Landform Degradation 48
   Rock Weathering and Soil Development 50
   Rock Varnish 50
   Calibrated Relative Dates 50

Numerical Dating Methods 52
   Dendrochronology 52
   Radiocarbon Dating 53
   K/Ar Dating 54
   U/Th Dating 56
   Luminescence Dating 56
   In-Situ Cosmogenic Nuclide Analysis 58

Measuring Rates of Geomorphic Processes 59
   Sediment Generation Versus Sediment Yield 59
   Landscape Change at Outcrop and Hillslope Scales 61
   Landscape Change at Basin Scales 63
   Erosion Rates over 10^6 to 10^8 Year Timescales 63

Experiments 63
   Field Experiments 63
   Laboratory Experiments 65
   Numerical Models 66

Proxy Records 66
Applications 67
Selected References and Further Reading 68
Digging Deeper: How Does a Dating Method Develop? 69
Worked Problem 72
Knowledge Assessment 72

PART II
Source to Sink 75

3 Weathering and Soils 77

Introduction 77

Physical Weathering 80
   Exfoliation 80
   Freeze–Thaw 82
   Thermal Expansion 82
   Wetting and Drying 83

Chemical Weathering 83
  Mineral Stability 84
  Oxidation and Reduction 85
  Solution 85
  Hydrolysis 86
  Clay Formation 87
  Hydration 88
  Chelation 89
  Cation Exchange 89

Soils 89
  Soil-Development Processes 89
  Factors Affecting Soil Development 90
  Processes and Rates of Soil Production 91
  Soil Profiles 91
  Soil Classification 95

Soils and Landscapes 99
  Soil Development over Time 99
  Soil Catenas 99
  Paleosols 100

Weathering-Dominated Landforms 101
  Inselbergs and Tors 102
  Duricrusts 103
Applications 104
Selected References and Further Reading 105
Digging Deeper: How Fast Do Soils Form? 106
Worked Problem 109
Knowledge Assessment 110

4 Geomorphic Hydrology 111

Introduction 111

Precipitation 112
  Duration and Intensity 113
  Recurrence Intervals 113
  Precipitation Delivery 114
  Climate Effects on Hydrology and Geomorphology 115

Evapotranspiration 116
  Evapotranspiration Rates 116
  Actual Versus Potential Evapotranspiration 116
  Geomorphic Importance of Evapotranspiration 116

Groundwater Hydrology 118
  Infiltration: Moving Water into the Ground 119
  Moving Water Through Earth Materials 120
  Hydrologic Flowpaths 122

Surface Water Hydrology 126
  Hydrographs 126
  Interactions Between Groundwater and Surface Flow 129
  Flood Frequency 130
  Water Budgets 132

Hydrologic Landforms 133
Applications 137
Selected References and Further Reading 137
Digging Deeper: Humans, Hydrology, and Landscape Change—What's the Connection? 138
Worked Problem 141
Knowledge Assessment 142

5 Hillslopes 145

Introduction 145

Slope-Forming Materials 146
  Strength of Rock and Soil 147
  Effects of Weathering on Rock Strength 149

Diffusive Processes 150
  Rainsplash 150
  Sheetwash 151
  Soil Creep 151

Mass Movements 153
  Slides 154
  Flows 156
  Falls 158

Slope Stability 159
  Driving and Resisting Stresses 159
  Infinite-Slope Model 160
  Environmental and Time-Dependent Effects 161

Slope Morphology 163
  Weathering-Limited (Bedrock) Slopes 163
  Transport-Limited (Soil-Mantled) Slopes 165
  Threshold Slopes 166
  Hillslope Evolution 166
  Drainage Density 166
  Box 5.1 Derivation of the Form of Convex Hillslope Profiles 167
  Channel Initiation 170
Applications 172
Selected References and Further Reading 173
Digging Deeper: How Much Do Roots Contribute to Slope Stability? 174
Worked Problem 177
Knowledge Assessment 178

6 Channels 179

Introduction 179

External Controls on Fluvial Processes and Form 180
  Discharge 181
  Sediment Supply 181

Bed and Bank Material   181
Vegetation   184

Fluvial Processes   185
Flow Velocity   185
Discharge Variability   187
Stream Power   189
Box 6.1 Derivation of Stream Power   189
Bedrock Incision   189
Channel Migration   191

Sediment Transport   194
Initiation of Transport   194
Sediment Loads   197
Bedforms   198

Channel Patterns   199
Straight and Sinuous Channels   200
Meandering Channels   201
Braided Channels   201
Anastomosing Channels   201

Channel-Reach Morphology   202
Colluvial Reaches   202
Bedrock Reaches   202
Alluvial Reaches   202
Large Organic Debris   204

Floodplains   205

Channel Response   207
Applications   208
Selected References and Further Reading   209
Digging Deeper: What Controls Rates of Bedrock River Incision?   211
Worked Problem   214
Knowledge Assessment   214

7 Drainage Basins   217

Introduction   217

Basin-Scale Processes   219
Sediment Budgets   219
Sediment Routing and Storage   222

Channel Networks and Basin Morphology   223
Drainage Patterns   223
Channel Ordering   225
Downstream Trends   225

Uplands to Lowlands   227
Process Domains and Valley Segments   228
Longitudinal Profiles   230
Channel Confinement and Floodplain Connectivity   231

Box 7.1 River Longitudinal Profiles   232
Downstream Trends   232

Drainage Basin Landforms   233
Knickpoints   233
Gorges   234
Terraces   235
Fans   237
Lakes   239
Applications   240
Selected References and Further Reading   244
Digging Deeper: When Erosion Happens, Where Does the Sediment Go?   245
Worked Problem   250
Knowledge Assessment   251

8 Coastal and Submarine Geomorphology   253

Introduction   253

Coastal Settings and Drivers   254
Tectonic Setting   254
Sea-Level Change   254
Salinity   256
Substrate and Sediment Supply   256
Tides   257
Waves   259

Coastal Processes and Landforms   264
Rocky Coasts   264
Beaches and Bars   266
Spits, Tidal Deltas, and Barrier Islands   268
Lagoons, Tidal Flats, and Marshes   270
Estuaries   271
Deltas   272
Coastal Rivers   274

Marine Settings and Drivers   274
Currents   275
Marine Sedimentation   276
Dissolved Load   276

Marine Landforms and Processes   276
Continental Margins   277
Abyssal Basins   278
Mid-Ocean Ridges   278
Trenches   279
Coral Reefs   279
Applications   281
Selected References and Further Reading   281
Digging Deeper: What Is Happening to the World's Deltas?   283
Worked Problem   286
Knowledge Assessment   287

PART III
Ice, Wind, and Fire — 289

9 Glacial and Periglacial Geomorphology — 291

Introduction 291

Glaciers 294
  Glacier Mass Balance 294
  Glacier Energy Balance 296
  Accumulation and Ablation of Glacial Ice 297
  Glacier Movement 299
  Thermal Character of Glaciers 302
  Glacial Hydrology 303

Subglacial Processes and Glacial Erosion 305

Glacial Sediment Transport and Deposition 309
  Subglacial Sediments and Landforms 309
  Ice-Marginal Sediments and Landforms 310
  Glacially Related Sediments and Landforms 311

Glacial Landscapes, Landforms, and Deposits 313
  Landforms of Alpine Glaciers 313
  Landforms of Ice Sheets 314
  Geomorphic Effects of Glaciation and Paraglacial Processes 315

Periglacial Environments and Landforms 316
  Permafrost 317
  Characteristic Periglacial Landforms and Processes 318
Applications 322
Selected References and Further Reading 323
Digging Deeper: How Much and Where Do Glaciers Erode? 324
Worked Problem 327
Knowledge Assessment 328

10 Wind as a Geomorphic Agent — 329

Introduction 329

Air as a Fluid 331
  Wind Patterns and Speeds 332
  Vertical Distribution of Wind Speed 333
  Settling Speed of Particles in Air 333

Spatial Distribution of Wind-Driven Geomorphic Processes 334

Aeolian Processes 335
  Disturbance 335
  Erosion 335
  Sediment Transport 337
  Deposition 341

Aeolian Features, Landforms, and Deposits 342
  Aeolian Erosional Features and Landforms 342
  Aeolian Transport Features and Landforms 343
  Aeolian Dust Deposits and Loess 347
Applications 350
Selected References and Further Reading 351
Digging Deeper: Desert Pavements—The Wind Connection 352
Worked Problem 354
Knowledge Assessment 354

11 Volcanic Geomorphology — 355

Introduction 355

Distribution and Styles of Volcanism 356
  Magma Chemistry and Volcano Morphology 359
  Tectonic Forcing and Volcanic Provinces 361

Eruptive Mechanisms and Products 363
  Lava Flows 363
  Pyroclastic Flows and Falls 365
  Volcanic Gases 366

Eruption Sizes and Types 368

Volcanic Landscapes 368
  Landscapes of Basaltic Volcanism 368
  Landscapes of Silicic Volcanism 371

Processes of Volcanic Landform Evolution 372
  Geomorphic Effects of Magma Intrusion 372
  Biologic Colonization 373
  Denudation and Aging 374
  Mass Movements 374
  Lahars 375
  Volcano-River Interaction 377
  Hydrologic Considerations 377
  Erosional Landforms 380
Applications 381
Selected References and Further Reading 382
Digging Deeper: Geomorphic Effects of Volcano Sector Collapse 383
Worked Problem 386
Knowledge Assessment 387

PART IV
The Bigger Picture — 389

12 Tectonic Geomorphology — 391

Introduction 391

Tectonic Processes 392
  Uplift and Isostasy 393
  Thermal and Density Contrasts 397

Tectonic Settings 397
  Extensional Margins and Landforms 399
  Compressional Margins and Landforms 401
  Transform Margins and Landforms 404
  Continental Interiors 404
  Structural Landforms 408
Landscape Response to Tectonics 411
  Coastal Uplift and Subsidence 412
  Rivers and Streams 413
  Hillslopes 413
  Box 12.1 Drainage Area-Slope Analysis 414
  Erosional Feedbacks 415
Applications 417
Selected References and Further Reading 417
Digging Deeper: When and Where Did that Fault Last Move? 419
Worked Problem 422
Knowledge Assessment 423

13 Geomorphology and Climate 425

Introduction 425
Records of a Changing Climate 427
  Landform Records of Climate Change 427
  Lake and Marine Sediment 429
  Ice Cores 432
  Windblown Terrestrial Sediment 433
Climate Cycles 434
  Glacial Cycles 434
  Orbital Forcing 436
  Local Events—Global Effects 436
  Climate Variability Within a Climate State 438
  Short-Term Climate Changes 439
Geomorphic Boundary Conditions 439
  Precipitation and Temperature 440
  Vegetation, Fire, and Geomorphic Response 440
  Base Level 442
Climatic Geomorphology 444
  Köppen Climate Classification 445
  Climate-Related Landforms and Processes 445
  Relict Landforms 446
Landscape Response to Climate 447
  Glacial-Interglacial Changes 447
  Isostatic Responses 448
  Climatic Control of Mountain Topography 449
  Climate Change Effects 449

Landscape Controls on Climate 452
  Regional Climate 452
  Earth's Energy Balance 452
  Hydrologic Cycling 452
  The Atmosphere 454
Applications 454
Selected References and Further Reading 455
Digging Deeper: Do Climate-Driven Giant Floods Do Significant Geomorphic Work? 457
Worked Problem 459
Knowledge Assessment 460

14 Landscape Evolution 461

Introduction 461
Factors of Landscape Evolution 462
  Tectonics 462
  Climate 463
  Topography 464
  Geology 465
  Biology 465
Models of Landscape Evolution 467
  Conceptual Models 467
  Physical Models 469
  Mathematical Models 469
Landscape Types 471
  Steady-State Landscapes 471
  Transient Landscapes 474
  Relict and Ancient Landscapes 478
  Basin Hypsometry and Landscape Form 479
Rates of Landscape Processes 480
  Uplift Rates 481
  Erosion Rates 481
  Spatial and Temporal Variability 483
Applications 487
Selected References and Further Reading 487
Digging Deeper: Is This Landscape in Steady State? 490
Worked Problem 493
Knowledge Assessment 494

Table of Variables T-1

Index I-1

Foreword

Key Concepts in Geomorphology is a book about our planet's dynamic surface, a place where Earth and atmosphere meet and life thrives. By its very nature, geomorphology, the study of Earth surface processes and history, is an integrative discipline. As geomorphologists, we strive to make sense of the complicated web of interactions shaping mountains and valleys, moving sediment and water downslope, and changing the shape of continents and ocean basins over millennia. A deep understanding of geomorphology requires not only expertise in solid-Earth geology but also the ability to use principles of physics, chemistry, biology, and mathematics to understand Earth surface processes and the evolution of topography over short and long timescales.

Geomorphology is also an applied science, one that solves problems important to people and societies the world over. Geomorphologists identify geologic hazards, provide information for effective land management, and are trained to understand the linked processes that produce and erode the soil on which agriculture, and thus civilization, depends. Without geomorphologists, who would identify areas where landslides could bury critical infrastructure? Who would help road builders find the most stable terrain? Who would understand river dynamics well enough to protect endangered salmon? Geomorphology is science that matters.

You will quickly notice that this textbook is concise and to the point, but it covers in one way or another the most important topics in previous geomorphology textbooks. Each chapter is focused specifically on key concepts rather than regional or local examples. We expect that professors will want to use examples well known to them and relevant to the locality in which they teach. The book is designed to present general background for the student both in the text and in the illustrations, which focus on underlying principles.

The design of this textbook is no accident. It reflects ideas developed by more than 50 scientists, engineers, and mathematicians at a National Science Foundation (NSF) supported workshop held at the National Academy of Sciences in 2006. Since then, with help from dozens of geomorphologists around the world, in small meetings and large, we have worked to identify the core concepts of our discipline, and we base this book on that knowledge. Truly a community effort, Key Concepts in Geomorphology is intended for use in a first geomorphology or physical geography class or as a readable reference book for experts in other allied fields, including forestry, engineering, and botany.

As with any textbook, we have made choices about what to include and what to leave out. We are aware that no two geomorphologists would agree on the same list of key concepts for chapters such as those on rivers or tectonic geomorphology. We have done our best to focus on areas of common interest to many people in the field—areas we and eight to twelve different reviewers of each chapter thought were those most important for aspiring geomorphologists.

In considering the history and evolution of landforms over time and space, each chapter of the book focuses on several similar themes—mass transport, energy transfer, and explicit linkages between the process that shape Earth's surface and the landforms and deposits those processes leave behind. To help readers find the information they need easily, each chapter begins with an outline. Because each chapter focuses on broadly accepted concepts in geomorphology, we have not placed citations in the text; rather, each chapter ends with a list of suggested readings that allows readers to delve more deeply into the chapter's subject matter. These readings include both classic and recent peer-reviewed papers as well as more specialist texts and reference books.

A Digging Deeper section and a Worked Problem come at the end of each chapter. Each Digging Deeper section poses an important question related to the chapter and then addresses the question by providing both factual material and a history of the geomorphic thinking related to the question. The Digging Deeper sections expose readers to more formal, academic ways of thinking and information delivery. In them, we use in-text citations to the most relevant literature and reproduce figures from important journal articles, adding interpretive and explanatory captions; these are very different in style than the figures used in the rest of the book and provide an introduction to the type of visual presentation common in peer-reviewed journal articles. The Worked Problems provide examples that readers can use, either qualitatively or quantitatively, to reflect on some of the material in the chapter.

By incorporating modern tools for learning, Key Concepts in Geomorphology builds on a tradition of excellent textbooks that have defined our field for decades. All the figures are new and were created by us working closely with a third geomorphologist, Luke Reusser, who is also an artist. The goal of the figures is to help readers understand how and why changes occur at Earth's surface; toward that end we use extensive annotation and have coordinated the use of colors and symbols. Thanks to our publisher, W. H. Freeman, all figures are available online for use in the classroom.

Each chapter also includes a Knowledge Assessment, which allows readers to assess for themselves how well they know the material. Our students find these Knowledge Assessments very useful as study guides. Faculty can use them as a source of questions for tests and quizzes.

Linked to the book are a pair of publicly available electronic resources, the creation of which was supported by the NSF. To add depth and breadth to the book, specific examples of geomorphology from around the world are given in several hundred web-based Vignettes (http://serc.carleton.edu/vignettes). These Vignettes are authored by an international cadre of geomorphologists and have been peer-reviewed and revised in working groups at a series of professional meetings. Vignettes allow readers and faculty to customize their study and lectures, delving deeper into topics of interest or localizing their study by focusing on examples germane to particular geographic settings. We thank Christine Massey for organizing the many Vignette workshops and checking all of the contributions before they went online.

Almost all of the photographs used in the book, plus many others, are available on the website "Imaging Earth's Surface," a collection of images for teaching and learning about geomorphology (http://www.uvm.edu/~geomorph/gallery/). The collection includes thousands of images selected from the authors' personal archives as well as public domain images and others donated by geomorphologists throughout the world. We thank Jamie Russell, who spent countless hours acquiring and organizing the images to make them accessible to the public, and Wes Wright for the programming expertise that makes the site work. Marli Bryant Miller provided many spectacular images for the book. She is credited as M. Miller and we thank her for the great perspective her images bring to understanding landscapes the world over.

There are many people and organizations without whom this book would not have happened. To our mentors in geomorphology—David Dethier, Bill Dietrich, Tom Dunne, Alan Gillespie, Bill Locke, and Reds Wolman—we owe particular thanks. We thank Russ Pimmel and others at NSF who supported the National Academy of Sciences workshop that catalyzed this project. We thank the NSF for providing the Course, Curriculum, and Laboratory Improvement grant that supported community involvement, Vignette creation, the initiation of "Imaging Earth's Surface," and detailed, specialist peer review of every chapter and the book in its entirety.

In spring 2008, eight of our colleagues, Missy Eppes, Eric Leonard, Pat McDowell, Milan Pavich, Mary Savina, Ray Torres, Cam Wobus, and Ellen Wohl, generously took several days out of their lives to come to NSF headquarters in Virginia and help determine the chapter list for the book as well as take the first stab at outlining 9 of the 14 chapters. In summer 2008, more than 60 geomorphologists at the Cutting Edge workshop "Teaching Geomorphology" toiled over two days and late one night refining all 14 chapter outlines and creating the first Vignettes. We thank the staff of the Whitely Center, Friday Harbor Laboratory, University of Washington, who provided us a quiet place to write many of the chapters of this book, as well as the University of Vermont and the University of Washington for providing us with sabbatical leaves during which much of the book was written.

Twenty-six scientists served as editorial experts for *Key Concepts in Geomorphology*. These people reviewed chapter outlines, revised initial drafts, marked up figures, and then read and edited one final draft of their chapter in page proofs after we pulled together input from other community reviewers. We thank all these expert reviewers for this effort: Paul Bishop, Derek Booth, Scott Burns, Kathy Cashman, Doug Clark, David Dethier, Lisa Ely, Missy Eppes, Gordon Grant, Arjun Heimsath, Nick Lancaster, Eric Leonard, Scott Linneman, Frank Magilligan, Ari Matmon, Leslie McFadden, Dorothy Merritts, Grant Meyer, Sara Mitchell, Milan Pavich, Frank Pazzaglia, Eric Steig, Ray Torres, Beverley Wemple, Cam Wobus, and Ellen Wohl. After the editorial experts were done, Alan Gillespie, Mary Savina, and Tom Dunne read and edited the entire manuscript. We owe them all great thanks for taking so much of their time to improve this book. Chandler Noyes and Molly Conroy, both geomorphology students at the University of Vermont, read proofs, catching typos and ensuring that the chapters contained all the information needed to answer the knowledge surveys.

Deepest thanks go to our families for all their time without us as we wrote, edited, and edited some more. We are indebted to the students at the Universities of Vermont and Washington, who over the past 20 years shaped our teaching and our understanding of geomorphology with their questions in the field and in the classroom.

This book is dedicated to everyone who cares about Earth's surface and how it works.

Paul R. Bierman, University of Vermont

David R. Montgomery, University of Washington

Electronic Textbook Options

For students interested in digital textbooks, W. H. Freeman offers *Key Concepts in Geomorphology* in an easy-to-use format.

The CourseSmart e-Textbook

The CourseSmart e-Textbook provides the full digital text, along with tools to take notes, search, and highlight passages. A free app allows access to CourseSmart e-Textbooks on Android and Apple devices, such as the iPad. Textbooks can also be downloaded to your computer and accessed without an Internet connection, removing any limitations for students when it comes to reading digital text. The CourseSmart e-Textbook can be purchased at www.coursesmart.com.

Instructor Ancillary Support

Whether you're teaching the course for the first time or the tenth time, the Instructor Resources to accompany *Key Concepts in Geomorphology* should provide you with the material you need to make the semester easy and efficient.

Textbook Images and Answers to Knowledge Assessment Questions

Instructors will have access to all images and photographs in the textbook, along with the answers to Knowledge Assessment questions, through www.whfreeman.com/geology. This password-protected resource is designed to enhance lecture presentations by providing all the illustrations from the textbook (in jpg format).

Imaging Earth's Surface

Imaging Earth's Surface is a publicly available collection of images for teaching and learning about geomorphology. Almost all of the photographs used in the book, plus many others, are available on the website. You can access *Imaging Earth's Surface* at http://www.uvm.edu/~geomorph/gallery/.

Web-Based *Vignettes*

To add depth and breadth to the book, specific examples of geomorphology are given in several hundred web-based *Vignettes* authored by geomorphologists around the world. *Vignettes* allow readers and instructors to customize their study and lectures, delving deeper into topics of interest or localizing their study by focusing on examples germane to particular geographic settings. You can access these web-based *Vignettes* at http://serc.carleton.edu/vignettes.

A Geomorphologist's View: Paul R. Bierman

Paul Bierman excited to fly above the margin of the Greenland Ice Sheet.

It is after dinner in Greenland, close to the Arctic Circle, and despite wearing only shorts and a t-shirt, I am uncomfortably warm. Our C-130 lumbered late into the afternoon, disembarking 30 or so scientists all thinking we had left the heat and humidity of an Albany, New York, summer behind. Well, one out of two is not so bad. It is dry here, desert dry. As I roller-ski along the only paved road in town, near-vertical walls of rock tower hundreds of meters above me on either side of the fjord. From up the fjord comes a thundering green-gray river, running bankfull all day and all night. Down the fjord, the brackish, still water is filled with glacial silt. There is no sign of the ice sheet for which this place is famous—only beat-up trucks, massive gray cargo planes, and lots of bright red Greenland Air Dash-7s, the Greyhound buses of the north. Once a day, a lumbering Airbus jet brings tourists from Denmark. They land in the morning, ride several bone-jarring hours over what might only charitably be called dirt roads, see the rapidly melting glacier margin a few tens of kilometers away, then bounce back, eat dinner, and fly home.

As geomorphologists, we are here to collect rock samples that we hope will reveal when the Greenland Ice Sheet last shrank significantly, as it is doing now. We are after rocks that just today are melting out of the icesheet—rocks that may have been entombed in ice for a hundred thousand years, for half a million years, or maybe a million years.

We know that the geochemical and isotopic records, preserved in the tiny skeletons of marine organisms buried deep in ocean-floor sediment, show that Earth's climate has warmed and cooled repeatedly over the past several million years. Glaciers have come and gone, sea level has yo-yoed up and down, and the Greenland Ice Sheet has alternately extended far beyond today's coast and then retreated to a mere shadow of its current dimension.

Analysis of rock samples, collected years ago at the base of the several kilometer-long ice core drilled at the summit of the Greenland Ice Sheet, provides tantalizing clues to how dramatic the retreats of the Greenland Ice Sheet have been. The rocks suggest that about half a million years ago, during one of Earth's warm spells, the ice sheet melted away, exposing bare rock at the center of Greenland. Since that time, the center of Greenland has been covered by ice continuously. But what does this imply for the future? Could retreat of the glaciers again be as drastic as it once was?

Looking to bring home evidence of that massive melting half a million years ago, we fly helicopters past 1000-meter cliffs, over sparkling blue water, and around towering white icebergs, and we hike to the edge of the ice, over recently uncovered rocky expanses where no human may have ever set foot. Forgetting that the Sun never sets, we work long into the night to find places we can safely pry rocks out of the ice. We wade through deep meltwater streams and walk over ice melting in place after a series of warm summers. There is water everywhere on the ice. It seems like everything is melting.

We are here to learn more about how our Earth has behaved in the past. Perhaps, we will also gain insight into how our world will respond in the future, as the greatest inadvertent scientific experiment ever alters the composition of our atmosphere and thus our planet's energy balance. The past provides clues to the future, and even if this analog is imperfect, it is the best we have. Geomorphologists have the skills to decipher the complex web of interactions that shaped our planet's surface, and with care we can read something of Earth's history. In a rapidly changing world, such insight into the past is critical to our future.

A Geomorphologist's View: David R. Montgomery

Few of us take the time to notice how topography influences our daily lives or how Earth surface processes shape the world we know. But you can see the general principles of geomorphology in operation anywhere—if you know how to look and take the time to see.

One unusually sunny March morning, some fundamental interrelationships between geology, climate, and biology suddenly came into focus on my way down a steep stretch of sidewalk where the slope drops off a glacially sculpted hill toward Seattle's University District. The street is not that well-traveled and is pretty icy at times; most people take the next street over, where the slope is gentler. But partway down the hill, right where it starts to steepen, little half-inch-high ribs of concrete running across the path break the slope into small terraced steps. For some reason, on this particular day I saw how these miniature terraces create distinct patterns of decaying organic matter, fresh soil, and thriving moss that told of a city showing the first signs of decay.

This being the Pacific Northwest, the autumn leaves piled up between the tiny steps had not quite rotted away over the damp winter. So, as spring arrived and worms started to reexcavate their burrows in the cracks between concrete slabs, they mixed mineral soil with rotting organic matter still on the ground. With little foot traffic to clear the surface, this rich, new soil supported thick, squishy mats of moss that trapped additional organic matter dropped from overhanging ferns, shrubs, and trees.

Dave Montgomery along the banks of the Siang River in northeast India.

This set up a feedback in which the subtle topographic influence of those little concrete ribs favored more life that more quickly broke down the concrete sidewalk into new soil. The sidewalk was evolving into a living carpet as more soil retained more water, which favored more life. There, right before my eyes, was a grand experiment in which life reshaped its environment in ways that promoted more life.

Looking down the street, it was obvious the moss had help. Trees growing in the parking strip next to the street had pried up large pieces of sidewalk. In some places, several one-by-two-meter concrete slabs were jackknifed and starting to break apart. But just downhill of the trees, the sidewalk turned back to a bare surface where no new soil was forming as direct sunlight turned the walkway into a miniature desert.

By the time I got to the bottom of the hill, I realized that the most basic principles of geomorphology were written right there on my way to work, hidden in plain sight for all to see but few to notice. This bit of sidewalk showed how interactions among climate, hydrology, geology, and biology shape topography and generate soil—and how these, in turn, shape ecological communities and sustain life. Given all the possible interactions among these factors, which ones are most important for shaping terrain in different regions around the world? To me, this question defines the intellectual domain of geomorphology—the science of physiography and landscape evolution, the study of Earth's dynamic surface.

KEY CONCEPTS IN GEOMORPHOLOGY

PART I

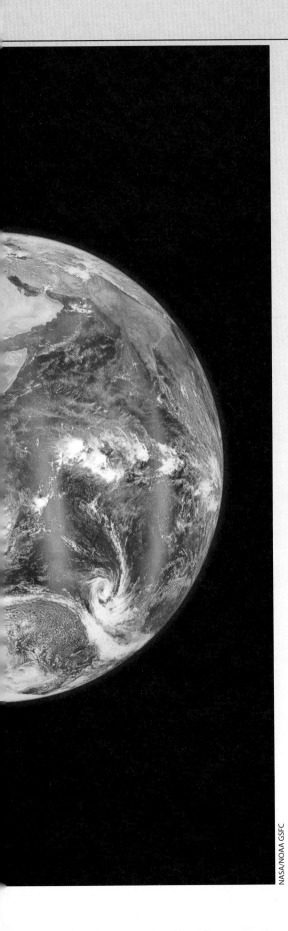

Geomorphology and Its Tools
(Chapters 1 and 2)

Geomorphology is the study of Earth's dynamic surface, its history, and its active processes. Geomorphology is a synthetic science drawing from many disciplines, including geology, physics, chemistry, and biology. Integral to Earth surface processes are the balances between forces, the transport of mass, and the distribution and redistribution of energy on our planet. Part I contains two chapters. Chapter 1 broadly introduces the field of geomorphology, laying out the interactions among the geosphere, hydrosphere, and biosphere that define Earth as a system and geomorphology as a discipline. It is followed by a short history of geomorphic thought. Chapter 2 introduces techniques that geomorphologists use to understand the form, changes, and history of Earth's surface. These techniques range from direct field observations to indirect chemical, mathematical, physical, and isotopic approaches. Once you have finished reading this part of the book, you should understand what geomorphology is and know the tools geomorphologists use to study Earth's surface.

Earth is a planet of water and land. Blue oceans dominate the planet. The continents range from brown in arid regions to green where vegetation thrives in areas that receive more precipitation. White clouds–some swirled in storms, others stretched out in linear weather fronts–hide some of Earth from view. Perspective is of a viewer looking down from almost 13,000 kilometers. NASA Portraits of Earth made from composite natural color images taken by the Visible Infrared Imaging Radiometer Suite (VIIRS) instrument aboard a polar orbiting satellite.

Earth's Dynamic Surface

Introduction

Geomorphology is the study of the processes shaping Earth's surface and the landforms and deposits that they produce. The word itself was introduced into the English language in the late nineteenth century, through the combination of the Greek word *geo* (earth) and the suffixes *-morphos* (form) and *-ology* (study of). Geomorphologists learn to observe and interpret landscapes systematically in order to understand how processes shape Earth's surface, decipher the history of a place, and recognize (and potentially mitigate or manage) the impacts of environmental hazards on societies. Whether one is on the way to work or exploring remote regions of the globe, geomorphology offers insights into how water and sediment move, how rocks break down to create soil, how tectonic forces raise mountains, and how uplift is locked in a geologic duel with erosion. Geomorphology provides a new way to see the world because landscapes tell stories that can be read if one has the proper knowledge and tools.

Until recently, geomorphology was a qualitative, interpretive science. Classification of landforms and topographic features informed speculation about the history of landscapes and provided information useful for the development of human settlements and the infrastructure they require. In the latter half of the twentieth century, geomorphology underwent a dramatic transformation with the advent of quantitative characterization and analysis of both landforms and the processes driving the development and evolution of topography. As the plate tectonics revolution illuminated the mystery of mountain

A braided, high-gradient stream roars down a canyon in the Tien Shan mountains, Kyrgyzstan. The large levees along the channel are studded with meter-size boulders, indicating that the channel was impacted by debris flows that most likely originated from landslides off the steep basin slopes.

IN THIS CHAPTER

Introduction
Geosphere
  Isostasy
  Tectonics
  Lithology and Structure
Hydrosphere
  Climate and Climate Zones
  Hydrologic Cycle
Biosphere
  Geographical Distribution of Ecosystems
  Humans

Landscapes
  Process and Form
  Spatial Scales
  Temporal Scales
Unifying Concepts
  Conservation of Mass
  Conservation of Energy
  Material Routing
  Force Balances and Thresholds
  Equilibrium and Steady State

  Recurrence Intervals and Magnitude-Frequency Relationships
Applications
Selected References and Further Reading
Digging Deeper: Why Is Earth Habitable?
Worked Problem
Knowledge Assessment

building and solved the problem of continental drift, radiometric dating techniques provided accurate landscape chronologies, and developments in electronics, sensors, and computational techniques allowed quantitative characterization of geomorphological processes. These developments allowed scientists to reliably connect process with the evolution of landforms. Today, geomorphology is an integrative science that brings a wide range of methods and techniques to bear on understanding the processes that shape Earth's dynamic surface.

Geomorphology is an inherently synthetic discipline that draws on geology, physics, chemistry, and biology to understand landscape-forming processes and to decipher how weathering, erosion, and deposition sculpt the land. It is a science with innate intellectual and aesthetic appeal. Geomorphology also has fundamental societal relevance because we live on Earth's surface, and the dynamic processes that shape topography influence human societies by controlling the spatial distribution of arable land, where and when floods and landslides occur, the geography of erosional uplands, the locus of coastal erosion, the retreat of glaciers, and the offset of land surfaces by earthquake-generating faults in different parts of the world [Photograph 1.1]. The study of geomorphology intersects with a variety of geologic and geographic disciplines because topography reflects the interaction of driving forces that elevate rocks above sea level, the erosional forces that wear rocks down, and the factors that determine resistance of the land surface to erosion—including how weathering progressively weakens slope-forming materials and how vegetation reinforces and stabilizes them.

Most people think of the land surface as stable and unchanging because we do not usually notice subtle, ongoing changes in our daily experience. Geomorphologists see landscapes as constantly changing and evolving, usually slowly but sometimes catastrophically shaping and remodeling our world. Knowing what to look for and knowing how to see is key to understanding landscapes. Recognizing landscapes as dynamic systems is central to understanding how natural forces shape Earth's surface, influence land use, and respond to human actions.

Tectonics (the movement of Earth's lithosphere) and **volcanism** (the eruption of molten rock) drive the **endogenic** (internal, or generated from within) processes that raise mountains and elevate topography above sea level. Tectonics determines the first-order topography, or megageomorphology, of Earth. In these ways, Earth's internal dynamics provide the raw material for erosional processes to shape landscapes and landforms. Endogenic processes are the result of Earth's internal heat, which drives plate tectonics through deep convection of the planet's mantle. The vast array of erosional processes that influence Earth's surface arise from **exogenic** (external, or generated from outside) processes imposed on landscapes through the action of wind, water, and ice. Exogenic processes are driven by the Sun's energy and the temperature gradient between the poles and the equator.

In framing how to study geomorphology, we consider three systems that intersect at and near Earth's surface. Specifically, we focus on the **geosphere**, which includes the rocks comprising Earth's **crust** and the global tectonic system that elevates the rocks that get sculpted into topography; the **hydrosphere**, which encompasses the oceans, atmosphere, and the surface and subsurface waters that erode, transport, and deposit sediment; and the **biosphere**, living things that are part of ecosystems that help shape and are in turn shaped by landscape dynamics. Global patterns within and among these three basic systems set the broad regional templates within which geomorphological processes shape landforms and landscapes evolve.

Geosphere

Planetary-scale processes drive the motion of Earth's **tectonic plates** and control the locations of mountains, plains, plateaus, and ocean basins. Earth's rigid lithosphere is broken into seven major and a number of minor plates that move independently from one another, creating **mountain belts** and **volcanic arcs** where plates converge and great **rift valleys** where the plates spread apart [**Figure 1.1**]. Many major surface features of the planet, including mountain ranges, earthquake-prone fault zones, and volcanoes, occur along tectonic boundaries, but plate boundaries do not necessarily coincide with the edges of the continents. Tectonic setting generally determines the distribution of different rock types with different degrees of resistance to erosion. Weakly cemented sedimentary rocks typically are found in depositional basins. Hard, erosion-resistant metamorphic and igneous rocks are exposed in the cores of most mountain ranges. The global distribution of rock types and rates and styles of crustal uplift, deformation, and extension reflect the history and patterns of tectonic motion that set the stage for topographic evolution and landscape development.

Isostasy

The most obvious feature of global topography is the difference between the most common land surface elevation of the **continents** (~1 km) and the ~4 km depth of the seafloor in **ocean basins** [**Figure 1.2**]. These two dominant elevations reflect the contrast in density between **continental crust** and **oceanic crust**, the result of recycling of Earth's crust over geologic time to produce continents rich in elements lighter than those that compose oceanic crust.

Earth's crust rides on the denser material of the planet's upper mantle, in much the same way that less dense ice floats in water [**Figure 1.3** on page 10]. This phenomenon, called **isostasy**, leads to a balance in regard to the total mass of rock above the depth of isostatic compensation, a reference level within the **asthenosphere** (a ductile or slowly deforming layer in the upper mantle) at which the pressure exerted by the overlying columns of rock is equal.

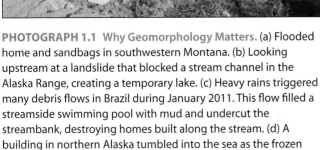

PHOTOGRAPH 1.1 Why Geomorphology Matters. (a) Flooded home and sandbags in southwestern Montana. (b) Looking upstream at a landslide that blocked a stream channel in the Alaska Range, creating a temporary lake. (c) Heavy rains triggered many debris flows in Brazil during January 2011. This flow filled a streamside swimming pool with mud and undercut the streambank, destroying homes built along the stream. (d) A building in northern Alaska tumbled into the sea as the frozen ground (permafrost) below melted and the coast became more susceptible to erosion. (e) Rocks from a 100,000 m³ rockfall in Yosemite Valley in 1982. The slide closed the highway for several months. The rockfall originated in jointed granite about 200 m above the highway. (f) Bouldery fault scarp at the site of an 1872 rupture on the Owens Valley Fault Zone, southern California. The land on the left moved up several meters, exposing the once-buried boulders.

Continents stand higher than ocean basins because continental crust is composed of less dense, silica-rich granitic rocks, whereas oceanic crust is made of denser rocks with a more iron-rich basaltic composition. Similar to icebergs, which lie mostly below the water surface, a deep root of low-density continental crust extends down to displace denser mantle material and support high-standing mountains. Some such roots, like those under the

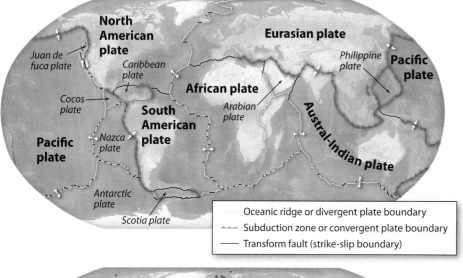

Plate boundaries govern the distribution of many distinctive geomorphic features, including mountain ranges and volcanoes. **Convergent boundaries** are associated with high mountain ranges and explosive volcanism. **Divergent boundaries** also generate high-standing topography, but volcanism there is less explosive and more effusive. **Strike-slip boundaries** (transform faults) offset landforms laterally and create local relief where faults bend.

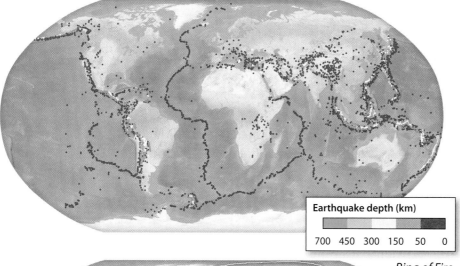

The global distribution of earthquakes follows closely the boundaries of tectonic plates. The largest and deepest earthquakes occur at and near convergent boundaries, especially **subduction zones.** Likewise, earthquakes are concentrated along mid-ocean spreading centers. Shallow earthquakes also occur in plate interiors.

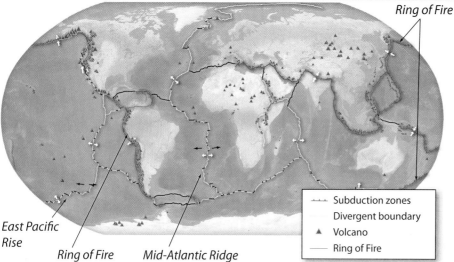

Many of the world's volcanoes are found on and near plate boundaries. The Pacific Ocean Basin is encircled by subduction zones and their associated volcanoes that together form the "Ring of Fire." The Mid-Atlantic Ridge and the East Pacific Rise are divergent plate boundaries where new oceanic crust is created by volcanism. Volcanoes are also common in terrestrial rift zones, such as East Africa.

FIGURE 1.1 Volcanoes, Earthquakes, and Plate Boundaries. The character of Earth's dynamic surface reflects the forces acting upon it. The movement of tectonic plates, shaking during earthquakes, and volcanic eruptions influence the development of regional topography.

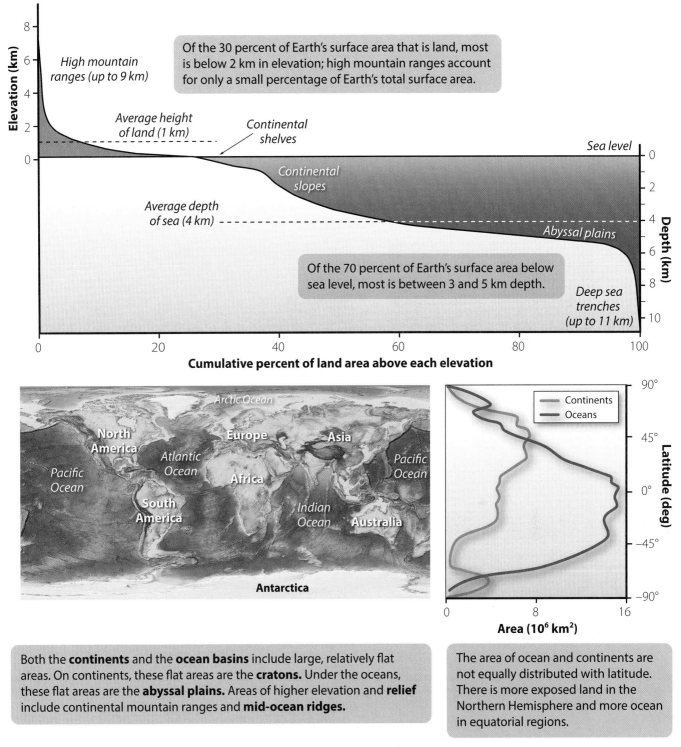

FIGURE 1.2 Global Topography. The surface of the Earth has two dominant elevations—the deep ocean basins at an average depth of about 4 km below sea level and the continental cratons, where the average elevation is about 1 km above sea level.

Himalaya, can be greater than 50 km thick—more than 6 times the height Mount Everest rises above sea level. Even the ancient, worn down Appalachian Mountains are supported by a crustal root tens of kilometers thick.

Due to isostasy, erosion that removes rock and soil from Earth's surface results in **isostatic compensation** or **uplift** of the remaining rock as denser material moves in at depth to compensate for eroded material. Consider how an iceberg continues to float even as it melts; the remaining ice rises to replace that lost off the top. With typical continental crust having a density of about 2700 kg/m^3, while typical mantle rocks have a density of more than 3300 kg/m^3, isostatic uplift can offset almost 82 percent of erosion (i.e., 2700/3300 = 0.82). Hence, for every meter

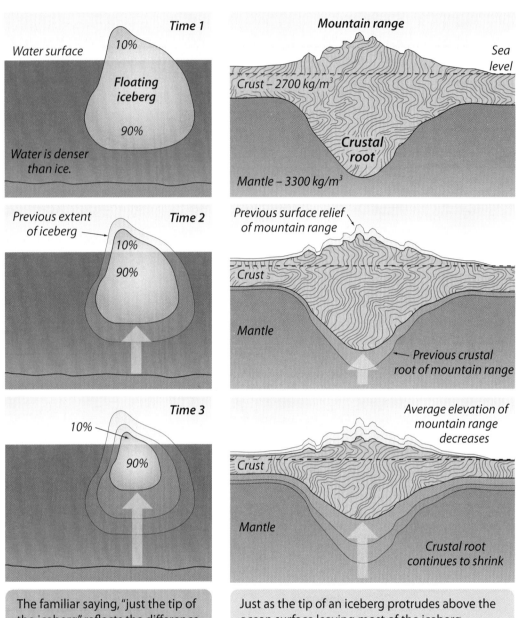

FIGURE 1.3 Introduction to Isostasy. Just as an iceberg of low density floats on water of higher density, mountains with thick crustal roots float on the denser mantle rocks. As ice above the water surface melts or mountains erode (Time 1, Time 2, Time 3 = time steps), the underlying material rises to take its place.

of rock stripped off by erosion, isostatic compensation triggers more than 80 cm of rock uplift, resulting in net surface lowering of a little less than 20 cm. Isostasy means that subduing or erasing a mountain range requires eroding through its root because it takes erosion of more than 5 times the thickness of rock to reduce average elevation by a comparable amount. Consequently, today's landscape can reflect tectonic events that occurred in the distant past, as it can take an extremely long time to erode mountains away because their roots rise almost as fast as the surface erodes. For example, peaks in the Appalachian Mountains stand more than 2000 m tall in a region where the last tectonic activity capable of producing large mountains (continental collision) occurred several hundred million years ago.

Tectonics

As tectonic plates move, the lithosphere is rafted along by thermal convection of Earth's deep mantle. Plate margins are classified as convergent, transform, and divergent boundaries [**Figure 1.4**]. **Convergent boundaries** are those where two plates move toward and collide with one another. **Transform boundaries** are those where two plates slide laterally past one another—for example, along California's earthquake-prone San Andreas Fault Zone. **Divergent boundaries** are those where two plates spread apart.

Mid-ocean ridges are divergent plate boundaries where mantle upwelling at **seafloor spreading centers** creates new oceanic crust. The older crust is pushed aside and moves away from the ridge as new material is extruded; this basaltic crust forms the floors of Earth's oceans. The new oceanic crust cools, grows more dense, and thus sinks. This thermal subsidence explains why the ocean floors in general get deeper the farther they are from mid-ocean ridges. Continental divergent boundaries are found where spreading centers extend onto land, resulting in extensional rift valleys that parallel the plate boundary. Such valleys typically host lakes, large alluvial fans, and volcanoes.

Subduction zones are convergent plate boundaries where old, denser oceanic crust sinks beneath less dense continental crust or younger, more buoyant oceanic crust. The plate boundary on the Pacific side of the Andean Mountains in South America and the trench seaward of the Japanese islands are examples of subduction zones. Less dense continents and islands do not subduct.

Continental margins are called **active margins** where they coincide with plate boundaries and **passive margins** where there is no relative motion between the continent and the seafloor, as for example along the east coasts of North America, South America, Africa, and Australia. Great mountain belts, like the Himalaya, form in **continental collision zones** where two continents made of low-density crustal rocks are squeezed between converging plates, shoving one beneath the other. In the case of the Himalaya, the Indian subcontinent has been ramming into Asia for the past 50 million years. The collision has piled up crust to form the Tibetan Plateau, much like how sand piles up in front of an advancing bulldozer blade. Continental collision zones build high mountains because crustal thickening gradually elevates the surface while building a deep crustal root.

Tectonically active mountain belts, like the Himalaya, the Andes, and the Cascades in the Pacific Northwest, are found along convergent margins (continental collision zones and subduction zones). Mountain belts also form above subduction zones where material scraped off the downgoing slab piles up (like the Olympic Mountains in Washington). Farther inland, linear chains of volcanoes are parallel to the subduction zone. The volcanoes result from partial melting at depth of the slab, overlying mantle, and moist, subducted oceanic sediments. Other ranges, such as the Appalachian and the Ural mountain chains, record ancient continental collisions at plate boundaries that are no longer tectonically active. Crustal thickening, when convergence is sustained for long enough and at a rate greater than the rate of erosion, can cause the height of mountain ranges and the stress they exert at depth to exceed the mechanical strength of the lithosphere. Where this happens, it leads to development of high plateaus, like Tibet and the Altiplano. Lithospheric strength limits plateau elevation and further convergence results in lateral extrusion of material that expands the area of high terrain without further surface uplift.

Continents are composed of mountain belts called **orogens** (sometimes referred to as orogenic belts), extensional rift zones, and ancient, tectonically stable and generally low-relief **cratons** [**Photograph 1.2**]. Active mountain belts typically consist of sedimentary and crystalline rocks that are highly deformed and erode rapidly, at rates up to several millimeters per year. Inactive mountain belts erode much more slowly, just a fraction of a millimeter per year. Tectonically stable continental landscapes, like the interior of Australia, erode even more slowly, at rates on the order of 1 meter per million years. Rift zones, like the East African Rift, are places where active divergent plate boundaries extend into continents, tearing landmasses apart, raising steep rift flanks, and emplacing relatively young volcanic rocks.

Continental shelves (see Figure 1.2) are submerged plains at the edges of the continents. During the last several million years, these areas were repeatedly exposed to subaerial erosion as cycles of glaciation periodically locked up sufficient ice on the continents to lower sea level by more than 100 m. Continental shelves extend beyond present-day shorelines, and they range from narrow features a few kilometers wide on tectonically active margins to broad areas of relatively shallow marine inundation that extend hundreds of kilometers offshore on passive margins.

Within tectonic plates, upwelling plumes of hot rock from deep in the mantle, called **hot spots**, dramatically influence topography because they produce prodigious and intense volcanism and can drive uplift thermally by heating rock and thus making it less dense. In oceanic settings, hot spots build chains of islands and seamounts like

Marine environments

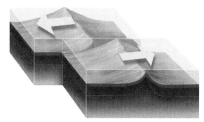

1. Where an oceanic plate **subducts** under another oceanic plate at a **convergent boundary**, a chain of steep, explosive volcanoes, known as an **island arc**, forms.

2. Marine transform faults offset spreading ridge segments. These strike slip faults are clearly visible in remotely sensed of sea-floor topography near spreading ridges.

3. Divergent boundaries are characterized by **basaltic volcanism** and shoulders uplifted by **thermal buoyancy** of the warm, less dense mantle rock that comes close to Earth's surface here.

Terrestrial environments

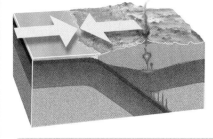

4. At **convergent** boundaries where oceanic plates subduct under continental plates, a chain of steep, explosive volcanoes forms inland of the boundary. An **accretionary wedge** of sediment scraped off the downgoing plate can form a coastal mountain belt. Continent–continent collisions result in less volcanism but greater uplift because buoyant continental crust does not subduct.

5. Transform plate boundaries generate both linear landscape elements related to the fault zone and the deformation of rocks along it as well as topography at bends along fault systems.

6. Continental divergent boundaries are areas where continental crust is moving apart and hot mantle material is moving toward the surface. Continental divergent boundaries have elevated rift shoulders bounded by steep normal fault planes. Alluvial fans form in the down-dropped basins and there can be basaltic volcanism.

FIGURE 1.4 Plate Margin Types. Each type of plate tectonic boundary (convergent, divergent, and transform) is associated with a particular suite of landforms that enables geomorphologists to predict landscapes and active surface processes in different parts of the world.

PHOTOGRAPH 1.2 Continental Landscapes. (a) The mountains of the Dolomite Range of Italy are steep and shed large cones of talus (broken rock). (b) The Dead Sea rift shoulder in Israel is steep, its topography controlled by a rift-bounding normal fault at its base. (c) The Australian craton is in large part flat-lying with little relief. The most weathering-resistant crystalline rocks are exposed as rounded core stones.

and the soil it produces, the durability of weathered material, and the strength of materials from which hillslopes form. Different rock types vary greatly in their chemical composition, texture, and material strength. For example, the stair-stepped walls of the Grand Canyon, in which hard units hold up steep cliffs and weak units form topographic ledges, provide a well-known example of how lithology affects topography [**Photograph 1.3**].

Rock structure also plays a key role in determining erosion resistance, because the degree to which rocks have been tectonically fractured, sheared, or deformed the Hawaiian Islands and the Emperor seamounts. In continental settings, hot spots produce areas of high elevation and volcanic activity, like the Yellowstone plateau and caldera in Wyoming. Because hot spots are rooted in the mantle below the crust, they remain stationary as plates move over them, and they result in topographic features that track the direction and rate of plate movement. For example, in the linear volcanic chain of the Hawaiian Islands, the younger islands lie progressively southeastward of older islands, a trend that translates into systematic differences in soil properties (e.g., soil fertility) and degree of topographic dissection (increased valley incision) with increasing island age.

Lithology and Structure

Lithology (rock type) and geologic structures like sedimentary layering, faults, and folds greatly influence erosional resistance. Lithology affects the style of weathering

PHOTOGRAPH 1.3 The Grand Canyon of the Colorado River in Arizona. Here, rock type controls topography with strong rocks, such as sandstones and well-cemented limestones, forming steep cliffs, and weak rocks, such as shale, holding gentler slopes and being buried beneath debris.

PHOTOGRAPH 1.4 Folded Manhattan Schist, Central Park, New York City. Weathering has etched out the foliation showing the folds.

influences their material strength. Sedimentary rocks are generally stratified and have bedding planes that create material discontinuities and zones of weakness. The degree of consolidation and the amount and type of interstitial cement also affect the strength and geomorphic expression of sedimentary rocks. Crystalline rocks, like granite, are typically more massive and stronger than many sedimentary rocks. Preferentially oriented cleavage planes and fabrics in metamorphic rocks often dominate their geomorphic expression. Geologic structures that expose rocks with different resistance to erosion can generate distinctive topographic signatures that allow geologists to recognize features like folds, faults, and fracture patterns solely from their geomorphic expression [Photograph 1.4].

Different rock types weather at different rates and produce weathering products with various material strengths and other properties that influence landform development. For example, weathering and erosion often lead to hill and valley terrain with high-standing topography capped by hard, erosion-resistant rock (like quartz-rich sandstone or quartzite) and areas of subdued topography that are underlain by relatively weak rock that is more susceptible to weathering and erosion (such as siltstone and shale). Rocks of a particular lithology may, however, behave differently in different climates. For example, limestone commonly holds vertical cliffs where it is exposed in the arid landscape of Arizona's Grand Canyon but supports only gentle slopes in the humid Appalachian Mountains. The patterns of rock types resulting from the geologic history of a region provide the raw material from which regional topography develops.

Hydrosphere

The distribution and movement of water over Earth's surface, as well as its form (rain, snow, or ice), greatly influence weathering, erosion, and landscape evolution. **Overland flow**, in which rainfall or snowmelt runs off across the land surface, is rare in well-vegetated, humid and temperate landscapes except near seasonally wet river courses. This rarity is the result of plants. Their roots and the organic-rich material that makes up the surface soil horizon (duff) lead to high **infiltration rates**. Depressions, such as the pits left when trees fall over, slow overland flow, allowing it to **infiltrate** (soak into the ground). Shallow subsurface flow is common as water moves through soils and shallow permeable rock on its way to rivers and streams. Subsurface flow also influences rock weathering and soil development. Overland flow is an important runoff mechanism in many arid regions with little vegetation and in disturbed areas where the soil has been burned, compacted, degraded, or paved over. The dynamics of glaciers and seasonally frozen ground dominate hydrologic processes in alpine and high-latitude regions.

Climate and Climate Zones

Climate describes long-term spatial and temporal patterns of precipitation and temperature on Earth's surface. Different climates are characterized not only by the average weather but how much the weather varies from day to day and month to month.

Solar radiation provides the energy that drives Earth's climates; it evaporates water from the oceans into the atmosphere, drives atmospheric circulation, and thereby produces wind and storms. The amount of incoming solar energy (**insolation**) varies with latitude, producing an air-temperature gradient between the poles (low insolation and therefore low temperatures) and the equator (high insolation and higher temperatures). The different **albedo** (reflectivity) of the ocean and the land surface also results in temperature gradients. About 30 percent of the energy our planet receives from the Sun is reflected back into space by clouds, particulate matter, and aerosols in the atmosphere and the ground surface.

Energy gradients drive atmospheric and oceanic circulation, which result in substantial poleward transfer of heat that delivers warm air and water to higher latitudes. Because Earth's total heat budget is relatively constant, well-defined temperature and precipitation patterns impart different, predictable climates to different regions. Over time, global and regional climates shift in response to changes in solar insolation, including those that arise from changes in Earth's orbit and from fluctuations in the concentrations of heat-trapping greenhouse gases (see Chapter 13).

Atmospheric circulation on our rotating planet gives rise to orderly circulation of rising and falling air masses as well as easterly and westerly zonal winds [**Figure 1.5**]. These systematic latitudinal variations in temperature, precipitation, and wind create a pole-to-equator sequence of climate zones that have distinctive ecological and geomorphological characteristics. Climate patterns in the Northern Hemisphere are generally mirrored in the Southern Hemisphere. Patterns of atmospheric circulation greatly influence the distribution of precipitation (and thus

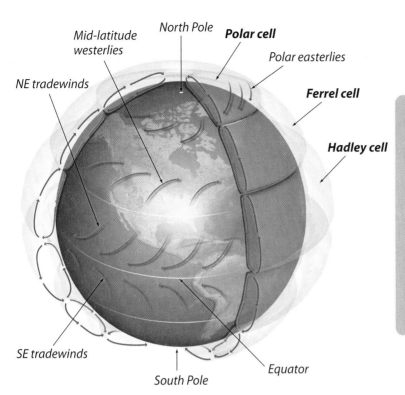

Large-scale **atmospheric circulation** is dominated by latitudinal circulation cells, defined by bands of rising warm, moist air along the equator (**Hadley cell**) with descending dry air in the mid-latitudes that defines the global belt of deserts. Similar temperate (**Ferrel**) and **Polar** circulation cells characterize atmospheric circulation at higher latitudes.

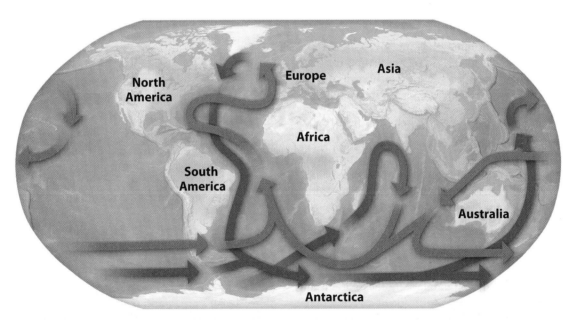

Ocean thermohaline circulation involves sinking of cold, salty water at the poles (shown in blue). This sinking water produces deep cold currents and shallow warm surface currents (shown in red).

FIGURE 1.5 Global Oceanic and Atmospheric Circulation. Air and water circulate through Earth's atmosphere and oceans, moving heat around Earth in response to unequal input of energy from the Sun over time and over the surface of our planet.

surface processes) by controlling the moisture content and trajectory of moving air masses.

Latitudinal temperature differences give rise to three great convective cells of atmospheric circulation. In the equatorial **Hadley cell**, moisture-laden air rises in the zone of high surface temperature at the equator. As the air ascends, it cools and loses its capacity to hold moisture, creating a zone of high precipitation in the tropical equatorial zone. Atmospheric circulation in this zone is dominated by the equatorial easterlies (trade winds) that consistently blow east to west. Between latitudes of about 30° and 35°, precipitation amounts are low beneath the dry air masses in the descending limb of the Hadley cell. Earth's great deserts, like the Sahara, the Arabian, the Sonoran, and the Kalahari, are concentrated in these high-pressure subtropical belts. Poleward of the Hadley zone, the prevailing westerlies that consistently blow west to east give rise to the temperate-zone **Ferrel cell**, in which rising air masses produce another belt of high rainfall between 40° and 60° latitude. In polar regions, the descending air of the third, or **polar cell**, delivers little moisture, resulting in cold deserts at high latitudes. These broad latitudinal patterns not only shape regional climates but they influence individual weather events as well. The trajectories of storms in the Northern Hemisphere temperate zone track the unstable, constantly shifting jet stream at the boundary between the Ferrel and polar cells. The short-term variability of this atmospheric interface defines weather patterns, whereas the long-term position controls regional climates.

Temperature and precipitation change with elevation. Within a mountain belt, the local gradient in temperature and precipitation with elevation can be as great as the difference between latitudinal climate zones. Differences in temperature and precipitation define distinct climate zones that greatly influence the ecosystems and styles of geomorphic processes in landscapes at different latitudes or elevations [**Figure 1.6**]. Geomorphic processes involving ice are important in cold, high-latitude, and high-elevation regions; rainfall predominates in temperate and tropical regions; and both wind and water shape arid-region landforms. Some climates can produce distinctive landscapes such as those formed by glaciers; however, some landforms (such as sand dunes) occur in a wide range of climates [**Photograph 1.5**].

At a regional scale, the distribution and orientation of mountains relative to prevailing winds and moisture sources affect the delivery of precipitation. For example, as moisture-laden winds approach a mountain range, air masses flow up and over the high topography. As wet air ascends the windward (upwind) slopes of a range, it cools, loses some of its ability to hold moisture, and the resulting condensation generates precipitation that nourishes lush vegetation on the upwind side of the mountains. By the time an air mass reaches the leeward (downwind) side of a high mountain range, it has lost substantial moisture. Warming as it descends, this low-humidity air continues on to form arid regions and deserts downwind of major mountain ranges.

The **orographic effect** of topography wringing moisture out of the atmosphere results in the development of so-called **rain shadows** on the leeward sides of mountain ranges (and substantial oceanic islands like Hawaii) that tend to be drier than their windward sides. Continental interiors are also drier than coastal areas. Such differences profoundly affect the local climate, ecological communities, and erosion rates and processes that influence the development of topography.

Ocean circulation plays a large role in determining terrestrial climate because ocean currents move large amounts of heat and water from low to high latitudes. Cold currents reduce evaporation and thus contribute to aridity in locations downwind. For example, upwelling of cold water off the west coasts of South America and Africa is a critical factor maintaining the aridity of the Atacama (Chile) and Namibian deserts. In contrast, the warm Gulf of Mexico delivers large amounts of atmospheric moisture to southeastern North America; its warm waters and high humidity nourish hurricanes that can cause large amounts of geomorphic change.

Hydrologic Cycle

The movement and exchange of water among the oceans, land, and atmosphere is known as the **hydrologic cycle** [**Figure 1.7**]. Water evaporated from the ocean surface rises together with warm air to become clouds that move with winds that blow air masses over continents. Precipitation that falls as rain or snow eventually either infiltrates into the ground, runs off over the ground surface to join bodies of surface water, evaporates, or is transpired by plants back to the atmosphere. Whether it is slow-moving groundwater or quickly flowing surface and near-surface runoff, water eventually flows into stream systems that ultimately discharge into the sea. The distillation of freshwater from the oceans, its delivery to the continents, and subsequent return to the oceans drives key geomorphological processes. Most of the water on Earth (97 percent) is held in the oceans. About 2 percent is stored as ice in glaciers and polar ice sheets, and about 0.6 percent is stored as groundwater. Freshwater rivers, lakes, and wetlands account for <0.5 percent of the water on Earth. Humanity depends on the continuous movement of water through the hydrologic cycle to maintain the freshwater supplies that sustain us, irrigate our crops, and support the terrestrial ecosystems upon which we depend.

Biosphere

The **biosphere** defines the realm of life within, upon, and above Earth's crust, and includes organisms ranging from small, short-lived soil microbes that enhance the weathering of rock to gigantic millennia-old redwoods that when they fall over move large amounts of soil downslope. Living systems depend on their physical environments and, in turn, influence the soils, rocks, air, and water in which they live. This interdependence is well illustrated by the complex ecosystems that exist in the soils and fractured and weathered rock at the interface between the abiotic, sterile geology of our planet's deep interior and the biology that covers much of its surface.

The dynamics of Earth's surface processes control the frequency and size of disturbances like floods and landslides that alter biologic and human communities. Disturbances may act to maintain stable ecological systems when

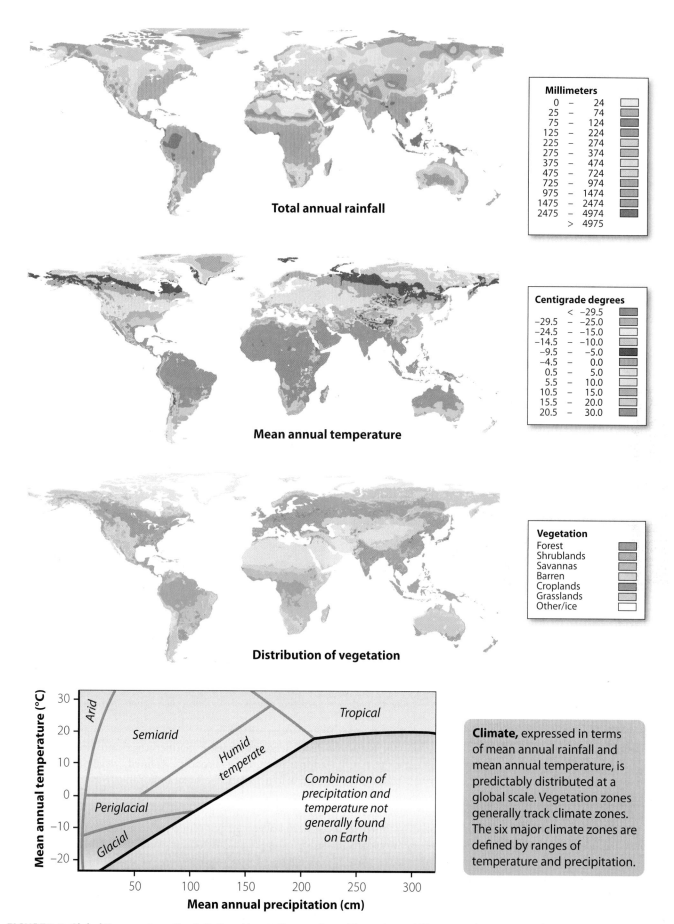

FIGURE 1.6 Global Temperature, Precipitation, Vegetation, and Climate. Earth has distinctive and predictable climate and vegetation zones determined by mean annual precipitation and temperature. These zones shift over geologic time as climate warms and cools.
[Adapted from Wilson (1968).]

(a)

(c)

(b)

(d)

PHOTOGRAPH 1.5 Climate, Geomorphic Processes, and the Appearance of Landscapes. (a) Largely devoid of vegetation, hillslopes in the Atacama Desert of northern Chile are gullied by rare but intense rainfalls. (b) Arctic landscapes are often dominated by ice and diminutive vegetation, stunted by cold and the short growing season, such as here, in Greenland near a recently drained glacial-margin lake. (c) In humid temperate zones, where much of Earth's population lives, the landscape is often tree-covered, such as in Shenandoah National Park, central Appalachian Mountains, Virginia. (d) Alpine landscapes are often steep, rocky, and cold with little vegetation. Such landscapes are characterized by the persistent effects of glaciation such as here in the Austrian Alps.

they predictably recur in the same place. For example, regular inundation of floodplains leads to characteristic riparian vegetation that slows flood flows along river corridors. Conversely, disturbances that are unpredictable and do not occur in the same place (for example, landslides) favor weedy plants that specialize in colonizing and stabilizing freshly disturbed land. Understanding the relationships between ecological communities and the landscapes they inhabit informs our understanding of both.

Geographical Distribution of Ecosystems

The global distribution of plant communities generally tracks global latitudinal climate zones and the elevation-dependent patterns of temperature and precipitation in mountains. Five broadly defined vegetation zones—semi-arid grasslands, temperate forests, tropical forests, arid deserts, and polar regions—characterize the global distribution of plant communities. Each zone has distinctive ground surface coverage, root reinforcement, soil types, and weathering properties that result from its plant communities.

Plants in grassland and forest zones not only generate different types of soils, but their root systems reinforce soils differently so the landscapes differ in their resilience to environmental disturbance as well. Grasslands generally have more biomass below ground than above ground, most of it in roots. Removal of root reinforcement when grasslands are impacted by overgrazing and/or plowing leaves them

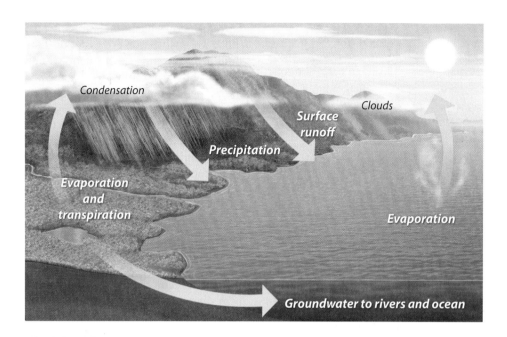

The **hydrologic cycle** drives many surface processes. Solar energy causes **evaporation** from lakes, rivers, wetlands, and the oceans as well as **transpiration** from plants. These processes move water vapor into the atmosphere where it condenses. Rainfall and snowmelt convert the **potential energy** of water vapor to the **kinetic energy** of raindrops and flowing water and do work, both as drops that impact the ground and as surface water runoff that collects and flows through channels. Groundwater, recharged by precipitation, weathers rock and destabilizes hillslopes.

FIGURE 1.7 Global Hydrologic Cycle. Water moves through the hydrosphere in vapor, liquid, and solid forms driving many geomorphic processes.

particularly vulnerable to erosion. In forests, extensive root networks form interlocking webs that mirror the extent of the forest canopy and reinforce hillside soils. Geomorphically important effects of forest type include the depth of root penetration, root strength, and the shape of fallen trees that enter rivers. In river channels, the difference between the long, pole shape typical of conifers and the branching structure typical of deciduous trees influences the stability and transportability of wood and thereby the propensity for logjam formation. Tropical forests tend to have little below-ground organic matter and extensively weathered mineral soils. Because most bio-available plant nutrients are held in the plants themselves, it can be hard to reestablish native forests after forest clearing, resulting in ongoing soil erosion.

In contrast to temperate and tropical regions where vegetation clearly plays a major role in the type, frequency, and intensity of geomorphological processes, the geomorphological role of vegetation in arid and polar landscapes can be more subtle. In desert landscapes where plant communities contain little biomass and plant cover is sparse, the presence or absence of even a little ground cover or a thin web of roots can greatly affect soil development and landscape stability. Fragile **cryptogamic crusts** made up of bacteria, mosses, and lichen hold together desert surfaces and prevent wind and water erosion. Biological respiration affects the deposition of impermeable, erosion-resistant calcium carbonate in desert soils—eventually changing subsurface water flow and influencing landform development.

In the arctic, plants, in particular small trees, influence the distribution of **permafrost**, the presence of ground that is frozen year-round. Areas with plants have deeper snow cover because the plants trap blowing snow. This snow insulates the ground beneath from frigid winter temperatures, limiting the freezing depth and keeping ground temperatures warmer over winter. Such areas are more likely to thaw in the summer and thawed permafrost tends to be weak and thus very active geomorphically.

The distribution of animal communities generally tracks climate zones, and these animals can influence the geomorphology of the landscapes they inhabit. For example, spawning salmon reshape bars and pools in Pacific Northwest rivers. Overgrazing by domestic animals accelerates soil erosion, leaves terracettes (small terraces) on hillslopes, and through soil compaction, can trigger gully development. Burrowing animals and mound-building ants and termites can displace and mix tremendous amounts of soil and weathered rock [Photograph 1.6]. Charles Darwin calculated that over the course of centuries, worms steadily plowed and thus mixed the hillside soils of England.

PHOTOGRAPH 1.6 Termite Mounds. Termite mounds in the Northern Territory of Australia indicate the amount of soil moved and stirred by insects.

Plants and animals influence geomorphological processes directly, as when burrowing activity and roots mechanically pry weathered rocks apart, and indirectly, as when plants protect soils from erosion during rainstorms, and roots mechanically reinforce slope-forming materials. Plants also are central to the chemical transformations that accompany the breakdown of rock-forming minerals into the clay minerals that hold nutrients essential to soil fertility. The size of an organism is not necessarily related to its importance or impact. Soil bacteria, for example, are important chemical weathering agents in many environments.

Humans

Humans move enough rock and soil to be counted among the primary geomorphic forces shaping Earth's modern surface. Coal and mineral mining operations move whole mountains and excavate great pits [Photograph 1.7].

PHOTOGRAPH 1.7 Open Pit Mining. This face of an open pit mine in Brazil shows how tropical landscapes, dominated by extreme weathering, are mined for residual ores, such as aluminum and iron oxides. The depth of weathering is shown by the red iron oxidation in rock tens of meters below the surface. Trees above the mine provide scale.

Farmers' plows push soil gradually but persistently downhill, and construction crews cut or fill the land to facilitate building or to suit our aesthetic whims. Human activities further influence geomorphological processes through the indirect effects of our resource management and land-use practices. In manipulating our world, we alter hydrological processes by changing surface runoff, stream flow, and flood flows. Clearing vegetation and changing water fluxes in the landscape affect slope stability and erosion rates. Construction of dams and coastal jetties interrupts the transport and storage of sediment. Human-induced changes often have unintended consequences far downstream, such as when upriver dam construction starves beaches and deltas of sand and mud by trapping the sediment that formerly flowed to coastal environments. Learning to recognize and understand such connections is central to applied geomorphology, whether to aid in the design of resilient communities, develop more sustainable land-use practices, or construct measures to protect critical infrastructure.

Human activities have resulted in changes to a wide range of local and landscape-scale geomorphological processes. For example, at a local scale, clearing of forests has triggered the erosion of slopes and the accumulation of sediment in rivers. On a global scale, the influence of human activity is great enough that geologists have proposed that we are entering a new period of geologic time that they call the **Anthropocene Epoch** (the human era). Over the next century, human-induced changes in the global climate are predicted to cause increasingly variable weather, more frequent hurricanes, rising sea levels, and a host of related regional impacts, like the loss of winter snow pack critical for maintaining spring and summer stream flows in mountainous areas. Predicting the ways that landscapes respond to such changes will be central to planning societal adaptation and mitigation efforts.

Landscapes

Landscapes are suites of landforms that share a common genesis, contiguous location, and related history. The study of landscapes involves investigations over a tremendous range of spatial and temporal scales, from the movement of a sand dune or gravel bar on a riverbed in a single storm to the rise of the Himalaya and growth of the Tibetan Plateau over tens of millions of years [Figure 1.8]. Consequently, the approach, methods, and scale of analysis involved in any particular study need to be tailored to the questions the geomorphologist seeks to address. Landscapes can be divided into distinct units that can be studied over discrete periods of time; the relationships between processes and landforms are fundamental in the modern approach to geomorphology. In seeking to understand landscape evolution, it is important to match the types of measurements and the understanding of geomorphological processes to the spatial and temporal scales over which relevant processes act.

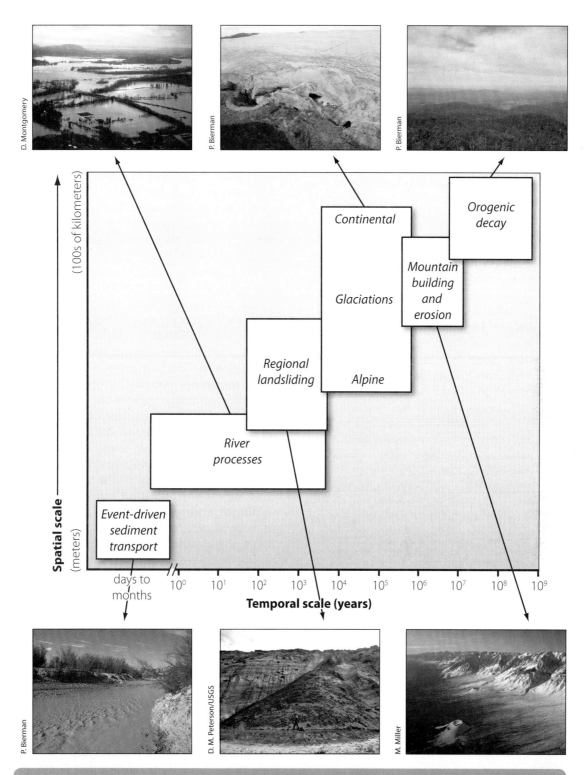

Geomorphically important processes occur on a variety of time and length scales, ranging from **event-driven** transport of sediment grains on a riverbed that happens in seconds and moves material meters, to the uplift and erosion of mountain ranges extending over hundreds of kilometers and taking millions to hundreds of millions of years.

FIGURE 1.8 Spatial-Temporal Range of Geomorphology. Scale is critical to understanding the geomorphology of Earth's surface. Some processes happen on short timescales and over small distances while others are imperceptibly slow and occur on the scale of continents.

Process and Form

The relationship between process and form lies at the heart of geomorphology. Stream flow, slope failure, moving glaciers, and blowing wind act to shape landscapes. At the same time, topography itself determines the style and rate of geomorphological processes. The shapes and orientations of large-scale landforms generally control the rates and distributions of the small-scale erosional and depositional processes that determine how landforms evolve over time. A geomorphologist can often read form to infer process, as in determining dominant wind directions from the shape and orientation of sand dunes. But in order to understand landforms and predict landscape response, we must understand the processes that form the landscape.

The geological and environmental history of a region can leave a lasting signature on landforms and on the processes operating on them. The physiographic signature of long-vanished glaciers still dominates the topography of formerly glaciated regions around the world. Thus, it should not be surprising that **process geomorphology** and **historical geomorphology** are complementary disciplines. Although many geomorphologists specialize in one or the other approach, no geomorphologist can afford to ignore either. Proposing testable hypotheses and interpreting field and laboratory data and the output of computer models in the evaluation of those hypotheses requires considering the action of surface processes over time because of the fundamental linkages between landscape history, process, and form.

Landscape dynamics and evolution result from the interaction of multiple systems that can be studied and understood in their own right. Considered broadly, the interaction of coupled tectonic, climatic, and erosional systems controls patterns of uplift, erosion, and sedimentation over geologic time. The topography of a mountain belt reflects the interaction of tectonic processes that raise rocks above sea level with the hydrologic and runoff processes that govern how erosion acts to shape slopes and incise valleys. At finer scales, many geomorphological processes reflect the interaction of ecological and hydrological systems, such as when the binding effect of tree roots helps to stabilize soils on landslide-prone slopes, or when logjams create dams that divert stream channels to new courses across their floodplains. Understanding landscapes often requires an appreciation of how processes interact in different regional contexts.

Spatial Scales

At global and continental scales, geomorphologists study major physiographic features like mountain belts, depositional basins, and great river systems like the Mississippi, Amazon, or Nile. At these scales, broad patterns in global climate and plate tectonics influence erosion and deposition that, in turn, influence the size and extent of mountains, plateaus, lowlands, coastal plains, and river basins. The resulting geomorphology reflects the relative importance of glacial, fluvial, aeolian (wind-driven), and coastal processes in shaping topography.

At a regional scale, tectonic forces have created distinct **physiographic provinces**, areas in which suites of geomorphological processes govern landscape formation and dynamics, producing similar landforms. In North America, examples of such provinces include mountain belts like the Sierra Nevada or Appalachian Mountains as well as features like the Great Plains, California's Central Valley, the Colorado Plateau in the Southwest, and the eastern Coastal Plain [**Figure 1.9**].

At more local scales, geomorphologists use a variety of techniques to define logical units for the analysis of landscape processes and history. In glaciated terrain, one might consider the area covered by an ice sheet or a valley glacier. Stretches of coastline with similar orientation or substrate (rocky or sandy) could be used for analysis of coastal landscapes. Sand source, wind direction, or the area covered in sand are logical units of analysis where aeolian processes dominate.

In areas where running water is the predominant agent of erosion and sediment transport on land, a **drainage basin** (the land surface area drained by a stream or river) is the logical unit for analysis of geomorphological processes. Small streams flow together to form larger rivers, so landscapes are naturally organized into smaller drainage basins nested within larger drainage basins. Drainage basins range in size from a headwater catchment that collects water from a single hillside to the Amazon River basin that drains more than half of South America. **Drainage divides** are topographic ridges that separate drainage basins.

At the scale of individual valley segments, the tectonic or climatic setting of a region often defines areas where different types of processes and/or histories have led to development of distinct landforms and dynamics. The difference in cross-sectional form between U-shaped valleys carved by glaciers and V-shaped valleys cut by streams is a classic example. There are systematic downstream changes in stream valley morphology, from erosional headwater streams with narrow valleys confined between bedrock walls, to the broad, unconfined valleys of depositional lowlands. Distinct suites of valley segment types are diagnostic of specific physiographic provinces.

At finer spatial scales, landscapes can be divided into distinct hillslopes, hollows, channels, floodplains, and estuaries. **Hillslopes** (including hilltops) are the undissected uplands between valleys that function as sources of sediment. **Hollows** are unchanneled valleys that typically occur at the head of channels in soil-mantled terrain. **Channels** are zones of concentrated water flow and sediment transport within well-defined banks. Valley bottoms are generally zones of sediment storage and include both active **floodplains** formed along river valleys by land that is inundated during times of high discharge under the present climate, and **terraces**, abandoned floodplains formed under conditions that differ from those at present. **Estuaries** are locations where streams entering coastal waters arrive at their ultimate destination—sea level. Each of these individual landscape

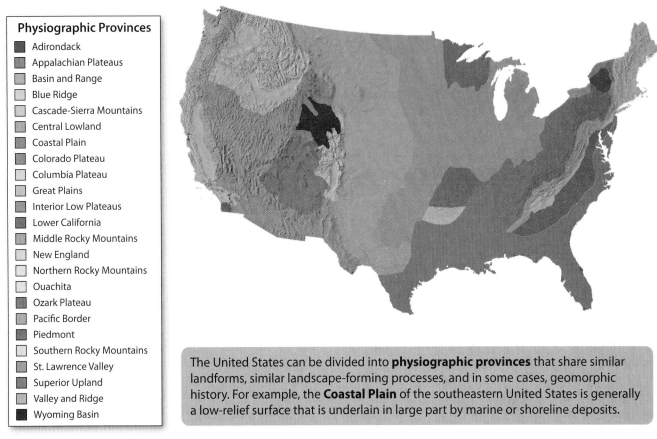

The United States can be divided into **physiographic provinces** that share similar landforms, similar landscape-forming processes, and in some cases, geomorphic history. For example, the **Coastal Plain** of the southeastern United States is generally a low-relief surface that is underlain in large part by marine or shoreline deposits.

FIGURE 1.9 Physiographic Provinces of North America. Physiographic provinces of the United States are defined on the basis of similar landforms, geomorphic histories, and active surface processes.

units exhibits a variety of specific landforms that are discussed in later chapters.

Temporal Scales

Interesting changes in the landscape occur over a variety of timescales. For example, topography evolves over periods of time that range from the few seconds, minutes, or days it takes for an earthquake-driven fault displacement, landslide, or flood to disrupt the land surface to the tens of millions of years required to erode away mountain belts. Climate cycles influence landscapes over millennia as glaciers advance, retreat, and scour out alpine valleys. Likewise, river profiles adjust to the sea-level changes that accompany glaciations. Large disturbances, like hurricanes and volcanic eruptions, often affect landscapes for centuries; it can take decades for landslide scars to revegetate and for river channels to process the sediment shed from slopes during intense, long-duration storms. River flow exhibits annual and seasonal variability that controls the timing of sediment movement.

Geomorphologists deal with a wide variety of measurements that represent process rates over very different timescales. For example, erosion rates directly measured in the field might be representative of months to years of geomorphic activity. In contrast, indirect estimates of rates of long-term landscape change inferred from isotopic analyses, or constrained by the volume of sediment preserved in depositional basins, can represent the average erosion rate over hundreds of thousands of years. Because of the disproportionate influence on erosion and sediment transport of infrequent extreme events like storms, landslides, and floods, rates of processes measured over short time spans may be very different from those measured over longer timescales. Since erosion and deposition include both chronic and episodic contributions to mass transport and landscape change, it is necessary to use a measuring scheme that samples over the magnitude and frequency of both types of contributions.

It is important to use measurements and analyses relevant to the timescale of interest because the timescale of observations can strongly influence geomorphic interpretations. For example, in the Blue Mountains of Australia, near Sydney, and in the mountains of Idaho, short-term rates of sediment export by streams are much lower than isotopic estimates that represent thousands of years of surface erosion and estimates of rock removal that consider millions of years of change. Geomorphologists reconciled this apparent difference by considering the impact of very large but very rare storms that move large amounts of sediment out of drainage basins. Because such storms recur hundreds of years apart, they are unlikely to influence short records, but they matter greatly over the long term.

Unifying Concepts

A number of core scientific concepts are central to contemporary geomorphology and guide our understanding of Earth's dynamic surface. Among these are the conservation of mass and energy, the routing of material through landscapes, force balances and thresholds in natural systems, equilibrium and steady state, and the relationship between the magnitude and frequency of geomorphic events. These concepts are fundamental to the discipline of geomorphology, and you will encounter them throughout this book.

Landscapes are created by the movement of matter through space. Gravity pulls water and sediment downhill. Thus, the routes that material takes from sources in eroding uplands to sinks in depositional lowlands and marine environments provide a common framework within which to consider the action of geomorphological processes across landscapes. The balance between driving and resisting forces allows us to quantify the way that stable landscape features may change state or become destabilized. The frequency with which events of a given magnitude occur is as important as the magnitude of events in considering their geomorphic significance. Together, these unifying concepts help structure a general, systematic approach to studying how Earth's surface processes interact to shape landscapes.

Conservation of Mass

Landscapes are dynamic systems through which mass is continually moving. Individual landforms can lose or gain mass over time; for example, rocky outcrops disaggregate and erode over time as they weather, but mass is conserved as it moves downslope. Eroded material is transformed by chemical reactions or physical abrasion as it moves, but it always ends up somewhere else. For example, a delta system at the mouth of a river grows because of input of sediment from upstream sources. Deltas erode away when gravity and ocean currents remove more sediment than the river supplies. Similarly, a mountain range erodes down to gentle hills over geologic time after tectonic uplift ceases because erosion removes mass from the mountain system. Tracking the flow of mass through geomorphic systems gives us a powerful framework for understanding Earth's surface processes.

In geomorphic systems, a simple yet very useful concept is that the difference between inputs (I) and outputs (O) during some time period equals the change in storage (ST):

$$I - O = \Delta ST \qquad \text{eq. 1.1}$$

where the Greek symbol delta, Δ, stands for change. This idea is simple and familiar. If you continually deposit money in the bank, your balance (storage) increases. If all you do is spend, you deplete your savings. Do this for long enough and you eventually run out of money. The same concept holds for soil, rock, and sediment. A mountain range eroding faster than tectonics raises it loses elevation. A river that receives more sediment than it can transport fills its valley with alluvial deposits.

Conservation of Energy

Energy is neither created nor destroyed but only transformed. Potential energy, the energy of position and configuration, like that of a boulder perched high on a cliff, may be converted into kinetic energy. Imagine that same boulder rolling off the cliff and crashing to the ground below. Its initial potential energy is transformed into kinetic energy as it falls and picks up speed (with a small loss to heat by friction in the air). When the boulder strikes the ground, the kinetic energy gained in the fall might shatter the rock or be imparted to anything it dislodges.

Rain or snowfall that is delivered to the land surface has potential energy of position. As water sinks into, collects upon, and runs off over the land surface, that potential energy is converted to kinetic energy. The kinetic energy is used to transport sediment and generates heat through friction. Water flows downhill under the influence of gravity, providing the energy for rivers and streams to maintain channels, transport sediment, and incise valleys. The frequency and magnitude of precipitation across a landscape—how much water falls where—determines the potential energy that is converted to kinetic energy. Most of the kinetic energy of the flow is dissipated by friction and turbulence generated as water moves across the channel bed and banks, leaving just a fraction to do the geomorphic work of transporting sediment and eroding bedrock.

Topographic slope sets the erosive potential of a landscape. Steep landscapes in general erode more quickly because the potential energy gradient (slope) controls the rate and magnitude of energy conversion, thereby linking process and form. The rugged Himalaya is eroding rapidly; the flat plains of central Australia are not.

Material Routing

Geomorphic systems route material from eroding sources to depositional sinks. In coastal systems, sand may move from beaches offshore to deeper water. In eolian systems, dust may originate in barren windy areas and be deposited in more heavily vegetated regions downwind. However, the one-way flow of water down river networks, of ice down glacial valley systems, and of soil and rock down hillslopes are the predominant agents of mass transport on the continents. Material eroded from the headwaters of a drainage basin eventually makes its way down through the river network and becomes the sediment supply for downstream channels before finding its way to depositional basins.

Tracking material from upland sources to lowland depositional sinks is particularly useful for considering how geomorphological processes interact to form and shape most landscapes. Headwater areas are generally steep, erosional regions where rocks weather and break down into transportable material. Sediment mobilized from upland slopes is delivered to streams and rivers and then transported to lowland depositional environments, to the coast, and ultimately to the deep ocean plains [Figure 1.10]. Because they are zones of erosion, steep headwaters are rarely preserved in the geologic record. Thus, to reconstruct the erosional history of continental highlands, we rely mainly on the geologic record of depositional environments where material stripped from erosional source areas was deposited near or below sea level. A source-to-sink framework shows how the erosion, transport, deposition, and storage of sediment drive landscape evolution and dynamics.

Force Balances and Thresholds

Force balances are fundamental to our understanding of geomorphic processes, particularly the balance between driving forces imparted by gravity and resisting forces offered by Earth materials. Force balance calculations help geomorphologists analyze a diverse set of processes, including, but not limited to, slope stability, the movement of sediment along the bed of a river, and the flow of ice down glacial valleys [Figure 1.11]. Generally, the ratio between driving and resisting forces is determined using vectors to resolve gravitational driving forces in the downslope direction; **constitutive equations** are used to quantify how earth materials like rock, soil, and ice resist deformation or failure.

In considering the relationship between an applied force and the surface it acts upon, it is useful to resolve the force into normal and shear components that respectively act orthogonal to (normal) and parallel to (shear) the land surface. We use trigonometry to resolve forces arising from the unit weight (the product of density, ρ, and gravitational acceleration, g) of a column of rock, soil, water, or ice of thickness z, on a surface, like a hillside or riverbed that slopes at an angle θ, into normal and shear components. The **normal force** per unit area, the stress, is given by

$$\rho \, g \, z \, \cos\theta \qquad \text{eq. 1.2}$$

The normal stress acts into the surface to hold material in place on slopes and resist downhill movement on slopes through solid friction. The shear stress is given by

$$\rho \, g \, z \, \sin\theta \qquad \text{eq. 1.3}$$

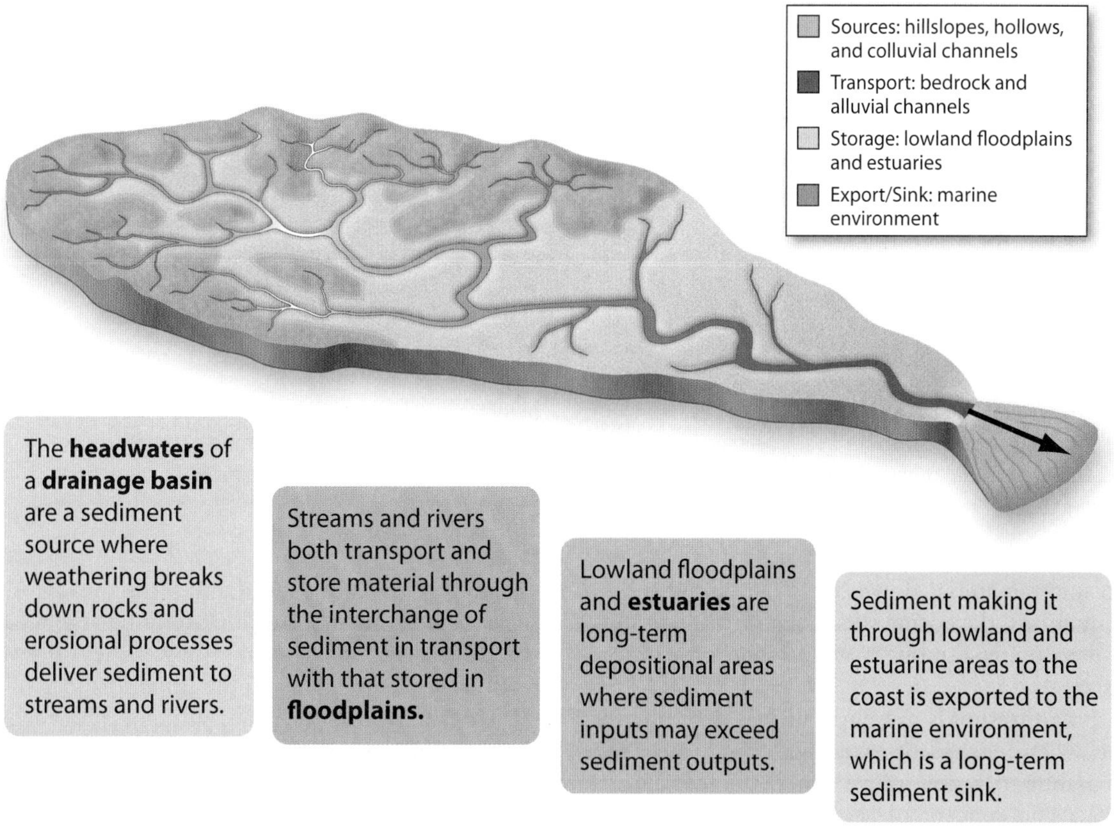

- Sources: hillslopes, hollows, and colluvial channels
- Transport: bedrock and alluvial channels
- Storage: lowland floodplains and estuaries
- Export/Sink: marine environment

The **headwaters** of a **drainage basin** are a sediment source where weathering breaks down rocks and erosional processes deliver sediment to streams and rivers.

Streams and rivers both transport and store material through the interchange of sediment in transport with that stored in **floodplains**.

Lowland floodplains and **estuaries** are long-term depositional areas where sediment inputs may exceed sediment outputs.

Sediment making it through lowland and estuarine areas to the coast is exported to the marine environment, which is a long-term sediment sink.

FIGURE 1.10 Drainage Basins: Source-to-Sink. It is useful to understand Earth's surface from the perspective of drainage basins—units of the landscape in which mass can be accounted and conserved. In general, sediment is sourced in the eroding uplands and deposited in lowland sinks, although it may be stored temporarily in floodplains along the way to long-term storage in marine environments.

Many surface processes are driven by gravity. To calculate force balances and determine whether a landscape element will be stable, it is helpful to resolve the downslope force into the components resisting and driving movement, respectively oriented perpendicular to the slope (**normal force,** ⟶) and parallel to the slope (**shearing force,** ⟶).

Particle of sediment on streambed

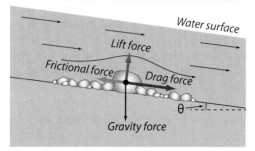

Sediment on a stream bed is subject to a variety of forces. The **gravity force** holds the grain on the bed opposing the **lift force** generated by the current. The current also applies a **drag force** to the grain. The drag force is resisted both by **frictional forces** and by the resistance offered by neighboring grains if the clast is embedded. When the lift and drag forces exceed the forces of gravity and friction, the grain of sediment moves.

Boulder on hillslope

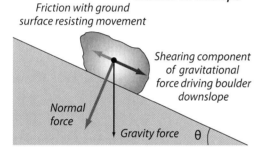

A boulder on a hillslope remains stable as long as the frictional force, holding the boulder in place, exceeds the **driving force,** in this case a **shearing force** parallel to the slope. Once the driving force exceeds the frictional force, down goes the boulder. Earthquakes, tectonic tilting, and slope undercutting can increase the driving force or reduce the normal force.

Flowing glacier

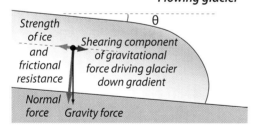

Glaciers flow by deforming and sliding along their bed. The driving force is the shear stress governed by the slope of the ice surface. The **resisting forces** include the strength of ice to resist internal deformation and the frictional resistance to sliding of material at the bed of the ice.

FIGURE 1.11 Force-Balance Diagrams. Force balances are a physical way of understanding Earth surface processes. The dynamic interaction between driving forces that tend to move objects and resisting forces that tend to prevent movement can be used to explain a diverse range of processes.

and acts to impel material downslope. The difference in these two equations, the cosine versus the sine function, is fundamental to understanding a wide range of geomorphic processes.

Geomorphic **thresholds** are physical or chemical conditions that, when reached or exceeded, trigger a change in state or a shift to a new range of average conditions [Figure 1.12]. The concept of geomorphic thresholds is important because many landscape processes and landforms are prone to the influence of individual events. The idea is that a geomorphic system may remain stable until an event of sufficient strength tips the balance, causing the system to cross a threshold and settle into another steady state. For example, a hillslope may be stable for decades until a very large rainstorm saturates the soil, reducing the resisting force and causing the slope to slide.

Thresholds are particularly important when considering a variety of geomorphic environments and processes including mass movements, sediment transport, and volcanic eruptions. A central implication of the concept of geomorphic thresholds is that small changes can trigger large responses. Changes triggered by crossing geomorphic thresholds sometimes cause a cascade of responses, referred to as a **complex response,** in which different parts of a

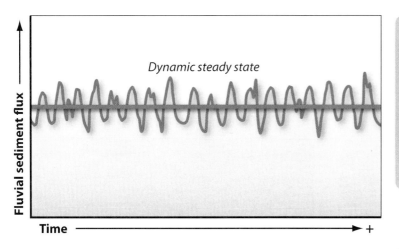

The flux of sediment from an undisturbed drainage basin changes over the short term as rainstorms come and go, individual hillslopes fail in mass movements, and riverbanks collapse. Over the long term, the flux of sediment from a drainage basin oscillates around a mean value, producing a **dynamic steady state,** unless there are significant changes in **boundary conditions,** such as climate, vegetation cover, or uplift rate.

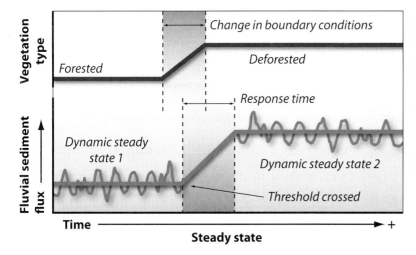

When boundary conditions change significantly, geomorphic systems adjust. Such adjustment does not happen instantaneously, rather it lags the change in boundary conditions, over a **response time.** In this case, deforestation and land conversion to agriculture increased the fluvial sediment flux to a new and higher dynamic steady state because soils are now disturbed by plowing and thus more vulnerable to erosion.

FIGURE 1.12 Steady State, Thresholds, and Response Time. Many geomorphic systems are in steady state, their central tendencies oscillating in equilibrium around a mean value. When external factors, such as climate or base level, change, the system can cross a threshold and, after a certain response time, change to a new and different state, in which the system may oscillate around a different mean.

system reach the threshold conditions at different times. The responses may be out of phase in different portions of the channel network, such as downstream channels that incise while upstream reaches remain sediment-choked.

Equilibrium and Steady State

The idea of balance, or **equilibrium**, between landforms and geomorphological processes provides a useful conceptual framework through which to study landscape evolution, as well as a reference frame for identifying and understanding nonequilibrium landforms and landscapes. It is often useful or convenient to assume an equilibrium landscape that does not change over time—a condition referred to as **steady state**. Equilibrium is, however, often not **static**, but is rather a **dynamic steady state** (see Figure 1.12), with landscape characteristics that vary over time around a central tendency. Steady state is strongly scale-dependent. The average slope of a mountain range, for example, remains constant if erosion and rock uplift rates are equal over time, even if individual erosional events greatly change local slopes in the short term. The timescales over which topography equilibrates to changes in landscape-forming processes range from seasonal resurfacing of gravel streambeds following winter storms to the tens of millions of years it can take to erode mountain ranges.

Landscapes may appear unchanging, but considered geologically, topography is dynamic because material is constantly being entrained, transported, and deposited. Over centuries to millennia, such changes result in a **dynamic equilibrium** that maintains topographic forms in an average sense even as individual slopes experience landslides; coastal landforms shift with currents, tides, and storms; and rivers migrate across their floodplains. Over longer timescales, sometimes referred to as **cyclic time**, landforms evolve in concert with tectonic and climatic changes. For example, once tectonically driven rock uplift ceases, mountains slowly erode away and average slopes decline. The **response time** of a landscape or landform to changes in driving or resisting forces varies greatly

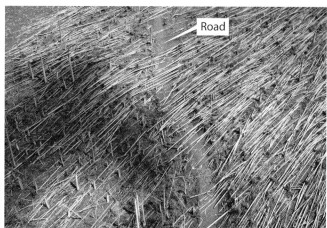

PHOTOGRAPH 1.8 Disturbances Come in Many Different Length Scales. (a) Blowdown of a single tree in the Adirondack Mountains of New York State disturbs several tens of square meters. Two-meter-high student for scale. (b) When Mount St. Helens erupted on May 18, 1980, much of the volcanic edifice collapsed and many square kilometers of trees were blown down by the volcanic explosions. Aerial view of timber blown down by eruption, with forest road for scale (Skamania County, Washington).

depending on the type of change, the rates of landscape-forming processes, and the resistance of the landscape to a particular change. All landscapes reflect the shifting balance between driving and resisting forces, and the response time to changes in either.

Recurrence Intervals and Magnitude-Frequency Relationships

Most changes on Earth's surface happen in response to discrete events that disrupt the land surface, move mass, and change surface forms. Geomorphic disturbances come in a wide range of scales, from massive volcanic eruptions and hurricanes with 200 km/hr winds and storm surges several meters high to a lone tree falling over, mixing the soil held by its roots [Photograph 1.8]. Both the frequency of disturbances and the duration of geomorphic events vary over many orders of magnitude.

Geomorphic events, while randomly distributed in time, generally have a characteristic **recurrence interval**, the average time between events of a similar magnitude. Take, for example, a flood that is just barely capable of overtopping the banks of a river, a level termed **bankfull flow**. A flood of this magnitude occurs, on average, once every year or two in most humid, temperate zone streams, while the recurrence interval of such bankfull flow may exceed 50 years in arid-region channels. Likewise, landslides on hillslopes have typical recurrence intervals of centuries to millennia. Many events that are important in shaping landscapes, including earthquakes, floods, and hurricanes, have discrete **magnitude-frequency relationships** that quantify the relationship between the size of an event and the chance that it will occur in a given time period [Figure 1.13].

It is important to note that different environments can have different recurrence intervals for the same process, a result of differing climate and tectonic setting. Furthermore, sometimes it is an infrequent event that shapes the landscape (for example, glaciation) and sometimes it is a frequent event, such as the annual flood.

Applications

The relationship between geomorphological processes and landforms has applications across all aspects of the human endeavor, from feeding a growing population sustainably to disaster preparedness, land-use planning, ecological restoration, and planetary exploration. Fundamental understanding of geomorphological processes lays the foundation for applied geomorphology. The dynamic nature of Earth's surface means that geomorphology has

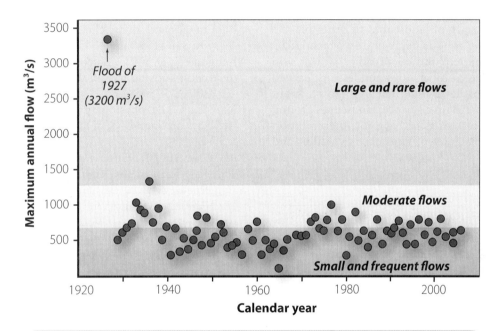

Even in humid, well-watered northeastern North America, there is great variability in maximum **annual flood** flows. On the Winooski River (~2700 km²) in northern Vermont, the largest annual flood (1927) was almost 20 times greater than the smallest annual flood (1963) and caused immense damage and channel change.

FIGURE 1.13 Magnitude-Frequency Relationships. At Earth's surface, small, low-intensity events are common, but rare, large-extent or high-intensity events may dramatically affect the landscape.

important ties to other disciplines, from civil engineering to agriculture and ecology.

Geomorphology plays a key role in natural hazard assessment, prevention, and recovery. Our ability to predict how landslides, floods, earthquakes, storms, and sinkholes will change Earth's surface is essential in efforts to minimize loss of life and property destruction, to mitigate or repair damage, and to modify our behavior to minimize future damage from these hazards. Geomorphology is an essential foundation for rational land-use planning and landscape management, particularly for landscape-scale natural resource management, like forestry and agriculture. In particular, understanding how natural processes shape landforms and river systems is a key part of developing strategies that mitigate human impacts and guide ecological restoration efforts.

The ability to read the landscape has proven invaluable for military planning for millennia. History is replete with examples of how knowledge of geomorphology led to military triumph, and how ignorance or neglect led to disaster. Thick layers of dust hidden just below rocky desert surfaces in Iran, Iraq, and Afghanistan have plagued helicopters that were originally designed to maneuver and deploy troops in the humid and muddy, but dust-free, jungles of Vietnam. Hannibal crossed the Alps in 218 BCE and lost thousands of troops and many of his elephants when they were overwhelmed by avalanches. History repeated itself during World War I, when many troops were buried by avalanches in northern Italy, some triggered intentionally on steep mountain slopes by the opposing army. Knowing what to expect from the terrain has proven invaluable time and again throughout history.

Spectacular advances in planetary geomorphology have extended the study of landscapes to other planets and moons, including robotic fieldwork on Mars. Exciting challenges in the exploration of other worlds will continue to be informed by analogs and lessons learned here on our home planet. While much is known about the nature of Earth's dynamic surface, landscapes around the world still harbor stories waiting for interpretation using the general principles of landscape evolution and dynamics.

Selected References and Further Reading

Bull, W. B. *Geomorphic Responses to Climate Change.* New York: Oxford University Press, 1991.

Butler, D. R. *Zoogeomorphology: Animals as Geomorphic Agents.* New York: Cambridge University Press, 1995.

Chorley, R. J., ed. *Water, Earth, and Man: A Synthesis of Hydrology, Geomorphology, and Socio-Economic Geography.* London: Methuen, 1969.

Chorley, R. J., A. J. Dunn, and R. P. Beckinsale. *The History of the Study of Landforms, or the Development of Geomorphology.* New York: Wiley, 1964.

Coates, D. R., and J. D. Vitek, eds. *Thresholds in Geomorphology.* London: Allen and Unwin, 1980.

Darwin, C. *The Formation of Vegetable Mould, Through the Action of Worms, With Observations on Their Habits.* London: John Murray, 1881.

Davies, G. L. *The Earth in Decay: A History of British Geomorphology, 1578–1878.* New York: American Elsevier, 1969.

Derbyshire, E., ed. *Geomorphology and Climate.* New York: Wiley, 1976.

Dietrich, W. E., C. J. Wilson, D. R. Montgomery, J. McKean, and R. Bauer. Erosion thresholds and land surface morphology. *Geology* 20 (1992): 675–679.

Gardner, T. W., D. W. Jorgensen, C. Shuman, and C. R. Lemieux. Geomorphic and tectonic process rates: Effects of measured time interval. *Geology* 15 (1987): 259–261.

Hack, J. T. Interpretation of erosional topography in humid temperate regions. *American Journal of Science* 258-A (1960, Bradley Volume): 80–97.

Hassan, M. A., A. S. Gottesfeld, D. R. Montgomery, J. F. Tunnicliffe, G. K. C. Clark, G. Wynn, H. Jones-Cox, R. Poirier, E. MacIsaac, H. Herunter, and S. J. Macdonald. Salmon-driven bed load transport and bed morphology in mountain streams. *Geophysical Research Letters* 35(2008), doi:10.1029/2007GL032997.

Hooke, R. L. On the history of humans as geomorphic agents. *Geology* 28 (2000): 843–846.

Kirchner, J. W., R. C. Finkel, C. S. Riebe, D. E. Granger, J. L. Clayton, J. G. King, and W. F. Megahan. Mountain erosion over 10 yr, 10 k.y., and 10 m.y. time scales. *Geology* 29 (2001): 591–594.

Molnar, P., and P. England. Late Cenozoic uplift of mountain ranges and global climate change: Chicken or egg? *Nature* 346 (1990): 29–34.

Montgomery, D. R., G. Balco, and S. D. Willett. Climate, tectonics, and the morphology of the Andes. *Geology* 29 (2001): 579–582.

Raisz, E. J. The physiographic method of representing scenery on maps. *Geographical Review* 21 (1931): 297–304.

Reid, L. M., and T. Dunne. *Rapid Evaluation of Sediment Budgets*, Reiskirchen, Germany: Catena-Verlag, 1996.

Summerfield, M. A., ed. *Geomorphology and Global Tectonics.* New York: Wiley, 2000.

Syvitski, J. P. M., C. J. Vörösmarty, A. J. Kettner, and P. Green. Impact of humans on the flux of terrestrial sediment to the global coastal ocean. *Science* 308 (2005): 376–380.

Thomas, W. L., Jr, ed. *Man's Role in Changing the Face of the Earth.* Chicago: University of Chicago Press, 1956.

Thornes, J. B., and D. Brunsden. *Geomorphology and Time.* New York: Wiley, 1977.

Whipple, K. X. The influence of climate on the tectonic evolution of mountain belts. *Nature Geoscience* 2 (2009): 97–104.

Wilkinson, B. H., and B. J. McElroy. The impact of humans on continental erosion and sedimentation. *Geological Society of America Bulletin* 119 (2007): 140–156.

Willett, S., C. Beaumont, and P. Fullsack. Mechanical model for the tectonics of doubly vergent compressional orogens. *Geology* 21 (1993): 371–374.

Willett, S. D., and M. T. Brandon. On steady states in mountain belts. *Geology* 30 (2002): 175–178.

Wilson, L. Morphogenetic classification. In R. W. Fairbridge, ed., *Encyclopedia of Geomorphology.* New York: Reinhold Book Corporation, 1968.

Wolman, M. G., and J. P. Miller. Magnitude and frequency of forces in geomorphic processes. *Journal of Geology* 68 (1960): 54–74.

DIGGING DEEPER Why Is Earth Habitable?

Earth is the only planet we know of that harbors life. So far, our planet appears to be a special place where the interactions and feedbacks among the solid Earth, the hydrosphere, the biosphere, and the atmosphere provide conditions favorable for living organisms.

What makes Earth so special? Earth is habitable because liquid water is stable on its surface; it is large enough to retain an atmosphere and some internal heat; and it has a composition of long-lived **radioactive elements**, along with the original accretionary heat, to warm the planet's interior even after 4.5 billion years. This heat warms and softens interior rocks, sustains mantle flow, and thereby drives plate tectonics. The same heat keeps the core partially molten, allowing for the generation of a strong magnetic field. Without our magnetic field and thick atmosphere, Earth's surface would be bombarded by life-threatening levels of **cosmic radiation**.

Together, these characteristics enable active tectonic and volcanic processes to continue. **Tectonism** and **volcanism** are critical for life because they recycle **volatile elements** and compounds critical for carbon-based life from the geosphere to the atmosphere and hydrosphere—making Earth's surface dynamic and life-supporting. Life is capable of reducing **entropy** (disorder) and driving chemical reactions that are not thermodynamically favorable. For example, without life, atmospheric oxygen is unlikely to occur in Earth's atmosphere because it is easily and rapidly consumed in **oxidation** (weathering) reactions with rocks.

Life as we know it requires the presence of liquid water and is thus limited to average temperatures in the range of approximately −15°C to 115°C. The **habitable zone** around a star is defined as the range of distances at which surface temperatures on a planet could potentially support liquid water. For our solar system, the habitable zone extends from about 0.84 to 1.7 times Earth's distance from the Sun, a range that includes Earth and Mars but not Venus.

The presence of liquid water is not only important for life itself, but it is also important for the **differentiation** of **continental crust** through the formation of granitoid rocks (Campbell and Taylor, 1983). In our solar system, Earth is the only inner planet with abundant water and the only known planet with continents. The formation of continents occurred as partial melting of oceanic crust released silicon- and potassium-rich, felsic magmas, a process that depends on water being carried (by **subduction**) into the mantle to initiate **partial melting**. Creation of extensive continents with a density low enough to stand above sea level requires the transport of large amounts of water into the upper mantle and thus requires tectonics. Continental rocks weather subaerially (under the atmosphere), consuming carbon dioxide (CO_2) and water, and thus in part control the composition of Earth's atmosphere [**Figure DD1.1**].

Planetary atmospheres are a key part of the habitability puzzle, in part because they buffer thermal swings that would otherwise occur between night and day and between seasons.

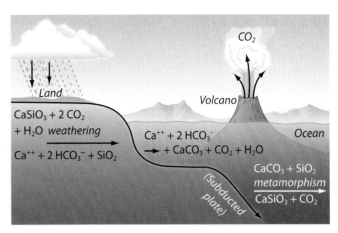

FIGURE DD1.1 This diagram shows the global cycle of carbonate and silicate, which is driven by surface weathering and plate tectonics. Silicate rocks exposed on land weather as they combine with water and carbon dioxide (CO_2). The weathering products, those that are not left on the land, enter the ocean and are eventually subducted by plate tectonics. Once subducted, CO_2 and water are released back to the atmosphere by volcanism and the cycle continues. Without plate tectonics, weathering on Earth's surface would slow and the atmospheric levels of CO_2 would be quite different. [From Kasting and Catling (2003).]

The mass of a planet influences its potential to retain an atmosphere, a seemingly critical ingredient for life; small planets and moons have insufficient gravity to hold an atmosphere. For example, the atmospheric density of Mars, which is one-third the size of Earth, is 1/100th that of Earth and does not provide much insulation (it gets very cold on Mars); nor does the Martian atmosphere provide much radiation shielding. In contrast, the atmosphere of Venus is 100 times thicker than that of Earth but is made of greenhouse gases, primarily CO_2, that make it too hot there for life. On Earth, the weathering of rocks and tectonic cycling of crustal materials absorb the greenhouse gas carbon dioxide from the atmosphere and thus stabilize global climate (Kasting and Catling, 2003; see Figure DD1.1), keeping Earth cooler than Venus. Most of the carbon on Earth is sequestered in limestones and organic fuels including coal and oil. Early life made Earth more habitable for later organisms by sequestering carbon and reducing atmospheric CO_2 content.

Humans have been inadvertently tinkering with the systems that stabilize our planet's environment for millennia. The dramatic rise of many different human civilizations occurred over the past 11,500 years, the **Holocene Epoch**, a period of generally stable climate and atmospheric composition [**Figure DD1.2**]. After sea level stabilized about 6000 years ago, agriculture spread into highly productive river deltas and preindustrial, human-induced changes to Earth's surface included regional deforestation and the erosion of soils that followed. Studying the record of gases preserved in ice cores and compiling historic data,

DIGGING DEEPER Why Is Earth Habitable? *(continued)*

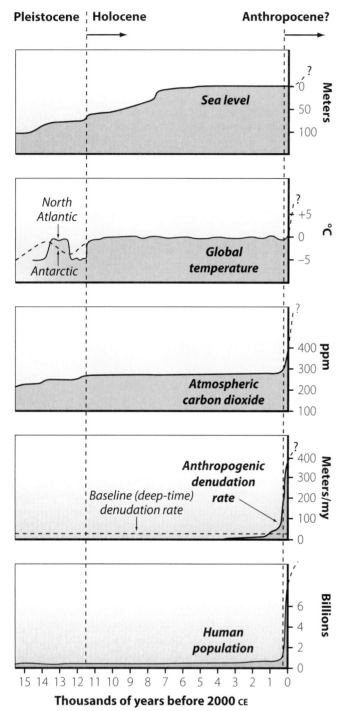

Ruddiman et al. (2008) and Ruddiman and Ellis (2009) hypothesized that flooding of rice paddies and domestication of livestock may have led to subtle changes in atmospheric composition thousands of years before the Industrial Revolution [**Figure DD1.3**]. Today, changes in atmospheric composition (primarily increases in CO_2 and CH_4) are directly tied to human activities, and the result is clear—unmistakable changes in global temperature, the intensity of the hydrologic cycle, and the distribution and intensity of storms such as hurricanes.

Over the past century, with the advent of mechanized agriculture, irrigation, and industrial fertilization, the human impact on global systems became more substantial and readily detectable. Compiling data about the rate at which humans, now powered by fossil fuels, move soil and sediment, Hooke (1994) suggested that humans are a major geomorphic agent on the planet, moving huge amounts of mass on and from uplands. At the same time, people have littered the planet with dams that are trapping much of this sediment. Compiling sediment yield data and land-use data, Syvitski and Milliman (2007) argued that more than one-quarter of the sediment coming off the continents is trapped in dams and not reaching the oceans. The consequences include sediment starvation for some of the world's deltas and beaches.

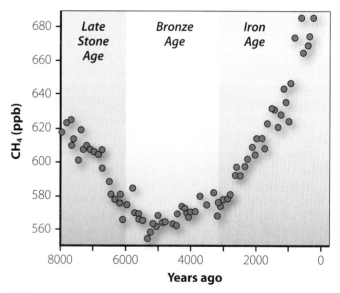

FIGURE DD1.2 The Holocene Epoch, the past 11,500 years, has until recently been a period of great stability on Earth. Global temperatures and atmospheric carbon dioxide levels have changed relatively little since the end of the last glaciation. Human population and global erosion rates were also steady and low until the last few thousand years, when both began to rise. During the twentieth century, after the Industrial Revolution, human population increased explosively and erosion rates have skyrocketed as people rapidly change our planet's surface. [From Zalasiewicz et al. (2008).]

FIGURE DD1.3 Ancient glacial ice in Antarctica and Greenland preserves a long record of Earth's atmospheric composition. Here, Ruddiman et al. (2008) present the trend of methane gas (CH_4) composition in the Dome C ice core from Antarctica. They suggest that the decline from 8000 to 5000 years ago represents a natural trend and that the reversal of this trend about 5000 years ago represents the addition of methane to the atmosphere coinciding with human activities, primarily agriculture, during the Bronze and Iron ages. If Ruddiman is right, humans have been affecting Earth's climate for several thousand years. [From Ruddiman et al. (2008).]

The impact of more than 7 billion people on Earth's surface, atmosphere, and geochemical systems threatens to destabilize the linked planetary systems and cycles on which society critically depends (Rockström et al., 2009). How much is too much impact on our planet? Rockström et al. proposed that there is a "safe operating space for humanity" on planet Earth [**Figure DD1.4**]. To define this safe space, they identified nine planetary systems and associated thresholds, which include geochemical cycles, the climate, the biosphere, the hydrosphere, and the atmosphere. If human impacts to Earth exceed reasonable limits, they argue that the stable, habitable planet on which humans evolved and societies have flourished may become unstable. In other words, the planet on which we depend may no longer be able to support us. In their view, we have already exceeded three boundaries: climate change, nutrient cycling, and losses from the biosphere. We are fast approaching other boundaries through acidification of the oceans, land-use changes, and freshwater use. Given that many Earth systems react abruptly to external changes, crossing thresholds could result in unexpected and potentially painful consequences like the rearrangement of ocean circulation (and thus regional climates) or rapid sea-level rise as ice sheets melt catastrophically.

There is solid evidence from past periods when Earth systems changed state radically that such thresholds exist. Probably the best known manifestation of a climate threshold in the recent geologic past is the abrupt Northern Hemisphere cooling that occurred just as the Earth was warming from the last glacial period. In this period, called the **Younger Dryas**, at least the area around the North Atlantic plunged back into near-glacial conditions for more than a millennium. The driving force may have been a change in ocean circulation, last triggered at the end of the Pleistocene Epoch by rapidly melting ice sheets (Alley, 2004).

Earth did not come with a user's manual. Yet to keep our planet habitable, we need to understand how surface processes act, and interact, to sustain or limit critical ecological, climatological, and hydrological systems on which we all depend. However one looks at it, a deeper and broader understanding of the linked processes, feedbacks, and history of Earth's dynamic surface is a useful prerequisite to learning how to live within our planetary means.

Alley, R. B. Abrupt climate change. *Scientific American* 291 (2004): 62–69.

Campbell, I. H., and S. R. Taylor. No water, no granites— No oceans, no continents. *Geophysical Research Letters* 10 (1983): 1061–1064.

Hooke, R. L. On the efficacy of humans as geomorphic agents. *GSA Today* 4 (1994): 217, 224–225.

Kasting, J. F., and D. Catling. Evolution of a habitable planet. *Annual Review of Astronomy and Astrophysics* 41 (2003): 429–463.

Rockström, J., et al. A safe operating space for humanity. *Nature* 461 (2009): 472–475.

Ruddiman, W. F., Z. Guo, X. Zhou, H. Wu, and Y. Yu. Early rice farming and anomalous methane trends. *Quaternary Science Reviews* 27 (2008): 1291–1295.

Ruddiman, W. F., and E. C. Ellis. Effect of per-capita land use changes on Holocene forest clearance and CO_2 emissions. *Quaternary Science Reviews* 28 (2009): 3011–3015.

Syvitski, J. P. M., and J. D. Milliman. Geology, geography, and humans battle for dominance over the delivery of fluvial sediment to the coastal ocean. *Journal of Geology* 115 (2007): 1–19.

Zalasiewicz, J., et al. Are we now living in the Anthropocene? *GSA Today* 18, no. 2 (2008): 4–8.

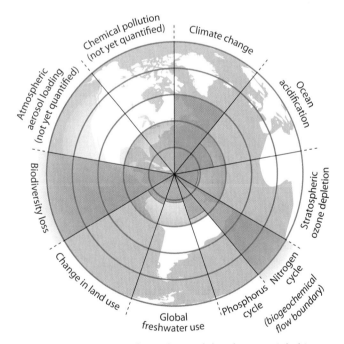

FIGURE DD1.4 Humans have changed the planet we inhabit. How do we know if this impact is significant and when it might become hazardous to the world's societies? Rockström et al. (2009) identified nine major effects that humans have on planetary systems and then sought consensus on the level of impact we could have without significantly disturbing the operation of our planet (the green zone). The orange shaded slices represent the current position of each system. Look carefully and you will see that human impact has taken the planet out of the safe operating zone for biodiversity, nitrogen cycling, and climate change. [From Rockström et al. (2009).]

WORKED PROBLEM

Question: Calculate and then describe how the normal force and the shear force per unit area (stress) vary as hillslope angle increases. Make two graphs that show the change in normal and shear stress as a function of slope angle.

Answer: The normal and shear stresses on a hillslope vary inversely. When one goes up, the other goes down. Equations 1.2 and 1.3 describe the change in shear and normal stresses as a function of slope angle. As slope increases, the normal stress declines and the shear stress increases. Eventually, this increase in shear stress and decrease in normal stress lead to hillslope failure in what could be a dramatic landslide or rock fall. The normal and shear stresses are the same on a 45° slope.

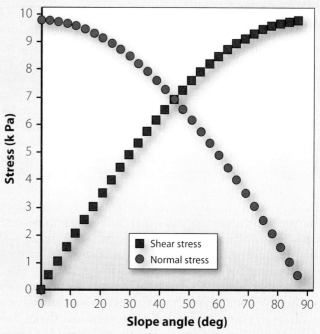

Shear and normal stresses calculated using equations 1.2 and 1.3 and assuming a 1-m-thick slab, a soil bulk density of 1000 kg/m³, and an acceleration of 9.8 m/s².

KNOWLEDGE ASSESSMENT Chapter 1

1. Define geomorphology.
2. Geomorphology draws on four disciplines to explain the behavior and history of Earth's surface. Name them.
3. What two contrasting forces does topography reflect?
4. Define the geosphere.
5. What controls the location of the first-order features of Earth's surface?
6. What are the most common elevations of Earth's topography?
7. What controls the location of mountain systems?
8. Define isostasy, explain how it works, and consider why it is important to geomorphology.
9. How thick are the roots of some large mountain ranges?
10. Explain how erosion causes uplift of rocks.
11. Why are mountain ranges long-lasting topographic features?
12. Compare and contrast the rate of erosion and the rate of land surface elevation change.
13. Describe a cratonic landscape.
14. What does a rift zone look like at Earth's surface?
15. Why are hot spots geomorphic agents, shaping topography?
16. How does rock type (lithology) influence the shape of Earth's surface?
17. What is the hydrosphere?
18. Sketch the hydrologic cycle.
19. How do plants influence overland flow?
20. How does the distribution of precipitation help to determine the shape of Earth's surface?
21. What is the primary driver of Earth's climate?
22. Describe a convective cell and predict how such cells affect Earth's large-scale climate and geomorphology.
23. Predict the distribution of rainfall across a mountain range.
24. Where do you find the biosphere?
25. What does the biosphere contain?

○ 26. Give three examples showing how the biosphere influences Earth's surface processes.
○ 27. In what ways have humans become geomorphic agents?
○ 28. Give an example of how geomorphic processes and landforms are related.
○ 29. At what temporal and spatial scales do geomorphologists work?
○ 30. How is the concept of steady state applied to landscapes?
○ 31. List the unifying concepts underlying modern geomorphology.
○ 32. Define a drainage basin.
○ 33. Describe the conservation of mass and why it is important for understanding Earth's surface processes.
○ 34. Why are thresholds important in geomorphology? Provide an example.
○ 35. Explain the idea of "source-to-sink."
○ 36. What is a force balance and why is the concept useful in geomorphology? Give an example.

A Brief History of Geomorphology

Since the establishment of geomorphology as a distinct scientific discipline in the late nineteenth century, geomorphic thinking has evolved from broad conceptual ideas ("conceptual models") of the origins and dynamics of landscapes and how fluvial and glacial processes shaped landforms to mathematical models of landscape evolution that allow formal evaluation of how day-to-day and rare extreme events can shape landscapes. Field observations, mapping of surficial deposits, and the evolution of measurement technology drove thinking and provided both a foundation for and a means to test theoretical models.

Since the 1950s, geomorphologists extended the qualitative insights of prior workers to develop a quantitative, physics-based understanding of how geomorphological processes erode, transport, and deposit sediment and in so doing shape landforms. Today, prediction of landscape response is now a common goal, although the push to develop predictive models faces significant challenges because of the number of underconstrained but important variables in many geomorphic systems. Since the time of the ancient Greeks, our view of landscapes has shifted from descriptions and imaginative theories regarding the origin of topography to the analysis of topographic contours, usually focusing on the profiles of individual hillslopes and rivers, to fully three-dimensional investigations and simulation models of topographic change over entire landscapes. Today, geomorphologists explore the interactions among climate, tectonics, and erosion in shaping Earth's dynamic surface.

Classical Knowledge

We recognize the geographers and philosophers of classical Greece as the first geomorphologists. They realized that Earth was a globe and that its surface evolved over unimaginably long time spans. For example, the historian and geographer Herodotus (c. 484 BCE–c. 430 BCE) described the striking contrast between the rich, black alluvial soils of the Nile River delta and the bare rocky soils of Libya and Syria. He concluded that the Egyptian coast advanced out into the Mediterranean Sea as the Nile deposited its load of silt, and calculated the age of the Nile delta from his estimated rate of sediment deposition. The annual flood of the Nile was so important to the prosperity of ancient Egypt that the rise and fall of the river was carefully monitored and recorded [**Figure A**]. The great philosopher Aristotle (384 BCE–322 BCE) argued that the land and sea constantly swapped places as rivers carried silt and sand to the sea, gradually filling it in, causing sea level to rise and submerge coastal land. Aristotle thought that an endless cycle in which land became sea, and then land again, happened so slowly as to escape observation as civilizations rose and fell in a world without beginning or end. His view of an ancient, eternally changing world did not appeal to those

FIGURE A Nilometer used to measure the stage of the Nile River in ancient Egypt. The chamber was connected to the river. As the river level rose, so would the level of water, covering stairs one by one as the flood crested. Flood heights were quantified by the number of stairs that were under water.

convinced that God created the world just a few thousand years ago.

Nicolaus Steno (1638–1686)

In the late seventeenth century, Danish physician Niels Stensen, better known by his Latinized name Nicolaus Steno, laid the foundation for modern geology, including geomorphology, while working for the Grand Duke of Tuscany in Florence. Steno recognized the organic nature of fossils and proposed a foundational principle of modern geology known as **Steno's law of superposition**, the idea that the oldest sedimentary layers are on the bottom and the youngest are on top. He also recognized that sedimentary rocks are deposited horizontally and used these principles to infer the geologic and physiographic history of the landscape of northern Italy, proposing a conceptual landscape-evolution model based, in part, on the biblical flood (Steno, 1669). Steno saw how layered rock containing marine fossils lay high above sea level, and that while some layers lay flat, others were contorted or lay pitched at steep angles.

His conceptual model to explain the modern geology and topography of the region involved two rounds of collapse in which continents fell into great subterranean caverns [**Figure B**]. Geologists respect Steno because he began an ongoing process of formalizing geological observations and interpreting evidence of Earth history through a set of guiding principles. Likewise, geomorphologists recognize

Charles Lyell (1797–1875)

In the early nineteenth century, it was well established that the world was significantly older than the 6000 years suggested by biblical chronology, but conventional wisdom still held that evidence for the biblical flood was preserved in the world's surficial deposits and in the incision of valleys. The creation of modern topography during the most recent of many grand catastrophes was thought to have ushered in the most recent era of geologic time.

Lyell hypothesized that today's surface processes, if they had acted over vast expanses of time, could shape landscapes. Touring the Auvergne region of France, he was intrigued by deep valleys carved into stacked lava flows topped by delicate cinder cones. A global flood powerful enough to have carved the rivers down through solid basalt would surely have swept away the loose cinder cones. Lyell recognized how explaining deposits of river gravel buried beneath lava flows required multiple episodes of valley incision, gravel deposition, and burial by lava flows [**Figure C**]. A single flood did not carve modern topography; rather, over time, rivers gradually carved their own valleys. Lyell's *Principles of Geology* (1830–1833) made the case for how the laws of nature governing geological processes remain constant, even though their effects vary through time. His view that present processes were the key to understanding processes in the geological past became known as **uniformitarianism** and stood in contrast to prior belief that Earth's surface was shaped by one or more grand catastrophes, a view known as **catastrophism**.

Steno as an intellectual forerunner because he attempted to explain the history and dynamic, changing nature of landforms based on a combination of field observations and general principles.

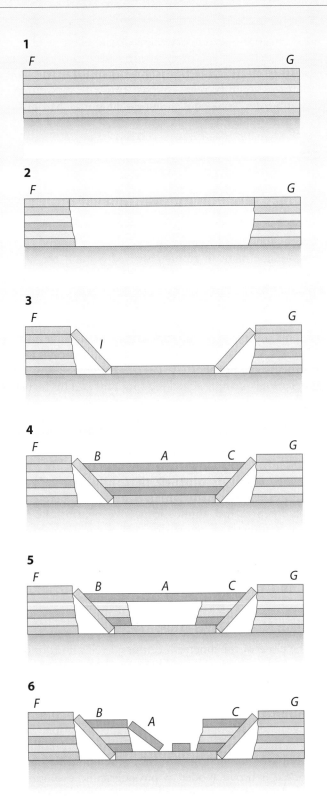

FIGURE B Steno's conceptual model for the geologic and physiographic history of northern Italy: (1) precipitation of fossil-free sedimentary rocks beneath a planetwide ocean; (2) the creation of subterranean caverns; (3) collapse of undermined continents and creation of drowned valleys by a great flood (Noah's flood); (4) new layered (sedimentary) rocks containing fossils form in inundated valleys; (5) continued undermining destabilizes younger rocks in valleys, resulting in (6) another round of collapse that created modern topography (from Steno, 1669).

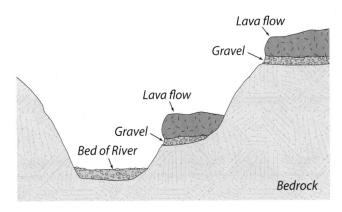

FIGURE C Lyell's illustration of river gravels buried by basaltic lava flows on the walls of valleys in the Auvergne region of France (from Lyell, 1830–1833).

Louis Agassiz (1807–1873)

In the 1830s, the origin of the great deposits of surficial gravel and debris that covered much of northern Europe remained enigmatic. What could have deposited the great blanket of geologic detritus that covered the lowlands across much of the continent? Contemporary processes active in the landscape were not doing so; there was no evidence of similar deposits forming anywhere at the time. Agassiz recognized the geomorphic role of glaciation and showed how different climates in the past changed the mix of geomorphic processes that shaped topography. Living in Neuchâtel in the Swiss Alps, he began to see how the landforms found around modern glaciers also occurred farther downvalley. Glaciers had been more extensive in the past, during an age of ice (Agassiz, 1840). Noting the stray, exotic boulders carried far from potential sources and the extent of ice-shaped landforms [Figure D] beyond the Swiss Alps, Agassiz established that glaciers had advanced not just short distances down their valleys but had overrun much of Europe in the past.

After emigrating to North America, he documented evidence for extensive continental glaciation in the surficial deposits and landforms of New England and the northern continental interior. His glacial theory was hotly debated before it was widely accepted in the 1860s; the discovery of glacial epochs introduced the idea that landforms could be relics of past eras with very different climates. One could not assume that things had always been as they are today.

FIGURE D Agassiz's illustrations of glacially polished and striated bedrock and boulders moved by glacier transport (from Agassiz, 1840).

Grove Karl Gilbert (1843–1918)

Exploration of western North America in the late nineteenth century illuminated the connection between the landscape and the geology below. The federal surveys commissioned to inventory the resource potential of the expanding American frontier proved a boon for understanding the relationship of topography to the type and structure of the underlying rocks. Such connections were apparent across grand vistas due to the excellent exposures in the arid west (Gilbert, 1877) [Figure E].

One of the most influential of these government scientists, Gilbert introduced ideas and methods central to **process geomorphology**. He revolutionized how geomorphologists approached reading the land by focusing on the morphologic implications of erosion, transport, and deposition processes. Gilbert also recognized the role of past climates in forming the now-vanished but once extensive lakes in the arid American West (Gilbert, 1890). Intrigued by the relationship of geology to topography, he began to formalize thinking about landscape evolution. Gilbert also initiated studies of active processes for practical applications, investigating the routing of sediment through drainage basins affected by hydraulic mining in California's gold rush, and conducting experimental studies of sediment transport in flumes (Gilbert, 1914).

William Morris Davis (1850–1934)

At the close of the nineteenth century, Lyell's uniformitarianism had become well established and the biological sciences were being revolutionized by Darwin's idea of natural selection. Evolutionary thinking was in vogue, and William Morris Davis adapted concepts of biological change and development to landscape evolution. Davis systematized thinking about landscape evolution as ordered around what he termed a **geographical cycle**. He proposed a broad model of physiographic evolution in which topography progressed in stages from youthful to mature to old age after an initial pulse of uplift (Davis, 1899, 1909). Relief increased during uplift, and then gradually decreased, ultimately approaching a beveled-off surface he termed a **peneplain** [Figure F].

Davis also wrote influential papers on the relation between geomorphic processes and landforms and the evolution of river systems, but it was his geographical cycle that dominated geomorphological thinking for the first half

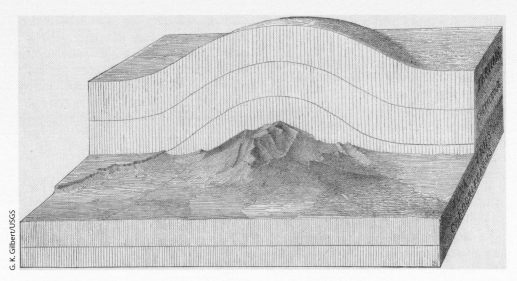

FIGURE E Frontispiece from *The Geology of the Henry Mountains* (Gilbert, 1877), showing the relationship between modern topography and the domal uplift inferred from the geologic structure of the range, in particular, the dip of the beds away from the core of the uplift. The mass of rock eroded off the landscape in the geologic past is shown above the current rugged landscape.

of the twentieth century. His ideas promoted qualitative interpretation of landscape history from broad aspects of landscape form. Davis's conception of peneplains became central to thinking about landscape evolution and motivated extensive searches for, and speculation about, suspected **erosion surfaces** thought to represent ancient peneplains. While Davis's conception of a life cycle of topography has been sidelined, the idea of topographic response to changes in rock uplift rate and the long time required to reduce the elevation of a mountain range remain pertinent in thinking about landscape evolution. Indeed, his thinking about general models of hillslope and river valley evolution led to modern mathematical models of landscape behavior.

J Harlen Bretz (1882–1981)

For the first half of the twentieth century, uniformitarianism dominated geological thinking to such an extent that few in the geologic establishment were comfortable entertaining the idea of grand catastrophes. In the 1920s, J Harlen Bretz, then a young geologist, uncovered evidence for an enormous ancient flood in eastern Washington State (Bretz, 1923, 1925). Through extensive fieldwork, Bretz pieced together the story of how a surging wave of water hundreds of meters high roared across eastern Washington, carving deep channels where today no rivers flow. It took most of the twentieth century for geologists to accept his radical hypothesis.

Bretz reintroduced the idea of grand catastrophes as effective geomorphic events. His work showed how ancient events and processes no longer active today could catastrophically affect landscapes. Over time, the intensity as well as type of geomorphic processes affecting an area could change greatly and leave a lasting impression on the landscape. In establishing that landforms could be the result of catastrophic floods, Bretz opened the door for recognizing the topographic signature of catastrophic events of many different kinds.

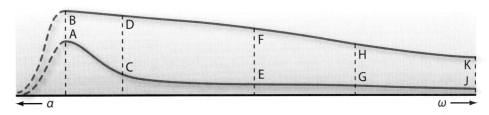

FIGURE F Davis's schematic diagram of the change in relief through time of the idealized cycle of erosion. Vertical dashed lines represent particular times; B-D-F-H-K represent the elevation of the ridge crest through time, whereas A-C-E-G-J represent the elevation of the valley bottom through time. The horizontal scale, α to ω, indicates the passage of time from the beginning (α) to the end (ω) of a geographical cycle (modified from Davis, 1899).

Ralph A. Bagnold (1896–1990) and Luna B. Leopold (1915–2006)

In the mid-twentieth century, most geomorphic studies were dominated by map analysis or Davisian thinking about grand cycles of uplift and response. But, by mid-century, things began to change as Bagnold and Leopold pioneered the application of physics to studies of landscape-forming processes and promoted the application of quantitative field measurements to test theories of aeolian and fluvial processes, respectively (Bagnold, 1941; Leopold et al., 1964). Bagnold explored the North African desert, where he studied the physics of blowing sand and the processes involved in the formation of sand dunes before putting his expertise to practical use as a major in the British army in Libya during World War II. In the 1950s, Leopold led a group of U.S. Geological Survey researchers who ushered in the modern era of process geomorphology with an aggressive campaign to quantify studies of river channels and explain the physics underlying fluvial processes.

Bagnold and Leopold introduced the practice of relating process to form. They did this by making rigorous field and laboratory measurements of geomorphological processes and the forms that resulted. Their approach to understanding the basic physics of aeolian and fluvial erosion, transport, and deposition began the quantification of landscape form, evolution, and response. The approach that Bagnold and Leopold adopted—coupling field and experimental observations and measurements with theoretical models to explain geomorphological processes—professionalized the practice of geomorphology and opened the door for development of process-oriented geomorphology in the late twentieth century.

Plate Tectonics

The gradual development of the concept of plate tectonics in the 1950s and 1960s revolutionized geology and elegantly tied Earth history together in a unifying framework that provides the basis for understanding the forces shaping Earth's dynamic surface. Plate tectonics was not the discovery of a single individual; its development was driven by data from many sources.

The theory of plate tectonics explained three independent mysteries that made sense only when considered together—magnetic stripes on the seafloor, high heat flow over the mid-ocean ridges, and the global distribution of earthquakes. The development of sonar during World War II revealed that linear submarine mountain chains circled the world (Heezen, 1960). Along the axes of these mountains were extensional zones, where the crust appeared to be pulling apart or extending. Mapping the magnetic field of the seafloor, to provide background values against which to hunt for submarines, revealed that away from these mid-ocean mountain chains the oceanic crust had alternating bands of normal and reverse magnetic polarity (Mason & Raff, 1961)—a natural strip-chart recorder of seafloor spreading over time because oceanic crust locked in the contemporary magnetic polarity as it cooled and moved away from seafloor spreading centers. Seismological networks set up to verify nuclear test ban treaties revealed that although most earthquakes occurred in the upper crust, mysterious deep earthquakes defined slabs of crust sinking hundreds of kilometers below ground, deep enough that rocks should be too hot and too soft to break (Benioff, 1954).

Considered together, these observations defined a cycle in which new crust formed from molten lava at mid-ocean ridges, moved away from the axial ridge, and dove back down to be recycled in the deep trenches at the edge of ocean basins. Here was a single, grand mechanism to explain the shapes of continents; how they moved and how mountains formed; why different rock types occur in different regions; and why earthquakes, volcanoes, and mountains all line up where plates split apart, collide, or slide past one another. Plate tectonics provided a unifying framework within which to understand the evolution of Earth's landscapes (e.g., Summerfield, 1991).

Modern Geomorphology

Today, geomorphologists work on a wide range of problems, using diverse methods and tools to understand the processes shaping individual landforms, and the history and evolution of global topography. Studies of the interactions of climate, tectonics, and surface processes of erosion and deposition frame new questions about the evolution of Earth's dynamic surface. Although rapid advances in the resolution, quality, and availability of digital topographic data have contributed to the quantification of geomorphology over the past several decades, there is still a significant role for insightful observation and fieldwork to advance our understanding of landscape processes and evolution, the geomorphology of particular regions, and the role of human actions in shaping the world we live on and the nature of the one our descendents will inherit. In addition, studies of Earth's landscapes provide the basis for interpreting evidence, gathered remotely, of the processes acting to shape surfaces of other planets and their moons.

Agassiz, L. *Etudes sur les Glaciers*. Neuchâtel, 1840.
Bagnold, R. A. *The Physics of Blown Sand and Desert Dunes*. London: Methuen, 1941.
Benioff, H. Orogenesis and deep crustal structure—Additional evidence from seismology. *Bulletin of the Geological Society of America* 65 (1954): 385–400.
Bretz, J. H. The channeled scabland of the Columbia Plateau. *Journal of Geology* 31 (1923): 617–649.
Bretz, J. H. The Spokane Flood beyond the channeled scabland. *Journal of Geology* 33 (1925): 97–115, 236–259.
Davis, W. M. The geographical cycle. *Geographical Journal* 14 (1899): 481–504.

Davis, W. M. *Geographical Essays*. Boston: Ginn and Co., 1909.

Gilbert, G. K. *Report on the Geology of the Henry Mountains*. U.S. Geographical and Geological Survey of the Rocky Mountain Region. Washington, DC: Government Printing Office, 1877.

Gilbert, G. K. *Lake Bonneville*. Monographs of the U.S. Geological Survey, vol. 1. Washington, DC: Government Printing Office, 1890.

Gilbert, G. K. *Transportation of Debris by Running Water*. Professional Paper 86, U.S. Geological Survey. Washington, DC: Government Printing Office, 1914.

Heezen, B. The rift in the ocean floor. *Scientific American* 203 (1960): 98–110.

Leopold, L. B., M. G. Wolman, and J. P. Miller. *Fluvial Processes in Geomorphology*. San Francisco: W. H. Freeman, 1964.

Lyell, C. *Principles of Geology: Being an Attempt to Explain the Former Changes of the Earth's Surface, by Reference to Causes Now in Operation*, 3 vols. London: John Murray, 1830–1833.

Mason, R. G., and A. D. Raff. Magnetic survey off the west coast of the United States between 32°N latitude and 42°N latitude. *Bulletin of the Geological Society of America* 72 (1961): 1259–1266.

Steno, N. *De solido intra solidum naturaliter contento dissertationis prodromus* (Outlines of a dissertation concerning solids naturally contained within solids). Florence, 1669.

Summerfield, M. A. *Global Geomorphology*. Chichester, UK: John Wiley, 1991.

Geomorphologist's Tool Kit

Introduction

Geomorphology has been revolutionized in the past several decades by rapidly advancing technologies. Today, geomorphologists routinely date landforms, measure rates of **erosion,** and quantify the **morphology** of Earth's surface with high precision. Techniques geomorphologists now take for granted (including the dating of rocks, sediments, and landforms; high-precision, satellite-based topographic surveys; multispectral remote sensing; and sophisticated computer modeling) have moved from the experimental realm to widespread application. These techniques and others provide fundamental data about the rates of some Earth surface processes and the distribution of earth materials. Such data allow geomorphologists to understand when, where, and how quickly our planet's surface changes.

Geomorphologists employ a variety of physical, chemical (elemental), isotopic (same element, different mass), biological, and remote-sensing methods to understand the rate, timing, and amount of change at and near Earth's surface. The certainty of these methods varies. Some are capable of generating data with precisions better than 1 percent (e.g., **radiocarbon dating** of organic material), while other methods provide only **relative age** information, like measuring the thickness of weathering-rinds on rock surfaces. All analytical methods have limitations,

Surveying at Iron Mountain, one of General George Patton's World War II training camps in the Mojave Desert of California. In the distance, a graduate student is using a GPS (Global Positioning System) total station to map the location of channels eroding a walkway built by soldiers in the early 1940s. In the foreground, an undergraduate uses a prism pole as part of a total station survey to measure the movement of painted pebbles placed at specific locations on the desert surface as tracers of sediment movement. These and other modern survey techniques allow precise quantification of landscape change over time.

IN THIS CHAPTER

Introduction
Characterizing Earth's Surface
  Field Surveys
  Active Remote Sensing
  Passive Remote Sensing
  Digital Topographic and Landscape
    Analysis
Relative Dating Methods
  Landform Degradation
  Rock Weathering and Soil
    Development
  Rock Varnish
  Calibrated Relative Dates

Numerical Dating Methods
  Dendrochronology
  Radiocarbon Dating
  K/Ar Dating
  U/Th Dating
  Luminescence Dating
  In-Situ Cosmogenic Nuclide Analysis
Measuring Rates of Geomorphic Processes
  Sediment Generation Versus Sediment
    Yield
  Landscape Change at Outcrop and
    Hillslope Scales
  Landscape Change at Basin Scales
  Erosion Rates over 10^6 to 10^8 Year
    Timescales

Experiments
  Field Experiments
  Laboratory Experiments
  Numerical Models
Proxy Records
Applications
**Selected References and Further
  Reading**
**Digging Deeper: How Does a Dating
  Method Develop?**
Worked Problem
Knowledge Assessment

both in their application and in their resolution. Usually, these limitations are related to assumptions underlying the method (e.g., the initial composition or origin of the sample).

The key to accurate application of any analytical method is in understanding how the method works as well as knowing its uncertainties, assumptions, and limitations. In general, getting accuracy (the right answer) comparable to the precision (the uncertainty of the reported measurement) of most dating methods requires careful consideration of the geologic and geomorphic context of a sample (e.g., the type of sediment from which the sample was taken or the relationship of the sample to the landform of interest). Therefore, the involvement of a geomorphologist in sample collection is often the key to obtaining meaningful geochronologic data about the rate of surface processes or the date of a landform.

It is worth stepping back in time a bit to appreciate how quickly geomorphology has changed from a dominantly descriptive science to one capable of testing hypotheses quantitatively and predicting the evolution of landscapes. Consider how we measure and map Earth's surface. Surveys of western North America throughout the 1800s and most of the 1900s painstakingly documented the elevation and form of mountain passes and river courses using crude and time-consuming optical and compass survey techniques. These surveys and later analysis of air photographs created topographic maps at a variety of scales and frequently known by the shorthand name "quadrangles." Today, satellites capture such information repeatedly each year using remote-sensing instruments.

Until the early nineteenth century, scholars thought that surficial gravel deposits, glacial striations, and even topography were the record of erosion and/or deposition by biblical floodwaters. It has been about two centuries since geologists realized that Earth history involved more than a single grand cataclysm, and only more recently that they recognized the essential roles of incremental everyday processes, ancient catastrophes, and past climate changes in shaping modern topography. Now, studies of landscape history are anchored in quantitative knowledge of process and chronologies based on a variety of chemical, physical, and isotopic measurements. Observational field studies are complemented by experimentation in the field, in the lab, or using computer models. Rates of processes, including erosion and deposition, are determined either by direct measurement or indirectly through physical and isotopic measurements.

Geomorphologists often require quantitative data to test their hypotheses regarding landscape process, form, and history. Such data may include simple field measurements, descriptions of topography at different scales, process rates integrated (averaged) over different time intervals, and landform ages measured using various **chronometers** (dating tools). New tools include high-resolution **mass spectrometers** and **radioactive-decay** counters that allow isotopic analysis and **radiometric dating.** These data give geomorphologists a means of measuring the age of landforms quantitatively, data useful for estimating the rates at which mass is removed, transported, and deposited on Earth's surface. Records of landscape change preserved in near-surface deposits are useful for understanding our planet's history and determining rates of change over time. The advent of powerful computers and the concurrent improvement in our ability to simulate complex physical and chemical interactions over time has further fueled both interest in and our ability to do process-based landscape studies.

This chapter introduces the wide range of techniques available to geomorphologists and reviews some of the most common methods of quantitatively describing topographic forms and of collecting relevant data on rates and dates of surface processes. In geomorphology, as in other sciences, the key to success is to select the best technique for the problem at hand—one that yields data at the needed resolution without excessive cost or acquisition time.

Characterizing Earth's Surface

Traditionally, geomorphologists studied and analyzed landforms using publicly available aerial photographs and paper topographic maps, supplemented by survey data gathered at higher spatial resolution using manual instruments like levels and transits. Recent technological advances, specifically remotely sensed data, now provide high-resolution topographic and compositional data over large areas of Earth's surface at a variety of spatial resolutions. The availability of digital data is a major advance because such data can be used in conjunction with other similarly formatted data or fed directly into landscape models [**Figure 2.1**].

Field Surveys

Despite rapid advances in technology, manual surveying of fine-scale landscape features remains an important tool for geomorphologists. For example, a **stadia rod** (essentially, a large ruler), a level, and a measuring tape can be used to quickly and accurately measure a series of points that define the dimensions of river channels in cross section. Hand levels can be used with lightweight stadia rods to rapidly collect the point data needed to characterize hillslope profiles in even the most remote locations. Adding a compass to the tool kit allows the collection of data in three dimensions, a prerequisite to making a topographic map. A variety of manual optical surveying instruments (including auto-levels, alidades/plane tables, and transits) generate similar elevation/distance/angle data more precisely but are heavier, bulkier, and require more setup time.

The advent of laser ranging and digital total stations with data loggers in the last several decades greatly streamlined the process of optical surveying and made it easier to collect larger amounts of high-precision data—hundreds to thousands of ground points in a day. These instruments

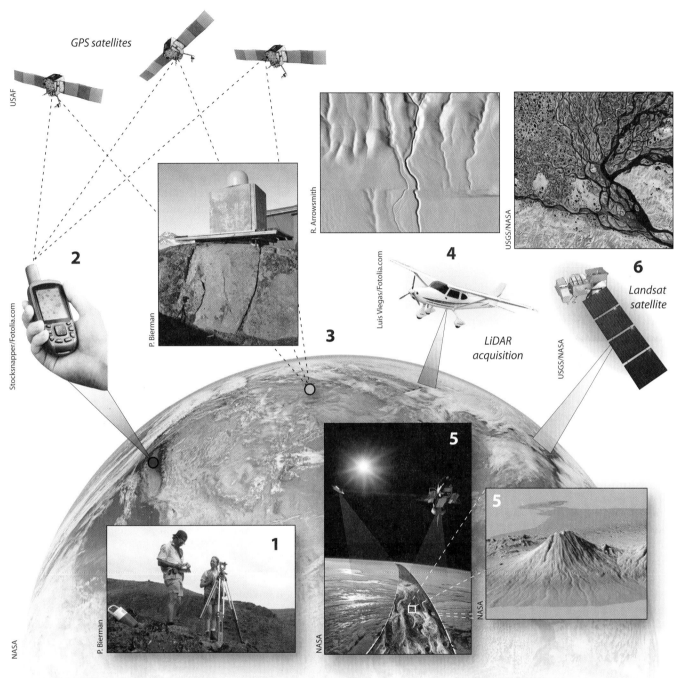

FIGURE 2.1 Characterizing Earth's Surface. Geomorphologists use a variety of techniques to characterize the shape and composition of Earth's surface. Some of these techniques are ground-based while others rely on airplanes and satellite platforms.

1. Optical surveys are done with a variety of instruments including levels, plane tables, and total stations; these provide fine-scale topographic data for hillslopes and river channels at relatively low cost.

2. Handheld **GPS** receivers are used routinely to locate positions and sample sites in the field. Vertical and horizontal locations are determined to better than tens of meters by trilateration to at least four GPS satellites.

3. Recording GPS receivers anchored to stable **benchmarks** are used to make high-precision (mm-scale) measurements of lateral and vertical position change caused by plate tectonics and the isostatic response to changing glacier mass over time.

4. LiDAR data are collected from airplanes carrying lasers capable of measuring precisely the travel time of laser light to Earth's surface. High-precision GPS determines the location of the airplane when each laser pulse is sent out and received.

5. Active remote sensing data collected by the Space Shuttle include Synthetic Aperture Radar (**SAR**), used both to map the topography of Earth's surface (**SRTM**) and to determine small changes in elevation through interferometry (**InSAR**).

6. Landsat and other satellites collect **passive remote sensing** data. Multi-spectral sensors on the satellites quantify the amount of light at different wavelengths reflected by the materials covering Earth's surface, such as differing vegetation or lithology.

determine distance by measuring the travel time of laser light between the instrument and a reflector. Typically, this reflector is a prism mounted on a pole and held by a rod person, with the rod person moving between points of interest. Thus, two people are required for this type of survey work.

Instruments are now available that allow laser ranging to natural surfaces (no laser-reflecting prism needed), and some total stations are automated so that only one operator, the person carrying the reflector from place to place, is needed. Despite technological advances, on-the-ground field collection of topographic data remains labor-intensive, time-consuming, and difficult in rough terrain. Laser-based systems require power, which means carrying and charging batteries—a difficulty in remote locations.

The development of the **Global Positioning System (GPS)** in the 1990s has been a boon for geomorphology. Locations of sampling sites or landforms anywhere in the world can now be easily and rapidly determined. The system relies on radio-based trilateration between a constellation of broadcasting satellites and passive receivers. The major limitation is terrain and forest cover. The GPS receiver must be able to lock onto the signal from at least four satellites simultaneously. The limited view of the sky in rough terrain and under thick forest patches precludes the use of at least high-precision GPS equipment in such settings. Uncorrected GPS data, easily collected in seconds by inexpensive handheld units, consist of locations and altitudes accurate to within a few meters.

A variety of correction schemes can greatly improve the precision and accuracy of GPS measurements. Post-processing of single-point GPS data can provide horizontal accuracies better than a meter. Differential GPS measurements, where one unit is placed on a benchmark and the other is moved to the survey point, can provide centimeter-scale accuracy in both the vertical and horizontal dimension in open terrain.

GPS units placed on secure **benchmarks** are sensitive and accurate enough that they can measure directly the slow movement (centimeters per year) of tectonic plates, glaciers, and landslides. Such data allow mapping of contemporary deformation rates in real time. For example, GPS measurements can detect the seasonal flexure of the lithosphere under Greenland as the ice mass grows heavier in the winter (snow accumulates) and lighter in the summer (ice melts away).

Active Remote Sensing

Advances in remote sensing allow geomorphologists to determine the composition of materials exposed at Earth's surface as well as topography and surface roughness. Remote sensing differs from manual data collection in that it can gather large amounts of information quickly; however, computers are needed to process the information and models are usually needed to interpret the resulting data, which are gathered pixel by pixel (a shorthand word for "picture element") over a large area, instead of in irregular transects like field data. The data collected from each pixel are averaged by the sensor over an area on the ground corresponding to the pixel size. For commonly available remote-sensing images, pixels range from a few meters to a kilometer or more on a side.

Active remote sensing involves an instrument that sends out a pulse of energy, receives that energy back, and analyzes the characteristics of the energy pulse as it returns to the detector—just like a total station, but on a coarser scale. In the past decade, several new active remote-sensing technologies have fundamentally changed the acquisition, quality, and global coverage of topographic data. **Interoferometric Synthetic Aperture Radar (InSAR)** can be used not only to measure topography over vast areas, but also in differential mode to measure topographic change over time and space. To do this, the SAR image is measured 2 times from the same point in orbit. If the ground surface has not changed between measurement times, the phase of the radar waves will be the same, but, if there has been deformation, there will be phase shifts proportional to the amount of elevation change.

Interferometry data are widely used by geomorphologists. They can reveal tectonic deformation occurring before, during, and after earthquakes or the deformation that accompanies slow, **aseismic creep** on faults. Volcanic deformation from magma intrusion or from incipient mass movements is easily detected, as is land subsidence from groundwater withdrawals. Interferometry reliably measures the surface elevation of ice sheets in response to climate change. InSAR is very sensitive. For example, it has been able to detect the swelling (~10 cm) due to hydration of clays after an alfalfa field was watered, to the delight of the scientists who first noticed it.

Synthetic Aperture Radar (SAR) data collected by the Space Shuttle during the **Shuttle Radar Topography Mission (SRTM)** in the year 2000 provided the first worldwide digital topographic data set, at a resolution of ~90 m per pixel and covering latitudes less than 55°. The topographic aspect of the data came from interferometry using the two shuttle antennas. Vertical precision of the data is on the order of a few meters. This mission mapped areas of Earth for which topographic data had never been gathered, most importantly, for areas where persistent cloud cover had previously prevented aerial photography and areas so remote that detailed land-based surveying was impractical. The SRTM data are available on line free of charge.

At finer spatial scales, a key advance in depicting Earth-surface topography has been the development and deployment of **LiDAR** systems (**Light Detection and Ranging**). This technology involves sweeping laser pulses across the landscape and detecting the timing of their returns. When operated from airplanes, LiDAR produces local data sets covering tens to thousands of square kilometers with meter-scale topographic resolution and elevation precision in the decimeter range. LiDAR maps, made after post-processing of the data to remove the effects of vegetation and buildings, have been used to identify and locate fault scarps, quantify channel dimensions, and detect and map landslides. Deployed on the ground, LiDAR can map outcrops and fine-scale landforms with centimeter-scale precision. Repeated LiDAR surveys can detect subtle changes over large areas and thus allow calculation of rates of mass loss or addition. LiDAR data are increasingly becoming publically available.

Radar data can also be used to estimate the roughness of Earth's surface remotely, because surface roughness controls the amount of radar energy that returns to the detector. Such measurements are particularly useful in arid regions where there is little vegetation. If the scale of the roughness elements (e.g., sand grains and boulders) is much less than the radar wavelength (typically <1/5), the surface appears smooth and reflects radar like a mirror, away from the detector. The resulting radar image looks black. However, if the roughness scale is large compared to radar wavelength, then the radar energy **backscatters** in all directions including back to the sensor and the rough surface appears bright in the image.

Measuring surface roughness is important to geomorphologists because surface roughness changes with surface age. In the desert, very young alluvial surfaces are rough and then smooth over time as desert pavements develop, the banks of abandoned channels erode, and exposed boulders weather; thus, radar roughness initially decreases with age. However, after a long period (tens to hundreds of thousands of years), the buildup of calcium carbonate in arid soils underlying smooth alluvial surfaces reduces porosity and limits the ability of rainwater to soak into the ground (**infiltrate**). Rainwater that cannot infiltrate runs over the surface and erodes channels. These channels increase relief so that radar roughness increases on old surfaces.

Passive Remote Sensing

Passive remote sensing uses a detector to measure the amount and wavelength of energy emitted from Earth's surface. Aerial photographs, taken from balloons and then airplanes, were the earliest passive remote-sensing images. Later, multispectral digital remote-sensing instruments were flown on both satellites and airplanes. Such instruments collect reflectance data for different wavelengths of electromagnetic radiation. **Multispectral remote sensing** works because different earth materials (vegetation, rocks, soil) have characteristic electromagnetic reflectivity **spectra**. Since most pixels contain a mixture of earth materials, the resulting spectra are also mixtures. One can mathematically unmix the pixel-scale spectrum and estimate the relative contributions of different materials to the total spectrum.

The first Landsat satellite was launched in 1972 and made a global data set of images available to the public for the first time. These images included four spectral bands of different wavelengths, from visible to near-infrared. Landsat images were the first multispectral images to cover most of Earth repeatedly at a resolution that has improved from ~80 m/pixel on the first satellite to ~15 m/pixel on the latest satellite. Most significantly, the continuous operation of the system since 1972 has meant that every 16 days an image is taken of a particular location, permitting monitoring of geomorphic change on timescales from weeks to decades. Using ground-based calibration, ratios of radiation intensity in selected bands can be used to reduce the sensitivity of images to topographic shading and infer the density of vegetation cover and the composition of earth materials exposed at the surface. Landsat data have been used extensively in geomorphology for mapping abandoned river channels, delineating the extent of floods, and monitoring deforestation over time.

Analysis of repeat photography can be a powerful geomorphic tool, allowing one to calculate the rate and distribution of surface processes. Aerial photographs are available for many locations, starting in the 1930s and 1940s with repeat photography usually every decade. Such photographs, once rectified to correct for distortion, are particularly useful for determining the migration rates of rivers, the retreat rates of cliffs, and changes in land use over time. Ground-level repeat photography has been used to document changes in glacier extent, vegetation response to climate change, and landscape response after disturbance.

Digital Topographic and Landscape Analysis

Beginning in the 1980s, collection and analysis of topographic data at all scales began the transition from analog (hand-contoured paper maps based on individually collected point data) to digital (computer analysis of information, pixel by pixel) formats. Early **Digital Elevation Models (DEMs)** were created primarily by digitizing printed topographic maps and, as a result, included a variety of artifacts such as elevation terracing—the appearance of terraces where there are none, the result of poor interpolation between contours. Later, remote-sensing data collected by sensors on satellites (for example, SPOT images) were manipulated to extract topographic information.

Today, LiDAR and SRTM data are routinely used as the basis for DEMs. The result is huge data sets in which millions of points on the landscape are assigned values for horizontal and vertical location. Each data point represents the average elevation for a pixel, the dimensions of which vary, depending on the technology used to collect and extract the data. If you do not need the high spatial resolution that LiDAR provides, the free, publically available SRTM data (90 meter pixels) may be a better tool as they are much easier to manipulate than LiDAR data, and because they are public domain, there is no cost.

There are limitations to DEM-based landscape analysis. Uneven distribution of the source topographic data can result in data gaps or faulty interpolation. The grid size or spacing between data points on a DEM will affect a variety of calculated landscape parameters. For example, because steep slopes have finite extents, a DEM-based slope map will show steeper slopes (closer to field-measured values) when based on high-resolution data (small pixels) than if based on low-resolution data (large pixels).

Digital landscape data are best managed by computer programs known as **Geographic Information Systems (GIS)**. These programs allow manipulation of digital data and automate many of the landscape analysis tasks that were once done manually, such as the calculation of watershed area, **hypsometry** (the proportion of the landscape at different elevations), and average slope. GIS can calculate flow paths based on topography and can be a platform for landscape models that simulate the movement of mass downslope over time. One of the more powerful aspects of GIS is the ability to layer information in a database format so that various landscape parameters such as soil type, infiltration capacity, and vegetation cover can be compared and considered together. Bringing these disparate spatial data sets together in the context of mapping is very important, because it highlights the relationships between surface processes, landscape form, and geologic substrates.

Relative Dating Methods

Relative dating methods are used to order landscapes, deposits, and geologic events according to their relative ages, whereas **numeric dating** methods provide estimates of chronologic age. For example, as streams and rivers incise into the landscape, the terraces they leave behind are oldest at high elevation and youngest at elevations just above the stream. Such a temporal sequence of land surfaces in a similar environment provides a **chronosequence,** a natural experiment where most landscape variables except time are constant.

The most straightforward relative dating techniques exploit the principles of **superposition** and **cross-cutting** relationships. Cross-cutting and superposition provide relative age information over a wide range of timescales from minutes to millions of years. For example, layered glacial lake sediments overlying **till** (unsorted sediment deposited from the base of glaciers) indicate that a proglacial lake formed *after* the glacier that deposited the till melted away [**Photograph 2.1**]. Such sequences are common in the northern Appalachian Mountains of New England.

PHOTOGRAPH 2.2 Moraines. Moraines of different ages are nested at Convict Creek on the eastern side of the Sierra Nevada in southern California.

There, till deposited under ice was covered by lake sediments as the shrinking Laurentide Ice Sheet dammed glacial lakes and impounded tributary valleys. Similarly, on the eastern flank of the Sierra Nevada of southern California, younger **moraines** (ice-bulldozed deposits of glacial sediment) lie within the perimeter of larger, older moraines, clearly indicating their relative age [**Photograph 2.2**].

Landform Degradation

Surface processes sometimes change landforms in predictable ways that allow age estimates. In some environments, steep, well-defined landforms with angular boundaries become rounded, less steep, and lower in elevation over time. For example, glacial moraines that are initially sharp-crested [**Photograph 2.3**] become smoother and broader over time as mass moves from the crest down

PHOTOGRAPH 2.1 Glacial Lake Sediment and Till. When a glacial lake formed next to a retreating ice margin in eastern New York State, rhythmically bedded fine-grain couplets of sand and silt were deposited over till. The underlying till is older than the overlying lake sediments. The fieldbook is about 20 cm long.

PHOTOGRAPH 2.3 Lateral Moraine. Young, sharp-crested lateral moraine with no soil development at the Ak-sai Glacier, Tien Shan Mountains, Kyrgyzstan.

PHOTOGRAPH 2.4 Colluvial Apron. Moraine cross section showing rounding of the crest and soil formation (reddish upper layer) indicative of age. On the left side of the moraine, toward the base, is a thickening wedge of brown soil, a colluvial apron. The apron is composed of material eroded from the moraine and transported downslope. White Mountain Range, California, is in the distance.

the side slopes to accumulate in **colluvial aprons** below [Photograph 2.4]. Similarly, fault scarps, across which one block is dropped down and the other uplifted, change predictably over time, rounding as they age [Figure 2.2]. Rough surfaces, like boulder-rich alluvial fans, smooth with age as chemical and physical weathering break

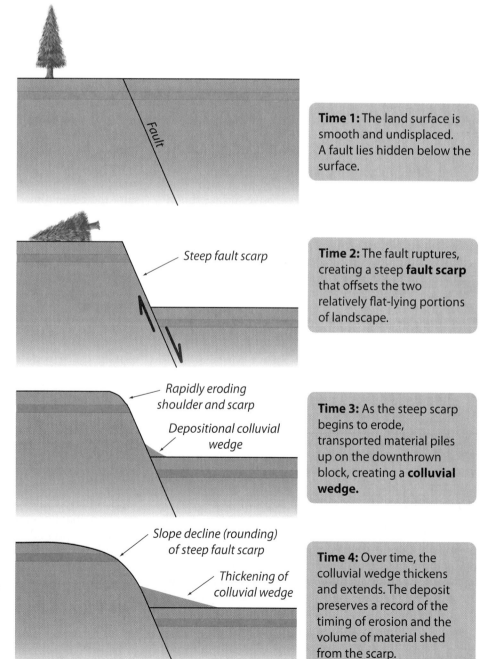

FIGURE 2.2 Fault Scarp Degradation. Over time, surface processes erode landforms such as fault scarps in unconsolidated materials, rounding sharp edges and moving mass from higher to lower elevations.

Time 1: The land surface is smooth and undisplaced. A fault lies hidden below the surface.

Time 2: The fault ruptures, creating a steep **fault scarp** that offsets the two relatively flat-lying portions of landscape.

Time 3: As the steep scarp begins to erode, transported material piles up on the downthrown block, creating a **colluvial wedge.**

Time 4: Over time, the colluvial wedge thickens and extends. The deposit preserves a record of the timing of erosion and the volume of material shed from the scarp.

PHOTOGRAPH 2.5 Cinder Cone. Old, weathered cinder cone on the wet, eastern side of the island of Hawaii is deeply gullied and dissected.

PHOTOGRAPH 2.6 Weathering Rind. Weathering rind in a granite pebble that is about 15 cm across. Weathered material is red and oxidized. Gray center of the clast is fresh granite.

down boulders, and wind deposits sediment in the low spots.

In some cases, landscape degradation results in roughening of initially smooth surfaces. For example, permeable basaltic cinder cones become more rugged as they age because basalt weathers to clay, increasing the content of fine material in the regolith and thus decreasing the ability of rainwater to soak into the surface. Since much of the rain can no longer infiltrate, it generates erosive surface runoff [Photograph 2.5]. Old cinder cones are often deeply incised by stream channels and rills. The degree of landform degradation provides useful relative age information over timescales of centuries to hundreds of thousands of years, depending on erosion rates.

Rock Weathering and Soil Development

Minerals that are exposed to weathering processes at Earth's surface chemically and physically transform, or weather, over time. The progress of weathering into a rock can be used as a relative dating tool in several different ways. By cracking open clasts and chipping into rock outcrops exposed on Earth's surface, a geologist can measure the thickness of the **weathering rind,** which for any particular rock type is related to the duration of time since the rock surface was last fractured or abraded [Photograph 2.6].

The degree of rock weathering can also be estimated based on the sound generated from blows of a rock hammer (crisp ping versus a dull thud) because weathered rock is generally less dense and more fractured than unweathered rock and transmits sound more slowly. This same principle applies when measuring the compressive strength or hardness of rock outcrops using a tool known as a Schmidt rebound hammer, or measuring the speed at which sound waves move through rocks using a sonometer. Rock weathering provides age information over millennia to at most a few hundred thousand years.

Soil properties change with time and are thus a tool that geomorphologists can use to estimate the age of some land surfaces. For example, on a sequence of sand and gravel river terraces in humid regions, soils on the older, upper terraces will likely be thicker, have better developed horizons, tend to be redder in color, and have more clay in the soil profile than soils developed on younger, lower terraces nearer the river. In arid regions, salts and compounds such as calcium carbonate, gypsum, and halite will accumulate over time, in some cases cementing older soils. Aeolian dust will also accumulate in desert soils as they age, building thick, fine-grain horizons near the surface. See Chapter 3 for more details about soil development.

Rock Varnish

In arid regions, a dark iron (Fe)- and manganese (Mn)-rich coating called **rock varnish** or **desert varnish** gradually forms on the surface of exposed rock. The concentration and buildup of Fe and Mn is thought to be the result of biological activity because these elements are only present at low levels in desert dust. The thickness, dark color, and extent of rock varnish appear to increase over time as long as the substrate remains stable. Rock varnish can be used as a relative dating tool over thousands to tens of thousands of years [Photograph 2.7]. In a higher-tech version of the same technique, multispectral remote sensing from satellites quantifies the chemical and physical characteristics of soil and rock, detecting the increase in Fe and Mn that results from more extensive varnish cover over time. This allows relative dating of desert land surfaces.

Calibrated Relative Dates

An approach known as **calibrated relative dating** uses observations of time-dependent geomorphic processes occurring on surfaces or deposits of known age to add

PHOTOGRAPH 2.7 Relative Ages. The Hanaupah Canyon alluvial fan extends from the Panamint Mountains to the floor of Death Valley. The different shades of brown reflect different intensities of rock varnish cover and thus the relative age of the abandoned fan surfaces.

Over decades to centuries, the maximum diameter of lichens growing on rock surfaces (**lichenometry**) increases with surface age [**Figure 2.3**]. Lichens growing on rock surfaces of known age, like tombstones or stone buildings, provide calibration. Applications of lichenometry include dating earthquake-induced rockfalls in New Zealand and in the Sierra Nevada, related to the activity on active faults, as well as dating young glacial deposits in Arctic Canada.

Where buried shell or bone material is present, a biological calibrated relative dating method known as **amino-acid racemization** can be used to estimate times of burial. This technique relies on the time and temperature dependence of a change in the structure of amino acids, the building blocks of proteins. Although this technique can be imprecise, it has provided useful age information in locations as disparate as central Australia and the Arctic over times ranging from thousands of years to several million years, depending on the climate and the amino acid analyzed.

Calibrated relative dating methods differ from isotopic methods in that the rate of change over time does not depend on a single fixed **rate constant** but rather on a

timing information to relative dating techniques. Soil development indices, which seek to quantify the intensity of soil-forming processes and relate them to time (see Chapter 3) are a calibrated relative dating technique.

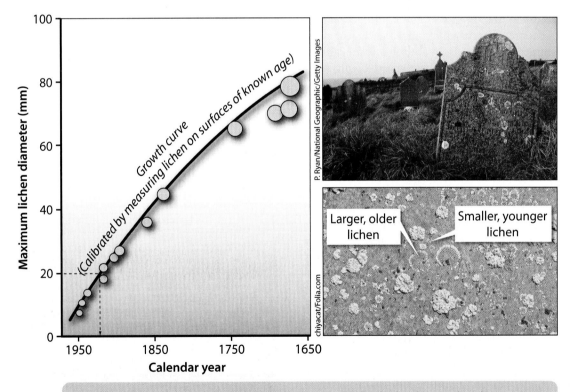

Lichenometry is a calibrated **relative dating method,** which relies on the observation that lichens of a single species have similar growth rates. Thus, by calibrating a **growth curve** on surfaces of known age, such as buildings and tombstones, the maximum width of a lichen found on a surface of unknown age (such as a glacial **moraine**) can be used for dating. Lichenometry can be used to date features between a few decades and a few centuries in age.

FIGURE 2.3 Lichenometry. Lichens get larger over time, providing a calibrated dating method for rock surfaces over the last few hundred years.

TABLE 2.1

Dating methods frequently used by geomorphologists

Method	Type	Age Range (years)	Requirements/Assumptions
Radiocarbon (^{14}C)	Numeric dating	10^2 to 5×10^4	Organic material present in interpretable geologic context
Cosmogenic nuclides	Numeric dating	10^2 to 10^6	Continuous exposure of noneroding surface that was free of cosmogenic nuclides before exposure
Luminescence	Numeric dating	10^3 to 10^6	Quartz or feldspar exposed to light or heat before burial
U/Th	Numeric dating	10^3 to 10^5	Carbonate minerals
Dendrochronology	Numeric dating	10^0 to 10^4	Wood from trees
K/Ar	Numeric dating	10^3 to 10^8	Potassium-bearing minerals
Lichenometry	Calibrated relative dating	10^1 to 10^3	Lichens on both unknown and dated calibration sites
Amino-acid racemization	Calibrated relative dating	10^3 to 10^5	Well-preserved shell material
Rock weathering	Relative dating	10^2 to 10^4	Dated surfaces for calibration
Soil development	Relative dating	10^2 to 10^6	Dated chronosequence for calibration

complex set of biological, chemical, and physical interactions. The accuracy and precision of calibrated relative dating methods are thus usually much lower than for many other numerical methods. In addition, local calibration curves are required because the speed of physical, chemical, and biological changes varies between different environments.

Numerical Dating Methods

Dating methods that give numerical ages have improved greatly over the past several decades and now offer the opportunity for geomorphologists to constrain the age of most materials and landforms at and near Earth's surface. **Numerical dating** methods rely on processes that cause measurable changes in the physical, chemical, or isotopic properties of materials in a predictable and regular fashion over time. Each method involves a set of assumptions that needs to be met in order for the resulting data and age assignments to have chronological meaning. Thus, in this book we refer to numerical dating rather than to absolute dating because all numerical ages depend on interpretive models.

For isotopically derived ages, the models include empirically derived constants and assumptions about initial and subsequent environmental conditions. For ages derived from counting layers (such as glacial lake varves, tree rings, or ice cores), the age model assumes each layer is, in fact, annual. Different dating methods are applicable over different time frames and for different landforms, Earth materials, and geological processes. The key to reliable numerical dating is understanding the method that best suits the problem at hand (**Table 2.1**).

Dendrochronology

Tree rings offer a robust, inexpensive method of numerical dating that is easy to apply, but their utility in geomorphology is limited to certain settings and relatively short time frames, typically centuries. Tree-ring dating (**dendrochronology**) is based on the observation that most trees add a discrete ring of wood to their stem each year that can be recognized and counted because early season wood is lighter in color and has larger fluid transport vessels than late season wood. Cores are collected from trees using sharp steel hand-drills called increment borers [**Photograph 2.8**]. Multiple samples are used to avoid errors from false or missing rings.

Dendrochronology is particularly useful for constraining the time of plant establishment and thus the minimum

PHOTOGRAPH 2.8 Collecting a Tree Core. This tree core and several others were used to date the stabilization time of the landslide scars in northern Vermont where the trees now grow.

age of river terraces, the timing of landslide stabilization, the recurrence intervals of wildfires, and the age of recent glacial activity [Figure 2.4]. Not only can trees be dated by ring counting, but the width of rings can be used to infer whether growing conditions were favorable, a reflection many times of climate, both temperature and precipitation. Incomplete or partial rings can indicate trauma to the tree such as flooding, debris impact, or fire.

Radiocarbon Dating

Since it was first measured in the 1940s, carbon-14 (^{14}C), the only **radioactive isotope** of carbon, has revolutionized geomorphology. Radiocarbon analyses are now commercially available for only several hundred dollars, and geomorphologists routinely use them to estimate the age of surficial deposits like river terraces, lake sediments, and landslide deposits. Analysis of organic debris in sediment shed from fault scarps is used to constrain the timing and recurrence interval of earthquakes on faults all over the world.

Radiocarbon dating is based on the observation that ^{14}C is produced in the atmosphere by the interaction of cosmic-ray neutrons with nitrogen. When a nitrogen nucleus absorbs a neutron, it emits a proton and an electron (beta particle), its atomic number decreases from 7 to 6,

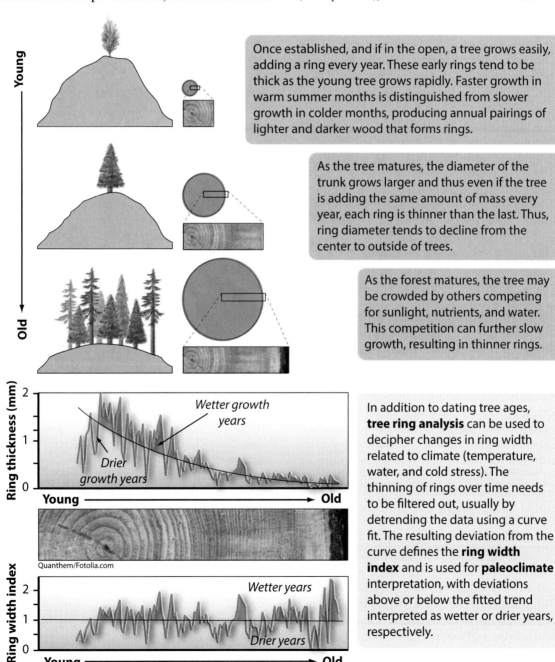

FIGURE 2.4 Dendrochronology. The analysis of tree-ring width patterns is a useful tool for understanding both the age of young landforms and for interpreting climate change over time.

and it becomes a carbon nucleus while maintaining the same atomic mass. Plants take up carbon dioxide from the atmosphere where ^{12}C, the stable common isotope of carbon, and radioactive ^{14}C are well-mixed, so all living tissue in plants contains the same ratio of $^{14}C/^{12}C$ as the atmosphere. After plant material (wood, leaves) no longer interacts with the atmosphere, the ^{14}C in the organic material begins to decay with a **half-life** of 5730 years while the concentration of ^{12}C remains unchanged. Buried organic matter recovered from an outcrop can thus be dated to determine the age of the deposit by measuring the ratio of $^{14}C/^{12}C$, assuming a known initial $^{14}C/^{12}C$ ratio [**Figure 2.5**].

Radiocarbon concentrations measured in samples are typically expressed as percent modern Carbon (pmC), where modern carbon is taken as the pre-1950 **activity** (radioactive decay rate) of radiocarbon in living plant material. For example, a piece of wood that had been dead for one radiocarbon half-life would have a ^{14}C activity, or pmC half that of living material. The age equation, which considers the half-life of ^{14}C and the pmC is:

$$\text{Radiocarbon age (years)} = 8033 \ln(1/pmC) \quad \text{eq. 2.1}$$

Note that radiocarbon age calculations use the same half-life (5568 years) assumed when the technique was first developed, although the actual half-life is now known to be 5730 years. Calibration, as discussed below and demonstrated in the Worked Problem at the end of the chapter, corrects for the half-life change as well as changes in the ^{14}C content of the atmosphere over time.

Carbon-14 is exceptionally rare in nature, so radiocarbon dating requires very sensitive instrumentation. In living tissue, one atom of ^{14}C is typically found for every 10^{12} other carbon atoms, the equivalent of a one-square-foot floor tile in a floor the size of the state of Indiana. Originally, ^{14}C content was measured by counting **beta particle** (electron) emissions as the isotope decayed. Since the 1980s, the development of **accelerator mass spectrometry** has meant that carbon isotope ratios can be measured directly, even in samples as small as a single spruce needle. Sample processing usually involves burning the material to be dated in an evacuated tube before capturing and purifying carbon dioxide [**Photograph 2.9**].

Radiocarbon dating is exceptionally useful in geomorphology, but it does have limitations. Because of its 5730-year half-life, radiocarbon becomes increasingly difficult to measure in older samples, particularly those older than 7 or 8 half-lives, in which there is little ^{14}C left to measure. The practical limit for dating is about 50,000 years. Coal, petroleum, and natural gas contain no original ^{14}C because of their great antiquity—the organic material in them has been isolated from the atmosphere for many hundreds to thousands of ^{14}C half-lives and so all original ^{14}C has long ago decayed away.

Radiocarbon years are similar to but not equal to calendar years because the ^{14}C concentration in the atmosphere has changed over time. The atmospheric concentration of ^{14}C changes in response to changes in the flux of cosmic radiation that produces ^{14}C, the rate and dynamics of ocean-atmosphere carbon exchange, the addition of ^{14}C to the atmosphere by nuclear weapons detonations, and the burning of fossil fuels that adds ^{12}C to the atmosphere, thereby changing the $^{14}C/^{12}C$ ratio. Careful measurement of past atmospheric ^{14}C concentration using dendrochronologically dated tree-ring samples and samples from well-dated corals allows calibration of radiocarbon ages, improving their accuracy. Samples of organic material less than several hundred years old usually give ambiguous ^{14}C ages because the ups and downs of atmospheric ^{14}C content on the scale of centuries interacts with decay to produce multiple apparent ages for a single ^{14}C concentration.

For geomorphologists, the most important uncertainty in radiocarbon dating is the relationship between the sampled organic material and the process or landform to be dated. For example, organic materials collected from glacial sediment could predate the glacial advance if they were overridden by ice and incorporated or could postdate the sediment if the dated organic material were bits of tree roots. In many arid settings, organic material simply cannot be found—either because it was rare at the time of deposition or because it has long since decayed away in the oxidizing near-surface environment. Organic material for ^{14}C dating is best preserved in cool, moist, oxygen-poor settings below the water table, such as in marsh and pond sediments or material buried below stream banks [**Photograph 2.10**].

K/Ar Dating

Potassium/argon (K/Ar) dating and its variant, argon/argon ($^{40}Ar/^{39}Ar$) dating, are widely used in geomorphology, especially to date the eruption of volcanic rocks. These methods rely on the buildup of ^{40}Ar gas (the **daughter isotope**) within mineral crystals as ^{40}K (the **parent isotope**) decays with a half-life of 1.25 billion years. As long as a crystal is hot, Ar gas formed by decay will leave the crystal structure. When the crystal cools below what is termed the **closure temperature** (in actuality, a temperature range), Ar gas is retained and the radiometric clock starts ticking.

Argon dating is simple and reliable for dating eruptions of potassium-rich rocks, like andesite and rhyolite, even if they are less than several million years old. Geologically young, low-potassium basalts are much more difficult to date because much of the argon in the rock was already present in the original magma and was not the result of radioactive decay since the lava cooled. In the best circumstances (very high-potassium rocks), K/Ar and Ar/Ar dating provide useful information about the age of volcanic rocks as young as several thousand years.

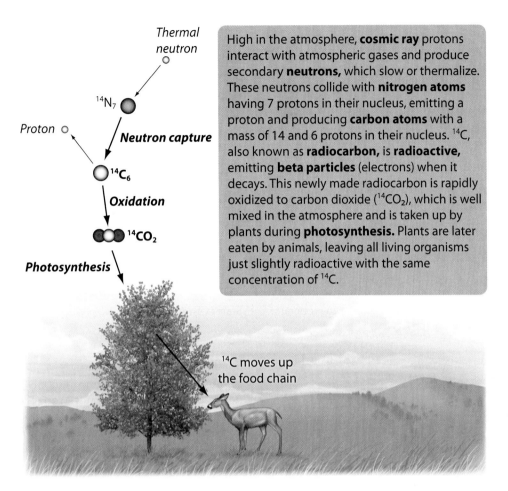

High in the atmosphere, **cosmic ray** protons interact with atmospheric gases and produce secondary **neutrons,** which slow or thermalize. These neutrons collide with **nitrogen atoms** having 7 protons in their nucleus, emitting a proton and producing **carbon atoms** with a mass of 14 and 6 protons in their nucleus. ^{14}C, also known as **radiocarbon,** is **radioactive,** emitting **beta particles** (electrons) when it decays. This newly made radiocarbon is rapidly oxidized to carbon dioxide ($^{14}CO_2$), which is well mixed in the atmosphere and is taken up by plants during **photosynthesis.** Plants are later eaten by animals, leaving all living organisms just slightly radioactive with the same concentration of ^{14}C.

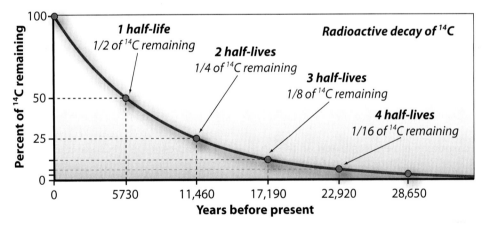

The time of death of organic material can be dated using the concentration of radiocarbon (^{14}C) remaining in the material. As the object ages, radiocarbon atoms decay back to nitrogen at a steady rate; thus, the concentration of ^{14}C is reduced over time. Because the **half-life** (decay rate) of radiocarbon is well known (5730 years), one can estimate an age from a radiocarbon concentration. In practice, radiocarbon dating is useful for 7 or 8 half-lives, about 50,000 years.

FIGURE 2.5 Radiocarbon Dating. Radiocarbon dating revolutionized geomorphology starting in the 1950s by providing detailed chronologies through the Holocene and much of the late Pleistocene epochs.

PHOTOGRAPH 2.9 Sample Preparation for Carbon Dating. Vacuum lines for the preparation of graphite from carbon dioxide. The $^{14}C/^{12}C$ ratio will be measured in graphite using an accelerator mass spectrometer (see Photograph 2.12).

PHOTOGRAPH 2.10 Dating Fluvial Sediments. Buried logjam, exposed along the banks of the Queets River on Washington State's Olympic Peninsula, is a perfect target for radiocarbon dating.

K/Ar ages of volcanic rocks have been used to calibrate other dating methods, to assess eruption frequency and thus volcanic hazards, and to determine erosion rates over time by measuring the mass lost from dated lava flows.

U/Th Dating

Uranium (U) and thorium (Th) are used primarily by geomorphologists to date calcium carbonate deposits, including limestone cave deposits and coral reef rocks. Uranium has three different radioactive isotopes, ^{234}U, ^{235}U, and ^{238}U, with half-lives of 250,000, 710 million, and 4.5 billion years, respectively. Although ^{234}U is rare, making up <0.01% of all natural uranium, it is the most useful uranium isotope geomorphically because of its relatively short half-life.

U/Th dating of carbonate minerals is based on the observation that soluble uranium, but not its daughter product insoluble ^{230}Th, is incorporated into calcium carbonate. Thus, the concentration of ^{230}Th will increase over time in newly formed carbonate minerals at a predictable rate as ^{234}U decays into ^{230}Th, which itself has a half-life of 80,000 years. The half-lives of ^{234}U and ^{230}Th indicate that U/Th dating will be useful for materials from thousands to a few hundred thousand years old.

The precision, accuracy, and application of U/Th dating in the solution of geomorphic problems has increased in the last decade, as mass spectrometers replaced decay counting as the method of choice for measuring these isotopes. Mass spectrometers allow measurement of much smaller samples. For example, small samples drilled from cave stalactites can be dated and used to constrain the timing of past climate changes, as indicated by changing stable oxygen isotope ratios. U/Th dates of uplifted coral reef rocks help to determine changes in local relative sea level over time.

Luminescence Dating

Luminescence methods are commonly applied in settings where organic material is not available for dating. These techniques, which include both **thermoluminescence (TL)** and **optically stimulated luminescence (OSL)**, rely on the observation that electrons become trapped over time in the crystal lattices of minerals like quartz and feldspar—the dominant minerals in sand derived from continental rocks. When the sample is heated or stimulated by light, the electrons are released and they give up their energy as photons (luminescence). Emitted photons can be counted and dates calculated if the intensity of irradiation and the radiation sensitivity of the minerals are known. Luminescence methods give dates that represent the last time the electron traps were emptied or "bleached," either by exposure to light during sediment transport or by heating, perhaps in a fire or volcanic eruption [**Figure 2.6**].

Luminescence methods have been widely effective in dating aeolian and fluvial sediments that lack organic material. They are useful over the time ranges of centuries to hundreds of thousands of years, depending on the characteristics of each sample. Improvements in OSL techniques are now enabling the use of this method for some samples, back to nearly 1 million years.

Limitations to these methods include overestimates of age because of incomplete bleaching, as occurs when transport times are short (just minutes) or transport is at night, and underestimates due to anomalous fading, the loss over time of trapped electrons in the absence of heat or light. Samples must not be exposed to light during collection, so blocks of material are typically removed from outcrops and handled in darkness in the lab to preserve the thermoluminescence signal [**Photograph 2.11**]. Recently, luminescence dating has been used in surface process studies to track the exposure history of individual grains as well as collections of grains. Such studies have measured transport times of sand along coastlines and down hillslopes.

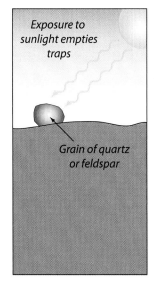
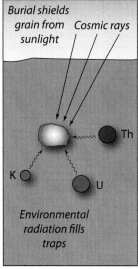

Luminescence dating relies on releasing energy stored in mineral crystal lattices. Over time, **environmental radiation** damages the crystal lattice, creating dislocations in the mineral structure when the energy is absorbed. When the lattice is heated (**thermoluminescence**) or exposed to light (**optically stimulated luminescence**), these **"traps"** empty and the stored energy is released as **photons** (light) that can be counted in the laboratory.

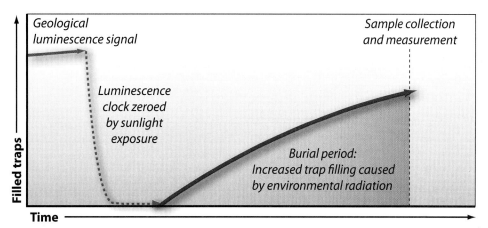

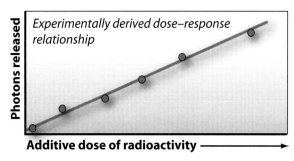

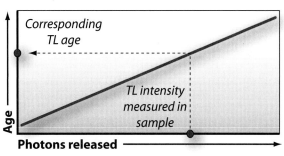

Every sample behaves somewhat differently and has a different sensitivity to radiation. **Additive dosing** is done to test how much luminescence is generated from a unit dose of radiation. Once the radiation sensitivity and the luminescence of the sample have been determined, an age can be calculated. Because ambient light resets the luminescence signal, all sampling needs to be done in the dark and all samples need to be protected from light before and after sampling.

FIGURE 2.6 Luminescence Dating. Luminescence is a useful tool for dating sediments that either contain no organic material, such as dune sands, or for sediments older than the limit of radiocarbon dating (about 50,000 years).

PHOTOGRAPH 2.11 Sampling for Luminescence Dating. The sample is collected in a metal or plastic tube so that it is not exposed to direct sunlight that would reset the luminescence signal.

In-Situ Cosmogenic Nuclide Analysis

In-situ (Latin for "in place") cosmogenic nuclide techniques have developed rapidly since the late 1980s. These techniques use measurements of a variety of stable and radioactive **isotopes** (nuclides) formed within mineral crystal lattices to estimate total cosmic-ray exposure of rock outcrops and sediments. Commonly measured nuclides include ^{3}He, ^{10}Be, ^{21}Ne, ^{26}Al, and ^{36}Cl. The technique relies on the observation that cosmic rays interact with the atoms in rocks at and near Earth's surface to produce new and novel nuclides that are otherwise extremely rare in nature [**Figure 2.7**]. These nuclides are produced at a rate that varies over time and across the Earth's surface because of variations in the strength of the magnetic field, which moderates the flux of cosmic rays from space to Earth's surface.

Once the inventory of cosmogenic nuclides is measured in a rock, then a model age can be calculated. Isolation and measurement of cosmogenic nuclides is done using extensive chemical processing and mass spectrometry. Researchers use accelerator mass spectrometers [**Photograph 2.12**] to measure concentrations of radioactive nuclides (^{10}Be, ^{26}Al, and ^{36}Cl) and noble gas mass spectrometers to count atoms of other, stable nuclides (^{3}He and ^{21}Ne).

There are several key assumptions that underlie the application of cosmogenic nuclides as a dating tool. First, dating assumes that rocks are exposed at the surface with no inventory of cosmogenic nuclides from a prior period of exposure—in reality, cosmic radiation penetrates several meters below Earth's surface, and rocks often contain at least low levels of "inherited" cosmogenic nuclides. Inherited nuclides lead to overestimates of rock exposure ages. Second, dating assumes that rocks do not erode after they are exposed—of course, rocks do erode, and when they do, cosmogenically generated nuclides are lost. Erosion of rock surfaces leads to age underestimates, as does intermittent or past burial by snow, soil, or windblown sediment.

Application of cosmogenic nuclides for geomorphic dating has become widespread. Perhaps the most common sampling targets are glacial boulders and glacially sculpted

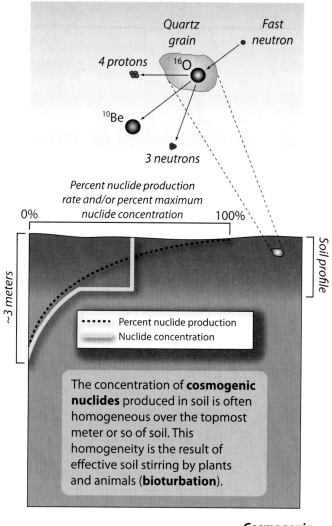

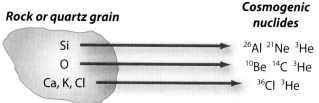

FIGURE 2.7 Cosmogenic Nuclide Production. In-situ produced cosmogenic nuclides, first applied to problems in geomorphology in the 1980s, are now providing rates and dates in many landscapes around the world.

bedrock surfaces [**Photograph 2.13**]. Dates from such surfaces have clarified long-standing controversies about the extent of glacial ice in high latitudes. Many studies, however, have found significant variability in ages of boulders sampled from single landforms, so these types of dates

PHOTOGRAPH 2.12 Accelerator Mass Spectrometer. The accelerator mass spectrometer at Lawrence Livermore National Laboratory occupies much of a football-field-long building and spends its days (and nights) counting atoms. The blow-up Emperor penguin, a little over a meter high, is for scale.

PHOTOGRAPH 2.13 Cosmogenic Nuclide Sampling. Drive-by cosmogenic nuclide sampling being done to date a glacially deposited boulder on Baffin Island, Canada.

are often considered minimum limiting ages. Cosmogenic nuclides have proved useful for dating a wide array of other geomorphic features, including river terraces, rock falls, fault scarps, marine platforms, archaeological artifacts, and desert pavements. All applications require knowledge of specific settings; some applications require depth profiles and advanced analytical models to decipher complex exposure and burial histories.

Cosmogenic nuclides can be used to date landforms over a wide range of ages. Recent advances in measurement techniques now allow dating of mountain glacial deposits formed during the Little Ice Age, only a few hundred years ago. Because erosion removes rock surfaces and the cosmogenic nuclides those rock surfaces contain, in most cases, erosion, rather than nuclide half-lives, constrain the upper age range of cosmogenic dating. Rock surfaces are generally not stable over hundreds of thousands or millions of years, with the exception of places with almost no liquid water, like Antarctica and the Atacama Desert, and in places where the rocks are exceptionally strong (e.g., massive quartzite, sandstone, and granite). Thus, cosmogenic exposure ages tend to be most reliable in the 1000 to 100,000 year range.

Measuring Rates of Geomorphic Processes

Over many years and in different environments, geomorphologists have employed a variety of techniques to understand rates of landscape change, specifically erosion of the land surface and transport of mass from one place to another. Some of these techniques measure rates of mass removal directly; others infer rates of change by measuring proxies like isotope abundance and surface elevation. It is always important to consider the **integration time** (time over which the resulting rate is averaged) inherent in each method, and to acknowledge the assumptions underlying the rate calculations.

Different techniques for estimating erosion provide data useful over a variety of spatial and temporal scales; thus, one needs to decide what integration time is most representative of the process being studied. For example, thermochronologic data provide a long-term view of the landscape but do not have the resolution to "see" climatically driven changes in denudation over the last few million years. In contrast, modern-day sediment yields that are based on just a few years or decades of data may grossly underrepresent or overrepresent the long-term flux of mass out of a drainage basin. In some cases, like the southern Appalachian Mountains, sediment yield, cosmogenic nuclide, and thermochronologic measurements all show similar erosion rates and/or sediment yields over widely different timescales.

Sediment Generation Versus Sediment Yield

There is a crucial distinction between sediment generation (the rate at which sediment forms and erodes from its parent rock) and sediment yield (the rate at which sediment is exported from the landscape). Sediment yields and sediment generation rates are often assumed to be equal, but that is only the case if sediment storage on the landscape is unchanging. If sediment is being mined by fluvial incision and carried out of the watershed by stream flow, sediment yield will exceed the rate of sediment generation. Conversely, if sediment is being trapped in the watershed, perhaps by deposition on footslopes or in river terraces and not exported, sediment yields will be less than erosion rates.

Sediment yields (mass/time) from a river basin are generally calculated for a series of time intervals by multiplying river discharges (volume/time) measured at gauging stations with suspended sediment concentrations (mass/volume) in samples from the same stations [**Figure 2.8**]; the results are then summed to give the annual load. Bedload

Estimating contemporary **sediment flux** out of a **drainage basin** is a complicated, multi-step process. First, measurements of water discharge must be made at a variety of **stages** (water surface elevations) and used to create a water **discharge rating curve.** Then, stage is measured over time and recorded. Together, stage measurements and the water discharge rating curve are used to estimate water discharge over time. A **sediment rating curve** is created by measuring **suspended sediment** load at different water discharges. Using the sediment rating curve and water discharge estimates, one can calculate sediment loads for each water discharge value. Summing daily sediment loads allows calculation of **sediment load** from a drainage basin on an annual basis.

Step 1. River stages can be measured manually by reading **staff gauges** or automatically by stage recorders. Some recorders use floats that measure water surface elevation directly; others use pressure transducers that measure the mass of water overlying the sensor.

Step 3. Convolving measurements of stage with the discharge rating curve allows for calculation of discharge over time. Data are collected at different time intervals and are often reported as mean daily discharges.

Step 5. Sediment loads (mass/time) are calculated by multiplying sediment concentrations (mass/volume) and water discharge (volume/time). Often, daily data are used for this calculation and then summed to give annual sediment fluxes out of a drainage basin. Since most sediment moves in large runoff events, there can be great variability in sediment yields year to year.

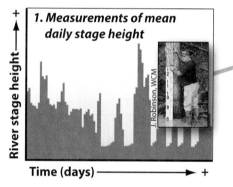

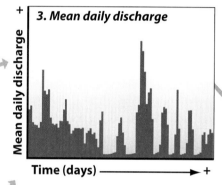

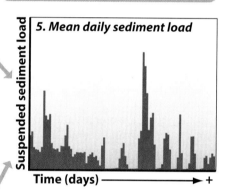

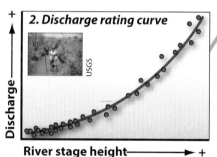

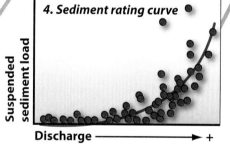

Step 2. Water discharge rating curves are created from numerous measurements of discharge at a gauging station. These measurements are made at different river stages using velocity meters and measured channel cross sections or by damming the river and forcing flow over a **weir** for which the stage-discharge relationship is known. Discharge rating curves are remeasured to ensure accuracy as channel dimensions and thus the rating curve can change over time.

Step 4. Sediment rating curves compare sediment concentration and water discharge and are often quite noisy, due to variations in sediment supply to the river. They are constructed by measuring sediment concentrations at different water discharges. Sediment concentrations can be measured discretely by sampling river water or continuously using recording **turbidity** meters.

Contemporary sediment yields, calculated from measurements of riverine sediment flux over only years to decades, may either over- or underestimate long-term erosion rates due to human impacts or climate variability.

FIGURE 2.8 Measuring Sediment Loads. Sediment is produced on hillslopes but is exported by river channels. Estimating the sediment load carried by a river or stream is a complex, multistep process.

PHOTOGRAPH 2.14 Measuring Channel Erosion. Geologists surveying channel erosion in Upper West Fork Muddy River near Mount Saint Helens after the 1980 eruption. Note instrument and geologist on bank and the rod person in the channel at the foot of the slope.

is usually either disregarded or assumed to be <10 percent of the total load. Until recently, sediment concentration was typically measured only episodically, and thus **sediment rating curves** (which show the relationship between water discharge and sediment load) had to be used to estimate loads. Now, automated turbidity meters can provide continuous estimates of suspended sediment loads, allowing improved estimates of mass export by streams over time.

Landscape Change at Outcrop and Hillslope Scales

Measuring the change in surface elevation at the scale of centimeters to meters provides a direct means of documenting landscape change. This can be done in the field using erosion pins (stakes that provide stationary reference points) as well as detailed, repeated topographic surveys [Photograph 2.14]. The exposure of tree roots, which are normally buried by soil, provides a natural datum by which to measure surface lowering [Photograph 2.15]. Similarly, the accumulation of soil upslope of dated stone walls or foundations allows calculation of mass flux rates down hillslopes. Such approaches are useful for understanding erosion over weeks to perhaps centuries on a farm field or a hillslope, but the meaning of these data at larger time and spatial scales is uncertain. However, such measurements can reveal much about processes that can be generalized and used at broader spatial and temporal scales.

Isotopic methods can also be used to estimate rates of erosion over small spatial scales. **Cesium-137** is a radioactive isotope (half-life = 30.1 years) that geomorphologists use to estimate rates of soil erosion and sediment deposition over the timescale of decades. Nuclear weapons tests in the 1950s and 1960s released large amounts of ^{137}Cs into the atmosphere. The ^{137}Cs rained down on the world's landscapes, and because cesium binds tightly to soil particles, subsequent erosion carried the ^{137}Cs off hillsides and into valley bottoms. ^{137}Cs deposition from the atmosphere has been well documented and one can predict the amount that should be present in a soil profile. Inventories of ^{137}Cs in soils can thus be used to determine

PHOTOGRAPH 2.15 Soil Erosion. The roots of this tree, in Utah's Bryce Canyon National Park, were completely undercut by soil erosion after it first grew from a seed. If the tree were cored and the rings counted, the depth of root exposure could be used to estimate a local erosion rate over the life of the tree.

whether an area is eroding (less ^{137}Cs than expected) or aggrading (more ^{137}Cs than expected) [**Figure 2.9**].

On timescales of thousands to hundreds of thousands of years, cosmogenic nuclides measured in samples collected from exposed rock surfaces are used to infer rates of erosion at that point, presuming that erosion rates have remained steady over time and there has been no soil cover. Cosmogenic estimates of point erosion rates integrate over time, but they only tell the erosion history of the sample site and say little about erosion on a larger spatial scale.

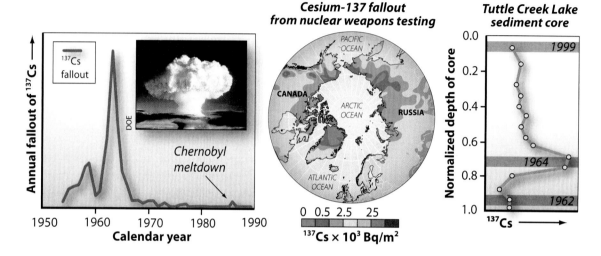

^{137}**Cs** was released into the environment primarily by the explosion of **nuclear weapons** in the atmosphere. This **radionuclide** is the product of **fission,** or the splitting of heavy elements such as uranium. The detonation of powerful hydrogen bombs (which have fission bombs as their core), starting in the 1950s, released large amounts of ^{137}Cs, with the peak release in 1964. After atmospheric testing of nuclear weapons was curtailed, the release of ^{137}Cs quickly decreased. The 1986 meltdown of the nuclear reactor at Chernobyl, in the former Soviet Union, released additional ^{137}Cs to the atmosphere.

^{137}Cs, which decays with a half-life of about 30 years, emits **gamma particles,** which are easily detected by sensitive **decay counters.** In **sediment cores,** the peak ^{137}Cs concentration is assigned to 1964, defining the age of the core at this horizon. ^{137}Cs concentration is described in units of becquerels (Bq), a unit of radioactivity, corresponding to one emission per second.

^{137}Cs is a useful monitor of erosion and deposition since the early 1960s, when the isotope was delivered to the soil surface by precipitation. ^{137}Cs sticks to soil and thus migrates no more than a few tens of centimeters below the ground surface. Using decay counting, geomorphologists measure the inventory of ^{137}Cs. If the measured inventory is greater than the amount delivered by nuclear weapons testing, then the measured profile is a depositional zone where eroded soil and the attached ^{137}Cs have accumulated. If the measured inventory is less than the amount delivered by nuclear weapons testing, the sampled profile is in an area that has eroded since the 1960s.

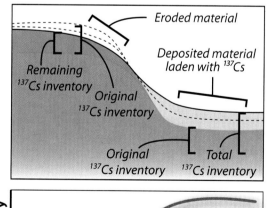

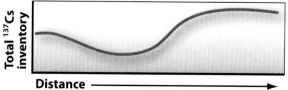

FIGURE 2.9 Cesium-137 Deposition and Interpretation. The ^{137}Cs produced by the explosion of atomic weapons and distributed around the world by atmospheric processes provides a tool for dating sediment and for determining where erosion and deposition have occurred over the past several decades.

Landscape Change at Basin Scales

Sediment yield over long periods of time (decades) and large areas (square kilometers) can be calculated by estimating the volume of sediment trapped in man-made reservoirs. By measuring the volume of retained sediment, the area of the drainage basin supplying the sediment, and the age of the dam, one can estimate the average sediment delivery rate to the impoundment from upstream and, assuming steady state, calculate a catchment average erosion rate integrated over the time of deposition. No information about the spatial variability in erosion rates is retained—only the catchment average value integrated over the age of the reservoir can be calculated.

In arid regions, such as the southern Negev Desert of Israel, dry streambeds have been dammed awaiting rare flows. The area behind the dam is carefully surveyed. When the flow finally comes, sediment and water are trapped behind the dam, the water infiltrates, the sediment settles out, and the area is resurveyed. The volume difference between the two surveys can be used to calculate the sediment yield for each flow event.

Similar approaches have been taken for natural sediment traps, including deep lakes and the ocean basins. Over longer timescales, up to millions of years, sediment volumes in depositional basins can be estimated using a combination of cores and remote sensing data such as seismic surveys. Calculations of erosion rates based on these types of data rely on a variety of assumptions, the most uncertain of which is steady state, the idea that the volume of sediment stored in the drainage basin has not changed appreciably over the time frame during which the sediment load has been captured. For sediment emplaced over millions of years, such as sediments deposited off the passive margin of eastern North America, the boundaries of the drainage basin itself may have changed, making such calculations inherently uncertain.

Measurements of the concentration of cosmogenic nuclides, such as ^{10}Be in river sand, can also be used to estimate the average rate at which drainage basins erode over time. Slowly eroding basins generate sediment with high concentrations of cosmogenic nuclides because the sand has resided within a meter or two of the land surface, where most cosmogenic nuclides are produced, for many thousands to tens of thousands of years. Sediment transported out of rapidly eroding basins will have low nuclide concentrations, reflecting its short residence time near Earth's surface. For rapidly eroding basins in tectonically active areas, cosmogenic nuclide measurements usually only record erosion rates over the Holocene Epoch, whereas in slowly eroding cratonic areas, this method averages erosion rates over 100,000 years or more.

Erosion Rates over 10^6 to 10^8 Year Timescales

Thermochronometry is a radiometric dating technique that allows geologists to measure erosion and rock uplift over time frames of millions to tens of millions of years. The accumulation of stable noble gas isotopes (helium and argon), as well as development of **fission tracks** (damage to the mineral lattice) in certain uranium-bearing rocks and minerals, both provide time/temperature histories of rocks [**Figure 2.10**]. Thermochronometric methods assume that noble gases and damage tracks begin to be retained in minerals when they cool below a mineral-specific range in **closure temperature,** and thus at a specific depth below the surface.

Because temperature generally increases with depth below Earth's surface, defining the **geothermal gradient,** thermochronologic methods provide a means of assessing depth/time relationships for rocks and can thus be used to interpret long-term erosion or "unroofing" rates. These interpretations, however, carry significant uncertainty because past geothermal gradients are difficult to estimate and because the closure temperature is actually a range rather than a single temperature, the extent of which depends on the rate of cooling. Changes in tectonic regime, changes in speed of erosion that can advect warm rock toward the surface, and cooling of rock once warmed by igneous activity may all complicate paleo-depth/time measurements. Even with these uncertainties, thermochronology provides a unique view into Earth's history and long-term, broad-scale changes at and near Earth's dynamic surface.

Experiments

Experiments, both physical and virtual, can provide insight into the rate, distribution, and importance of various surface processes. Some experiments are conducted in the field, while others are run under controlled laboratory conditions. Recent advances in computational power and efficiency have led to increasingly sophisticated computer models that run numerical experiments addressing systems, processes, and timescales that defy analysis by conventional field and laboratory methods.

Field Experiments

Field experiments manipulate or investigate natural systems. For example, geomorphologists conduct sprinkling experiments at various time and spatial scales to better understand the hydrology and sediment dynamics of hillslopes. The response of instrumented slopes and even whole drainage basins to natural and applied rainfall is monitored to understand the rate at which water infiltrates into and moves through different soil types and soils that have been subjected to different land uses. Sprinkling tests on natural slopes test predictions of water-table rise after precipitation events and help determine the influence of rainfall on slope stability. Such experiments have revealed that a surprising amount of storm runoff can move through weathered, near-surface bedrock, beneath the soil.

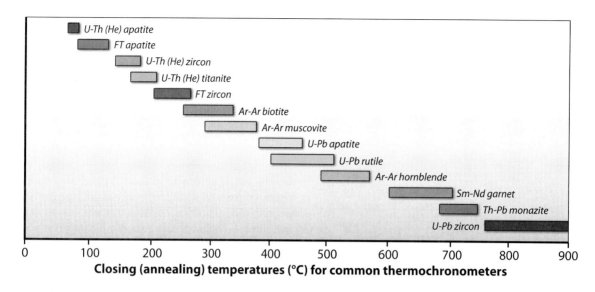

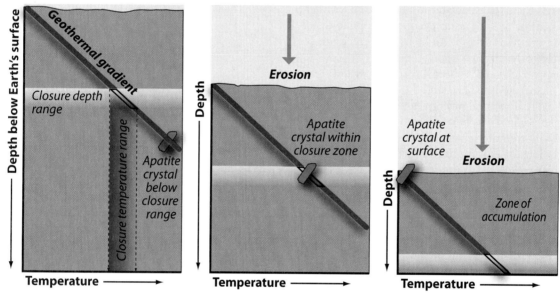

FIGURE 2.10 Thermochronometry. Thermochronologic analyses of different elements retained in different minerals provide a means by which to understand how quickly erosion and tectonic processes move material toward Earth's surface over timescales of millions of years.

Grain-size analysis is used to characterize the size distribution of material carried down hillslopes and by streams and rivers. Such grain-size data can be used to infer the power of sediment transport processes. The grain size of fine sediments is inferred from their settling rate or light scattering behavior in a fluid. For sandy sediments, grain size is determined using sieves with openings of known diameter. For coarser sediments—pebbles, gravel, and boulders—geomorphologists perform pebble counts. Such pebble counts are a statistical analysis of a natural system and involve random sampling of a gravel bar or hillslope often using a grid system or a blind reach to collect samples so that the diameter of each pebble can be measured. Usually, 100 or more pebbles are counted to ensure a representative median particle size. If compositional data are important, perhaps to identify the sources of sediment, then the lithology of each pebble can be identified.

Marked clasts have long been used to understand the movement of sediment in rivers and down hillslopes [Photograph 2.16]. Painted or magnetically tagged pebbles and cobbles are introduced into streams and used to measure average velocities of clasts moved as bedload. For example, researchers deployed and tracked 1600 painted pebbles at four sites in the Mojave Desert in order to calculate short-term rates of sediment movement. In another experiment, painted clasts were placed on steep, arid hillslopes. Repeated surveying over 20 years allowed measurement of both the timing and amount of downslope movement. A higher-tech approach involves drilling the pebbles and inserting a small magnet or radiotag, allowing them to be identified even when buried in the stream bottom. Similarly, the introduction of geochemically unique waste into a river system such as the Rio Grande, as happened at the Los Alamos lab during development of nuclear weapons in the 1940s, allows tracing of sediment movement.

In other field experiments, landforms have been instrumented with automated sensors in order to understand better how they are changing over time. For example, geomorphologists used soil temperature, moisture, and tilt meters to study intriguing landforms called sorted circles that are found in permafrost regions like the Canadian and Scandinavian arctic (see Chapter 9). After instrumenting the circles, they determined that the soil within them slowly convects during freeze–thaw cycles. Over decades to centuries, this movement sorts fine material into the center of the circle; coarse material accumulates on the edges. In humid and arid regions, instrumentation of small channels and slopes has contributed greatly to understanding the rates of and factors controlling geomorphic processes.

Laboratory Experiments

Physical models of natural systems are widely used to understand geomorphic systems better. In the lab, these physical models are manipulated to see how they respond to changes in boundary conditions. Such an approach considers the model as an analog of a natural system. Such **analog experiments** include elements of both field and lab science. Results can be quite useful, but there are limitations related to **scaling** issues due to differences in size between models and natural systems; that is, most physical models of rivers must use smaller than natural channels but cannot alter the **viscosity**, or resistance to deformation, of water. When constructing physical models of hillslopes, boundary conditions, such as the edges of the experimental device, can have unforeseen consequences on the results. Nevertheless, analog techniques are useful for posing and answering questions at time and length scales that are impractical to address otherwise.

A **flume** (an artificial stream channel) is perhaps the most common physical model of a geomorphic system [Photograph 2.17]. Flumes of different sizes and shapes are used to understand the behavior of streams and rivers, including the dynamics of fluid flow and its effect on fluvial channel banks and bank materials. Flumes with shallow slopes and more water than sediment are used to understand how rivers work. Steeply sloping flumes charged with large amounts of sediment are used to examine debris flow behavior. Hillslopes and drainage basins are also modeled using physical analogs. Experiments range from those that simulate development of a single hillside, to larger, more complicated experiments designed to simulate the development of large basins as they fill with sediment and subside over geologic time [Photograph 2.18].

Laboratory experiments are particularly useful for understanding processes such as rock weathering that occur on timescales so long that direct observation of change is difficult. For example, early laboratory heating experiments

PHOTOGRAPH 2.16 Tracking Sediment Movement. Placing painted pebbles at General Patton's former Camp Iron Mountain in the Mojave Desert to trace sediment movement over time.

PHOTOGRAPH 2.17 Experimental Flume. This experimental flume at the University of California's Richmond Field Station is designed to mimic the flow in a gravel-bedded stream.

simulated rock weathering both by fire and by diurnal temperature fluctuations. Such experiments confirmed field observations that fire is an effective rock-weathering agent. Other lab experiments include studying chemical weathering rates by leaching rocks in laboratory beakers and simu-

PHOTOGRAPH 2.18 The Jurassic Tank. An analog model at the National Center for Earth-Surface Dynamics, University of Minnesota, is set up to simulate deltaic sedimentation.

lating abrasion of clasts during downstream transport using rock tumblers.

Numerical Models

Over the past several decades, the advent of faster computers and a more quantitative approach to geomorphology have stimulated the development of **numerical models.** Such models simulate the behavior of natural systems over time in either two or three dimensions. They require an initial condition (the shape of the landscape), a set of mathematical equations that describe change over time quantitatively, and the establishment of boundary conditions such as the fate of sediment leaving the system or the rate of uplift and erosion.

Numerical models have been used at all scales in geomorphology. Some describe the movement of bedforms on a streambed; others simulate the dissection and retreat of great escarpments after continental rifting. Numerical models are particularly useful for testing the sensitivity of geomorphic systems to changes in boundary conditions imposed by tectonics and climate. Model results often provide hypotheses that can be tested in the field by using other analytical techniques.

Proxy Records

A variety of natural archives record past changes in the characteristics and behavior of Earth's biosphere, hydrosphere, and atmosphere. Some archives are sedimentary; others are biologic. Most archive data are referred to as **proxy records** because they cannot be interpreted directly as changes in climate or surface characteristics. Rather, these records provide information that can be interpreted through the use of **transfer functions** (quantitative relationships between a physical/chemical/isotopic/biologic property and the characteristic of interest). For example, in tropical regions, the chemistry of corals (e.g., calcium-magnesium ratio) changes over time, reflecting sea-surface temperature. In arid and semi-arid areas, the width of tree rings is a proxy for water availability and thus drought status and climate. In more humid environments, tree rings (and the scars they preserve) are used to estimate forest-fire frequency, and in the right setting, the impact frequency of land-altering debris flows.

Sedimentary deposits, such as those preserved in the depths of lakes and oceans, hold many important proxy records that can be interpreted both in terms of changing climate and geomorphic response (see Chapter 13). Such deposits are routinely sampled by coring, although samples from other sedimentary archives, such as deposits of wind-blown loess, are recovered through surface excavation. Sediment samples not only provide physical proxy records, but they also host biological proxy records including pollen, other plant parts, and the remains of diagnostic insect species. Glaciers and ice sheets are a different sort of sedimentary system, but they too preserve chemical, isotopic, and physical records of changes in climate and ice accumulation over time. Ice cores have been collected in many places on the Greenland and Antarctic ice sheets and document atmospheric and climatic conditions over much of the past million years.

Applications

As geomorphologists, we now have the ability to measure and to quantify accurately the rates of surface processes as well as the dates at which sediments were deposited and landforms were created. The powerful analytical tool kit at our disposal allows geomorphologists to understand and predict the behavior of Earth's surface in ways that are directly relevant to society. A variety of different visualization tools allows us to communicate broadly and clearly our understanding of geomorphic processes and landscape change over time [**Figure 2.11**].

Erosion and the sediment it releases are fundamental to human survival. Eroding topsoil decreases fertility and the ability of societies to feed their people. Once eroded, that same soil clogs waterways with sediment and impacts aquatic organisms. Fifty years ago, geomorphologists could do little more than measure suspended sediment load and survey discrete locations along changing channels and slopes—extrapolating their results to the rest of the landscape. Today, the game has changed. Cosmogenic ^{10}Be can be used to estimate background rates of soil formation. Recording turbidity meters can accurately estimate sediment yields. ^{137}Cs can be used to estimate erosion rates, depths, and timing, and LiDAR, GPS, and InSAR can be used to quantify changes in topography over time across the entire landscape.

Geomorphologists and the analytic tools they employ are playing a key role in understanding the timing, magnitude, cause, and effects of past climate changes. For example, a broad range of field and analytic tools have been used to understand how sea level has changed over time in relation to warming and cooling of the climate. The role of ice sheets and their response to changing climate is key to understanding sea-level change. Ice-sheet behavior in the past is now routinely quantified by carbon and cosmogenic dating of glacial deposits. Sea-level change is quantified by dating and surveying both submerged and raised beaches, and the contemporary behavior of ice sheets is closely monitored with GPS and InSAR.

Earthquakes and the seismic shaking and land movement they cause are a major geologic hazard. Geomorphologists and their tool kit of analytical and field techniques play an important role in defining earthquake hazard levels and the probability of future earthquakes. Fieldwork,

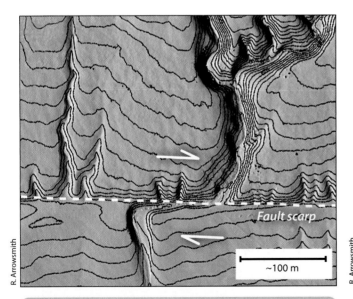

A detailed topographic map with a 2-meter contour interval shows the offset of Wallace Creek by the San Andreas Fault.

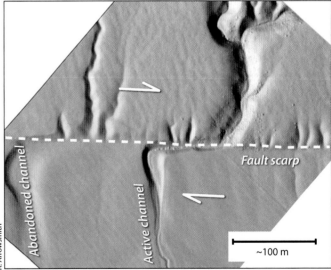

High-resolution **LiDAR** image shows both the current (active) and abandoned channels of Wallace Creek as well as the San Andreas Fault **scarp.**

FIGURE 2.11 Visualizing Geomorphology. There are many different ways of looking at Earth's surface, each of which communicates information differently. Here are two different ways of viewing tectonic landforms in the Carrizo Plain of California, where the San Andreas strike-slip fault offset the channel of Wallace Creek about 3700 years ago. Recent earthquakes have offset the stream in approximately 5-m increments every few centuries on average.

along with remote sensing such as LiDAR and high-precision GPS, locates fault scarps. Dating techniques such as carbon-14, luminescence, and cosmogenic nuclides allow estimates of the timing of paleo-earthquakes and thus the probability of future seismic events. After an earthquake, GPS and InSAR are used to quantify deformation.

Not only do data collection techniques allow geomorphologists to understand change over time, but they allow us to better understand fine-scale process-response relationships. For example, remote-sensing data such as aerial photography allow land managers to track channel changes remotely and to quantify changes in fish habitat (such as the size and locations of pools) over time and after significant disturbance events. LiDAR can be used to estimate both the top of the forest canopy and "bare-earth" elevation beneath the trees; thus, while foresters might find the data useful to characterize stand composition and tree height, geomorphologists can use the same data set to find fault scarps and earthflows hidden under the forest cover.

Although many technologies useful for understanding Earth's surface are readily available, their application is not always straightforward. Producing meaningful and accurate results requires an understanding not only of the field situation in which the technique will be used, but also of the limitations of each technique and the biases introduced if underlying assumptions are not properly understood or met. In short, the challenge facing today's geomorphologist is choosing the right tool for the job and understanding the geomorphic context in which the tool is used.

Selected References and Further Reading

Bada, J. L., and R. A. Schroeder. Racemization of isoleucinen in calcareous marine sediments: Kinetics and mechanism. *Earth and Planetary Science Letters* 15 (1972): 1–11.

Bierman, P. R., and M. Caffee. Cosmogenic exposure and erosion history of ancient Australian bedrock landforms. *Geological Society of America Bulletin* 114 (2002): 787–803.

Bierman, P. R., and K. Nichols. Rock to sediment—Slope to sea with ^{10}Be—Rates of landscape change, *Annual Review of Earth and Planetary Sciences* 32 (2004): 215–255.

Bull, W. B., and M. T. Brandon. Lichen dating of earthquake-generated regional rockfall events, Southern Alps, New Zealand, *Geological Society of America Bulletin* 110 (1998): 60–84.

Colman, S. M., K. L. Pierce, and P. W. Birkeland. Suggested terminology for Quaternary dating methods. *Quaternary Research* 28 (1987): 314–319.

Crook, R. J., and A. R. Gillespie. "Weathering rates in granitic boulders measured by P-wave speeds." In S. M. Colman and D. F. Dethier, eds., *Rates of Chemical Weathering of Rocks and Minerals*, Orlando: Academic Press, 1986.

Ehlers, T. A., and K. A. Farley. Apatite (U-Th)/He thermochronometry: Methods and applications to problems in tectonic and surface processes. *Earth and Planetary Science Letters* 206 (2003): 1–14.

Gardiner, V., and R. V. Dackombe. *Geomorphological Field Manual.* London: Allen & Unwin, 1983.

Gosse, J. C., and F. M. Phillips. Terrestrial in situ cosmogenic nuclides: Theory and application. *Quaternary Science Reviews* 20 (2001): 1475–1560.

Kirchner, J. W., R. C. Finkel, C. S. Riebe, et al. Mountain erosion over 10 yr, 10 k.y., and 10 m.y. time scales. *Geology* 29 (2001): 591–594.

Lal, D. Cosmic ray labeling of erosion surfaces: *In situ* nuclide production rates and erosion models, *Earth and Planetary Science Letters* 104 (1991): 424–439.

Libby, W. F. Radiocarbon dating. *Philosophical Transactions of the Royal Society of London. Series A, Mathematical and Physical Sciences* 269 (1970): 1–10.

Lingenfelter, R. E. Production of carbon 14 by cosmic-ray neutrons. *Reviews of Geophysics* 1 (1963): 35–55.

Miller, G. H., and S. J. Clarke. "Amino-Acid Dating." In S. A. Elias, ed., *Encyclopedia of Quaternary Sciences*, Boston: Elsevier, 2007.

Montgomery, D. R., W. E. Dietrich, R. Torres, et al. Hydrologic response of a steep, unchanneled valley to natural and applied rainfall. *Water Resources Research* 33 (1997): 91–109.

Nichols, K. K., and P. R. Bierman. Fifty-four years of ephemeral channel response to two years of intense World War II military activity, Camp Iron Mountain, Mojave Desert, California. In J. Ehlen and P. Harmon, eds., *The Environmental Legacy of Military Operations*, Boulder, CO: Geological Society of America, 2001.

Persico, L. P., K. K. Nichols, and P. R. Bierman. Tracking painted pebbles: Short-term rates of sediment movement on four Mojave Desert piedmont surfaces. *Water Resources Research* 41 (2005): W07004, doi:10.1029/2005WR003990

Pierce, K. L., and S. M. Colman. Effect of height and orientation (microclimate) on geomorphic degradation rates and processes, late-glacial terrace scarps in central Idaho. *Geological Society of America Bulletin* 97 (1986): 869–885.

Portenga, E., and P. R. Bierman. Understanding Earth's eroding surface with ^{10}Be. *GSA Today* 21, no. 8 (2011): 4–10.

Staley, J. T., J. B. Adams, and F. E. Palmer. "Desert varnish: A biological perspective." In G. Stotzky and J.-M. Bollag, eds., *Soil Biochemistry*, Volume 7, 1992.

Tarolli, P., J. R. Arrowsmith, and E. R. Vivoni. Understanding earth surface processes from remotely sensed digital terrain models. *Geomorphology* 113 (2009): 1–3.

Von Blanckenburg, F. The control mechanisms of erosion and weathering at basin scale from cosmogenic nuclides in river sediment. *Earth and Planetary Science Letters* 237 (2005): 462–479.

Wintle, A. G. Luminescence dating: Where it has been and where it is going, *Boreas* 37 (2008): 471–482.

Wintle, A. G., and A. S. Murray. A review of quartz optically stimulated luminescence characteristics and their relevance in single-aliquot regeneration dating protocols. *Radiation Measurements* 41 (2006): 369–391.

Woldenberg, M. J., ed. *Models in Geomorphology*. Boston: Allen & Unwin, 1985.

Zhang, W., and D. R. Montgomery. Digital elevation model grid size, landscape representation, and hydrologic simulations. *Water Resources Research* 30 (1994): 1019–1028.

DIGGING DEEPER How Does a Dating Method Develop?

For geomorphologists, radiocarbon (^{14}C) dating is the most powerful and widely used dating method. Dozens of commercial and academic labs routinely measure thousands of samples every year, some smaller than the size of a flea. Geomorphologists use radiocarbon measurements to constrain the timing of climate swings, the recurrence interval of earthquakes, and the age of volcanic eruptions. But the technique has only existed since about 1950.

Libby and colleagues at the University of Chicago laid the foundation for radiocarbon dating by developing sensitive analytical equipment (decay counters) capable of measuring the radioactivity given off by decaying ^{14}C atoms (half-life of 5730 years). Sample preparation involved combustion of samples to create carbon dioxide (CO_2) gas, waiting weeks for radioactive radon gas to decay, and then days to weeks of counting emissions as ^{14}C decayed. Using the equipment they developed, Libby and others measured the **activity** (decay rate per unit time) of ^{14}C in plant and animal material from around the world. They found that living organisms had similar concentrations of ^{14}C, whether expressed as a ratio of $^{14}C/^{12}C$ ($1.4 \cdot 10^{-12}$) or in terms of their disintegration rate (13.6 disintegrations per minute per gram C). By measuring **beta emissions** (electrons) given off as ^{14}C decayed, early workers were actually counting ^{14}C atoms indirectly.

In 1949, Arnold and Libby published a paper that revolutionized geomorphology (Arnold and Libby, 1949). To establish the validity of radiocarbon as a dating method, they analyzed samples of wood that had been dated by other means [**Figure DD2.1**]. Their data were convincing: Radiocarbon ages matched the other ages within the uncertainty of the measurements. The fledgling dating method was now verified, other labs were set up, and the method was improved (Suess, 1954). Soon, measurements were being made around the world, including the first numerical dating of marine cores (Suess, 1956). Dating organic matter in these cores with radiocarbon showed that the coldest temperatures were reached ~15,000 years ago; extrapolation back in time from the dated sections of the cores suggested that the last glacial episode began ~100,000 years before the present. At last, there was a numerical chronology stretching into and beyond the last glacial maximum, now taken to be about 20,000 years ago.

Quickly, scientists began investigating the assumptions underlying the radiocarbon method and learning more about how Earth's surface, biota, oceans, and atmosphere interact. Suess (1955) noted that modern wood had lower $^{14}C/^{12}C$ ratios than nineteenth-century wood. He attributed this difference to recent plant uptake of ^{14}C-depleted CO_2, added to the atmosphere by the burning of coal old enough that it contained no radiocarbon. Now known as the **Suess effect**, the presence of ^{14}C-depleted CO_2 was critical evidence that humans were indeed changing atmospheric composition and suggested that atmospheric ^{14}C concentration

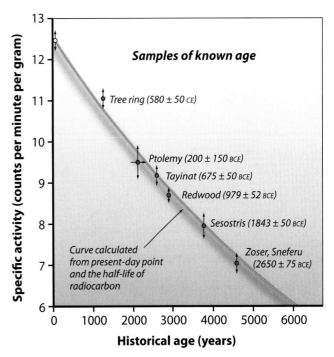

FIGURE DD2.1 Working closely with other scientists, including archaeologists and dendrochronologists, Arnold and Libby measured radioactivity generated by the decay of ^{14}C in material that had been well-dated by other means. Their samples included wood from trees that had been dated by ring counting and wooden artifacts from Egyptian tombs and Syrian palaces. [From Arnold and Libby (1949).]

DIGGING DEEPER How Does a Dating Method Develop? *(continued)*

responded to changes in carbon cycling. Just a few years later, Broecker and Walton (1959) detected ^{14}C produced by atmospheric testing of nuclear weapons. They used these data to show that the CO_2 in the atmosphere was well-mixed between the hemispheres in only a year or two—another principal assumption of the radiocarbon method.

Accurate radiocarbon dating depends on knowledge of the initial concentration of ^{14}C in the sample. Arnold and Libby's 1949 work showed, to a first approximation, that the initial concentration of ^{14}C in living organisms varied little over at least the last 5000 years. Later work by Stuiver (1978a) extended the first-order comparison of radiocarbon dates to other dates back over <30,000 years [**Figure DD2.2**]. However, changes in cosmic-ray flux, the strength of Earth's magnetic field, which moderates the cosmic-ray flux, and ocean/atmosphere CO_2 exchange all suggest that atmospheric ^{14}C content changes over time. Several scientists tested the temporal variation in initial ^{14}C concentration by measuring ^{14}C in independently well-dated materials (Suess, 1970; Stuiver, 1978a), principally wood that had been accurately dated by dendrochronology [**Figure DD2.3**]. They found that radiocarbon years were not equal to calendar years, showing convincingly that the concentration of radiocarbon in the atmosphere had changed over time.

Today, radiocarbon ages are routinely calibrated before comparison to age data generated by other chronometers. This calibration is done iteratively, considering the measured activity of ^{14}C in the sample, the decay constant of ^{14}C, and the atmosphere ^{14}C content estimates painstakingly worked out by Stuiver and others over the past several decades. The development of web-based calibration programs, such as the widely used, CALIB, makes it easy to account for changing initial concentrations and to produce calibrated radiocarbon ages (Stuiver and Reimer, 1993).

In the late 1970s, a revolution in measurement technology produced a major advance in radiocarbon dating (Bennet et al., 1977; Muller et al., 1978; Nelson et al., 1977). Instead of counting the decays of radiocarbon atoms, several research groups used both cyclotrons and linear particle accelerators to count atoms of different

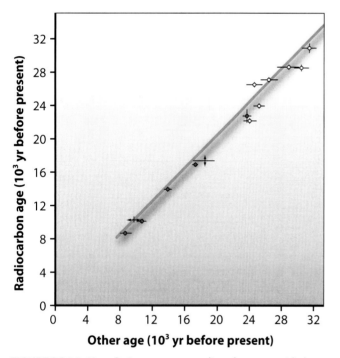

FIGURE DD2.2 Here, Stuiver compares radiocarbon ages with those determined by other methods, including U/Th on lake sediments (open circles), luminescence dates on basalt (double-headed arrows), and dates based on measurements of Earth's magnetic field inclination preserved in a lake core (solid circles), to demonstrate that inaccuracies in radiocarbon dating over the past 32,000 years amounted to, at most, a few thousand years. [From Stuiver (1978a).]

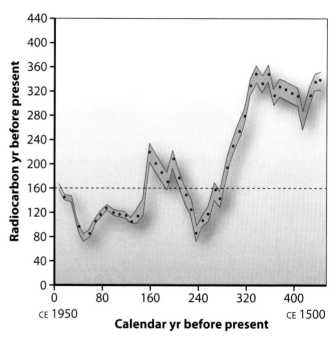

FIGURE DD2.3 If the concentration of radiocarbon in the atmosphere were constant over time, radiocarbon ages would directly match calendar ages. To test the consistency of atmospheric radiocarbon concentration, Stuiver precisely measured the radiocarbon content of wood taken from slabs of trees for which ring counts established each sample's exact calendar age. The result is that radiocarbon concentration has changed over time; thus, one radiocarbon age can correspond to multiple calendar ages. For example, a radiocarbon age of 160 years (dashed line) intercepts the curve in four places, giving calendar ages of 0, 150, 220, and 290 years. Over the past few hundred years, only radiocarbon ages between 230 and 260 years correspond to unique calendar ages. The blue band around the data represents the uncertainty of the laboratory measurements. [From Stuiver (1978a).]

mass rather than radioactive decays. The accelerators turned out to work best, and a new field, **Accelerator Mass Spectrometry**, or **AMS** was born.

Just as in the evaluation of the original method, the new method was established by blind tests of old wood of known age [**Figure DD2.4**], followed by a period of comparison between established decay-counting methods and the new atom-counting methods (Stuiver, 1978b). Applications soon followed. Tucker et al. (1983) used AMS ^{14}C analysis of tiny flecks of charcoal to date soils associated with earthquakes on Utah's Wasatch Fault, demonstrating that paleoseismic studies, which often rely on dating small amounts of organic matter in buried soils, were possible. Rather than waiting months for decay-counting results, AMS data could be available in days, allowing fault trenches to remain open while geomorphic and faulting hypotheses were evaluated. Such real-time feedback between the field and the lab proved invaluable.

Today, almost all ^{14}C measurements are made by AMS and the results are impressive. Radiocarbon dating by decay counting required at least a gram of carbon. Atom counting by AMS today requires much less than a milligram. Decay counting that took weeks has been supplanted by AMS measurements made in tens of minutes. It is now possible to date virtually any organic matter younger than about 50,000 years.

Arnold, J. R., and W. F. Libby. Age determinations by radiocarbon content: Checks with samples of known age. *Science* 110 (1949): 678–680.

Bennet, C. L., R. P. Beukens, M. R. Clover, et al. Radiocarbon dating using electrostatic accelerators: Negative ions provide the key. *Science* 198 (1977): 508–510.

Broecker, W. S., and A. Walton. Radiocarbon from nuclear tests. *Science* 130 (1959): 309–314.

Muller, R. A., E. J. Stephenson, and T. S. Mast. Radioisotope dating with an accelerator: A blind measurement. *Science* 201 (1978): 347–348.

Nelson, D. E., R. G. Korteling, and W. R. Stott. Carbon-14: Direct detection at natural concentrations. *Science* 198 (1977): 507–508.

Stuiver, M. Radiocarbon timescale tested against magnetic and other dating methods. *Nature* 273 (1978a): 271–274.

Stuiver, M. Carbon-14 dating: A comparison of beta and ion counting. *Science* 202 (1978b): 881–883.

Stuiver, M., and P. J. Reimer. Extended ^{14}C database and revised CALIB ^{14}C calibration program. *Radiocarbon* 35 (1993): 215–230.

Suess, H. E. Natural radiocarbon measurements by acetylene counting. *Science* 120 (1954): 5–7.

Suess, H. E. Radiocarbon concentration in modern wood. *Science* 122 (1955): 415–417.

Suess, H. E. Absolute chronology of the last glaciation. *Science* 123 (1956): 355–357.

Suess, H. E. "Bristle-cone pine calibration of the radio-carbon timescale 5200 BC to the present." In I. U. Olsson, ed., "Radiocarbon variations and absolute chronology." *Proceedings of the Twelfth Nobel Symposium*, Stockholm: Almqvist & Wicksell, 1970.

Tucker, A. B., W. Woefli, G. Bonani, and M. Suter. Earthquake dating: An application of carbon-14 atom counting. *Science* 219 (1983): 1320–1321.

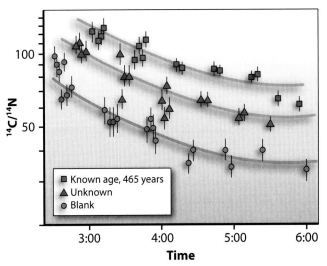

FIGURE DD2.4 Muller et al. (1978) demonstrated that ^{14}C atoms could be counted using a cyclotron, a circular particle accelerator. The graph shows the ratio of ^{14}C/^{14}N reported for three different samples over a 3.5-hour period that the machine was running. The ^{14}C originated in the sample. The ^{14}N is present at background levels. The ^{14}C/^{14}N ratio lowers over time as samples are consumed. The three samples include 465-year-old wood, a wood sample of unknown age, and a blank containing no ^{14}C. Although ^{14}C was measured in all samples (probably the result of contamination in the cyclotron), ^{14}C levels were highest in the youngest material and lowest in the oldest material (the blank). The unknown has a ^{14}C concentration consistent with being 5900 ± 800 years old, statistically consistent with the 5080 ± 60 year age measured by traditional ^{14}C decay counting methods. The data proved that ^{14}C could be measured by particle accelerators and opened a new chapter in the application of radiocarbon dating. Over the past few decades, AMS ^{14}C dating has become increasingly precise. Uncertainties today are as good as or better than those derived from counting techniques. [From Muller et al. (1978).]

WORKED PROBLEM

Question: A piece of wood is removed from a depth of 560 cm in a lake core collected from northern Vermont and prepared for radiocarbon dating by chemical cleaning, combustion to CO_2 gas, conversion to graphite, and atom counting by AMS. The resulting data indicate that the wood has 38.47 percent of the ^{14}C content of modern wood. Using the Libby half-life of 5568 years for ^{14}C (as is the accepted tradition for reporting radiocarbon ages, although the actual half-life is now known to be 5730 years), what is the radiocarbon date for the wood? Using the program CALIB (http://calib.qub.ac.uk/calib/), what is the calibrated age range for the wood sample? The lab reported a one standard deviation age error of ±90 years.

Answer: The radiocarbon age of the sample is between one (50 percent) and two (25 percent) Libby half-lives of ^{14}C (5568 years). Using eq. 2.1, 7670 yr = 8033 ln (1/0.3847), and the reported lab uncertainty, the radiocarbon age of the sample is 7670±90 radiocarbon years. Entering these data in CALIB and using the *intcal* global calibration database indicates that the wood most likely formed between 8396 and 8543 years ago. The offset between radiocarbon and calendar ages is almost 1000 years.

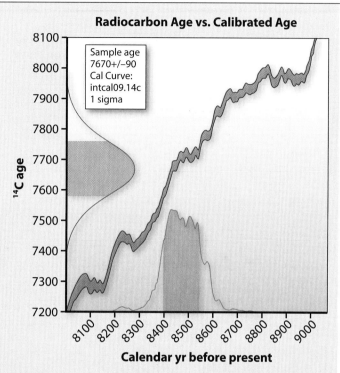

Radiocarbon Age versus Calibrated Age This graph, created by the CALIB program, shows the radiocarbon age as a probability distribution of the vertical axis, the calibration curve as the wiggly line that goes diagonally across the graph, and the calibrated radiocarbon age as the probability distribution across the horizontal axis.

KNOWLEDGE ASSESSMENT Chapter 2

1. Explain why the involvement of a geomorphologist is so important for dating Quaternary materials and events.
2. Describe several field surveying techniques.
3. List several uses of GPS that are important for geomorphologists.
4. Explain the difference between active remote sensing and passive remote sensing.
5. What are SRTM data and why are they important to geomorphology?
6. How does LiDAR work and what is it useful for geomorphically?
7. Explain what a DEM is and how it can be made.
8. How have changes in mapping technology over the past decades changed the field of geomorphology?
9. Explain the difference between relative dating methods and numerical dating methods.
10. List three relative dating methods.
11. How does a calibrated relative dating method work?
12. Explain how the shape of landforms relates to their age.
13. What is a weathering rind and how does it change over time?
14. What is rock varnish, where would you find it, and how does it probably form?
15. How are lichens used to date geomorphic features?
16. Explain how calibrated relative dating methods differ from isotopic dating methods and predict which one is likely to give more precise and accurate ages.
17. What is dendrochronology and how does it work?
18. What happens to the radiocarbon content of wood after the tree dies and the wood is buried in a river terrace?
19. How is most radiocarbon (^{14}C) produced?

20. Where and over what time frame is radiocarbon most useful as a dating method?

21. Explain the physics underlying luminescence dating methods.

22. What is the most important assumption underlying luminescence dating methods?

23. Explain the two primary uses of cosmogenic nuclides in geomorphology.

24. What dating technique is frequently used to quantify the age of young volcanic rocks?

25. Clearly state the difference between sediment generation and sediment yield.

26. List four ways to measure rates of geomorphic processes.

27. Define integration time and explain why it is important to geomorphology.

28. Explain how the atmospheric testing of nuclear weapons in the past helps today's geomorphologists.

29. How does thermochronometry work and why is it important to geomorphology?

30. Explain how laboratory and field experiments can help us to better understand surface processes.

31. What is a numerical model and how would it be useful to a geomorphologist?

32. Explain what proxy records are and give an example of how they are used in geomorphology.

PART II

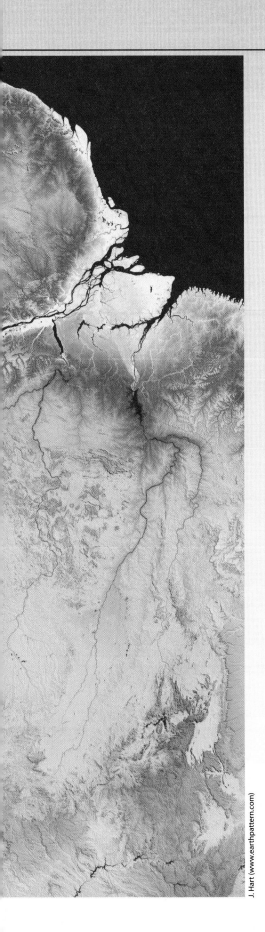

Source to Sink
(Chapters 3 to 8)

Landscapes are shaped by both subtle and dramatic geomorphic processes that move mass from uplands to lowlands and eventually to the coast and offshore to the deep ocean basins. In the next six chapters, we examine the processes that move earth materials and generate landforms. Chapter 3 considers how rocks weather to produce soil and sediment. In Chapter 4, we trace the movement of water at and near Earth's surface. Chapter 5 introduces the processes active on hillslopes and the landforms that these processes produce. Flow through channels is covered in Chapter 6; drainage basins, the fundamental unit of landscape analysis, are the topic of Chapter 7. In Chapter 8, we consider coastal and marine environments, the depositional archives that store material removed from the land surface. After reading these six chapters, you should understand how material moves across Earth's surface from source to sink and how surface processes drive this cycle and shape our planet's landscapes.

The Amazon is Earth's largest river system. As recently as 15 million years ago, the river flowed westward, until the rise of the Andes forced a new eastward flow path to the Atlantic. Today the material eroded from the Andes is transported east to the river mouth and the long-term sink of the Atlantic Ocean. This image shows a digital elevation model of the Amazon River Basin from Shuttle Radar Topography Mission (SRTM) data. By assigning color and brightness to the elevation values (from light gray at sea level to cool pink at the highest elevations), the image shows both the local terrain detail and the major regional landscape elements.

Weathering and Soils

Introduction

A thin layer of mechanically broken and chemically altered rock mixed with living and dead organic material defines the transition from Earth's rocky interior to its gaseous atmosphere. This layer is produced by **weathering,** the chemical or physical alteration of **primary minerals** (unaltered by weathering) that make up the **parent material,** rock or sediment. From a geomorphologist's perspective, this thin layer plays a key role in defining the shapes and processes of Earth's surface because erosion is much more effective at altering landscapes once solid rock breaks down into transportable material—a process involving both physical changes and chemical transformations that create **secondary minerals** and **dissolved ions.**

Three types of weathering are generally recognized and closely related: **physical weathering, chemical weathering,** and **biological weathering.** The greater surface area that results from physically breaking rocks into smaller fragments accelerates chemical weathering. Changes in mass or volume accompanying chemical decomposition can promote physical weathering. Even though different environmental conditions tend to favor physical versus chemical weathering, the processes are complementary and often strongly influence each other. Biological activity catalyzes both physical and chemical weathering. The weathering effects of biological activity depend on the active species, substrate, and weathering processes in question. Tree roots, for example, slowly pry rocks apart as they grow [**Photograph 3.1**], and the organic acids they

Rock weathering can produce spectacular landscapes. In Monument Valley, Arizona, differential rock weathering and erosion of sandstone formations result in stair-stepped topography and distinctive towers of rock. Weathering etches less resistant layers so that bedding becomes visible. The pattern of jointing and overall rock strength varies with rock type. Debris shed from rock faces forms angle-of-repose slopes.

IN THIS CHAPTER

Introduction
Physical Weathering
   Exfoliation
   Freeze–Thaw
   Thermal Expansion
   Wetting and Drying
Chemical Weathering
   Mineral Stability
   Oxidation and Reduction
   Solution
   Hydrolysis
   Clay Formation
   Hydration
   Chelation
   Cation Exchange
Soils
   Soil-Development Processes
   Factors Affecting Soil Development
   Processes and Rates of Soil Production
   Soil Profiles
   Soil Classification
Soils and Landscapes
   Soil Development over Time
   Soil Catenas
   Paleosols
Weathering-Dominated Landforms
   Inselbergs and Tors
   Duricrusts
Applications
Selected References and Further Reading
Digging Deeper: How Fast Do Soils Form?
Worked Problem
Knowledge Assessment

secrete, as well as the carbon dioxide (CO_2) they respire into the soil, increase rates of chemical weathering.

Physical, chemical, and biological processes all act as weathering agents, but their rates and relative importance vary dramatically among different landscapes around the world. Weathering processes create and influence the physical and chemical properties of the layer of **regolith** (unconsolidated and weathered material covering fresh bedrock) that provides the foundation for terrestrial life. Not all regolith is produced in situ—for example, some of the glacial debris covering New England was imported by glaciers from Canada and some regolith on hillslopes is derived from upslope. Such mobile hillslope regolith integrates the characteristics of the rocks upslope, from which it was derived. Regolith properties—such as clay content, the speed with which water moves through it, and its grain size—directly influence the distribution of geomorphological processes and the resulting landforms.

Biological activity affects both the rate and style of physical and chemical weathering. Organisms speed up mineral weathering by mechanically breaking down rocks and by catalyzing chemical reactions and causing changes in environmental conditions (such as soil pH). The growth of plant roots helps disintegrate rocks by gradually prying apart cracks, fractures, and other openings. Burrowing animals, like gophers, ants, and termites, excavate rock fragments and mix them into overlying soils [**Photograph 3.2**]. The process of biologically mediated mixing, called **bioturbation,** disrupts original structures or fabric in the **parent material.**

Biological activity also increases the potential for chemical weathering by increasing surface area and creating and enlarging pathways for subsurface water flow. Because oxygenated water is reactive, the more water that moves through a soil, the more weathering takes place. Decaying organic matter and respiration of soil microorganisms and plant roots indirectly influence chemical weathering by increasing the concentration of CO_2 in soil gases, thereby promoting acidification of soilwater and enhancing chemical weathering in the fractures and pores through which soilwater flows. Plant roots release organic acids that enter soilwater and either attack fresh mineral surfaces directly or exchange hydrogen ions for nutrient cations that plant roots absorb as they take up water. Decaying organic matter also releases humic acids that facilitate further weathering. Bacteria can mediate weathering reactions, and lichen colonization is the first step in weathering of many rocky surfaces. In short, more life leads to more rapid rock breakdown and thus more weathering.

Weathering and erosion do not necessarily progress at the same pace. Where weathering outpaces erosion, landscapes develop thick mantles of **saprolite,** chemically altered but in-place rock that has lost mass and strength but not volume during weathering. Saprolite can be hundreds of meters deep in flat-lying, slowly eroding areas of tropical Africa and South America. It can erode like loose sand, yet still retain

PHOTOGRAPH 3.1 Biological Activity and Weathering. Plant roots can pry apart rock, such as this small tree wedging open a glacial erratic in Ledyard, Connecticut.

PHOTOGRAPH 3.2 Bioturbation. Funnel ants live in the semi-arid woodlands of eastern Australia and make volcano-shaped ant hills that dominate the forest floor. Their burrowing activity turns over the entire soil profile down to a depth of 30 cm every 200 years or so.

structures or features of the original parent material, such as igneous dikes, calcite veins, or sedimentary bedding planes [Photograph 3.3]. Saprolite density can be as low as half that of intact rock due to mass lost by solution.

Not everyone defines soil similarly. Some consider soil to be any unconsolidated material at Earth's surface. Others consider soil to be only the material that has been affected by pedogenesis. In this book, we define soil broadly as the unconsolidated mineral and organic material on the surface of the Earth that serves as a natural medium for the growth of land plants and animals and that differs from the material from which it is derived in its physical, chemical, biological, and morphological properties. Where it is present, soil forms the upper part of the regolith.

Soil forms on Earth's outer surface and provides the substrate in which plants root, terrestrial life derives sustenance, and erosion acts to shape many landscapes. Soils not only harbor and sustain life, they are themselves partly composed of organic matter. Plants and animals both depend on soils and, in turn, influence the rate of **regolith production** (rock weathering) and the rate of soil development, or **pedogenesis**—the result of physical, chemical, and biological processes that alter the appearance and properties of the parent material upon which the soil is developed. In this sense, soil development may be considered a top-down process through which the properties of surficial materials change in response to the influence of environmental factors (like climate and vegetation), whereas soil production may be viewed as a bottom-up process through which rocks break down into surficial materials.

The presence, amount, and phase of water (i.e., vapor, liquid, or ice) strongly influence patterns and rates of both chemical and physical weathering. It is not surprising, then, that global patterns in the style and intensity of weathering generally track regional climate. Differences in temperature and precipitation control both the magnitude and relative importance of physical and chemical weathering processes [Figure 3.1]. Chemical weathering is most important in regions with high year-round temperatures and abundant precipitation, like the equatorial regions, and least important in cold, dry areas, like the poles. Higher temperatures promote much faster reactions because of the nonlinear dependence of chemical reaction rates on temperature; thus, temperature extremes can control chemical weathering rates. A short time spent at high temperature can produce as much chemical weathering as a long time at the average temperature. Relative to other environments, physical weathering is thought to be least active in dry environments and most active in regions where temperatures repeatedly cycle through freezing for part of the year. Consequently, tropical regions are generally dominated by chemical weathering and high latitudes by physical weathering.

Topography also greatly influences weathering. Mechanical weathering dominates in high mountains with steep slope gradients that create stresses in rock, extensive fracturing from tectonic forces, and periodic freezing temperatures. Low-relief and especially low-elevation environments generally favor chemical weathering. Rates of physical and chemical weathering influence one another because high rates of mechanical breakdown promote high rates of chemical decomposition by exposing

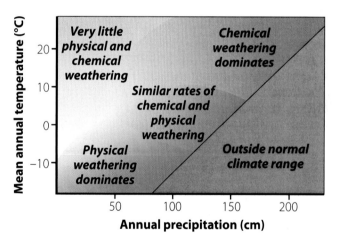

PHOTOGRAPH 3.3 Saprolite. Saprolite developed on nickel-rich weathered serpentinite in Greece. Veins of calcite stand above the weathered rock surface in the lower left.

Environmental controls on **physical weathering** and **chemical weathering** vary depending on the temperature and the amount of moisture. Chemical weathering dominates in hot, wet climates, and physical weathering dominates in cold, dry climates, where chemical weathering proceeds slowly. Chemical weathering tends to decrease with increasing elevation and at higher latitude, whereas physical weathering tends to increase at higher elevation and higher latitude.

FIGURE 3.1 Physical and Chemical Weathering. The intensity and thus importance of physical and chemical weathering vary with climate (precipitation and temperature) and degree of prior weathering.

greater surface area to chemical attack and vice versa (through reduced material strength). Consequently, the highest total rates of rock weathering typically occur in environments conducive to both physical and chemical weathering, such as steep terrain in tropical regions.

Styles and rates of weathering vary among rock types because of differences in chemical (mineralogical) characteristics and physical attributes (strength and fracturing). In particular, the ease with which water is able to penetrate rocks strongly influences weathering rates. As discussed earlier, mineral composition is a primary factor in chemical weathering. Some rock types, such as limestone and marble, are particularly vulnerable to chemical weathering in humid climates because they can be dissolved by weakly acidic soilwater or rainwater. In contrast, these rock types can be quite erosion resistant in arid regions where water is not sufficiently abundant to carry away dissolved ions. Compared to limestone, quartz is far more resistant to both chemical and physical weathering, and rocks like quartz sandstone and quartzite weather more slowly than most other rocks, regardless of environmental conditions.

This chapter reviews the processes that act to weather rocks, sediment, and minerals at and near Earth's surface and introduces the physical, chemical, and biological processes that produce soils. We explore the dominant controls on the transformation of primary rock-forming minerals into secondary minerals and consider global and lithologic controls on weathering and the resulting influences on landforms. We also address how soil scientists recognize diagnostic characteristics and classify soil types.

Physical Weathering

Physical processes mechanically break rocks into smaller pieces (**disaggregation**). During physical weathering, rocks and rock-forming minerals break apart without changes in composition. In some cases, physical weathering reduces parent material into fragments of individual mineral grains. Zones of weakness in the original material, such as cleavage or bedding planes, metamorphic foliation, and even mineral boundaries, often determine the size and shape of rock fragments. Cracks form where stresses imparted by expansion, contraction, or shearing exceed the strength of rocks or minerals. The most pronounced physical weathering in bedrock occurs where there is a strong directional contrast in pressure. This is the case at and near the land surface where overburden is minimal and where open fractures are more abundant than at depth, where confining pressures are greater and more uniform.

Important general mechanisms of physical weathering include release of confining pressure that allows rock to expand and fracture, thermal expansion from **insolation** (heating by the sun) or forest fires, and cyclic expansion and contraction from freeze–thaw action in cold environments. Disintegration as a result of wetting and drying is important in rocks with minerals susceptible to shrinking and swelling. Expansion by growth of salt crystals can crack rocks in arid and coastal environments.

Exfoliation

Exfoliation describes the processes by which exposed outcrops of rock break into sheets oriented parallel to the land surface. These **exfoliation sheets** are typically 0.5 to 10 m thick and result in onionlike fracture patterns [**Photograph 3.4**]. Because exfoliation sheets are more readily eroded than underlying unfractured rock, slope-parallel fractures tend to mimic topographic forms as subsequent sheets form and erode.

The process of exfoliation commonly produces dome-shaped surfaces, like Yosemite Valley's famous granitic Half Dome. Exfoliation sheets are most apparent on bare rock surfaces, but slope-parallel fractures also develop within soil-mantled slopes, creating networks of discontinuities that can greatly influence near-surface groundwater flow paths and slope stability. Exfoliation is typically more common and better developed in igneous and metamorphic rocks that formed deep (>10 km) within Earth's crust than in sedimentary and volcanic rocks formed at shallower depths.

Exfoliation occurs as erosion brings rock closer to Earth's surface, increasing the contrast between the stresses locked in during crystallization or metamorphism and the stress imposed by adjacent rock [**Figure 3.2**]. When the difference becomes greater than the strength of the rock, the rock cracks. Even in sedimentary rocks, upward expansion from unloading separates bedding planes, allowing outward expansion where bedding tilt matches the topographic slope. Fracturing due to stress release is an important mechanism that allows water, oxygen, and plant roots to penetrate into a rock mass.

Rock masses subject to tectonic stresses or rocks that have shrunk upon cooling develop patterns of **joints** (fractures along which no significant movement has occurred) that provide planes of weakness along which weathering and erosion

PHOTOGRAPH 3.4 Exfoliation. Exfoliation (sheet jointing) in granite in Yosemite National Park, California, north of May Lake.

Physical Weathering   81

Granitic rocks cropping out at Earth's surface often weather into characteristic forms, rounded domes termed **bornhardts** that shed mass from their surface in sheets. Sheet thickness is set by the spacing between **exfoliation joints** that form as erosion brings rock close to the surface and the distribution of stresses changes.

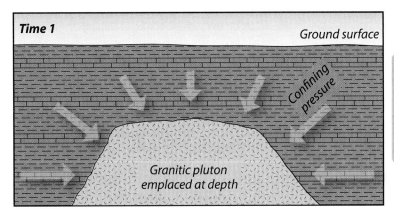

Kilometers below the surface, granitic plutons are emplaced by igneous activity. As the plutons slowly cool, minerals in the rock crystallize under uniform confining pressure.

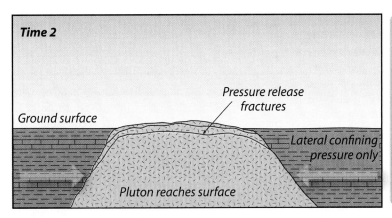

As erosion removes rock from Earth's surface, burial depth of the plutonic rocks decreases. The stress field on the rock is no longer equant; the least confining stress is at the surface because the rock that once covered the pluton is gone. The rock mass responds by cracking; resulting joints are oriented parallel and subparallel to the ground surface.

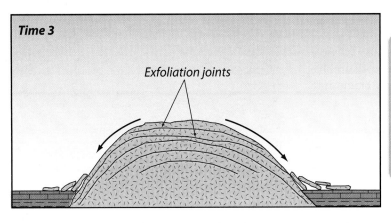

Once exposed at the surface, the rock mass develops fractures (joints) parallel to the land surface. Known as **exfoliation joints,** these sheets of rock peel off the exposed surface. Their debris builds up at the base of the resulting landform.

FIGURE 3.2 Exfoliation Sheets. Exfoliation occurs when the confining pressure on a rock mass is not equal on all sides. As erosion brings rock closer to the surface, the pressure from above is reduced and joints open.

can penetrate and preferentially remove material. Because the tensile (extensional) strength of rock is typically many times less than its compressive and shear strength, jointing is often well developed even in relatively strong, erosion-resistant rocks. Parallel sets of extensional joints develop orthogonally to the direction of maximum stress as a result of either crustal extension or cooling of igneous rock.

Joint systems can have strong topographic expression in arid and semi-arid landscapes where regolith is thin or absent and bedrock dominates hillslope morphology. Joints provide avenues for water and plant roots that focus erosion and facilitate infiltration and groundwater flow into rock that, in turn, promote more aggressive weathering. Erosion along intersecting sets of joints can produce isolated columns of rock separated by weathered-out joints. Joints also form parallel to the strike of bedding as a result of bending or folding of brittle rocks. Lithology greatly influences the degree of jointing, with joints typically being better developed in more brittle rocks such as sandstone and granite and less well-developed in more ductile rocks like shale. In addition, joint patterns developed in igneous rocks tend to be less linear and more irregularly spaced than those developed in sedimentary rocks.

Freeze–Thaw

The expansion of water confined in fractures and pore spaces as it freezes into ice is particularly effective at breaking rocks apart. Consequently, **frost shattering** is a primary weathering process in alpine and polar environments that are subject to frequent freeze–thaw cycles. Water expands by almost 10 percent when it freezes, so rock generally will not shatter unless 90 percent or more of the available pore space is saturated. If only some pore space is filled by water, ice simply expands into partially saturated voids without generating high pressures on the surrounding rock.

Hydrofracturing, a process by which freezing of water proceeds from the outside of the rock inward, forces water into the tiny ends of fractures, producing an effect like a hydraulic jack. In addition, as ice lenses form, water and vapor flow toward them and freeze, imparting enough force to cause growing ice crystals to crack rocks. Many alpine slopes in environments subject to frost shattering are covered by a blanket of rock blocks called **felsenmeer** (German for "rock sea") [**Photograph 3.5**], for example, the rubble-covered summit of Mount Washington in New Hampshire.

Thermal Expansion

Rocks and minerals expand when subjected to heat, but rock's low **thermal conductivity** generally prevents the zone of heating from penetrating more than a few centimeters into rocks over the course of a forest fire or a day of exposure to intense sunlight. Extreme temperature differences between a rock's surface and its cool interior can produce large differential stresses that exceed rock strength and result in **spalling**

PHOTOGRAPH 3.5 Felsenmeer. Felsenmeer and frost-shattered bedrock, on the summit of Mount Darling, Marie Byrd Land, Antarctica. The geologist in the image is between two much less weathered glacial erratics deposited on the surface by glacial ice frozen to its weathered bed.

(breaking away) of a thin outer layer that can be up to several centimeters thick [**Photograph 3.6**]. Such differences are particularly acute during range and forest fires and can result in extensive loss of mass from rock surfaces over time. Mass loss from rock surfaces has implications for surface exposure dating techniques examined in Chapter 2.

The weathering effect of daily temperature fluctuations due to heating by sunlight has long been debated, but recent field work provides convincing data that daily heating and cooling can lead to rock fracture. The majority of cracks in desert rocks that are not related to rock heterogeneities, like bedding, are oriented north–south. Thermal stresses, due to differential heating when the sun crosses the sky from east to west, are thought to produce these oriented cracks. The large **coefficient of thermal expansion** of certain minerals, such as calcite, can cause a rock to break apart along mineral boundaries as individual

PHOTOGRAPH 3.6 Fire Spalling. Fire-spalled granitic boulder in Pingree Park, Colorado, with accumulation of thin chips of rock around the base of the boulder.

grains expand and contract differently with small changes in temperature. The weathering product of this type of granular disintegration, known as **grus** from the German for grit, fine gravel or debris, consists of loose, unconsolidated, individual mineral grains. Grus is most commonly formed by the disintegration of coarse-grained intrusive igneous rocks, such as granite.

Wetting and Drying

The addition and removal of water from minerals, processes known as **hydration** and **dehydration,** can cause swelling or contraction capable of fracturing rocks or disaggregating them into individual mineral grains. The volume increase in the anhydrite to gypsum transformation, in particular, has been ruinous to archaeological monuments. Most clay minerals shrink when they dry and swell when they absorb water. Some can expand to twice their original volume, and certain particularly susceptible clays, like smectite (including montmorillonite), swell severalfold when they get wet. **Expansive soils,** containing minerals with such shrink-swell behavior, cause substantial engineering problems and extensive damage (e.g., cracked foundations) in the southern United States outside of the glacial limits and in the western plains where such soils are common. Swelling of smectite-rich soils in Montana, Wyoming, and Colorado can make travel on dirt roads nearly impossible after rainstorms, because sticky mud bogs down even vehicles with four-wheel drive.

In some rocks, like granite, physical expansion of biotite micas during weathering can pry apart grain boundaries and produce grus. Similarly, cycles of wetting and drying can lead to repeated expansion and contraction that disaggregate micaceous sandstone and turn hard,

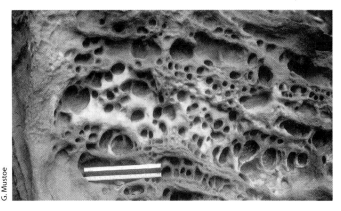

PHOTOGRAPH 3.8 Salt Residue. White salt residue on honeycomb weathering in Capitol Reef National Park, Utah. The ruler is 15 cm long.

erosion-resistant rock into a loose pile of sand in just a few years or decades. In rock types susceptible to this process, exposure to seasonal cycles of wetting and drying leads to rapid rates of bedrock erosion through the annual formation of a loose crust or outer covering of highly erodible material easily removed by subsequent heavy rainfalls or high river flows.

Hydration and expansion of salts such as gypsum or halite (table salt) within pore spaces also cause spalling and rock disintegration [**Photograph 3.7**]. Salts can expand severalfold when hydrated. Repeated wetting and drying, as well as the growth of salt crystals due to evaporation of fluids in near-surface fractures and pore spaces, can gradually pry rocks apart in arid climates and in coastal settings where spray covers rocks in salt [**Photograph 3.8**]. Salt weathering may be an important mechanism of rock disintegration in the dry desert landscapes on Mars.

Chemical Weathering

Chemical weathering involves breaking of chemical bonds—metallic, ionic, and covalent. The corresponding principal weathering processes are electron exchange (oxidation-reduction), solution (ionization), and ion exchange (as in acid attack). Because many rocks are dominated by a mix of ionic and covalent bonds, solution and acid attack are major weathering processes. Hydration and dehydration are also important weathering mechanisms for certain rock types and in certain environments. Chemical weathering is essential for the biosphere; vegetation needs calcium, magnesium, potassium, and phosphorus to thrive. Nutrients derived from minerals are cycled through ecological systems because of the slow pace of weathering or the depleted nutrient status of surficial materials (especially in elements such as phosphorus, which, if not present in sufficient concentrations, can limit plant growth). Such cycling is of particular importance in areas with deeply weathered, nutrient-depleted soils.

PHOTOGRAPH 3.7 Limestone Weathering. Honeycomb weathering of limestone blocks in Malta results in the formation of gypsum at the surface. The weathering pattern on the rock surface here is controlled by the fossil pattern of bioturbation from marine animal burrows in the original sediment. The width of the image is approximately 70 cm.

Chemical weathering occurs because the minerals in rocks form in equilibrium with deep-Earth conditions—high pressure, high temperature, and low oxygenation—quite different from those at Earth's surface. When these minerals are exposed to the relatively cool, wet, and oxygen-rich surface conditions, they are vulnerable to chemical decomposition and transformation. Chemical weathering involves reactions that change primary rock-forming minerals into **secondary minerals,** such as clays. In the process, some elements are lost in solution to surface water and groundwater.

The primary weathering agent is rainwater that percolates into the ground and promotes chemical weathering because it contains dissolved ions gained from the atmosphere and from the soils through which it moves. The bipolar water molecule can be a potent solvent, given time. The metabolism of soil microorganisms and decay of organic matter enhance weathering as they add organic acids to water moving through soils. Root respiration and microbial oxidation enrich the soil atmosphere in CO_2, which makes carbonic acid when dissolved in water, thus lowering pH and increasing the aggressiveness of soilwaters. Sulfuric and nitric-acid weathering are important in some areas.

Processes of chemical weathering include **solution, oxidation, reduction, hydrolysis, ion exchange,** and the formation of new, secondary minerals like clays and hydrous oxides, which are more stable near Earth's surface than are their parent minerals. The end results of chemical weathering depend on a variety of interacting factors, including the composition and texture of the parent material and the chemical, physical, and biochemical processes acting in a particular environment. The mobility and stability of the secondary minerals and solutions produced depend on environmental conditions such as pH, redox potential, and temperature.

Mineral Stability

The general susceptibility of rock-forming minerals to weathering is the inverse of the sequence in which they form deep within the Earth. Minerals that formed at the highest temperatures and pressures are farthest from equilibrium at surface conditions and are therefore most susceptible to weathering when exposed to the elements. Among the common silicate rock-forming minerals, olivine and pyroxene are most susceptible to weathering, followed in order of decreasing susceptibility to breakdown by amphibole, biotite, muscovite, and quartz [**Figure 3.3**]. This progression, known as **Goldich's Weathering Series,** is the opposite of the order of crystallization as magma cools, familiar to geologists as **Bowen's Reaction Series.**

Under similar environmental conditions, rocks composed of more **mafic** iron- and magnesium-rich minerals (olivine, pyroxene, amphibole, and biotite) will weather faster than those composed of muscovite and quartz. But

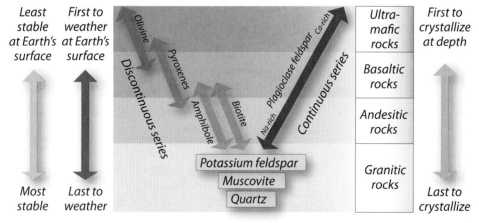

FIGURE 3.3 Goldich's Weathering Series. Minerals formed at high temperatures tend to be less stable at and near Earth's surface than minerals formed at low temperatures. This relationship, known as Goldich's Weathering Series, is the inverse of Bowen's Reaction Series, used to describe the order in which minerals crystallize from rock.

it is not always as simple as this. Feldspars have a range of compositions and thus a range of weathering susceptibilities. Due to the relative weakness of ionic bonds, silicates with complicated mineral structures break down more readily than do those with simpler covalent-bond structures like quartz (SiO_2) or zircon ($ZrSiO_4$), a very stable mineral even though it has a very high melting temperature. Both crystal complexity and formation conditions are central to mineral stability.

The mobility of cations in rock-forming minerals varies greatly and influences the relative ease and order in which weathering strips cations from rocks and secondary minerals, with the sequence from most to least mobile proceeding as Ca^{2+}, Na^+, Mg^{2+} > K^+ > Fe^{2+} > Si^{4+} > Fe^{3+} > Al^{3+}. The most mobile cations (Ca^{2+}, Na^+, Mg^{2+}) are readily stripped from mineral surfaces, tend to remain in solution, and are the first to be lost from rocks as they weather. The least mobile cations (Si^{4+}, Fe^{3+}, and Al^{3+}) are relatively insoluble and become concentrated in residual soils over time as weathering strips away more mobile elements.

Oxidation and Reduction

In oxidation, an element such as iron loses an electron to a receptor, often an oxygen ion—for example, when iron rusts. Conversely, reduction is defined as gaining an electron. Free oxygen is rare at crustal depths where rocks form, but abundant at Earth's surface. Certain rocks and rock-forming minerals oxidize when they are exposed to well-oxygenated soilwater, directly to the atmosphere, or to gases in soil pores. Reducing conditions in oxygen-poor waters with lots of organic matter, found for example in swamps and peat bogs with high seasonal water tables, generally prevent oxidation, retard organic decay, and slow weathering. When soils alternate between saturated and unsaturated conditions, a blotchy color pattern known as **mottling** develops—with gray colors due to reduced iron alongside reddish colors due to oxidation.

Oxidizing potential is expressed in terms of **redox potential (Eh)**, the availability of free oxygen, which is greatly influenced by the amount of dissolved organic matter in pore fluids. Soils typically have Eh high enough to oxidize most common elements, but iron, manganese, and sulfur are especially prone to rapid oxidation and typically occur as red, black, and yellow coatings in soils. Redox potential exerts a substantial influence on ion mobility, and oxidation is often the first form of weathering to alter freshly exposed rock surfaces. Most oxidized elements form hydroxides and are thus relatively insoluble and immobile, whereas most reduced elements are far more soluble and thus mobile.

Over time, rinds of oxidized material form on the surfaces of outcrops, boulders, and cobbles (see Photograph 2.6). When rocks containing common iron-bearing carbonates, sulfides, and silicates (such as olivine and biotite) oxidize, they become susceptible to additional physical weathering. Oxidation produces relatively insoluble ferric oxides like hematite (Fe_2O_3) and oxyhydroxides like goethite ($FeO(OH)$), which color soils and weathered rock various shades of reddish or yellowish brown. By the same token, oxygen-starved conditions (such as those under stagnant water rich in decomposing organic matter) reduce iron and manganese, which allows them to be dissolved and leached if the water drains or is flushed from the soil. Portions of the soil that are frequently subject to reduction processes are recognizable by their gray-blue colors.

Solution

The flow rate and acidity of pore water are two of the most important factors influencing the amount of dissolution from soil, sediment, or rock in a given weathering environment. In particular, pH strongly affects the solubility of most elements [**Figure 3.4**]. Rainwater is slightly acidic (pH = 5.7) from dissolved atmospheric CO_2 (which forms carbonic acid, H_2CO_3), and chemical weathering and biologic processes (respiration and decomposition) further alter the pH of water moving through soil and regolith

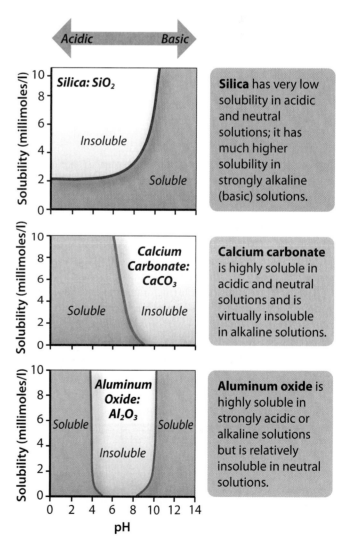

FIGURE 3.4 Solubility Diagrams. Solubility relationships with pH are different for each mineral. The three geomorphically important minerals shown here have very different solubility-pH relationships.

(Table 3.1). In weathering zones with active groundwater circulation, freshwater comes in contact with parent material, and weathering continues as leaching removes dissolved constituents. Slowly circulating pore waters retard dissolution as the amount of dissolved ions in solution approaches an equilibrium or saturated concentration.

Natural soilwater tends to be slightly acidic, due to dissolution of soil carbon dioxide in water to produce carbonic acid (H_2CO_3). Carbonic acid is not a strong acid, but it is extremely abundant because it forms through the carbonation reaction wherever water encounters CO_2:

$$H_2O + CO_2 \longleftrightarrow H_2CO_3 \qquad \text{eq. 3.1}$$

Decay of organic matter together with respiration of soil invertebrates, bacteria, and root systems can elevate CO_2 concentrations in soil pores so that they are 10 to 100 times greater than atmospheric concentrations. Thus, carbonation is a particularly important factor in heavily vegetated areas. Cold temperatures also favor formation of carbonic acid in soilwater because the solubility of CO_2 is inversely proportional to temperature, as is true of most gases.

In solution, carbonic acid partially disassociates into hydrogen (H^+) and bicarbonate ions (HCO_3^-):

$$H_2CO_3 \longleftrightarrow H^+ + HCO_3^- \qquad \text{eq. 3.2}$$

Consequently, bicarbonate is the most common anion in natural groundwater.

Atoms exposed on the mineral surfaces of rock and soil particles tend to have surfaces with net positive or negative charges and react with dissociated hydrogen (H^+) and hydroxide (OH^-) ions in water. These interactions break bonds, effectively disassociating individual mineral molecules and causing exchanges that release cations from the mineral surface into solution. Mineral structures become unstable and vulnerable to further weathering when they lose cations, so initial weathering promotes more weathering. **Congruent dissolution** occurs when all the constituents of an individual molecule are separated and remain in solution. During **incongruent dissolution**, some of the released ions recombine to create new compounds and secondary minerals (as discussed later).

Dissolved material may remain in solution and move along with flowing groundwater, may reprecipitate elsewhere, or may enter streams and rivers and eventually reach the ocean. Ocean salinity results from the long-term delivery of dissolved material in stream water. Most common elements are soluble to some degree in both rainwater and soilwater. Consequently, water circulation promotes solution by introducing fresh water that removes dissolved ions from mineral surfaces.

The dissolution of calcite ($CaCO_3$, calcium carbonate) is a particularly important chemical weathering reaction that occurs in the presence of carbon dioxide dissolved in water (carbonic acid, see equation 3.1) and introduces bicarbonate ions (HCO_3^-) into solution. The resulting reaction is expressed as

$$CaCO_3 + H_2CO_3 \longleftrightarrow Ca^{2+} + 2(HCO_3^-) \qquad \text{eq. 3.3}$$

The carbonate dissolution reaction is reversible. An increase in CO_2 concentration within soil gases, a decrease in pH, or dilution will drive the reaction to the right (as written above); carbonate dissolution will increase, and the bicarbonate concentration in groundwater will go up. This effect helps percolating water erode fractures and form extensive cave systems typical of regions underlain by carbonate rocks (limestone or dolomite). Conversely, decreased CO_2 concentration, increased pH, or evaporation will drive the reaction to the left and favor precipitation of calcium carbonate ($CaCO_3$). It is this reaction that deposits stalagmites and stalactites in caves, as well as calcite in desert soils.

Mineral dissolution can affect near-surface process rates by increasing pore space and thus increasing the volume of percolating water, soil acids, oxygen, and bacteria moving into and through the regolith. Calcite and salts are readily dissolved in water, so carbonate rocks and evaporites are particularly susceptible to dissolution, especially in regions with abundant precipitation. In contrast, quartz and most other rock-forming silicate minerals are not very soluble at typical Earth surface conditions, leading to slow rates of dissolution in most environments.

Iron and aluminum oxides are virtually insoluble under oxygenated soilwater conditions, so these compounds are typically left behind while more soluble, mobile material is depleted. Consequently, the abundance of iron and aluminum oxides increases over time as rocks and sediment are exposed to weathering. The red soils of unglaciated, stable continental regions are an example of how these oxides can accumulate over time.

Hydrolysis

Hydrolysis is a chemical reaction in which water molecules (H_2O) are split into protons (H^+) and hydroxide anions (OH^-) that react with primary rock-forming minerals to form new compounds (secondary minerals). Hydrolysis is an important chemical weathering process that acts to break rocks apart and transform silicate minerals (the most widespread minerals in Earth's crust) into

TABLE 3.1

Common pH Values	
Normal rainfall	5.7 (mostly due to carbonic acid)
Acid rain	< 5.6
Normal soil (moist climate)	≈ 4.6
Vinegar	2.4
Coca-Cola	≈ 2.3

weathering products. The process is critical in generating regolith and making clay minerals that are important for soil fertility. In hydrolysis reactions, mineral cations are released into solution and replaced by hydrogen (H^+), producing a new mineral. This process results in incongruent dissolution, the irreversible transformation of aluminosilicate minerals, like feldspars and micas, into various clay minerals or oxides.

When carbonic acid dissociates to form an "acid" or protons, the resulting weathering of aluminosilicate minerals consumes CO_2, drawing down atmospheric CO_2 levels and thus helping to cool global climate through the general reaction:

$$\text{aluminosilicate} + H_2CO_3 + H_2O \rightarrow \text{clay mineral} + \text{cations} + HCO_3^- + H_4SiO_4 \quad \text{eq. 3.4}$$

Earth's long-term climate is thus mediated by organic matter burial (which sequesters carbon in geologic materials), and silicate weathering, which consumes CO_2 (producing bicarbonate). Glaciations and the anthropogenic contribution to atmospheric CO_2 are short-term perturbations of these long-term geologic controls on global atmospheric composition and thus climate.

Once secondary minerals (clays) are formed, further weathering can strip additional cations and can convert secondary aluminosilicates into other, more cation-depleted clays or oxides. Each step in the weathering of clay minerals strips additional cations, sequentially reducing the complexity of mineral structures.

Clay Formation

Clay minerals are both a product and a player in processes of hydrolysis and hydration. Unlike most primary minerals, with the exception of quartz, secondary minerals such as clays and hydrous oxides are chemically stable under Earth surface conditions. They become a major constituent in soils because their relative stability and immobility leaves them as common in situ by-products of weathering.

Most clay minerals are layer silicates composed of sheets of alumina octahedra (an atom of aluminum bonded to six atoms of oxygen) or silica tetrahedra (an atom of silica bonded to four atoms of oxygen) [Figure 3.5]. Aluminum (and, rarely, other elements) may substitute for silica in the tetrahedral layers. Iron and magnesium commonly substitute for aluminum in the octahedral layers. These sheets generally are bonded together in either a 1:1 structure (T-O), in which each layer of silica tetrahedra (T) is paired with a layer of alumina octahedra (O), or in a 2:1 structure (T-O-T), in which each octahedral layer is sandwiched between two tetrahedral layers. These building blocks are themselves interlayered and bound together by shared ions between the sheets. Layer architecture (1:1 versus 2:1) and ionic substitution within and between the sheets determine the physical properties of different clay minerals. For example, substitution within the layers makes the difference between illite and smectite clays.

Adjacent layers in **kaolinite,** a clay mineral with a 1:1 structure, are held together by hydrogen bonds that are strong enough to prevent cations or water from entering the spaces between the sheets. Because it has few exchangeable cations held between its layers, kaolinite does not swell much when wetted, and it has low plasticity (and thus little capacity to be molded). Clays with 2:1 layer structures exhibit much more variability in the chemical composition of their octahedral sheets (typically due to substitution of Fe^{2+} and Mg^{2+} for Al^{3+}) and in the abundance and type of ions present between layers. **Smectite** clays, like **montmorillonite,** have weak bonds between the silica layers, which allow water and ions to penetrate the crystal structure readily. Also known as swelling clays, smectites expand readily upon wetting and are a main component of expansive soils. **Illite,** another common clay mineral in soils, has a strongly bonded 2:1 structure. The cations between its layers are tightly held, so it has less swelling potential than smectite.

Weathering of secondary minerals involves stripping off layers of the silicate structure. The modification of muscovite (mica) to illite (clay), both of which consist of T-O-T "sandwiches," involves removal of the interlayer cations. Extreme weathering conditions can go beyond leaching of the intermediary cations and remove one of the two T layers, leaving a T-O sequence of silicate layers, resulting in kaolinite clay. Stripping the remaining tetrahedral layer leaves a basic octahedral layer of aluminum hydroxide, gibbsite ($Al(OH)_3$). In general, smectites weather to illites and ultimately become kaolinite. Deeply weathered soils generally have high concentrations of kaolinite and oxides.

The particular minerals produced by weathering processes depend on the parent materials and environmental conditions. But the sequence of weathering reactions, starting with the conversion of the rock mineral potassium feldspar (orthoclase) to illite clay, illustrates how progressive weathering changes clay minerals from 2:1 mineral structure to 1:1 mineral structure and then eventually to oxides. Orthoclase incongruently reacts with carbonic acid to produce the 2:1 clay mineral illite:

$$3KAlSi_3O_8 + 2H_2CO_3 + 12H_2O \rightarrow KAl_3Si_3O_{10}(OH)_2 + 2K^+ + 2HCO_3^- + 6H_4SiO_4$$

Orthoclase (feldspar)   *Illite (clay)*   eq. 3.5

Further intensive weathering of illite produces the 1:1 clay mineral kaolinite, which consists of just hydrogen, aluminum, silica, and oxygen:

$$2KAl_3Si_3O_{10}(OH)_2 + 2H_2CO_3 + 3H_2O \rightarrow 3Al_2Si_2O_5(OH)_4 + 2K^+ + 2HCO_3^-$$

Illite (clay)   *Kaolinite (clay)*   eq. 3.6

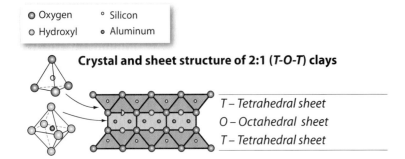
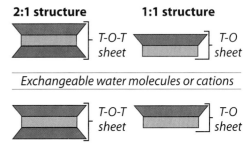

Because silicate tetrahedra are linked in sheets, the **phyllosilicate** group of minerals have sheetlike properties. Muscovite and biotite are two common phyllosilicate minerals that exist as layers. The **clay minerals** are part of the phyllosilicate group and form from surface and near-surface weathering of common rock-forming minerals. The clay minerals are "sandwiches" of silicate tetrahedral layers and octahedral layers, the latter of which are commonly of $Al_2(OH)_6$ composition.

A 2:1 clay mineral has two tetrahedral sheets and one octahedral sheet, and a 1:1 clay mineral has one of each type of layer. Various cations or H_2O molecules can be incorporated between the composite layers in some types of clays, particularly those termed **expandable clays.**

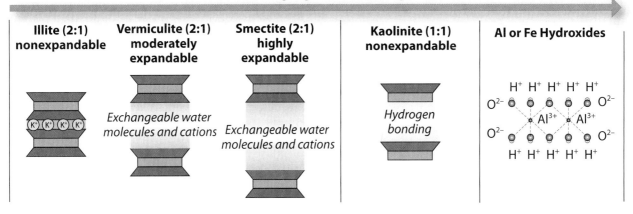

Illite and smectite have a 2:1 structure with alumina sheets sandwiched between silica sheets. They exhibit a wide range of properties, chemical compositions, and **cation exchange capacities** due to substitution of ions between layers.

Kaolinite has a 1:1 structure and low **cation exchange capacity** with alumina and silica sheets held together by strong ionic bonds.

Extremely weathered soils may result in a lateritic residuum of weathering-resistant Fe and Al oxides.

FIGURE 3.5 Clay Weathering Sequence. Increasing degrees of weathering result in different, indicative types of secondary clay minerals as interlayer cations are removed.

Weathering that strips the silica from kaolinite leaves a residue of aluminum oxide, or gibbsite ($Al(OH)_3$):

Kaolinite (clay)   *Gibbsite (oxide)*

$Al_2Si_2O_5(OH)_4 + 5H_2O \rightarrow 2Al(OH)_3 + 2H_4SiO_4$ eq. 3.7

Secondary minerals, like kaolinite and gibbsite, can also form directly from primary minerals, depending on the parent material and specific environmental conditions.

Hydration

Hydration describes the process by which minerals combine with water or hydroxide ions (OH^-) to form hydrated compounds. Hydration is another way that primary minerals are converted to secondary minerals. Common hydration reactions include the conversion of anhydrite ($CaSO_4$) to gypsum ($CaSO_4 \cdot 2H_2O$), and the formation of relatively insoluble iron and aluminum hydrous oxides, such as limonite (FeO(OH) · nH$_2$O) in regions of intense tropical weathering such as the Amazon basin.

Chelation

Chelation is the process through which relatively immobile metal ions, such as iron and aluminum in soils, are mobilized by soluble organic compounds that form ring structures around metal ions. Chelation is facilitated by organic acids (particularly fulvic acid) produced by the breakdown of leaf litter, soil organic matter, and lichens. Iron and aluminum mobilized by chelating agents may be carried along in solution with soilwater flow until concentration changes or microbial actions break down the chelating agent, causing the metal to reprecipitate. Conifer needle decay in cool, moist environments is a common source of chelating agents, leading to the stripping of iron and other elements from the upper portions of forest soils.

Cation Exchange

An important outcome of chemical weathering is the ability of secondary minerals and organic matter to exchange cations with soilwater, thereby making vital nutrients (Ca^{2+}, Mg^{2+}, K$^+$) available to plants. Clays and organic compounds vary in their ability to adsorb and release cations, a property called **cation-exchange capacity**. Rates of ion exchange are controlled by soil cation-exchange capacity as well as by the ionic composition and pH of soilwater. Strongly acidic (low pH) pore fluids allow H$^+$ to substitute for and replace metal cations. As hydrogen ions exchange places with nutrient cations held on a clay surface, the number of potentially exchangeable cations decreases.

The degree to which the exchange sites are occupied by exchangeable cations other than H$^+$ and Al^{3+} is called **base saturation**. Repeated cation exchange progressively lowers base saturation in clay minerals by removing cations from between clay sheets. A clay with a high cation-exchange capacity but a low base saturation has had its balancing cations stripped out and replaced by hydrogen ions through substantial chemical weathering. Such clays are typically found in tropical regions with high temperature and rainfall, such as Hawaii. The progressive loss of exchangeable cations, which are needed for plant growth, reduces soil fertility. Older, more intensively weathered soils have low base saturation and are thus proportionately less fertile.

Soils

To a geomorphologist, soil is a layered residue of decomposed rock and organic matter left by weathering over an extended period of time. Soils have mineralogical, biological, and morphological characteristics that are distinct from those of its parent material. Other disciplines have their own definitions of soil—engineers generally consider soil to be any loose material above bedrock (if it can be dug, it is soil), and agronomists often consider soil anything in which one can grow a plant.

A tremendous variety of soils has resulted from the wide array of parent materials, climates, vegetation, and weathering environments around the world. Due to differences in soil-forming processes, patterns in the distribution of soil types mirror different climate zones and geologic settings. Soils also reflect differences in landscape history, and in some cases their characteristics can be used to infer past climate changes, periods of landscape stability, or the nature of ancient environments. Soils are generally unconsolidated, but calcite (CaCO$_3$), silica (SiO$_2$), and iron oxide (Fe$_2$O$_3$) cements that develop in some soils add strength and allow formation of distinct erosion-resistant landforms.

Soil-Development Processes

Soil development occurs by a variety of processes through the addition, loss, transformation, and translocation (movement) of material within a soil profile. These processes include the accumulation of organic matter at the ground surface and its subsequent integration down into the soil, decomposition of primary minerals into secondary minerals, the leaching of particular organic and mineral constituents from the parent material, and the movement of material leached or washed from upper soil layers deeper into the soil. Soil profiles lose material by either erosion or dissolution. Erosion by surface water flow and wind removes material from the top of the soil profile, and dissolution removes mineral material from within the profile.

New material may be added to a soil profile from sources below, within, or above the ground surface. Dustfall adds material to the top of a soil profile. The breakdown of bedrock adds mineral matter to soil from below. Organic matter is added to soil profiles from above when leaves and other organic debris fall to the ground and the remains of once-living plants and animals become mixed into the soil, or roots die within the soil. In addition, floods can deposit fluvial sediment on floodplains, wind can add dust to the top of a soil profile, and volcanic eruptions contribute ash to soils.

Soil scientists refer to the movement of material from one part of a soil profile to another as translocation. Dissolved matter may move within the soil or leave the soil in groundwater, a process known as **leaching**. Water percolating down through the upper soil may also physically entrain clays, silt, and colloids and redeposit them deeper in the soil. Precipitation infiltrating into a soil dissolves material from the upper layers. Soilwater flow carries the dissolved material downward and may deposit it in lower layers as water evaporates and solute concentration increases or as redox, pH, or gas contents change. The process of transporting clays and mobile ions out of a layer of soil is called **eluviation;** washing of material into a layer of soil is called **illuviation**.

Soils can be residual, colluvial, or cumulic. **Residual soils** develop in place on stable Earth materials including weathered rock and alluvial deposits, for example, river terraces. **Colluvial soils** move downslope and thus are typically well stirred (also known as turbated) by both biological and physical processes. **Cumulic soils** accumulate slowly enough, or episodically enough, that material added at the top of the soil profile weathers to some degree before being buried. Examples of such inputs include wind-blown dust or volcanic ash and flood-deposited silts on floodplains. Repeated, episodic deposition results in stacked soils, with multiple buried A and B horizons evident in soil pits.

Factors Affecting Soil Development

The factors that control soil development are climate, organisms (biological activity), topography (including slope orientation and steepness), parent material, and time. Climate and time are the predominant factors in soil development at a regional scale, but geologic factors greatly influence local soil characteristics. Other than time, soil-development factors are to some degree interdependent. Climate, for example, influences vegetation, and topography and parent material influence one another. Still, considering these factors independently helps to organize thinking about soil development.

Temperature and moisture conditions strongly influence rates and styles of weathering and thus soil development, because weathering reactions proceed faster at higher temperatures and where more water is available. In hot, wet tropical regions, intense weathering creates thick soils depleted of most of their original constituents, leaving behind relatively stable weathering products like kaolinite clay and aluminum and iron oxides and hydroxides. In contrast, soils formed slowly in cold, dry polar regions tend to be relatively shallow (thin) and have only minimal chemical alteration. In mid-latitude temperate regions that have moderate temperatures, precipitation, and evaporation, the resulting soils typically have intermediate depths and development intensities. In arid regions, low precipitation rates and high evaporation rates cause development of carbonate- or salt-enriched soils because dissolved constituents (usually derived from weathered, wind-blown dust) percolate into the soil, reprecipitate, and accumulate over time.

Climatically driven differences in soil are most pronounced on level, stable slopes where soil **residence time** (the time between a typical mineral grain's initial weathering from the underlying rock and erosion from the slope) is sufficiently long to reveal the influence of climate.

Organisms, including microbes, plants, and animals, affect the composition and behavior of soil. Vegetation influences soil composition directly when the growth of roots and tree-throws break up bedrock and mix it into the soil [**Photograph 3.9**]. The type and amount of vegetation growing in a soil influences erosion resistance and the ratio of surface water runoff to infiltration after precipitation

PHOTOGRAPH 3.9 Tree-Throw. Tree-throw stirs soils, breaking up underlying rock and moving it downslope. This tree-throw has ripped up rock held within the tree's roots.

events. Root size, density, and depth can greatly influence soil development through the location of organic inputs, whether at the surface or at depth. Vegetation also helps keep soil in place and shades the soil, helping it retain moisture and increasing rates of chemical transformation. Animals contribute to soil development and mixing through burrowing activity. In tropical regions, ants and termites can build great mounds that move soil vertically. In more temperate regions, burrowing mammals, such as gophers, churn the regolith and mix weathered rock fragments into the soil. Microbes, fungi, and other decomposers are central to the breakdown of organic matter and the cycling and recycling of key nutrients derived from rock weathering.

Topography greatly influences rates and styles of soil development through its control on rates of soil erosion and patterns of soil moisture. Gentler slopes may foster greater plant growth and accelerated soil development because they retain soil moisture better than steeper slopes. In upland regions, soils are better developed on gentle slopes, while shallow soils and even bare rock characterize steep slopes. Steeper slopes also promote faster rates of downslope soil movement, leading to higher soil turnover rates and therefore younger, less well-developed soils.

Slope aspect (the direction a slope faces) can affect microclimate and local soil moisture. In some landscapes, different types and densities of vegetation and soils develop on north-facing versus south-facing slopes because systematic differences in the amount of direct sunlight translate into differences in soil moisture. North-facing slopes in the Northern Hemisphere have lower evaporation rates, retain snow cover longer in the spring, and tend to hold soil moisture longer into the summer growing season. Such differences may result in local differences in soil thickness, moisture, pH, and organic matter content that may influence vegetation patterns, such as the presence of grasses versus trees.

Rocks or unconsolidated sediments provide the raw parent material for soil, and the mineralogy, porosity, and permeability (e.g., fracturing) of the parent material influence the style of weathering and types of secondary minerals that form from weathering. In particular, the structure, hydraulic conductivity, and fracturing of rock and sediments greatly influence the movement of soilwater and groundwater and thus the progress of weathering into parent material.

Differences in parent material generally exert the greatest influence during initial soil development and determine clay mineralogy, chemistry, and certain physical characteristics of the soils that first develop from them. For example, in the Piedmont of the eastern United States, smectite clays in soils, developed from calcium- and magnesium-rich rocks like gabbro and dolomite, tend to have thicker, more organic-rich topsoil and redder, clay-rich subsoil than well-drained soils developed on granite; the reason is that expansive clays seal up soil structures and impede soil drainage. In tropical regions like Brazil, deeply weathered residual soils developed on silica-rich granite tend to consist mainly of kaolinite and quartz (SiO_2), whereas those formed from silica-poor basalt typically consist of aluminum oxides, like gibbsite ($Al(OH)_3$). Young desert soils are strongly influenced by parent material, particularly the contrast between soils developing in carbonates (limestone) versus aluminosilicates (like granite or basalt). As these desert soils age, the high influx of dust in many arid regions over time blurs the parent material–related differences.

Soils develop over time as pedogenic (soil-forming) processes mechanically and chemically transform parent material. Young surfaces have weakly developed soils, which are thin and show little evidence for pedogenesis. Old surfaces typically have more strongly developed, thicker soils with distinct evidence for pedogenesis, including color and soil structure. On relatively flat-lying surfaces, where erosion and deposition rates are low, soils generally remain in place and chemically evolve over long periods of time (tens of thousands to hundreds of thousands of years). Given enough time, a soil will progressively mature as weathering proceeds until it approaches equilibrium between the rate at which weathering products are removed and the rate at which they are produced. The timescale to reach an equilibrium soil thickness varies greatly. In general, rates of weathering and ion loss from soils tend to decrease over time because the supply of fresh minerals declines as pedogenesis progresses. Soil development can be interrupted and equilibrium conditions reset by a changing regional climate or by environmental changes like glaciation that scrape away some or all of the soil.

Most soils on steep slopes are poorly developed because high rates of erosion and downslope transport move soil before pedogenesis can alter parent material significantly. Another way to think about this is to consider **soil residence time**. In sloping, actively eroding landscapes where regolith production and soil erosion are in equilibrium, soil residence time is equal to the ratio of the soil thickness to the rate of erosion. For example, one meter of soil developed on a hillside that is eroding at an average rate of 0.1 mm/yr has a residence time of 10,000 years, whereas soil 0.1 m thick developed on a slope that is eroding at 1 mm/yr has an average residence time of just a century. In general, soils with longer residence times will be better developed than those with shorter residence times.

Processes and Rates of Soil Production

Soil production is the rate at which bedrock is broken down into erodible material. Because soil erosion occurs everywhere, albeit at different rates, soil can be maintained only if it is replaced by soil production. Soil production is critically important for maintaining soil and soil fertility in the form of fresh minerals and exchangeable cations.

Geomorphologists have proposed two general models for the relationship between soil depth and soil production from unweathered bedrock [**Figure 3.6**]. In one model, soil production declines with increasing soil depth because soil-forming processes have less access to fresh bedrock beneath deeper soils. For example, burrowing activity in thick soils breaks down less rock to produce fresh soil than does burrowing in thin soils where organisms are more likely to entrain weathered rock fragments in their digging. Likewise, once soils become deeper than the average depth of tree rooting, the roots of falling trees simply rework soil rather than ripping up bedrock. In addition, chemically percolating water becomes less aggressive as it runs out of reactive oxygen and CO_2. For soils produced by such processes, soil production rates are highest in shallow soils and decline with increasing soil thickness. In addition, development of a thick regolith can impede weathering and soil production by keeping the base of the soil profile cooler than the surficial soil where daily temperatures may vary greatly. Thus, given time, soil thickness will approach a steady state that reflects local environmental conditions in which soil production and soil erosion are matched.

In the other model, soils promote rock weathering because they hold more water and support more vegetation than bare rock outcrops. This model predicts that soil production rates are highest at some intermediate soil thickness but lower in both thin soils that hold less water and support less vegetation and in thick soils that act to protect the underlying bedrock. In landscapes that have this relationship between soil production and thickness, soil that thins to the point that soil production declines substantially will erode to expose bare rock. In this case, bedrock weathers and erodes slower than surrounding terrain, giving rise to isolated, high-standing bedrock outcrops.

Soil Profiles

Soils exhibit tremendous variability, but almost all soils have an upper zone of leaching that is composed of inorganic primary minerals mixed with organic matter in

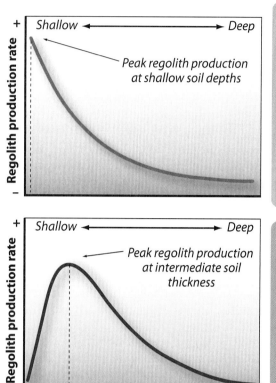

FIGURE 3.6 Soil Production Functions. The rate at which soil is produced from weathered rock is a function of soil depth. Field observations and data suggest two somewhat different relationships that reflect soil-forming processes and subsurface water content.

temperate climates and silt-sized dust in dry climates. Most soils also have a lower zone of accumulation, the nature of which is determined by climate. In temperate and tropical climates, clays and iron oxides dominate the accumulation zone. In dry climates, calcium carbonate ($CaCO_3$) and salts dominate. The upper zone, where losses dominate except in very dusty regions, is called the **zone of eluviation;** the lower zone where particles accumulate and chemical compounds precipitate out of solution is called the **zone of illuviation.**

As soils develop, they become differentiated into distinct **soil horizons.** These horizons are pedogenically created zones in the soil profile with unique characteristics typically found at different depths. Horizons have compositions that reflect the action and interaction of different soil-forming processes. Letters are used to identify master horizons within soil profiles reflecting major differences in physical and chemical characteristics and soil-forming processes. Soil profiles typically consist of certain associations and combinations of soil horizons that are typical of common climatic zones with distinctive environmental conditions [Figure 3.7; Photograph 3.10].

The layer of decomposing organic matter at the ground surface is known as the **O horizon.** Organic matter like leaves, twigs, pine needles, fallen logs, and animal remains dominate the volume of solids in an O horizon; inorganic minerals make up the remainder. In some soil profiles, the O horizon is a thick layer that acts as organic mulch in various stages of decomposition, for example, in wetlands. In other profiles, such as those in deserts, the O horizon is thin or missing altogether.

The next layer down, the **A horizon,** is a mixture of decomposed organic matter and mineral grains. In humid regions, the A horizon is typically dark brown to black in color because of high concentration of decomposed organic matter. Humic acids and high soil CO_2 concentrations promote decomposition of fresh mineral grains within the A horizon, and percolating water leaches material downward in the soil profile. Many A horizons have a relatively loose, friable (crumbly) texture because of soil aggregates, relatively high organic matter concentration, and pervasive disruption by plant roots. In soils developed in dry climates, the A horizons are often light tan in color and have accumulated dust rather than organic matter. Desert A horizons often have visible pores, termed vesicles. In cool, moist, acidic soils, extreme leaching removes organic matter and iron oxides, and a nutrient-poor, bleached gray **E horizon** develops below either the O or A horizon.

The **B horizon** lies below the O, A, or E horizon, and consists predominately of inorganic material in which leached elements and minerals have accumulated after

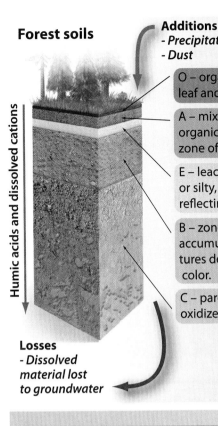

Forest soils

Additions
- Precipitation
- Flood Deposits
- Dust
- Organic Matter

- **O** – organic material, primarily leaf and needle litter, dark color.
- **A** – mixture of decomposed organic material and mineral soil, zone of leaching, dark color.
- **E** – leached horizon, often sandy or silty, white or gray in color reflecting lack of grain coatings.
- **B** – zone of clay and iron oxide accumulation, blocky soil structures developed here, reddish color.
- **C** – parent material, may be oxidized if not saturated.

Humic acids and dissolved cations →

Losses
- Dissolved material lost to groundwater

Forest soils have strong gradients in elemental concentrations as organic acids, generated from decaying organic material, leach material from the A horizon and deposit it into the B horizon. Where forest soils develop under coniferous forests and on sandy soils, bright white, highly leached E horizons are often present.

Grassland soils

Additions
- Precipitation
- Flood Deposits
- Dust
- Organic Matter

- **O** – organic material, primarily grass litter, dark color.
- **A** – mixture of decomposed organic material and mineral soil, zone of leaching, dark color.
- **B** – zone of clay and iron oxide accumulation, blocky soil structures developed here, reddish color.
- **C** – parent material, may be oxidized if not saturated.

Humic acids and dissolved cations →

Losses
- Dissolved material lost to groundwater

Grassland soils typically have large amounts of decaying organic matter in dark and deep A horizons because alkaline grass litter and dry summers slow the decay of organic material. These soils have high **base saturation** and thus are fertile, producing much of the world's grains.

Arid soils

Additions
- Dust
- Precipitation

- **A** – often dominated by aeolian dust, may have vesicular structure (especially at the top), little organic material, tan color.
- **B** – zone of clay, iron oxide accumulation soluble mineral accumulation, and blocky soil structure. In arid environments, salts such as calcium carbonate may precipitate, forming horizons such as white Bk horizon.
- **C** – parent material, likely oxidized.

Dissolved cations →

CaCO$_3$

Losses
- Wind erosion
- Water erosion

Arid-region soils are characterized by small amounts of organic matter and significant accumulation of salts and minerals, such as calcium carbonate. These soils are usually dry, allowing organic carbon to oxidize and preventing soluble elements from leaching away in groundwater. Many arid region soils have large inputs of dust, trapped by surface roughness elements including plants and stones. This dust forms fine-grained A horizons, some with visible pores or vesicles termed Av horizons, that are typical of many arid-region soils.

FIGURE 3.7 Soil Profiles. Soil profiles and the soil horizons that typify them reflect climate and vegetation. Surface, atmospheric, and biospheric processes add mass to soil profiles. Mass is lost through groundwater transport of soluble materials and by wind- and water-induced erosion.

(a)

(c)

(b)

(d)

PHOTOGRAPH 3.10 Soil Variability. Soils are extremely variable in appearance. (a) Forest soil profile, Cape Cod, Massachusetts, formed in sandy glacial fluvial deposits. This soil has surface layers of organic and mineral sandy material (O and A horizon), a white E horizon of uncoated quartz sand grains, and a reddish B horizon. (b) Calcium carbonate collecting in a light-colored K horizon below the red B horizon in this aridisol, Escalante, Utah. (c) Red soil of Martha's Vineyard, interpreted to be developed on an older interglacial deposit, has well-developed O/A horizon over reddened B horizon. (d) Soil section in the hyperarid Paran region of the Negev Desert, Israel. At the top there is a desert pavement, below is the Av soil horizon, a clast-free horizon, the result of ~1.8 million years of dust accumulation.

percolating down from the overlying horizons. Depending on the environment, these accumulations may consist of clay minerals (Bt), carbonates (Bk), salts, or iron and aluminum oxides. In general, clay content is usually higher in B horizons than in the overlying or underlying horizons. B horizons have textures that are distinct from overlying soil horizons as well as from underlying parent material, and often are redder than the overlying A horizon because they contain less organic matter and more iron oxides. B horizons retain little to no evidence of original rock or sediment structure such as bedding or foliation. In many soils, clay films are deposited by water seeping down into the soil. These films coat grains or aggregates and line fractures and voids within the B horizon. In arid and hyperarid regions, evaporation of pore water concentrates calcium carbonate ($CaCO_3$) or increases soil CO_2, causing precipitation and development of a petrocalcic horizon, a thick erosion-resistant layer of calcium carbonate (K horizon).

The **C horizon** consists of unweathered parent material and weathered material that retains some original rock or depositional structure, including sedimentary bedding planes, depositional fabrics, and core stones surrounded by altered weathering rinds. Note that the parent material for a soil can be lithified rock or unconsolidated sediment like dune sand or river gravel. Thoroughly weathered saprolite is generally noted as a Cr horizon and consists of rock material (r) that has weathered in place. Deeply weathered soils and saprolite extending down hundreds of meters into weathered rock are common in the flat, hot, wet tropical landscapes typical of the cratons of Africa and South America.

Biological activity can disrupt the development of soil horizons and mix soil by moving material upward or downward. Bioturbation by plant roots and burrowing animals acts to disrupt development of soil horizons. In particular, the burrowing activity of worms, termites, and gophers may thoroughly mix soil profiles, dragging rock fragments to the surface and organic matter down into the soil. Significant material mixing and net downslope transport also occur when a tree falls and the material pulled up by roots gradually settles back into or near the resulting hole. The persistence of strong soil horizonation in the presence of episodic disturbance indicates the pervasive nature of soil-forming processes, bioturbation rates that are slower than pedogenic processes, and the confinement of the most rapid bioturbation to the A horizon.

Soil Classification

Soils are distinguishable by differences in one or more soil properties that reflect the degree of soil development and provide both indications of process and measures of time and relative landform stability. The most important soil properties observable in the field—horizonation, color, texture, and structure—reflect a soil's clay and organic matter content, its development, and its ability to hold moisture. For the most part, soils are classified based on such field observable characteristics.

Soil color indicates both mineral composition and organic matter content. Soils with high organic matter content are typically dark brown to black. Soils are red to yellow-brown in color in oxidizing environments and blue to black or gray in reducing environments such as wetlands. Light colors, like white, gray, and beige are generally associated with accumulations of calcite ($CaCO_3$) or leached horizons consisting mostly of residual silica (SiO_2). In the field, soil scientists use a book of color charts resembling those used at paint stores to classify soil colors based on a palette of standardized colors.

Soil texture is measured in terms of the proportions of different particle sizes in the soil [**Figure 3.8**]. The U.S. Department of Agriculture defines 12 standard soil textures based on the relative proportions of clay, silt, and sand (e.g., clay loam, silt loam, silty clay). Although soils may contain larger clasts (like gravel), soil properties measured in the field generally are described based on particles less than 2 mm in diameter (i.e., sand and smaller) and then, if appropriate, the modifier gravelly is added to the classification. Soil texture greatly influences the **cohesion** (consistency) of a soil as well as its **plasticity** (ability to resist deformation). A simple field test for soil composition is to moisten a sample of soil and try to make a ball in one's hand. Clay can be rolled into a "worm"; silt and sand cannot. Another simple field test—one that is not always advisable to employ—is that medium to coarse silt feels gritty between one's teeth, while clay has a smooth, creamy texture. Sand grains are visible to the naked eye.

Soil structure describes the shapes in which soil particles cluster together. Most A horizons have granular or crumb structures in which individual clumps, or aggregates, are more or less spheroidal. In contrast, B horizons often have structures, called **peds,** that are shaped like blocks, prisms, or plates with rounded or angular edges. Peds can range from pea size to fist size and reflect the deposition of clay enhanced by wetting and drying along gaps or discontinuities in the soil structure.

Soil organic matter includes undecomposed material like leaves, branches, and bones (**litter**) as well as amorphous decomposed material (**humus**). Soil microorganisms that decompose litter into humus become active at temperatures slightly above freezing (generally $>5°C$), and rates of microbial decomposition increase as temperatures rise. At temperatures above about 25°C, little humus accumulates because it decomposes as fast as it is produced. Consequently, soil organic matter content tends to be greatest in mid-latitude regions with mean annual temperatures between 5° and 25°C.

Soil taxonomy is one way in which scientists classify soils. **Soil orders** are the highest level of this soil classification scheme. The U.S. Natural Resource Conservation Service (formerly the Soil Conservation Service) uses a system of soil taxonomy that recognizes 12 soil orders, distinguished by diagnostic horizons and characteristics that reflect different environments, processes, and soil residence time on the landscape [**Figure 3.9**]. Soils can be further subdivided into 64 common suborders and then into more specific soil types, but the major soil orders are sufficient to characterize soil characteristics, parent material, and degree of weathering at a regional scale.

Young, incompletely developed soils are either **entisols** that have a faint A horizon, or **inceptisols** with a weak B horizon. These soils do not have enough horizon development to qualify as one of the more developed soil orders. Soils formed in seasonally submerged or frozen environments have distinctive characteristics. **Histosols** are organic-rich soils that form in wetlands where reducing

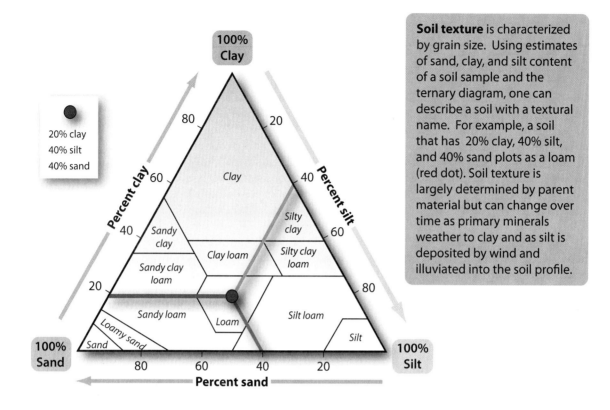

FIGURE 3.8 Soil Structure and Texture. Soil texture reflects the relative percentages of sand, silt, and clay in the soil. Pedogenic processes, soil materials, and water interact to create soil structure, which can be recognized when examining soil profiles in the field.

conditions restrict rates of organic decomposition. Histosols are identified by their black color, high organic matter content (more than 25 percent), thick organic-rich O and A horizons, and lack of evidence for oxidation. **Gelisols** form in polar and subpolar environments where the ground at some depth stays frozen all year. These cold climate soils include, by definition, a layer of permanently frozen soil (**permafrost**) within 2 meters of the land surface. They typically show little evidence of chemical weathering and have only an A horizon developed over permafrost. Gelisols are often structurally disrupted and mixed by seasonal freeze–thaw processes called **cryoturbation**,

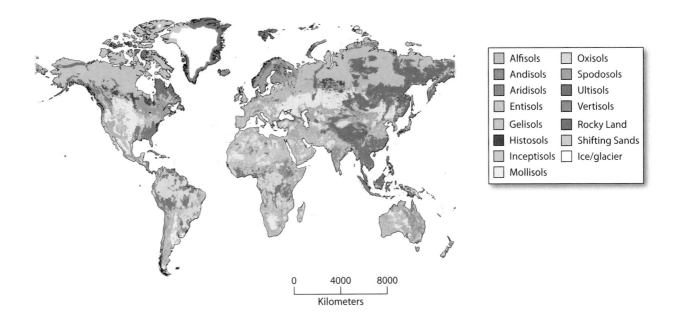

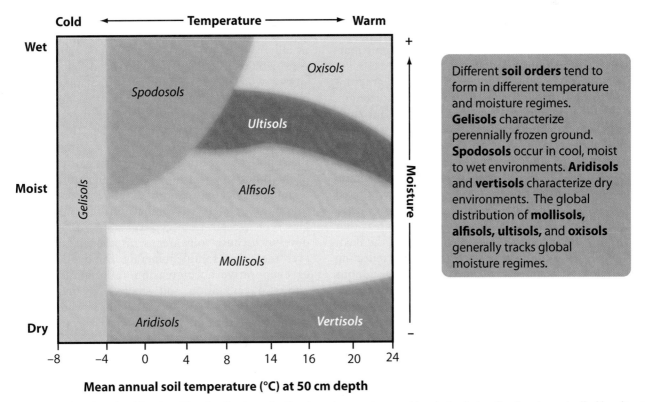

FIGURE 3.9 Soil Order Classification. The distribution of soil orders, the most general level of soil classification, is controlled by climate, predominantly temperature and precipitation. [Adapted from Birkeland (1999).]

which complicates foundation engineering in periglacial environments.

Several types of soil reflect distinctive parent materials. **Andisols** are formed from volcanic parent material. Many andisols contain amorphous colloids from weathering of glassy volcanic ash and have high cation-exchange capacities. **Vertisols** are soils that contain high concentrations of swelling clay (smectite) and exhibit significant shrink-swell behavior; thus, vertisols do not have distinctive horizons because of the physical mixing. They exhibit large desiccation cracks when dry and tend to become quite sticky when moist. The other six soil orders are associated with particular climate or environmental settings and form under certain combinations of temperature and relative wetness.

Mollisols typically develop in temperate regions with grassland vegetation, like the prairies of the Great Plains of North America. Mollisols have well-developed, black, organic-rich A horizons that are often more than a meter thick and B horizons that are enriched in clay. Grassland soils accumulate abundant organic matter because of the dense and deep root structures of most grasses. In many grasslands, the soils are too cold for decomposition in the winter and are too dry for much decomposition in the summer. Such environments permit incomplete decay of organic material, producing organic colloids that leach down into the soil, darkening both A and B horizons. The abundant organic matter and clays with high base saturation in mollisols generally make fertile agricultural soils. The fertility of the grassland soils of the U.S. Midwest is provided by A horizons so thick and rich that they remain productive even after more than a century of agriculturally induced erosion.

Surface litter produced by forests, particularly coniferous forests, tends to be more acidic than organic matter produced by grasses. Consequently, forest soils are generally more acidic and have A horizons that are thinner and more leached of soluble ions (notably calcium, sodium, and magnesium) than grassland soil horizons. Soils that develop beneath temperate forests are called **alfisols** and **ultisols**. Alfisols and ultisols have thin A horizons. Alfisols form under deciduous forests where decaying leaves provide organic acids, and as such are common in the eastern United States. Alfisols are less weathered than ultisols; alfisols have higher base saturation. Ultisols tend to form in old and highly weathered areas that are warm and wet and were not glaciated (e.g., the Piedmont region of southeastern North America), reflecting more prolonged or intense weathering under deciduous or coniferous forests. In both alfisols and ultisols, chelating agents (organic acids from leaf and needle decomposition) leach iron, aluminum, and organic matter from the A horizon and deposit these materials in the B horizon. A distinctive, light-colored, leached E horizon between the A and B horizon characterizes many forest soils (see Photograph 3.10a).

Spodosols are forest soils that form under cooler temperature regimes and a range of moisture conditions in environments characterized by coniferous or boreal forests. Spodosols exhibit severe leaching due to acidic conditions under coniferous forests resulting from organic acids produced by decaying needles. This results in translocation (elluviation) of clays, organic matter, iron, and aluminum from the A and E horizons. Thus, the B horizon in spodosols is enriched in iron and organic matter, typically forming a hardpan horizon (**fragipan**) within the soil. Spodosols are most common in parent materials with high permeability, such as sands on a coastal plain.

Oxisols are the most deeply weathered of all soil orders and are characterized by highly oxidized soil horizons that form in hot, wet (tropical) forest environments. Abundant rain falling through a surficial organic layer leads to acidified water that promotes intensive leaching, which strips nutrients from the soil. Oxisols generally have distinctive red coloration and little organic matter accumulation because plant and animal remains decay rapidly in such conditions. Red-colored oxisols have an abundance of iron oxides (hematite and goethite) and tend to be depleted in exchangeable nutrient cations. The most heavily weathered tropical soils (**laterites**) consist of little more than residual iron and aluminum oxides such as gibbsite and kaolinite. The components, once heated and dried by the sun, harden into bricklike peds. Organic matter and nutrients in oxisols are contained in the thin O or A horizon or in the standing vegetation, posing impressive challenges to sustaining agricultural productivity when these surficial horizons or the vegetation are removed by erosion.

Aridisols are soils that develop in dry regions. They usually have low organic matter content and a minimal A horizon overlying a B horizon, the development intensity of which varies by soil age. In semi-arid and arid climates, infiltrating precipitation typically evaporates from the ground surface or from within the soil profile before reaching the water table. When this occurs, ions and particles in percolating rainwater or mobilized from the A horizon precipitate and deposit calcium carbonate in the B horizon, coating rock and mineral surfaces and forming a calcic horizon. In extremely arid settings like the Negev Desert in southern Israel, salt-rich (salic, NaCl) or gypsum-rich (gypsic, $CaSO_4 \cdot 2H_2O$) horizons develop because soilwater is insufficient to leach away even these very soluble minerals. The depth to this zone of evaporative accumulation reflects the amount of annual precipitation and shallow groundwater flowpaths. In arid and semi-arid regions, if water containing silica dissolved from fine dust and rock evaporates, silica can reprecipitate. Increasing aridity generally results in less well-developed A horizons and shallower zones of evaporite accumulation. Given enough time, however, red, clay-rich B horizons can develop in some of Earth's driest deserts.

Dustfall is important in the development of many soils. In particular, dust-influenced desert soils typically have a stone-free, silt-rich horizon just below the surface. This accumulation of fine, wind-delivered sediment is known as an Av horizon because of the presence of small vesicles (air pockets) that form when the horizon wets from rain and then dries. Sometimes, Av horizons are capped by a layer of coarse clasts known as a desert pavement (see Photograph 3.10d).

Soils and Landscapes

Soil properties and orders vary among climate zones and physiographic regions [**Figure 3.10**]. Soil orders and the degree of soil development, particularly soil thickness and organic matter content, generally track latitudinal patterns in temperature and precipitation. Deeply weathered oxisols are typical of the equatorial tropics. Aridisols are typical of mid-latitude deserts. Organic-rich mollisols and forest soils are typical of temperate latitudes. However, soils also differ within single landscapes because of variations in environmental factors, rates of erosion, and landscape history.

Soil Development over Time

As soils develop over time, they gradually lose the physical and chemical characteristics of their parent materials and increasingly take on characteristics that reflect soil-forming processes and the four other factors affecting soil formation, especially climate and vegetation. Consequently, differences in parent material are best expressed in poorly developed entisols and inceptisols. Soil properties like the amount of organic matter, the degree and depth of oxidation, the removal of cations or minerals, development of clay minerals, and clay or $CaCO_3$ content in the B horizon change as the soil weathers and develops. Thus, the degree of weathering can serve as a proxy for the stage of development and relative age of a soil.

The parts of a soil profile develop at different rates. As a soil ages, the A horizon generally achieves steady state or equilibrium first, followed by the B horizon, and then such features as carbonate accumulation and oxidation. Different soil orders take different times to reach equilibrium. Mature spodosols and alfisols are capable of forming in millennia. Mature ultisols may take tens of thousands of years to form. Mature oxisols may require hundreds of thousands of years to form.

A **chronosequence** is a series of soils of different ages that formed on the same parent material under similar conditions of vegetation, topography, and climate—for example, a sequence of soils formed on alluvial terraces at different heights above an incising river [**Figure 3.11**]. Chronosequences of soils that have been dated by radiometric or other means are used to evaluate rates of change in soil properties. Calibrated soil development rates can then be used to infer the ages of landforms from soil properties or to estimate the amount of time that a soil was exposed at the surface before being buried.

Soil Catenas

Different kinds of soils can develop on different parts of a single slope because of local variability in soil-forming

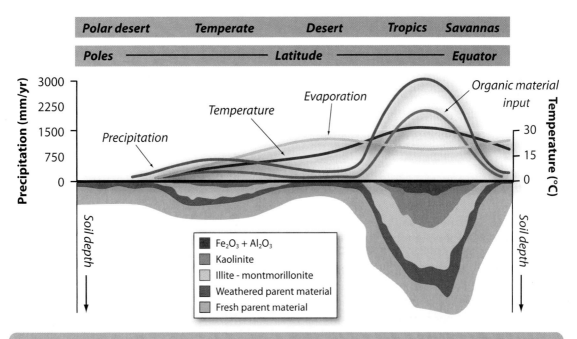

FIGURE 3.10 Soil Properties and Depths with Latitude. Soil properties, including average soil depth and characteristic weathering minerals, vary by latitude because climate and vegetation (important for soil formation) are latitude-dependent. [Adapted from Strakov (1967).]

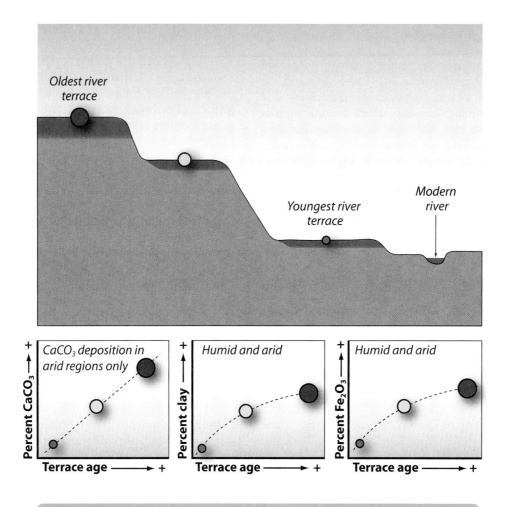

FIGURE 3.11 Soil Chronosequences. Soil characteristics, such as horizon development, color, clay content, and chemical composition, change predictably with age in a soil chronosequence when other soil-forming variables such as parent material are constant.

processes, erosion, and deposition. A **soil catena** (Latin for "chain or series") is a suite of soils from different positions in a landscape—each of which has different characteristic soil moisture conditions, sediment transport rates, slope steepness, and chemical weathering rates [**Figure 3.12**]. Catenas reflect local differences in soil-forming processes that result from topographic position and the hydrological and geomorphological processes that govern infiltration, runoff, and soil erosion. For example, soils at the top of slopes where water readily washes over or through the soil profile may be well drained, oxidized, and reddish in color, whereas those soils at the base of slopes tend to have a higher water table and reducing conditions that produce more gray to blue colors. Soils developed on the steep slopes around a plateau may be quite different from those developed on the flat plateau surface. In many environments, sloping surfaces are the rule and distinct soils develop on hilltops at mid-slope, and on the base of slopes as erosion thins the soils on the upper slopes and accumulation stacks the soils a the base of slopes.

Paleosols

Unlike soils actively developing at Earth's surface, **paleosols** are ancient soils in which soil development processes have ceased. Some paleosols are buried and thus preserved in the geologic record. **Relict soils** are those that have remained at the surface since their formation but may have formed under different climatic or hydrologic conditions. **Buried soils** are readily recognized, typically by an abrupt change in color, texture, or mineralogy, and can be found in geologic materials of almost any age [**Photograph 3.11**].

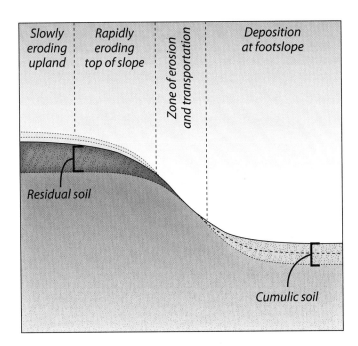

The suite of soils associated with landforms in a particular region form a soil **catena** in which soil properties vary systematically with landscape position. The slowly eroding upland consists of soils developed on in situ parent material. Steeper hillslopes may display truncated soil profiles and represent a zone of translation and more rapid erosion (transport slopes). Lower gradient slopes may have soils developed on deposits of material transported from upland environments.

FIGURE 3.12 Soil Catena. Soils are generally eroded from upland areas, transported down hillslopes, and deposited at low-gradient areas below slopes. A sequence of soils with properties that vary as a function of hillslope position is known as a catena.

When found in the geologic record, paleosols indicate a land surface on which a soil was able to form. They thus represent a period during which soil production and development outpaced both erosion and burial by deposition. Because the characteristics of paleosols reflect the soil-forming factors of the landscape in which they developed, they can be useful indicators of past climatic or environmental conditions. Some soil features, such as organic matter, do not survive extended burial and are poorly preserved, whereas other characteristics, like soil structure and texture of B horizons, are geologically more robust.

Weathering-Dominated Landforms

Weathering processes produce a variety of distinctive coarse- and fine-scale landforms. At fine scales, physical and chemical processes wear away the edges and corners of rock outcrops, joint-bounded blocks, and boulders fastest; they thus promote **spheroidal weathering**, which

PHOTOGRAPH 3.11 Paleosol. Wall of trench cut into historic alluvial fan that covers a paleosol. Well-preserved buried A horizon (black) and B horizon (brown) buried by tan and gray post-settlement sediment resulting from clearance of hillslopes above. Meter stick for scale.

converts angular blocks of rock that result from tectonic and unloading fractures into rounded cobbles, boulders, and monoliths. Outward expansion of clay minerals on chemically altered rock surfaces enhances spheroidal weathering and produces weathering rinds that peel off like onion skins [Photograph 3.12].

PHOTOGRAPH 3.12 Spheroidal Weathering. Spheroidal weathering of granitic rock at Goldstone in the Mojave Desert, was largely brought on by hydrolysis of feldspar and by wetting-drying and oxidation of biotite and expandable clay. Rock is surrounded by grus, weathered, disarticulated mineral grains that once made up the granite rock.

PHOTOGRAPH 3.13 **Weathering Pit.** A weathering pit on the top of granitic Pildappa Rock, on the Eyre Peninsula in south central Australia, is more than a meter wide.

PHOTOGRAPH 3.14 **Tafoni.** Tafoni developed in granite near Caldera, Chile.

Wetting and drying or freezing of water causes localized spalling and granular disintegration that creates **weathering pits** on bare rock surfaces of certain rock types [Photograph 3.13]. Extreme cases of weathering pit development result in cavernous or honeycomb textures known as **tafoni**, from the Greek for tomb or the French for window [Photograph 3.14]. Dissolution pits and cavities often characterize highly soluble carbonate outcrops. At coarser scales, weathering-dominated landforms include karst topography (discussed in later chapters), inselbergs, tors, and duricrusts. Differential weathering can also operate on a much larger scale. For example, in the valley and ridge province of the eastern United States, differential weathering and subsequent erosion results in quartzite ridge tops and limestone valley bottoms.

Inselbergs and Tors

Inselbergs (German for "island mountains") and **tors** are high-standing bodies of exposed rock that rise above surrounding terrain and result from spatial variability in the rates of rock weathering and erosion [Photographs 3.15 and 3.16]. Inselbergs are large residual rock masses still attached to bedrock after episodes of deep weathering and

PHOTOGRAPH 3.15 **Inselbergs.** The Olgas of central Australia are sandstone inselbergs, also known as domes or bornhardts.

PHOTOGRAPH 3.16 Tor. Devil's Marbles, a tor with boulders, is the residuum of spheroidal weathering in central Australia.

subsequent erosion removed the surrounding material. Such stripping can reveal **etchplains,** extensive bedrock surfaces formed under a mantle of weathered material. Tors are smaller features consisting of bedrock on which may sit multiple, smaller exhumed core stones not necessarily still attached to the rock. In some places, field evidence suggests that inselbergs and tors weathered more slowly in the subsurface than did surrounding areas because of differences in either mineral composition or fracture density that rendered them less susceptible to chemical weathering than the surrounding rock [**Figure 3.13**]. In other places, a chance exposure of one part of the rock mass at the surface renders it less susceptible to further weathering, providing a positive feedback to amplify the topography (until the outcrop is destroyed by erosion from the sides).

Variability in the progress of deep weathering produces an uneven **weathering front,** the boundary between fresh rock below and that altered by weathering closer to Earth's surface. Inselbergs and tors represent exhumation of less-weathered material protruding above the weathering front. These exhumed landforms record a changing balance between weathering and erosion. Specifically, they represent the change from an erosion-limited landscape to one where the pace of weathering limits erosion rates.

Tors are common landforms in weathered landscapes around the world; the rock towers of southwest England and Wales are some of the best-known examples of tors. Australia's famous Uluru (also known as Ayers Rock), a massive sandstone outcrop that rises hundreds of meters above the desert plain of the central outback, is perhaps the largest and most dramatic example of an inselberg in an arid setting. In a more humid climate, Rio de Janiero's famous Sugarloaf, a granitic inselberg, towers above deeply weathered surroundings [**Photograph 3.17**]. Along Brazil's coast, mechanically weak fault zones along which weathering is enhanced and where shattered rock is easily eroded have been mapped between the towering inselbergs. It appears that the steep sides of these inselbergs likely result from enhanced weathering along these near-vertical fault zones.

Duricrusts

Duricrusts are erosion-resistant soils hardened and turned back into rock by cementation within the pedogenic zone. In arid regions, enough calcium carbonate ($CaCO_3$) may accumulate within a soil profile to form a cementlike layer of caliche (a Spanish word for "porous material cemented by calcium carbonate"), or **calcrete,** that is erosion resistant when it is exposed at the land surface. **Silcrete,** a hard silica-rich (SiO_2) layer, behaves similarly [**Photograph 3.18**] and also forms in arid and semi-arid regions as silica in groundwater reprecipitates. Duricrusts also form by evaporation of solute-rich groundwater along stream valleys. These protective shells can even be tough enough to cause **topographic inversions,** which occur when formerly low-lying terrain becomes more resistant to erosion than the neighboring uplands and eventually becomes elevated as the surrounding terrain erodes away.

Extreme oxidation, intense weathering, drying, and leaching produce iron-rich residual soils known as **laterite** and **ferricrete.** Ferricrete is an iron-rich duricrust made of highly concentrated iron oxides as a result of intensive weathering. Laterite is clay-rich soil containing large proportions of iron and aluminum oxides. These erosion-resistant crusts form where intense weathering removes all but the least mobile elements, leaving behind just aluminum and iron oxides. Laterite and ferricrete are often found on slowly eroding topographic highs in the tropics and form the caprock for mesas in central Australia and the Amazon.

Time 3

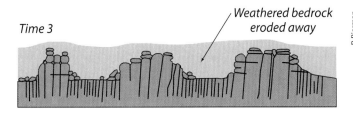

Weathered bedrock eroded away

Time 2

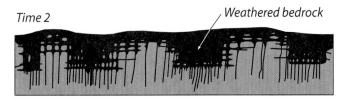

Weathered bedrock

Time 1

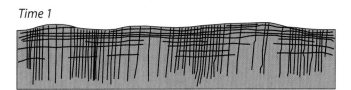

The relief developed on **etchplains**, regional bedrock surfaces characterized by **tors** and **inselbergs**, reflects differential weathering and subsequent erosion of an irregular **regolith**. As weathering proceeds faster in more highly fractured portions of a landscape, the interface between regolith and bedrock may develop substantial relief. Subsequent erosion of weaker, weathered material, perhaps triggered by a change in climate or base level, can leave less-highly fractured and weathered bedrock standing above surrounding terrain as **core stones**, tors, and inselbergs.

FIGURE 3.13 Etchplain Formation. Erosion of deeply weathered regolith exposes an irregular bedrock surface of tors and inselbergs that was shaped in the subsurface by physical and chemical alteration of bedrock along the weathering front. [Adapted from Linton (1955).]

Applications

Weathering is of fundamental geomorphological and societal importance because it transforms hard rock into material that can be moved by surface processes and used by

PHOTOGRAPH 3.17 The Sugarloaf. Rio de Janiero's famous Sugarloaf is a granitic inselberg towering above deeply weathered surroundings. Its steep margins are defined by near-vertical fault zones along which chemical weathering proceeded more quickly, isolating the more resistant rock.

living organisms. The physical breakdown of rock into smaller pieces and the chemical transformation of rock into secondary minerals and dissolved elements influence the types and rates of geomorphological processes that shape topography.

Soils are the frontier between geology and biology. The thin skin of weathered rock and decomposing organic matter provides the nutrients that nourish the plants on which all terrestrial life depends. Cation-exchange capacity and base saturation of soils are the basis for soil fertility because plants can readily extract nutrients from soils with high base saturation. Sustained cultivation without replenishing soil nutrients leads to declining crop yields. This is one reason why floodplains that receive annual deposits of fresh minerals and volcanic soils that are periodically replenished by ash fall are prized and highly productive agricultural lands around the world. Modern

PHOTOGRAPH 3.18 Silcrete. Silica-cemented rocks at Coober Pedy in central Australia. Such silica-cemented layers (silcrete) hold up the mesa in the distance, acting as caprocks.

industrial agriculture uses tremendous amounts of chemical fertilizers (principally nitrogen, phosphorus, and potassium) to supplement native soil fertility in place of traditional crop rotations, applications of manure, and organic farming techniques that are based on soil ecology. Understanding soil-forming processes and soil fertility are important for evaluating options for maintaining soil fertility and sustaining agricultural productivity.

Maintenance of soil fertility depends not only on sustaining soil nutrient levels, but also on conserving the soil itself. Extensive soil loss to erosion can follow deforestation, tillage, and destruction of vegetative cover by fire or overgrazing. On some now-barren Caribbean islands, sugarcane cultivation on steep slopes sent most of the topsoil into the ocean in less than a century. Recent Earth history is rife with truncated and thinned soil profiles that record examples of ancient societies (such as classical Greece, Rome, and Easter Island) that failed to prevent rates of soil erosion from exceeding rates of soil production. Globally, the average rate of net soil loss from agricultural fields has increased 10- to 20-fold as a result of tillage and exposure of bare soil to the effects of wind, rainfall, and runoff. Basic training in geomorphology can help practitioners tailor sustainable agricultural practices to particular soils and landforms.

Farming practices have reduced soil organic matter across vast areas of the continents, particularly in mollisols. About one-third of the carbon dioxide added to the atmosphere by human activity since the Industrial Revolution came from the decay of soil organic matter, the result of plowing fertile grassland soils. But people can also improve soil fertility. The recently discovered organic-rich, incredibly fertile "terra preta" soils in the Amazon jungle formed over millennia as indigenous people burned trash and broken pottery in their fields. Today these soils form islands of fertility in otherwise infertile tropical soils.

Weathering also provides minerals critical for modern life. Deep weathering on ancient land surfaces produced iron and aluminum ores through pervasive leaching and removal of other common elements, thereby concentrating relatively immobile elements in residual soils. Aluminum is a common element in terms of its distribution in Earth's crust, but it is dispersed in most rocks such that it is not economically extractable. It takes millions of years to dissolve away everything else and make laterite soils that are enriched enough in aluminum that they constitute aluminum ores. Few people realize that we wrap our soda and beer in metal mined from ancient soils.

Selected References and Further Reading

Amit, R., R. Gerson, and D. Yaalon. Stages and rate of the gravel shattering process by salts in desert Reg soils. *Geoderma* 57 (1993): 295–324.

Armson, K. A. *Forest Soils: Properties and Processes.* Toronto, Buffalo: University of Toronto Press, 1977.

Berner, R. A. "Chemical weathering and its effects on atmospheric CO_2 and climate." In A. F. White and S. B. Brantley, eds., *Chemical Weathering Rates of Silicate Minerals.* Washington, DC: Mineralogical Society of America, 1995.

Bierman, P. R., and A. R. Gillespie. Range fires: A significant factor in exposure-age determination and geomorphic surface evolution. *Geology* 19 (1991): 641–644.

Birkeland, P. W. *Soils and Geomorphology.* New York: Oxford University Press, 1999.

Birkeland, P. W., R. R. Shroba, S. F. Burns, et al. Integrating soils and geomorphology in mountains—An example from the Front Range of Colorado. *Geomorphology* 55 (2003): 329–344.

Brady, N. C., and R. R. Weil. *The Nature and Properties of Soils*, 14th ed. Upper Saddle River, NJ: Prentice Hall, 2008.

Butler, D. R. *Zoogeomorphology: Animals as Geomorphic Agents.* New York: Cambridge University Press, 1995.

Colman, S. M. Rock-weathering rates as function of time. *Quaternary Research* 15 (1981): 250–264.

Colman, S. M., and D. P. Dethier, eds. *Rates of Chemical Weathering of Rocks and Minerals.* Orlando: Academic Press, 1986.

Eppes, M. C., L. D. McFadden, J. Matti, and R. Powell. Influence of soil development on the geomorphic evolution of landscapes: An example from the Transverse Ranges of California. *Geology* 30 (2002): 195–198.

Gile, L. H., F. F. Peterson, and R. B. Grossman. Morphological and genetic sequences of carbonate accumulation in desert soils. *Soil Science* 101 (1966): 347–360.

Harden, J. W. A quantitative index of soil development from field descriptions: Examples from a chronosequence in central California. *Geoderma* 28 (1982): 1–28.

Harrison, J. B. J., L. D. McFadden, and R. J. Weldon III. Spatial soil variability in the Cajon Pass chronosequence: Implications for the use of soils as a geochronological tool. *Geomorphology* 3 (1990): 399–416.

Heimsath, A. M., W. E. Dietrich, K. Nishiizumi, and R. C. Finkel. The soil production function and landscape equilibrium. *Nature* 388 (1997): 358–361.

Heimsath, A. M., W. E. Dietrich, K. Nishiizumi, and R. C. Finkel. Cosmogenic nuclides, topography, and the spatial variation of soil depth. *Geomorphology* 27 (1999): 151–172.

Jenny, H. *Factors of Soil Formation.* New York: McGraw-Hill, 1941.

Linton, D. L. The problem of tors. *Geographical Journal* 121 (1955): 470–486.

Loughnan, F. C. *Chemical Weathering of the Silicate Minerals.* New York: American Elsevier, 1969.

Machette, M. N. "Calcic Soils of the Southwestern United States." In D. L. Weide, ed., *Soils and Quaternary Geology of the Southwestern United States.* Geological Society of America Special Paper 203. Boulder, CO: Geological Society of America, 1985.

Markewich, H. W., and M. J. Pavich. Soil chronosequence studies in temperate to subtropical, low-latitude, low-relief terrain with data from the eastern United States. *Geoderma* 51 (1991): 213–239.

McAuliffe, J. R., E. P. Hamerlynck, and M. C. Eppes. Landscape dynamics fostering the development and persistence of long-lived creosotebush (*Larrea tridentata*) clones in the Mojave Desert. *Journal of Arid Environments* 69 (2007): 96–126.

McFadden, L. D., M. C. Eppes, A. R. Gillespie, and B. Hallet. Physical weathering in arid landscapes due to diurnal variation in the direction of solar heating. *Geological Society of America Bulletin* 117 (2005): 161–173.

Montgomery, D. R. *Dirt: The Erosion of Civilizations*. Berkeley: University of California Press, 2007.

Mustoe, G. E. Biogenic origin of coastal honeycomb weathering. *Earth Surface Processes and Landforms* 35 (2010): 424–434.

Ollier, C. D. *Weathering*, 2nd ed.. New York, London: Longman, 1984.

Reheis, M. C, R. R. Shroba, J. W. Harden, and L. D. McFadden. Development rates of Late Quaternary soils, Silver Lake Playa, California. *Soil Science Society of America Journal* 53 (1989): 1127–1140.

Richter, D. D., and D. Markewitz. How deep is soil? *BioScience* 45 (1995): 600–609.

Riebe, C. S., J. W. Kirchner, and R. C. Finkel. Erosional and climatic effects on long-term chemical weathering rates in granitic landscapes spanning diverse climate regimes. *Earth and Planetary Science Letters* 224 (2004): 547–562.

Robinson, D. A., and R. B. G. Williams, eds. *Rock Weathering and Landform Evolution*. New York: Wiley, 1994.

Schaetzl, R., and A. Anderson. *Soils: Genesis and Geomorphology*. Cambridge: Cambridge University Press, 2005.

Selby, M. J. Form and origin of some bornhardts of the Namib Desert. *Zeitschrift für Geomorphologie* 26 (1982), 1–15.

Soil Survey Staff. *Soil Taxonomy*, 2nd ed. Natural Resource Conservation Service, Agriculture Handbook 436. Washington, DC: U.S. Department of Agriculture, 1999.

Sposito, G. *The Chemistry of Soils*, 2nd ed. New York: Oxford University Press, 2008.

Strakov, N. *Principles of Lithogenesis*. London: Olive & Boyd, 1967.

Stuiver, M. Atmospheric carbon dioxide and carbon reservoir changes: Reduction in terrestrial carbon reservoirs since 1850 has resulted in atmospheric carbon dioxide increases. *Science* 199 (1978): 253–258.

Swoboda-Colberg, N. G., and J. L. Drever. Mineral dissolution rates in plot-scale field and laboratory experiments. *Chemical Geology* 105 (1993): 51–69.

Twidale, R. *Granite Landforms*. New York: Elsevier Scientific 1982.

Zimmerman, S. G., E. B. Evenson, J. C. Gosse, and C. P. Erskine. Extensive boulder erosion resulting from a range fire on the type-Pinedale moraines, Fremont Lake, Wyoming. *Quaternary Research* 42 (1994): 255–265.

DIGGING DEEPER How Fast Do Soils Form?

How quickly does soil, the discontinuous mantle of weathered rock and organic material that covers large parts of Earth's surface, form? This is an important and practical question, because soil and its fertility are so important to agriculture and thus to human survival. It is also a difficult question to answer because the answer varies not only by where one looks but also by how one defines soil formation.

There are many different ways of approaching the concept of soil formation, the processes and rates by which soil is created from rock and sediment. In many landscapes, soils develop on stable deposits of unconsolidated material such as blankets of loess, bouldery colluvium, or the sand and gravel of river terraces. In these cases, the rate of **soil development** describes the speed at which pedogenic processes alter the parent material. In such settings, it is the pedogenic alteration of primary to secondary minerals and the accumulation of organic matter that makes soils fertile and useful for agriculture.

This rate of soil development (pedogenesis) has been quantified in many different environments by geomorphologists studying **chronosequences,** soils formed in similar parent materials but having different ages [**Figure DD3.1**].

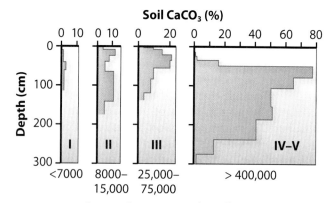

FIGURE DD3.1 Soil chronosequences are constructed from observations of soils that developed in similar climates and on similar parent materials but differ in age. Here, the soil chronosequence (soils I through V) is used to indicate that as arid soils develop over time, the amount of calcium carbonate ($CaCO_3$) in the soil increases predictably. This increase reflects both in situ weathering of calcium-bearing minerals and the addition by wind of carbonate-bearing dust. Over time, carbonate is carried more deeply into the soil profile by infiltrating water. [From Birkeland, (1984).]

In some areas, the rate of soil development has been quantified well enough that soil ages are estimated using a **soil development index** that takes into account soil characteristics that change over time, such as reddening, structure, and clay content (Harden, 1982).

In the last work of his long, stimulating, and controversial career, Darwin (1881) concluded that soils formed at a pace of less than a millimeter per year. He arrived at this conclusion by measuring the thickness of new soil that lay on top of the foundations and floors of Roman buildings in England. Darwin attributed the development of new soil to the action of worms gradually bringing up material to the ground surface; thus, Darwin's soil was cumulic—it was not formed by the weathering of bedrock from below but rather by the addition of material from above.

In contrast to soil development on unconsolidated parent material, **soil production,** or the production of regolith, refers to the in situ development of soil by the loosening of individual mineral grains from bedrock (Heimsath et al., 1997). Such loosening is the result of both physical and chemical weathering, the rate of which depends on many factors including soil moisture, rock type, and biological disturbance by burrowing animals and tree roots. The concept of soil production from rock is applicable to upland, often hilly, soil-mantled bedrock landscapes where the limiting factor for soil formation is the weathering of the underlying rock. The concept of soil production does not apply to agricultural lowlands where soils develop on unconsolidated deposits.

Two centuries ago, Playfair (1802) recognized that the long-term rates of soil production and soil erosion must balance for soil-mantled, upland landscapes to persist over time. If soil erosion occurred faster than soil production, the soil would thin until bare rock was exposed at the surface, but soil blankets slopes in most humid and temperate landscapes. More than a century and a half later, a thorough review of the literature found no reliable, quantitative estimates of how fast soil formed from bedrock (Smith and Stamey, 1965). Since then, a variety of geochemical measurements have been used to quantify rates of soil production from rock.

Measurements of the bulk composition of rocks, soils, and the **dissolved load** of runoff collected in studies of small watersheds, along with the assumption of steady state, allows estimation of rates of soil production from a **mass balance** of major rock-forming elements (Alexander, 1985; 1988). Inherent in this approach is the assumption that cations are released from minerals as rock chemically weathers into regolith.

Elemental mass fluxes and thus rates of chemical weathering and soil production in 18 North American, European, Australian, and African watersheds, each underlain by a single rock type, were positively correlated with runoff; that is, rock weathered faster in wetter climates. Assuming all cations came from weathering of bedrock, rates of rock weathering and thus soil production varied from 0.002 to 0.158 mm/yr. Similar application of a global geochemical mass balance, based on the mean compositions of major elements (aluminum, calcium, iron, potassium, magnesium, sodium, silicon) in Earth's crust, soils, and the dissolved load of rivers, found that the global average rate of soil production from rock weathering was between 0.03 and 0.10 mm/yr with a mean rate of 0.06 mm/yr (Wakatsuki and Rasyidin, 1992). Not all of this weathering was conversion of bedrock to regolith and eventually soil. Some of the cations may have been released as soils developed on unconsolidated parent materials.

Analyses of cosmogenic isotopes (see Chapter 2) in subsoil bedrock allows direct measurement of rates of soil production (Heimsath et al., 1997) on upland landscapes where hillslope processes dominate the movement of material downslope. This approach assumes steady-state soil production and steady soil thickness; that is, the rate of bedrock weathering and new soil production matches the rate of downslope transport of soil [Figure DD3.2].

Investigations at several different field sites show that ^{10}Be concentrations vary systematically with the thickness of soil cover. Applications of this approach to soil profiles in California and Australia support an exponential decline of soil production from rock with increasing soil depth

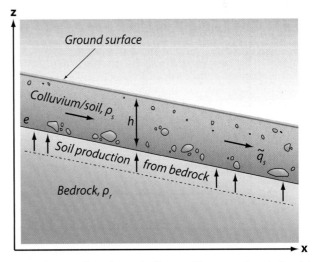

FIGURE DD3.2 This schematic diagram illustrates the relationship between soil production and soil erosion on an idealized cross section of a hillslope. The soil thickness h will be maintained if soil production from bedrock matches mass transport, q_s, of the active layer of soil. The area shown between the base of the soil at elevation e and the dash-dot line is the amount of bedrock that would be converted to soil over some specified time interval. Here ρ_s and ρ_r are the densities of soil and bedrock, respectively.
[From Heimsath et al. (1997).]

DIGGING DEEPER How Fast Do Soils Form? *(continued)*

[**Figure DD3.3**]. Heimsath et al. (1997) attribute this decline in soil production rates with increasing depth to decreased effectiveness of mechanical (physical) weathering processes such as freeze–thaw, tree-throw, and biogenic disturbance in breaking up bedrock as soil cover thickens. The cosmogenic approach allows direct determination of soil production rates from bedrock and provides a means to map their variability across soil-mantled, upland landscapes around the world.

A global compilation of soil formation rate estimates, which includes both rates of soil production and soil development made by various methods, finds they range widely between 0.0001 to 0.5 mm/yr, with global median and mean values of 0.017 and 0.036 mm/yr, respectively (Montgomery, 2007). The same study found that short-term rates of soil formation, short-term rates of soil erosion under native vegetation, and long-term geological erosion rates were similar. Before people got involved, it appeared as though soils formed at about the same pace as they eroded [**Figure DD3.4**].

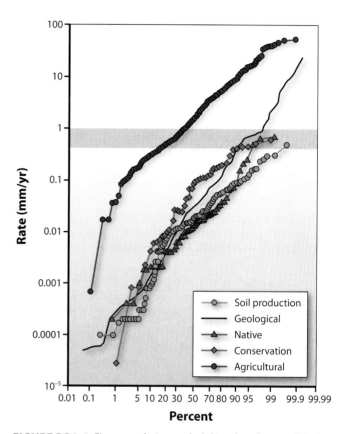

FIGURE DD3.4 Five cumulative probability plots from a global compilation show that rates of soil erosion from agricultural fields under conventional tillage regimes (red circles; sample size, n = 448) are far higher than other rates of erosion: where soil conservation techniques are used (green diamonds, n = 47), where plots are under native vegetation (blue triangles, n = 65), long-term, geological rates of erosion (solid line, n = 925), and rates of soil production (orange circle, n = 188). The shaded area represents the range of soil erosion rates that the U.S. Department of Agriculture defines as tolerable soil loss from agricultural fields. Even these "tolerable" rates of soil erosion are unsustainable in terms of geological rates of soil production and erosion. To read the graph, note that the 50 percent value is the median; the lowest rate plots on the far left and the highest rate plots on the far right. [From Montgomery (2007).]

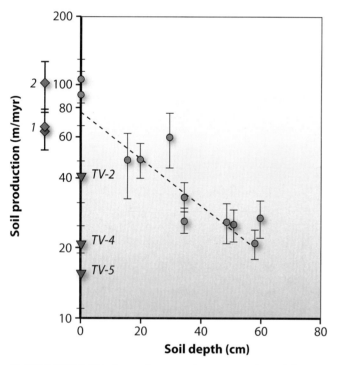

FIGURE DD3.3 This figure shows erosion rates on three different parts of the same landscape, Tennessee Valley, in northern California. Left of the vertical axis are diamonds that represent basin-scale erosion rates estimated by measuring in situ cosmogenic ^{10}Be in river sediment. These rates are similar to the rates of soil production measured in rock outcropping at the soil surface (filled circles plotted at zero soil depth). As the soil thickens, soil production rates decline. Outcrops standing well above the soil surface had the lowest erosion rates (plotted as upside-down triangles, labeled TV). [From Heimsath et al. (1997).]

The maximum reported pace of soil production (0.5 mm/yr) is less than the average rate of soil erosion reported for U.S. and global croplands (0.89 to >1 mm/yr) (Montgomery, 2007; Wilkinson and McElroy, 2007). This discrepancy between geological rates of soil production and contemporary agricultural erosion rates leads to the unsettling conclusion that conventional farming practices are not sustainable. Fortunately, however, soil erosion rates under conservation agriculture, such as "no till" farming, are much closer to soil production rates, showing that agriculture need not result in accelerated rates of soil loss.

Unless better soil management practices are adopted worldwide, the gradual, ongoing loss of fertile topsoil and degradation of agricultural lands present a growing problem for the world, especially as population continues to rise. Not only does the loss of soil impact agricultural productivity, but some of the eroded, nutrient-rich topsoil ends up in rivers and lakes, impacting water quality and encouraging the growth of nuisance organisms. Although soil erosion (coupled with deforestation and climate shifts) plagued many ancient societies, soil conservation measures rooted in geomorphological principles as well as agronomy can be employed to make agriculture sustainable. Because even unsustainable rates of soil erosion appear slow on human timescales, there is little urgency surrounding the issue of soil loss. Reducing the rate of soil erosion is important because healthy soil and fertile land are arguably society's most overlooked strategic resource.

Alexander, E. B. Rates of soil formation from bedrock or consolidated sediments. *Physical Geography* 6 (1985): 25–42.

Alexander, E. B. Rates of soil formation: Implications for soil-loss tolerance. *Soil Science* 145 (1988): 37–45.

Birkeland, P. W. *Soils and Geomorphology*. New York: Oxford University Press, 1984.

Darwin, C. *The Formation of Vegetable Mould, Through the Action of Worms, With Observations on Their Habits*. London: John Murray, 1881.

Harden, J. W. A quantitative index of soil development from field descriptions: Examples from a chronosequence in central California. *Geoderma* 28 (1982): 1–28.

Heimsath, A. M., W. E. Dietrich, K. Nishiizumi, and R. C. Finkel. The soil production function and landscape equilibrium. *Nature* 388 (1997): 358–361.

Montgomery, D. R. Soil erosion and agricultural sustainability. *Proceedings of the National Academy of Sciences* 104 (2007): 13,268–13,272.

Playfair, J. *Illustrations of the Huttonian Theory of the Earth*. London: Cadell and Davies, 1802.

Smith, R. M., and W. L. Stamey. Determining the range of tolerable erosion. *Soil Science* 100 (1965): 414–424.

Wakatsuki, T., and A. Rasyidin. Rates of weathering and soil formation. *Geoderma* 52 (1992): 251–263.

Wilkinson, B. H., and B. J. McElroy. The impact of humans on continental erosion and sedimentation. *Geological Society of America Bulletin* 119 (2007): 140–156.

WORKED PROBLEM

Question: For an arid landscape and for a humid landscape, developed on the same type of granitic rock, contrast rock weathering on the respective hillslopes and soil development in the valleys. Consider the germane rock-weathering processes and describe the resulting soil profiles.

Answer: In the humid landscape, hillslopes will be soil-mantled and much of the weathering will be chemical, occurring below the cover of soil. There will be some physical weathering, much of it accomplished by tree roots. In the arid landscape, there will be less weathering overall, chemical weathering will be less intensive, the slopes will be rocky, and a greater proportion of the weathering will be physical, occurring on bare rock surfaces. In both cases, the granite will decompose to its constituent minerals; however, in the humid landscape more of the minerals will have been chemically altered, with the feldspars weathering to clay minerals and the iron-rich minerals oxidizing. The chemical alteration of the granite will assist physical decomposition to grus, probably through the alteration of biotite.

Soils in the two landscapes will be distinctive. In the arid landscape, soils will likely be capped with a vesicular A horizon, reflecting the accumulation of aeolian dust and the formation of small vesicles from wetting and drying. This will be underlain by a B horizon, characterized by reddening and perhaps the accumulation of calcium carbonate. If the soil is old enough, the carbonate will coat B-horizon clasts and create a durable and impermeable petrocalcic horizon (calcrete). The humid soil may have an organic-rich O horizon at the surface, which is likely to be underlain by an A horizon and possibly an E horizon if the soil is permeable and if there are trees growing on the surface that produce acidic leaf litter. Below the upper horizons, there will be a B horizon typified by reddening and an increase in iron and clay content.

KNOWLEDGE ASSESSMENT Chapter 3

1. Define and differentiate saprolite, soil, and regolith.
2. Describe how physical, chemical, and biological weathering differ.
3. How does chemical weathering influence physical weathering and vice versa?
4. List three mechanisms of physical weathering and describe how each works.
5. What are borndhardts, tors, and inselbergs?
6. Propose a series of physical processes and rock history that could lead to exfoliation.
7. What process likely created felsenmeer?
8. Predict what will happen to boulders or outcrops of rock exposed to a forest or range fire or to extended heating in the desert sun.
9. Explain how salts can physically weather rock.
10. Why does chemical weathering occur?
11. Explain how silica and most rock-forming minerals differ from calcium carbonate in terms of chemical weathering?
12. Why are iron and aluminum oxides characteristic of materials left as residue after extensive chemical weathering?
13. List the most important factors influencing the amount of leaching by solution of material from soil or rock.
14. Discuss how various types of biological activity affect the physical and chemical weathering rates of rocks.
15. What controls the swelling potential of clay minerals?
16. Define cation exchange, identify which ions are most likely to be exchanged, and explain why cation exchange is an important process.
17. What is oxidation of minerals and where can you see it occur?
18. Sketch and explain Goldich's Weathering Series.
19. Starting with feldspar, explain how it weathers over time to eventually become gibbsite.
20. What is chelation and why is it important in soil formation?
21. Why is the density of saprolite less than that of the rock from which it was derived?
22. Describe the patterns of and balance between chemical and physical weathering on a global scale.
23. Explain how vegetation type and density affect soil formation rates and the type of soils that result.
24. Explain how topography (slope steepness) affects soil thickness.
25. How is soil development related to the age of a soil?
26. List three ways by which material can be added to and removed from soils.
27. Define translocation and explain how it can occur in soils.
28. Explain how bioturbation occurs and why it is important.
29. List the most common soil horizons and explain how each forms.
30. Describe the two general models relating soil production rates to soil depth.
31. Compare and contrast pedogenesis and soil production.
32. Make a table listing the 12 soil orders, describing their salient characteristics, and suggesting in what environment each might be found.
33. List the five factors thought to control rates of pedogenesis (soil development).
34. Define a soil catena and explain differences observed across a landscape in terms of soil-forming processes.
35. Suggest in what geomorphic environment you might find a chronosequence and explain why defining a chronosequence could be useful to you as a geomorphologist.
36. Compare a paleosol and a cumulic soil.
37. List and describe several examples of weathering-induced landforms at both small and large scales.
38. List several types of duricrusts and explain how they help shape landscapes.
39. Compare rates of soil production to contemporary rates of soil loss and explain why this comparison is important.

Geomorphic Hydrology

4

Introduction

Without water, Earth's surface would be a vastly different, inhospitable place—rather like present-day Mars. Sediment would move only when meteorites struck, stirring broken rock at the planet's surface, or when sand-charged winds blasted the land. There would be little weathering, no rivers, no lakes, and no streams. Our planet would be dry and barren, devoid of plants and animal life. But Earth is not a dry planet; we live on a planet of water—about 70 percent of its surface is covered by oceans. The temperature range on Earth appears to be unique in our solar system, allowing water to exist at the surface in three phases (liquid, solid, and vapor) simultaneously. All three phases are geomorphically important. Ice and liquid water are effective mass transport agents. Water vapor is also a critical part of the **hydrologic cycle.**

Much of the water that falls on Earth's surface as precipitation eventually ends up in streams and rivers as it moves toward the ocean. Water is constantly on the move, evaporating from the ocean, entering the atmosphere, and then precipitating back onto the land. Once on land, water moves under the influence of gravity (and inertia), flowing down rivers and percolating through the ground until, once again, it enters the atmosphere or the sea. Plants short-circuit this cycle, using solar energy to pump water directly back into the atmosphere through the process of **transpiration,** the uptake of water through roots and the release of water vapor to the atmosphere through leaves.

Most landscapes are shaped either directly or indirectly by water. Understanding how water moves and where it is

Turbid floodwaters fill the bedrock channel of the Potomac River just upstream of Washington, DC, after the passage of Hurricane Isabel in 2003. More than 4500 cubic meters per second of water are flowing over Great Falls, which is completely submerged.

IN THIS CHAPTER

Introduction
Precipitation
  Duration and Intensity
  Recurrence Intervals
  Precipitation Delivery
  Climate Effects on Hydrology and Geomorphology
Evapotranspiration
  Evapotranspiration Rates
  Actual Versus Potential Evapotranspiration

  Geomorphic Importance of Evapotranspiration
Groundwater Hydrology
  Infiltration: Moving Water into the Ground
  Moving Water Through Earth Materials
  Hydrologic Flowpaths
Surface Water Hydrology
  Hydrographs
  Interactions Between Groundwater and Surface Flow

  Flood Frequency
  Water Budgets
Hydrologic Landforms
Applications
Selected References and Further Reading
Digging Deeper: Humans, Hydrology, and Landscape Change—What's the Connection?
Worked Problem
Knowledge Assessment

found are fundamental to understanding how Earth's surface changes. Raindrops dislodge soil grains, sending material downslope in hillslope runoff. Runoff eventually collects, slowly incising valleys or rapidly eroding disturbed surfaces like construction sites and plowed agricultural fields. **Groundwater,** flowing slowly through surficial deposits or deeper in porous or fractured rock, dissolves elements while leaving pore space, caves, and weathered rock behind. Rising groundwater can destabilize hillslopes, triggering landslides that can degrade mountain slopes, take lives, and devastate communities. Surface water, flowing through streams and rivers, transports sediment and shapes channels used for both recreation and transport.

Meltwater moving through and under glaciers sustains river flow through dry summer months, providing clean and abundant drinking water. Yet, when glacial dams burst, areas downstream can be devastated and cubic kilometers of freshwater pour into the world's oceans. Ancient catastrophic megafloods have raised sea level and altered global climate many times in the past. Hydrologic processes have shaped landscapes around the world and the hydrologic response to climate change affects society and may do so with increasing intensity over the coming century.

The geomorphic influence of water on Earth's surface is by no means uniform. In hyperarid regions, such as the deserts of southern Africa, rain may fall only once a decade and channels may flow only several times a century. When rain does fall, massive amounts of sediment can move because most of the rainfall generates runoff and there is little vegetation to hold the landscape together. Conversely, in warm humid climates, abundant water supports dense forests that largely shield the ground surface from the immediate hydrologic effects of heavy precipitation [**Photograph 4.1**]. At the same time, beneath the surface, continually moist conditions and organic acids generated by decaying vegetation dismember rocks and weaken underlying slopes. Reduced erosion and greater weathering promote soil development.

Both the amount of water and the rate at which it is supplied to the landscape are important when trying to understand the geomorphological behavior and ecological diversity of Earth's surface. In addition, temperature is of critical importance because it determines where and when liquid and frozen precipitation will fall and greatly affects the rate at which water-mineral weathering reactions take place. The same amount of water will do more weathering in the hot tropics than if it falls as precipitation in a cooler climate. In this chapter, we follow the water as it moves across, into, and through the landscape and consider how it shapes the planet on which we live.

Precipitation

Precipitation is a key link in the **hydrologic cycle,** as it returns water from the atmosphere to the land surface (see Figure 1.7). Water can fall from the sky either as rain or as various forms of frozen precipitation, including snow. Frozen precipitation is mostly a winter phenomenon (except at high latitudes and altitudes), and in spring and summer, the ice and snow melt and move as liquid water over, into, and through the ground.

Precipitation has long been measured using rain gauges, but such gauges are rare; on average, there is about one official gauge per 1000 km^2 in the United States. Furthermore, dense networks of such gauges are required for accurate delivery estimates because precipitation amounts can vary greatly over the scale of kilometers. In the 1990s,

(a)

(b)

PHOTOGRAPH 4.1 Water Shapes Earth's Surface. The amount of water on the landscape dramatically affects its appearance. (a) Water-sculpted bedrock channel flowing off the wet tropical escarpment in Queensland, Australia. This area receives around 2 m of rain yearly. (b) The Gaub River in Namibia receives only several centimeters of precipitation a year and both the channel and surrounding area are largely devoid of vegetation. Most of the year, the stream channel is dry.

advanced weather radars, called NEXRAD (Next Generation Weather Radar), were installed across the United States. These systems, when used in concert with interpretive mathematical algorithms, allow forecasters to estimate rainfall intensity remotely and at high resolution (km^2). Other satellite-based observations are being used to estimate precipitation amounts around the world and, in some cases, drive geomorphic models.

Duration and Intensity

Precipitation events have two fundamental characteristics: **duration** (how long the storm lasted) and **intensity** (the volume of water that came down, expressed as length per unit time, such as mm/hr). Both of these characteristics are geomorphically important. Long-duration events can saturate the ground, thereby triggering landslides and raising river levels. High-intensity events also trigger landslides and can deliver large volumes of water to landscapes so quickly that some of it runs off, eroding soils and carving rills or deep gullies.

Commonly, precipitation is only geomorphically effective when an event exceeds a combined intensity-duration **threshold** that triggers a discrete change—like a landslide. For example, landslides that produce downstream mudflows can be triggered by specific combinations of rainfall intensity and duration [**Figure 4.1**]—heavy rains lasting a short time or less intense rains that last a long time. If the combination of intensity and duration does not exceed the threshold, slopes will remain stable. Note that the data underlying these threshold curves can be quite variable—the curves describe the average behavior of a landscape. If the threshold values are exceeded, some hillslopes will fail and others will not. While it is important to consider that individual hillslopes behave quite differently, once the threshold is exceeded, more rainfall would be expected to cause more landslides in more locations across the landscape.

Recurrence Intervals

Experience tells us that extreme events, such as long-duration or intense storms, are rare and that short, low-intensity events are more common. This relationship can be quantified as a **recurrence interval**, the probability of an event occurring (e.g., rainfall over a 24-hour period) based on the average amount of time that passes between events of at least a certain magnitude. Calculating a recurrence interval (RI) for an event of a particular **magnitude** (size) requires many observations (N) of different annual maximum events (the largest event in each year) collected over time. Each annual maximum event in the record is assigned a rank (r) based on its magnitude. The largest event is given a rank of 1. The smallest event is given a rank of N. Rank and recurrence interval (in years) are inversely related:

$$RI = (N + 1)/r \qquad \text{eq. 4.1}$$

Large, rare events have long recurrence intervals and small, common events have short recurrence intervals. In other words, events with long recurrence intervals have a low annual probability of recurring while those with short recurrence intervals occur much more frequently [**Figure 4.2**]. The probability, p, of an event (a rainstorm, flood, or landslide) occurring in any one year is the inverse of its recurrence interval:

$$p = 1/RI \qquad \text{eq. 4.2}$$

Although the recurrence interval formulation is simple, its application is fraught with assumptions and misconceptions. Chief among the latter is that a 100-year event is something that only happens once every hundred years. The 100-year event actually describes an event the size of which has a 1-in-100 chance of occurring in any given year (1 percent chance per year). Similarly, the 50-year event has a 1-in-50 chance of occurring in any year (2 percent chance per year), and so on for the 1-in-25-year (4 percent chance per year) and 1-in-10-year events (10 percent chance per year).

Estimates of event likelihood often require extrapolating beyond the period of record because most stream or precipitation gauges have only a few decades of data. Although estimating recurrence intervals based on short records results in great uncertainty, it is routinely done. Anyone reporting that an area has been devastated by the 500-year storm (i.e., a storm which on average occurs about once every 500 years), when only 30 years of weather data

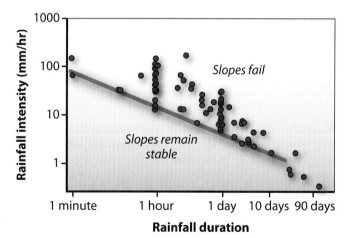

> **Landslides** that trigger debris flows are a geomorphic process that occurs when a **threshold** is exceeded. In this case, rainfall is the trigger, saturating the ground and causing slopes to fail. Both rainfall intensity and duration, not just cumulative rainfall, are important in determining slope stability. The combinations of rainfall intensities and durations that trigger slope failure vary regionally, reflecting differences in slope strength, topography, and near-surface hydrology.

FIGURE 4.1 Intensity-Duration Thresholds for Geomorphic Processes. Many geomorphic processes are triggered only when a force or process, such as rainfall intensity and duration, exceeds certain limits. [Adapted from Caine (1980).]

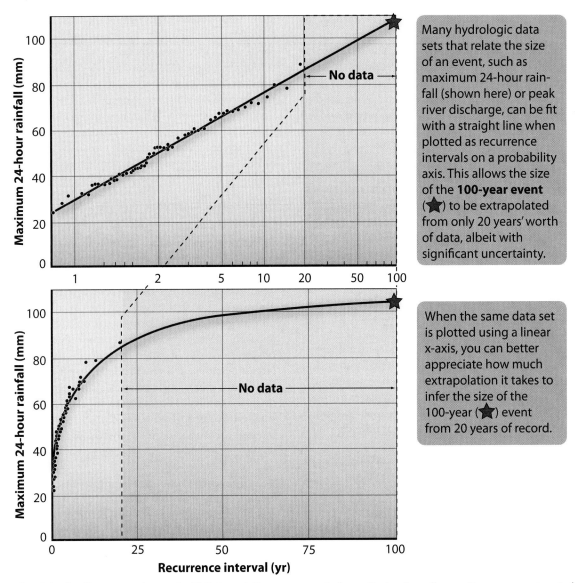

FIGURE 4.2 Precipitation Recurrence Interval with Extrapolation. Extreme hydrologic events are rare. Small events are common. Probability plots are a useful means of estimating how often an event of a particular size will recur. However, most periods of record are short, only a few decades, and extrapolation beyond the period of record is both commonplace and uncertain.

exist, is extrapolating. Fundamental to the application of recurrence intervals is the idea that weather events are random, implying there is no long-term trend in the weather and thus no trend in climate over time. This assumption, termed **stationarity**, is growing less valid given the reality of climate change. Although the forecasts of global climate models differ in details, all predict significant rises in global average temperatures and changes in the distribution, frequency, and amount of precipitation over the next century—a time frame relevant both to geomorphology and human interactions with Earth's dynamic surface.

Precipitation Delivery

Precipitation is delivered to Earth's surface in discrete weather events—each having unique geomorphic impacts because of differing spatial extent and intensity [**Photograph 4.2**].

For example, many areas near the coast at low and middle latitudes can be affected in summer and fall by intense storms born in the tropics; they are called **hurricanes** in the Atlantic Ocean, **typhoons** in the Pacific and Indian oceans, and **cyclones** in Australasia. These tropical low-pressure systems can rapidly release large amounts of precipitation and trigger floods over thousands of square kilometers.

Some storms are large and move slowly. In northwestern North America, cold, mid-latitude winter storms moving in from the Pacific Ocean are often long-lasting, gently soaking temperate rainforests with low-intensity rainfall for days on end but leaving little geomorphic evidence of their passage. In eastern North America, sprawling nor'easters draw in Atlantic Ocean moisture and can pile snow meters deep in the Appalachian Mountains while battering the Atlantic coast with rain

PHOTOGRAPH 4.2 Storms. Storms come in a variety of shapes and sizes. (a) Hurricanes impact southern and eastern North America every summer and fall; this is Hurricane Katrina, August 2005. (b) Small but intense convective thunderstorms pour down rain on isolated desert valleys in Namibia. (c) Heavy rain from large, mid-latitude, low-pressure system triggered small landslide along a road near Nelson, New Zealand.

and high surf for days on end. Compare such large, regional weather systems with the intense summer thunderstorms common in arid and semi-arid regions. These convective storms commonly affect only small areas, perhaps a few square kilometers in a single mountain drainage basin, but deliver intense downpours capable of quickly triggering debris flows and devastating flash floods. At an intermediate scale, frontal lifting of moist air can cause condensation and precipitation over large areas. **Warm fronts** tend to produce soaking rains of low intensity that can last many hours. **Cold fronts** are sharper boundaries, producing high rainfall rates and strong winds that can be geomorphically effective, even though they last for only a few hours.

Mountains and large lakes can create local weather. Lake effect snows caused by cold air moving over relatively warm water can quickly bury downwind locations, such as Buffalo, New York, on the eastern shore of Lake Erie, under a meter of snow. Mountains force air masses to rise, wringing out precipitation as the air cools and water condenses. This **orographic** effect can be significant. Burlington, Vermont, and Seattle, Washington, both receive on average a little less than a meter of precipitation each year. The Green and Cascade Mountains, ranges just to the east of these cities, receive 2 and 3 times the mean annual precipitation of the adjacent lowlands, respectively.

In some areas, fog and clouds contribute significant amounts of moisture to the water budget, water that is critical for rock weathering and the survival of life. For example, in hyperarid coastal Namibia, more water is delivered to the land surface by frequent and persistent coastal fogs than by rainfall, which occurs only rarely. The large amount of fog-derived condensation in the Namib Desert may be responsible for enhanced salt weathering of rock surfaces, resulting in rock erosion rates comparable to those observed in topographically similar areas of Australia that receive 10 times more rainfall. Similarly, at high elevation and along humid, temperate coastal zones, large amounts of water may be delivered to trees by condensation from clouds and fog and then flow down stems and tree trunks to the ground. Conversely, leaves, particularly the broad leaves of deciduous trees, can intercept precipitation. If the precipitation is of low intensity and short duration, it will never reach the ground but rather will evaporate directly from the leaves. The effect is greatly lessened for deciduous trees during winter when leaves are off; however, conifers can intercept and retain snowfall on their leaves and branches. Some of this snowfall can sublimate and never reach the ground.

Climate Effects on Hydrology and Geomorphology

Climate affects hydrology, and thus the processes that shape Earth's dynamic surface, in a variety of ways. The primary climatic control on hydrology is the amount and seasonal distribution of precipitation. But the primary hydrologic control on geomorphology is the frequency and timing of high-magnitude events such as spring floods or hurricanes. The rest of the year, precipitation and the weathering that it catalyzes prepares material on hillslopes for evacuation by the big storm events.

In arid regions, rainfall is sporadic and spatially disjunct, often affecting only small parts of a drainage basin as discrete storm cells move across the landscape. Many arid-region storm cells are small relative to the size of drainage basins. Thus, specific geomorphic processes (flooding, landslides, debris flows) may be active in one subcatchment or even one part of a subcatchment while neighboring subcatchments remain unaffected. In humid regions, precipitation events are more frequent, but rain gauge data also show significant spatial variability that helps explain observed spatial differences in active geomorphic processes and process rates. Rainfall data clearly show that the highest rainfall intensities typically cover only small portions of the landscape.

Some climates are strongly seasonal and that seasonality is reflected in regional hydrology and geomorphology. **Monsoonal** climates have warm, wet summers and cooler, drier winters. These climates result from the geographic juxtaposition of an elevated landmass and an adjacent tropical ocean. The landmass is heated by the Sun during the summer, causing the air above it to rise and entrain moisture from the ocean. Strong lift causes the moist air to condense and produce particularly intense rainfall for several months each year. River and stream channel dimensions are adjusted to this regime of heavy rain and significant runoff. The months after monsoon rains pass may be quite dry. During the dry season, channels appear grossly oversized for the small flows they carry. The Himalaya, northern Australia, and the Colorado Plateau are all affected by summer monsoons.

Cold regions, such as the Arctic and Antarctic, are characterized by long periods of hydrologic inactivity punctuated by periods where water comes out of frozen storage and runoff is dramatic. Spring snowmelt quickly releases water from storage in the snowpack and 30 to 50 percent of the yearly runoff can leave the landscape in just a few weeks. Similar to monsoonal climates, channels in cold regions are shaped by and sized for flows that occur during only part of the year.

Evapotranspiration

Much of the water delivered to Earth's surface by precipitation returns directly to the atmosphere without ever entering a stream or river. The portion of the precipitation that is evaporated and transpired does not contribute to runoff.

Evapotranspiration Rates

Evapotranspiration is a broad term describing the flux of moisture into the atmosphere both through the physical process of **evaporation** and the biological process of **transpiration** [Figure 4.3]. **Evaporation** removes water directly from Earth's surface and from the stems and leaves of vegetation. **Transpiration** is distinguished from evaporation by the active role that plants play in both facilitating water transfer to the atmosphere and pulling water from the soil. The rate of evaporation is controlled by temperature, the relative humidity of the air, wind speed, and ultimately by the intensity of solar radiation. These same factors, although biologically driven, control transpiration. The effects of temperature on both evaporation and transpiration are indirect, through temperature's effect on relative humidity.

Evapotranspiration rates vary over space and time. Rates are high during the summer and during daytime hours when the input of solar radiation is greatest. Rates of evapotranspiration drop at night in the summer and for deciduous trees in winter, when the leaves are absent and photosynthesis is minimal. In contrast, coniferous trees, which keep their leaves all winter, continue to transpire water all year long. The importance of winter transpiration by conifers is confirmed by dozens of studies showing that the increase in **water yield** (the amount of water leaving the basin) following removal of conifer forests substantially exceeds that when hardwood forests are removed.

Actual Versus Potential Evapotranspiration

It is critical to make the distinction between actual and potential rates of evapotranspiration (ET) [Figure 4.4]. Low and mid-latitude deserts have very high potential rates of ET because of low relative humidity and large inputs of solar energy. However, actual rates of ET are low in arid regions because little water is available to evaporate or transpire. Conversely, in humid regions, where rainfall is common and soils are often moist, actual and potential rates of ET are more closely matched. Actual ET is difficult to measure and is often calculated on a yearly basis as the difference between water supplied to a watershed by precipitation and that running off through rivers. The implicit assumption underlying this calculation—that the flux of groundwater is small by comparison—is reasonable in most terrain.

Geomorphic Importance of Evapotranspiration

Evapotranspiration can be a geomorphically important process, removing large amounts of water from near Earth's surface. In many North American drainage basins, more than half the precipitation is returned to the atmosphere by ET. Movement of water from the soil to the atmosphere by transpiring plants is sufficient to reduce soil moisture levels and draw down **water tables** (the level below the surface where void spaces within soil, in rock fractures, or between sediment particles are completely filled with extractable water, also known as the **phreatic zone**) during the growing season. Trees exert a direct control on water levels in streams by reducing stream flow

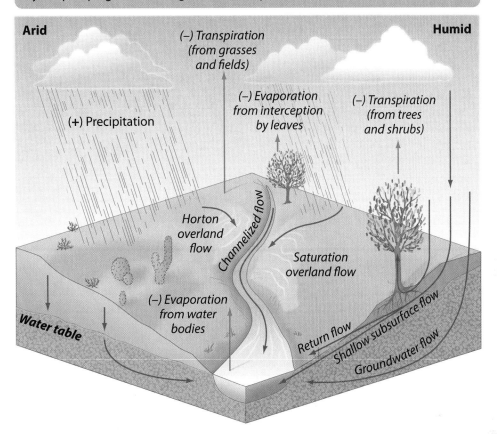

Water, falling as precipitation, takes various paths on, below, and away from Earth's surface. Some precipitation never reaches the ground because it is **intercepted** and evaporates from vegetation. Some water **infiltrates** and flows through the ground, some runs off the land surface, and some is returned to the atmosphere by the pumping action of vegetation, **transpiration.**

In arid regions, the groundwater table beneath hillslopes is deep, and precipitation mainly comes as high-intensity, short-duration storms. Because the dry climate supports scant vegetation, little precipitation is intercepted. **Infiltration rates** can be low, especially for A-horizons composed of wind-blown dust and clay. Large amounts of **Horton overland flow** occur because the rate of precipitation exceeds that of infiltration. Runoff also occurs through direct precipitation onto stream channels.

In humid regions, the water table can be close to the surface. Large amounts of vegetation intercept precipitation, make soil more permeable, and transpire water. Where and when the **water table** intercepts the ground surface, there is **return flow** and rainfall generates **saturation overland flow.** Some precipitation infiltrates and moves through the shallow, permeable, near-surface soil as **shallow subsurface flow** before emerging as return flow and entering streams. Precipitation that infiltrates deeply becomes **groundwater.**

FIGURE 4.3 Schematic Diagram of Evapotranspiration and Flowpaths. Water falling as precipitation can take many different paths. Some water is intercepted by vegetation and evaporates before it ever reaches the ground. Other water is taken up by plants and transpired back to the atmosphere. Much of the remaining water either runs off or enters the groundwater system.

during storm events (**peak flow**) and flow between storms (**baseflow**). Diminished baseflow results from water table lowering, which occurs during extended periods of evapotranspiration. Reduced flood peaks result from reduced soil moisture levels because tree roots remove water held by **capillary** forces, or surface tension, from the unsaturated zone. Stream flow records in forested, temperate regions show baseflow discharge and groundwater table

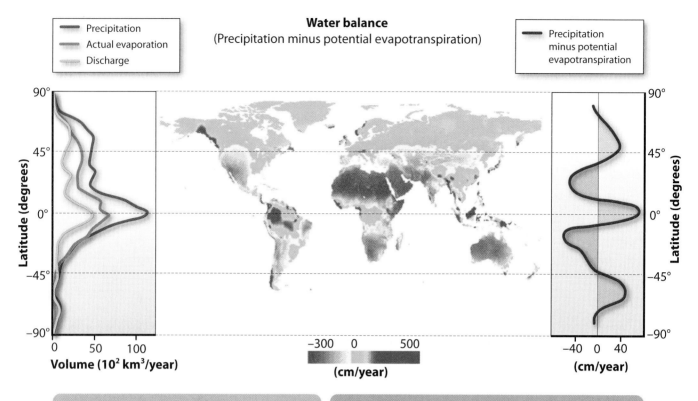

FIGURE 4.4 Water Balance by Latitude. Water balance varies greatly over Earth's surface. Over deserts, potential rates of evapotranspiration far exceed available precipitation. In temperate high and low latitudes, there is an excess of water. The volume of precipitation, evaporation, and discharge are controlled by the location of land areas and by high temperatures near the equator. Differences in land area lead to asymmetry between the northern and southern hemispheres.

elevations increasing in the fall when deciduous trees stop pumping water.

The power of trees to pump water from the ground has led people to deforest watersheds in regions where water is scarce so that water previously transpired by trees would be available for human use. Such modifications have been done in New Zealand and South Africa, as well as in arid southwestern, humid southeastern, and wet northwestern North America. Indeed, **water yields** went up by hundreds of millimeters a year in some cleared watersheds.

There are significant geomorphic implications to such ecosystem manipulation. Erosion, landsliding, and the flux of sediment tend to increase in deforested watersheds because along with removal of the trees comes loss of the binding effect of roots that otherwise help hold hillslope soils in place. Rising water tables on cleared hillslopes have the potential to further exacerbate landslide and erosion hazards. Virtually every paired-watershed study with more than a decade of data postharvesting shows that increases in water yields following vegetation removal are ephemeral, declining to pretreatment levels generally within a decade after cutting as vegetation regrows. Thus, to maintain the increased water yield, the landscape must be kept open and unforested, increasing the likelihood and intensity of erosion.

Groundwater Hydrology

Groundwater, although unseen, maintains baseflows in streams and rivers and has important geomorphic effects, including weathering of rocks. Soils saturated with groundwater are more likely to fail in mass movements. The degree and extent of soil saturation by groundwater influence surface-water flowpaths and the type and distribution of vegetation, fundamental controls on the rate and distribution of surface erosion.

Infiltration: Moving Water into the Ground

Soil, because it covers much of Earth's surface, is the critical link between precipitation, runoff, and geomorphology. The linkage is expressed by a soil hydrologic characteristic termed the **infiltration rate**. Infiltration rates describe how quickly water moves into soils and are calculated as a volume infiltrated over an area per unit time; often, they are expressed in units of length per unit time such as mm/hr (do the unit analysis and you will see that the two units are equivalent). Soil infiltration rates are critical values because they determine the fate of precipitation—how much rainfall or snowmelt soaks into the ground and how much runs over the land surface.

Not all soils are similar, and infiltration rates vary widely among different materials (**Table 4.1**). Compacted, clay-rich soils might infiltrate only a millimeter or two per hour while loose, sandy soils could easily pass 30 times more water in the same time interval. Initially, dry soils infiltrate water rapidly as water fills empty pores [**Figure 4.5**]. As a rainstorm continues and the soil wets up or saturates, the rate of infiltration falls and then levels off at a steady value. In some clay-rich soils, apparent infiltration rates decline as clays hydrate (take on water) and swell, closing cracks and reducing the rate at which water can enter the soil.

Biological activity dramatically affects soil infiltration rates. Earthworms and burrowing animals loosen soil and create large open spaces in the soil, known as **macropores**, some of which are centimeters to decimeters across. Similarly, plant roots disturb soils. When plants die and their roots begin to decompose, the cavities left behind provide additional conduits through which water can easily move. Macropores intersecting the surface increase infiltration rates.

Human activity tends to decrease the rate of infiltration. Both agriculture and urbanization typically reduce or remove the permeable, organic-rich cover typical of naturally vegetated areas. Soil is easily and quickly compacted by only a few passes of foot or vehicle traffic, both of which can reduce infiltration capacity by an order of

TABLE 4.1

Typical infiltration rates

Material	Infiltration rate (mm/hr)
Clay soil	1–5
Clay loam soil	5–10
Loam soil	10–20
Sandy loam soil	20–30
Sandy soil	>30

From United Nations Food and Agriculture Organization, http://www.fao.org/docrep/S8684E/S8684E00.htm

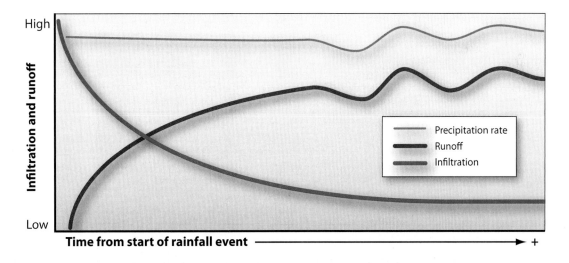

FIGURE 4.5 Infiltration and Runoff over Time. When rainfall begins, infiltration rates are high. As the ground saturates, infiltration rates drop off, and as soon as the rate of precipitation exceeds the rate of infiltration, runoff begins.

Infiltration trends: **Infiltration** rate decreases over time as dry **pore spaces** between soil particles fill with water. Rates of infiltration drop and become steady as the soil saturates. After soil saturation, the infiltration rate is a function of soil **hydraulic conductivity**.

Runoff trends: **Runoff** increases over time because soil pores fill with water, allowing less of the rainfall to soak into the ground. Once the soil is saturated, runoff varies with the **precipitation intensity** and the runoff rate equals precipitation intensity minus hydraulic conductivity.

magnitude or more. One only needs to hike in a rainstorm to see water running down a compacted trail; less than a meter away, precipitation will be infiltrating into the leaf-covered mineral soil [Photograph 4.3].

Moving Water Through Earth Materials

Infiltrating water begins what could be a long or a short journey through the solid Earth and its cover of loose, **unconsolidated** materials. As water enters the ground, it moves first through the unsaturated or **vadose zone** defined by the presence of both air and water in the pore spaces between grains. There, water movement is controlled by the force of gravity, which promotes soil drainage, and the **capillary force** that holds water between grain surfaces. After sufficient water infiltrates the soil, the gravitational force overcomes the capillary force and the water moves deeper into the soil. At some depth below the surface, all the pore spaces become filled with water. If a well were drilled and lined with a perforated pipe, water would flow freely from the soil into the pipe and come to rest at a certain level; this is the **water table** [Figure 4.6]. The area just above the water table, termed the **capillary fringe**, contains enough water held in tension between grains to saturate pores. But in this zone just above the water table,

PHOTOGRAPH 4.3 Overland Flow. Rainwater, unable to infiltrate into the compacted soil of an informal footpath, runs off the University of Vermont campus in Burlington, Vermont. This is an example of Horton Overland Flow—rainfall intensity exceeding the infiltration rate of the compacted soil.

the water is held tightly enough by capillary forces that it is not free to drain. The capillary effect is most significant in fine-grain sediments.

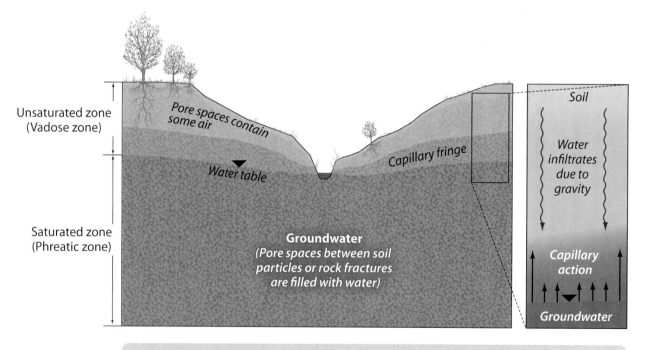

Below the surface, Earth materials can be classified by the degree to which they are saturated by water. In the **unsaturated** or **vadose** zone, pore spaces between grains contain at least some air. In the **saturated** or **phreatic** zone, all pore spaces are water-filled. Pore spaces in the **capillary fringe** are water-filled; however, the water is held tightly to and between grains by capillary forces and cannot freely drain. This diagram shows a typical situation in a humid region where rainfall is frequent.

FIGURE 4.6 Water Table and Definitions. The water table marks the boundary between zones where the pores in rock and sediment are saturated with water that can freely drain (the phreatic zone) and the vadose or unsaturated zone. The water level in the stream reflects the local water table.

The concept of **hydraulic head,** a measure of the total energy of water at a point (the sum of its potential and pressure energy), is particularly important in geomorphology, because it defines the energy gradient and thus determines the flow direction of groundwater. Head can be measured in reference to a variety of different datums, including the local ground surface, sea level, and geologic contacts [**Figure 4.7**]. Water moves from locations of high head to locations of lower head, down the energy gradient. A familiar example is how water flows into a well when the well is pumped and the level of water in the well (its head) drops beneath that of the adjacent groundwater, causing water to flow into and refill the well.

The ability of Earth materials to pass water depends on two physical properties, **porosity** and **permeability** (**Tables 4.2** and **4.3**). Porosity quantifies the amount of a material that is void space, open volumes that water could potentially occupy. **Primary porosity** is a function of the pore sizes within the material itself. Porosity can also be secondary, the result of biological, chemical, and physical processes altering soil and sediment. Biologically-induced, **secondary porosity** includes soil openings created by roots and animal burrows. Joints and fractures create secondary porosity through physical means; chemical dissolution can enlarge fractures, opening water-bearing conduits, particularly in soluble rocks such as limestone. The amount of groundwater moved in primary and secondary porosity varies widely and depends on the type of material. In coarse gravel, most flow passes through the primary pores. In well-developed karst terrain, almost all flow is through secondary conduits, fractures widened by dissolution of bedrock.

Permeability describes the ability of a fluid to move through any porous medium, including soil, sediment, and rock, and is a way to quantify the integrated effect of the size of pores through which the fluid moves and the degree to which these pores are interconnected. When the fluid is water, the permeability is expressed in terms of **hydraulic conductivity** (see Table 4.3), a value that incorporates both the **density** and **viscosity,** or resistance to flow, of water. Hydraulically conductive or permeable soils allow large amounts of rainfall and/or snowmelt to quickly infiltrate or soak into and then move through the ground, sustaining groundwater flow and minimizing surface runoff. Low permeability soils generate more rapid runoff at or near

TABLE 4.2

Typical porosity of selected Earth materials

Material	Porosity (%)
Soil	30–50
Gravel and sand	20–35
Clay	45–55
Sandstone	15
Shale or limestone	5
Granite	1

From http://geology.er.usgs.gov/eespteam/brass/aquifers/aquifersintro2.htm

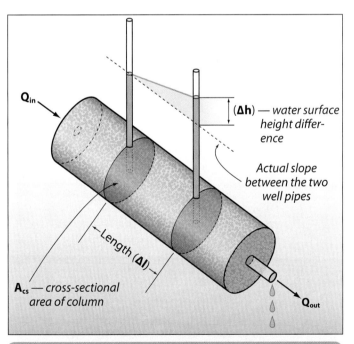

$$Q_{out} = KA_{cs}(\Delta h/\Delta l)$$

K = Hydraulic conductivity of the soil
A_{cs} = Cross-sectional area of the soil column
$\Delta h/\Delta l$ = Hydraulic gradient or head difference

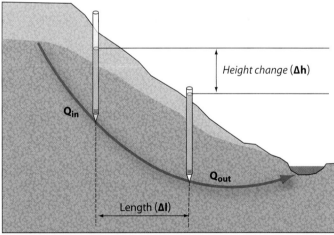

Darcy's Law describes the flow, Q, of a fluid, in this case water, through **porous material** as a function of the energy or **hydraulic gradient** and the ability of the material to pass water, the **hydraulic conductivity.**

FIGURE 4.7 Darcy's Law. Darcy's Law explains the flux of groundwater as a function of energy gradient (change in head), the cross-sectional area through which the groundwater flows, and the ability of earth materials to pass water (their hydraulic conductivity).

TABLE 4.3

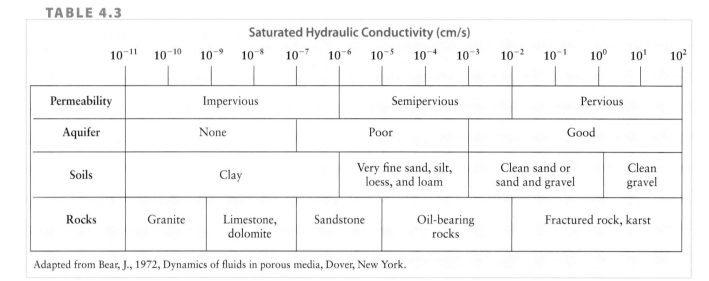

Adapted from Bear, J., 1972, Dynamics of fluids in porous media, Dover, New York.

the surface than do high permeability soils, promoting erosion and starting the movement of sediment downslope. Permeability and porosity are not always directly related. For example, some soils or rocks have high porosity but low permeability because the pores, although numerous, are not well interconnected.

Once infiltrating water reaches the water table, it becomes part of the groundwater system and its movement can be described quantitatively by **Darcy's Law**, a mathematical formulation proposed by a French engineer in the nineteenth century (see Figure 4.7). Based on a series of experiments, this expression relates the volumetric rate of fluid flow (Q) through a porous medium (like sand) to the cross-sectional area (A_{cs}), the material's hydraulic conductivity (K), and the energy or head gradient ($\Delta h/\Delta l$) available to move the fluid (the hydraulic head difference divided by the distance between the observation points):

$$Q = A_{cs} K \Delta h/\Delta l \qquad \text{eq. 4.3}$$

In other words, the groundwater discharge (Q) equals the product of the cross-sectional area (A_{cs}), the ability to pass water (K, the hydraulic conductivity), and the hydraulic gradient ($\Delta h/\Delta l$) driving the flow.

Examining Darcy's Law, it is clear that large volumes of groundwater can move rapidly through highly permeable material (with large hydraulic conductivity) since K and Q are directly related. Steep hydraulic gradients (large change in h for a unit change in l), such as those formed when the groundwater table is steepened just after a large storm, will also move large amounts of water through the ground. Note that the rate at which rainwater steadily infiltrates into surface soil after the initial wetting period is also governed by the saturated hydraulic conductivity of the subsurface material (see Figure 4.5). Hydraulic conductivities are derived experimentally either in the lab, or more usefully in the field; they range over many orders of magnitude (see Table 4.3) and depend on void size, including joint dimensions, the space between grains, and the size and connectivity of **macropores** (large voids) [**Photograph 4.4**]. Head gradients are measured in the field by comparing the level to which water rises in standpipes and monitoring wells (see Figure 4.7).

Hydrologic Flowpaths

Water reaching Earth's surface can follow a variety of **flowpaths**—the route water takes down the hydraulic gradient. Different flowpaths have different geomorphic implications (see Figure 4.3). Water flowing over the ground

PHOTOGRAPH 4.4 • Macropores. Macropore flow through otherwise low-permeability soil in Tennessee Valley, California.

surface moves rapidly, offering the potential to erode and deposit sediment, thereby doing geomorphic work and shaping landforms. Subsurface flow is considerably slower but can help weather subsurface materials, weakening regolith over much longer time frames. Rock weathering lowers hillslope strength over time and can change hillslope hydrology by concentrating flow in what are already high permeability zones of weakness. There are substantial and important feedbacks here. For example, more effective subsurface flow can increase rates of rock weathering, further increasing the permeability of preferential flowpaths.

Much of what we know about flowpaths has been learned using a combination of physical field experiments, chemical and isotopic tracers, and numerical models. Physically based field experiments typically involve placing instruments on a hillslope to understand the distribution of flow over time at various depths and locations. Chemical tracing of flowpaths can involve both natural and artificially introduced tracers, including major elements such as chlorine; noble gases including helium, neon, argon, and krypton; and chlorofluorocarbons (CFCs) introduced into the atmosphere by industrial activity. Ratios of stable isotopes ($^{16}O/^{18}O$, $^{1}H/^{2}H$) and the abundance of radioactive isotopes such as ^{3}H (tritium) in water have been used to trace the source, flowpaths, and age of groundwater.

When rainfall intensity exceeds the soil infiltration rate, water runs over the land surface. Runoff generated in this fashion is termed **Horton overland flow (HOF)**, after the hydrologist Robert Horton who first studied this process. HOF is common in arid regions where precipitation intensity often exceeds the infiltration capacity of low permeability, clay-rich soils, and on disturbed ground where the pounding of boots, hooves, or vehicles have lowered infiltration rates. In humid temperate regions, HOF is rare because dense vegetation and its root zones promote high infiltration rates. HOF in humid regions is largely restricted to areas where development or disturbance have compacted soils (see Photograph 4.3).

Most precipitation infiltrates and begins to move through the shallow subsurface in response to gravity. In the unsaturated zone, water can flow through connected pores such as animal burrows or root casts, spilling out farther downslope if the pore encounters the surface (see Photograph 4.4). Much water moves laterally in the subsurface along flowpaths largely controlled by contrasts in hydraulic conductivity, such as the interface between the looser, more permeable soil near the ground surface and more clay-rich, less permeable lower soil horizons or the interface between the soil or surficial deposits and underlying, solid bedrock. This **shallow subsurface flow** or **interflow** is largely disconnected from deep groundwater flow and is often ephemeral because the wetting front moves downward only temporarily after storms as water percolates into the soil.

Some of the near-surface groundwater moves downward and recharges deep groundwater flow systems where it can remain isolated from Earth's surface for days to millennia. Groundwater can discharge into rivers, which func-

PHOTOGRAPH 4.5 Seaonally Saturated Ground. Snowmelt saturates the ground in Greensboro, Vermont. A rapid April melt due to strong, moist southerly winds followed heavy warm-frontal rains, saturated the ground and caused water to run over the pastures. The wet areas, which are low sections of the landscape, will be dry by late spring.

tion as drains if they are lower than the water table along their banks. This groundwater discharge sets the **baseflow** that supports aquatic life between rainstorms. Fossil groundwater, recharged during wetter glacial times before the last glacial maximum more than 20,000 years ago, is today mined in the central United States, pumped from **aquifers** such as the famous Ogallala, and is used to irrigate the dry lands of the western Great Plains. The rate of water removal is much greater than the current recharge rate. Continued pumping of this and other aquifers has led to land-level **subsidence** as pore spaces dewatered and collapsed.

Saturation overland flow is generated when shallow subsurface flow returns to the surface, such as at the base of a hillslope, or where rainfall on already saturated soil cannot infiltrate and runs off. Saturation overland flow is common just after snowmelt or winter rainstorms, when water tables are high [**Photographs 4.5** and **4.6**]. The

PHOTOGRAPH 4.6 Saturation Overland Flow. Saturation overland flow on grassy surface after a March rainstorm in Tennessee Valley, California.

observation that saturated areas change seasonally led to the **variable source area** concept, the idea that source areas for rapid runoff expand and contract over time. In wet seasons, large amounts of runoff may reach the channel by saturation overland flow whereas during drier times of the year, most rainfall infiltrates and moves to channels through subsurface flowpaths [**Figure 4.8**].

Not only do flowpaths depend on time (variable source area), but they are also controlled by surface and subsurface heterogeneities. Transitions in slope topography, such as from convex to concave portions of the landscape, influence both surface and subsurface flow and the resulting sediment transport. For example, **hollows** or **swales**, which are concave portions of the landscape, concentrate flow and sediment transport [**Photographs 4.7** and **4.8**] and are often the sites where stream channels begin. In contrast, **noses** (topographic convexities) cause flow and sediment to diverge or spread out. Longitudinal, lateral, and vertical (with depth) heterogeneities in soil properties (such as hydraulic conductivity and grain size) control hillslope hydrology and sediment movement by influencing the amount of water moving at and near the soil surface.

Not all the water associated with a stream enters it over the ground surface. The subsurface component of flow along stream corridors, called **hyporheic flow**, sometimes accounts for a substantial portion of stream flow and can be ecologically important for riparian ecosystems [**Photograph 4.9**]. Channels often exchange substantial flow with permeable sediments within stream valleys, a

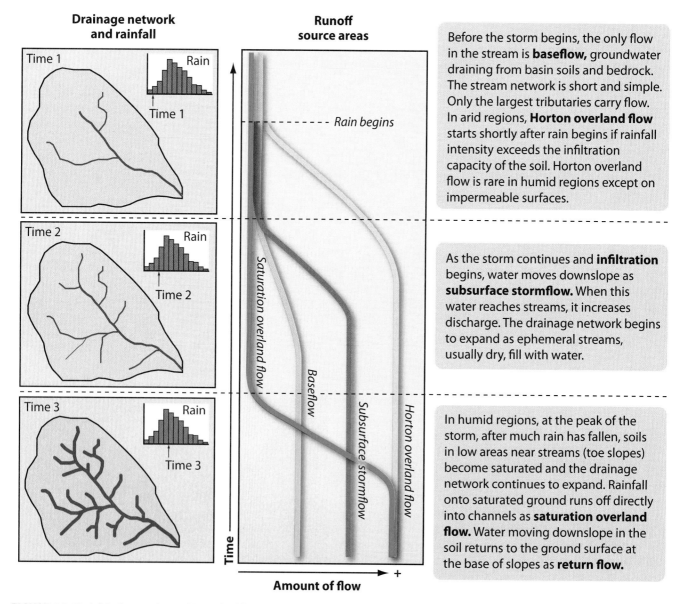

FIGURE 4.8 Variable Source Areas. Watershed flowpaths change during a storm. Areas of saturated ground expand away from channels, and the channel network itself grows larger as ephemeral streams fill with water. Baseflow increases as water tables rise. If rainfall intensity exceeds the infiltration rate or when the ground saturates near stream channels, overland flow will begin.

PHOTOGRAPH 4.7 Unchanneled Valley. Hollow or unchanneled valley, in southern Sierra Nevada, California. Here, the convergent topography focuses sediment transport, groundwater movement, and surface water flow.

PHOTOGRAPH 4.9 Hyporheic Flow. Rio Puerco (New Mexico) has incised its channel. The deep channel acts as a drain, lowering the local water table. Along the incised channel, tamarisk trees thrive as their deep roots tap groundwater below this ephemeral river.

common feature of gravel-rich mountain streams and floodplains in general.

Soil characteristics can be used to identify areas of the landscape that are alternately wet and dry, seasonally saturated zones (see Photograph 4.5). Soils that are always wet are gray-green, the result of reducing conditions (the lack of oxygen). Soils that are well-drained and usually dry are well-oxygenated, allowing minerals to oxidize and leaving the soil reddish-brown. Topography also has a large effect on drainage, including the water table position and its seasonal variation, and thereby the location of perennially saturated **gley** soils (gray-green from reduction of iron).

When a geomorphologist finds red **mottles** (evidence of oxygen) in an otherwise gray (anoxic) soil, that's a hint that the water table (and the amount of oxygen) fluctuates over time—red reflecting iron oxidation during the dry season and gray reflecting reducing conditions during the wet season [Photograph 4.10]. Such seasonally varying saturation of the ground is often used as a criterion for wetlands

PHOTOGRAPH 4.8 Channel Head. Channel head and initiation of surface flow at the base of a hollow in Tennessee Valley, California.

PHOTOGRAPH 4.10 Soil Mottling due to Variable Water Table Position. The red and gray features shown in this soil are mottles that form at and below the seasonal high water table, indicating alternation between oxidizing and reducing conditions.

delineation and subsequent protection. Soil mottles imply a varying water table. Finding mottled soils suggests that the ground has been saturated during wet periods, and thus saturation overland flow likely occurred during rainfall events.

Surface-Water Hydrology

Water moving over Earth's surface is a potent geomorphic agent, eroding slopes and riverbanks, moving sediment, and carrying dissolved loads. Surface water and groundwater flows are closely linked, with water moving rapidly and continually between the two systems.

Hydrographs

A fundamental means of characterizing surface water flow is to measure the volume of water passing a point in the channel over time. A common visualization tool used for describing this flux is the **hydrograph,** a graph that charts the volume of water moving through a channel (termed **discharge**) through time [Figure 4.9]. Hydrographs are

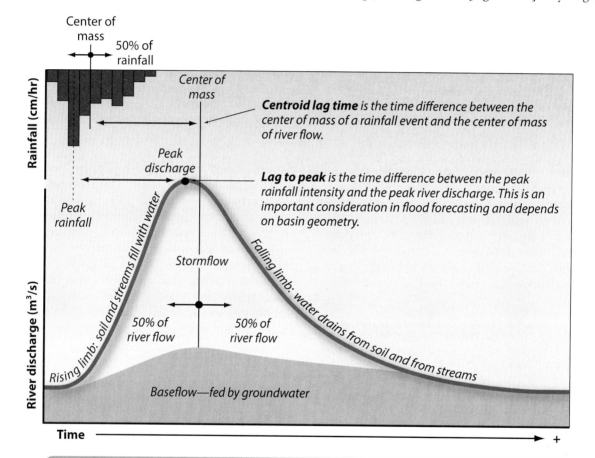

During a rainfall event, the intensity at which precipitation falls on a landscape is often variable. The **peak rainfall,** a measure of the greatest intensity of rainfall, does not necessarily occur in the middle of a rainfall event. Therefore, the peak and the center of mass of rainfall (the point at which half of the total rainfall has fallen) often do not correspond.

As precipitation falls across a landscape, some of it infiltrates and some runs off and enters river channels causing **discharge** and **river stage** to rise. As the soil fills with water, more precipitation enters river channels as **runoff,** groundwater flow, and **subsurface stormflow.** This flow moves downstream, entering progressively larger channels, where the water level, or stage, rises until reaching the peak, or maximum discharge resulting from a rainfall event. As water drains from the landscape and from channels, the river stage begins to fall, eventually returning to **baseflow,** which reflects normal groundwater discharge to rivers. Baseflow does not occur in most arid-region streams, because arid-region streams tend to flow ephemerally.

FIGURE 4.9 Hydrograph Definitions. Hydrographs describe flow through streams over time. Stream flow reflects the timing and volume of precipitation. The hydrograph can thus be interpreted through the lens of relevant runoff processes and pathways.

important because they can be interpreted to infer when geomorphically important events happen, such as when a river overflows its banks.

All hydrographs share common characteristics. **Baseflow,** derived from both deep and shallow groundwater drainage, is the amount of water flowing into and moving through a stream between storms. During a runoff event, flow rises from baseflow to a **peak discharge** defined by the intensity and duration of the precipitation or snowmelt that is providing water to the system. The **rising limb,** during which flow is increasing, is typically steep as water moves quickly to the channel over land and through shallow subsurface flowpaths. The **falling limb,** during which flow is decreasing, is generally less steep and persists longer than the rising limb as water slowly drains from the subsurface into the stream.

There is a delay between rainfall and runoff because it takes time for water to move across and through the landscape and into and down channels. This delay is referred to as the **lag-to-peak** (L_p) and is a function of basin shape and size [Figure 4.10]. L_p typically increases with basin size and varies by runoff generation mechanism, with short lag times between rainfall and peak runoff for overland flow and longer lag times for shallow subsurface storm flow and groundwater flow. Small, urbanized basins can have lag-to-peak times of minutes whereas the lag-to-peak in a large drainage basin, such as the Mississippi River Basin, may span weeks to months as water moves downstream through the river system. The lag-to-peak can be defined either as the time difference between the peak rainfall and peak runoff or as the difference between the center of mass of rainfall and the center of mass of runoff.

Hydrograph shapes reflect both the geomorphology and hydrology of the drainage basin as well as the antecedent weather and hydrologic conditions that preceded the runoff event. Steep narrow watersheds, saturated ground, and urbanization all generate hydrographs with especially steep rising limbs; basins with such characteristics are known as **flashy.** Headwater streams have steeper rises and more peaked hydrographs than main-stem, lowland rivers because these large rivers are fed by many tributaries, all contributing water at different rates and at different times. For small basins, the shape of the

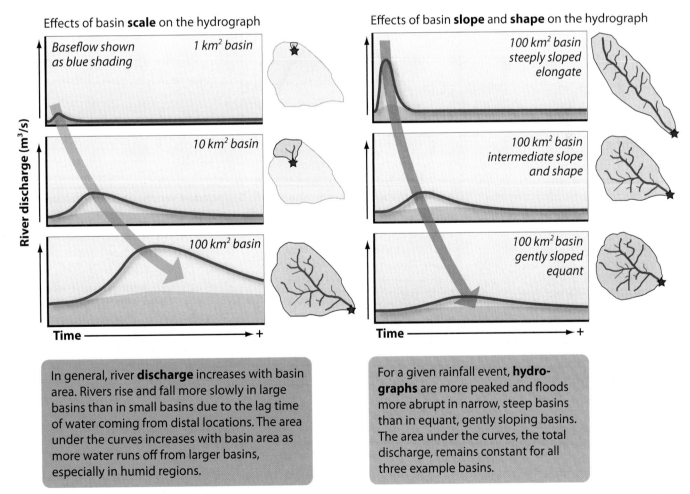

FIGURE 4.10 Hydrograph: Basin Scale and Shape. The shape, size, and timing of the hydrograph are directly related to the size and the geomorphology of the drainage basin.

hydrograph's receding limb is controlled by the rate at which both surface and subsurface processes deliver water to the channel after the storm ceases.

Elevated groundwater tables and long drainage networks lead to long, gradual **flow recession**. In contrast, hydrographs from small urban streams fall quickly because these basins are largely paved and thus little water infiltrates to sustain baseflow. In large basins, the shape and duration of the receding limb predominantly reflect the time it takes for water to move through the drainage network.

The **recession constant**, K_r, empirically describes the rate at which discharge decreases after the hydrograph peaks

$$Q_t = Q_p K_r^t = Q_p e^{-\alpha t} \qquad \text{eq. 4.4}$$

where Q_t is the discharge at time t, and Q_p is the peak discharge at the start of uninterrupted periods of declining discharge. By defining $\alpha = \ln K_r$, equation 4.4 may be expressed as

$$\ln Q_t = \ln Q_p - \alpha t \qquad \text{eq. 4.5}$$

and α can be calculated from the slope of semilogarithmic plots of discharge recession, where time is on the x-axis and $\ln Q$ is on the y-axis. K_r values cluster in different ranges for different runoff generation mechanisms, with low K_r values for Horton Overland Flow (<0.3), reflecting rapid discharge recession, whereas higher K_r values for subsurface stormflow (>0.3) imply sustained drainage and prolonged discharge after the storm ends. Saturation overland flow hydrographs exhibit a wide range of K_r values due to the influence of subsurface stormflow on sustaining return flow during discharge recession.

Hydrograph shape integrates the effects of rainfall patterns, runoff generation processes, and hydrologic and topographic properties of the catchment. The lag-to-peak (L_p) and discharge recession constant (K_r) provide simple measures of the timescale of runoff response. These two measures, L_p and K_r, quantify differences in drainage basin hydrologic processes, can change through a storm event or hydrologic year as different runoff generation mechanisms become active, and can be measured for basins from the scale of individual headwater channels to large rivers.

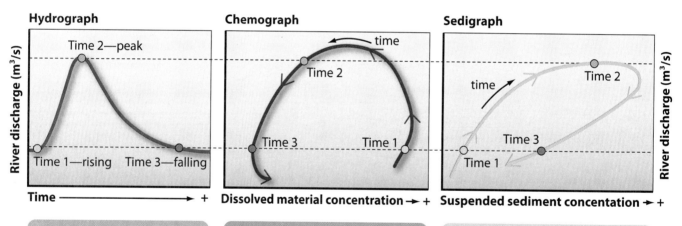

FIGURE 4.11 Hydrographs, Chemographs, and Sedigraphs. Streams and rivers move mass across Earth's surface, both as sediment and as material dissolved in solution. Due to the complex nature of sediment transport and flushing of pore water from soil by rainstorms, sediment and solute concentrations vary through the hydrograph and are not directly related to discharge—the phenomenon referred to as hysteresis.

Closely related to the hydrograph are the **chemograph** and the **sedigraph.** The chemograph relates the dissolved mass loading of an element to discharge; the sedigraph plots the concentration of suspended sediment in transport as a function of discharge. Both chemographs (dissolved material) and sedigraphs (solid material) tend to exhibit a phenomenon known as **hysteresis,** in which dissolved load and sediment load values are path- and time-dependent rather than depending only on discharge. For example, sediment and dissolved load concentrations at a discharge value on the rising limb of a hydrograph are different from those at the same discharge value on the falling limb [Figure 4.11].

Thinking about the physical processes active in and near the channel can explain hysteresis. For both sediment and dissolved load, hysteresis reflects the mobilization and flushing of material from the system. As the discharge increases, geomorphic action in the channel begins. As the velocity and the depth of flow rise, so does the shear stress on the bed. When bed material begins to move, the concentration of sediment in the flow increases. Larger amounts of sediment move after the flow becomes strong enough to mobilize the coarse layer of clasts that armors the surface of most channels, exposing the more easily moved, finer-grained material below.

Once liberated from the bed, sediment moves downstream rapidly so that sediment concentration is often higher during the receding limb than during the rising limb of the sedigraph. Interpreting chemographs is also complex. They reflect the addition to the channel both of rainfall precipitating onto the basin and the addition of "old" water previously stored in the groundwater system (and thus enriched in dissolved mineral constituents such as calcium and sodium) but now displaced and forced out by newly infiltrated stormwater.

Interactions Between Groundwater and Surface Flow

Although many people and many legal theories consider surface flow and groundwater to be separate systems, they are intrinsically connected. Water moves freely from streams, rivers, and lakes into the ground and then comes to the surface again in other locations. An informative way to understand this linkage and interaction is to consider the connections between streams and groundwater.

Streams and rivers can be characterized in different ways. One taxonomy places streams in one of two categories, **gaining streams** and **losing streams** [Figure 4.12]. Discharge in gaining streams tends to increase downstream because those streams act as drains for groundwater. This occurs because the water level or head in the stream is lower than that of the adjacent groundwater table, establishing a head or energy gradient so that groundwater moves into the stream. The result is baseflow. Losing

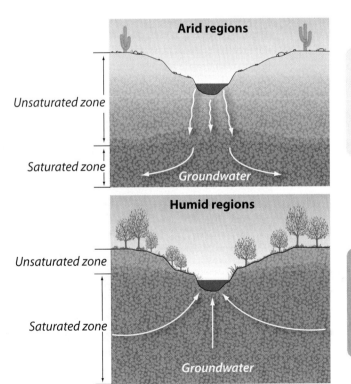

FIGURE 4.12 Gaining and Losing Streams. Streams and groundwater interact. In arid regions, streams lose flow through their beds, recharging aquifers; thus, discharge decreases downstream. In humid regions, streams usually function as drains, gaining flow from groundwater; thus, discharge increases downstream.

streams are most common in arid regions where groundwater tables are typically well below the bottom of the stream channel; thus, discharge in losing streams tends to decrease downstream. In the desert, these losing streams serve as intermittent sources of recharge to the groundwater system operating whenever water from storms or snowmelt in adjacent highlands flows through their channels. Losing streams often run dry between storm events, all their flow having infiltrated.

A single stream may gain or lose water at different times and at different places along its channel. When groundwater levels are high, such as during the winter or immediately after storms and snowmelt events, the head difference is great and large amounts of water flow into streams. However, in large channels that receive substantial discharge from upstream, river stage can rise more rapidly than adjacent groundwater levels. When the head in the stream exceeds that in the adjacent groundwater, river water flows into the bank material and the stream loses water. If flood levels drop quickly, pore pressures in the bank can remain elevated because soil takes longer to drain than river stage takes to drop. The resulting gradient in pore pressures can trigger bank collapse.

Another taxonomy classifies streams as perennial, ephemeral, seasonal, and intermittent. **Perennial streams** always contain flowing water and are typically found in humid regions where precipitation is distributed throughout the year and where groundwater tables are at or near the land surface during all seasons. The size of the drainage basin needed to support a perennial stream depends on climate and lithology; in moist climates and on generally impermeable lithologies, small drainage basins can support perennial streams. **Ephemeral streams** have flow only during large or intense precipitation events or during snowmelt. Ephemeral streams are common in arid regions and in upland drainage basins with small catchment areas. In many cases, such streams are disconnected from the regional groundwater table. During storms, shallow subsurface flow enters these upland streams and they fill with water. As a storm subsides, groundwater levels lower and water in ephemeral streams infiltrates into the bed.

Seasonal streams have flow only during certain, predictable parts of the year, for example, during spring snowmelt. Flow in **intermittent streams** is discontinuous in space, usually as a result of changing hydraulic conductivity in the subsurface and thus differences in the ability of the subsurface material to transmit water. For example, where a stream flowing over rock enters a valley bottom reach filled with large cobbles, water may sink into the subsurface and flow between the cobbles [**Photograph 4.11**].

Flood Frequency

Floods occur when rivers overtop their banks. Floods are natural and normal events punctuating the otherwise tranquil baseflow of many rivers and streams [**Photographs 4.12 and 4.13**]. Because of the geomorphic influence

PHOTOGRAPH 4.11 Intermittent Stream Flow. This intermittent stream drains the west side of the Appalachian Mountains in Shenandoah National Park, Virginia. The streambed, being highly permeable (because it is composed of large, quartzite clasts), carries flow beneath the surface in most conditions—except during floods, when the volume of water increases so much that some water flows on the surface. Upstream and downstream, where the channel bed is finer and less permeable, water flows on the surface.

of regular high flows, contemporary restoration efforts on large river systems often aim to recreate at least some of the effects of regular, high flows associated with snowmelt or other hydroclimatic drivers. The level or discharge of the annual high flow can have a great deal of interannual variability (which tends to increase as precipitation decreases—i.e., dry regions have the greatest interannual variability). Thus, the word "flood" can refer to a regularly predictable yearly event as well as to unusual, less frequently occurring, storm-driven high flows.

PHOTOGRAPH 4.12 Suspended Sediment. The Kuiseb River in Namibia rarely floods, but when it does, it carries large loads of suspended sediment. The blue sky suggests that the silty water pouring down the river likely fell on the more humid highlands rather than on the arid lowland region where this image was taken.

PHOTOGRAPH 4.13 Connecticut River in Flood. In 1927, the largest flood on record hit the Connecticut River. Heavy rain fell for two days in early November onto ground saturated by record October rainfall. The rain quickly ran off, driving rivers and streams out of their banks. At St. Johnsbury, Vermont, this bridge was swept away as the townspeople watched.

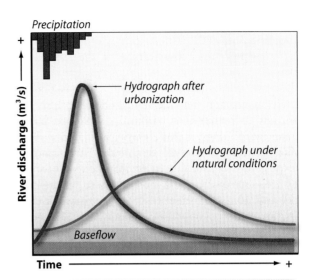

Under natural conditions, rainfall follows convoluted paths through the landscape. Water is held in **detention storage** by irregular pit and mound topography, infiltrating into organic-rich forest soil, and moving slowly to the channel. The infiltrating water feeds **baseflow** during times when it is not raining. **Flood peaks** are delayed because natural landscape characteristics slow the rising limb of the hydrograph, lower the peak flow, and extend the flood duration.

After urbanization, rainfall moves rapidly to the channel with little chance to infiltrate; thus, baseflow is reduced. Flowing directly off **impervious surfaces,** such as parking lots, and into storm sewers, runoff rapidly enters streams, raising their level quickly. Flood peaks now come sooner and are higher, increasing flood hazards and the tempo of geomorphic change. In the example below, the natural 25-year flow becomes the much more frequent urbanized 2-year flow.

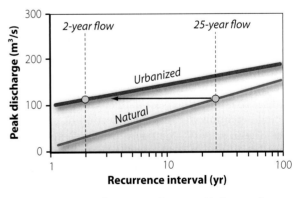

FIGURE 4.13 Pre- and Post-Development Hydrographs. Development, and the increase in impermeable surfaces including pavement and roofs, changes runoff processes and the shape of the hydrograph. As development increases, rising limbs steepen, peak flow increases, and baseflow diminishes. Return intervals shorten for floods of the same magnitude. In the case illustrated, a flow of ~125 m³/s occurred under natural conditions on average once every 25 years, but under urbanized conditions the same flow now occurs about once every 2 years.

Similar to precipitation events, floods can be described in terms of recurrence intervals using equation 4.1 and substituting peak annual **flood stage** (height) or discharge for precipitation intensity [Figure 4.13]. Flood recurrence intervals are established by analyzing discharge records typically maintained by government agencies. In the United States, many such records are kept by the Water Resources Division of the U.S. Geological Survey and are available online. Other records are kept by the Army Corps of Engineers, National Oceanic and Atmospheric Administration (NOAA), and state agencies.

Additional flood data, stretching back hundreds to thousands of years, can sometimes be determined geologically. **Paleoflood hydrology,** the study of ancient floods predating gauge records, relies on evidence for past flood heights by dating flood debris preserved above the modern-day channel. Often such material is preserved in caves or bedrock alcoves. Such an approach has been of particular utility for identifying the size of extreme floods in arid regions.

Bankfull flow is a commonly used datum to describe the size of floods that fill the channel but are not high enough to inundate the floodplain. In humid regions, many streams fill their banks every year or two, on average. Geomorphologists tend to pay particular attention to the bankfull flow, based on the idea that this flow shapes alluvial channels because larger floods may do more work but are so rare that they are relatively unimportant in setting channel form. Lower flows, although more common, do not have the energy to cause significant changes to the channel. The recurrence interval for bankfull flows varies depending on the environment. In arid regions, bankfull flows may recur as infrequently as every 50 years, on average. In bedrock canyons and channels without well-developed floodplains, defining the bankfull stage is not possible. In channels responding to disturbance, especially those that are incising, the bankfull flow level may be nothing more than a discontinuous low terrace at the base of steep banks.

Because the concept of bankfull flow is commonly applied in stream management and restoration, it is important to realize that bankfull, as defined based on channel morphology, is not always associated with a particular recurrence interval discharge. Furthermore, the concept is most applicable to environments where a relatively frequent discharge is most likely to fill the channel banks and do a great deal of geomorphic work (humid-temperate regions, snowmelt systems). In arid environments, where infrequent discharges shape the channel and transport most sediment, bankfull flow is not likely to be the annual flood flow but rather a flood that occurs more rarely. Still, the bankfull flow is important for setting the scale of alluvial channels even in arid regions.

Of broad societal importance is the **100-year flood**, the flood with a yearly return probability of 1 percent. There is nothing scientifically special about the event that has a 1 percent chance of occurring each year. Rather, the 100-year flood was an arbitrary risk level chosen by U.S. government agencies and Congress when they wrote the 1968 Flood Control Act. Even though the annual probability of a 100-year flood is very low, there is no physical reason preventing a 100-year flood from recurring in two successive years. Characteristics of the 100-year flood (discharge and stage) are used extensively for planning and zoning purposes and often form the basis for defining floodplains in terms of insurance and development regulation. In addition to planning that considers hazards of inundation from rising floodwaters, "smart" river corridor planning also needs to consider lateral movement of the channel driven by bank erosion. Such lateral migration is typical of most streams with erodible banks and can be particularly rapid in arid-region streams where riparian vegetation is sparse and there is little, if any, root reinforcement.

Land-use and climate changes (both natural and human-induced) can alter flood recurrence intervals. Clearing forest or grassland to build houses or businesses smooths the land surface so that lawns can be planted (reducing **detention storage**, the collection of rainwater in surface irregularities), compacts the land surface, and reduces infiltration, thereby speeding the movement of water to channels (see Figure 4.13). The addition of impermeable surfaces and storm sewers further increases the speed of runoff and thus the peak discharge, stage (height), and erosion potential of floods downstream. It is common to find that peak discharges that occurred, on average, once every decade before development, occur every year or two after development (see Digging Deeper). The geomorphic and societal effects of development, and its attendant hydrologic alterations, can be wide ranging and include channel incision, flooding, and reduced baseflow between runoff-producing events.

Water Budgets

Basin-scale **water budgets** are important tools for understanding basin behavior and managing water resources. They can be used to answer such practical questions as, "How much water can be withdrawn from a stream for snowmaking in the winter before flow will drop below levels needed to support overwintering aquatic life?" or "How much water can be withdrawn from a stream for irrigation before withdrawals exceed the annual water supply?" Water budgets are useful for predicting how much water might be available for uses and for partitioning that usage when demand exceeds supply.

Such budgets consider inputs from precipitation [**Photograph 4.14a**] and direct condensation from fog or clouds in locations where this is important (mountaintops, hyper-arid seacoasts). Outputs of water from drainage basins include direct evaporation from water surfaces and soils, transpiration by plants, groundwater flow, and channelized surface flow [**Photograph 4.14b**]. Constructing accurate water budgets, especially for small basins, is not straightforward because basin-specific data are usually hard to come by.

Precipitation, the primary input of water to most watersheds, can be estimated in a variety of ways. Established weather stations record precipitation directly, but such stations are few and far between. Where dense networks of precipitation collectors have been installed, the resulting data clearly show large variability in precipitation amounts and rates over small areas, especially in rugged terrain where topography affects precipitation patterns.

Recently, the application of basin-scale precipitation models [**Photograph 4.14c**] has become more widespread; these models rely both on the observational record and known relationships between elevation and rainfall/snowmelt to produce spatial estimates of precipitation. Weather radar systems have been calibrated to provide real-time precipitation estimates during storms. Such data can be integrated to estimate precipitation inputs over time. Flood forecasting combines water balances with computer models that include precipitation estimates and route stormwater downstream to predict the timing and magnitude of high flows. In areas where the winter precipitation falls as snow, transfer of snow between watersheds by wind can complicate water budgeting. In small, high-elevation catchments, large amounts of water (in the form of blown snow) can move from one basin to another.

Losses of water from a watershed can be measured or estimated. Evaporation and transpiration are usually modeled using established approximations that consider temperature, wind speed, and vegetation type. Because most watersheds, even those in developed countries, do not have stream gauges that provide continuous discharge records, generalized regional water discharge/basin area relationships are useful. Such relationships are created by regressing basin area and mean annual discharge for basins of different sizes that are gauged. These relationships are

PHOTOGRAPH 4.14 Instrumentation Needed for Water Budgeting (a) An early weather station (1880) with rain gauge on Pike's Peak, Colorado, at an elevation of ~4200 meters. (b) Dam and rectangular weir for measuring stream flow in the Los Angeles River in southern California, 1904. (c) Geography strongly controls the distribution of annual precipitation on the island of Hawaii. This map shows how wet the east side of the island is due to orographic lifting of moist tropical ocean air carried by northeasterly trade winds. The data are from the PRISM model, an algorithm that takes scarce observational data and combines these with terrain models to estimate the spatial variability of precipitation. [Source: 4 × 4 km PRISM Model Grid. Based on 1961–90 annual precipitation data from NOAA Cooperative Stations. Hillshade relief derived from U.S. Geological Survey DEMs.]

most useful when the ungauged basin has similar shape, topography, and orientation as the calibration basins.

Groundwater losses are even more difficult to quantify and are usually calculated by differencing after the other terms in the water budget have been estimated. While this approach has been very useful in the past, temporal variability, especially the nonstationary behavior of the climate and thus the hydrological system in response to human-induced climate change, may introduce greater uncertainties and render the approach less useful in the future.

Surface water moves rapidly through basins; in-channel storage times are short (minutes to days) and scale with channel length and inversely with slope and velocity. Groundwater **residence times** are much longer. Shallow subsurface groundwater might reside in the basin for days to months. Deeper groundwater can remain in large basins over glacial/interglacial timescales and is essentially fossil; the residence time of water in such deep aquifers can be many millennia. Indeed, groundwater extracted from the major irrigation aquifers in the central United States, such as the Ogallala, and in the Great Artesian Basin of Australia, was likely recharged during the Pleistocene Epoch under a completely different climate regime. Once withdrawn, such ancient water will not be replaced on human timescales, making fossil groundwater a nonrenewable resource.

Hydrologic Landforms

Weathering and erosion of carbonate rocks and evaporites produce unique topography dominated by dissolution and groundwater hydrologic processes; landscapes produced this way are referred to as **karst**, a German word for a limestone region in Slovenia. Karst landscapes are both morphologically and hydrologically distinct systems in which large amounts of water move between Earth's surface and interconnected cavities below ground [**Photograph 4.15**]. Chemical weathering and dissolution produce distinctive types of karst landforms, and drainage of karst terrain primarily occurs below ground, producing unique and often very limited

PHOTOGRAPH 4.15 Karst Landscapes. Limestone, and the karst landscapes developed on it, can be spectacular. (a) Touring watery passages of Mammoth Cave, Kentucky, by boat. The National Park Service discontinued tours such as this in the 1990s. (b) Sinkholes form as rock below collapses from dissolution. Here is a sinkhole in the Minnehakta limestone southeast of Boyd, Weston County, Wyoming, circa 1900. (c) Dramatic view of tower karst topography at river's edge, taken from a boat on a trip down the Li River south of Guilin, China. (d) Pillars of dripstone or flowstone deposited from calcium carbonate–saturated solutions at Carlsbad Caverns National Park, New Mexico.

surface drainage patterns. Most of the world's great cave systems are developed in limestone and occur in karst terrain.

The solubility of limestone is primarily regulated by groundwater flux, acidity (pH), and CO_2 concentration (eq. 3.3). Consequently, abundant vegetation cover and rainfall favor karst development. As calcite dissolution occurs, the largest, fastest-growing voids tend to capture flow and grow at the expense of smaller, slower-growing voids, resulting in preferential development of a limited number of conduits. This favors development of distinct caverns and subsurface drainage. Karst topography in silicate rocks, referred to as **silicate karst,** is rare, can take many millions of years to develop, and is found only on unusually stable, ancient land surfaces in some parts of the equatorial tropics, such as the stable cratons within the Amazon basin.

Although karst landforms may be found in any environment, they are most common in humid-temperate and tropical regions. In the United States, karst terrain is concentrated in the eastern and southern regions, principally in Florida, the Appalachian Mountains, southern Indiana, New Mexico, and western Kentucky. Other important karst landscapes include areas in China, Australia, Malaysia, Jamaica, and Spain. This distribution reflects the location of both carbonate rocks and sufficient precipitation to dissolve those rocks.

Most extensive regions of karst topography are developed on limestone consisting of more than 50 percent $CaCO_3$, although dolomite (magnesium-rich carbonate) and evaporites such as gypsum may also develop karst topography. In general, karst formation potential increases with limestone purity. In particular, the development and nature of porosity along joints, faults, fractures, and bedding planes promote dissolution if coupled with high permeability (connectivity) that allows free circulation of water.

Tufa and **travertine** are deposits that result from the evaporation of or degassing of CO_2-enriched waters, which causes precipitation of calcium carbonate that can coat silicate rocks or accumulate to form thick deposits [**Photographs 4.16** and **4.17**]. These deposits mark the prior location of springs, groundwater flow, and surface water flow; thus, mapping their locations allows for the reconstruction of past flowpaths.

The most common karst landforms are **sinkholes**, also known as **dolines**. Typically wider than they are deep, dolines are generally circular to elliptical and narrow with depth, producing a funnellike shape. Dolines can reach sizes of up to 1 km across and hundreds of meters deep. Areas with abundant dolines can have a distinctive pitted appearance and lack typical valley development. Dolines can form by either solution or collapse into subsurface caverns that themselves result from solution [**Figure 4.14**].

Hydrologically, karst is characterized by disrupted surface flow with interrupted stream valleys and closed depressions, some of which are filled by groundwater, indicating that they intersect the groundwater table. There are frequent diversions of surface water underground as streams disappear into subterranean conduits only to reemerge as **springs** (places where groundwater emerges at the surface) of all sizes. Springs are often found at topographic, stratigraphic, and structural discontinuities.

Karst landscapes have particularly high secondary porosity, with most groundwater moving through conduits of differing dimensions dissolved into the rock. Differences in nearby flowpaths (distinct conduits) can lead to significant groundwater level changes over short length scales. Groundwater moving through karst may move underground between watersheds (crossing below surface watershed boundaries), because the flow of groundwater responds to head gradients rather than topography, which controls the flow of surface water.

Caves and springs are common in karst terrain. Cave formation appears to be controlled by rock structure, fracture patterns, and lithology with caves developing as conduits along zones of preferential groundwater flow and dissolution. The extensive cave networks common in karst terrain exhibit a variety of morphologies based on their pattern as viewed from above, their branching characteristics, and their cross-sectional shape. In a positive feedback loop, fractures that capture the most flow grow most rapidly into caves at the expense of others.

Caves that form at the groundwater table can be used to estimate regional river incision rates if deposits in a vertical series of caves can be dated. Such dating can be done using U/Th on carbonate dripstone deposits in the cave; cosmogenic methods (using two isotopes with different half-lives) can be used to date how long clasts, washed into the cave, have been isolated from cosmic radiation.

PHOTOGRAPH 4.16 Tufa Deposits. Stream near Ruapehu volcano on the North Island of New Zealand, encrusting a waterfall with tufa, a deposit of calcium carbonate.

In some karst terrain, rivers flow into **blind valleys** with no outlet, from which the only exit for water is underground. **Dry valleys** are common in karst terrain; these are a kind of intermittent stream that results when a river, flowing across a valley floor, disappears into a sinkhole and flows below ground, leaving its former valley high and dry. Karst terrain often consists of closed depressions that are not connected by an integrated surface drainage network.

Tower karst is found most often in the tropics and describes a landscape where steep-sided hills of limestone rise above a low-lying alluvial plain. Some suggest that these towers are the more resistant rock and that recent stripping has removed the intervening weaker material, leaving the rugged, high-relief landscape.

PHOTOGRAPH 4.17 Tufa Towers. The "man and woman" tufa towers were deposited by springs entering Mono Lake in southern California when it was deeper, before the streams feeding the lake were diverted as water supplies for the city of Los Angeles. Now, they stand as dry monuments to the lowering of the lake level so that the Los Angeles population and farm fields could have more water.

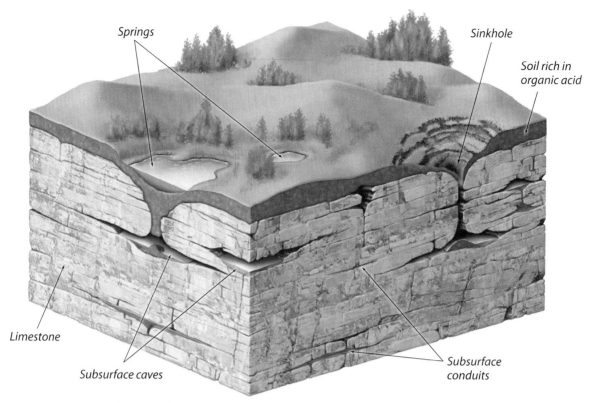

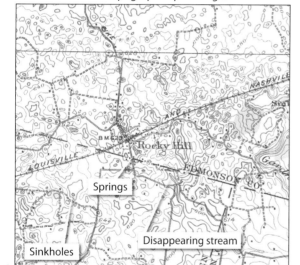

Section of Mammoth Cave, KY
USGS topographic quadrangle

Corresponding region from Google Earth

Karst landscapes are very distinctive; topographic maps show many closed contours and airphotos show pits, some filled with water. Karst landforms are the result of rock dissolution and include subsurface caves and conduits through which groundwater flows before emerging in springs and flowing into rivers. There are **sinkholes,** which can be caused by cave collapse. **Disappearing streams** appear to vanish; in fact they flow for some distance underground in connected passages. On the surface, karst terrain is often rough, pockmarked by sinkholes. There can be large areas with no surface drainage. The lack of streams in karst terrain, even in humid, temperate environments with high mean annual rates of precipitation, reflects the large amount of water moving underground.

FIGURE 4.14 Karst Landscape. Karst landscapes, such as the one shown on the Mammoth Cave, Kentucky, topographic quadrangle, are complex, with numerous closed depressions and intermittent surface drainage as water moves between the surface and the ground. Much of the subsurface flow in karst is carried by enlarged fractures, secondary porosity in otherwise nearly impermeable limestone. [Upper image adapted from Panno and Wolf (1997).]

Applications

There are complex and ongoing interactions and feedbacks between water and the landscape. Although surface water and groundwater shape our planet's surface by eroding earth materials physically and chemically, it is the landscape that determines hydrologic flowpaths. Topography controls the direction of surface water flow and in many cases the direction of subsurface, groundwater flow. Water flowing over slopes does geomorphic work, moving sediment downslope both in diffuse overland flow on surfaces with low infiltration rates and in concentrated flow in rills and small channels. As water erodes the landscape, it reshapes topography, which in turn controls future flows and the flow of groundwater. Together, surface water and groundwater are geomorphic agents important in shaping landscapes.

Vegetation plays a role in surface and subsurface hydrology and thus in the geomorphology of many landscapes. By transpiring water, trees change the hydrologic balance of watersheds, reducing the water available for transporting materials and reducing the frequency and extent of saturated soils during growing-season rainstorms. People have tried various schemes to increase water yield by harvesting trees, but unless the slopes are kept clear, young trees regrow and use water. In addition, when slopes are cleared of vegetation, the root systems rot, weakening the slope and eventually closing off macropores that can rapidly transmit large amounts of water through the subsurface.

Not all landscapes were once forested and changes in vegetation other than deforestation, such as the conversion of grassland to agriculture, also affect the hydrologic cycle, erosion, and sedimentation. One of the most important hydrologic effects of modern, mechanized agriculture is the installation of tile drains to lower groundwater levels in flat-lying fields, particularly in the U.S. Midwest. These drains improve crop yields and extend the growing season by drying the fields earlier in the spring and preventing waterlogging during rainy spells, but the drains speed the transport of both water and nutrients to streams and rivers. The results are increased flood peaks and eutrophication of water bodies (including the Gulf of Mexico) as nutrients, applied as fertilizers, move downstream.

Humans both affect and are affected by the hydrologic system. Landscape modification, including removal of forests, agriculture, and urbanization, change the rate and means by which water leaves the landscape through their effects on infiltration rate and detention storage. Not only do such changes affect groundwater and surface water response to precipitation at the local or hillslope scale, but they change the size and frequency of surface water discharges downstream. Floods that used to occur only rarely can become commonplace, and their effects can ripple through the landscape, destabilizing stream banks, changing the sediment transport capacity of streams, and degrading habitat for valuable aquatic organisms, such as salmon.

The intimate connection between groundwater, surface water, and plants is clear to geomorphologists. Extracting large amounts of groundwater can reduce baseflow, drying out streams and changing their ability to transport sediment and provide habitat. Groundwater levels go up and down with the seasons, in part reflecting seasonal changes in precipitation, but these levels are also driven by the ability of trees and other vegetation to pump water from the ground by transpiration. Measuring shallow groundwater levels in the winter often gives very different results than measuring them in the summer. Such changing levels could be key when modeling slope stability—low summer groundwater levels might suggest a slope is stable, whereas high winter groundwater levels might indicate that the slope is near failure.

Climate change will affect the hydrologic system, dramatically in some places. With a warming climate and increasingly active hydrologic cycle, recurrence intervals for precipitation and stream flows are changing. Over the next century, the warming climate will likely affect the distribution of vegetation and the frequency of droughts, forest fires, and storms. Together, these changes will affect slope stability, runoff, and the amount of water and sediment shed from landscapes. One of the most dramatic hydrologic changes will be in streams issuing from high-elevation, glaciated mountain basins. Many of these streams are critical water supplies for villages in the valleys below. If the climate warms sufficiently that glaciers melt away, then the natural reservoirs of ice and snow, which provide water all summer from melting ice, will vanish. Many towns in the Andes and the Himalaya will face the need to develop alternative water supplies they can ill afford.

Geomorphic hydrology and natural hazards are closely related. In karst terrain, sinkholes can open unexpectedly, swallowing homes and, in one famous case, a Porsche dealership—cars and all. Around the world, floods cause billions of dollars of damage yearly, often killing or displacing large numbers of people as channel locations change and water and sediment flow overbank. In the United States, floods are the number one geologic hazard in terms of dollars lost. Water supply has been and continues to be a critical geopolitical issue. Geomorphologists, with their long-term, broad-scale view on the landscape, are well equipped to advise society about issues related to reducing the hazard from large-magnitude, infrequent natural events.

Selected References and Further Reading

Baker, V. R., R. C. Kochel, and P. C. Patton, eds. *Flood Geomorphology*. New York: Wiley, 1988.

Bonnell, M. Progress in the understanding of runoff generation in forests. *Journal of Hydrology* 150 (1993): 217–275.

Brooks, K. N., P. F. Folliott, H. M. Gregersen, and L. F. DeBano. *Hydrology and the Management of Watersheds*, 3rd ed. New York: Wiley-Blackwell, 2003.

Brown, A. E., L. Zhang, T. A. McMahon, et al. A review of paired catchment studies for determining changes in water yield resulting from alterations in vegetation. *Journal of Hydrology* 310 (2005): 28–61.

Caine, N. The rainfall-intensity duration control of shallow landslides and debris flows. *Geografiska Annaler*, 62A (1980): 23–27.

Costa, J. E., and J. E. O'Connor. "Geomorphically effective floods." In J. E. Costa, A. J. Miller, K. W. Potter, and P. R. Wilcock, eds. *Natural and Anthropogenic Influences in Fluvial Geomorphology*. Washington, DC: American Geophysical Union Press, 1995.

Dunne, T. "Field studies of hillslope flow processes." In M. J. Kirkby, ed., *Hillslope Hydrology*. New York: Wiley-Interscience, 1978.

Dunne, T., and R. D. Black. Partial area contributions to storm runoff in a small New England watershed. *Water Resources Research* 6 (1970): 1296–1311.

Dunne, T., and L. B. Leopold. *Water in Environmental Planning*. New York: W. H. Freeman, 1978.

Granger, D. E., J. W. Kirchner, and R. C. Finkel. Quaternary downcutting rate of the New River, Virginia, measured from differential decay of cosmogenic ^{26}Al and ^{10}Be in cave-deposited alluvium. *Geology* 25 (1997): 107–110.

Hirschboeck, K. K. "Flood hydroclimatology." In V. R. Baker, R. C. Kochel, and P. C. Patton, eds., *Flood Geomorphology*. New York: Wiley, 1988.

Hombeck, J. W., M. B. Adams, E. S. Corbett, et al. Long-term impacts of forest treatments on water yield: A summary for northeastern United States. *Journal of Hydrology* 150 (1993): 323–344.

Horton, R. E. Erosional development of streams and their drainage basins: Hydrophysical approach to quantitative morphology. *Geological Society of America Bulletin* 56 (1945): 275–370.

Jones, J. A., and D. A. Post. Seasonal and successional streamflow response to forest cutting and regrowth in the northwest and eastern United States. *Water Resources Research* 40, W05203 (2004): doi: 10.1029/2003WR002952.

Kochel, R. C., and V. R. Baker. Paleoflood hydrology. *Science* 215 (1982): 353–361.

McDonnell, J. J. Where does water go when it rains? Moving beyond the variable source area concept of rainfall-runoff response. *Hydrological Processes* 17 (2003): 1869–1875.

McDonnell, J. J., M. Sivapalan, K. Vaché, et al. Moving beyond heterogeneity and process complexity: A new vision for watershed hydrology. *Water Resources Research* 43, W07031 (2007): doi: 10.1029/ 2006WR005467.

Milly, P. C. D., J. Betancourt, M. Falkenmark, et al. Stationarity is dead: Whither water management? *Science* 319 (2008): 573–574.

Montgomery, D. R., and W. E. Dietrich. Hydrologic processes in a low-gradient source area. *Water Resources Research* 31 (1995): 1–10.

Montgomery, D. R., and W. E. Dietrich. Runoff generation in a steep, soil-mantled landscape. *Water Resources Research* 38 (2002): 1168, doi: 10.1029/2001WR000822.

National Research Council. Hydrologic Effects of a Changing Forest Landscape. Washington, DC: National Academies Press, 2008.

O'Loughlin, E. M. Prediction of surface saturation zones in natural catchments by topographic analysis. *Water Resources Research* 22 (1986): 794–804.

Panno, S. V., and E. M. Wolf. "Karst land in Illinois: Hills, hollows, and honeycomb rock." Illinois State Geological Survey poster, 1997.

Singh, P., and L. Bengtsson. Impact of warmer climate on melt and evaporation for the rainfed, snowfed, and glacierfed basins in the Himalayan region. *Journal of Hydrology* 300 (2005): 140–154.

Stanford, J. A., and J. V. Ward. The hyporheic habitat of river ecosystems. *Nature* 335 (1988): 64–66.

Tallaksen, L. M. A review of baseflow recession analysis. *Journal of Hydrology* 165 (1995): 349–370.

Thornthwaite, C. W. An approach toward a rational classification of climate. *Geographical Review* 38 (1948): 55–94.

U.S. Geological Survey. *Tracing and Dating Young Ground Water*, Fact Sheet 134–99. Washington, DC: U.S. Department of the Interior, 1999.

Ward, J. V. The four-dimensional nature of lotic ecosystems, *Journal of North American Benthological Society* 8 (1989): 2–8.

Wolman, M. G. A cycle of sedimentation and erosion in urban river channels. *Geografiska Annaler Series A* 49 (1967): 385–395.

Wolman M. G., and R. Gerson. Relative scales of time and effectiveness of climate in watershed geomorphology. *Earth Surface Processes* 3 (1978): 189–208.

Wolman, M. G., and J. P. Miller. Magnitude and frequency of forces in geomorphic processes. *Journal of Geology* 68 (1960): 54–74.

Yair, A., and H. Lavee. Runoff generation in arid and semi-arid zones. In M. G. Anderson and T. P. Burt, eds., *Hydrological Forecasting*. New York: Wiley, 1985.

DIGGING DEEPER   Humans, Hydrology, and Landscape Change—What's the Connection?

Seven billion people occupy Earth and their actions are changing our planet's landscapes [**Figure DD4.1**] (Hooke, 2000). Using a series of model-based calculations and some data, Hooke (1994) argued that humans have become the most effective geomorphic agent on Earth. Not only are people active geomorphic agents, but their actions alter both hydrology and sediment transport at local and global scales. Clearing native vegetation for agriculture and

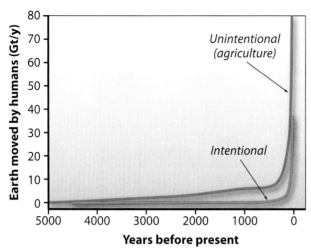

FIGURE DD4.1 Using a variety of assumptions and different data sets, Hooke estimated the amount of earth moved by humans over the past several thousand years in units of gigatons per year (a gigaton is 10^{12} kg). He differentiated movement of soil for agriculture (tillage) with that moved for construction (intentional). The dramatic increase in earth moving corresponds to the Industrial Revolution and the advent of cheap, easily available energy from fossil fuels. [From Hooke (2000).]

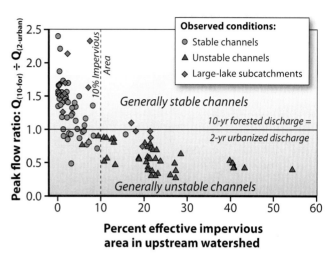

FIGURE DD4.2 Examining stream morphology, Booth et al. (2002) found that in watersheds where more than 10 percent of the land had been converted to impervious cover (such as roads, buildings, and parking lots), stream channels were unstable, and streams were incising, eroding their banks, and migrating laterally. Channel instability increased dramatically when the 2-year recurrence interval peak discharge after development (2-urban) exceeded the 10-year recurrence interval discharge (10-for) before development. [From Booth et al. (2002).]

through urbanization reduces infiltration rates and changes sediment yields as well as the locus of erosion, all in a predictable fashion. Damming rivers influences the distribution and timing of water and sediment discharge. Changing climate affects the volume, intensity, and spatial distribution of storms, precipitation, and runoff (IPCC, 2007).

How do we know the effects of landscape change on rivers, their channels, and the sediment loads they carry? Wolman (1967) documented the hydrologic and resulting geomorphic effects of urban land-use change by surveying channels and collecting sediment yield data before, during, and after development. In a now-classic schematic diagram, he showed how the yield of sediment changes over time as progressive development of a humid, vegetated landscape alters hydrologic flowpaths and the availability of sediment for transport (see Figure 7.12).

Initially, sediment yields are low under native forest cover as rainfall infiltrates and stream banks are stable. Sediment yields rise as agriculture disturbs the land surface and reduces infiltration. More sediment pours off the landscape during urbanization as construction exposes easily eroded soil and paving decreases infiltration and increases runoff. Finally, with the landscape urbanized, most sediment sources are paved and impermeable or otherwise stabilized with plantings; stormwater flows quickly into remaining channels, which, carrying lots of fast-moving water, rapidly scour their banks, deepening and widening as the channel cross section enlarges (Trimble, 1997). In many cases, the flood that occurred on average once every 10 years under native vegetation occurs after urbanization at least every 2 years, if not more frequently [**Figure DD4.2**] (Booth et al., 2002).

Damming rivers changes the frequency and magnitude of water flows, and thus the movement of sediment through river systems, by attenuating flood flows and, in some cases, augmenting baseflows. In arid regions, the most obvious hydrologic and geomorphic effect of dams on rivers is the creation of lakes in the desert, but there are other, more subtle effects. Doing fieldwork and collecting flow data, Grams et al. (2007) documented major changes that occurred downstream after the Glen Canyon Dam on the Colorado River was closed in 1963; the size and duration of spring floods decreased, baseflow increased, and sediment was trapped behind the dam [**Figure DD4.3**]. There were major geomorphic changes downstream. Starved for sediment, the alluvial channel downstream incised and narrowed; gravel and sandbars disappeared from the channel and the average size of riverbed material increased 80-fold. Floodplains were abandoned and became terraces, providing ideal habitat for nonnative species.

The hydrologic and geomorphic effects of dams in humid regions can also be substantial. Magilligan and Nislow (2001) used flow records from seven rivers in northeastern North America that had been dammed and created two flood frequency-magnitude curves for each flow record, one before and one after damming. The pre-dam record included, on average, 30 years of flow data, and the post-dam record included about 40 years. They found that peak flows decreased an average of 32 percent on impounded rivers, which was not a surprise since most of these were flood control dams. They found major

DIGGING DEEPER Humans, Hydrology, and Landscape Change—What's the Connection? *(continued)*

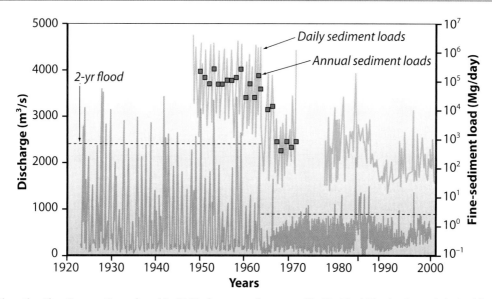

FIGURE DD4.3 When the Glen Canyon Dam closed in 1963, the flow regime and the sediment load of the Colorado River changed dramatically. Flow variability (the brown curve) dropped and sediment load (the light green curve) was diminished by orders of magnitude. The green boxes are the annual fine sediment loads (expressed in Mg/day). The horizontal dashed lines show that the magnitude of the 2-year recurrence flood dropped from 2400 m³/s to about 800 m³/s after damming. Look carefully at the two vertical axes (left and right). They are different—discharge is linear, and sediment load is logarithmic. [From Grams et al. (2007).]

changes in the frequency of bankfull discharge, the geomorphically important, channel-shaping flow. In the 40 years since damming, four of the seven channels did not experience a single bankfull flow; on these rivers, water and sediment have not reached the floodplain for decades.

There is overwhelming scientific consensus that humans have changed global climate and that climate change has altered the hydrologic cycle (IPCC, 2007). How might these changes in hydrology affect the rate and distribution of Earth surface processes? Over the past 150 years, data show that rivers and streams are freezing later and that the ice is breaking up sooner [**Figure DD4.4**] (Magnuson et al., 2000). In New England, the timing of spring runoff has advanced between one and two weeks in the last 30 years and is predicted to become even earlier (Hayhoe et al., 2007) as snowpacks thin and melt earlier than they used to. Hayhoe et al. (2007) predict that if warming continues, by the end of this century there will no longer be winter snowpacks in southern New England and therefore spring snowmelt floods will be a thing of the past. Warmer temperatures increase the saturation vapor pressure of water, putting more water into the atmosphere and making for more

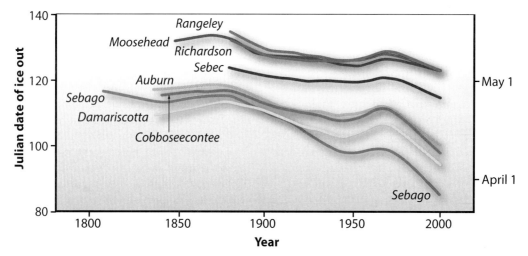

FIGURE DD4.4 Today, ice on lakes in New England and in other regions in the Northern Hemisphere (Magnuson et al., 2000) is melting and breaking up earlier than it did a century ago. These curves show the number of days after January 1 (the Julian date) that ice left each of these lakes in Maine. The yearly data are shown as smooth curves. [From Hodgkins et al. (2002).]

powerful storms. Indeed, Emanuel (2005) compiled meteorological data and showed that over the past 30 years, hurricanes are releasing increasingly more energy (wind speed integrated over time) and have greater maximum power release (the cube of the storms' maximum wind speeds). These enhanced storms now have the potential to cause more coastal geomorphic change; in particular, the increased power of tropical storms may increase the amount of beach erosion and longshore sediment transport.

Booth, D. B., D. Hartley, and R. Jackson. Forest cover, impervious-surface area, and the mitigation of stormwater impacts. *Journal of the American Water Resources Association* 38 (2002): 835–846.

Emanuel, K. Increasing destructiveness of tropical cyclones over the past 30 years. *Nature* 436 (2005): 686–688.

Grams, P. E., J. C. Schmidt, and D. J. Topping. The rate and pattern of bed incision and bank adjustment on the Colorado River in Glen Canyon downstream from Glen Canyon Dam, 1956–2000. *Geological Society of America Bulletin* 119 (2007): 556–575.

Hayhoe K., C. P. Wake, T. Huntington, et al. Past and future changes in climate and hydrological indicators in the U.S. Northeast. *Climate Dynamics* 28 (2007): 381–407.

Hodgkins, G. A., I. C. James, and T. G. Huntington. Historical changes in lake ice-out dates as indicators of climate change in New England, 1850–2000. *International Journal of Climatology* 22 (2002): 1819–1827.

Hooke, R. L. On the efficacy of humans as geomorphic agents. *GSA Today* 4 (1994): 217, 224–225.

Hooke, R. L. On the history of humans as geomorphic agents. *Geology* 28 (2000): 843–846.

IPCC (Intergovernmental Panel on Climate Change). *Climate Change 2007: The Physical Science Basis—Summary for Policymakers.* Paris, 2007.

Magilligan, F. J., and K. H. Nislow. Long-term changes in regional hydrologic regime following impoundment in a humid-climate watershed. *Journal of the American Water Resources Association* 37 (2001): 1551–1569.

Magnuson, J. J., D. M. Robertson, B. J. Benson, et al. Historical trends in lake and river ice cover in the Northern Hemisphere. *Science* 289 (2000): 1743–1746.

Trimble, S. W. Contribution of stream channel erosion to sediment yield from an urbanizing watershed. *Science* 278 (1997): 1442–1444.

Wolman, G. M. A cycle of sedimentation and erosion in urban river channels, *Geografiska Annaler Series A* 49 (1967): 385–395.

WORKED PROBLEM

Question: The table below lists the mean annual flood discharges between 1970 and 1989 for the Winooski River, which drains 2700 km² of northern Vermont. Based on these data, how large is the flood that occurs on average every 10 years (the flood with the 10 percent annual chance of recurrence)? The 1927 flood, the flood of record for the Winooski River, had an estimated discharge of 3200 cubic meters per second (m³/s). Extrapolating the short flood record, what is the estimated discharge of the 100-year flood (1 percent recurrence probability) and how does it compare to the discharge recorded in 1927? In what season does the annual maximum flood most often occur? Why?

Annual maximum flood discharges, Winooski River at Essex, Vermont

Year	Date	Discharge (m³/s)
1970	25 Feb	581
1971	4 May	586
1972	5 May	776
1973	1 Jul	813
1974	22 Dec	651
1975	2 Apr	629
1976	2 Apr	790
1977	14 Mar	906
1978	17 Oct	612
1979	6 Mar	702
1980	27 Nov	331
1981	25 Feb	521
1982	18 Apr	858
1983	25 Apr	493
1984	14 Dec	603
1985	3 Dec	396
1986	31 Mar	572
1987	1 Apr	790
1988	29 Apr	473
1989	7 May	575

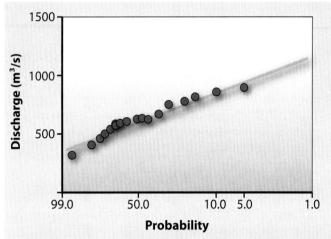

For the Winooski River at Essex, Vermont, the annual maximum flood series for 1970 to 1989 is plotted on probability axes that make the record linear and allows for extrapolation. Extrapolating from the 19-year record to the 1:100 event (the 100-year return flood; 1 percent annual probability) suggests that flood would be about 1200 m³/second.

Answer: First, rank the flood data. Then calculate a probability and recurrence interval for each flood using equations 4.1 and 4.2. The discharge of the 10 percent annual probability flood is ~860 m³/s and it occurred in 1982. Use probability graph paper and plot the flood data, connecting them with a line. If you do this, you find that the 100-year flood (probability of 1 percent) would be estimated, on the basis of this short record, to have a flow of about 1200 m³/s. The 1927 flood, with its discharge of 3200 m³/s, is more than 3 times higher than any flood experienced in the short, 10-year record. This analysis clearly demonstrates the uncertainty of extrapolating a short discharge record to estimate the magnitude of extreme events. The annual maximum floods on the Winooski River tend to happen in late winter and early spring; this is typical for rivers draining areas that receive significant winter snowfall. These winter and spring floods are often driven by rain-on-snow events. The runoff is a mix of snowmelt and rainfall onto saturated ground.

KNOWLEDGE ASSESSMENT Chapter 4

1. Explain why each of water's three phases is geomorphically important.
2. Explain the difference between the intensity and duration of precipitation.
3. Give an example of how thresholds are important in geomorphology.
4. Explain a recurrence interval in words stating the underlying assumptions.
5. Give the formula for calculating recurrence intervals.
6. What is stationarity?
7. List several types of weather systems that can deliver precipitation.
8. Explain why the geomorphic effects differ depending on the type of weather system delivering the precipitation.
9. Define evapotranspiration.
10. Provide examples of how climate affects the geomorphology of streams and rivers.
11. Explain why the seasonal distribution of runoff differs between monsoonal climates and cold region climates.
12. Predict how evapotranspiration will change with climate.
13. Explain why evapotranspiration is geomorphically important.
14. Explain the difference between actual and potential rates of evapotranspiration.
15. Give an example of a place where potential and actual rates of evapotranspiration are different.
16. Explain the hydrologic effects of removing trees from a drainage basin.
17. List factors that can affect the infiltration rate.
18. Explain how macropores affect the movement of water across a landscape.
19. How do the vadose zone and phreatic zone differ?
20. Explain why some very porous materials have low permeability.
21. Contrast the causes of primary and secondary porosity.
22. How is Darcy's Law used?
23. Provide the formula for Darcy's Law.
24. Define hydraulic conductivity and explain how it varies between different Earth materials.
25. Explain how Horton overland flow differs from saturation overland flow.
26. Explain the variable source area concept.
27. Using the variable source area concept, explain how and why flowpaths will differ by season.

28. What are mottles and what do they tell geomorphologists about local hydrology?
29. Draw a labeled diagram of a hydrograph.
30. Predict how lag-to-peak will differ as a function of basin geomorphology.
31. Predict how the hydrograph will change as a basin is urbanized.
32. What is hysteresis?
33. What is a chemograph?
34. Why do sedigraphs exhibit hysteresis?
35. Where would you go to find a losing stream?
36. How does detention storage influence flood peaks?
37. Define bankfull flow and explain why it is important in shaping channels.
38. What is a water budget, how would you create one, and why could it be geomorphically important?
39. How do surface water and groundwater systems interact?
40. How does the flow system in a karst terrain differ from the flow system in a terrain underlain by granite rock?
41. List four common karst landforms and explain how each forms.

Hillslopes 5

Introduction

Hilltops and hillslopes, the elevated land between valley bottoms, account for much of the landscape; thus, understanding hillslope processes is central to understanding landscape evolution. In general, hillslope topography reflects the nature of the slope-forming materials, the environmental factors that govern processes on inclined surfaces, and the history of specific landscapes.

From a geomorphological perspective, the pace of soil production and sediment delivery determine the sediment supply to channels at the base of slopes and influence fluvial processes far downstream. Eventually, this sediment is transported to depositional basins, coastal plains, and continental margins. Many processes that transport material down hillslopes do so without flowing water. For example, **mass wasting** is hillslope sediment transport during which soil and rock move downslope when the gravitational stress acting on a slope exceeds the slope's ability to resist that stress (its strength). Rates of downslope transport range from the slow movement of soil displaced by burrowing animals, trees falling over, and gravitational creep to catastrophic landslides that can move a neighborhood in an afternoon or destroy a house in less than a minute.

For society, properly assessing the nature of geologic hazards and the environmental impacts of upland land use depends on understanding how our actions influence hillslope processes and the places, styles, and rates at which such processes occur. Hillslope geomorphology has practical implications because upland land use influences the

Steep, threshold slopes in the mountains of Alaska. In the center of the image, a gray rockslide partially covered with tundra vegetation descends from a ridgeline, entering the river below.

IN THIS CHAPTER

Introduction
Slope-Forming Materials
  Strength of Rock and Soil
  Effects of Weathering on Rock Strength
Diffusive Processes
  Rainsplash
  Sheetwash
  Soil Creep
Mass Movements
  Slides
  Flows
  Falls

Slope Stability
  Driving and Resisting Stresses
  Infinite-Slope Model
  Environmental and Time-Dependent Effects
Slope Morphology
  Weathering-Limited (Bedrock) Slopes
  Transport-Limited (Soil-Mantled) Slopes
  Threshold Slopes
  Hillslope Evolution
  Drainage Density
  Channel Initiation

Applications
Selected References and Further Reading
Digging Deeper: How Much Do Roots Contribute to Slope Stability?
Worked Problem
Knowledge Assessment

stability of steep hillslopes and the supply of sediment to river systems. From a biological perspective, the style of hillslope sediment transport defines the geomorphic disturbance regime; rapid catastrophic delivery of material has impacts different from slow, steady sediment movement.

This chapter discusses the properties of slope-forming materials and their influence on hillslope processes, topography, and landforms. We emphasize the key distinction between soil-mantled and bare rock slopes. The chapter also explains how erosion is related to slope steepness, climate, tectonics, and lithology. We begin with the nature of slope-forming materials before discussing controls on hillslope processes and slope form.

Slope-Forming Materials

The properties of slope-forming materials exert a profound influence on both hillslope processes and form. Slopes made of loose, unconsolidated sediment, slopes mantled by soil reinforced by plant roots, and slopes that expose bedrock each offer substantially different resistance to erosion and gravity-induced failure. Because of this, the material that makes up a slope strongly influences the processes that determine that slope's morphology and evolution [**Figure 5.1**].

Rock is made of mineral grains that are bound together by interlocking crystal structures or interstitial cement. Bedrock properties directly influence the morphology of bare rock slopes. Underlying bedrock also influences the stability, soil properties, and topography of soil-mantled slopes. **Saprolite**, deeply weathered bedrock that is still in place, is usually weaker and more prone to failure than unweathered bedrock, because the transformation of primary minerals into secondary minerals involves volume and chemical changes that reduce material strength [**Photograph 5.1**].

Bulk material properties measured on samples brought into the lab are typically used to characterize soil and rock strength, but it is widely recognized that the strength of rock masses is typically determined more by the frequency and orientation of discontinuities than by the properties of intact, unfractured rock. For example, slopes underlain by rock layers that dip parallel to the slope surface are more prone to instability and will adopt a lower gradient

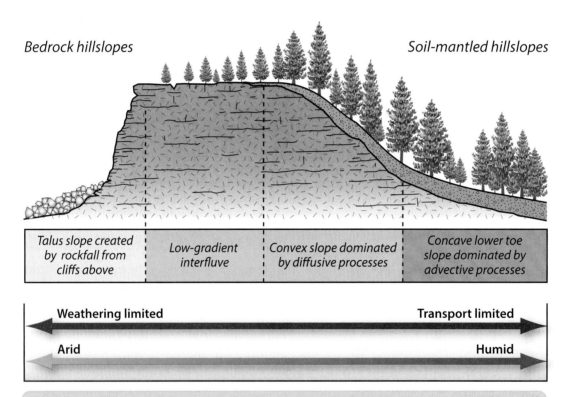

FIGURE 5.1 Hillslope Geometry and Nomenclature. Hillslopes can be categorized by whether they are mantled by soil or regolith, in which case they are known as transport limited, or whether they are characterized by extensive bedrock outcrops (weathering limited slopes). Each type of slope is shaped by a characteristic suite of surface processes.

PHOTOGRAPH 5.1 Saprolite. Weathered biotite gneiss saprolite in Clemson, South Carolina, is bright red from iron oxidation. The original foliation is preserved, although many of the original primary minerals have weathered away and been replaced by secondary minerals.

than slopes underlain by rock with fractures that dip into the slope [Photograph 5.2].

Colluvium is the unsorted, mobile or potentially mobile hillslope material that overlies a more stable substrate, including bedrock and consolidated sediments. It generally

PHOTOGRAPH 5.2 Dip Slopes. Dipping quartzite beds at Keurbooms along the southern coast of South Africa. The outcrops fail along bedding plane dip slopes when the toe of the outcrop is undercut by waves.

forms by weathering of underlying material and moves downslope as the result of processes that involve gravity-driven mass wasting, frost wedging, and animal burrowing. Even with bedrock parent material, hillslope soils tend to be colluvial because they are transported downhill by the force of gravity.

In some places, hillslopes are formed of unconsolidated sediments that were deposited by rivers (**alluvium**), glaciers (**till**), or wind (**loess**). In such cases, the material properties of the slope and surficial materials may be similar to those of a soil-mantled bedrock slope, even if little to no soil development has taken place. Surficial deposits that have been overridden, compacted, and thus strengthened by glacial ice are an important exception, because they can form extremely resistant vertical cliffs that behave more like rock outcrops than loose sediment.

Strength of Rock and Soil

The rock and soil that make up Earth's surface vary greatly in material strength and ability to resist erosion. Granite is difficult to break with a sledgehammer because the material is cohesive and has high **compressive strength** (resistance to squeezing) and high **tensile strength** (resistance to pulling apart). A loose pile of sand at the base of a slope of weathered granite can be scooped up with a spoon (see Photograph 3.12) because it has no cohesive strength. Deeply weathered saprolite in tectonically stable continental interiors or tectonically shattered bedrock in rapidly uplifting mountains may have less strength than an overlying clay-rich soil B horizon. The strength of slope-forming materials is a dominant influence on slope processes and topography.

The ability of material to resist shearing stress (the sliding of one body over another) is called its **shear strength**, a property that is quantified by three components, two of which are intrinsic material properties—the **angle of internal friction** (a measure of frictional strength commonly referred to as the **friction angle**) and **cohesion** (the tendency of material to stick to itself). The third component, called the effective **normal stress**, is the material's unit weight (the product of bulk density and gravitational acceleration) perpendicular to the slope (a force) considered per unit area of the slope less the **buoyant force** due to any interstitial water. The way that these factors together determine the shear strength of a material is described by the Coulomb equation,

$$SS = C + \sigma' \tan \phi \qquad \text{eq. 5.1}$$

where SS is shear strength, C is cohesion, σ' is effective normal stress, and ϕ is the angle of internal friction [Figure 5.2].

The frictional strength of rock and soil (ϕ) arises from the resistance to shear between mineral grains that are in contact across potential failure surfaces. Frictional strength increases in direct proportion to the normal stress holding grain surfaces in contact. The friction angle, or the angle of internal friction (ϕ), corresponds to the slope of

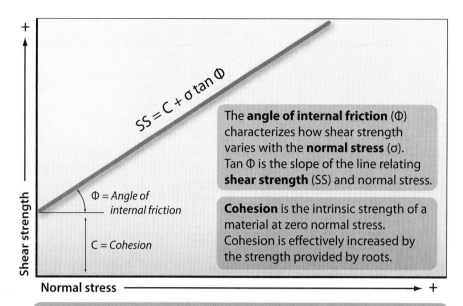

FIGURE 5.2 Shear Strength (Coulomb Criteria). The strength of earth materials has two components: a frictional component that is linearly related to the applied normal stress and a cohesive component that is unrelated to normal stress. Together, these are known as the Coulomb criteria.

the line that describes the relationship between shear strength and the confining stress (which on a hillslope is equal to the effective normal stress).

In other words, as the effective normal stress increases, the shear strength of a material increases at a rate set by the friction angle. In loose, granular materials, the friction angle is usually close to the **angle of repose,** the maximum angle at which a slope of dry, cohesionless material can stand. Rock masses and granular material like sand typically have friction angles of about 30 to 40 degrees; clay has a lower friction angle of 20 to 30 degrees. The great difference in strength between loose sand and rock is due to cohesion; in contrast to mineral grains making up a rock, grains of sand are not bonded to each other and thus have no cohesion.

Friction angles for both soil and rock generally fall in the range of 10 to 40 degrees, but cohesion values of rock are typically many orders of magnitude greater (100,000 kPa) than those of most soils (10–100 kPa, **Table 5.1**). (*Note:* 1 Pa = kilograms per meter per second squared, kg/ms^2.) Consequently, there is a profound difference in strength on many slopes where weaker soil and weathered rock lie above much stronger bedrock. Because of the disaggregation of soil and rock particles once slope failure occurs, the postfailure strength of earth materials often is less than the peak strength before failure. For most hillslope processes, it takes less force to maintain downslope transport than to initiate it. Once it is moving, slope-forming material tends to keep going until it spreads out onto a gentler slope or dissipates its kinetic energy as friction when flowing over, through, or around whatever was in its way.

Cohesion is a measure of the intrinsic strength of a material when there is no normal stress. This corresponds to a material's y-intercept value in Figure 5.2 and varies greatly among slope-forming materials (see Table 5.1). Cohesive strength arises from various types of internal bonding, including the chemical bonds and interlocking fabric between mineral grains that provide substantial strength to crystalline rocks. Electrostatic bonding between the charged surfaces of clay particles and ions in interstitial water enhances the cohesion of clay-rich soils, but these electrostatic forces are much weaker than chemical bonds in rock. Sediments compacted by the weight of now-melted glacial ice often have high cohesion.

Interstitial cements like calcium carbonate also greatly increase rock and soil cohesion. Plant roots can impart an **apparent cohesion** to soils in a manner similar to the way steel rods (rebar) contribute tensile strength to concrete. Root strength can make the difference between slope stability and failure for thin soils on steep slopes. But the apparent cohesion from roots changes over time as trees grow, mature, and die—whether from natural disturbance, succession, fire, and senescence or because of root decay after timber harvesting.

Normal stresses—those oriented into the slope—help hold soil on hillslopes. On a dry slope, the normal stress (σ) from the weight of dry rock or soil is supported by the contacts between grains. The normal stress is greater on gentle slopes than on steep slopes because it is proportional to the component of the weight of the soil ($\rho_s\, g\, z_s$, where ρ_s is soil density, g is the acceleration of gravity, and z_s is soil depth) oriented into the slope, and thus is a function of the cosine of the slope angle (θ):

$$\sigma = \rho_s\, g\, z_s\, \cos\theta \qquad \text{eq. 5.2}$$

However, if water fills the void spaces between rock or soil particles, it reduces the normal stress (σ) by an amount equal to the pore-water pressure (μ). One way to understand this effect is to consider that some of the weight of the overlying material is supported by pressurized water rather than by solid material contacts. The remaining portion of the normal stress that is supported by a rigid network of grain-to-grain contacts is called the **effective normal stress** (σ') when considered per unit area, and this is the normal stress that matters for slope stability:

$$\sigma' = \sigma - \mu \qquad \text{eq. 5.3}$$

In dry soil where $\mu = 0$, the effective normal stress is equal to the applied normal stress (i.e., $\sigma' = \sigma$). In partially saturated soils, the surface tension produced by capillary stresses can increase stability through negative pore pressures, but only to a point. Negative pore pressures are significantly reduced as a soil approaches saturation. Thus, they do not contribute much, if at all, to soil strength at the time it is needed most—in the middle of a soaking rainstorm.

Below the water table, positive pore pressures ($\mu > 0$) reduce the effective normal stress and lower the shear strength of the soil. The higher μ becomes, the greater the reduction in the effective normal stress. Thus, landslides tend to happen during and after rainstorms because even partially saturated soils are much weaker than dry soils. A slope does not, however, need to be completely saturated to fail. Slopes fail when they become saturated enough that material strength is less than the shearing stress, a condition that requires lower pore pressures (μ) on steeper slopes, as we will see below.

Effects of Weathering on Rock Strength

Weathering lowers rock strength over time through physical and chemical alteration of rock properties and by changing slope hydrology. The cohesion of weathered rock, soil, and unconsolidated sediment is generally much lower than that of intact rock, and the development of

TABLE 5.1

Typical strength of Earth materials

Material	Friction angle (degrees)	Cohesion (kPa*)
Soil		
Sandy soil	30–40	0
Soft organic clay	22–27	5–20
Stiff glacial clay	30–32	70–150
Rock		
Intact sandstone (lab)	35–45	>10,000
Intact shale (lab)	25–35	>1,000
Sandstone (field)	17–21	120–150
Shale (field)	15–25	40–100

*1 Pa = 1 kg/ms^2; 1 kPa = 1000 Pa
lab = laboratory data on small samples
field = data collected from field measurements

zones of weakness as weathering proceeds also greatly reduces rock strength. As fresh rock weathers to become soil, porosity and permeability increase, sometimes by orders of magnitude. Such changes proceed preferentially along fissures and fracture zones and produce patterns of variable weathering intensity within the rock (see Figure 3.13).

Many slope failures involve sliding of the surficial soil and sediment mantle over underlying bedrock, because unweathered or lightly weathered rock has much higher cohesion and is much stronger than soil. In landscapes with intense chemical weathering, as is common in humid-tropical regions, a zone of pervasively weathered, virtually cohesionless saprolite that extends deep beneath the soil may slide off a slope as pore pressures rise during storms. Such saprolite may erode rapidly if gullied by surface water drainage or seepage pressures generated by high groundwater tables. The **gullies** of Madagascar are prime examples of hillslope erosion in heavily weathered, saprolitized bedrock [Photograph 5.3].

Calculations based on strength values measured in hand specimens of rock suggest that some rock types should be capable of supporting vertical cliffs far taller than any that exist in nature. Why the discrepancy? Over time, weathering and tectonic stresses reduce the strength of even the most resistant rocks at or near Earth's surface and produce the deep-seated bedrock landslides that are common in pervasively fractured, tectonically active upland landscapes. Laboratory measurements of shear strength use small intact samples of rock and thus generally overestimate rock strength because lab work cannot take into account the localized but critical zones of weakness, such as fractures and bedding surfaces that control the actual strength of slopes. Because of the tremendous range of slope-forming materials and their relationship to topography, methods for characterizing rock strength have been developed based on classification systems that describe the qualities of rock outcrops in the field, including the orientation and density of bedding planes, faults, and fractures.

The incision of valleys by flowing water and ice concentrates compressional stresses in valley bottoms and causes extensional stresses along ridgetops and valley walls that help break up intact bedrock. Mechanical unloading at the ground surface commonly results in a zone of fractured rock that extends to some depth beneath the land surface. This creates planes of weakness parallel to the ground surface that promote landslides and accelerate rock weathering by focusing groundwater flow. Chemical weathering by groundwater flow along such fractures further reduces bedrock strength by destroying or weakening the bonds between mineral grains and smoothing asperities (small protrusions) that roughen joint surfaces.

Diffusive Processes

Sediment-transport processes on hillslopes include **diffusion-like processes** (generally called **diffusive processes**) in which the transport rate is proportional to hillslope gradient. Rainsplash, sheetwash, and soil creep are diffusion-like processes that reduce relief and fill in depressions. Diffusive hillslope processes involve sediment movement without entrainment by concentrated flow of water, wind, or ice and are distinguished by basic differences in transport style.

Rainsplash

Raindrops strike bare ground with substantial force during intense rainfall events and the resulting impacts can move a significant amount of sediment by **rainsplash** [Photograph 5.4]. Such splash is geomorphically important in regions with scant vegetation cover (such as deserts and disturbed landscapes), because it loosens sediment from soil surfaces, allowing transport by moving water. The thick vegetation cover of humid temperate regions shields many soil surfaces from rainsplash impacts.

On bare sloping ground, rainsplash produces net downslope sediment transport. To understand why this is the case, consider the trajectory of particles ejected from a sloping surface. If the raindrop falls vertically, the same number of particles are directed uphill as downhill; however, those headed uphill run into the slope at a shorter distance than those headed downhill, because the slope, like a ski jump, drops out from beneath the particles transported downslope.

If rainfall rates exceed infiltration rates, as in many desert thunderstorms, then water will begin to flow over the land surface. Raindrops can penetrate shallow overland flow and kick sediment up into suspension. However, as surface water flow deepens, less sediment is entrained because the water shields the soil surface from raindrop impacts. Consequently, the contribution of rainsplash to

PHOTOGRAPH 5.3 **Madagascar Gully.** This large gully near Amparafaravola, Madagascar, is 155 m wide and almost 40 m deep. These features are locally known as lavaka, the Malagasy word for "hole." The gully has a large flat floor and an external debris apron that is partly vegetated. Its headwall has cut back through the ridge crest.

PHOTOGRAPH 5.4 Rainsplash. Rainsplash impacting shallow flow dislodges both water and sediment.

hillslope sediment transport is largely limited to thinly vegetated areas relatively close to drainage divides, where overland flow is absent or shallow.

Sheetwash

Overland flow that is not concentrated into discrete channels and spreads across the ground surface is called **sheetwash** [Photograph 5.5]. Overland flow is rare on heavily vegetated soil-mantled slopes because the surficial organic layer and porosity caused by root cavities ensure that infiltration capacity generally exceeds rainfall rates and thus that rainwater sinks into the ground. Consequently, sheetwash does not transport much sediment in humid or temperate environments, except where soil has been disturbed or compacted, for example, on construction sites and walking trails (see Photograph 4.3).

Unchannelized sheetwash does, however, transport significant amounts of material down undissected slopes in the arid and semi-arid environments that make up large portions of continental land masses. Sheetwash processes are instrumental in shaping the morphology of hillslopes in these areas, moving mass downslope from slowly eroding hillslopes to depositional basins. Experiments using painted pebbles over the short term and cosmogenic nuclides over the long term showed that sheetwash moved sediment over gently sloping Mojave Desert slopes at rates of at most a few tens of centimeters per year. Sheetwash, while pervasive in these regions, does not change landforms rapidly.

Rainsplash and sheetwash are both diffusion-like processes that fill in topographic depressions and smooth over relief. Where enough overland flow concentrates to incise the ground surface, channels form and the flow then acts to incise and enhance relief at least temporarily. Because sheetwash events can be few and far between in arid regions, other surface processes, such as wind erosion and **bioturbation** (soil-stirring by animals), may erase small channels before the next flow event.

Soil Creep

Soil creep is the incremental downslope movement of soil and sediment. Soil creep processes are too slow to see without instrumental measurement or other indicators of long-term movement. The term incorporates a wide range of processes that include seasonal heave from ice or expanding clays, downslope movement of soil from the burrowing activity of animals, and displacement of material downhill by uprooted trees [Photograph 5.6]. In cold regions with permafrost, seasonal thawing of the surficial layers can result in **solifluction**, which involves creep of weak, saturated soils over stronger, impermeable, frozen ground. Seasonally frozen ground also experiences **heave,** cyclic

PHOTOGRAPH 5.5 Sheetwash. Sheetwash covers a gentle slope on the coastal plain outside Okambahe, Namibia, during a very heavy thunderstorm. Flow resulted from a high-intensity, short-duration (<1 hour) summer (February) rainfall event. Runoff lasted for about an hour.

PHOTOGRAPH 5.6 Tree-throw. Tree-throw can move large amounts of material downslope. Here, in the Nahanni National Park in Canada's Northwest Territories, a tipped up rootwad moves the shallow layer of soil downslope. (Lens cap on the top of the rootwad provides scale.)

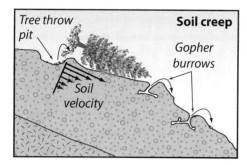
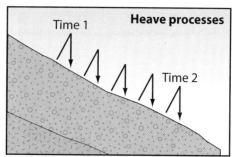

Soil creep describes the suite of processes that move soil and regolith downslope at a velocity proportionate to the to slope angle. Processes contributing to soil creep include tree-throw, animal burrowing, and deformation of fine-grained soil.

Heave contributes to soil creep. Heaving soil rises up perpendicular to the slope through the wetting and expansion of clays or the freezing of interstitial water. When the soil thaws or dries and shrinks, the material drops vertically under the influence of gravity, causing net downslope movement of soil.

FIGURE 5.3 Soil creep. Soil creep is the movement of material downslope at a rate proportional to slope. Heave is one creep process.

expansion and contraction of material that produces net downslope transport because of a differential bias in the direction of movement. On sloping surfaces, soils expand perpendicular to the ground surface, but gravity-induced contraction occurs vertically, resulting in a net downslope movement with each freeze–thaw cycle (or season) [Figure 5.3].

Downslope creep rates are variable and depend on slope angle, climate, soil moisture content, and particle size, but they rarely exceed a few millimeters per year. Creep rates typically decrease with depth below the ground surface, and most displacement happens within about a half meter of the surface. Evidence of active soil creep (specifically, the higher creep velocity at the surface than at depth) includes such indicators of net downslope movement as tilted fence posts, curved lower tree trunks (known as pistol-butt trees, in honor of their resemblance to antique dueling pistols), cracked building foundations, and accumulation of soil on the upslope side of fixed obstructions [Photograph 5.7]. Slow, gravity-driven deformation of rocky slopes, from repeated freeze–thaw or wetting-and-drying-induced expansion and contraction of near-surface regolith, can lead to downslope bending of fractured and weakened near-surface bedrock [Photograph 5.8].

PHOTOGRAPH 5.7 • Pistol-Butt Trees. Bent tree trunks resulting from soil creep in Nevada.

PHOTOGRAPH 5.8 Creep. Creep moves fractured rock downslope near Marathon, Texas. Creep is faster nearer the surface where the rock is less competent.

Mass Movements

The material displaced and the style and rate of deformation distinguish different types of **mass movements** [Figure 5.4]. In general, mass movements involve translation of partially to fully saturated soil and/or rock along a well-defined **failure surface**, or **shear plane**. Generally, the failure surface is either approximately parallel to the land surface (**planar landslides**) or extends to some depth as a concave surface along which rotational slippage or slumping occurs (**rotational landslides**). Factors that influence the occurrence of mass movements include ground-shaking, intense or long-duration rainfall, undercutting that removes material buttressing the toe of a slope, and the progressive weathering of hillslope materials.

There are three general types of mass movements: **slides, flows,** and **falls**. These processes can act alone or in combination to form landslide complexes. Mass movements are usually described using a prefix that identifies whether the material involved is rock, coarse soil, and sediment (debris) or fine-grained material (earth). Different types of mass movements can be either wet or dry and slow or fast. Complex mass movements consist of several failure styles.

Shallow mass movements generally involve surficial materials like soil and saprolite. Because there is typically a large strength discontinuity between soil and the underlying bedrock, shallow planar landslides often detach and slide along the soil-bedrock contact. Shallow planar slides are common on steep, soil-mantled slopes, such as the wet, tectonically active Oregon Coast Range.

Deep-seated bedrock failures may involve fresh bedrock as well as weathered surficial material. Bedrock landslides typically involve either failure along a discrete plane of weakness—like a bedding plane or a fault—or slippage along a rotational failure surface. Rockslides, for example, are usually associated with structures like faults, fractures, or joint sets [**Photograph 5.9**]. The orientation of bedding planes can also enhance or decrease slope stability. Bedding that dips back into a slope promotes slope stability; beds that dip in the same direction as the slope promote instability. Hillslopes where underlying strata have an inclination close to parallel with that of the topographic surface are especially prone to sliding (see Photograph 5.2).

Mass movements may be active or dormant, and some can be readily reactivated because the material within landslides and along failure planes generally loses strength after it fails. Recently active landslides can be distinguished from older, inactive landslides based on a variety of criteria. Sharply defined scarps, tilted trees, **sag ponds** (closed depressions that are unconnected to external

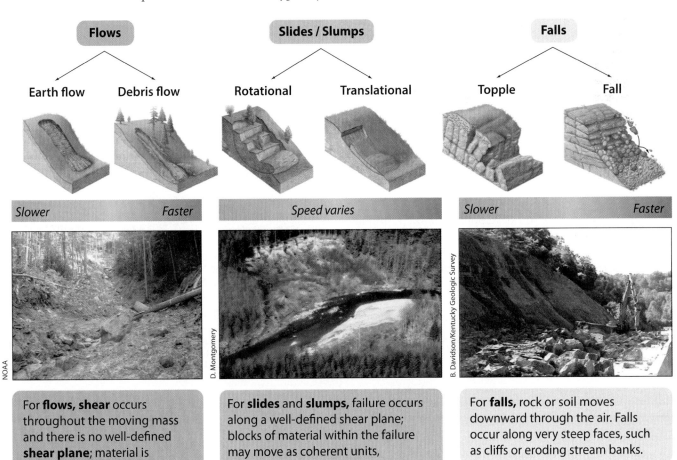

For **flows**, **shear** occurs throughout the moving mass and there is no well-defined **shear plane**; material is disrupted throughout the flow.

For **slides** and **slumps**, failure occurs along a well-defined shear plane; blocks of material within the failure may move as coherent units, preserving relict structures.

For **falls,** rock or soil moves downward through the air. Falls occur along very steep faces, such as cliffs or eroding stream banks.

FIGURE 5.4 Taxonomy of Mass Movements. Mass movements are typically classified by both their shape and by the speed at which they move.

PHOTOGRAPH 5.9 Rockslides. (a) Rockslide on Elephant Rock, Yosemite National Park. The debris spread down the slope, knocking down trees, and the dust cloud covered a wide area. (b) The large boulder failed along the bedding plane in a dip slope rockslide in Death Valley National Park, California.

drainage), and lack of soil development on depositional areas all may indicate recent activity [Photograph 5.10a]. Rounded scarps, unaffected trees, well-integrated drainage with few ponds, and soil development on depositional areas generally indicate passage of significant time since landslide activity [Photograph 5.10b]. Such features can help determine the relative age of landslides, but the amount of time it takes for a landslide to appear inactive depends on a number of local and regional factors, like lithology and climate, that complicate inferences of landslide age.

Slides

Slides involve downslope movement of cohesive blocks of soil or rock along a relatively thin and well-defined **shear plane**, a zone of intense **shear strain** (deformation). There is little internal shearing within the sliding block or within discrete blocks that move together in the slide. Resistance to movement drops after initial failure, as material weakens during downslope transport and deformation; thus, movement generally continues until the sliding block(s) encounters sufficient resistance to halt further movement, usually because of decreased slope. Slides may have planar or rotational failure surfaces and may be fast or slow moving once initiated. Some slides become flows after initial failure.

Translational slides are typical of many small landslides with shallow planar failure surfaces on which the failed material moves. During shallow, planar slides, coherent slabs of soil move downhill over more solid rock or consolidated soil. Such slides typically occur on planar slopes below ridgelines and in concavities where groundwater flow converges and raises pore pressure, reduces normal stress, and thus lowers the frictional strength of the material. The margins of shallow planar landslides are defined by **scarps,** vertical faces along the top and edges of

PHOTOGRAPH 5.10 Active and Inactive Landslides. Recently active and older inactive landslides can easily be distinguished. (a) An active shallow planar landslide, near Florence, Oregon, has recently disrupted a steep forested slope. The slide was likely triggered by wave erosion at the toe of the slide. (b) Inactive, deep-seated bedrock landslide with well-defined scarp in the Bolivian Andes. Landslide is currently stable enough that it is crossed by roads and covered by agricultural fields.

the detached slide block. Downslope of their **initiation zone,** translational slides can move out over the original ground surface. Translational slides may be soil slides in which blocks of soil detach but retain some internal structure while moving as coherent slabs [**Photograph 5.11a**], or they can be rock slides in which rock slabs detach along planar failure surfaces, such as bedding planes. Translational slides may be rapid or slow, depending on their moisture content and the nature of the failed material. Translational slides can involve whole mountainsides sliding along bedding planes, as occurred in the Frank Slide in Alberta, Canada, and the Love Creek Slide near Santa Cruz, California [**Photograph 5.11b**].

(a)

(b)

PHOTOGRAPH 5.11 Translational Slides. (a) Shallow planar failure in disturbed glacial sediment that occurred during a wet spring, the year after the slope was graded as part of construction of a new highway near Essex, Vermont. (b) The Frank rockslide in Alberta, Canada, failed early on the morning of April 29, 1903, burying almost 100 people beneath about 30 million cubic meters of limestone rock debris.

PHOTOGRAPH 5.12 Rotational Landslide. A magnitude 9.2 subduction earthquake on March 27, 1964, in Anchorage, Alaska, triggered a rotational landslide. Note the large upper scarp, the back-tilting of blocks, and the bulging toe of the slide. The scar of an older landslide is offset by the recent slide.

Rotational landslides, commonly called **slumps,** involve the movement of soil or rock along a curved, concave failure surface [**Photograph 5.12**]. They typically exhibit a pronounced head scarp, as well as secondary scarps at the heads of back-tilted, rotated blocks within the slide mass and an elevated bulge where material accumulated at the toe of the failure [**Figure 5.5**]. Rotational slides may consist of either a single slump or have multiple nested failure surfaces that create multiple scarps and merge at depth within a single larger complex feature. Rotational slides are common in thick, cohesive, relatively homogenous deposits such as glacial lake clays. They can move rapidly (in seconds) or slowly (over days) and are frequently reactivated, particularly if a stream erodes the toe of the slide.

Spreads are landslides that involve extension of cohesive or hard rock masses due to lateral movement of softer, weaker underlying material. Development of extensional fractures in the overlying rock mass accompanies general subsidence of the fractured cohesive material into the softer underlying material. The dominant mode of deformation in spreads is lateral extension that can break the surficial material into a maze of high-standing **horsts** and downdropped **grabens.** The intricate maze of grabens in Canyonlands National Park in southern Utah represents large-scale lateral spreading of the brittle capping rock (or **caprock**) above underlying, readily deformed, viscous salt that was exposed and remobilized when downcutting of

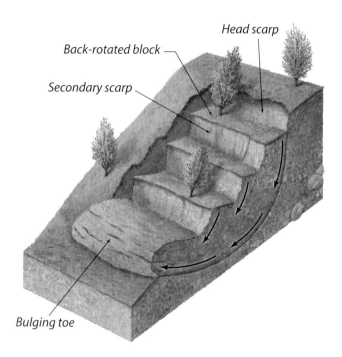

Rotational landslides generate characteristic landforms that are indicative of the underlying physical processes. Near the **head** of the slide there are often multiple back-rotated blocks, each bordered by **scarps** and each having a back-tilted top. At the **toe** of the slide, material may pile up, increasing ground elevation.

FIGURE 5.5 Rotational Slump. Rotational slumps have specific features that make them easy to identify in the field, including back-rotated blocks, multiple scarps, and bulging toes.

the Colorado River removed the lateral confining support during the past half million years.

Formations of fine-grained sediment deposited in seawater and now elevated above sea level, such as glacial marine clay, are susceptible to spreading for two reasons. After hundreds to thousands of years of subaerial exposure and percolation of freshwater from rain and melting snow, many of the ions that helped to hold the loose clay together have been washed away. If shaken strongly by an earthquake, these already weakened clays collapse, increasing interstitial pore-water pressures and allowing them to flow. A prime example is the spreading that affected the Turnagain Heights neighborhood of Anchorage, Alaska, during the 1964 earthquake.

Flows

Flows move by differential shearing within the sliding material and have no well-defined internal shear planes. Mass flows resemble the flow of viscous fluids in which the maximum shearing occurs at the base and flow velocity decreases with increasing depth from the top of the flow. Most mass flows involve some amount of water, but large rock slides and falls sometimes transform into dry flows that run out long distances at the base of slopes. In such massive, high-energy flows, fluidity can be maintained by grain-to-grain collisions that promote efficient energy transfer. Such dry fluidization is invoked as the explanation for large flows on Mars, where liquid water is currently absent. Flows may move slowly (cm/day) to quite rapidly (m/second) and may involve soil or rock.

A **debris flow** is a rapid movement of saturated material, often channelized, down a steep slope. Debris flows are a slurry of soil, rock, and water that can travel far downslope from the point of failure initiation [**Photograph 5.13**]. Too much water, and the flow will separate into two phases, water and sediment, with the coarser component of the sediment settling to the bed. Too little water, and the flow will be too strong to move downslope. Fine-grained debris flows are often called mudflows. Some geomorphologists use the term **debris avalanche** to describe a very rapid flow of partially or fully saturated debris down a steep slope without confinement in an established channel or valley. These rapidly moving slope failures typically originate on steep bedrock slopes with thin soil cover and are thus common in formerly glaciated, alpine terrain where melting ice left debris on steep slopes.

Initiation of a debris flow generally requires at least a small landslide; thus, abundant groundwater, a steep slope,

PHOTOGRAPH 5.13 Debris Flow. The San Francisco earthquake of April 18, 1906, after heavy winter rains raised soil moisture levels, triggered this debris flow in Marin County, California.

and soil that is susceptible to shear failure (landsliding) are necessary ingredients. Soil properties greatly influence the potential for landslides to mobilize into debris flows because shearing or shaking within loose, sandy soil results in consolidation of the soil matrix that raises the pore pressure of interstitial fluids, triggers fluidization, and enhances mobility. In contrast, deformation of **dilative materials** that expand when sheared, like nonmarine clays, reduces pore pressures by expelling fluid, thereby limiting mobility because such deformation promotes drainage. Debris flows thus tend to form most frequently from landslides that occur in sandy soils with just enough fine material (5 to 20 percent) to retard drainage of interstitial fluids.

As they move downslope, debris flows generally entrain surficial materials like soil, unconsolidated sediment, and saprolite; they are a major sediment-transport process in mountainous landscapes. They typically begin on slopes between 26 and 45 degrees and have flowpaths that consist of a source area defined by the zone of initial slope failure, a scour and transport zone, and a depositional zone [**Figure 5.6**]. Debris flows can grow to more than 100 times the initial failure volume by scouring material from the base and edges of their runout paths—in many cases, steep headwater channels. Debris flows slow down and can then form depositional fans once they reach slopes of about 3 to 6 degrees.

Debris flows typically have a coarse, boulder-rich snout and sometimes deposit substantial levees along their runout path. Flowing material deforms rapidly and then deposits material abruptly when the driving stress (**shear stress**) falls below the yield strength of the flow. This abrupt change in deformation rate with applied stress is referred to as **plastic** behavior. Perfectly plastic materials are considered to have a finite yield strength. At shear stresses below the yield strength, they do not deform. If shear stress exceeds yield strength, plastic materials deform rapidly. A debris flow stops moving once it thins sufficiently or encounters a slope low enough that the yield strength exceeds the driving stress. The main bodies and tails of debris flows are often more fluid slurries than the coarse fronts, and debris flows commonly exhibit pulses because of variations in the fluid content within the flowing mass.

Debris-flow deposits are **matrix-supported**; clasts float in a finer-grain matrix and rock fragments are isolated from each other [**Photograph 5.14**]. A good analogy is chocolate chip cookie dough. Debris-flow deposits are readily distinguishable from those laid down by flowing water because fluvial deposits are **clast-supported**; individual particles rest in contact with one another, as would a pile of beans. Debris flows often come to rest on fans at the bottom of steep slopes where gradients are lower [**Photograph 5.15**].

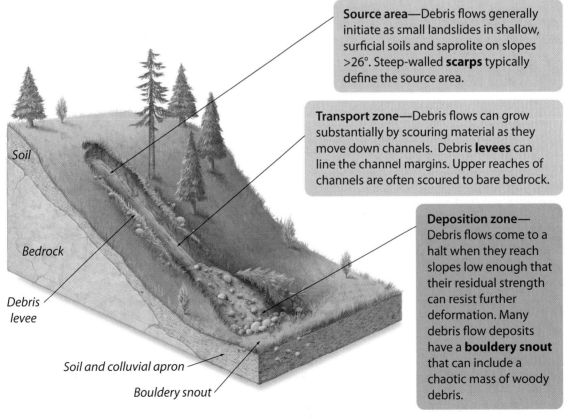

Source area—Debris flows generally initiate as small landslides in shallow, surficial soils and saprolite on slopes >26°. Steep-walled **scarps** typically define the source area.

Transport zone—Debris flows can grow substantially by scouring material as they move down channels. Debris **levees** can line the channel margins. Upper reaches of channels are often scoured to bare bedrock.

Deposition zone—Debris flows come to a halt when they reach slopes low enough that their residual strength can resist further deformation. Many debris flow deposits have a **bouldery snout** that can include a chaotic mass of woody debris.

Soil

Bedrock

Debris levee

Soil and colluvial apron

Bouldery snout

FIGURE 5.6 Debris-Flow Landforms. Debris flows leave characteristic clues on the landscape, including a source area (typically a landslide scar), a transport zone (an incised channel sometimes surrounded by levees), and a deposition zone that is often in the form of a fan.

PHOTOGRAPH 5.14 Debris-Flow Deposit. Bouldery debris flow deposit in the Jiangjia River Valley of China, showing the matrix-supported nature of material deposited by the flow. The large boulder at the top of the photo was carried by the flow.

PHOTOGRAPH 5.15 Debris Fan. This debris fan of unconsolidated granitic sediment resulted from a debris flow that swept down a narrow canyon in the area denuded by a 2008 fire in Sequoia National Forest, California.

Earthflows are a type of landslide common in clay-rich earth materials; they are often deep-seated, and many originate as slumps involving masses of material that, unlike debris flows, move as rigid blocks within the failed mass. Although earthflows typically occur on slopes of 5 to 25 degrees, they can start on a variety of slope angles and exhibit highly variable mobility and speed, depending on the nature of the material involved (i.e., its unit weight and shear strength), the pore pressure, and geometry of the failure. Earthflows often consist of a complex of multiple flow lobes that may move at different speeds and at different times. They can move over periods of months to years and can episodically show distinct pulses of movement, with individual zones within a larger earthflow reactivating or moving at different paces [Photograph 5.16]. Excavation of material from their toes, ground shaking, or excessive rainfall can reactivate earthflows and accelerate their movement.

Rock flows involve deformation distributed among many through-going fractures within a rock mass. They may be extremely slow, deforming over long time spans, or quite rapid, as in the case of **rock avalanches**—massive, extremely rapid flows of fragmented rock that typically originate from a large rock slide or rock fall. Also commonly termed debris avalanches, these slope failures may run out long distances. One of the best-known examples of such a rock avalanche is the Blackhawk landslide in the Mojave Desert of southern California, which dropped more than a kilometer in elevation over its 9-kilometer runout [Photograph 5.17].

Falls

Falls begin with the detachment of soil or rock. They involve the downward motion of rock or soil through the air by falling, bouncing, or rolling, with little to no initial interaction with other materials. Falls typically occur on slopes that are steeper than the internal friction angle of the slope-forming material. Thus, falls are the most common mass movements on slopes that range from 45 to 90 degrees; in contrast, slides and flows are most common on slopes gentler than the friction angle (i.e., those < 45 degrees).

Falls tend to be rapid (m/s). Rock and soil falls typically occur when material is dislodged from very steep faces, like cliffs or streambanks. Rock falls are common on steep slopes in arid and alpine landscapes and are often triggered by earthquakes. Soil falls typically occur in places where flowing water or progressive slope failure undercuts cohesive soil or sediment, such as at gully heads or along incised streambanks. Slopes formed of tall blocks are especially prone to toppling, as the forward rotation of a soil or rock mass pivots about a point below its center of mass. **Topples** can range in speed from extremely slow (mm/yr) to extremely rapid (m/s).

PHOTOGRAPH 5.16 **Slump and Earthflow.** The scarp and bulging toe are clearly visible in this central California slump and earthflow.

PHOTOGRAPH 5.17 **The Blackhawk Landslide.** This landslide in a Mojave Desert valley originated as a rock avalanche from the distant ridge and spread pulverized rock over the valley. The road in the foreground provides scale.

Slope Stability

Slope stability is analyzed by considering the balance between shear stress and the strength of earth materials. An examination of the ratio of driving forces to resisting forces per unit area reveals the relationships among the factors that resist and those that promote slope instability.

Driving and Resisting Stresses

The shear stress generated by the weight of the soil and rock overlying a unit area of a potential failure plane drives slope instability. Shear stress is defined as the downslope component (force per unit area) of the weight of the slope-forming material. If we consider the stresses acting on a hillslope, the shear stress (τ) acting on a planar surface covered by unconsolidated soil is given by

$$\tau = \rho_s \, g \, z_s \sin\theta \qquad \text{eq. 5.4}$$

where ρ_s is the average density of the soil, g is gravitational acceleration, z_s is the soil thickness measured perpendicular to (into) the slope, and θ is the slope angle [**Figure 5.7**]. In order to simplify the development of the infinite-slope model, we do not explicitly consider the changing density of the soil as it saturates with water during storms.

The shear strength of slope-forming materials (as expressed in eq. 5.1) resists the shear stress acting to move material downhill. Factors that decrease shear strength include weathering processes that weaken earth materials and increase pore pressure (μ), which increases buoyancy and thus acts to reduce the effective normal

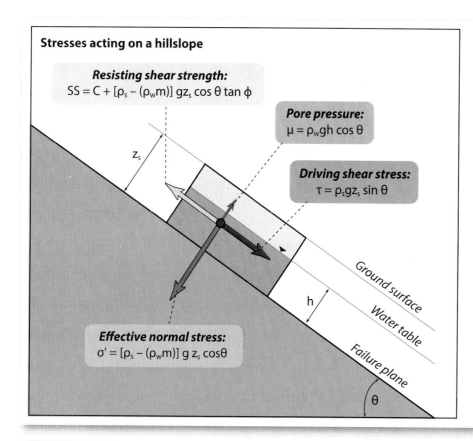

FIGURE 5.7 Infinite-Slope Model. The stability of planar landslides is commonly estimated using a force-balance approach (the infinite-slope model) to determine the factor of safety, the ratio between the resisting and driving forces per unit area. The analysis considers soil density, soil saturation, slide-block thickness, and the slope of the failure plane.

stress ($\sigma' = \sigma - \mu$). The hydrostatic pore pressure is expressed in terms of the height of the water table above the slide plane (h):

$$\mu = \rho_w \, g \, h \, \cos\theta \qquad \text{eq. 5.5}$$

It is worth noting that if we cast this equation (and subsequent equations in this section) in terms of vertical soil thickness, we would introduce an additional $\cos\theta$ term on the right-hand side of eq. 5.4 and 5.5 (and thus on the trigonometric terms in the numerator and denominator of eq. 5.8). Substituting eq. 5.2 and 5.5 into eq. 5.3 yields the effective normal stress for a partially saturated slope:

$$\sigma' = (\rho_s \, g \, z_s \, \cos\theta) - (\rho_w \, g \, h \, \cos\theta) \qquad \text{eq. 5.6}$$

After collecting and simplifying terms this reduces to

$$\sigma' = [\rho_s - (\rho_w \, m)] \, g \, z_s \, \cos\theta \qquad \text{eq. 5.7}$$

where m is the ratio of the saturated soil thickness (h) to the total soil depth (z_s) above the slide plane that is saturated ($m = h/z_s$). Expressing the relative soil saturation as the ratio m allows for consideration of the effect of variations in moisture on the potential for slope failure. If the soil is fully saturated, and the water table is at the land surface, m = 1. If the water table is below the potential failure surface, or if the soil is dry, then m = 0. Slope stability models typically portray relative soil saturation as a single value, but direct measurements of pore-water pressures during landslide-inducing storms show that variations in soil properties, the presence of macropores, and groundwater recharge and discharge (springs) at the base of the soil profile impart significant spatial variability to soil moisture.

Infinite-Slope Model

The simplest physical model of slope stability is the **infinite-slope model**. It considers the balance of forces at a point on a potential failure plane that is assumed to have a constant slope (θ) and extend infinitely at its margins to avoid edge effects. The infinite-slope model effectively reduces the complex three-dimensional problem of slope failure to a two-dimensional problem by ignoring the effects of lateral slide boundaries or along-slope variations in soil depth and slope angle. This formulation greatly simplifies the analysis of the factors that contribute to slope instability, while still allowing for meaningful quantitative assessment of slope properties. In particular, the infinite-slope model allows stability to be assessed for a unit area of the slope as representative of all the unit areas around it.

At the core of the infinite-slope model is the **factor of safety** (FS), which is the ratio of resisting shear strength to the driving shear stress (FS = SS/τ), and it equals 1 at slope failure. Stable slopes have FS > 1, and unstable or failed slopes have (or had) FS ≤ 1. The infinite-slope model is derived by substituting the definitions of the shear strength (SS = C + σ' tanϕ: eq. 5.1) and shear stress (τ = ρ_s g z_s sinθ: eq. 5.4) into the definition of the factor of safety (FS = SS/τ). From there, we further substitute the expression for the effective normal stress (eq. 5.7) in the numerator and arrive at:

$$FS = \frac{C + [\rho_s - (\rho_w m)]\, g\, z_s \cos\theta \tan\phi}{\rho_s\, g\, z_s \sin\theta} \quad \text{eq. 5.8}$$

This equation illustrates the basic controls on slope stability and shows all the key factors involved in slope stability: the material properties (ρ_s, ρ_w, C, and ϕ), the geomorphic properties of soil depth and ground slope (z_s and θ), and the time-varying environmental factor, that is, the proportion of the soil thickness that is saturated (m), or the pore-water pressure (μ; see eq. 5.5), and gravitational acceleration (g), which is important in the case of earthquake-induced landslides.

Not only is the infinite-slope model useful for predicting slope failures, it can also be used to understand landscapes. Given the characteristic strength parameters (C and ϕ) for the earth materials underlying a region and assuming groundwater levels, one can predict the maximum stable angle of hillslopes. In the simplest example, the maximum stable slope angle (FS = 1.0) for dry cohesionless material like sand (C = 0 and m = 0) equals the friction angle of the slope-forming material (tanθ = tanϕ). As cohesion values increase, so will the steepness of slopes the material can support. Conversely, the wetter the slopes (higher water table), the less steep the slopes will be.

Rotational failures can be modeled using a simple adaptation of the infinite-slope approach. In the **method of slices**, the failure surface is defined as a series of chords that collectively approximate the arc of a circle. The analysis then divides the circular arc of the failure surface into a series of slices, each a segment of the arc and each having an average vertical soil thickness. Individual stability analyses for each slice can then be summed to assess the overall stability of the slope based on the geometry and material properties (ρ_s, C, and ϕ) of the slope [**Figure 5.8**]. For rotational slides, this method improves on the infinite-slope model because it includes the geometry of the toe of the slide, where the rotation surface curves upward and acts to resist stresses generated upslope.

Environmental and Time-Dependent Effects

Slope failure is a good example of a geomorphic threshold. When the material on a formerly stable slope slides downhill, it means that something changed to upset the preexisting relationship between driving stress and the resisting material strength. A variety of environmental factors that change over short and long timescales influence slope stability. In the short term, slope failure can be triggered by response to rainfall (change in pore pressure, μ) and seismically driven forcing (change in g, gravitational acceleration). Over decadal time frames, vegetation succession following disturbance (and the associated decline in apparent cohesion from root reinforcement) influences slope stability. Over millennial timescales, the evolution of soil thickness and slope morphology change normal and shear stresses and thus the stability of slopes.

Landslides on potentially unstable slopes can be triggered by earthquake shaking. Ground acceleration during earthquakes destabilizes slope material by changing the effective value of g. Both the intensity of ground shaking and the horizontal and vertical acceleration depend strongly on material properties of the soil. The greatest number of slides typically occurs near the earthquake epicenter and in weak materials (such as alluvium, marsh deposits, or artificial fill) where the intensity of ground shaking is greatest.

Studies of the rainfall conditions associated with slope failures have shown that shallow failures in hillside soils tend to occur once rainfall exceeds a quantifiable intensity-duration threshold that varies by region (see Figure 4.1). During storms, pore pressures increase as soil below the hillside becomes saturated, thereby reducing the effective normal stress and decreasing the shear strength. More intense or longer-duration rainfall translates into a greater saturated thickness and therefore greater pore pressures and less stability. In other words, slides tend to occur once it rains hard enough for long enough to sufficiently saturate hillslopes. Deep-seated landslides may move long after rainfall ceases, or in response to seasonal changes in the water table because of the time required for water to infiltrate to and thus raise the water table.

The apparent cohesion that plant roots contribute to soils differs greatly among species and changes as vegetation grows, matures, and dies [**Table 5.2**]. Consequently, plant succession or a change in community structure, like conversion of forest to grassland, can cause substantial changes in slope stability. Grassland soils are also stabilized by root reinforcement, and the stability of grass-covered hillslopes has been negatively impacted by the replacement of deep-rooted native grasses with invasive shallow-rooted grasses (as happened both in California and in the Great Plains of North America).

Among trees, conifers generally contribute the most apparent cohesion to soils. Following forest fires or timber harvest, roots die and it takes about a decade for them to rot completely and lose all strength. Thus, there is usually a period of at least several years during which root strength

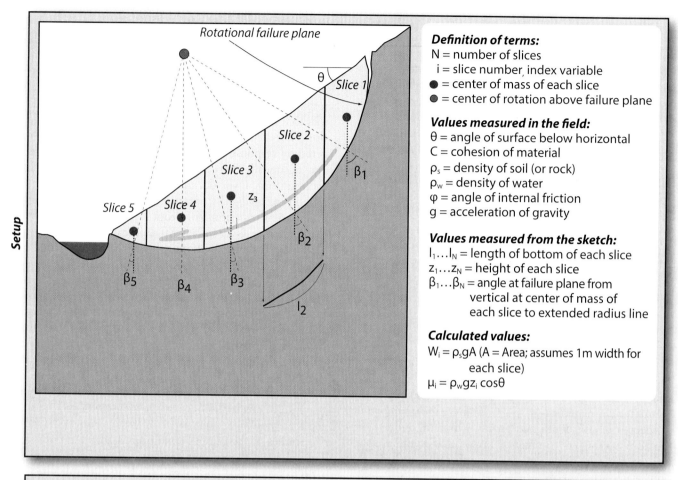

FIGURE 5.8 Method of Slices. The stability of rotational failures is modeled using a permutation of the infinite-slope model, referred to as the method of slices. This approach considers the slide surface as a series of slices.

remains low as root networks of newly planted or established trees regrow. During this window of low root strength, potentially unstable slopes are particularly vulnerable to slope failure (see Figure DD5.4). If slopes are kept clear, as they were in New England during most of the 1800s, they remain susceptible to failure during soaking rain and spring snowmelt. During periods of low root strength, more landslides occur in response to rainfall events than would have prior to vegetation clearing, or under mature forest.

In addition to altering or clearing vegetation, human land use affects slope stability by changing hydrologic pathways, such as when urbanization or road construction concentrates and delivers runoff to steep slopes [**Photograph 5.18**]. The effect is even more dramatic when rainwater runoff is routed to topographic concavities, known as **hollows**, where colluvial soil accumulation and both surface and shallow subsurface runoff are focused (see Photograph 7.9). Shallow, planar failures often start in hillslope hollows

TABLE 5.2

Typical values of apparent cohesion associated with vegetation

Vegetation	Apparent cohesion (kPa*)
Grass	0–1
Stumps	0–2
Chaparral	0–3
Hardwoods	2–13
Conifers	3–20

*kPa = kg/ms^2

because these parts of the landscape favor both the accumulation of thicker soils and greater soil saturation.

Although landslides tend to be triggered by destabilizing events, the potential for slope failure changes gradually over centuries to millennia as the hydrology and strength properties of a slope change. Once-stable slopes can become unstable and slopes near the stability limit can become more stable. Over time, soil thicknesses (z_s) and slope angles (θ) change. For example, when processes like tectonic tilting and stream incision cause a slope to steepen, the slope's factor of safety decreases. Soil depth increases over time as colluvium fills in topographic hollows, resulting first in decreased stability as soil depth exceeds tree-root depth; then, strength grows as clay content in older soils increases, and the thickness of fill prevents the hollow from saturating completely during storms. Consequently, the slope grows stronger and landslides are less likely.

Simple slope-stability models like eq. 5.8 can generally predict failure of steep slopes with wet, thick soils, but the spatial variability in cohesion (C), soil thickness (z_s), and soil saturation (m) complicates predictions of landslide initiation in particular locations in response to specific storm events. Regional landslide hazard assessments and inventories typically find that only a fraction of potentially unstable slopes fail during any single storm. The number of slope failures generally increases with storm size and duration, but accurately identifying individual slopes that will fail in a given storm is not yet possible. Model-based assessments of slope stability are thus best viewed as assessing the generalized probability of slope failure.

Slope Morphology

Slope morphologies reflect slope-shaping processes. In profile, hillslopes may be convex, straight, or concave [**Figure 5.9**]. In humid and temperate regions, soil-mantled slopes generally steepen downslope from convex ridgetops at drainage divides to a planar midslope segment with a constant angle and a concave, less steep basal zone at the toe of the slope. Straight portions of slope profiles are typically more pronounced in steep terrain where frequent landslides plane off topography in the midslope zone. In contrast, arid slopes in areas of high relief often have a vertical cliff face downslope of the ridgetop and exhibit slope variations that are controlled by the relative strength of the bedrock, with harder rocks holding up steeper slopes and weaker rocks supporting gentler slopes. Such bedrock differences are apparent when the rock is not mantled by soil or regolith.

In map view, slopes are described as convergent, divergent, or planar. In general, ridgelines are divergent, and valleys are convergent. Planar slopes are neither convergent nor divergent. On convergent slopes, flow lines converge downslope. On divergent slopes, flow lines diverge downslope. Two adjacent balls rolling down a convergent slope would be at risk of colliding. They would move farther apart rolling down a divergent slope. Likewise, sediment moving down slopes tends to accumulate in convergent valleys over time.

Slope processes generate divergent, planar, or convergent topography, depending on the predominant erosional and depositional mechanisms. The diffusion-like action of creep and sheetwash creates zones of hilltop convexity and divergent noses on soil-covered hillsides. Divergent slopes are farthest from streams, and broad, rolling hills generally reflect slow uplift and erosion rates. In contrast, steep slopes and deeply incised valleys suggest more rapid uplift and erosion. Long, planar slope segments with only short hilltop convexities are common in rapidly eroding terrain and reflect the influence of landslides where the relief and hillslope gradient are high.

Weathering-Limited (Bedrock) Slopes

Weathering and sediment transport both affect slope morphology. At one extreme, **weathering-limited slopes** (also called production-limited slopes) have net rates of soil transport that are determined by the rate at which weathering

PHOTOGRAPH 5.18 Debris Flows from Forest Roads. Debris flows originate on a series of forest roads and carry sediment downslope to the Tolt River in the Cascade Mountains, Washington State.

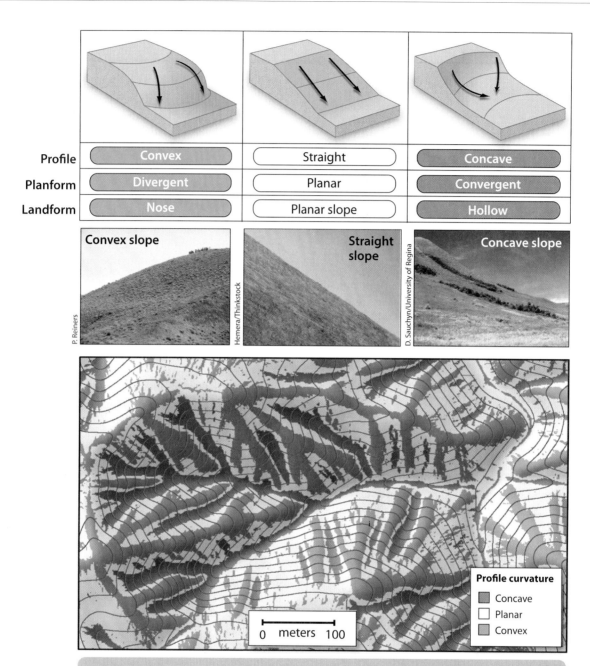

Hillslope geometry can be characterized by profile and planform shape. Patterns of mass flux down slopes are dependent on hillslope geometries. For example, on convex **noses,** flow diverges. Conversely, in concave **hollows,** flow converges. Most landscapes are composed of different slope geometries. The lower panel shows how areas of profile curvature (colors) generally correspond with areas of planform curvature (as indicated by purple topographic contours); hollows are typically concave and convergent, whereas noses are typically convex and divergent.

FIGURE 5.9 Hillslope Geometries. Hillslopes can be grouped into three distinct geometries. Convex slopes are divergent and found on noses or ridgelines. Concave slopes are convergent and make up valleys. Planar slopes are straight and often reflect slopes steep enough to fail by mass movements.

provides new material [**Photograph 5.19**]. On these slopes, rates of mass removal by erosion generally match rates of regolith production, so weathered material is transported downslope about as rapidly as it is produced. These slopes tend to be bedrock with thin, discontinuous soils. The steepness of weathering-limited slopes is generally controlled by rock mass strength, and slope morphology is often closely related to the underlying rock type. Weathering-limited slopes are commonly found in arid regions where chemical weathering and rock detachment rates are

PHOTOGRAPH 5.19 Bedrock Hillslope. Bare rock hillslope (weathering limited) at the Nahal Yael experimental watershed in the Negev Desert, Israel. Here, runoff from the bare granite hillslope is funneled into a weir so that it can be measured.

PHOTOGRAPH 5.20 Talus Slopes. These talus slopes, which originate from the cliffs above, line a canyon in the Judean Desert, Israel.

slow, as well as in arctic and high alpine areas where strong, bare rock slopes were steepened and stripped of previously weathered rock by glaciers.

The properties of bedrock slopes with little to no soil are strongly influenced by the material and strength properties of the bedrock itself. On many bedrock hillslopes, the angle of the slope depends on the resistance of the underlying bedrock to collapse or disintegration. Harder rocks form steeper slopes; weaker rocks form gentler slopes. Where slopes are formed by interbedded strong and weak rocks (sandstone and shale, for example), variability in erosion resistance commonly produces the stair-stepped morphology typical of arid and semi-arid landscapes. The walls of the Grand Canyon provide a classic illustration of this morphology (see Photograph 1.3). Rock slopes are typical of arid and semi-arid landscapes but generally are found only in the steeper parts of temperate and humid landscapes because weathering maintains a soil mantle on most lower-gradient slopes.

In places where the sediment supply from a bedrock cliff face exceeds the rate of removal by streams, angular fragments of rock debris, called **scree**, accumulate and pile up at the base of the cliff. Eventually, such material builds up a **talus slope** or ramp that progressively buries the cliff under its own debris [**Photograph 5.20**]. Because scree falls from the cliff face above, the size of individual blocks primarily reflects the structural characteristics and fracture density of the parent rock.

Transport-Limited (Soil-Mantled) Slopes

Transport-limited slopes have a supply of readily transportable material at the surface and generally have soil production rates that equal or exceed rates of downslope soil transport, a condition that results in the development of a persistent, continuous soil mantle. There is enough transportable material available on these slopes that the transport capacity of hillslope processes governs the rate at which soil leaves the slope. The morphology of transport-limited slopes is strongly controlled by soil properties. The smoothly convex to planar form of soil-mantled slopes typically masks variations in the underlying rock type as well as in bedrock structures like folds, joint patterns, and fault scarps.

Soil-mantled slopes generally steepen downhill, with a greater sediment flux corresponding to the increasing slope. On soil-mantled, transport-limited slopes, rates of soil movement are generally considered to be a simple linear function of slope steepness, θ, so the volumetric flux rate, q_s, increases with increasing slope, θ, as governed by a hillslope erosion (or diffusivity) rate constant, k, the value of which reflects soil type, vegetation, and climate through

$$q_s = k \tan\theta \qquad \text{eq. 5.9}$$

The flux of sediment moving downslope grows as one moves from the ridge crest to the valley bottom. Increasing soil flux downslope is the inevitable outcome of mass conservation, as each bit of hillslope above the valley bottom contributes mass that moves downslope under the influence of gravity. Hence, unless the shape of the slope is changing with time (from aggradation), slope angle must increase downslope in order to accommodate the downslope increase of soil flux by diffusive processes (as θ increases in eq. 5.9, so does q_s). Increasing slope with distance downslope by definition leads to hillslope convexity [**Photograph 5.21**].

If rock-uplift rates are uniform over time, soil-mantled slopes can reach an equilibrium profile that reflects a balance between soil-formation rates and soil-transport rates.

PHOTOGRAPH 5.21 Convex Hillslopes. Grass-covered convex hillslopes in Brazil with a cow in foreground and hillslope hollow (unchanneled valley) in distance.

PHOTOGRAPH 5.22 Bedrock Landslide. Large bedrock landslide on a steep, threshold slope in the Tien Shan mountains, Kyrgyzstan.

In such a case, slopes will maintain a convex profile. Indeed, the convexity of drainage divides can be used to define the zone of diffusion-dominated erosion and sediment transport near those divides. For the case of an equilibrium hillslope with a uniform soil thickness eroding at the same rate, the convex hillslope form can be derived analytically [Box 5.1].

Threshold Slopes

On slopes that have steepened to the point where landslides are the dominant erosional mechanism, river incision, rather than further slope steepening, triggers additional landsliding. Consequently, once slopes reach an upper, limiting angle between about 30 and 40 degrees (depending on soil and rock strength), they become **threshold slopes** on which landslide frequency, instead of slope angle, controls slope erosion rates [Photograph 5.22]. At slope angles above about 26 degrees, the relationship between slope angle and hillslope erosion rate thus changes from the linear relationship common on low-gradient slopes to a nonlinear, asymptotic relationship [Figure 5.10].

Slopes with gradients less than about 26 degrees, a lower limit for the initiation of most debris flows and shallow soil landslides, tend to erode and evolve through slope-dependent sediment transport in which the pace of soil erosion increases linearly with slope. Development of threshold slopes as hillslopes approach a **critical angle**, the angle at which landsliding is both frequent and the dominant means of mass transport, indicates that in steep terrain, erosion rates can increase greatly with little change in slope. Thus, it becomes increasingly difficult to infer erosion rates from slope angles as slope steepness approaches the upper limiting angle set by the soil or rock strength. The development of threshold slopes at angles close to the upper limiting angle is a major reason why mountainous topography looks remarkably similar across a wide range of climates and tectonic settings worldwide.

Hillslope Evolution

Geomorphologists have developed three general models of slope-profile evolution to explain how slope morphologies change over time: slope replacement, parallel retreat, and slope decline [Figure 5.11]. In **slope replacement**, which occurs in all climates, the gentler profile of the lower slope segment extends uphill as the steeper upper angle (or cliff face) retreats and a talus apron accumulates. This scenario occurs when material shed from a cliff face is not removed from its base. During **parallel retreat**, a slope maintains a constant angle as erosional retreat of the ground surface cuts back into the bedrock. **Slope decline** occurs as steep slopes gradually become shallower, flattening the overall slope profile. The end result of slope decline is a profile with a convex upper slope segment above a concave lower segment that develops where material is deposited near the base of the slope. All three modes of slope retreat occur in nature. Slope decline generally characterizes profile evolution in temperate and humid regions with soil-mantled slopes, and slope replacement and parallel retreat are more common in arid and semi-arid regions with bedrock slopes.

Drainage Density

Drainage density quantitatively describes the degree to which a landscape is dissected by stream channels and their valleys [Photograph 5.23]. The degree of topographic dissection of a landscape is related to the distance that channels begin downslope from drainage divides. Drainage density (DD) equals the total channel length (ΣL) within a given drainage area (A):

$$DD = \Sigma L/A \qquad \text{eq. 5.10}$$

BOX 5.1 Derivation of the Form of Convex Hillslope Profiles

Combining the concepts of conservation of mass and a simple, slope-dependent soil transport rate, we can derive the expected form of a soil-mantled hillslope. Based on the assumption of conservation of mass, the relationship between the change in soil thickness z_s over time (expressed as the derivative of soil thickness with respect to time, dz_s/dt) and the rate of bedrock weathering (W_b), the density of rock and soil (ρ_r and ρ_s, respectively), and the downslope change in the transport of soil (q_s) may be expressed as a function of the distance downslope (x):

$$dz_s/dt = (\rho_r/\rho_s) W_b - (1/\rho_s)(dq_s/dx) \quad \text{eq. 5.A}$$

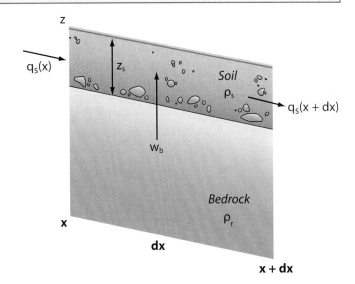

In other words, the rate at which the soil thickness changes (dz_s/dt) will be equal to the difference between the rate of bedrock weathering (adjusted for the difference in bulk density between rock and soil; the first term on the right-hand side of the equation) and the downslope change in the rate of soil transport (dq_s/dx, the spatial derivative of the sediment flux; the second term on the right-hand side of the equation).

For the case of steady-state soil thickness for which the soil thickness does not change over time (i.e., $dz_s/dt = 0$), eq. 5.A reduces to

$$\rho_r W_b = dq_s/dx \quad \text{eq. 5.B}$$

which indicates that the local soil production ($\rho_r W_b$) is balanced by the downslope change in soil transport (dq_s/dx). Integrating eq. 5.B with respect to x results in an expression for the soil flux as a function of position on the hillslope:

$$q_s = \rho_r W_b x \quad \text{eq. 5.C}$$

which states that the soil flux increases linearly with distance downslope.

Recasting the equation for slope-dependent hillslope transport (eq. 5.9) with a negative sign to reflect the convention of using the ridgetop elevation as a benchmark yields

$$q_s = -k \, dz/dx \quad \text{eq. 5.D}$$

where q_s is the soil transport (soil flux), k scales the transport efficiency, and dz/dx is the ground slope, the change in elevation with distance downslope. Substituting eq. 5.D into eq. 5.C reveals that the slope (dz/dx) increases with distance (x), resulting in a convex slope profile.

An expression for the form of the slope profile may be derived by substituting the expression for q_s in eq. 5.D into eq. 5.A, which yields a form of the standard one-dimensional diffusion equation:

$$dz_s/dt = (\rho_r/\rho_s)W_b + (k/\rho_s)(d^2z/dx^2) \quad \text{eq. 5.E}$$

where the ratio k/ρ_s is the landscape diffusivity, which may be expressed as $K_d = k/\rho_s$. For the case of a spatially uniform hillslope lowering rate and constant soil thickness (i.e., a steady state topographic form where $dz_s/dt = 0$), eq. 5.E reduces to

$$(d^2z/dx^2) = -(\rho_r W_b/K_d \rho_s) \quad \text{eq. 5.F}$$

This expression shows that the hillslope curvature, the second derivative of elevation as a function of distance, is a constant set by the rate of soil production (W_b) and the landscape diffusivity (K_d). Integrating this expression once yields the topographic slope (d_z/d_x) as a function of position along the slope (x):

$$(d_z/d_x) = -(\rho_r W_b/K_d \rho_s)x + c_1 \quad \text{eq. 5.G}$$

where c_1 is a constant of integration, which equals zero because the slope (d_z/d_x) is zero at the hilltop (x = 0). Integrating this expression once again yields the shape of the hillslope profile expressed through the relationship between hillslope elevation (z) and distance downslope (x):

$$z = -(\rho_r W_b/2 K_d \rho_s)x^2 + c_2 \quad \text{eq. 5.H}$$

in which c_2 (another constant of integration) is equal to the hillslope elevation at the ridgecrest, z_{max} (the value of z at x = 0) from where elevation decreases as a function of the square of the distance downslope. This yields the functional form of an inverted (convex-up) parabola:

$$z = z_{max} - (\rho_r W_b/2 K_d \rho_s)x^2 \quad \text{eq. 5.I}$$

While the functional form of this equation closely fits the convex profile of diffusion-dominated hillslopes, the hillslope curvature and the width of the zone of convexity vary with differences in soil production (and thus the bedrock erosion) rate and landscape diffusivity.

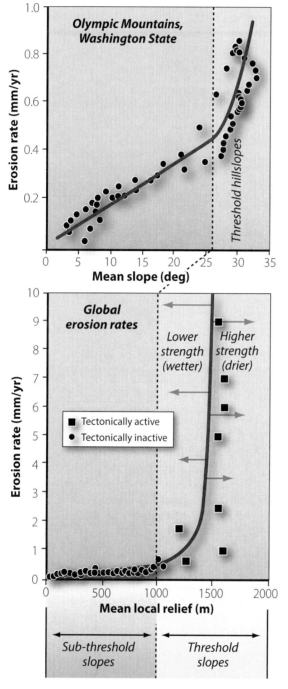

FIGURE 5.10 Threshold Hillslopes. Threshold slopes are steep and transport of mass from them is dominated by landsliding. The slope angle of the transition to threshold slopes reflects the strength of the underlying rock. [Adapted from Montgomery and Brandon (2002).]

Terrestrial drainage densities typically range from 1 to 100 km/km², and the length of unchanneled slopes upslope of well-defined stream channels ranges from less than a meter on some intensively rilled badlands to more than a kilometer on broad, low-gradient hillsides in semi-arid landscapes developed on cratons (such as in Kenya, eastern Wyoming, and the Amazon Basin). In disturbed areas and sparsely vegetated arid terrain, channel heads may extend as rills above unchanneled valleys onto the surrounding side slopes.

The drainage density within a watershed depends on the underlying lithology and the length of time over which dissection has occurred. In highly dissected watersheds, typical of shales and other easily erodible bedrock, drainage density can be quite high. This degree of dissection has important hydrologic manifestations, as it is one determinant

Slope Evolution

Humid regions	Arid regions	Steep terrain
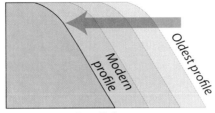	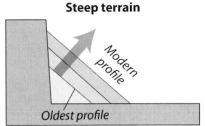	
Slope decline	**Parallel retreat**	**Slope replacement**
Hillslopes in humid regions tend to be **soil-mantled** and erode by slope-dependent **diffusive processes,** resulting in progressive rounding and lowering of hillslope profiles. Over time, this results in the decay of hillslope profiles as gradients gradually decline.	Hillslopes in arid regions tend to form bedrock slopes that erode by progressive back-wearing. Processes such as **rock fall** preserve the initial shape and gradient of hillslopes, thereby promoting **parallel slope retreat.**	**Slope replacement** occurs in very steep terrain where surface processes cannot remove debris that accumulates below a cliff. In this case, a **talus slope** builds up as the base of the cliff. The cliff is "replaced" by a more gently sloping pile of broken rock (**scree**).

FIGURE 5.11 Slope Evolution. Slopes can evolve over time by parallel retreat, common in arid regions, or by declining steepness, more common in humid regions.

(a)

(b)

PHOTOGRAPH 5.23 Drainage Density. (a) High drainage density of the weak, nonvegetated badlands landscape eroded into poorly consolidated lakebed deposits, at Zabriskie Point, Death Valley, California. Spacing between the finest valleys shown is about 10 m.

(b) Lower drainage density shown in the tree-covered, humid-temperate landscape of the southern Appalachian Mountains in West Virginia, an area underlain by strong, metamorphic rocks.

of the shape of the hydrograph. High drainage densities efficiently move water off slopes and into channels; therefore, where drainage density is high, the hydrograph tends to have a higher peak discharge and shorter lag times relative to watersheds that have lower drainage densities.

Channel Initiation

The initiation of stream channels on hillslopes marks an important process transition. Above the site of **channel initiation,** hillslope processes dominate sediment transport and smooth relief. Downstream of where channels initiate, fluvial processes cause incision, increase local relief, and dominate sediment transport through channel networks [**Figure 5.12**]. This transition is often characterized as a contrast between diffusion-like hillslope processes driven by gravity and dominated by the local slope angle and channel processes dominated by advective transport involving flowing water and thus strongly dependent on discharge as well as slope.

Soil strength and vegetation act to resist channel initiation. The cohesive strength of soil helps to retard entrainment of material from the ground surface by overland flow and the detachment of material by seepage erosion in humid terrain; vegetation shields the ground surface and provides roughness that helps reduce the velocity of moving water, thereby reducing its erosive capacity. Roots also help bind soil together and retard channel incision. As discharge increases downslope from a drainage divide and overland flow deepens, the basal shear stress that flow imparts on the slope surface also increases. Because of the importance of discharge and soil moisture in channel initiation, channels typically begin in topographically convergent source areas at the head of the channel network in humid, soil-mantled landscapes.

The **channel head** is the most upslope part of the channel and is defined by evidence of sediment transport by flowing water confined between identifiable banks. The channel head may be a gradual or abrupt change in shape. Gradual channel

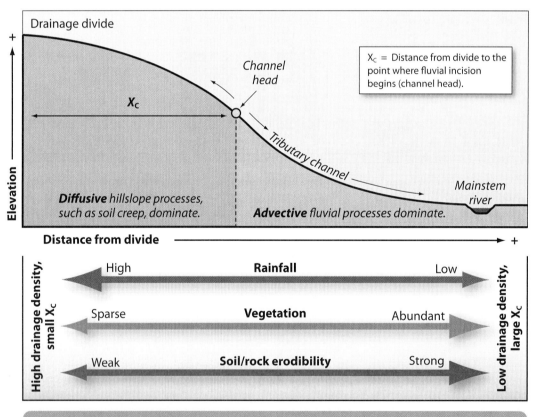

FIGURE 5.12 Channel Head Location. The channel head defines the location where advective processes moving mass out of the drainage basin become more effective than diffusive processes delivering sediment from the hillslopes to the channel. Drainage density is inversely related to the distance of the channel head from the drainage divide.

heads tend to form where sufficient overland flow accumulates to overcome the erosion resistance of the ground surface (see Photograph 4.8). On gentler slopes, abrupt channel heads often occur at the upslope end of incised gullies where subsurface water converges into unchanneled valleys. Landsliding is a primary channel initiation process in steep terrain.

Channels typically initiate where drainage areas become large enough to support a discharge capable of generating flow with sufficient shear stress to carve a channel. In many landscapes, the drainage area upslope of channel heads is inversely related to the slope angle, giving rise to smaller source areas, and thus shorter hillslopes, on steeper slopes and larger source areas, and broader hillslopes in low-gradient terrain [Figure 5.13].

Hollows are unchanneled valleys that typically lie at the head of the channel network. Hollows concentrate subsurface flow and gradually fill with colluvium derived from adjacent slopes. Over time, channel head extension

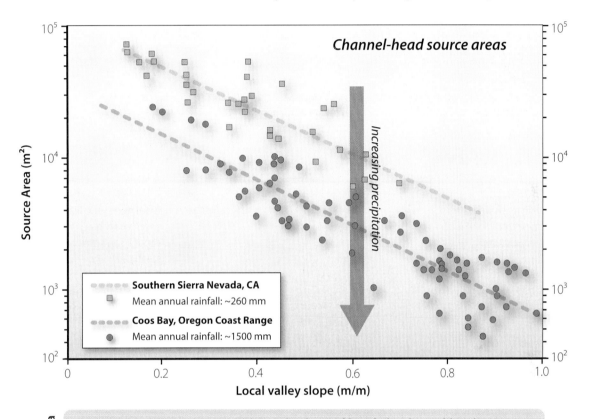

Drainage Area	The size of the drainage area upslope of a **channel head,** the channel-head source area, reflects the balance between the processes acting to incise channels and the erosional resistance of the ground surface. Consequently, steeper slopes, in weak rocks subject to higher precipitation, need smaller drainage areas to generate enough runoff to incise a channel.
Slope	The size of channel-head source areas is governed by near-surface runoff and typically decreases with local valley slope. Steeper slopes with smaller source areas, and thus more highly dissected topography, have greater **drainage density.** In regions where near-surface hydrologic response is controlled by fracture flow and subsurface processes, channel-head source area size may be slope independent.
Precipitation	Drier regions typically have larger channel-head source areas than wetter regions for the same slope. Consequently, dry regions tend to have broader hillslopes and lower drainage density, whereas wetter regions tend to have more highly dissected terrain with higher drainage density. However, complexities introduced by rock strength and vegetation cover also influence the degree of landscape dissection.

FIGURE 5.13 Drainage Area, Slope, and Climate Control on Channel Initiation. Channels begin at the downstream end of source areas, the size of which is a function of both slope and climate. In humid climates, where water is plentiful, smaller source areas are required to initiate channels than in arid climates. [Adapted from Montgomery and Dietrich (1988).]

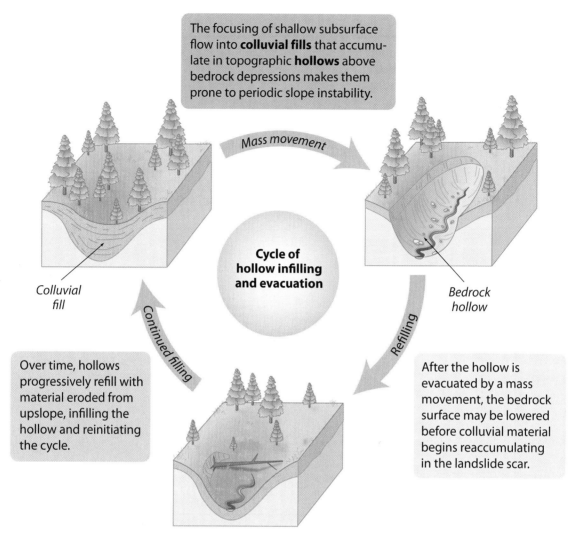

FIGURE 5.14 Bedrock Hollows. Hollows, depressions along bedrock ridgelines, cyclically fill with colluvium from bordering slopes and then episodically empty in landslides. Sliding is usually triggered by heavy rains and/or the removal of tree roots that otherwise stabilize the weak, granular colluvium on steep slopes. [Adapted from Dietrich et al. (1982).]

by landsliding or fluvial erosion is required to excavate and maintain hollows (see Photographs 4.7 and 4.8). Radiocarbon dating of charcoal collected from colluvial soils in hollows indicates gradual infilling between excavation events. In steep, landslide-prone terrain, hollows undergo a cycle of infilling, excavation, and refilling over the course of hundreds to thousands of years [**Figure 5.14**].

Applications

An understanding of slope processes and dynamics is important for engineering, hazard assessment, and land-management applications. Recognition of landslide hazard zones and ancient landslides is essential for upland development and provides important context for geotechnical engineering analyses. Building a house on an ancient slow-moving earthflow or in the path of future debris flows can invite disaster if such potential hazards are not recognized and accounted for during siting of infrastructure and in project design. Ancient landslides sometimes have residual strength lower than their original, prefailure strength and thus can be reactivated by changes in slope hydrology that accompany development.

For example, increased soil moisture from lawn irrigation in northern California communities has reactivated enormous ancient landslides. By reading terrain well enough to identify an upslope head scarp or downslope landslide toe, one can better identify ancient landslides and assess hazards to communities or new developments. Landslide hazard mapping of potentially unstable slopes—those that are prone to failure but that have not yet failed—can be based on slope stability calculations like eq. 5.8, extrapolation based on statistical characterization of observed failure locations, or the propensity for landslides to initiate in steep, convergent topography.

The increasing accessibility of LiDAR topographic data over the past decade has made slope morphology much easier to portray and analyze. Such high-resolution topographic data provide new ways to assess topography in order to identify the locations of past landslides that pose a risk to development on their surfaces or locations downslope. Similarly, advances in GIS, radar-based precipitation estimates, and landslide modeling have made it possible to produce real-time landslide hazard assessments based on digital topography at high spatial resolution over even large areas.

Human actions influence slope processes by changing the properties of hillslope materials, altering slope configuration, changing the hydrology, or adding and removing vegetation. "Follow the water" is the first rule of forensic landslide investigations in rural and urban environments where many slope failures can be traced to concentration of road runoff, leaky underground pipes, or places where homeowners and developers have directed storm runoff from roof drains and gutters onto steep slopes. Excavating the toe of a slope to make a road cut, to construct a building pad, or to mine gravel can remove basal support and trigger slope failure. Routing street or gutter runoff onto steep, slide-prone slopes can increase soil moisture levels and cause slopes to fail. The risks associated with clear-cutting steep, forested hillsides are significant and can be assessed using slope-stability analysis to delineate areas that are especially susceptible to postharvest slope failures. Slope stability is increased either by reducing the driving stress, for example, by building terraces, or by increasing the stability of the slope by building retaining walls or replanting denuded areas.

Changes to the erosion resistance of low-gradient slopes associated with agricultural practices like conventional tillage or overgrazing cause gully development, soil loss, and other types of damage to farm lands, as well as sedimentation problems for locations downslope and downstream. Effective erosion control practices begin with an understanding of the nature of the slope processes. Generally, it is far less expensive and more effective to address sediment problems in streams by reducing erosion at its source on upland hillslopes than it is to deal with an overload of sediment and its environmental consequences downstream.

Selected References and Further Reading

Abrahams, A. D., ed. *Hillslope Processes*, Boston: Allen & Unwin, 1986.

Black, T. A., and D. R. Montgomery. Sediment transport by burrowing mammals, Marin County, California. *Earth Surface Processes and Landforms* 16 (1991): 163–172.

Binnie, S., W. Phillips, M. Summerfield, and L. Fifield. Tectonic uplift, threshold hillslopes, and denudation rates in a developing mountain range. *Geology* 35 (2007): 743–746.

Burbank, D. W., J. Leland, E. Fielding, et al. Bedrock incision, rock uplift and threshold hillslopes in the northwestern Himalayas. *Nature* 379 (1996): 505–510.

Caine, N. The rainfall intensity-duration control of shallow landslides and debris flows. *Geografiska Annaler Series A* 62 (1980): 23–27.

Carson, M. A., and M. J. Kirkby. *Hillslope Form and Process*. London: Cambridge University Press, 1972.

Cruden, D. M., and D. J. Varnes. "Landslide types and processes." In A. K. Turner and L. R. Schuster, eds., *Landslides—Investigation and Mitigation*, National Research Council Transportation Research Board Special Report no. 247. Washington, DC: National Academies Press, 1996.

DiBiase, R., K. Whipple, A. M. Heimsath, and W. B. Ouimet. Landscape form and millennial erosion rates in the San Gabriel Mountains, CA. *Earth and Planetary Science Letters* 289 (2009): 134–144.

Dietrich, W. E., and T. Dunne. Sediment budget for a small catchment in mountainous terrain. *Zeitschrift für Geomorphologie, Supplementbände* 29 (1978): 191–206.

Dietrich, W. E., T. Dunne, N. F. Humphrey, and L. M. Reid. Construction of sediment budgets for drainage basins. In F. J. Swanson, R. J. Janda, T. Dunne, and D. N. Swanston, eds. *Sediment Budgets and Routing in Forested Drainage Basins*, General Technical Report PNW-141, Forest Service, U.S. Department of Agriculture, 1982.

Gilbert, G. K. The convexity of hilltops. *Journal of Geology* 17 (1909): 344–350.

Horton, R. E. Erosional development of streams and their drainage basins: Hydrophysical approach to quantitative morphology. *Geological Society of America Bulletin* 56 (1945): 275–370.

Iverson, R. M. The physics of debris flows. *Reviews of Geophysics* 35 (1997): 245–296.

Iverson, R. M., M. E. Reid, and R. G. LaHusen. Debris-flow mobilization from landslides. *Annual Review of Earth and Planetary Sciences* 25 (1997): 85–138.

Kirkby, M. J. Measurement and theory of soil creep. *Journal of Geology* 75 (1967): 359–378.

Kirkby, M. J. "Hillslope process-response models based on the continuity equation." In D. Brunsden, ed., *Slopes: Form and Process*. London: Institute of British Geographers Special Publication 3, 1971.

Korup, O. Rock type leaves topographic signature in landslide-dominated mountain ranges. *Geophysical Research Letters* 35 (2008): L11402.

Larson, I. J., D. R. Montgomery, and O. Korup. Landslide erosion controlled by hillslope material. *Nature Geoscience* 3 (2010): 247–251.

Matsuoka, N., and H. Sakai. Rockfall activity from an alpine cliff during thawing periods. *Geomorphology* 28 (1999): 309–328.

McKean, J. A., W. E. Dietrich, R. C. Finkel, et al. Quantification of soil production and downslope creep rates from cosmogenic ^{10}Be accumulations on a hillslope profile. *Geology* 21 (1993): 343–346.

Montgomery, D. R. Road surface drainage, channel initiation, and slope stability. *Water Resources Research* 30 (1994): 1925–1932.

Montgomery, D. R., and M. T. Brandon. Nonlinear topographic controls on erosion rates in tectonically active mountain ranges. *Earth and Planetary Science Letters* 201 (2002): 481–489.

Montgomery, D. R., and W. E. Dietrich. Where do channels begin? *Nature* 336 (1988): 232–234.

Montgomery, D. R., K. M. Schmidt, W. E. Dietrich, and J. McKean. Instrumental record of debris flow initiation during natural rainfall: Implications for modeling slope stability. *Journal of Geophysical Research—Earth Surface* 114 (2009): F01031, doi: 10.1029/2008JF001078.

Reneau, S. L., W. E. Dietrich, M. Rubin, et al. Analysis of hillslope erosion rates using colluvial deposits. *Journal of Geology* 97 (1989): 45–63.

Roering, J. J., J. W. Kirchner, and W. E. Dietrich. Evidence for nonlinear, diffusive sediment transport on hillslopes and implications for landscape morphology. *Water Resources Research* 35 (1999): 853–870.

Schmidt, K. M., and D. R. Montgomery. Limits to relief. *Science* 270 (1995): 617–620.

Selby, M. J. A rock-mass strength classification for geomorphic purposes: With tests from Antarctica and New Zealand. *Zeitschrift für Geomorphologie* 24, (1980): 31–51.

Selby, M. J. *Hillslope Materials and Processes*, 2nd ed. New York: Oxford University Press, 1993.

Sidle, R. C., and H. Ochiai. *Landslides: Processes, Prediction, and Land Use*. AGU Water Resources Monograph 18. Washington, DC: American Geophysical Union, 2006.

Terzaghi, K. Stability of steep slopes on hard unweathered rock. *Géotechnique* 12 (1962): 251–270.

DIGGING DEEPER How Much Do Roots Contribute to Slope Stability?

Keen observers have long recognized that trees help stabilize soils on steep mountain slopes. Lyell (1853) and Marsh (1864) interpreted associations between forest cutting and mass wasting as evidence that forest clearing accelerated erosion in mountainous terrain. Since Lyell's day, the influence of root reinforcement on shallow landsliding has been well established by studies of landslide erosion under mature forest and in harvested plots, mechanistic studies of root reinforcement, and theoretical analyses based on the infinite-slope stability equation (eq. 5.8), where root strength is considered as part of the cohesion term (Sidle et al., 1985). Although roots contribute to soil strength by providing apparent cohesion and holding the soil mass together, they have a negligible effect on frictional strength. Studies from the western United States, Japan, and New Zealand all indicate that the stability of the soil mantle on steep, soil-mantled slopes depends in part on reinforcement by tree roots and that after the loss of forest cover (either by timber harvest or fire), the decay of tree roots increases the potential for slope instability, especially when soils are partly or completely saturated (Sidle et al., 1985; Bierman et al., 2005).

Root reinforcement may occur through the base of a potential landslide as roots grow into the underlying bedrock or more stable surface materials. Dense, interwoven root networks both reinforce soil and provide lateral reinforcement across potential failure scarps. Burroughs and Thomas (1977) demonstrated a rapid decline in the tensile strength of Douglas-fir roots following timber harvest in western Oregon and central Idaho and indicated the increased potential for landslides when trees were removed. Building on the Burroughs and Thomas approach, Sidle (1992) developed a quantitative model of root-strength reinforcement that combined the decay of roots after timber harvest with the regrowth of new roots [**Figure DD5.1**]. Although the decay and regrowth times vary for different tree species, a period of low root strength occurs some time between 3 and 20 years following timber harvest or fire. If a big storm occurs in

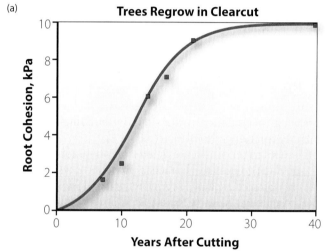

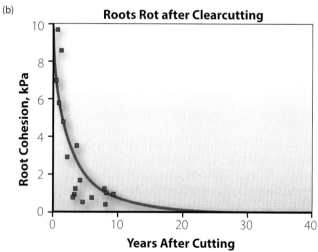

FIGURE DD5.1 Root strength changes over time as (a) trees grow in clearcuts and (b) as roots decay after trees are clear-cut. It takes about a decade after cutting for the dead roots of coastal Douglas fir trees to lose all of their strength and about 20 years for new trees to take root and develop full root strength. Planting seedlings right after harvest is a land-management strategy that reduces the chance of landsliding because new roots are growing as the old ones are decaying. [From Sidle (1992).]

this window and saturates the soil, landslides will likely follow.

Studies comparing the rate of landsliding on forested versus clear-cut slopes have reported a range of effects, from no detectable increase in landslide frequency to more than a ten-fold increase following timber harvest (Sidle et al., 1985). In a study that both analyzed a regional data set of >3200 landslides and intensively monitored a study area, Montgomery et al. (2000) found that storms with 24-hour rainfall recurrence intervals of less than 4 years (common storms) triggered landslides in the decade after timber harvesting in the Oregon Coast Range [Figure DD5.2]. Comparison of these postharvest rates of landsliding with the estimated background rate implied that clear-cutting of slopes increased landsliding rates by 3 to 9 times over the natural background. This increase reflected reduced root strength as the dead roots of the cut trees rotted and weakened. Without strong roots, less soil saturation was required to induce slope failure, and thus smaller storms could trigger landslides.

Schmidt et al. (2001) measured root cohesion in soil pits and scarps of landslides triggered during large storms in February and November of 1996 in the Oregon Coast Range. They found a preponderance of broken roots in the margins of recent landslide scarps, indicating that root tensile strength contributed to stabilizing the soil (until the roots snapped) in most locations. They also found that root density, root penetration depth, and the tensile strength varied among species; the tensile strength increased nonlinearly with root diameter. The median lateral cohesion provided by roots in mature natural forest ranged from 26 to 94 kPa. It was much lower in planted, industrial forest stands, ranging from 7 to 23 kPa. In clear-cuts, the lateral root reinforcement was uniformly low, under 10 kPa [Figure DD5.3].

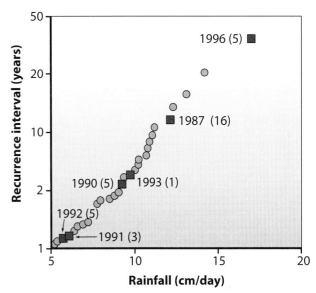

FIGURE DD5.2 Plot of recurrence intervals for 24-hour rainfall events from 1931 through 1996 (yellow circles) in a steep 0.43 km^2 study area that was clear-cut in the 1980s. Storms that occurred after clear-cutting and are known to have generated landslides are shown as blue squares. Numbers in parentheses after years indicate how many landslides occurred. Note that eight landslides occurred in this area during storms having less than 2-year recurrence intervals, all after clear-cutting. Vertical axis is logarithmic. [From Montgomery et al. (2000).]

Similar to Montgomery et al. (2000), Schmidt et al. (2001) found that a persistent reduction in root strength

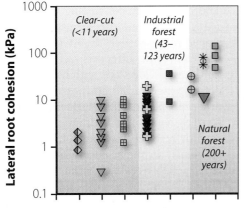 

FIGURE DD5.3 In the Oregon Coast Range, not all roots provide the same amount of lateral root cohesion. Roots in clear-cuts do little to stabilize slopes. Industrial forests, those planted and managed for wood products, have roots that provide some stabilization, but the highest apparent root-cohesion values are found in mature, natural forests. [From Schmidt et al. (2001).]

DIGGING DEEPER How Much Do Roots Contribute to Slope Stability? *(continued)*

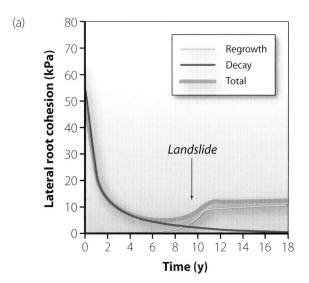

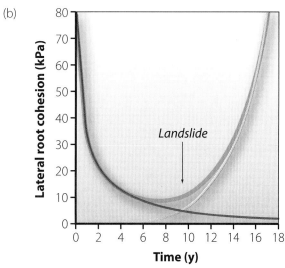

FIGURE DD5.4 Predicted total lateral root cohesion considering contributions from tree regrowth and decay of old roots for two sites that were clear-cut in 1986 and yielded landslides in 1996. Figure (a) represents a site where understory regrowth dominates vegetation. Figure (b) is a site where growth consists of abundant conifers and deciduous trees. [From Schmidt et al. (2001).]

resulting from timber harvest significantly reduced the soil moisture (m in eq. 5.8) required to trigger slope instability. They modeled root decay and regrowth for two sites that were clear-cut in 1986, and then slid in 1996. Both failures occurred close to the predicted root-strength minima, about 10 years after clear-cutting [**Figure DD5.4**].

Root strength varies spatially in a forest, complicating slope-stability modeling. Roering et al. (2003) docu-

mented the distribution and characteristics of trees adjacent to 32 shallow landslides in the Oregon Coast Range. Not surprisingly, bigger trees had larger root systems. The diameter of the tree crown and the root network was a function of the tree diameter (and thus tree age), and Roering et al. (2003) quantified root strength in landslide scarps by pulling on roots and measuring the tensile strength at which they broke. Summing the total root strength in each landslide perimeter, they found that root strength correlated with the size, species, condition, and spacing of trees around the landslide scarps; bigger, healthier trees spaced more closely together gave greater root strength. They also found that landslides tended to occur in areas of low root strength and thus that the potential for shallow slope instability was a function of the diversity and distribution of vegetation on potentially unstable slopes. Well-vegetated slopes were more stable.

Root strength can also vary with topographic position. Hales et al. (2009) investigated the spatial variability of root network density and strength in the southern Appalachian Mountains in North Carolina by measuring the distribution and tensile strength of roots from soil pits on topographic noses and hollows. They found that roots from trees on noses had greater tensile strength than those found in hollows, a pattern suggesting that not only does vegetation help stabilize topography but that topography affects vegetation, specifically, root strength (presumably due to differences in soil moisture). Trees on noses provided more effective root cohesion than those in hollows, a pattern that would increase further the propensity for landslides to occur in hollows.

The variability of root reinforcement with tree species, root diameter, tree diameter, topographic position, and time after timber harvest complicates quantitatively predicting the effect of root reinforcement on slope stability. The evidence is convincing that taking trees off slopes reduces root reinforcement and allows soils to fail on slopes more easily, i.e., in smaller precipitation events; however, this effect is difficult to incorporate into landscape-scale slope stability models due to the tremendous spatial variability not only in root strength but in other properties that influence slope stability, such as regolith depth and hydraulic conductivity, and the influence of bedrock fractures on soil saturation. There is no ambiguity in the science indicating that clear-cut slopes, from which trees have been removed, are more likely to fail than similar slopes under mature forest. However, managing timber-harvest-related slope instability is difficult because it is impossible to identify with certainty which potentially unstable slopes will actually fail in a particular storm. [**Figure DD5.5**].

FIGURE DD5.5 Debris flows off a steep, clear-cut slope, Stillman Creek, Washington. The timber company's application to the State Department of Natural Resources before harvest reported that the site had been inspected and was found to have no potentially unstable slopes. [Photograph by S. Ringman, from *Seattle Times*.]

Bierman, P. R, J. Howe, E. Stanley-Mann, et al. Old images record landscape change through time. *GSA Today* 15, no. 4 (2005): 4–10.

Burroughs, E. R., and B. R. Thomas. *Declining root strength in Douglas fir after falling as a factor in slope stability*. U.S. Forest Service Research Paper INT-190, Ogden, UT: U.S. Department of Agriculture, 1977.

Hales, T. C., C. R. Ford, T. Hwang, et al. Topographic and ecologic controls on root reinforcement. *Journal of Geophysical Research* 114 (2009): F03013, doi: 10.1029/2008JF001168.

Lyell, C. *Principles of Geology; Or, the Modern Changes of the Earth and Its Inhabitants, Considered as Illustrative of Geology*, 9th ed. London: J. Murray, 1853.

Marsh, G. P. *Man and Nature; or, Physical Geography as Modified by Human Action*. New York: Charles Scribner's Sons, 1864.

Montgomery, D. R., K. M. Schmidt, H. Greenberg, and W. E. Dietrich. Forest clearing and regional landsliding. *Geology* 28 (2000): 311–314.

Roering, J. J., K. M. Schmidt, J. D. Stock, et al. Shallow landsliding, root reinforcement, and the spatial distribution of trees in the Oregon Coast Range. *Canadian Geotechnical Journal* 40 (2003): 237–253.

Schmidt, K. M., J. J. Roering, J. D. Stock, et al. The variability of root cohesion as an influence on shallow landslide susceptibility in the Oregon Coast Range. *Canadian Geotechnical Journal* 38 (2001): 995–1024.

Sidle, R. C. A theoretical model of the effects of timber harvesting on slope stability. *Water Resources Research* 28 (1992): 1897–1910.

Sidle, R. C., A. J. Pearce, and C. L. O'Loughlin, C. L. *Hillslope Stability and Land Use*. Water Resources Monograph 11, Washington, DC: American Geophysical Union, 1985.

WORKED PROBLEM

Question: Using the infinite-slope model, what is the maximum stable angle for both dry and saturated sand with no cohesion and a friction angle of 37 degrees? How does this stable angle compare to that of more cohesive material such as till or clay?

Answer: For dry cohesionless materials, the maximum stable angle is the friction angle, ϕ, in this case, 37 degrees. For the failure of a fully saturated, cohesionless soil like coarse sand (FS = 1.0, C = 0, and m = 1.0), eq. 5.8 reduces to $\tan\theta = [[(\rho_s - \rho_w)/\rho_s]\tan\phi]$, which may be approximated by $\tan\theta = 1/2 \tan\phi$ (since for most soils $\rho_s \approx 2\rho_w$). This indicates that sandy slopes steeper than about half the friction angle tend to fail if saturated. Thus, when saturated, cohesionless sand with a friction angle of 37 degrees will fail when the slope is about 23.5 degrees. At higher slopes where $\theta \geq \phi$, cohesionless soils tend to slide even when dry; the soil mantle rarely stays on such steep slopes unless there is significant root reinforcement. Soils with even modest amounts of cohesion can stand at much steeper angles over length scales shorter than typical hillslope lengths. For example, excavations in clay (and other cohesive materials like glacial till) can hold vertical faces of up to several meters in height, as can riverbanks, especially if reinforced by roots that provide apparent cohesion.

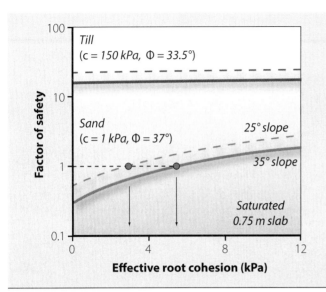

Infinite-slope models are useful for evaluating slope stability and determining the effect of both cohesion and friction angle on slope stability. Here, we plot factors of safety for saturated earth materials—till with high cohesion and virtually cohesionless sand—as a function of saturation, root cohesion, and slope. The dashed lines are model results for a 25-degree slope. The solid lines are model results for a 35-degree slope. The red circles indicate the amount of root cohesion needed to prevent slope failure (factor of safety = 1). Clearly, small amounts of root cohesion can stabilize saturated sandy hillslopes. For glacial till, the effect of roots is unimportant as the material already has large amounts of cohesive strength. [From Bierman et al. (2005).]

KNOWLEDGE ASSESSMENT Chapter 5

1. List three factors that control rock strength.
2. Explain how saprolite and colluvium differ.
3. Explain how the angle of internal friction relates to soil strength.
4. What is cohesion and why is it important geomorphically?
5. List three properties that control the cohesion of soils.
6. Predict the effect of a heavy rainstorm on slope stability and explain the reasoning behind your prediction.
7. Define shear stress and explain why and how it varies with slope angle.
8. Explain two biologic effects on soil strength and erosion.
9. Contrast the form of arid and humid region slopes and explain why they differ.
10. There are straight, convex, and concave slopes. Where is each type most likely to be found?
11. How does weathering affect hillslope form and behavior?
12. Explain how weathering-limited slopes differ from transport-limited slopes and where you might typically find each.
13. Explain how talus and stair-step topography form and why they are linked.
14. Where would you be most likely to find threshold slopes?
15. List and explain the three general models of slope profile evolution.
16. Explain the importance of raindrops in shaping hillslopes.
17. Where is sheetwash likely to occur and why?
18. List the three main categories of mass movements.
19. List four specific physical processes that cause soil creep.
20. What specific observations would allow you to distinguish between active and inactive landslides?
21. Describe the physical characteristics of debris flows and the deposits they leave.
22. Predict the behavior of low-yield strength debris flows.
23. Describe three distinguishing characteristics of rotational landslides.
24. Explain the primary controls on rockslides.
25. Describe how slope stability is modeled for planar and rotational slides.
26. Explain why landslides are random in time or clustered.
27. How and where are channels initiated on slopes?
28. Explain what a hollow is and where it might be found.
29. Provide the equations for resisting and driving stresses on slopes.
30. What is the effective normal stress and how does it relate to slope stability and climate?
31. How does climate affect the shape and behavior of hillslopes?
32. Explain why the strength of rock samples measured in hand samples differs from the strength of the same rock at the scale of hillslopes.
33. Give three examples of diffusive hillslope processes and explain how the rate of such processes relates to slope steepness and slope length.
34. Describe the infinite-slope model and explain why it is useful.
35. What characteristics control drainage density and how does drainage density affect landscape response to storms?

Channels

6

Introduction

Stream channels drain landscapes and carry material from erosional uplands to depositional lowlands, estuaries, and coastal environments. Channels range in size from small, headwater rivulets to large, continent-draining rivers. Channels are shaped by the flows and sediment loads they carry, the cohesive strength of the material in their banks, the slopes they flow down, and the rocks they cut into. Without rivers and streams to carry debris away, upland valleys would gradually become choked with sediment eroded off hillslopes, reducing ridgetop to valley-bottom relief and leveling the land. Fluvial processes control local deposition and erosion of sediment, allowing rivers to migrate across valley bottoms and form floodplains. These processes also govern the way that rivers incise into bedrock and generate the topography of upland valleys. An understanding of **fluvial** (river) processes and dynamics is thus central to studies of landscape evolution.

The interaction of flowing water and sediment shapes channels, so the supply of water, sediment, and large organic debris (such as logs) greatly influences channel morphology and dynamics. These factors vary within drainage basins, across regions, and over time. Channels exhibit tremendous variability in morphology and can respond quickly and significantly to changes in discharge and sediment load. The physical processes that determine channel morphology and dynamics are similar across different regions, but the importance of local controls and the influence of the downstream routing of water and sediment lead to a wide variety of channel

A braided river in central Alaska cuts through boreal forest and tundra; its gravel-rich, light-colored active channel stands out from the dark vegetation.

IN THIS CHAPTER

Introduction
External Controls on Fluvial Processes and Form
  Discharge
  Sediment Supply
  Bed and Bank Material
  Vegetation
Fluvial Processes
  Flow Velocity
  Discharge Variability
  Stream Power
  Bedrock Incision
  Channel Migration

Sediment Transport
  Initiation of Transport
  Sediment Loads
  Bedforms
Channel Patterns
  Straight and Sinuous Channels
  Meandering Channels
  Braided Channels
  Anastomosing Channels
Channel-Reach Morphology
  Colluvial Reaches
  Bedrock Reaches

  Alluvial Reaches
  Large Organic Debris
Floodplains
Channel Response
Applications
Selected References and Further Reading
Digging Deeper: What Controls Rates of Bedrock River Incision?
Worked Problem
Knowledge Assessment

types. Over long periods of geologic time, channels respond to tectonic uplift and subsidence, sea-level rise and fall, erosion of the landscape, and changes in climate and vegetation. Over shorter timescales, channels adjust to seasonal, annual, and decadal cycles of discharge and sediment supply as well as to changes in land use and extreme events like droughts and floods.

Regional climate and tectonic setting together broadly determine whether streams flow through steep or gentle terrain, and whether channels carry low or high loads of fine or coarse sediment. Climate establishes the amount, type, and seasonal pattern of precipitation and runoff, and controls the dominant runoff-generating mechanisms and thereby streamflow magnitude and variability. The depth to the water table influences whether channels carry flow only during storms or whether they also carry **baseflow** during periods between storms. Channels in arid and semi-arid regions may be **ephemeral** and flow for only part of the year, while channels typical of humid or temperate regions are **perennial** and flow year-round.

The potential for intense thunderstorms over small areas makes flash floods an important (and dangerous) fluvial process not only in semi-arid landscapes like much of Australia and the U.S. Southwest, but also in subhumid landscapes of the North American Midwest and in the humid southern United States and northeastern Australia, where hurricanes and cyclones strike. Ice jams that form in floods during spring ice breakup are an important fluvial process (eroding banks, removing riparian vegetation, and spreading sediment over floodplains) in places where rivers annually freeze over, like New England, subarctic Canada, and the northern Great Plains of North America [**Photograph 6.1**]. In general, regional climate is the dominant influence on the timing, magnitude, and regularity of the streamflows that control channel processes.

Regional tectonic forces generate uplift that elevates the land surface as well as subsidence that lowers the topography over which streams flow and into which they incise channels. Over geologic time, streams **grade** their channels, adjusting channel slope in response to changes in uplift and erosion and the balance between sediment supply and transport capacity. Over shorter time frames, however, channel slope is an externally imposed factor that controls channel processes, particularly the flow of water and the transport of sediment. Regional geologic and tectonic history determine the rock types (lithologies) over which streams flow. Rock type influences the processes of bedrock erosion, as well as the durability of the sediment that streamflow carries.

Steep channels incised into bedrock in tectonically active mountains differ greatly from lowland rivers that flow across depositional basins where sediment accumulates and can be stored over long periods of time. Upland channels tend to have little sediment cover, typically have rocky beds and/or banks, and have little **alluvial** storage in their valley bottoms. In contrast, lowland alluvial channels typically have bed and banks composed of material transported by the channel and store substantial amounts of sediment in their valley bottoms. Mountain (bedrock) streams generally have a **transport capacity** that exceeds their sediment supply, whereas lowland (alluvial) streams have a sediment supply that equals or exceeds their transport capacity.

Long-term and short-term controls on fluvial morphology and processes both act to shape habitats for in-channel and near-channel organisms. River dynamics influence the distribution of species along river corridors, and **riparian** (near-stream) areas are the most biologically diverse parts of many landscapes. Not only do rivers shape riparian zones, but riparian zones also shape rivers by contributing woody debris and slowing flood flows.

This chapter addresses the influences of regional landscape context on stream systems, explains the flow and transport processes that move mass through channel networks, introduces the fluvial landforms that result from the action and interaction of these processes, and describes channel response to common changes in these processes. Rivers around the world have been dammed, straightened, pinned between levees, and cleared of logs and logjams. Modern channel restoration, rehabilitation, and flood-control efforts are based on an understanding of the ways in which environmental controls, the history of human modifications, and physics interact in particular rivers. This chapter provides context for such work by presenting the key concepts in fluvial geomorphology.

External Controls on Fluvial Processes and Form

Some factors that influence streams, including regional climate and tectonics, are externally imposed controls to which channel processes must respond. Other factors, like

PHOTOGRAPH 6.1 Cleared Ice Jam. In the late 1800s, a spring ice jam along the White River in Sharon, Vermont, had cleared—but not before rising water levels, sufficient to move large ice blocks, spilled out of the channel and onto the floodplain, covering a road and damaging several buildings. Such ice jams recur every spring as river ice breaks up.

the pattern of erosion and deposition within a stream valley, are themselves influenced by channel processes and response. As an example of the feedback between external controls and channel processes, consider how net differences in erosion or deposition lead to changes in channel slope. If sediment supply exceeds a stream's transport capacity, the additional deposition will cause channel aggradation that increases channel slope downstream of the point of deposition, thereby increasing the local transport capacity, which will facilitate moving the deposited material downstream. Conversely, if a channel's transport capacity exceeds its sediment supply, erosion will scour the channel bed (potentially down to bedrock), thereby reducing the channel bed slope. In this context, a **graded stream** has a profile that is adjusted to carry its sediment load. A graded stream profile is concave up, with steeper channels in headwaters declining progressively downstream toward the outlet of the river network.

Upland slopes deliver water and sediment to stream channel networks in varying ways and amounts and over different periods of time. Streamflow sorts, breaks down, and transports this material, and thus it shapes channels in response to external and self-limiting controls. Fluctuations in water flow and sediment supply give rise to spatial and temporal variability in channel morphology, processes, and response. The shape and behavior of a channel is governed primarily by its sediments, its discharge, the composition of its bed and banks, and the vegetation growing in and immediately adjacent to the stream. Many of these factors interact in complex ways with human modifications like dams and levees that impose new external controls on a channel or stream system.

Discharge

Discharge is the volume of water flowing past a point on a stream per unit time [**Figure 6.1**]. Stream discharge (Q) is typically measured in cubic meters per second (m^3/s) and is equal to the product of the channel's cross-sectional area (A_{cs}) and the flow velocity (U), or to the product of the average stream width (W), flow depth (D), and flow velocity:

$$Q = A_{cs}U = WDU \qquad \text{eq. 6.1}$$

Discharge stays the same downstream unless water is added or lost, an expression of continuity and the conservation of mass. In humid-temperate regions, discharge systematically increases downstream within channel networks because small channels converge to form larger ones, and because perennial streams in wet climates typically gain water from groundwater aquifers. In arid regions, discharge often decreases downstream as water infiltrates through the streambed.

The discharge in a channel varies over both event and seasonal timescales as individual storms and annual weather patterns deliver precipitation. High-discharge events mobilize large volumes of sediment and typically govern the processes that dramatically reshape channels. The frequency, magnitude, and duration of high flows vary with seasonal patterns of precipitation and temperature. Some stream systems experience regular, predictable flood events, including those in regions with monsoon climates, streams dominated by seasonal melting of snowpacks, and high-latitude rivers that flow poleward and experience massive ice jams when they thaw upstream while downstream reaches remain frozen. In most regions, however, irregular storm-driven rainfall events like hurricanes and local intense thunderstorms produce flood flows.

The way stream channels and their valleys carry high discharges greatly affects channel morphology and valley-bottom landforms. High-discharge events in mountain channels confined between bedrock valley walls have greater flow depths, for the same flood discharge, than do channels flowing through wide valley bottoms across which floodwaters disperse. This fundamental difference in how channels convey high flows leads to the common observation that very little sediment is stored in mountain valleys and there is generally extensive sediment deposition in broad alluvial lowlands.

Sediment Supply

The sediment supply to a channel is imposed by the rate at which material is delivered from upstream channel reaches, from neighboring hillsides, and from bed and bank erosion during channel incision and migration. The volume, grain size, and degree of sorting of sediment supplied to a channel influence how much of the material subsequent streamflow is able to transport and sort, and how much the stream is unable to move and must flow around. Sediment supply often varies tremendously over time. Hillslope processes generally deliver much greater volumes of sediment during abrupt events, such as landslides and intense erosion by overland flow during large storms, than they do through slower, steadier processes like soil creep. In mountain drainage basins, sediment supplied by catastrophic mass wasting typically dominates the sediment supply to stream channels [**Photograph 6.2**]. In contrast, the sediment supplied to lowland channels generally comes from upstream channels and local bank erosion [**Photograph 6.3**].

Grain size and degree of sorting of the sediment supplied to a channel often profoundly influence sediment transport patterns, channel dynamics, and channel morphology of the stream. For example, steep channels that receive only sand from upstream bank erosion may rapidly transport their full sediment load and expose the underlying bedrock, while comparable channels that receive a wide range of particle sizes from a landslide off of a valley wall may develop a coarse surface layer of large clasts that move only during the highest flows, if ever.

Bed and Bank Material

The materials that form the bed and banks of a river control the frequency and rate at which the channel mobilizes sediment, erodes into its banks, and incises its bed. The

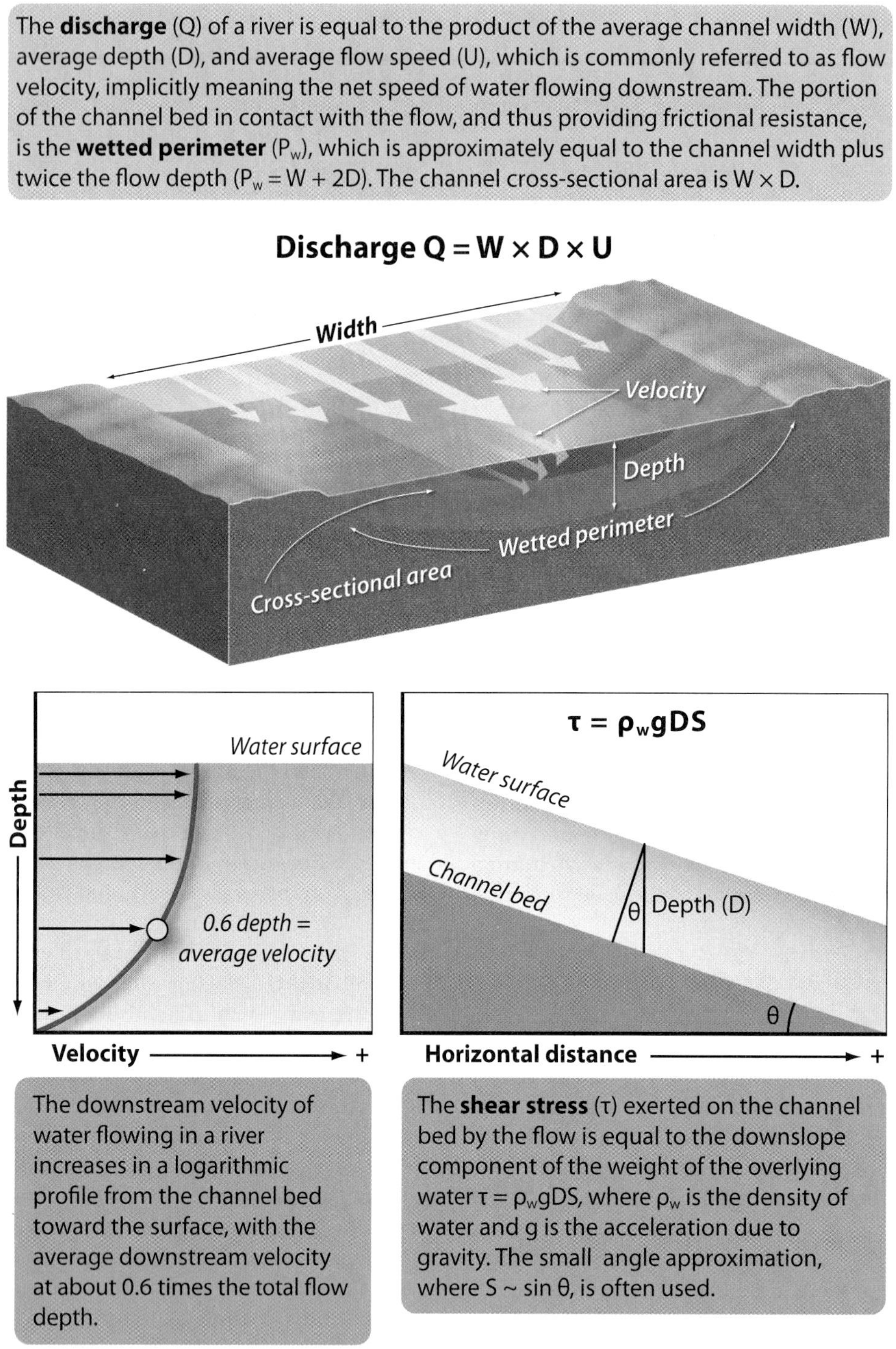

The **discharge** (Q) of a river is equal to the product of the average channel width (W), average depth (D), and average flow speed (U), which is commonly referred to as flow velocity, implicitly meaning the net speed of water flowing downstream. The portion of the channel bed in contact with the flow, and thus providing frictional resistance, is the **wetted perimeter** (P_w), which is approximately equal to the channel width plus twice the flow depth ($P_w = W + 2D$). The channel cross-sectional area is W × D.
Discharge $Q = W \times D \times U$
Width
Velocity
Depth
Wetted perimeter
Cross-sectional area
Water surface
Depth
0.6 depth = average velocity
Velocity +
$\tau = \rho_w gDS$
Water surface
Channel bed
θ
Depth (D)
θ
Horizontal distance +
The downstream velocity of water flowing in a river increases in a logarithmic profile from the channel bed toward the surface, with the average downstream velocity at about 0.6 times the total flow depth.
The **shear stress** (τ) exerted on the channel bed by the flow is equal to the downslope component of the weight of the overlying water $\tau = \rho_w gDS$, where ρ_w is the density of water and g is the acceleration due to gravity. The small angle approximation, where $S \sim \sin\theta$, is often used.

FIGURE 6.1 Channel Cross Section. Channel characteristics, including depth, width, slope, and water velocity, are measured for a variety of reasons, including calculating discharge (volume/time) and shear stress on the bed and bed sediments.

character of the bed and bank material also helps to set the tempo and style of channel migration. Some channels are composed of cohesive materials that resist bank erosion, such as bedrock or clay, whereas some are made of more erodible, noncohesive materials, such as sand and gravel, that are easily and frequently remobilized by streamflows.

Alluvial channels, most common in lower-gradient stream valleys that are not confined by valley walls, have a thick cover of sediment that shields the bedrock from active channel incision and allows channels to migrate laterally across valley bottoms. The beds and banks of alluvial channels are predominantly made of **alluvium,** unconsolidated material that is transported and sorted by the

PHOTOGRAPH 6.2 Landslide and Debris Flow Sediment. Channel eroded into debris flow sediment issuing from the massive Tarndale landslide in the steep uplands of the Waipaoa drainage basin, located in northern New Zealand.

PHOTOGRAPH 6.4 Alluvial Channel. The Rio Puerco flows through banks composed of unconsolidated alluvial sediments in northern New Mexico.

flow [Photograph 6.4]. In contrast, streams with **bedrock channels** actively cut into rock and flow directly over bedrock or over a thin layer of alluvium [Photograph 6.5]. They typically occupy narrow valleys with rocky walls. Most bedrock channels are found in uplands and hilly or mountainous terrain, but some occur in relatively gentle landscapes that have undergone deglaciation, recent uplift or downcutting, or in places that have dramatic lithologic contrasts or a limited sediment supply.

The grain-size distribution of material carried by a stream is an important control on channel morphology, but it is often not reflected in the sediments that accumulate temporarily in and form the channel bed. Perennial streams generally sort the material they carry; coarser sediment collects in the channel bed beneath fast-moving

PHOTOGRAPH 6.3 Bank Failure. Rotational landslide and bank erosion on a small stream in the Waipaoa basin, North Island, New Zealand. Note the termination of the slump where the bank is stabilized by the large root system of the tree at upper left. The right-hand bank shows older slumping.

PHOTOGRAPH 6.5 Bedrock Channel. An ephemeral stream in flood is confined to a bedrock channel as it flows through Red Canyon in the southern Negev Desert of Israel.

currents, and finer grains travel farther downstream until deposited in calmer water. Further sorting of the channel bed occurs during high-discharge flows that mobilize larger sediment grains. Channel beds can be armored by coarse clasts that overlie finer material. Only when high flows mobilize the armor layer, can the fine sediment beneath become entrained in the flow.

Boulders introduced into channels by landslides can be too large for even rare high flows to transport. Accumulations of boulders sometimes form persistent obstructions to the flow, such as the unmovable lag deposits from debris flows where tributaries enter the Colorado River in the Grand Canyon. Debris from these flows creates the steep rapids for which the river is famous. Since the damming of the Colorado River, reduced flood flows have allowed additional coarse debris to accumulate at such tributary junctions. Streambed material in the ephemeral streams common in arid regions and at the upstream tips of most channel networks is generally poorly sorted, lacks a surface coarse layer, and closely matches the composition of the sediment supply. In arid regions, flows are usually of such short duration and are so uncommon that there is little transit time for grain sorting.

Different sizes of sediment move at different rates and have different fates in fluvial systems. Coarse materials moving slowly as **bedload** (which rolls along the channel bed) and **saltating** load (which leaves the bed in short hops before settling back to the bed) remain in the channel and can be incorporated into floodplains and terraces by channel migration. Finer material, suspended in the flow, moves downstream at the speed of the water. Some of the fine material is evacuated from the basin, some is deposited on floodplains if the flow overtops the bank, and some is deposited in the channel as flow wanes and velocities decrease.

Vegetation

Riparian vegetation that grows along the banks of stream channels and **large woody debris** (logs) within channels influence channel processes and dynamics. Bank vegetation strongly affects channel shape and bank stability [Photograph 6.6]. Large trees growing in floodplains or along channels provide substantial root reinforcement (apparent cohesion) that stabilizes channel banks and reduces bank erosion. The stabilizing effects are most apparent along small, shallow channels where roots penetrate the full depth of the channel bank. Even in larger channels, where roots only penetrate a portion of the bank depth, root reinforcement still slows the pace of bank erosion. Deep rivers that undercut their banks can overwhelm the stabilizing contribution of roots.

Riparian vegetation also provides a source of in-channel woody debris (logs or mats of sticks, leaves, and roots) that acts as both mobile sediment and forms stable logjams. Accumulations of woody debris that slow, block, or divert streamflow influence patterns of streambed

PHOTOGRAPH 6.6 • Woody Debris. (a) Tangled root mats from silver maple trees protect the upstream margin of an alluvial mid-channel island in the Winooski River, Vermont. Flood debris (mostly wood) is jammed on the upstream side of the island. (b) Large woody debris in a small stream in Cobalt, Connecticut, creates a log step, changing the local stream gradient and impounding water and sediment.

scour, deposition, and sediment transport at scales that range from a single pool scoured from a streambed where flow is forced around a logjam, to long-term storage of alluvium behind log dams in steep, bedrock-floored mountain channels. Fallen trees may redirect stream currents into channel banks, setting off a new round of channel shifting and sediment remobilization. In forested regions, the amount of organic debris supplied to a channel can be as important a control on channel morphology, processes, and dynamics as the stream's sediment or discharge regime. For example, forest stream channels vary greatly in width locally along their courses as a result of flow deflection from wood debris. In contrast, grassland stream channels are more uniform in width.

Fluvial Processes

Understanding the discharge of water and the transport of sediment is crucial for understanding river dynamics because these are the most important processes that interact to form channels. The ability of a stream to erode, transport, and deposit sediment reflects the balance between driving and resisting forces. The shear stress (drag force per unit area) exerted by water flowing over the channel bed provides the impetus for sediment entrainment and transport. The shear strength of the bed and bank-forming materials resists erosion. The total load of a stream includes fine-grained sediment suspended within the flow; coarser sediment that slides, rolls, and bounces along the channel bed; and the material dissolved in the flow itself. Together, these characteristics influence the ability of channels to incise into bedrock and migrate laterally.

Flow Velocity

The flow velocity (U) in a channel depends on the gravity-impelled fluid driving force that is controlled by the flow depth and the slope of the channel (S), as well as the resisting force generated by the frictional resistance provided by the channel bed and banks. This frictional resistance is characterized by the channel **roughness**. Flow velocity is measured in meters per second, but channel roughness cannot be measured directly because it includes the integrated resistance to flow from the viscosity of the water and the irregularities of the channel bed and banks. The overall roughness of a channel reflects the resistance caused by obstructions that protrude into and impede the flow, including individual sedimentary particles, clusters of particles, and larger-scale **bedforms**.

Manning's equation relates the velocity of flow in a stream (U) to characteristics of the channel,

$$U = [R^{2/3} S^{1/2}]/n \qquad \text{eq. 6.2}$$

where S is water surface slope, R is called the **hydraulic radius** and is defined as the cross-sectional area of the flow (A_{cs}) divided by the wetted perimeter (P_w), and n is the **Manning roughness coefficient**. As the cross-sectional area of a rectangular channel is the product of width and depth ($A_{cs} = WD$) and the wetted perimeter is approximately equal to $W + 2D$, the hydraulic radius R may be approximated as the flow depth ($R \approx D$) in wide channels, where W is much greater than D (i.e., $W >> D$).

The Manning roughness coefficient (n) is one commonly used empirical assessment of frictional resistance to flow. The coefficient can be back-calculated if the depth, water-surface slope, and downstream water velocity are known, but it is usually estimated from channel characteristics or using comparisons with channels of known roughness value. It is typical to assign a single value of n to a channel reach, but roughness values actually change with discharge because obstacles become submerged at high flow. For natural channels, high-flow values of the Manning roughness coefficient range from 0.01 in smooth, sandy channels to ~0.2 in wood-clogged, bouldery channels. Note that when the equation is in feet rather than meters, the numerator on the right-hand side needs to be multiplied by 1.49 to account for unit conversion, because the equation is not dimensionless.

Channel roughness generally decreases as the flow depth increases in channels. At greater flow depths, friction on channel boundaries influences a smaller proportion of the flow, and deeper flows submerge large roughness elements, such as boulders and logs. Flow velocities in channels thus typically increase during floods. However, when the flood grows large enough that water spills over the bank, roughness greatly increases as shallow floodwaters move through valley bottom vegetation, such as riparian forests where trees and thick underbrush impede flow, slowing the water, and causing sediment to drop out on the floodplain. For example, Manning n values for overbank flows through woody vegetation commonly range from 0.07 to 0.2, many times higher than in most channels.

Hydrologists have developed a number of methods to estimate flow roughness from channel characteristics. One popular technique is to compare a channel with pictures of channels of known flow depth, slope, and velocity for which the roughness has already been calculated. For example, U.S. Geological Survey Water Supply Paper 1849 shows an array of channels with measured roughness values to which a channel of interest can be compared in order to arrive at a quick estimate of an appropriate roughness coefficient [**Photograph 6.7**]. Another method is to estimate roughness by consulting a table of channel characteristics (**Table 6.1**).

Bed and bank roughness that slows flow along the margins of a channel usually results in maximum flow velocity near the surface in the middle of the channel (see Figure 6.1). Flow velocity increases from zero at the bed to maximum velocity directly below the water surface. Flow at the surface is somewhat slower because of the drag associated with the air–water interface and because of vortices of more slowly flowing water shed from the channel margins. Flow immediately above the channel bed is typically characterized by a thin layer of water, sometimes called the **laminar sublayer,** where water molecules travel smoothly along parallel flowpaths. Higher in the water column, the velocity increases logarithmically toward the maximum near the surface. Because of this nonlinear increase, average flow velocity generally occurs at about 60 percent of the flow depth, and the average speed of the flow is usually about 80 to 90 percent of the surface velocity.

Discharge is not uniform across a channel, so determining an average velocity in the channel is best done by measuring flow depth and velocity at a number of locations (typically at least 10) across a channel and then calculating an average discharge value. This can be done either by averaging

(a)

(b)

(c)

(d)

PHOTOGRAPH 6.7 Channel Roughness. Channels have different roughness values, expressed as Manning n values. Images taken from the USGS Water Supply Paper 1849, a compendium of channel roughness values with photographs and cross sections of channels in which the n values were measured. (a) n = 0.026, Indian Fork below Atwood Dam, near New Cumberland, Ohio. (b) n = 0.036, West Fork Bitterroot River near Conner, Montana. (c) n = 0.057, Mission Creek near Cashmere, Washington. (d) n = 0.073, Boundary Creek near Porthill, Idaho.

many measurements of depth and the downstream component of the flow velocity and multiplying the averages by the stream width, or it can be done making a series of discharge measurements across the channel. The latter method requires sectioning the channel into virtual flow tubes. Each flow tube defines a width increment of stream that appears to have similar depth and velocity. Starting at either the right or left bank, as seen when facing downstream looking in the direction of the flow, one measures the width, average depth, and downstream water velocity (using a current meter) in the first increment and calculates a discharge. This process is continued

TABLE 6.1

Typical Manning coefficient values (n)

Channel-bed material	n =
Straight canal/concrete banks	0.01–0.02
Straight canal/earthen banks	0.02–0.03
Sand	0.01–0.03
Sand/gravel	0.03–0.05
Cobble/boulder	0.04–0.08
Timber/vegetation-choked	0.07–0.16

across the entire channel width and the resulting incremental discharges are then summed.

Measurements of discharge and **stage** (water-level elevation) taken at a range of flows over time define a **rating curve** (a plot of discharge versus stage) that allows for estimating discharge at different water levels (see Figure 2.8). Flow stage is readily converted to flow depth if one also knows the elevation of the streambed. Once a rating curve has been constructed for a channel, discharge can then be estimated by simply measuring the flow stage. This can be done manually by reading a **staff gauge** or can be automated using digital pressure transducers.

Most natural channels exhibit **turbulent flow** in which the velocity continuously fluctuates, producing eddies that mix the flow and greatly increase the flow resistance. Different types of turbulent flow (defined below) determine what kind of bedforms develop on the channel bed. The specific type of flow that occurs can be calculated using the **Froude number** (Fr), defined as the ratio of the flow velocity (U) to the speed at which a surface wave will propagate, which is given by the square root of the product of flow depth (D) and gravitational acceleration (g):

$$Fr = U/(Dg)^{0.5} \qquad \text{eq. 6.3}$$

Tranquil (or subcritical) flow happens when Fr < 1, during **critical** flow Fr ≈ 1, and **supercritical** flow is when Fr > 1. The Froude number indicates whether the water is moving faster or slower than its own wake, so a simple test for the difference between subcritical and supercritical flow is to toss a rock into the stream. If the ripple from the splash travels upstream against the current, then the current is subcritical. But if the current sweeps the expanding splash downstream, then the flow is supercritical. At critical flow, there are standing (or stationary) waves. These three flow regimes are geomorphically important because they transport and deposit sediment differently and are thus related to development of distinctive bedforms on channel beds.

Discharge Variability

Flows that fill a channel to the point of overflowing are called **bankfull flows** [**Photograph 6.8**]. The frequency with which a channel experiences bankfull flow varies from several times a year in humid environments to once every few decades in arid regions. When discharge exceeds bankfull stage, flood flows spill out over the channel banks and inundate valley bottoms, as discussed in Chapter 4.

Flows of different size have different velocities and thus transport differing amounts and sizes of sediment. The discharge that transports the most sediment over a period of years to decades is called the **effective discharge**. More extreme events (such as the flow that occurs on average once every 50 years) move more sediment per event than flows that recur more frequently. However, such extreme events are too rare to dominate alluvial channel development, although extremely large flows may cause **avulsions** or cutoffs, which change the path of channels, altering their form and the direction and location of flow.

PHOTOGRAPH 6.8 Bankfull Flow. Bankfull flow of high turbidity water after a very heavy rain, shown at Small Creek near Nelson, North Island, New Zealand.

In perennial humid-temperate streams, effective discharge often corresponds to the bankfull flow with a one-to-two-year recurrence interval. In monsoon-driven stream systems, the effective discharge corresponds to the monsoon-season baseflow because that flow is large and lasts long enough to carry a lot of sediment. In arid channel systems, the effective discharge usually has a long recurrence interval, reflecting the rarity of precipitation events long enough and intense enough to generate significant runoff.

In channels with alluvial banks, the bankfull flow is important geomorphically because it generally represents the discharge to which channel width and depth are adjusted. In humid regions, channel dimensions thus reflect flows that recur on average every year or two. In arid regions, channels are shaped by longer recurrence interval events because bankfull flows are much less common than in humid regions.

A channel's **"at-a-station hydraulic geometry"** describes how its width (W), depth (D), and velocity (U) vary as discharge (Q) rises and falls [**Figure 6.2**] over the course of many different flow events. Hydraulic geometry is defined using three linked equations that describe how these factors increase with discharge through a given channel cross section:

$$W = aQ^b \qquad D = cQ^f \qquad U = kQ^m \qquad \text{eq. 6.4}$$

where a, c, and k are empirically determined constants, and b, f, and m are empirically determined exponents. Because discharge is the product of the flow width, depth, and velocity (Q = WDU), the product of the constants must equal 1 (ack = 1), and the sum of the exponents must also equal 1 (b + f + m = 1). The wide range of b, f, and m values empirically determined for at-a-station hydraulic discharge relationships (0–0.5, 0.3–0.6, and 0.2–0.6, respectively) reflects differences in channel cross-sectional form that reflect, for example, whether the stream flows between resistant bedrock outcrops or

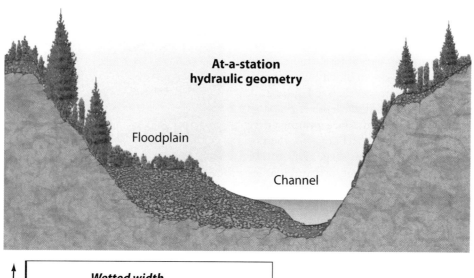

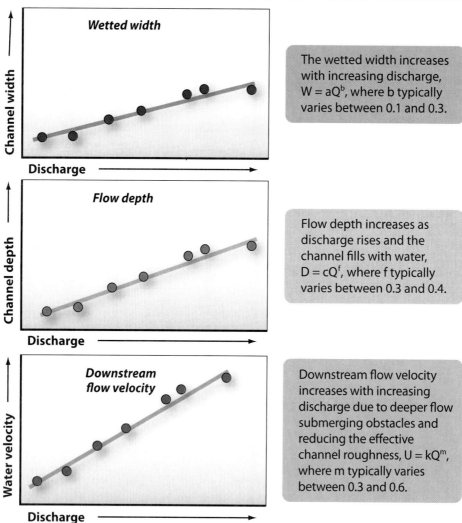

The wetted width increases with increasing discharge, $W = aQ^b$, where b typically varies between 0.1 and 0.3.

Flow depth increases as discharge rises and the channel fills with water, $D = cQ^f$, where f typically varies between 0.3 and 0.4.

Downstream flow velocity increases with increasing discharge due to deeper flow submerging obstacles and reducing the effective channel roughness, $U = kQ^m$, where m typically varies between 0.3 and 0.6.

The particular relationships between width, depth, and downstream flow velocity depend on the geometry and **roughness** of the channel. Width, depth, and downstream flow velocity (W, D, and U) typically increase as power-law functions of discharge. At a given channel cross section, flow velocity typically increases faster than depth or width, as deeper flow reduces the effective bed roughness.

FIGURE 6.2 At-A-Station Hydraulic Geometry. The hydraulic geometry of a channel cross section (at a station) describes the relationships between increasing discharge and the flow width, depth, and velocity.

erodible alluvial banks, which may be scoured back during high flows.

Flow depth at a particular channel cross section (at a station) typically exhibits significant proportional change as discharge increases, but channel widths are generally constrained by the streambanks. Consequently, increased discharge is mostly accommodated by increased flow depth and velocity. Knowing the hydraulic discharge relationships for a location allows predictions of flow depths and velocities at a range of discharge values.

Stream Power

The ability of a stream to transport sediment and carve into the underlying bedrock is related to its **stream power** (Ω), the rate of potential energy loss per unit channel length, which is defined as

$$\Omega = \rho_w g Q S \qquad \text{eq. 6.5a}$$

where ρ_w is water density, g is gravitational acceleration, Q is discharge, and S is channel slope ($\tan\theta$); see **Box 6.1** for derivation.

Unit stream power (ω) is the rate per unit area of channel bed at which the river expends or dissipates its potential energy in the process of flowing downstream,

$$\omega = \Omega/W \qquad \text{eq. 6.5b}$$

where W is the channel width. Unit stream power is equivalent to the product of shear stress and flow velocity (i.e., $\omega = \tau U$). If the same discharge flows through a narrower channel, the deeper and likely faster flow will have more stream power per unit channel bed and will therefore be capable of greater sediment transport or bedrock incision. Similarly, steep channels will have higher unit stream power than channels with low slope. The geomorphic implications of unit stream power are significant. For example, mountain stream channels, steepened by uplift, can have sufficient stream power to incise at high rates and thus keep up with uplift.

Bedrock Incision

Channel incision into bedrock requires mobilization of the alluvial cover to expose the underlying bedrock to erosion. In general, it takes more stream power (slope, discharge) to cut down into bedrock than to move sediment, so channel incision into bedrock occurs only after sediment is in motion [**Figure 6.3**]. In many cases, non-flood flows and frequent, smaller floods drive sediment transport processes, whereas larger, less frequent events control channel incision into bedrock. Thick sedimentary cover serves to protect the bedrock beneath a streambed from erosion. In contrast, a very low sediment load provides few clasts that can act as abrasive tools entrained in the flow to erode exposed bedrock. The highest rates of bedrock incision are thus expected to occur in channels with intermediate sediment loads (see Digging Deeper).

Streamflow incises into bedrock through a combination of **abrasion, plucking,** and **dissolution,** and the relative importance of each process depends on the bedrock type and stream morphology. Abrasion sandblasts bedrock with material

---

BOX 6.1 Derivation of Stream Power

Water flowing down a river channel loses gravitational potential energy as it drops in elevation. Conservation of energy requires that potential energy is transformed into other forms of energy, including frictional heating and performing work to transport sediment or erode the channel bed. The rate of potential energy loss, defined as the stream power, is a measure of the ability of the channel to move sediment and incise into rock. The potential energy (PE) of the water per unit channel length is given by the product of the mass of the volume of water, itself the product of water density (ρ_w) and gravitational acceleration (g), and the channel width (W), flow depth (D), and elevation (z):

$$PE = \rho_w g W D z \qquad \text{eq. 6.A}$$

The downstream rate of potential energy loss (dPE/dx) is given by how fast the water drops in elevation, the channel slope (S). Thus, stream power (Ω) is given by:

$$\Omega = \rho_w g W D U S \qquad \text{eq. 6.B}$$

where U is the flow velocity and S is the channel slope ($\tan\theta$ or dz/dx). Noting that the product of WDU is equal to the flow discharge (Q), this reduces to:

$$\Omega = \rho_w g Q S \qquad \text{eq. 6.C}$$

If we consider the amount of work that the channel is able to do on its bed, it is reasonable to formulate the downstream rate of power loss in terms of the rate of power loss per unit bed area, termed the unit stream power (ω). In this case, we divide eq. 6.C by the channel width, yielding

$$\omega = \rho_w g Q S / W \qquad \text{eq. 6.D}$$

an expression that suggests that the rate of channel incision in upland bedrock channels is a function of the local water discharge, slope of the channel, and channel width.

River channels are underlain by both rock and sediment. In order for the river to incise into the underlying bedrock, it must first entrain sediment from the channel bed. Bedrock cannot be eroded until its sediment cover is in motion.

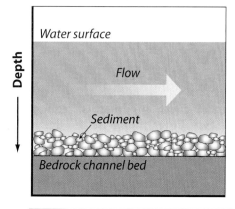

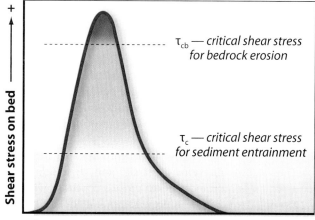

During a flood, shear stress on the stream bed increases as the depth and velocity of water increase. The **critical shear stress** above which bedrock erosion occurs (τ_{cb}) is higher than that required to entrain sediment from the channel bed (τ_c). Thus, rare floods, high-stream power events that generate large shear stresses on the bed, are required to erode rock. Often bedrock only erodes significantly during the highest flood peaks.

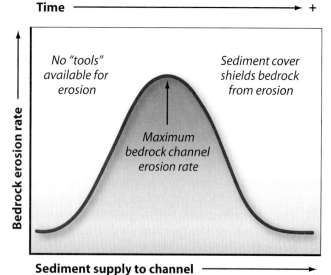

The rate of **bedrock incision** also depends on the sediment supply to the channel. The maximum bedrock erosion rate occurs when there are enough "tools" (clasts) entrained in the flow to erode the bed, but not enough sedimentary cover to protect the underlying bedrock from erosion.

FIGURE 6.3 Bedrock Erosion Thresholds—Geomorphically Effective Events. Bedrock erosion requires that shear stress exceeds a critical threshold. The rate of bedrock erosion depends on the competing effects of the availability of sediment to provide tools for erosion and its effect on shielding the bedrock surface from impacting grains.

entrained in the flow, carving channels and sculpting rock into polished bedforms like potholes and flutes on cohesive, erosion-resistant channel beds [**Photograph 6.9**]. **Potholes** are cylindrical forms eroded into rock by rapidly moving vortices carrying abrasive, sand-sized sediment. Potholes can form vertically or horizontally. Once formed, potholes significantly lower rock-mass strength and therefore catalyze plucking, which is the removal of rock slabs bedrock and make it susceptible to plucking. Streams also dissolve their way down through bedrock in regions with soluble rock, particularly limestone landscapes in humid climates where extensive dissolution leads to development of karst terrain (see Chapter 4). Dissolution often preferentially enlarges fractures or jointing in channel-bed rocks, which in turn promotes plucking.

Channel Migration

The flow of water around channel bends determines in-channel patterns of sediment erosion and deposition and drives channel migration in alluvial channels. As flow enters a bend, the centrifugal force elevates the water surface on the outside of the curve. This superelevation sets up a cross-channel component to the water surface slope that drives flow down the outer bank and back across the channel, creating helical flow that spirals downstream in a roughly corkscrewlike pattern [**Figure 6.4**]. This secondary circulation results in a zone of converging flow that scours out pools along the outer bank of a bend, erodes the bank, and deflects flow back across the channel. Divergent flow on the inside of the channel bend likewise results in local deposition that builds up a crescent-shaped **point bar** from deposition of bedload, creating a topographic obstruction that steers flow back across the channel into the outer bank. **Cutbank** erosion on the outside of channel bends and point-bar deposition on the inner bank drive progressive channel migration laterally and downstream [**Photograph 6.11**].

In almost all channels (not just meandering ones), the deepest part of the flow, called the **thalweg** (from the German, meaning "valley way"), follows a path through shallow riffles (areas of fast, shallow, turbulent flow) that connect deeper pools. During periods of low

PHOTOGRAPH 6.9 Bedrock Incision. (a) Fluted bedrock surface on a former bedrock riverbed above the Watson River near Kangerlussuaq, western Greenland. (b) Large, round symmetrical pothole next to swimmers on the Ammonoosuc River, New Hampshire.

along joints or other fractures in the channel walls and bed. Adjacent potholes can erode into one another, undercutting whole sections of gorge walls.

Flowing water also loosens and plucks material from the beds of channels. Loose or broken rocks along fractures, joints, or bedding planes are particularly susceptible to plucking during floods [**Photograph 6.10**]. **Hydraulic wedging** of sand and gravel into openings helps loosen

PHOTOGRAPH 6.10 Plucking Jointed Blocks. Joints in schist on the bed of the Potomac River, northern Virginia, provide weaknesses that allow tractive and lift forces, developed during flood flows, to pluck and move large pieces of rock. Note the pothole drilled on a gently dipping joint face in center of the image.

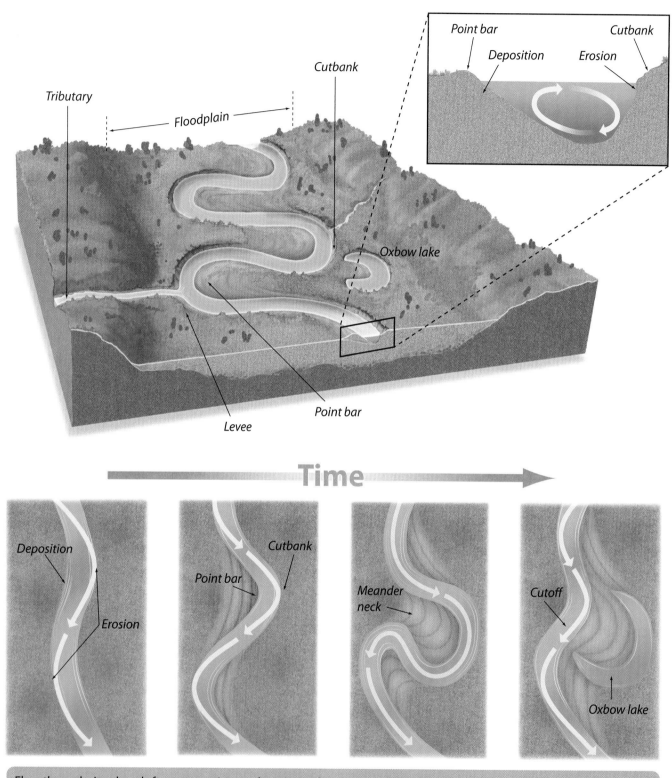

FIGURE 6.4 Meander Migration—Erosion and Deposition. Flow through meanders results in a predictable pattern of erosion and deposition as well as downstream translation of channel bends over time.

deposition rate on point bars, which then become buried by overbank deposition over time. The pace of bank erosion is controlled by bank erodibility, stream size, and flow velocity. In many channels, one can readily see evidence for this process of gradual channel migration in the progression of vegetation height (and age) away from the channel on the inner (point bar) side of meander bends [Photograph 6.12].

The point of maximum bank erosion is often just downstream of the meander apex, which leads to the downstream migration of meanders. However, obstructions like bedrock outcrops in alluvial floodplains or along valley walls can block downstream migration and force upstream meanders to bunch up. In-channel obstructions like boulders, ice dams, and logjams also influence meander patterns and locations.

In addition to gradual channel migration by cutbank erosion and point-bar deposition, channels may suddenly move to a new position, a process called **avulsion**. Typically, an avulsion happens in places where ongoing lateral migration brings the outer banks of two adjacent meanders so close that continued erosion allows them to intersect. When the upstream side of a meander captures the downstream side, it cuts off the intervening loop and creates a loop-shaped slough, called an **oxbow lake** [Photograph 6.13]. In addition to meander cutoff events, channels sometimes avulse when a flow obstruction like a logjam blocks the channel and causes flow to spill over the streambank and carve a new channel across the floodplain or shift flow into a secondary side channel or to an inactive, abandoned channel. Streams with multiple side channels are thus common in forested terrain where woody debris is large enough to create logjams capable of diverting flow. Stable logjams at the heads of diverted channels are often porous enough to allow some flow to enter, creating high-quality, low-disturbance fish habitat.

PHOTOGRAPH 6.11 **Point Bar and Cutbank.** Paired point bar and cutbank along a well-developed meander bend in Tuxedni Bay, Alaska.

flow, riffles have steep water-surface slopes and rapid flow, whereas pools have relatively flat water-surface slopes and slower flow. So how are pools scoured out? In contrast to conditions at low flow, high-flow velocities and water-surface slopes in the pools increase more rapidly than in the riffles, leading to a velocity reversal in which the water speed at high flow is greater in the pools than in the riffles. At high flows, shear stress increases toward the downstream ends of the pools, scouring and transporting coarse bed material to be deposited in lower velocity riffles where shear stresses decline downstream.

The combination of focused bank erosion at cutbanks and deposition of point bars on the inner bank causes channel migration. **Meanders**, the winding curves or bends in a river, migrate toward the eroding outer banks, and point-bar deposition fills the inside of the curve. In most cases, the rate of erosion on cutbanks matches the

PHOTOGRAPH 6.12 **Vegetation Height and Channel Migration.** Point bar on the Winooski River, Vermont, with "layers" of vegetation indicating time since disturbance. Closest to the river, only annual grasses are present; farther back are shrubs, then small trees, and, at a distance, mature trees.

PHOTOGRAPH 6.13 **Oxbow Lake.** This oxbow lake formed from an abandoned river meander and is now isolated from the channel of the Rio Aquidauana, Mato Grosso do Sul, Brazil.

Sediment Transport

Streams dissipate most of their energy by friction and turbulence. Only a small fraction is available to erode and transport sediment, but even so, streamflow is a key agent of sediment erosion, transport, and deposition on the continents. The physics involved in rigorously analyzing the movement of sediment in rivers is complex enough to have reportedly discouraged a young Albert Einstein from pursuing a career in physical geography. Consideration of the forces acting on particles within the flow and on the streambed provides substantial insight into how rivers transport sediment. Such insight is geomorphically important because the processes of sediment entrainment, transport, and deposition govern the sorting of material in transport, the development of systematic patterns of channel morphology, and the formation of channel bedforms.

Initiation of Transport

Water flows in a channel under the influence of gravity (see Figure 6.1). The ability of streamflow to displace or erode the material in its bed, whether to transport sediment or incise bedrock, is due to the **shear stress** (τ), the force per unit area that a river exerts on its bed:

$$\tau = \rho_w g D \sin\theta \qquad \text{eq. 6.6}$$

where ρ_w is the density of water, g is gravitational acceleration, D is flow depth, and θ is the water surface (channel) slope in degrees. The flow depth is location-specific and is typically an average value for the area over which the channel slope is determined. For the small slopes typical of riverbeds, $\sin\theta \approx \tan\theta$, and geomorphologists commonly adopt this small-angle approximation for applications in low-gradient channels. This means that $\tau \approx \rho_w g D \tan\theta$, where $\tan\theta$ is the drop in water surface elevation, divided by downstream distance for a given channel segment. A second common assumption is that the water-surface slope may be approximated by the channel slope.

The flow of water over sediment particles exerts both lift and drag forces that act to drive sediment transport (see Figure 1.11). The submerged weight of the particles and the frictional resistance between neighboring grains act to resist particle movement. The degree to which a grain protrudes into the flow and its surface exposure to the

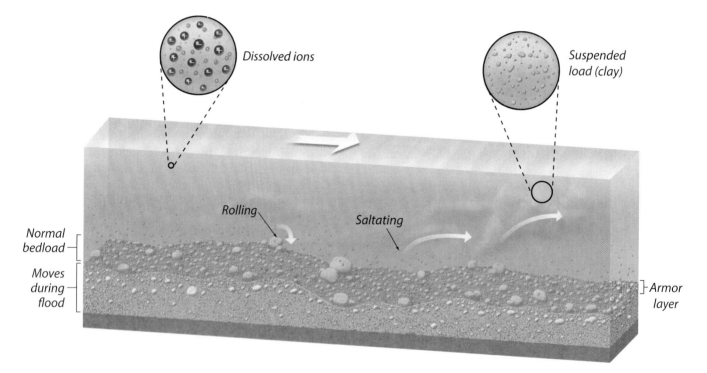

FIGURE 6.5 **Stream Load.** Streams and rivers carry sediment as bedload, suspended load, and dissolved load.

Streams carry material as **dissolved load, suspended load,** and **bedload.** Dissolved load is composed of ions in solution that travel at the speed of the flow. Suspended load (typically silt and clay) is composed of material suspended by turbulence in the flow and moving at the speed of the flow. Bedload moves by rolling or sliding along the channel bed and is typically composed of gravel and cobbles. Sand may travel as either suspended load or bedload, depending on the flow velocity. **Saltating** sediment is swept from the bed, then travels some distance while settling back to the channel bottom. Bedload moves intermittently and thus more slowly than the flow. Streambeds are often **armored** by a layer of large clasts due to winnowing of finer material from the bed.

flow influence the drag forces acting on the grain. When the driving forces exceed the resisting forces, sediment begins to roll along the streambed or is thrust up into the flow (**saltation**), where it is carried along and settles at a velocity that depends on clast size and density. Upward-directed velocity fluctuations keep small particles suspended in the flow while larger sediment rolls, slides along the bed, or bounces before it settles back to the bed [**Figure 6.5**]. Flow velocity increases rapidly above the bed (see Figure 6.1), so sediment tends to move downstream once it is entrained in the flow.

Entrainment of material from the streambed is necessary to initiate bedload transport [**Figure 6.6**]. Resistance to motion depends on the size, shape, and density of sediment particles; their interlocking relationship with neighboring grains; and their exposure to the flow. Although not all particles on a streambed are mobilized at the same time, streambed gravels generally mobilize at flow velocities that exceed a **critical shear stress** (τ_c), characterized by

$$\tau_c = \tau_c^* \, g (\rho_s - \rho_w) \, d_{50} \qquad \text{eq. 6.7}$$

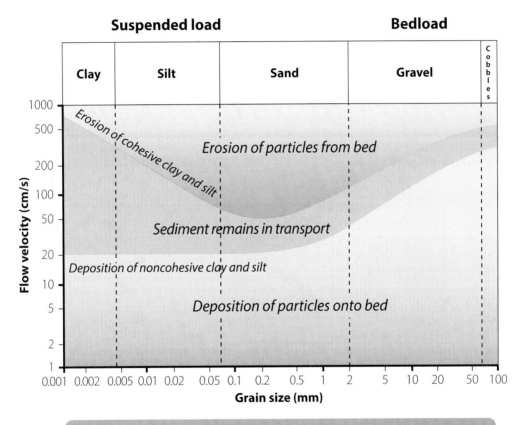

Erosion: The flow velocity required to erode material from a channel bed is a function of grain size. Sand is eroded at lower flow velocities than both coarser material (gravel and cobbles) and finer-grained material (silt and clay). To erode silt and clay, water must be moving quickly enough to overcome the **cohesive** strength of the material. The greater velocity needed to erode larger particles reflects their greater mass.

Transport: For large grain sizes that travel as **bedload,** there is little difference in the flow velocity required for erosion and deposition. In contrast, smaller particles that travel as **suspended load** can remain in motion at flow velocities well below those required to erode them.

Deposition: Fine-grained material (silt and clay) settles out in very still water, whereas coarse-grained material settles out even in swift water.

FIGURE 6.6 Hjulström Curve. Different sediment grain sizes are entrained, transported, and deposited in streams at different but characteristic flow velocities.

where τ_c^* is the **Shields parameter** (also known as the dimensionless critical shear stress), ρ_s is the density of the sediment, ρ_w is the density of water, and d_{50} is the median diameter of the bed-forming grains. Because individual particles are locally mobilized when bursts of high-velocity, turbulent flow sweep the bed and kick clasts up into the flow, initial motion is generally defined not by the motion of a single clast but rather by a less precise observation of general bedload mobility.

Streambed mobility is usually analyzed using the median grain size because larger clasts that protrude above the bed protect their smaller-than-average neighbors from the flow. Shields parameter values are generally about 0.06 for gravel streambeds, but the value varies among different channels, a reflection of the relative degree of sorting and packing of the streambed material, and ranges from 0.03 for loosely packed gravel to 0.1 for coarse, well-packed cobbles. Shields parameter values are generally high for cohesive, fine-grained sediments, like clay, that require large shear stress (velocity) to initiate bed mobility. Once the critical shear stress threshold is reached and streambed sediment is mobilized, bedload transport rates generally increase with increasing shear stress or stream power [**Figure 6.7**].

Sediment grain size determines the flow velocity at which sediment erodes and is deposited, in part because grain mass scales nonlinearly with grain diameter and in part because cohesive forces between grains become important at small grain sizes. For grain sizes larger than 1 to 2 mm, the critical-flow velocity required to erode material from a riverbed increases with grain size and is similar to the velocity below which sediment will be deposited. For cohesive materials smaller than about 0.2 mm in diameter, the flow velocity required to erode material from the bed increases with decreasing grain size because of the cohesion characteristic of fine-grained material like silt and clay (see Figure 6.6) and because fine particles do not extend above the laminar sublayer into the turbulent flow above.

Sediment mobility is a function of grain size. Once mobilized, fine-grained particles tend to remain suspended in the flow, even at very low flow velocities. Coarse grains rapidly fall back to the stream bottom and transport of coarse material ceases at flow velocities close to those at which it was entrained. Sediment finer than sand tends to travel in suspension, while gravel and coarser sediment travels along the bed, as bedload. Sand, however, can

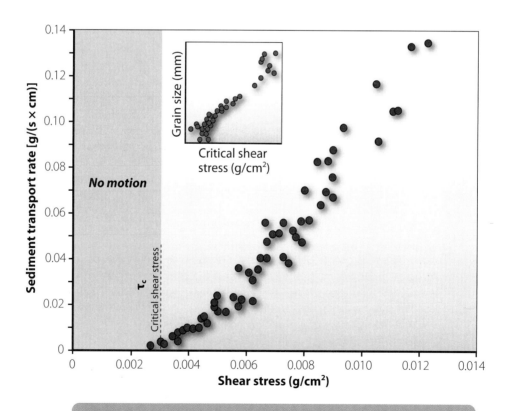

Bedload transport (**entrainment**) typically begins at a **critical shear stress (τ_c)**, below which there is no motion and above which sediment transport rates increase with increasing shear stress. The critical shear stress needed to initiate motion increases with grain size (see inset).

FIGURE 6.7 Shear Stress and Transport Rates. A critical shear stress must be exerted on the channel bed by flowing water before sediment can move. The critical shear stress needed to move sediment increases with sediment grain size. As shear stress increases, the sediment transport rate also increases. [Adapted from Leopold et al. (1964).]

travel either as suspended load or bedload in many river systems.

Sediment Loads

The total load of a stream consists of material dissolved in, carried within, or pushed along by the water flowing in a channel (see Figure 6.5). Loads vary between watersheds and over time. Most mass is carried by streams when they are in flood. Stream load can be partitioned into the **clastic load** (grains of sediment) and the **dissolved load** (ions and molecules in solution). Stream systems that drain steep catchments in rapidly eroding, tectonically active settings, like the Himalaya, carry most of their load as sediment grains, the clastic load. The relative amount of clastic load varies greatly among rivers, from less than 10 percent to more than 90 percent of the total load and averages about 75 percent for the world's largest rivers. The Saint Lawrence River, which drains to the North Atlantic, is a notable example of a river that carries a high percentage of its total load as dissolved material. Most of the clastic sediment is trapped by the Great Lakes, which are in the basin's headwaters.

The dissolved load consists of material contributed by chemical weathering and therefore reflects the weathering regime and solubility of the rocks within the drainage basin. Dissolved load concentration varies greatly, from ~40 parts per million (ppm) for the Amazon to ~850 ppm for the Colorado River, with a global average of about 120 ppm. A stream's total dissolved load is directly related to the discharge volume and the average concentration of dissolved material. The percentage of the total sediment load that is carried as dissolved matter varies greatly among streams in different environments. Channels that drain low-gradient catchments dominated by chemical weathering, or that have large lakes acting as sediment traps, carry most of their loads as dissolved matter. The dissolved load has little effect on fluvial processes and channel morphology; however, it is often critical for stream organisms and the health of downstream water bodies.

The **suspended load** is the part of the clastic load that consists of material fine enough to remain suspended by turbulence; it is carried along at a velocity similar to that of the water. For a grain to be held in suspension, its **settling velocity** (the rate at which it moves downward through still water) must be lower than the upward component of the velocity field created by turbulent eddies. The suspended load of most rivers consists of silt and clay because the settling velocity of a particle depends on its density and the square of its radius—large particles settle out faster than finer ones. The finer component of the suspended load is likely to stay suspended until it either ends up on a floodplain, river delta, or lake floor or settles out during waning discharge as a readily remobilized layer that blankets the streambed until the next high flow. This can be seen in patches of sand on bar tops along many gravel riverbeds and in clay drapes on sand ripples. Material fine enough that it never settles to the bed, even during low flow, is called the **washload**. Like the dissolved load, washload is transported downstream at the velocity of the flow with no geomorphic effect.

Bedload is clastic material that is transported by rolling, **saltating** (bouncing), and sliding along the channel bed. The material that makes up a streambed generally consists of bedload material that is periodically remobilized and deposited by fluctuating streamflow. Although bedload generally accounts for a minor portion of the sediment moved by rivers, typically amounting to 10 percent (and up to as much as 30 percent) of the total load, it is an important factor in determining channel morphology as it forms the bed and some banks.

The size of material transported as suspended load or bedload changes as the discharge volume and flow velocity rise and fall. For example, in many channels, sand moves as bedload at low flows but becomes part of the saltating and then suspended load during high flows. Material moving as suspended load during high-discharge events likewise settles to the streambed as a flood recedes. Gravel and coarser material generally travel as bedload in most rivers and streams, but even boulders can be temporarily suspended and moved downstream in large enough floods on big, steep rivers.

In contrast to the dependence of washload and suspended load on the supply of material delivered to the channel (most channels are capable of carrying far more suspendable material than they actually receive), bedload conveyance in alluvial channels is limited by the transport capacity of the channel. Bedload sediment transport rates (Q_b) generally increase nonlinearly with increasing discharge above the flow required to initiate bed mobility. However, prediction of bedload transport rates from channel characteristics like slope, depth, or stream power is complicated by the observation that rates of bedload transport vary greatly for similar hydraulic conditions in different channels. Consequently, the accurate prediction of bedload transport rates often requires calibration against field measurements.

Direct measurement of representative bedload transport rates is difficult because of temporal and spatial variability and because placing a sample collection device on the streambed disturbs stream flow and bedload transport. Also, it is usually neither safe nor feasible to collect samples during flood flows that mobilize coarse riverbeds. Clever ways to measure bedload transport include recessing open troughs in small streambeds and tracking the movement of magnetically tagged particles as they move downstream. Where conditions and access allow, bedload can also be measured directly with a sampling device that resembles a sturdy butterfly net, tipped on its side at the end of a long pole (called an Elwha sampler). Placed flush with the streambed, material rolling into the net is retained in a trailing mesh bag.

Wood and other floating organic matter sometimes constitute a substantial component of the overall load carried by a stream. Unsaturated wood is less dense than water and floats downstream as washload. Saturated logs can sink and become part of the bed material. In forested regions, extensive logjams can partially obstruct, dam, and even divert streams. The presence of a root ball enhances the stability of a log because the heavy, waterlogged wad of roots acts like an anchor as a log grounds out and begins to deflect flow. Logs that are large and waterlogged enough form stable **key members**, which are obstructions to flow that can capture additional wood and anchor logjams. The transport and storage of logs and organic debris within stream systems can greatly influence channel morphology and dynamics in forested regions.

Bedforms

Streambeds are rarely smooth because erosion, transport, and deposition of sediment shape a variety of **bedforms** on channel floors that result from feedback between flow and sediment transport. **Bars** are large-scale, elongate bedforms that are often longer than the channel width and occur in many shapes and positions within channels. Some bars are relatively permanent features that form in local sediment storage zones where sediment accumulates during high flows [Photograph 6.14]. Persistent bars commonly develop at channel bends, confluences, or logjams along mobile alluvial channel beds. In floods, bedload particles generally move from one sediment bar to the next and are temporarily stored there between bed-mobilizing flows. Gravel-bed and sand-bed channels have similar types of bars, but their suites of finer-scale bedforms differ. Some bedforms are stable features while others are mobile and progressively advance up- or downstream.

Flow over a channel floor of readily deformable material, like loose sand, produces a range of bedforms [Photograph 6.15] that change with the flow depth and contribute substantial flow resistance. In such channels, the Froude number is geomorphically important because it determines the bed geometry. For example, plane beds and low-amplitude ripples commonly form in sandy channels at low-flow velocities or when deep flow (Fr << 1) generates a flow regime near the channel bed with minimal bedform roughness. Dunes with wavelengths of 4 to 8 times the flow depth and heights up to one-third of the flow depth form at higher flow velocities and/or shallower flow depths (Fr < 1). When flow approaches critical velocities (Fr ≈ 1), dunes wash out and form an upper flow-regime plane bed devoid of bedforms. Supercritical flow (Fr > 1) at even higher velocities and/or shallower depths produces anti-dunes that are like dunes but face upstream and their form migrates upstream if the current is not too fast. Similar suites of bedforms develop in coarser-grained channels at high discharges.

In gravel-bed channels, bars are the dominant bedform, although other distinctive bedforms and erosional features can develop. Streamlined particle tails and clusters of coarse clasts aligned with the flow direction accumulate downstream of flow obstructions. Transverse-to-flow

PHOTOGRAPH 6.14 **Bars.** Bars are common bedforms along channels carrying sediment. (a) Alternating point bars along the gravel-bedded, meandering Tolt River in western Washington State. (b) Mid-channel bars form in a sandy stream reach, outside of Colorado Springs, that has been artificially straightened.

PHOTOGRAPH 6.15 **Dunes.** These bedforms in the South Platte River, Colorado, are large sediment waves called dunes. Dunes form in subcritical flow with high sediment-transport rates.

ribs, which consist of repeated ridges of coarse clasts at a spacing determined by the channel width and the largest clasts, develop in areas of supercritical flow. Larger, stair-like gravel steps sometimes floor steep gravel channels, and channel-spanning steps of coarse clasts are common in cobble-boulder channels.

In perennial channels, a coarse-grained surface layer generally forms from winnowing of small particles from the bed surface during nonflood flows between full bed mobilizing events. This coarse surface layer (armor) typically extends down about twice the median surface particle diameter and overlies finer material that is more characteristic of the total bedload grain-size distribution once the bed surface is mobilized. Ephemeral channel beds in semi-arid regions generally do not have this coarse surface layer because they lack day-to-day flows that are capable of sorting and winnowing particles between flood events, and because abundant sand precludes formation of a gravel armor layer in most locations.

Channel Patterns

Channel morphology reflects the interplay of fluvial processes and the routing of material through drainage basins, stream-valley segments, and individual channel reaches. Distinct channel patterns and the morphologies of individual reaches arise from different balances between sediment supply and transport capacity, as well as from the bedrock structure, climate, and supply of large organic debris along the stream valley.

Stream channels are either single-thread or multi-thread and follow either relatively straight or tortuous paths. The **sinuosity** of a river can be defined in various ways, but a simple definition is the ratio of the channel length measured along the center of the channel to the straight-line distance measured down the valley axis [**Figure 6.8**]. High-sinuosity channels follow convoluted, twisting paths, and low-sinuosity channels follow relatively straight paths. Sinuosity and channel pattern change

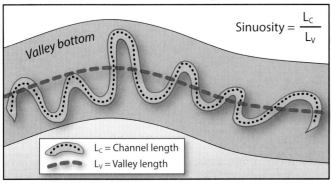

Channel sinuosity

Sinuosity = $\dfrac{L_C}{L_V}$

L_C = Channel length
L_V = Valley length

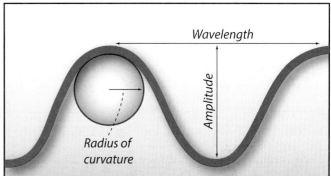

Geometry of a meander

Wavelength

Amplitude

Radius of curvature

Straight
sinuosity < 1.3

Meandering
sinuosity > 1.5

Braided
unvegetated bars

Anastomosing
vegetated islands

Single-thread channels exhibit either **straight** or **meandering** patterns. Straight channels are typically confined by valley walls and have **sinuosity** < 1.3, whereas meandering channels typically flow across broad floodplains and have sinuosity > 1.5. The meander belt is the zone of active meandering.

Multi-thread channels exhibit either **braided** or **anastomosing** patterns. Braided channels exhibit multiple unvegetated, frequently shifting channels that converge and diverge within a larger channelway. Anastomosing channels divide into multiple channels that flow around vegetated islands.

FIGURE 6.8 Channel Patterns. Channels can have a variety of forms, which are classified as single-thread or multi-thread. Both sinuosity and meander wavelength can be quantified.

from one reach to another downstream within a channel network as a result of local influences on sediment supply and discharge. Natural channels vary within a wide spectrum of patterns, but the simple distinction of **straight, meandering, braided,** and **anastomosing** channel patterns provides a useful framework for understanding the processes that control channel morphology (see Figure 6.8).

Different channel patterns arise from differences in bankfull discharge, gradient, sediment supply, and bank material. Braided channels occur on steeper slopes and with greater discharges and sediment loads than single-thread meandering channels [**Figure 6.9**]. Cohesive banks favor development of meandering channels, whereas weak, noncohesive banks favor development of braided channels. Local stabilization by vegetation or large organic debris leads to development of anastomosing channels that are split into multiple individual channels separated by stable, vegetated islands. Streams with highly erosion-resistant banks, such as bedrock canyons, often follow structural weaknesses like faults, joints, or sedimentary bedding planes. Although it has long been argued that meandering channels within deeply incised bedrock valleys inherited their courses from past channels cut in higher (now eroded) alluvial surfaces, it now appears that meandering channels can form over long periods of time as bedrock channels migrate laterally when carving down into competent bedrock.

Straight and Sinuous Channels

Straight channels with sinuosities < 1.3 are relatively rare in natural streams because even very slight flow irregularities lead to deposition and accumulation of sediment in **alternate bars** that are successively positioned on opposite sides of the channel. Sinuous channels (those with sinuosities of 1.3–1.5) are quite common because as soon as a subtle bar forms on one side of a channel, it steers flow coming from upstream toward the opposite bank, where the flow begins to excavate a pool as it impinges on and erodes the far bank. Flow returning across the channel is, in turn, directed into the opposite bank, leading to the development of another pool and bar downstream, and so on down the channel. This process leads to the development of a sequence of alternate bars and pools that promotes lateral channel migration and the growth of meanders through erosion of the cutbank. Most straight channels have erosion-resistant banks, follow structural controls like faults, or have been confined by engineered levees that prevent natural lateral migration (see Photograph 6.14b).

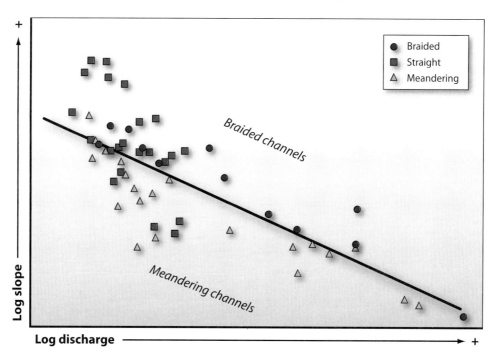

Discharge and slope influence channel planform patterns. At a particular slope, higher discharges are likely to produce **braided channels**. Likewise, for a particular discharge (or stream size), **meandering channels** tend to have lower slopes than do braided channels. **Straight channels** occur at low discharges over a variety of slopes. Sediment supply and the variability in discharge also play a role in determining channel form.

FIGURE 6.9 Channel Patterns as a Function of Discharge and Slope. Channel patterns are determined by slope and discharge. Braided channels are more common at higher slopes and higher discharges. [Adapted from Leopold and Wolman (1957).]

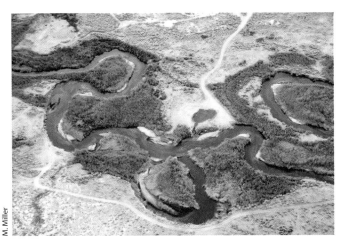

PHOTOGRAPH 6.16 **Meandering Channel.** Aerial view of meanders on floodplain of the Owens River, California.

Meandering Channels

Meandering channels have sinuosity values ≥ 1.5 and generally have a single, deep, narrow channel with few islands [**Photograph 6.16**]. Unconfined meandering channels typically develop a meander wavelength (the distance from the apex of one bend to the next on the same side of the channel) equivalent to about 10 to 12 channel widths, a relationship that is related to the physics of thalweg oscillation as water flows along an undulating channel. Pools are generally located on the outer banks of each bend and are thus typically separated by an average distance of 5 to 6 times the channel width.

The radius of curvature, the radius of a circle that fills the arc of a meander, describes the tightness of a river bend (see Figure 6.8). A small radius of curvature describes a tight bend. Feedback between flow through the bend and bank erosion typically leads to a radius of curvature between 2 and 3 times the channel width in meandering channels. In broad bends with a larger radius of curvature, the greatest shear stress on the outer bank occurs upstream of the meander apex, leading to preferential erosion that tightens the bend and decreases the radius of curvature. Conversely, in tight bends, where the radius of curvature is small relative to the channel width, the greatest shear stress and most intense bank erosion are downstream of the meander apex, which causes lateral migration that broadens the bend and increases the radius of curvature. Through such feedbacks, the typical form of meandering rivers represents an equilibrium adjustment to coupled flow and erosion through channel bends.

Meandering streams are common today, but we have found no evidence in the rock record that they existed before the evolution of land plants about 400 million years ago. Fluvial sediments from older periods of geologic time record braided channel morphologies. Conventional wisdom holds that this is because of the effect of root strength on bank stability, and recent flume experiments that demonstrate the role of streamside vegetation in stabilizing meander formation support this hypothesis. One would not expect meandering bedrock channels to be preserved in the geologic record for the simple reason that they occur in eroding upland environments.

Braided Channels

Braided channels are made of multiple, active threads within a broad, low-sinuosity, high-flow channel. A series of shallow, wide, low-flow channel strands that branch, diverge, and converge again form a distinctive braided pattern within the banks of a typical braided channel [**Photograph 6.17**]. The sediment bars that divide the flow into multiple strands are called **braid bars**. Braided channels generally are quite dynamic, and individual strands shift positions within the main channel, sometimes on a daily basis, as ephemeral patterns of deposition and erosion shift the sediment that makes up the braid bars. Erodible banks and a sediment load that exceeds the stream's carrying capacity favor formation of braided channels because these factors force the stream to flow around its own sediment at low flow. High slope, frequent variations in discharge, lack of bank-stabilizing vegetation, and a high load of coarse sediment promote the development of braided channels. Braided channels are commonly found downstream of glaciers and at mountain fronts with high sediment loads and steep channels.

Anastomosing Channels

Anastomosing (or anabranching) channels exhibit a complex pattern of individual channels that bifurcate and rejoin to flow around relatively stable, typically vegetated, islands [**Photograph 6.18**]. Anastomosing channels are generally narrower and deeper, have lower gradients than braided channels, and migrate by discrete avulsions instead of by steady lateral channel migration. Cohesive banks that limit lateral migration, flood-prone discharge regimes, and mechanisms that promote local overbank

PHOTOGRAPH 6.17 **Braided Channel.** The braided Resurrection River in Alaska flows between steep valley walls.

PHOTOGRAPH 6.18 Anastomosing Channel. Some of the sandbars in this anastomosing channel in the Alaska Range are stabilized by substantial vegetation growth.

flooding or channel blockage promote formation of anastomosing channels. In forested environments, the presence of large organic debris capable of forming stable logjams that locally split flow into multiple channels can result in an anastomosing channel pattern. Similarly, blockage of individual channels by logjams can trigger avulsions that shift flow from one channel to another or that cause flow to spill overbank and form a new channel.

Channel-Reach Morphology

A **channel reach** is a stretch of a channel that exhibits similar characteristics; reach types often reflect similarities in bed and bank material and position in the landscape. The distribution of typical channel reach types within a stream system reflects differences in relative transport capacity, as determined by the ratio of sediment supply to a stream's transport capacity. At a fine scale, within individual channel reaches, the channel consists of groups of morphologically distinct forms, called **channel units**, which are typically up to several channel widths in length. Unique suites of channel units—bars, steps, pools, and riffles—define different types of channel reaches [**Figure 6.10**].

Colluvial Reaches

Colluvial reaches typically occur in stream valley segments in the headwater portions of channel networks. Hillslope processes of mass wasting, soil creep, tree-throw, and burrowing activity introduce sediment into upland channel reaches and shape these channels. Flow in the channel does not govern the formation of the valley fill because shallow flow and limited fluvial transport capacity are insufficient to alter significantly the patterns of gravity-driven deposition, especially because of the stabilizing role of large woody debris.

Bedload sediment in colluvial stream reaches is typically poorly sorted and includes finer grain sizes than downstream alluvial channels. Average bed-surface grain size generally increases downstream as flow begins to winnow the finest grains; most colluvial reaches exhibit downstream coarsening. Average grain size typically reaches a maximum at or near the downslope transition from colluvial to alluvial reaches. In steep colluvial channels, debris flows that deposit clasts too large for normal flows to move are a dominant sediment transport process, and coarse-bed channels are common. The downstream terminus of a substantial debris flow often coincides with a grain-size maximum in a mountain stream profile because of the role that debris flows play in the delivery of coarse clasts. Because water flows in colluvial channels are insufficient to move large rocks, episodic debris flows are the primary means by which steep headwater channels are cleared of accumulated large debris.

Bedrock Reaches

Bedrock channel reaches are cut mostly into rock. They have little, if any, alluvial bed material or valley fill, generally lack floodplains, and are typically confined by narrow valley walls. Bedrock reaches typically occur on steeper slopes than alluvial reaches within the same drainage basin. In general, bedrock reaches lack an alluvial bed because the stream's transport capacity is greater than its sediment supply, a discrepancy that results from a high slope, and thus high transport capacity, a low sediment supply, or a combination of both.

Channels that are subject to scouring and/or deposition by periodic debris flows may alternate between bedrock, colluvial, and alluvial morphologies as they recover following disturbance; for example, a channel scoured to bedrock by a debris flow may slowly accumulate sediment delivered by creep down steep adjacent slopes. Channels in mountain drainage basins often exhibit mixed alluvial-bedrock reach morphologies that arise from fluctuations in local controls on sediment delivery, accumulation, and storage. For example, softer, more erodible rocks wear down faster and create wider valleys with more local **accommodation space** for storage of alluvium in the valley bottom than do more erosion-resistant rock types.

Alluvial Reaches

Alluvial channel reaches have morphologies that are predominantly formed by the interaction of flowing water and the sediment it carries. Several distinct channel reach morphologies can be identified in a natural continuum of alluvial channel types that reflect variations in relative transport capacity.

Cascade reaches are characterized by longitudinally and laterally disorganized bed material, and their bed typically consists of cobbles and boulders. Flow in cascade reaches diverges and converges around individual large clasts that protrude into the flow, generating large vortices and waves that dissipate a substantial portion of the total

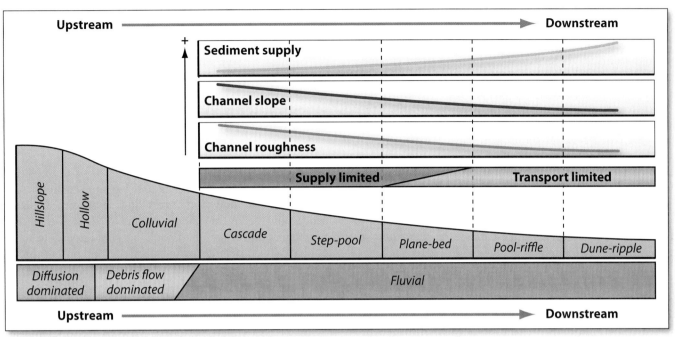

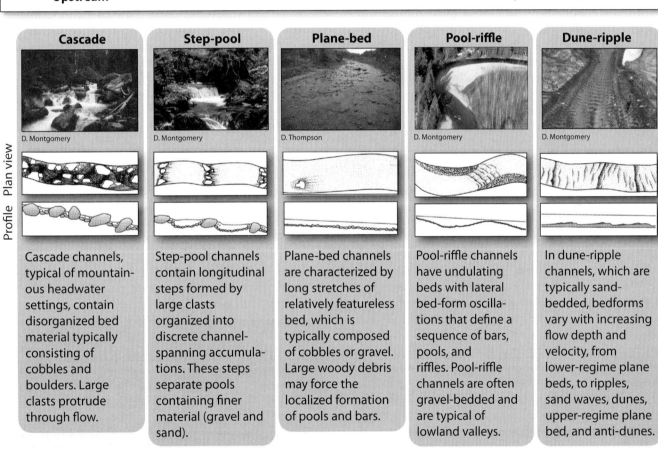

FIGURE 6.10 Downstream Trends in Mountain Stream Channel Types and Characteristics. Schematic illustration of typical downstream trends in sediment supply, channel slope, channel roughness, and channel types on a longitudinal profile from ridge crest to basin outlet. The particular sequence of channel types along any given river system will reflect both local and systematic downstream trends in channel slope.

flow energy; flow tumbles over obstructions. The largest clasts may move only every few decades; smaller clasts are rapidly transported during more frequent flows. Cascade reaches are steep (between 8 and 20 degrees), have large bed-material grain sizes, and have relatively shallow flow depths. Sediment transport in cascade channels tends to be supply limited because the transport capacity generally exceeds the sediment supply.

Step-pool reaches are characterized by a series of discrete channel-spanning accumulations of large clasts that form coarse-grained steps between pools floored by finer-grained sediment. The steps account for much of the elevation drop within step-pool reaches, and the steps provide bed roughness, causing areas of supercritical flow that alternate downstream with tranquil flow through pools. Steps nucleate where large clasts accumulate in congested zones with high local flow resistance, and they grow by trapping additional large clasts. Infrequent flood events that move large clasts typically form the framework of step-pool reaches; finer sediment is transported over the steps and deposited in pools during more frequent, lower-velocity flows. Step-pool reaches typically form at channel slopes between 4 and 8 degrees. Like cascade reaches, they are generally supply limited.

Plane-bed reaches have relatively featureless channel beds that are mainly composed of riffles. Although they share the same name, they are not to be confused with the upper and lower regime plane-bed morphology in sand-bed channels. Plane-bed reaches are distinguished from steeper cascade reaches by the absence of tumbling flow and by smaller grain sizes relative to flow depth. The largest clasts are submerged at all but the lowest flow in plane-bed channels. The few pools in plane-bed channels typically form in eddies where flow has been forced around obstructions. Plane-bed channels are typically straight, have slopes of 1 to 4 degrees, and are floored by a coarse surface layer that becomes mobile at or near bankfull flow. Plane-bed channels represent a transitional morphology between steeper, supply-limited channels and lower-gradient, transport-limited channels. Consequently, they are often found between step-pool reaches and lower-gradient meandering channels. Plane-bed reaches are rare in forested mountain drainage basins because stable obstructions, such as large woody debris, force local pool and bar formation.

Pool-riffle reaches consist of a sequence of bars, pools, and riffles and are typical of meandering channels. The pools and riffles themselves are often relatively stationary features, even though the bed-forming material that composes them is under constant flux. Development of alluvial bars requires a large channel width-to-depth ratio and small grains (relative to flow depths) that are readily mobilized by the flow. Pool-riffle reaches typically have gravel-sized to cobble-sized bedload, slopes less than 1.5 degrees, a coarse surface layer, and general bed-surface mobility at flows approaching bankfull. Pool-riffle reaches tend to be transport limited. Fining of the bed surface and changes in bedform size or amplitude adjust transport capacity to the sediment supply.

Sand-bedded channel segments may have a pool-riffle morphology but are referred to as **dune-ripple** reaches where they exhibit a succession of mobile bedforms that provide the primary flow resistance. Bedform type is determined by flow depth and velocity as expressed by the Froude number (eq. 6.3). Sediment transport in sand-bedded reaches occurs at almost all discharges, and transport rates are strongly dependent on discharge. Sand-bedded reaches are thus transport-limited channels.

Braided channels fit into the lower gradient parts of the classification scheme. The individual threads of most braided channels have pool-riffle morphology, although some consist of individual channel threads that are plane-bed and/or dune-ripple.

Large Organic Debris

Large organic debris, like logs and logjams, creates stable obstructions to flow and forces flow convergence, divergence, and sediment impoundment in stream channels, resulting in the formation of pools, bars, and steps. The morphologic effects of large organic debris depend on its volume, position, orientation, and size relative to channel dimensions. Large logs in small channels can be stable for decades, even centuries. These obstructions force local, long-term flow convergence that scours out pools and flow divergence where sediment accumulates to form bars. In larger channels, individual logs are more mobile and less likely to obstruct flow, but groups of logs and other organic debris form stable logjams that influence pool formation, channel patterns, and sediment storage [**Photograph 6.19; Figure 6.11**].

Stable logjams obstruct and divert flow and can split flow into multiple channels, forming anastomosing channel patterns. In forest channels, large organic debris can force the formation of pool-riffle or step-pool morphologies by retaining sediment in reaches that would otherwise have bedrock or plane-bed morphologies. In such reaches, the abrupt failure or removal of a logjam, or a gradual reduction (due to deforestation) in the supply of wood large enough to form jams, can lead to changes in channel morphology, allowing large volumes of sediment to be scoured and flushed downstream.

The type of trees that supply organic debris to a stream determines the effect of plant material on stream morphology.

PHOTOGRAPH 6.19 Logjam. Logjam on the cutbank of a meander on the Queets River, Washington.

Deposition of a large, stable log (shown here with an attached rootwad) causes local flow **convergence** and **divergence** that result in bed scour (erosion) upstream and sediment deposition downstream. Faster flow around the rootwad causes scour.

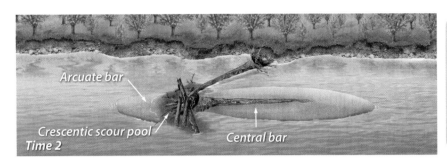

Scour around a stable, **key member** log creates a crescentic **scour pool** on the upstream side of the obstruction and deposition builds a **central bar** that buries the tree trunk on the downstream side. An arcuate bar forms as flow diverges upstream of the obstruction.

Continued scour and deposition occurs as additional debris racks up on the logjam, enlarging the scour pool. Continued deposition can build up the central bar into an island, and flow deflection can result in localized channel widening due to bank scour.

Eventually the logjam can become partially buried. It then protects the associated island from erosion, providing stable habitat where trees large enough to produce key member logs can grow even in disturbance-prone valley bottoms. The bar can eventually attach to the channel bank and become integrated into the floodplain.

FIGURE 6.11 Effects of Large Woody Debris. Large woody debris in channels changes the distribution of water flow and velocity. The resulting sediment scour and deposition can eventually form a new patch of floodplain. [Adapted from Abbe and Montgomery (1996).]

The highly branched trees typical of temperate deciduous forests and tropical rainforests form individual snags. Such snags plagued navigation on the Mississippi River historically and still interfere with boat traffic on tributaries of the Amazon River today. In contrast, the more readily transported, telephone pole-like morphology of coniferous trees leads to extensive development of logjams, where material routed downriver hangs up on large key members, as is common in the Pacific Northwest and in other evergreen forests around the world.

Floodplains

Floodplains, the low ground adjacent to stream channels, are built up by sediment deposited during floods. Alluvial channels typically have laterally extensive floodplains. Bedrock channels generally do not have floodplains, which makes defining bankfull flow difficult in many bedrock channels because they often have no discernible banktop. Floodplains are built by lateral and vertical accretion of sediment that results from (1) deposition of suspended load that settles out from overbank flow, (2) bedload deposition from lateral channel migration, and (3) amalgamation of local surfaces formed by alluvium trapped by blockages like landslides, channel-choking plant growth, and logjams [**Figure 6.12**]. These three processes lead to formation of different types of floodplains, composed of different materials and with different suites of floodplain landforms.

When sediment-laden discharge spills out over channel banks during floods, the velocity of the unconfined flow decreases due to decreased depth and increased roughness. This causes material carried by the flow to settle out onto the floodplain. Floodplains built up by the accumulation of suspended sediment are composed of landforms derived from differential settling of material leaving the river. As flow slows upon spilling out of the channel, the coarsest material settles out close to the channel, building berms parallel to the channel banks known as **natural levees** [**Photograph 6.20**]—sandy deposits decimeters to meters high along channel margins. These levees are breached by **crevasses**, which allow water and sediment out to create **crevasse splays**, fanlike deposits of sediment on the floodplain just outside the crevasse.

Farther from the channel, floodplains built primarily of overbank deposition are composed of fine-grained sediment, typically silts and clays that were carried as suspended

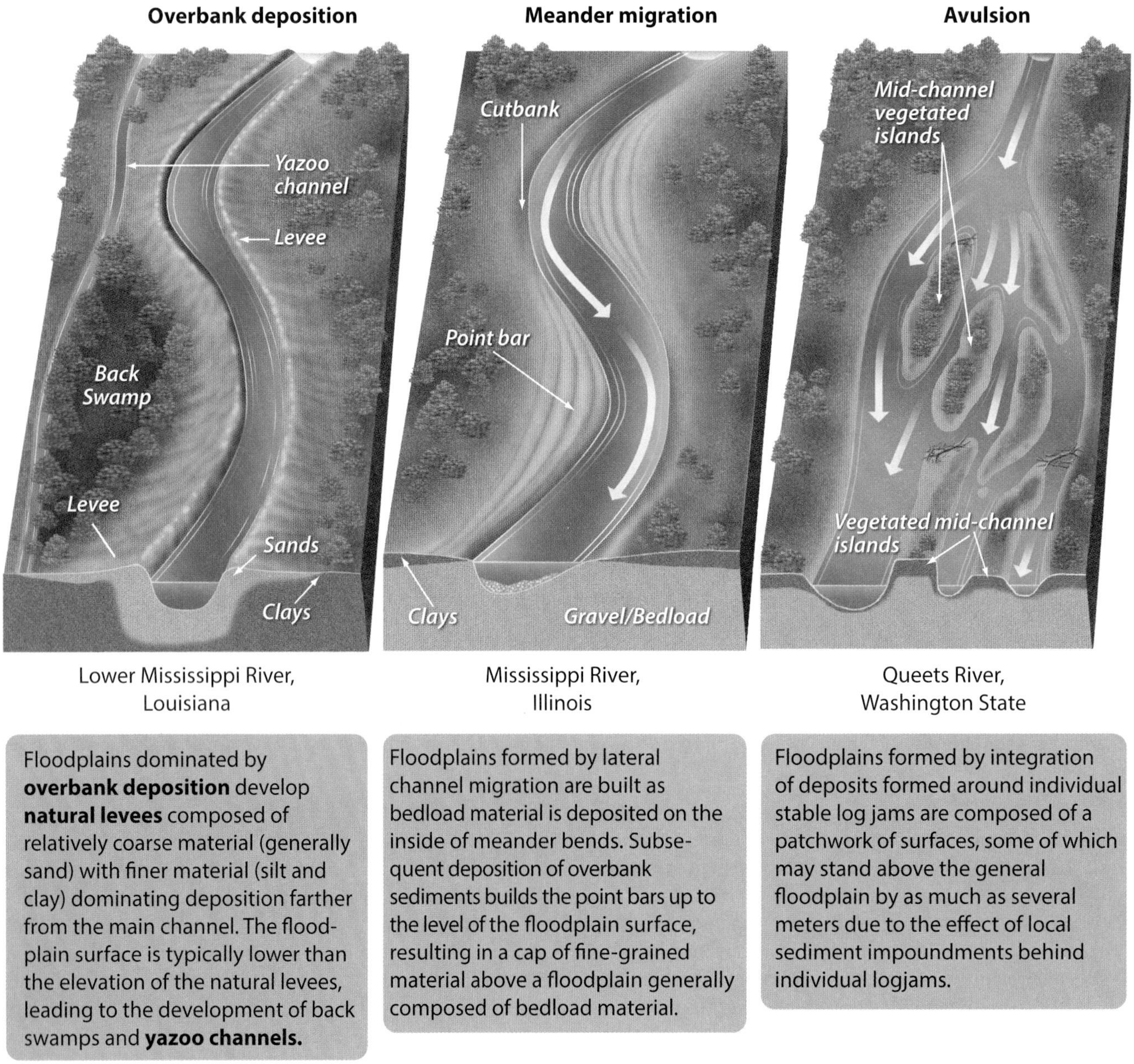

Overbank deposition

Lower Mississippi River, Louisiana

Floodplains dominated by **overbank deposition** develop **natural levees** composed of relatively coarse material (generally sand) with finer material (silt and clay) dominating deposition farther from the main channel. The floodplain surface is typically lower than the elevation of the natural levees, leading to the development of back swamps and **yazoo channels.**

Meander migration

Mississippi River, Illinois

Floodplains formed by lateral channel migration are built as bedload material is deposited on the inside of meander bends. Subsequent deposition of overbank sediments builds the point bars up to the level of the floodplain surface, resulting in a cap of fine-grained material above a floodplain generally composed of bedload material.

Avulsion

Queets River, Washington State

Floodplains formed by integration of deposits formed around individual stable log jams are composed of a patchwork of surfaces, some of which may stand above the general floodplain by as much as several meters due to the effect of local sediment impoundments behind individual logjams.

FIGURE 6.12 Floodplain Development. Floodplain formation processes and landforms differ depending on active channel processes. Floodplains can be built by overbank deposition, channel migration, and avulsion.

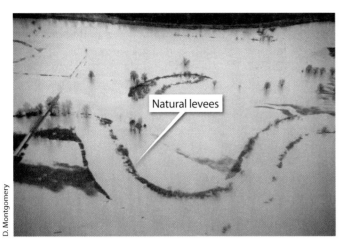

PHOTOGRAPH 6.20 Natural Levees. Natural levees with trees growing on their crests define the channel of the Snohomish River in Washington during a 1996 flood.

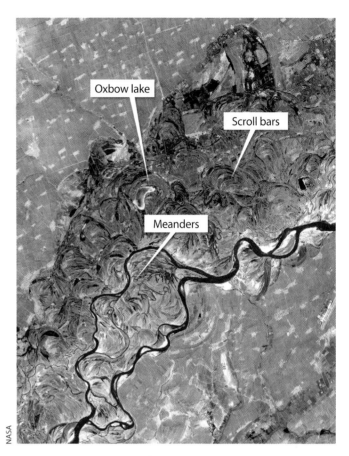

PHOTOGRAPH 6.21 Scroll Bars and Meanders. Satellite photograph of the Songhua River, just west of Haerbin, northeast China, showing meanders, scroll bars, and oxbow lakes. The area shown is about 30 km wide.

load. Extensive flooding can occur when rivers overtop their levees because the surrounding floodplain typically sits at a lower elevation; **backswamps** commonly occupy low-lying ground on valley margins along suspended sediment-dominated floodplains. **Yazoo channels** are tributary streams that flow down floodplains parallel to and outside the main channel levees and that serve to drain floodwaters back into the main channel at some distance downstream.

Lateral channel migration by cutbank erosion on the outer side of meanders and point-bar deposition on the inner bank builds floodplains composed of relatively coarse-grained bedload material at depth (old point bars and channel lags) overlain by fine-grained overbank deposits. **Scroll bars** are laterally stacked, abandoned point-bar deposits associated with former positions of the inner margin of channel meanders [Photograph 6.21]. They record past positions of the stream channel and can be used to track the history of its trajectory across a floodplain. Channel avulsions leave abandoned meanders that sometimes form **oxbow lakes,** narrow looping lakes that may remain partially connected to the main channel. Oxbow lakes slowly fill with fine-grained material deposited from suspension during subsequent floods. Once filled with relatively strong, silt- and clay-rich cohesive sediment, old oxbows become erosion-resistant masses that can obstruct lateral migration of meanders and modify groundwater flowpaths.

In forested terrain, floodplains are topographically varied and hydraulically rough. Logjams cause local channel damming and avulsions, creating networks of side channels. Local accumulations of sediment are trapped behind stable logjams and other obstacles to flow. These can coalesce into a patchwork floodplain composed of deposits at elevations determined by the depositional contexts of individual logjams. Unlike floodplain surfaces that have slopes similar to the valley slope, in forested floodplains, individual patches of bedload trapped by logjams can form terracelike surfaces at elevations up to several meters above the riverbank. These flat surfaces rise up to twice the diameter of the key member logs and are discontinuous both laterally and longitudinally.

Channel Response

Concern over impacts on human communities that rely on rivers and streams, as well as on aquatic and riparian ecosystems, motivates a desire to understand the ways that stream channel systems respond to disturbances, including both natural events and those that result from land use or climate change. The wide variety of channel types, the complex ways that channels adjust to local and regional factors, and the potential time lags between perturbation and channel response complicate interpretation and make prediction difficult.

Alluvial channel morphology adjusts to variations in sediment supply and discharge. In general, channels respond to fluctuations in (1) the delivery rate, volume, and grain size of supplied sediment; (2) transport capacity as affected by changes in bankfull discharge as well as the frequency, magnitude, and duration of high-discharge events; and (3) vegetation that influences bank stability or the size, amount, and stability of in-channel woody debris. Channels alter their width, depth, bed slope, grain size, and plan view patterns to accommodate changes in the three boundary conditions listed above.

Changes in discharge or sediment supply are often offset by corresponding changes in channel-bed grain size and slope, which can be formalized as a relationship that states that the product of the bedload supply (Q_b) and the bed-surface grain size (d_{50}) is proportional to the product of the water discharge (Q) and channel slope (S):

$$Q_b d_{50} \approx QS \qquad \text{eq. 6.8}$$

The sediment load (Q_b) and discharge regime (Q) are imposed on the channel by upstream conditions. The channel is free to adjust grain size (d_{50}) by sorting sediment and to adjust channel bed slope (S) by incising or aggrading. Different streams and stream segments within a drainage basin may be out of phase in their responses to the same initial disturbance because of the time lags that arise from routing sediment through the channel network.

Changes in the sediment supply to a channel occur naturally when mass movements (such as landslides, rock falls, and debris flows) suddenly deliver sediment from hillslopes into stream channels. Anthropogenic changes, like those that accompany mining, logging, urban development, or land clearing for agriculture can also dramatically alter the amount and grain size of sediment supplied to stream channels (see Figure 7.12). Channels with a high sediment load tend to have finer beds lacking a distinct, coarse surface layer because flows are insufficient to transport and sort all the material delivered to the channel. Conversely, channels with low sediment loads generally have coarser beds and a well-developed coarse surface layer because there is enough energy to flush fine material downstream.

Aggradation occurs when an increased sediment supply overwhelms stream transport capacity and the channel bed builds up. Rapid addition of a large amount of sediment can result in significant channel infilling and loss of flow conveyance, which increases the frequency of overbank flood flows. Channel responses to an abrupt increase in sediment supply may also include fining of the channel bedload, pool infilling, channel widening, and development of channel braiding and narrow braid bars.

Pulses of sediment introduced into channel systems may move down through channel networks as coherent slugs, causing a wave of progressive aggradation and re-incision as the sediment moves downstream. Patterns of sediment storage that are out of equilibrium with the stream system as a whole can persist for decades within a channel network as streamflow gradually mobilizes and redistributes sediment from the reaches where it accumulated. The impact of increased sediment loads can be long-lasting. Channels in parts of the Sierra Nevada are still adjusting to large inputs of mining debris that occurred during California's Gold Rush in the mid-1800s.

Mass movements can introduce material into the channel that is too large for moving water to transport. Rock falls, valley-wall rock slides, and debris flows all convey large volumes of sediment into headwater channels in mountain drainage basins. Like dams engineered by humans, large landslides block rivers and impound lakes, but these natural dams typically fail and erode away over time. As rivers erode down through landslide debris, clasts that are too large to transport sometimes become concentrated and form an immobile lag on the channel bed, retarding river incision.

The effects of debris flows vary with slope and position along the channel network. Although debris flows can scour steep headwater channels to bedrock, they generally deposit material when they reach lower slopes of 3 to 6 degrees or when they lose substantial momentum traversing sharp bends at tributary junctions. Debris flow deposition results in local aggradation. Sometimes debris-flow deposits completely fill in and obliterate a channel. Recovery following debris-flow disturbance differs between steep and low-gradient channels. Steep, high-energy channels (bedrock, cascade, or step-pool reaches) recover relatively quickly and rapidly transmit most debris-flow material to downstream reaches. In contrast, low-gradient channels (pool-riffle and plane-bed reaches) generally take longer to recover from debris-flow impacts because of their lower transport capacity. The cycle of debris flow disturbance and recovery can take decades to centuries and varies with position in the channel network.

Applications

Human actions greatly affect fluvial systems. In addition to direct impacts like channelization by levees and impoundment by dams, human modifications to the land surrounding a stream system (e.g., deforestation, agricultural development, and urbanization) profoundly affect stream systems by changing discharge, sediment supply, and the caliber and amount of woody debris supplied to the stream.

Construction and paving of land in urban and suburban areas increases the area of impervious surfaces, which alters the volume and timing of runoff delivered to stream channels, often triggering channel adjustment. Increased sediment delivery to channels from agriculture and forestry practices, as typically accompanies plowing and clear-cutting, alters sediment loads and causes streams to adjust. Channel responses associated with reduced sediment supply include both changes in the grain-size distribution and sorting of the channel-bed sediments and channel incision, or **entrenchment**, as occurs when people mine substantial amounts of gravel from a channel or when denuded catchments are revegetated.

Alteration of channel-margin vegetation changes the size and species of wood entering a channel, thereby influencing the abundance of pools, bars, and steps generated by accumulation of woody debris. The conversion of channel-margin vegetation from forest to grassland species (e.g., when forest is cleared for pasture land) can lead to systemic channel widening or narrowing, depending on the relative amount of bank reinforcement by roots. In general, however, in-channel logs and logjams recruited from streamside forests tend to promote variability in channel width by locally deflecting flow toward channel banks and by trapping sediment. In forest channels, depletion or reduced size of large organic debris causes pools to be lost, alluvial reaches to erode down to bedrock, and anastomosing channel patterns to simplify and become braided, straight, or meandering. In small channels, where logs and logjams provide significant sediment storage, a decreased supply of large woody debris accelerates sediment transport. Channels in which large wood provides a dominant control on pool formation and sediment storage, like those with a pool-riffle or step-pool morphology forced by logjams, are particularly responsive to changes in the size and amount of woody debris.

Dam construction and large natural stream obstructions like landslide deposits, logjams, lava flows, and glaciers, change both the discharge regime and the sediment supply to channels downstream. An abrupt reduction in sediment supply, such as occurs when a dam is constructed across a stream, typically results in channel incision and coarsening of bedload sediments downstream because the sediment formerly delivered from upstream is impounded behind the dam. Decreased discharge or frequency of high-flow events below a dam cause channel narrowing in downstream reaches. Conversely, dam removal causes channel widening and delivers a pulse of sediment that moves through downstream reaches as the stream redistributes the material that accumulated in the reservoir.

The construction of artificial levees along a river can exacerbate downstream flooding by preventing the spread of floodwaters across the floodplain and speeding flow to downstream reaches. When we look back at the history of many stream systems, it is common to see that once levees were constructed in one part of a basin, levees or other flood-control structures were soon required on stream reaches throughout the basin. In locations where channel gradients decrease downslope, levee construction also results in deposition of material inside the channel because flow is confined within the levees. Such confinement prevents the stream from carrying sediment over the banks and onto the floodplain. This leads to aggradation of the channel bed and deflation of the floodplain surface as loose sediments compact under their own weight. Ongoing levee reinforcement over the past several thousand years along China's Yellow River (Huang He) has elevated the channel more than 30 meters above its floodplain. Today, this ensures a catastrophic disaster every time a levee fails.

Structures in streams, including dams, levees, and other flood-control and flow-modification structures, affect river behavior. The direct effects of such changes are easy to appreciate. Levees prevent channel migration and dams store floodwater and trap sediment. The indirect geomorphic effects of changes to riparian and in-stream vegetation can be more difficult to identify but are no less profound. Consequences include changes in stream gradient, redistribution and destruction of aquatic habitats, and variations in the flux of sediment and water through the stream corridor over time and space.

Understanding how such changes influence particular river systems is fundamental to assessing and mitigating flood hazards, as well as to understanding human impacts on aquatic resources and to designing or evaluating river restoration and rehabilitation projects.

Selected References and Further Reading

Abbe, T. B., and D. R. Montgomery. Large woody debris jams, channel hydraulics, and habitat formation in large rivers. *Regulated Rivers: Research and Management* 12 (1996): 201–221.

Abbe, T. B., and D. R. Montgomery. Patterns and process of wood debris accumulation in the Queets River Basin. *Geomorphology* 51 (2003): 81–107.

Bagnold, R. A. The nature of saltation and of "bedload" transport in water. *Proceedings of the Royal Society of London* 332 (1973): 473–504.

Barnes, H. H., Jr. *Roughness Characteristics of Natural Channels.* U.S. Geological Survey Water-Supply Paper 1849. Washington, DC: Government Printing Office, 1967.

Brummer, C. J., and D. R. Montgomery. Downstream coarsening in headwater channels. *Water Resources Research* 39 (2003): 1294, doi: 10.1029/2003WR001981.

Buffington, J. M., and D. R. Montgomery. A systematic analysis of eight decades of incipient motion studies, with special reference to gravel-bedded rivers. *Water Resources Research* 33 (1997): 1993–2029.

Calow, P., and G. E. Petts, eds. *The Rivers Handbook: Hydrological and Ecological Principles.* Oxford: Blackwell Scientific, 1992–1994.

Carling, P. A., and G. E. Petts. *Lowland Floodplain Rivers: Geomorphological Perspectives.* New York: Wiley, 1992.

Constantine, J. A., and T. Dunne. Meander cutoff and the controls on the production of oxbow lakes. *Geology* 36 (2008): 23–26.

Costa, J. E., A. J. Miller, K. W. Potter, and P. Wilcock, eds. *Natural and Anthropogenic Influences in Fluvial Geomorphology.* Geophysical Monograph 89. Washington, DC: American Geophysical Union, 1995.

De Waal, L. C., A. R. G. Large, and P. M. Wade, eds. *Rehabilitation of Rivers: Principles and Implementation.* New York: Wiley, 1998.

Dietrich, W. E. "Mechanics of flow and sediment transport in river bends." In K. Richards, ed., *River Channels:*

Environment and Process. Oxford: Basil Blackwell, 1987.

Dietrich, W. E., and J. D. Smith. Influence of the point bar on flow through curved channels. *Water Resources Research* 19 (1983): 1173–1192.

Doyle, M. W., E. H. Stanley, and J. M. Harbor. Geomorphic analogies for assessing probable channel response to dam removal. *Journal of the American Water Resources Association* 38 (2002): 1567–1579.

Dunne, T., L., A. K. Mertes, R. H. Meade, et al. Exchanges of sediment between the flood plain and channel of the Amazon River in Brazil. *Geological Society of America Bulletin* 110 (1998): 450–467.

Emmett, W. W., and M. G. Wolman. Effective discharge and gravel-bed rivers, *Earth Surface Processes and Landforms* 26 (2001): 1369–1380.

Ferguson, R. I. Emergence of abrupt gravel to sand transitions along rivers through sorting processes. *Geology* 31 (2003): 159–162.

Finnegan, N. J., and W. E. Dietrich. Episodic bedrock strath terrace formation due to meander migration and cutoff. *Geology* 39 (2011): 143–146.

Gilbert, G. K. *Hydraulic-Mining Debris in the Sierra Nevada.* U.S. Geological Survey Professional Paper 105. Washington, DC: Government Printing Office, 1917.

Gomez, B., and M. Church. An assessment of bed load sediment transport formulae for gravel bed rivers. *Water Resources Research* 25 (1989): 1161–1186.

Graf, W. L. *Fluvial Processes in Dryland Rivers.* New York: Springer-Verlag, 1988.

Henck, A. C., D. R. Montgomery, K. W. Huntington, and C. Liang. Monsoon control of effective discharge, Yunnan and Tibet. *Geology* 38 (2010): 975–978.

James, L. A. Sustained storage and transport of hydraulic gold mining sediment in the Bear River, California. *Annals of the Association of American Geographers* 79 (1989): 570–592.

Knighton, D. *Fluvial Forms and Processes: A New Perspective.* New York: Arnold, 1998.

Kodama, Y. Experimental study of abrasion and its role in producing downstream fining in gravel-bed rivers. *Journal of Sedimentary Research* 64 (1994): 76–85.

Lane, E. W. The importance of fluvial morphology in hydraulic engineering. *Proceedings of the American Society of Civil Engineers* 81 (1955): 745–761.

Leopold, L. B., and T. Maddock Jr. The Hydraulic Geometry of Stream Channels and Some Physiographic Implications. *U.S. Geological Survey Professional Paper 252.* Washington, DC: Government Printing Office, 1953.

Leopold, L. B., and M. G. Wolman. River channel patterns—braided, meandering and straight. *U.S. Geological Survey Professional Paper 282A.* Washington, D.C.: Government Printing Office, 1957.

Leopold, L. B., and M. G. Wolman. River meanders. *Geological Society of America Bulletin* 71 (1960): 769–794.

Leopold, L. B., M. G. Wolman, and J. P. Miller. *Fluvial Processes in Geomorphology.* San Francisco: W. H. Freeman, 1964.

Mackin, J. H. Concept of the graded river. *Geological Society of America Bulletin* 59 (1948): 463–512.

Madej, M. A., and V. Ozaki. Channel response to sediment wave propagation and movement, Redwood Creek, California, USA. *Earth Surface Processes and Landforms* 21 (1996): 911–927.

Milliman, J. D., Q. Yun-Shan, R. Mei-E, and Y. Saito. Man's influence on the erosion and transport of sediment by Asian rivers: The Yellow River (Huanghe) example. *Journal of Geology* 95 (1987): 751–762.

Montgomery, D. R., and T. B. Abbe. Influence of logjam-formed hard points on the formation of valley-bottom landforms in an old-growth forest valley, Queets River, Washington, USA. *Quaternary Research* 65 (2006): 147–155.

Montgomery, D. R., T. B. Abbe, J. M. Buffington, et al. Distribution of bedrock and alluvial channels in forested mountain drainage basins. *Nature* 381 (1996): 587–589.

Montgomery, D. R., and J. M. Buffington. Channel-reach morphology in mountain drainage basins. *Geological Society of America Bulletin* 109 (1997): 596–611.

Murray, A. B., and C. Paola. A cellular model of braided rivers. *Nature* 371 (1994): 54–57.

Nanson, G. C., and J. C. Croke. A genetic classification of floodplains. *Geomorphology* 4 (1992): 459–486.

Parker, G. On the cause and characteristic scales of meandering and braiding in rivers. *Journal of Fluid Mechanics* 76 (1976): 457–480.

Pizzuto, J. E. Effects of dam removal on river form and process. *BioScience* 52 (2002): 683–691.

Richards, K. S. *Rivers: Form and Process in Alluvial Channels.* London: Methuen, 1982.

Schumm, S. A. *The Fluvial System.* New York: Wiley, 1977.

Schumm, S. A. *River Variability and Complexity.* New York: Cambridge University Press, 2005.

Schumm, S. A., M. P. Mosley, and W. E. Weaver. *Experimental Fluvial Geomorphology.* New York: Wiley, 1987.

Stover, S. C., and D. R. Montgomery. Channel change and flooding, Skokomish River, Washington. *Journal of Hydrology* 243 (2001): 272–286.

Summerfield, M. A., and N. J. Hulton. Natural controls of fluvial denudation rates in major world drainage basins. *Journal of Geophysical Research* 99 (1994): 13,871–13,883.

Tal, M., and C. Paola. Dynamic single-thread channels maintained by the interaction of flow and vegetation. *Geology* 35 (2007): 347–350.

Thorne, C. R., R. D. Hey, and M. D. Newson. *Applied Fluvial Geomorphology for River Engineering and Management.* New York: Wiley, 1997.

Vörösmarty, C. J., M. Meybeck, B. Fekete, et al. Anthropogenic sediment retention: major global impact from registered river impoundments. *Global and Planetary Change* 39 (2003): 169–190.

Whipple, K. X., G. S. Hancock, and R. S. Anderson. River incision into bedrock: Mechanics and relative efficacy of plucking, abrasion, and cavitation. *Geological Society of America Bulletin* 112 (2000): 490–503.

Wiberg, P. L., and J. D. Smith. Model for calculating bed load transport of sediment. *Journal of Hydraulic Engineering* 115 (1989): 101–123.

Wohl, E. *Mountain Rivers Revisited*. Water Resources Monograph 19. Washington, DC: American Geophysical Union, 2010.

Wohl, E., P. L. Angermeier, B. Bledsoe, et al. River restoration, *Water Resources Research* 41 (2005): W10301, doi: 10.1029/2005WR003985.

Wohl, E. E., and D. M. Merritt. Bedrock channel morphology. *Geological Society of America Bulletin* 113 (2001): 1205–1212.

Wolman, M. G., and L. B. Leopold. *River Flood Plains: Some Observations on Their Formation*. U.S. Geological Survey Professional Paper 282-C. Washington, DC: Government Printing Office, 1957.

Wolman, M. G., and J. P. Miller. Magnitude and frequency of forces in geomorphic processes. *Journal of Geology* 68 (1960): 54–74.

Zen, E., and K. L. Prestegaard. Possible hydraulic significance of two kinds of potholes: Examples from the paleo-Potomac River. *Geology* 22 (1994): 47–50.

DIGGING DEEPER What Controls Rates of Bedrock River Incision?

Upland bedrock landscapes are dramatic. Their form, dynamics, and evolution are greatly influenced by the processes and pace of bedrock channel incision. Early workers hypothesized that the rate at which rivers incised into bedrock was a function of rock resistance to erosion, river discharge, and slope (Gilbert, 1877). Indeed, discharge can be a critical variable in the rate of bedrock erosion. Although bedrock channels are typically the result of slow erosion over millennia, megafloods, caused by the release of large amounts of ponded water over short periods, can carve canyons rapidly. When a spillway on the Guadalupe River in Texas was first used to convey floodwaters threatening to overwhelm a dam, the flow cut a 7-meter-deep, several-kilometer-long canyon into limestone in just 3 days (Lamb and Fonstad, 2010). Studying streams in this same area of Texas, where intermittent high flows interrupt typically semi-arid conditions, Baker (1977) concluded that much, perhaps most, geomorphic change including bedrock incision was occasioned by rare, storm-driven, high flows.

Geomorphologists understand the effects of lithology on rock resistance to erosion using a variety of approaches. Geological constraints on rates of long-term channel incision provide context while field observations of bedrock channels are useful for characterizing erosion processes. Field experiments that monitor channel bed changes can measure short-term erosion rates, and computer modeling is useful for predicting future evolution of channels and the landscape of which they are part.

Formal expressions used to predict the rate at which mountain rivers incise into bedrock include terms not only for the discharge, flow depth, and channel slope, but also include the supply of sediment to the channel (Whipple and Tucker, 1999). Such sediment is key because it provides tools for erosion when it is mobile but, if thick enough, can shield the bed from erosion.

Stock and Montgomery (1999) used paleoriver profiles of known age to evaluate rates of bedrock river incision. They found that bedrock erodibility varies by more than five orders of magnitude among different lithologies. Not surprisingly, hard crystalline rocks erode more slowly than softer volcaniclastic rocks, and poorly indurated mudstones erode fastest among the five rock types studied. The wide range of bedrock erodibility between different lithologies implies that rates of river incision into bedrock could vary greatly in different geological settings.

Whipple et al. (2000) studied field evidence from a wide range of environments and concluded that the distribution of fluvial erosion processes (plucking, abrasion, and solution) is strongly influenced by bedrock lithology. Although lithology influences the pace of solution (more soluble rocks will erode faster), field evidence led Whipple et al. to conclude that the spacing and orientation of joints, fractures, and bedding planes exert direct control on whether plucking or abrasion dominates bedrock incision. Rocks that were well jointed or fractured on a submeter scale were most often eroded by plucking [**Figure DD6.1**]. Abrasion by suspended sediment controlled rates of incision into more massive, less fractured rocks. Weathering-limited erosion (such as spalling from disintegration by wetting and drying) tends to control erosion of weak, friable rocks.

Lab experiments can reveal a lot about bedrock erosion. Remember your rock tumbler? Sklar and Dietrich (2001) conducted experiments on the pace of bedrock erosion by loading 6-mm-diameter gravel into a rotating drum with a floor made of different types of rock for each of many experimental runs. They, too, found that the rate of bedrock erosion varied by more than five orders of magnitude (rapid = mudstone, slow = quartzite). Rocks with high tensile strength eroded more slowly than those

DIGGING DEEPER What Controls Rates of Bedrock River Incision? *(continued)*

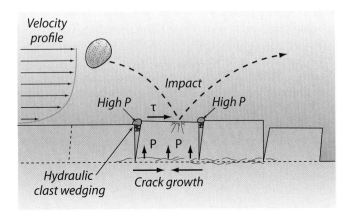

FIGURE DD6.1 Schematic illustration of the processes and forces contributing to bedrock channel erosion by plucking. On the left is the velocity distribution with depth in the channel (arrow length scales relative velocity). Impacts by large saltating grains (shaded) generate stresses that drive the crack propagation necessary to loosen joint blocks. High pressures (P) that develop across crack openings in the bed help drive hydraulic wedging of clasts into fractures that work to further open cracks. Surface drag forces due to shear stress (τ) and differential pressures (high P at the surface versus low P at depth) act to lift loosened blocks from the bed. Where the downstream neighbor of a block has been removed, both rotation and sliding become possible, and it is then much easier for rocks to be plucked out of the riverbed. [From Whipple et al. (2000).]

with low tensile strength [**Figure DD6.2**]. By varying the amount of gravel used in their experiments, they found that the pace of erosion increased initially as more sediment was added. Erosion rates decreased once the sediment covered the majority of the bed; the rate of erosion fell more than an order of magnitude once the bed was fully covered with sediment [**Figure DD6.3**].

The implication was clear. Sediment-covered channels erode slowly because impacting rocks cannot reach the bedrock. Increasing amounts of sediment in transport acted to shield the riverbed from erosion, whereas increasing numbers of sediment clasts (tools) acting as abraders increased the rate of erosion (Sklar and Dietrich, 2006). The combined effect of these two processes makes the pace of bedrock incision greatest when the bed is partially or lightly covered by sediment [**Figure DD6.4**].

Results of landscape evolution models show that bedrock channel incision plays a critical role in the development of mountain landscapes (Whipple and Tucker, 2002). In particular, the rate at which channels erode bedrock communicates changes in uplift rate and base-level lowering through mountain landscapes and governs the timescale of landscape response to such perturbations. The wide range of rates of bedrock channel incision means that the potential pace of landscape response to disturbance also varies widely.

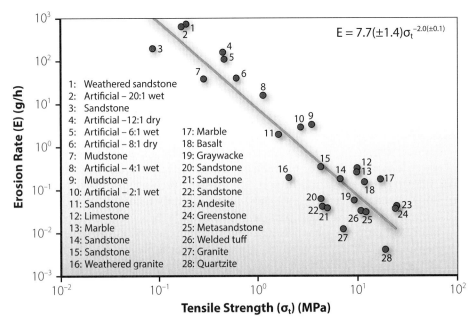

FIGURE DD6.2 Plot of experimental data from the rock-tumbling experiment described in the text. It shows that the variation in measured erosion rate (expressed in grams per hour) decreases with the square of rock tensile strength. Strong rocks (e.g., quartzite, #28) erode more slowly than weak rocks (e.g., sandstone, #3). For this set of runs, 150 g of 6-mm-diameter gravel sediment were loaded into the tumbler. [From Sklar and Dietrich (2001).]

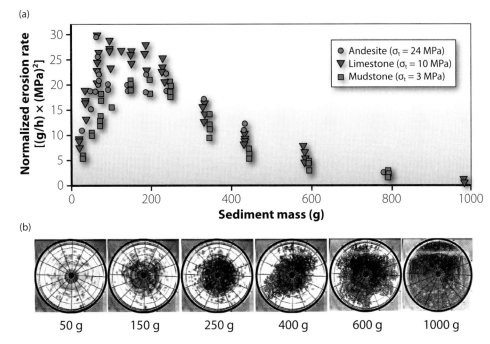

FIGURE DD6.3 (a) The variation in experimental erosion rate (the tumbler, again) with increasing sediment load (mass). Data for three rock types are normalized for rock strength by multiplying erosion rate (expressed in grams per hour) by the square of the rock tensile strength (expressed in megapascals, MPa). (b) A series of photographs taken from below the abrasion mill looking up through the glass bottom. They show how alluvial bed cover increases with increasing sediment loading. [From Sklar and Dietrich (2001).]

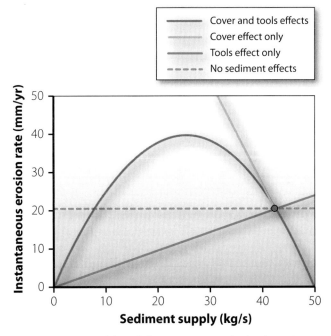

FIGURE DD6.4 Model result for instantaneous bedrock erosion rate as a function of sediment supply. The sloping (yellow) line indicates the effect of bed cover, a reduction in erosion rate with increasing sediment supply. The rising (blue) line indicates the increasing erosion rate caused by more tools abrading the bed. The arched (green) curve represents the combined effect of cover and tools and indicates that the erosion rate peaks when there is some but not too much sediment on the bed. The circle at the intersection of the lines indicates the instantaneous incision rate that corresponds to the assumed long-term bedrock erosion rate at a site on the South Fork Eel River in northern California. The dashed (orange) line represents the case of no sediment effects. [From Sklar and Dietrich (2006).]

Baker, V. R. Stream-channel response to floods, with examples from central Texas. *Geological Society of America Bulletin* 88 (1977): 1057–1071.

Gilbert, G. K. *Report on the Geology of the Henry Mountains*, U.S. Geographical and Geological Survey of the Rocky Mountain Region. Washington, DC: Government Printing Office, 1877.

Lamb, M. P., and M. A. Fonstad. Rapid formation of a modern bedrock canyon during a single flood event. *Nature Geoscience* 3 (2010): 477–481.

Sklar, L. S., and W. E. Dietrich. Sediment and rock strength controls on river incision into bedrock. *Geology* 29 (2001): 1087–1090.

Sklar, L. S., and W. E. Dietrich. The role of sediment in controlling steady-state bedrock channel slope: Implications of the saltation-abrasion incision model. *Geomorphology* 82 (2006): 58–83.

Stock, J. D., and D. R. Montgomery. Geologic constraints on bedrock river incision using the stream power law. *Journal of Geophysical Research* 104 (1999): 4983–4993.

Whipple, K. X., G. Hancock, and R. S. Anderson. River incision into bedrock: Mechanics and relative efficacy of plucking, abrasion, and cavitation. *Geological Society of America Bulletin* 112 (2000): 490–503.

Whipple, K. X., and G. E. Tucker. Dynamics of the stream-power river incision model: Implications for height limits of mountain ranges, landscape response timescales, and

DIGGING DEEPER What Controls Rates of Bedrock River Incision? *(continued)*

research needs. *Journal of Geophysical Research* 104 (1999): 17, 661–671, 674.

Whipple, K. X., and G. E. Tucker. Implications of sediment-flux-dependent river incision models for landscape evolution. *Journal of Geophysical Research* 107 (2002): 2039, doi: 10.1029/2000JB000044.

WORKED PROBLEM

Question: Manning's equation (eq. 6.2) describes the relationship between flow velocity and channel hydraulic radius (R), water surface slope (S, or $\tan\theta$), and the roughness or energy dissipation characteristics of an open channel (n). Do a sensitivity test so that you can describe how velocity of flow changes if (1) channel roughness increases through the range of Manning n values found in natural channels while holding R and S constant, (2) R (the wetted perimeter) increases while n and S are held constant, and (3) slope increases while R and n are constant. Express your results in words and graphically.

Answer: All three relationships are nonlinear. As n increases, flow velocities decrease. As the wetted perimeter, R, grows larger, flow velocities increase. As the water-surface slope increases, velocities also rise. Note that the difference in the exponents for R and S (2/3 versus 1/2) in Manning's equation control the rate of change in velocity for a unit change in the variables R and S.

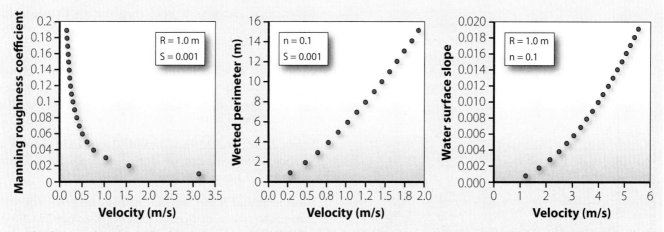

Graphs showing the relationship between velocity and changing values of the Manning roughness coefficient, channel wetted perimeter, and water surface slope.

KNOWLEDGE ASSESSMENT Chapter 6

○ 1. Explain why understanding channels and their behavior is fundamental to geomorphology.
○ 2. Define a graded stream.
○ 3. List three factors controlling the shape of channels.
○ 4. Describe several ways in which climate can influence channels.
○ 5. Contrast the change in discharge downstream in humid-region and arid-region channels.
○ 6. List three ways in which sediment reaches river channels.
○ 7. Describe how bed and bank material affect the behavior of channels over time.

8. Predict where you are most likely to find alluvial channels and where you are most likely to find bedrock channels.
9. Explain two very different ways in which riparian vegetation influences channel behavior.
10. The balance between what two physical constraints determines flow velocity?
11. Write out Manning's equation and explain each of the terms.
12. Write the formula for discharge and for the hydraulic radius of a channel.
13. Explain what the Manning roughness coefficient represents.
14. List several physical properties that determine the roughness of a channel.
15. Explain how flow depth affects channel roughness.
16. What is the Froude number for streamflow and why is this an important value to know?
17. Define effective discharge and explain how its frequency is controlled by climate.
18. Explain the velocity distribution along a channel cross section and identify where you would find the highest flow velocity.
19. Define bankfull flow and explain why it is important to geomorphologists.
20. Sketch graphs of at-a-station hydraulic geometry and explain why they are useful.
21. Describe how fluid forces can move and entrain sediment along the bed of a stream.
22. List the three types of sediment load carried by a stream. Which type generally represents the greatest amount of transported mass?
23. Define critical shear stress and explain why it is relevant to bedload transport.
24. Explain the concept of stream power and, in the context of stream power, explain why deeper, quickly flowing water can transport more sediment or incise rock more rapidly than slowly flowing, shallow water.
25. Explain why different discharges are required to move sediment than to incise bedrock.
26. List the three ways by which rivers incise into rock.
27. From the perspective of flow dynamics, explain why alluvial channels migrate across floodplains.
28. Contrast the characteristics of pools and riffles at high and low flows.
29. Explain why and how channels meander.
30. Define river sinuosity.
31. Sketch straight, meandering, braided, and anastomosing channel patterns and suggest in what settings each might be found.
32. Explain the difference between channel reach types and channel units by providing examples of each.
33. Sketch the usual sequence of channel reach types as one moves downstream from mountainous headwaters to lowland river valleys.
34. Predict the effects when large woody debris enters a channel by landsliding or bank collapse.
35. Describe three different ways in which rivers build floodplains.
36. Give three different examples of channel response to external perturbations.
37. Predict how alluvial channels will respond to both increases and decreases in sediment supply and discharge.

Drainage Basins

Introduction

A **drainage basin** is the entire area drained by a stream and its tributaries. Drainage basins, also known as **watersheds**, are the fundamental unit for geomorphic analysis of fluvial systems because they are the land surfaces over which water and sediment move down topographic gradients. Drainage basins are separated by the elevated land, or ridges, between them that define topographic **drainage divides.** Drainage basins are thus defined by both hydrology and topography [**Figure 7.1**].

Drainage basins come in all shapes and sizes, from the upper reaches of a small headwater valley to the Amazon River that drains more than one-third of South America. Most humid regions have **perennial streams** and rivers that flow continually. In arid regions, stream channels may stay dry for long periods of time, carrying water perhaps only once a year or even once a decade. These **ephemeral streams** flow only in response to rainfall events. Drainage basins are composed of nested subbasins that geomorphologists analyze as a hierarchical system in which the smallest basins (most of which lie at the head of the channel network) are located at higher elevations along the basin's drainage divide and drain into successively larger watersheds.

Most drainage basins contain a predictable set of components. The upper reaches of the basin generally have steeper slopes and supply sediment to the lower basin. Moving down from the watershed-bounding slopes and ridges, sediment and water flows begin to concentrate in channels until eventually, where slope and/or lateral confinement have diminished sufficiently, there are areas

Rakaia River, South Island, New Zealand, meanders through spectacular flat-topped alluvial terraces, former floodplains of the river. Along the river's course are well-defined gravel point bars and steep cutbanks. In the distance are the rapidly rising Southern Alps.

IN THIS CHAPTER

Introduction
Basin-Scale Processes
  Sediment Budgets
  Sediment Routing and Storage
Channel Networks and Basin Morphology
  Drainage Patterns
  Channel Ordering
  Downstream Trends

Uplands to Lowlands
  Process Domains and Valley Segments
  Longitudinal Profiles
  Channel Confinement and Floodplain Connectivity
  Downstream Trends
Drainage Basin Landforms
  Knickpoints
  Gorges

  Terraces
  Fans
  Lakes
Applications
Selected References and Further Reading
Digging Deeper: When Erosion Happens, Where Does the Sediment Go?
Worked Problem
Knowledge Assessment

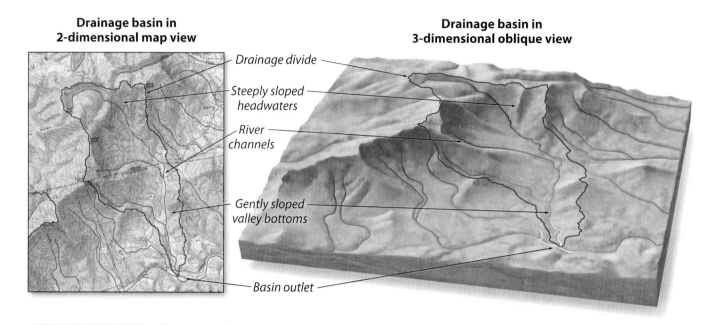

Drainage basins are the upslope area draining to a point along a stream and are a primary way by which geomorphologists subdivide landscapes. Separated by **drainage divides,** rivers and streams in drainage basins convey sediment from generally steep uplands to generally less steep lowlands and then onto an outlet defined as the end of the basin. Drainage basins contain streams of various sizes as well as smaller tributary drainage basins.

FIGURE 7.1 The Drainage Basin. Drainage basins, separated by drainage divides, are a fundamental unit of geomorphic analysis.

where sediment is deposited and stored at least temporarily alongside lowland streams and rivers. In the largest drainage basins, such as the Amazon or the Ganges-Brahmaputra, much of the sediment delivered from the mountains never makes it to the ocean; rather, it is deposited and trapped in isostatically and tectonically subsiding basins where rivers leave upland valleys to flow across broad lowlands (see Chapter 12). In general, steep uplands are zones of erosion, and relatively flat lowlands are zones of sediment deposition. Most sediment that reaches the ocean originates in mountainous headwaters, although it may have spent thousands of years in storage along the way.

Major physiographic, climatic, and geologic features impose broadly comparable controls on channels and their watersheds within a geomorphic (or physiographic) province. Neighboring watersheds in a geomorphic province tend to have similar relief, climate, and rock types. Consequently, broad relationships between drainage area, discharge, sediment supply, and bed-material grain size characterize channels and watersheds in a geomorphic province. In addition, regional similarities in vegetation, climate, and climate history (such as legacies of past glaciation) may impose similar general constraints on channels and thus watersheds.

Drainage basins may be closed or open. Most drainage basins are termed open because mass (water, sediment, and the dissolved load) is transported through and then out of the watershed either to larger watersheds or to the ocean. Drainage basins that are closed (also known as internally drained) terminate in lowlands, where water is lost only by evaporation and seepage into the groundwater system. Closed drainage basins are most commonly found in arid regions where drainage networks are not well developed, in areas where the local tectonic regime is extensional, and in glaciated areas where glacial erosion has overdeepened valley bottoms. Arid lowlands are often occupied by ephemeral lakes that dry out and become salt flats or **playas** (from the Spanish for "beach") between runoff events [Photograph 7.1]. These lakes are salty because the

PHOTOGRAPH 7.1 A Salty Playa. The white area is covered by salt left as water evaporates from this closed basin in Deep Springs Valley, California.

dissolved load delivered to them is left behind and concentrated by evaporation. Famous examples of closed drainage basins include those supplying water and sediment to the Great Salt Lake in Utah and to the Dead Sea in the Middle East.

Basin-Scale Processes

At the basin scale, one can investigate the production, transport, storage, and delivery of material. The specific process will vary from place to place; however, there is a generalized framework (sediment budgets, routing, and storage) one can apply to gain a better understanding of landscape behavior.

Sediment Budgets

The fluxes of sediment through a drainage basin can be described in terms of a **sediment budget** that considers **sources** from which mass enters the system (sediment derived from eroding slopes), and **sinks** where mass leaves the system. Sediment budgets allow one to understand where sediment comes from and where it goes. Such budgets are useful for predicting the probability that mass will move into and out of storage, as well as the response of watersheds to disturbances such as the installation or removal of dams that retain sediment. A sediment budget must also account for **storage,** the retention of mass in the system as sediment is deposited in floodplains and other depositional environments. The line between storage and sinks can be a blurry one and often depends on the timescale under consideration. For example, on the timescale of decades, floodplains can be considered as sinks, but over millennia, floodplains are eroded and reworked, functioning as temporary storage for sediment.

A generalized equation for water and sediment budgets considers the flux of mass in (Q_{in}), the flux of mass out (Q_{out}), and the change in storage (ΔST) all considered per unit time:

$$Q_{in} - Q_{out} = \Delta ST \qquad \text{eq. 7.1}$$

Often, the change in storage is assumed to be zero, meaning that $Q_{in} = Q_{out}$ and the basin is in steady state in terms of mass flux.

Sediment budgets can be simple or complex. A simple sediment budget might consider only the erosion rate of the sediment-producing hillslopes and the export rate of sediment from the watershed. More complex sediment budgets include terms describing the likelihood of, and thus the flux of material delivered by, dynamic exchanges between sediment storage and active transport—for example, the reworking of floodplain sediment as rivers erode their banks. These kinds of interactions can be assigned probabilities. Sediment stored in higher **terraces** (former floodplains of the river abandoned when the river incised) [**Photograph 7.2**] along valley walls is assigned a

PHOTOGRAPH 7.2 River Terraces. Sequence of river terraces storing large amounts of sediment in a wide, intermontane valley, Kadjerte River, Kyrgyzstan.

lower probability of entering the active channel than sediment stored in lower terraces abutting a river, which are more likely to be reworked by bank erosion. Sediment budgets can be constructed on a variety of different timescales (years, centuries, millennia), and the results can be very different because processes moving and storing sediment can be quite episodic.

Until recently, constructing **sediment budgets** was far more challenging than constructing water budgets, because there were typically few relevant data available [**Figure 7.2**]. The increased use of cosmogenic nuclides (see Chapter 2) to determine basin-scale rates of erosion (sediment supply) over the long term as well as repeat flights of high-resolution LiDAR to estimate short-term rates of surface change have greatly facilitated sediment budget construction. The development of optical sensors for measuring stream-water turbidity, and thus better assessment of suspended sediment load, has improved estimates of mass export from watersheds.

Sediment sources are dominated by the erosion of hillslopes through various processes, all of which can be characterized in terms of rates of mass delivery to the river network. The complication is that many slope processes, such as landslides and gullying, are episodic, and short-term and long-term mass flux rates often differ dramatically. Several different approaches have been used to address temporal variability. It is common to measure the mass of sediment accumulated and stored in depositional landforms in order to estimate average rates of sediment delivery over time and thus constrain rates of sediment generation (and erosion rates) for parts of basins. For example, alluvial fans or colluvial deposits, such as those in hollows, can be dated and the volume of the deposit measured. Defining the area contributing sediment allows calculation of the average rate of erosion and sediment supply from upland slopes. Sediment production rates can also be established through cosmogenic nuclide analysis of soil and sediment and the use of fallout nuclides such as ^{137}Cs (see Chapter 2).

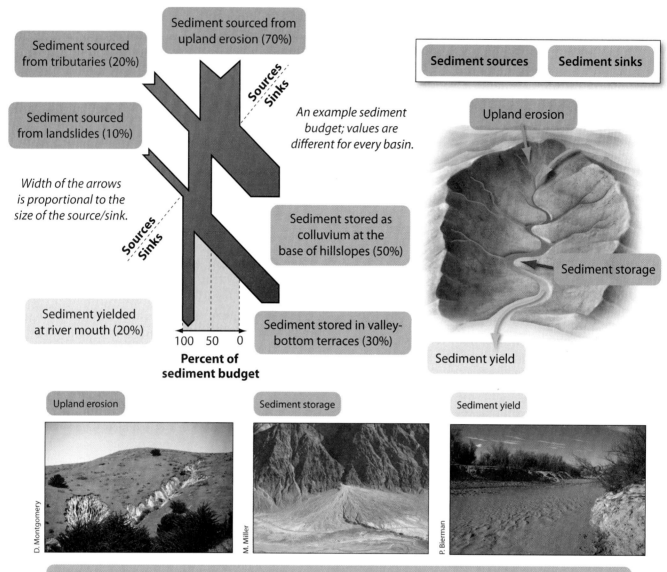

Sediment budgets describe where sediment comes from, where it goes, and how quickly it moves through the landscape. In many basins, sediment is sourced primarily in the uplands where steep slopes erode rapidly. That sediment is often stored for varying amounts of time at the base of slopes as **colluvium** and along stream and river channels as alluvium. Over time, for small basins in equilibrium, sediment generation in the uplands will match sediment yield coming out of the lowlands. However, large amounts of sediment can be stored for decades to millennia in disturbed landscapes. In basins large enough to contain extensive depositional lowlands (like the Amazon), sediment generation rates will exceed sediment yields as some sediment is trapped in the subsiding basin.

FIGURE 7.2 Sediment Budget. Sediment budgets quantify sources and sinks of sediment at the scale of drainage basins. [Adapted from Trimble (1999).]

Sediment evacuation rates are usually established by estimating flow over time and measuring suspended load during a variety of different discharge events, creating what is known as a suspended **sediment rating curve** [Figure 7.3]. This approach accounts for most of the sediment leaving the basin in suspension; however, in some rivers, dissolved load and bedload can be significant and need to be accounted for separately in the total mass budget. Some sediment rating curves display large amounts of scatter due to a phenomenon known as **hysteresis** or path dependence (see Figure 4.11).

Consider what happens to a streambed as flow rises during a storm. Until the flow rises sufficiently to mobilize the bed, sediment transport rates are low because large clasts generally protect smaller material from erosion. Once flow is sufficient to mobilize the bed, then large amounts of sediment move. After the flood peak passes, the flow begins to wane and the large clasts drop out, but the fine material

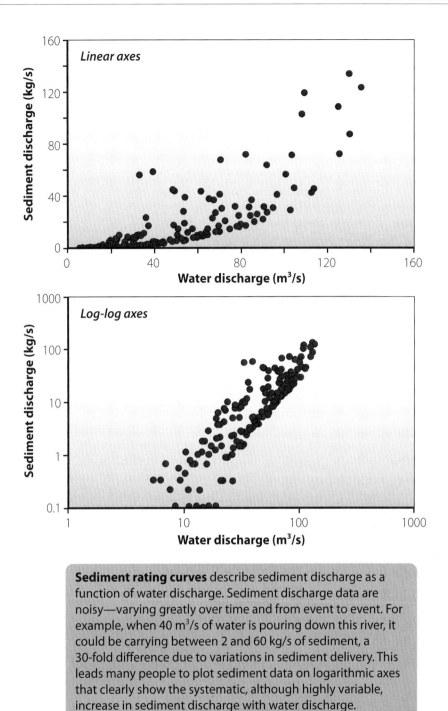

Sediment rating curves describe sediment discharge as a function of water discharge. Sediment discharge data are noisy—varying greatly over time and from event to event. For example, when 40 m³/s of water is pouring down this river, it could be carrying between 2 and 60 kg/s of sediment, a 30-fold difference due to variations in sediment delivery. This leads many people to plot sediment data on logarithmic axes that clearly show the systematic, although highly variable, increase in sediment discharge with water discharge.

FIGURE 7.3 Sediment Rating Curves. Sediment rating curves quantify the relationship between the flux of water and the flux of sediment through drainage basins.

continues to move in suspension. Thus, at the same discharge level during flood rise and subsequent recession, very different amounts of sediment may be moving. Measurements of sediment accumulation in lakes and reservoirs can be used to understand sediment flux over a wide range of basin areas and timescales. For example, this approach has been applied to small watersheds feeding farm ponds as well as the entire eastern United States by calculating the volume of sediment deposited offshore over the past 100 million years.

Understanding the magnitude and duration of sediment storage on the landscape is a particularly difficult and time-consuming part of constructing basin-scale sediment budgets. Quantifying storage requires both estimating the volume of sediment stored in depositional landforms (floodplains, terraces, fans, and colluvial deposits) as well as estimating the ages of these landforms and the sediment they contain. Both fieldwork and remote sensing data (LiDAR) are used to identify deposits and quantify their volume. Dating deposits requires techniques introduced in Chapter 2 and often relies on radiocarbon analyses of buried organic material in humid-temperate regions and luminescence or cosmogenic techniques in arid regions (due to the lack of preserved organic matter).

Sediment Routing and Storage

Sediment residence time in drainage basins varies widely. Numerous studies suggest that sediment eroded off slopes can be trapped for centuries to millennia before entering the fluvial system. For example, much **paraglacial** sediment (sediment deposited as the result of nonglacial processes in areas altered by glaciation) still remains on the landscape, and most sediment eroded by postcolonial land clearance and agriculture has yet to enter a stream channel. In both of these cases, the majority of recently eroded sediment remains in colluvial storage at the base of the hillslopes from which it eroded—isolated from the channel network [Photograph 7.3].

Once sediment enters stream channels, its residence time is controlled by grain size and valley morphology. In narrow, steep upland valleys, there are few places for sediment to be stored; there is little space for floodplain deposition, and most sediment is rapidly reworked during the next flood [Photograph 7.4]. In small, lower gradient valleys, typical of midwestern North America, rapid, human-induced hillslope erosion has caused massive aggradation, filling narrow valleys with sediment on which present-day streams now flow.

In large, lowland alluvial river valleys, floodplains, once formed, can persist for millennia and longer, isolating at least some sediment from fluvial reworking [Photograph 7.5]. For example, in the Amazon River

PHOTOGRAPH 7.4 **Lack of Sediment Storage in Narrow Bedrock Valleys.** The turbid Salween River stores little sediment in its narrow bedrock valley, Three Rivers Region, Eastern Tibet. The river is running between steep, strength-limited slopes with little soil cover.

Basin, sediment resides on the floodplain for an average of thousands of years. Floodplain sediment residence times along the Amazon increase downstream, and there is more sediment moving between the channel and the floodplain at any given time than there is sediment moving downstream. Sediment stored in terraces is even less accessible to the active channel and can be stored for many millennia, unless terraces are being actively undercut and eroded by the channel.

For sediment in humid-temperate river channels, transit times are grain-size dependent. The smallest grain sizes,

PHOTOGRAPH 7.3 **Colluvial Storage.** Here colluvium is stored at the base of a slope. This small fan-form feature, in the Tien Shan mountains, Kyrgyzstan, is fed by rock falls and debris flows from steep, bare rock slopes above.

PHOTOGRAPH 7.5 **Sediment Storage on Floodplains.** In the rapidly eroding North Island of New Zealand, large amounts of sediment are stored along the flat floodplains of the main stem Waipaoa River to the left side of the image and its tributary to the right. The headwaters of the main stem have been heavily logged, triggering erosion and filling the channel and much of the floodplain with sediment.

PHOTOGRAPH 7.6 **Suspended Silt and Sand.** The very turbid Colorado River carries large amounts of silt and sand as suspended load through the Grand Canyon, Arizona. Here, steep hillsides are incised into the Vishnu Schist.

those carried in suspension, generally move through drainage basins at the pace of river flow, except for those grains that settle out on the floodplain as overbank deposits [Photograph 7.6]. These grains will eventually be reintroduced to the channel by point bar/cutbank migration. Larger particles move episodically during higher flows. During low-flow periods, coarse particles are stored in the channel bed and on point bars where they reside until the next significant flood [Photograph 7.7]. In areas where flows are episodic, such as arid regions and parts of the tropics where precipitation amounts are strongly seasonal, bed material often moves all together, suggesting much less grain-size dependence of residence times.

PHOTOGRAPH 7.7 **Gravel Bar.** Deposition during a flood created a large gravel bar on the floodplain of the Swift River, New Hampshire. The bar may eventually become vegetated and incorporated into the floodplain.

Channel Networks and Basin Morphology

Channel networks carry water and sediment through drainage basins and are integral both to basin form and function. The form of the drainage network is controlled in part by the earth materials underlying the basin, in part by the tectonic setting of the watershed, and in part by the history of the channel network itself. Over the past century, several different approaches have been developed to describe and classify the pattern and organization of channel networks.

Drainage Patterns

Drainage network development reflects the integrated action of erosional processes as rivers incise over geologic time. The common idea that river courses develop on the blank slate of an initial surface is usually an oversimplification, although this does occur where newly exposed marine terraces that lack channels are first elevated above sea level or on extensive low-relief plains of volcanic flows and ash deposits. More typically, river networks inherit features from preexisting topography and develop on rocks that vary spatially in their resistance to erosion. Tectonic processes can also modify channel network patterns over time.

Channel networks in regions with little structural control on topography typically exhibit **dendritic** patterns consisting of a system of branching tributaries that form a treelike pattern fanning out toward the basin headwaters. There is no dominant orientation of channels in dendritic channel networks, and **junction angles** at which tributaries come together are typically much less than 90°. Dendritic channel networks tend to form in the absence of structural or lithologic controls; thus, they generally occur in relatively flat-lying sediments or in homogenous crystalline rocks. Nondendritic channel networks suggest that geology, topography, or tectonics are influencing network shape.

Qualitative analysis of drainage patterns can reveal much about the materials underlying the basin and about channel history, including structure, rock type, landform shape, and human impact [Figure 7.4]. For example, naturally straight channels are rare in nature. Where they are found, they often follow fault lines or joints in rock, exploiting brecciated (broken) zones of weakness or areas of preferential groundwater flow and enhanced chemical weathering. Sharp bends in channels can reflect structural control or they can reflect channels offset by strike-slip faulting (see Figure 2.11).

Geological structures can influence drainage patterns so much that the courses of rivers and streams can be used to interpret or map the underlying structural geology. Such structurally controlled drainage exhibits **trellis** and **rectangular drainage patterns.** Trellis drainage is characterized by two dominant channel orientations in which primary tributaries join main channels at roughly right angles, with secondary tributaries running parallel to main

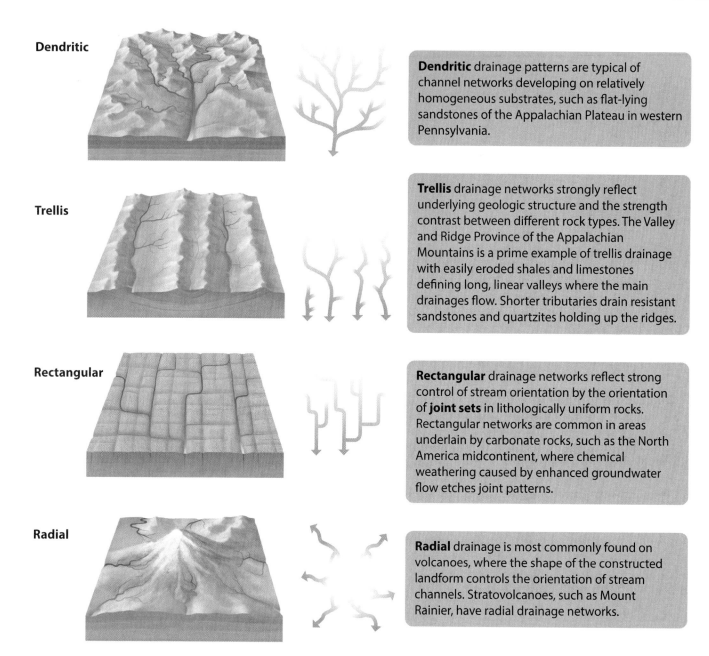

FIGURE 7.4 Drainage Patterns. Qualitative analysis of drainage patterns reveals the influence of underlying earth materials on the orientation of stream channels.

channels. Trellis drainage patterns form in areas underlain by tilted or folded beds of alternately weak and resistant sedimentary rocks, where preferential erosion along weak beds results in development of bedding-parallel strike valleys with short, steep, orthogonally oriented dip and antidip streams incised into the more resistant strata. Rectangular drainage is similar to, but more symmetric than, trellis drainage, with the two dominant drainage directions more equally developed. Rectangular drainage characterizes areas in which jointing or faults govern drainage patterns by producing linear zones that are more susceptible to weathering and erosion; such weaknesses are particularly common in landscapes underlain by carbonate rocks, such as limestones.

Topography can also influence channel networks. **Radial drainage patterns** consist of channel networks that flow away from or toward a central point. Radial drainage networks typically reflect flow off symmetrical landforms such as **volcanoes** or **structural domes**. Drainage patterns that consist of channels flowing toward a common point typically form in closed depressions, like volcanic calderas, craters, and down-dropped tectonic basins in extensional terrain.

The relationship of channel courses to the underlying geologic structure is another fundamental attribute of drainage patterns. In areas where rocks have been folded (for example, the Valley and Ridge Province in Pennsylvania), tributaries follow long linear courses between ridges controlled by the underlying structural (geologic) fabric (see Photograph 12.13). Rivers and streams that follow geologic structure, as described above, are readily explained by preferential erosion focused on zones of weakness and therefore greater erodibility. **Consequent** streams are those that generally follow the regional geologic structure, such as by flowing down tilted beds.

More curious is the development of **transverse drainage** that cuts across geological structures. The most common explanation for transverse drainage is that the rivers predate the deformation that accompanied development of a mountain range. In this view, erosion by the **antecedent,** or preexisting, drainage was able to keep pace with uplift and maintain the river's prior course as the mountains rose around it. This is the favored explanation for the development of transverse drainage in many mountain ranges, such as the Himalaya and the Cascade Range in the Pacific Northwest. Another explanation for transverse drainage is **drainage superposition** during which an alluvial drainage pattern or river course developed on sedimentary cover erodes into and is imposed on the underlying bedrock as the cover gradually erodes. Major rivers crossing the Appalachian Mountains are thought to be **superposed.** However, the superposition mechanism is difficult to test because it inherently involves removal of the evidence for it (the initial overlying deposit).

Basin shape and the associated network pattern also affect the likelihood of important confluence effects, changes that occur where a tributary enters the main stem stream. Tributary–main stem confluences are often places where bed-particle size or water geochemistry changes as material from two different basins mixes [**Photograph 7.8**]. Stream junctions frequently represent important sites of repeated disturbance and sedimentological heterogeneity, often leading to greater ecological diversity. The relative size difference between a tributary and the main stem determines whether a geomorphically effective confluence effect occurs. In general, major geomorphic change occurs at confluences where tributary basin size is ≥ 0.6 times the size of the main stem basin. These confluence effects are most likely to occur, for example, in oval basins having dendritic network patterns because such networks have large tributary basins. Rectangular basins with trellis network patterns have the least number of geomorphically important tributary inputs because, in general, trunk streams are large and tributaries are small.

Channel Ordering

Because river channels form networks, a variety of schemes have been developed that describe a channel's position and rank in the drainage network. Key to any network analysis is definition of the lowest-order or smallest channel. First-order streams are typically defined as a channel that regularly carries flow. While such a definition makes sense, implementation is more difficult because stream ordering is usually done from maps and is not based on extensive fieldwork; thus, the scale at which the basin is considered is critical. Stream orders calculated from ultra-high resolution LiDAR data will be quite different from those generated using large-scale topographic maps.

The dominant schemes used for channel ordering are those of Strahler and Shreve [**Figure 7.5**]. In the Strahler **stream ordering** scheme, when two first-order channels join, the channel downstream is designated as a second-order channel. When two second-order channels join, the result is a third-order channel, and so it goes down the network. Lower-order streams entering higher-order streams do not influence stream order in Strahler's approach. In contrast, Shreve's alternative approach defines **stream magnitude** as the total number of first-order streams contributing to the reach in question. Both stream order and magnitude are related positively to discharge, stream length, channel width, depth, and cross-sectional area, as well as sinuosity. Usually, stream order is inversely related to channel and basin slope. Some researchers argue that such relationships are the inevitable result of analyzing unidirectional branching hierarchical networks. The Strahler system is more commonly used to describe channel networks.

Downstream Trends

In addition to stream order, one can consider downstream changes in stream channel characteristics as a function of drainage basin area. Such area-based relationships have the advantage of being reproducible, independent of ordering scheme, and easily implemented in **Geographic Information Systems (GIS)**. Relationships between upstream basin area and channel characteristics depend critically on climate. In humid basins, discharge increases downbasin. In arid

PHOTOGRAPH 7.8 Stream Junction. Cloudy, turbid stream water, originating from a glaciated catchment (right), merges with a clear stream (left) at a confluence in the Noatak River drainage, at the Gates of the Arctic National Park and Preserve, northern Alaska.

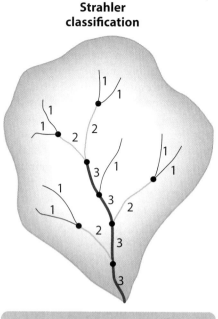
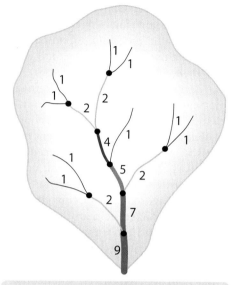

In the **Strahler** classification of stream ordering, **stream order** increases only when two streams of the same order come together. The addition of lower-order streams does not influence the order of the main stream.

In the **Shreve** classification of stream ordering, **stream magnitude** increases every time a tributary stream enters the main stem. Thus, stream magnitude increases more rapidly in the Shreve classification than does stream order in the Strahler classification.

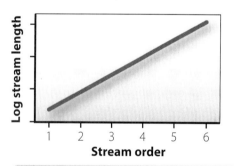
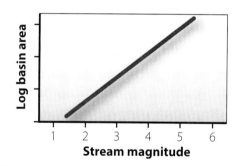

The ordering of streams leads to a series of relationships known as **Horton's laws** after the hydrologist, Robert Horton, who first developed such ordering schemes. When stream length and drainage basin area are plotted against stream order, there is a positive relationship. Streams of larger order or magnitude are systematically longer and have larger catchments.

FIGURE 7.5 Stream-Order Classifications. Stream-order classification is a quantitative analysis procedure that assigns a value to a stream reach based on its place in the stream network. The two most common ordering schemes, those of Strahler and of Shreve, use different classification algorithms.

basins and in areas underlain by soluble rocks and karst, the relationships are often more complex because losing streams are common (see Chapter 4); thus, channels need not convey increasing discharges downstream.

In humid regions, there is usually a robust, positive correlation between basin area and both bankfull and mean annual discharge. Indeed, numerous studies have found that in humid-region drainage basins, basin area is positively related to many different hydraulic geometry variables, including channel width, depth, velocity, and cross-sectional area [Figure 7.6]. In general, basin area is inversely related to both channel and basin average slope.

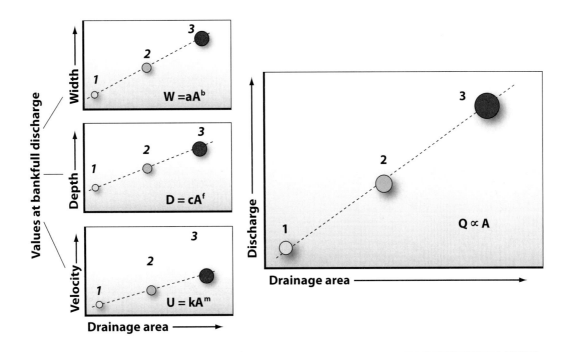

In humid-temperate regions, rivers and streams typically gain water over their length; thus, **discharge** increases downstream (with increasing basin area) and so do channel width, depth, and the average downstream velocity of water moving through the channel at any particular flow condition. Usually, comparisons are made at bankfull flow. Coefficients (a, c, and k) are determined empirically. Because basin area and discharge are positively related, basin area (which is easily measured) is used as a proxy for discharge when predicting downstream trends in width, depth, and downstream velocity, with typical values of b = 0.3 to 0.5, f = 0.3 to 0.4, and m = 0.1 to 0.2.

FIGURE 7.6 Downstream Changes in Channel Geometry. Downstream hydraulic geometry relationships quantify the observation that in humid-temperate drainage basins, channel width, depth, and water velocity all increase with basin area for a specified river level, such as bankfull.

Such correlation between basin area and geomorphically important channel characteristics means that by amalgamating hydraulic geometry data for multiple stations along a drainage, one can predict channel characteristics downstream at specific, geomorphically meaningful discharges, such as bankfull, as a function of basin area. Equations of the same form as at-a-station relationships (see Chapter 6, eq. 6.4) are used for downstream (basin area) relationships. When downstream relationships are plotted, the exponents represent the downstream rate of change in width, depth, and cross-sectional area with increasing basin area.

Finding that flow velocity generally increases downstream comes as a surprise to many people accustomed to thinking that the rushing, turbulent waters of mountain streams flow faster than the apparently calm water of large lowland rivers. However, the roughness of mountain streams and their shallow depth together lead to energy dissipation that more than makes up for their steeper slopes. Deep, wide lowland rivers flowing through relatively smooth channels have little friction along the bed and banks, allowing the water to move quickly despite the low slope (see Chapter 6).

Uplands to Lowlands

Drainage basins encompass a continuum of elevation, climate, and biota from their headwaters in the uplands to their outlets downslope. In large drainage basins, such as the Amazon, this continuum extends from frigid, lofty mountain peaks to sweltering tropical lowlands. Moving down the main stem river in a drainage basin, one can make predictions about changes in the dominant geomorphic process and form, changes indicative of slope, temperature, and weathering intensity [**Figure 7.7**]. In general, upland areas are sediment sources and lowland areas are sediment sinks.

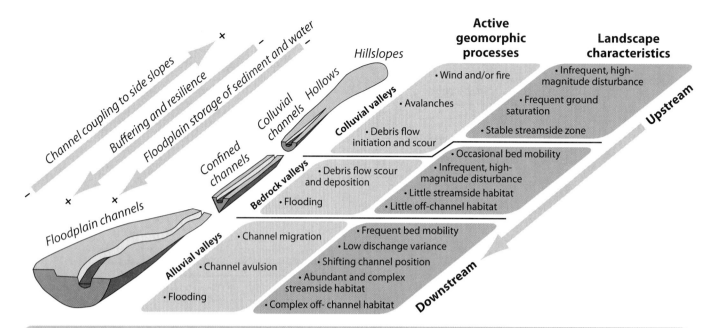

FIGURE 7.7 Hillslope-Channel Continuum. Active geomorphic processes, and thus the landscape character and landform distribution, change predictably from hillslopes to downslope valley bottoms. [Adapted from Montgomery (1999).]

Process Domains and Valley Segments

Different portions of a drainage basin are shaped by different processes. Hillslopes, hollows, channels, and valley bottoms define different process domains that may be distinguished on the basis of different relationships between drainage area and slope.

Near drainage divides on convex hillslopes (see Chapter 5), slope angle increases with distance downslope and thus with drainage area [**Figure 7.8**, middle right panel]. In contrast, drainage area-slope relations along the valley network typically exhibit an inverse relationship between slope and area:

$$S = K_s A^{-\theta_c} \qquad \text{eq. 7.2}$$

where K_s is the steepness index and θ_c is the concavity index. Spatial variations of the steepness index (K_s) have been attributed to regional and along-channel differences in precipitation, rock strength, sediment supply, and uplift rates. Different values of θ_c are typical for hollows, debris flow–dominated channels, and fluvial channels. Concavity index values range from 0.1 to >1 and are thought to reflect the influence of different surface processes, thus allowing recognition of different process domains within a landscape.

In mountain drainage basins, values of $\theta_c < 0.3$ characterize steep headwaters influenced by debris flows and channels with downstream increases in incision rate or rock strength. Bedrock channels dominated by fluvial incision typically exhibit $0.3 < \theta_c < 0.7$, and high concavities ($\theta_c > 0.7$) are associated with alluvial rivers. These downstream variations in river profile concavity are associated with identifiable **valley segments** defining portions of the channel network that have similar valley-scale morphologies and in which specific geomorphic processes operate (see Figure 7.8).

Colluvial, bedrock, and **alluvial** valley segments reflect different relative balances between sediment transport capacity and sediment supply along fluvial systems. Depositional **estuarine** valley segments define the transition from terrestrial to marine environments. In many landscapes, and mountain drainage basins in particular, fluvial processes in headwater valleys are relatively ineffective at transporting sediment delivered from surrounding hillslopes. Consequently, colluvial valley fills, composed of material that has undergone little to no transport by flowing water, accumulate in valley bottoms. Material stored in **colluvial valleys,** including the unchanneled valleys or hollows at the head of the channel network, is then transported to downstream channels when entrained by debris flows or rare high flow events [**Photograph 7.9**]. Colluvial valleys are net sediment sources. Colluvial channels that dominate the headwaters of steep, landslide-prone drainage basins typically give way to fluvial channels, such as those discussed below, at drainage areas of 1–10 km².

Colluvial valleys. Found in the headwaters of drainage networks, where hillslope processes deliver sediment to valley bottoms by soil creep and rare episodic processes like **debris flows** and **landslides**. In upland colluvial valleys, sediment is delivered more rapidly than day-to-day fluvial processes can remove it.

Bedrock valleys. Narrow, steep valley bottoms with little sediment storage and a high capacity for sediment transport. Here the fluvial system can efficiently move the sediment supplied from hillslopes and from upstream.

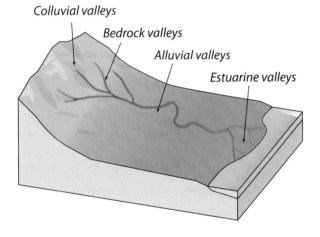

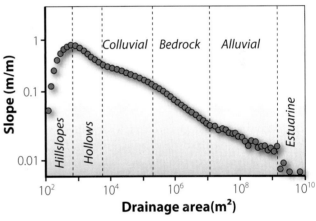

Alluvial valleys. Low gradient and filled with sediment, streams are usually unable to scour to bedrock. **Floodplains** and **terraces** are common and can store significant amounts of sediment.

Estuarine valleys. The interface between the terrestrial and marine realm, these low-gradient valleys are filled with fine-grained sediment, are wide, and are heavily vegetated.

FIGURE 7.8 Valley Segment Types. Characteristic valley segment types are found in different parts of the drainage network: colluvial valleys in the uplands, bedrock valleys where sediment transport capability exceeds supply, alluvial valleys in low-gradient lowlands, and estuarine valley segments where a river meets the ocean, or a local base level. [Adapted from Montgomery (2001).]

Bedrock valleys generally lack significant valley fill and have bedrock valley walls that typically confine channel migration to a narrow, often V-shaped valley bottom (see Photograph 7.4). Narrow valley bottoms favor development of relatively straight channels, although some regions have deeply entrenched bedrock meanders. Channel floors in bedrock valleys generally consist of either patchy accumulations of alluvium, exposed bedrock, or a mixed morphology of alternating alluvial patches and bedrock outcrops. The geologically insignificant sediment storage

PHOTOGRAPH 7.9 Hollows. Hillslope hollows (unchanneled colluvial valleys) along Mettman Ridge, near Coos Bay, Oregon. Area in the foreground was forested but has been clear-cut.

PHOTOGRAPH 7.10 Estuary. Low-energy estuary of the Salmon River in Oregon stands in stark contrast to the raging surf off the rocky headland.

in bedrock valleys, which can amount to just a few years' worth of sediment in active transport through the channel network, indicates rapid downstream export of the sediment supplied from hillslopes. Rates of bedrock valley incision limit the lowering of adjacent hillslopes and thus set the pace of landscape evolution in many unglaciated upland settings.

Thick, unconsolidated deposits characterize **alluvial valleys** where channels are able to sort and transport their loads but are unable to routinely scour to bedrock. The transport and deposition of sediment in alluvial valleys leads to the formation of floodplains and valley bottoms across which channels can migrate unconfined by bedrock valley walls. In mountainous terrain, postglacial fluvial sedimentation can fill in deeply excavated glacial or preglacial valleys. Channels in alluvial valley segments exhibit a wide range of channel patterns that reflect differences in sediment transport capacity, sediment supply, and vegetation type.

Low gradients and extensive deposits of fine-grained sediment characterize estuaries where saltwater and freshwater mix, and there can be a strong tidal influence moving sediment both up and down the estuary [Photograph 7.10]. Valley bottoms are typically wide and heavily vegetated with extensive areas of land that are occasionally submerged, either by exceptionally high tides or flood flows. During generally rising Holocene sea levels, estuaries were usually net sediment sinks, but during times of lower sea level in the past (such as at glacial maxima around 21,000 years ago when sea level was ~130 m lower), former estuaries would have become sediment sources as they were incised by rivers responding to lower sea level.

Valleys are formed in different ways and their shapes reflect both how they formed and how they were later modified. The formation of some valleys is tectonically influenced, such as those valleys running along the trend of faults where weak, brecciated (broken) rock is common. The shape of other valleys is directly controlled by the underlying structure, such as the linear valleys of the Valley and Ridge Province of the Appalachian Mountains. The shape of valley segments in cross section can be indicative of the processes that both formed the valley and subsequently modified it. For example, valley segments dominated by fluvial erosion tend to be V-shaped in cross section with relatively planar slopes. Valley segments shaped by glacial erosion tend to be U-shaped with wide, flat valley bottoms abutting steep, sometimes near-vertical valley walls.

Longitudinal Profiles

River **longitudinal profiles,** plots of channel elevation versus downstream distance, are generally concave-up, reflecting a systematic decrease in slope downstream from their headwaters [Figure 7.9]. When such a profile is smooth and concave-up, the river and its drainage basin are termed **graded**, implying that the river's slope has adjusted to a steepness that allows all the sediment supplied to the river to be transported [Box 7.1]. At the downstream end of the profile is the **base level**. The ocean, a lake, another stream, or a bedrock obstruction that controls the elevation of the channel can provide base-level control. Over geologic time, base level often moves up and down as the effects of climate change cause sea level and the water surface of closed-basin lakes to rise and fall. Uplift driven by tectonics and isostatic rebound alters the relative elevation of different parts of the landscape and can change base level. In general, rivers respond to falling base levels by incising their bed and to rising base levels by depositing sediment and aggrading.

Relief is a widely used description of drainage basins and can be considered and calculated in various ways. Total **basin relief** is simply the lowest elevation in the basin subtracted from the highest elevation. **Fluvial relief** is the elevation difference between the highest and lowest points on the river network and is less than the basin relief because of the slopes that connect the highest points in the basin to the channel heads where streams initiate. In mountainous terrain, wider mountain ranges tend to have

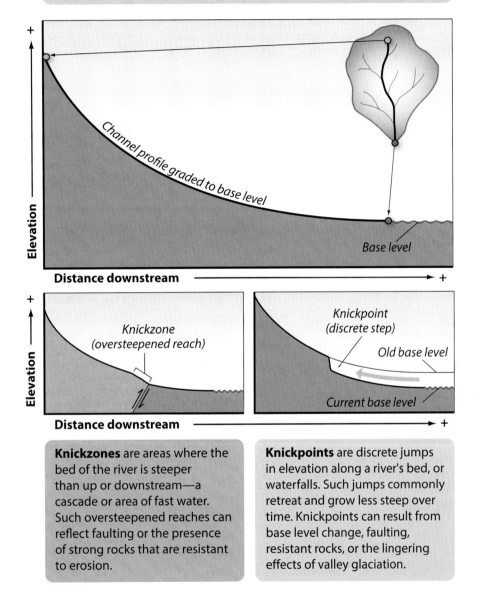

FIGURE 7.9 Longitudinal Profiles. Departures of river longitudinal profiles from the ideal, smoothly graded case reflect the influence of rock type, structure, climate/glaciation, tectonic forcing, and/or the effects of changing base level. Discontinuities in longitudinal profiles are referred to as knickzones and knickpoints.

greater relief because the main stem rivers cutting through them are longer and thus have a greater channel length, generating greater fluvial relief. In part to normalize for this effect, the **local relief** can also be calculated as the minimum and maximum elevations within an area, generally a circle of set radius, typically 5 or 10 km, or a grid cell of comparable dimensions.

Channel Confinement and Floodplain Connectivity

From the headwaters downstream to the channel mouth, there are significant differences in the lateral extent of floodplains and their connectivity with the channel. In the uplands, floodplains, if they exist, are narrow and channel

> **BOX 7.1** River Longitudinal Profiles

A simple prediction for the shape of a river longitudinal profile may be arrived at by assuming that the rate of energy loss, or stream power ($\Omega = \rho_w g Q S$), is uniform along the length of a river system. Because discharge generally increases with distance downslope along river systems, we can consider the predicted form of a river profile if there is a constant value of the stream gradient index (SL) defined by the product of channel slope (S, dz/dx) and distance from the drainage divide (x_d):

$$SL = (d_z/d_x)\, x \qquad \text{eq. 7.A}$$

Rearranging this expression and adopting a sign convention consistent with elevation dropping downslope results in an expression for channel slope as a function of SL and x:

$$d_z/d_x = -SL/x_d \qquad \text{eq. 7.B}$$

Integrating this expression with respect to x yields an expression for the riverbed elevation as a function of distance downstream:

$$z = -SL\, \ln x + c_1 \qquad \text{eq. 7.C}$$

where c_1 is a constant of integration. Considering the case where elevation declines to a base level of zero at a distance L from the drainage divide implies that $c_1 = SL\, \ln L$, in which case the full profile of the river may be described as a logarithmic function of distance downstream:

$$z = -SL\, \ln(x_d/L) \qquad \text{eq. 7.D}$$

Different values of the slope-gradient index (SL) lead to differing degrees of concavity in the overall river profile.

---

migration is confined by valley walls. Many steep upland basins have no floodplains at all; steep rocky hillslopes directly border the channels. In the lowlands, floodplains are wide and provide extensive areas for both temporary and long-term sediment storage. Wide floodplains, and the riparian vegetation communities they support, provide significant roughness that slows flood flows, dissipating energy and encouraging the deposition of additional fine sediment.

Floodplain connectivity, the ability of floodwaters to reach the floodplain, is easily changed by human activity; for example, dredging channels or building levees prevents floodwater from leaving a river. This increases the velocity and the depth of floodwaters in the river channel, thereby exacerbating the potential for damage downstream. Flow regulation by dams can also have significant effects on channel/floodplain connectivity by preventing or limiting the occurrence of overbank flows.

Downstream Trends

The characteristics of fluvial sediment (grain size, shape, and composition) systematically change downstream in channel networks. Bed material grain sizes typically decrease downstream in a drainage basin [**Photograph 7.11**]. Some of this fining is due to abrasion of grains as they are transported, some is due to weathering and breakdown of clasts during storage in terraces and bars, and some is due to selective transport through which finer grains are preferentially conveyed to downstream reaches.

In the small headwater channels of mountain drainage basins, mean grain size typically increases downstream from the channel head to a maximum before the onset of downstream fining that is typical for most of the channel network. The initial downstream increase is the result of slope processes (including rock fall, debris flows, and soil creep) delivering a range of grain sizes, including clasts large enough that flows in the relatively small channel are unable to move them. As the channel grows larger downstream and discharge increases, a larger fraction of the finer material is removed and fluvial processes become increasingly effective at transporting larger material. In addition, as floodplains widen downstream, it becomes less likely that a valley-margin landslide will deliver coarse material directly to the stream. Downstream of the zone where hillslope processes dominate sediment delivery to channels, bed grain size begins to fine downstream as particles abrade or shatter during fluvial transport and weathering during periods of storage along river systems.

Local inputs of coarse material to the channel, such as from steep tributaries, rock fall, and landslides, can reset downstream grain-size patterns. Similarly, downstream patterns of bed surface fining can be reversed where channel slopes steepen. Moving downstream in a drainage basin, one typically encounters boulder-bed, gravel-bed, and then sand-bed channels. Although the downstream transition from boulder-bed to gravel-bed channels is gradual, the transition from gravel-bed to sand-bed channels is typically rather abrupt. Sand-bed channels are actively mobile at most (if not all) flows, whereas gravel-bed and boulder-bed channels exhibit threshold mobility in which the bed is stable at flows lower than a critical discharge.

Particle size and shape change downstream as abrasion increasingly rounds clasts during transport. In basins with heterogeneous geology, the composition of particles in fluvial sediment also changes with distance downstream, as more readily weathered grains are altered. Weathering-resistant grains, such as feldspar and quartz, form an increasing proportion of the clast sediment population with distance downstream.

PHOTOGRAPH 7.11 Channel Bed Material. (a) Boulder-bed mountain channel with a step-pool morphology, North Saint Vrain Creek, Colorado. (b) Gravel stream-bed, Washington State. (c) Dry sand-bed channel of the Rio Puerco, New Mexico.

Drainage Basin Landforms

Drainage basins contain a variety of landforms that result from the complex interaction between rivers and both solid and unconsolidated earth materials. These landforms can be used in some cases to interpret landscape history and in others to infer tectonic and climatic change. The distribution of drainage basin landforms is closely tied to the history and location of base-level controls (see Longitudinal Profiles section above).

Knickpoints

Deviations from smooth concavity in a river's longitudinal profile are often interpreted as the influence of tectonics, climate history, and/or lithology. For example, uplift along a fault over which a river is flowing will often result in an unusually steep reach along a channel, known as a **knickzone** (see Figure 7.9). Some steepened stream reaches reflect the presence of immobile debris deposited into the channel by landslides and debris flows, and are not related to base-level change. If the steepened area is a discrete jump in bed elevation (a waterfall), it is known as a **knickpoint** [Photograph 7.12a]. Longitudinal profiles of streams in glaciated landscapes often have steps related to hanging valleys inherited from glacial times, steps that fluvial erosion over the Holocene Epoch has not been able to erase [**Photograph 7.12b**]. Steep channel reaches can also reflect the presence of strong rocks, such as quartzite or volcanic dikes, in the midst of otherwise weaker lithologies such as shale or mudstone [**Photograph 7.12c**]. The change in slope downstream affects important geomorphic processes, especially sediment transport and the energy available for channel bed incision.

Some knickpoints, once established, retreat upstream over time, leaving in their wake incised channels and abandoned floodplains, which then become terraces. Knickpoint retreat is easily demonstrated in some locations, for example, at Niagara Falls, on the border between Ontario, Canada, and upstate New York and in many small catchments affected by postglacial isostatic response and relative sea-level drop such as in Scotland. In other areas, such as the Great Falls of the Potomac River, cosmogenic dating of exposed bedrock terraces is more consistent with spatially uniform incision and persistence of the knickzone at about the same location. At Great Falls, it appears that the river incised downward into rock along a several-kilometers-long reach of the channel at the same time.

Gorges

Gorges are deeply incised reaches along rivers, the result of river channels cutting into competent material, usually rock, that can hold steep to near-vertical slopes [**Photograph 7.13**]. Gorges form in steep channel reaches

PHOTOGRAPH 7.12 Knickpoints. Knickpoints come in different sizes and reflect different conditions and processes. (a) Skogafoss, a large waterfall fed by glacial meltwater on the southern coast of Iceland, acts as a migrating knickpoint in response to changes in base level imposed by the eruption of basaltic lava flows. (b) Hanging valleys are common in once-glaciated terrain. Here, Bridalveil Creek hangs above the deeper Yosemite Valley, California, forming the famous falls. (c) This waterfall in Bryce Canyon National Park, Utah, reflects a resistant layer of bedrock. Undermining at the base of the falls allows upstream migration of the knickpoint.

where sufficient energy is available to erode rock and prevent sediment accumulation. Oversteepening can be the result of tectonic uplift; for example, deep gorges are found on Himalayan rivers that flow across localized areas of active rock uplift. Diversion of rivers by glaciers or large rock slides can steepen gradients and lead to localized incision. Many gorges in northeastern North America are the result of postglacial rivers reestablishing their courses over bedrock ridges and then cutting down through these obstacles.

Gorge incision occurs through a variety of bedrock incision processes including plucking, pothole formation, and abrasion. Because flow in gorges is deeper and narrower than in reaches that typify streams outside of gorges, unit stream power exerted on the bed is higher. This provides a positive feedback that enhances incision within gorges.

Very narrow, steep-sided **inner gorges** are often found incised into the bottom of wider gorges or valleys. Inner gorges in nonglaciated regions of the Himalaya and northern California have been attributed to recently increased uplift rates and/or concomitant base-level fall, leading to renewed and focused incision. In some locations, inner gorge formation has been attributed to greater soil moisture, producing a higher landslide frequency on the lower portions of valley walls. The origin of V-shaped gorges at the base of glaciated U-shaped valleys invites a different question, especially in tectonically quiescent landscapes. Do these gorges record postglacial incision in response to lowered base level or have they survived under ice through multiple glaciations?

Terraces

Terraces are important drainage-basin landforms that preserve sediment and provide a record of the past behavior of the fluvial network, as well as of the supply and nature of sediment delivery to the river system. Over time, as rivers shift position and erode their banks, terrace remnants are removed so that eventually only isolated parts of any originally continuous terraces remain.

The existence of terraces provides strong evidence that thresholds are important in geomorphology. For terraces to form, floodplain abandonment must be an episodic rather than gradual process. Terrace formation can be caused by changes in many different boundary conditions. Changes in sediment supply, often linked to changes in climate over the Quaternary (the Pleistocene and Holocene Epochs), have caused many rivers to aggrade and then abandon and incise their floodplains, leaving flights of terraces behind. In areas that have been uplifted, terraces, if they can be dated, are a means by which to characterize rates of uplift and surface deformation over time.

Terraces can be divided into erosional and depositional terraces [**Figure 7.10**]. Terraces eroded into stable substrates are known as **strath terraces** [**Photographs 7.14 and 7.15**]. Aggradational terraces [**Photograph 7.16**] that represent filling of an already incised valley are known as **fill terraces** and consist of fluvially deposited unconsolidated sand and gravel. They form when, after a period of incision and valley formation, a river does not have the transport capacity to move the sediment load supplied to it from drainage basin slopes. When formed, fill terraces were laterally continuous floodplains across the valley and downstream. Fill terraces are common in valleys just outside glacial margins. For example, in the Teton Range near the Grand Teton in Wyoming, aggradational terraces reflect the very large sediment loads supplied to nearby rivers by glacial meltwater during glacial periods. Fill terraces can also reflect increasing sediment supply off hillslopes when the climate changes. Isolated terrace remnants perched high on gorge walls indicate that thick stacks of sediment once filled deep valleys in the Himalayan Mountains. The terraces, of which these sediments are the last surviving evidence, are thought to reflect increased monsoonal activity and sediment delivery in the mid-Holocene, about 7000 years ago.

Strath terraces have thin deposits of unconsolidated alluvium overlying beveled and better consolidated material beneath. The beveled material can be bedrock or consolidated

PHOTOGRAPH 7.13 River Gorge. Steep river gorge cut into rock along a tributary to the Salween River in eastern Tibet. Note sediment deposition and temporary storage in the foreground where the canyon widens (along with the washed-out pier of a bridge).

236 Chapter 7 • Drainage Basins

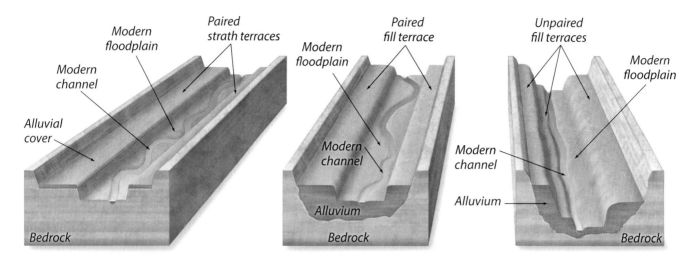

Erosional fluvial terraces are indicative of a geomorphic regime in which the river has sufficient energy not only to move the sediment load supplied to it, but to cut into the material that makes up the channel bed. Terraces formed by erosion are referred to as **straths.** Such terraces are frequently found in areas where active uplift or tilting provides the potential energy for incision. Straths can be covered by a thin layer of alluvial sediment.

Depositional fluvial terraces are indicative of river systems where sediment supply once exceeded the capacity of the river to transport sediment. The excess sediment was deposited in valley bottoms, filling them. At later times, if the sediment transport capacity increases (from more water, or steepening of the river gradient by tectonic tilting) or the sediment supply decreases, then the depositional surface can be incised.

Paired terraces are found at the same elevation across the width of a valley. Paired terraces form when river migration rates across the valley are rapid in comparison to incision rates. **Unpaired terraces** are found on only one side of the valley and result when rivers incise much more rapidly than they migrate across the valley bottom. Unpaired terraces also can result from stream erosion removing the terrace from one side of the valley but not the other.

FIGURE 7.10 Terrace Types. Depositional (fill) and erosional (strath) terraces form in response to different forcings. Erosional terraces imply incision capable of eroding underlying rock or sediment. Depositional terraces indicate changing sediment-transport capacity and/or sediment supply for the channel.

PHOTOGRAPH 7.14 Strath Terrace. Series of bedrock strath terraces along the Susquehanna River in Holtwood Gorge, Pennsylvania. Any alluvium that once capped these terraces has been removed by later river flows.

PHOTOGRAPH 7.15 Strath Terrace and Alluvial Mantle. Strath terrace eroded across steeply dipping sedimentary rock and covered with a thin mantle of alluvium, Nenana River, Alaska.

PHOTOGRAPH 7.16 Fill Terrace. This fill terrace along the Copper River in Alaska shows that the river once aggraded significantly before incising.

PHOTOGRAPH 7.17 Alluvial Fan. Dark alluvial fan abutting white playa deposits in Death Valley, California. Note the road around the toe of the fan.

sediment. Current hypotheses suggest that strath terraces form when a river with a bedrock-floored channel meanders across a valley bottom and that strath formation is favored by, but not restricted to, weak rocks—especially those that lose strength when desiccated (such as micaceous siltstone). Beveling bedrock is widely thought to require a thin, mobile layer of sediment, which provides the tools for cutting the bed (see Chapter 6, Figure 6.3, and Figure DD6.4). If the river carries too much sediment, it will aggrade rather then bevel the rock beneath; if the river carries too little sediment, it will incise. Strath terraces are abandoned when boundary conditions change, allowing the river to incise. The incision leaves the strath terrace and its thin blanket of alluvial sediment high and dry (see Photograph 7.15).

The spatial distribution of terraces reflects the ratio between vertical and lateral rates of channel migration. Rapid downcutting, with little lateral channel migration, leads to **paired terraces** at similar elevations on both sides of the channel. More rapid lateral migration over greater distances provides the opportunity for **unpaired terraces** to form in close proximity to one another but at different elevations. Such unpaired terraces form because by the time a laterally migrating river returns to its original position, the river has incised enough that it forms a floodplain at a new, lower level (see Figure 7.10).

There are numerous reasons why rivers form terraces. In rapidly uplifting, tectonically active areas, strath terraces are left behind as rivers incise into rock. Changes in base level, such as those resulting from the draining of ice marginal lakes in the mountainous terrain of New England, have resulted in strath terrace sequences as newly reestablished rivers cut through large accumulations of glacial and postglacial sediment. These are not bedrock-floored strath terraces but rather terraces cut into cohesive glacial lake sediments and the underlying glacial till.

Fans

Fans are cone-shaped depositional features that form in both arid and humid regions where generally steep channels emerge from confined canyons and narrow valleys onto lower gradient surfaces where channels and flows are less confined. Sediment supplied from the highlands builds up on fans from which it cannot easily be removed. Fans are a dominant and conspicuous landform in arid regions where fluvial sediment transport processes in valley bottoms are limited by the paucity of water. Fans are also present in humid temperate zones, although they are often hard to find under forest cover. Large fans are commonly found at the base of slopes in tectonically active regions where rates of hillslope erosion are very high and sediments fill the valleys of the Basin and Range in western North America [**Photograph 7.17**]. Small fans can form in places that receive more sediment from upslope than can be removed by surface processes [**Photograph 7.18**].

PHOTOGRAPH 7.18 Historical Alluvial Fan. A horse stands at the apex of a small alluvial fan near Tunbridge, Vermont, in the late 1800s. This fan, an accumulation of material eroded from the river terrace above, testifies to the ease with which soil moves after trees are stripped from the landscape.

PHOTOGRAPH 7.19 **Bajada.** Fault-controlled mountain front and recently erupted cinder cone on a bajada (the surface of coalescing fans sloping gently away from the mountain front) draped in snow. Fans here, in the Owens Valley below the Sierra Nevada, California, are dominated by debris-flow deposits that accumulated primarily during glacial periods.

Sediment transport capacity is lower on fans than in the channels of drainage basins that supply fan sediment for several reasons. When source streams cross from the mountain front onto a fan, channel hydraulic geometry changes as the streams are no longer confined in bedrock channels but rather can adjust their geometries in unconsolidated fan materials. In arid regions, where fans are a common landform, there can be significant infiltration losses from the channel, reducing discharge and sediment transport capacity down fan. In addition, channel slope declines down fan, reducing stream power. Streams flowing on fans commonly split into several channels resulting in less stream power per unit area (see eq. 6.5) on the broad fan toe than at the fan apex.

Geomorphologists distinguish **alluvial fans** and **debris fans** on the basis of the dominant sediment-transport process, as inferred from surface morphology and the sedimentology of fan deposits. Although some fans are composed entirely of fluvially transported material and others are entirely composed of debris flow–delivered sediment, many fans are composed of a mix of water-borne sediments and debris-flow deposits. Nevertheless, the distinction persists in the literature based on the observation of depositional processes and deduction from fan stratigraphy that debris fans are built primarily by debris flows, while alluvial fans are built primarily by stream flows [Photograph 7.20].

The largest fans generally form where channels cross a mountain front and flow out onto a structural valley. Along many mountains fronts, drainage basins are spaced closely enough and sediment loads are sufficiently high that individual fans merge into a broad, low-gradient piedmont termed a **bajada**, from the Spanish for "slope" [Photograph 7.19]. Fan surfaces typically decrease in slope, and fan sediments typically decrease in grain size, from the **fan head** to the distal end of the fan, the **toe**. Changes in sediment load due to climate change and/or tectonic tilting can result in sediment redistribution on fans with **trenching** (incision) at the head of fans and deposition at the toe.

Alluvial sediments tend to be better sorted than debris-flow deposits, and many alluvial sediments are **clast-supported** (one clast resting on another with no intervening fine-grain matrix). Active alluvial fan surfaces are commonly dominated by channels with bars and swales formed by flowing water. In cross section, the alluvial sediments are sorted by grain size and stratified, indicating

(a)

(b)

PHOTOGRAPH 7.20 **Alluvial and Debris Fan Formation.** Fans can result from both alluvial and debris flow deposition. (a) Trench into small fan in Huntington, Vermont, reveals stratified layers of water-lain, alluvial sediment demarcated here by flags. A standard soil color chart (known by its trade name, Munsell) is being used to describe the color of the fan sediments. (b) Nearly 2-meter-high geologist attempts to climb an exposure in a fan dominated by debris-flow deposition, as indicated by the unstratified, clast-supported nature of the deposit, in Coso Range, southern California.

fluvial transport. Alluvial deposits typically lack very fine material, at least close to the range front.

In contrast, debris-flow deposits are typically unsorted and **matrix supported** (there is a fine-grained matrix between the clasts). The surface morphology of fans dominated by debris flows is varied but can be distinctive. Common landforms include channels bordered by debris-flow levees, channels plugged by debris flows that ran out far enough that the shear stress imparted by gravity acting on the fan slope was no longer sufficient to overcome the **yield strength** of the flowing debris, "freezing" it in place [**Photograph 7.21**], and lobes of debris that spilled out of channels and onto the broader fan surfaces.

Road cuts into such fans, which are common along the White Mountains and Sierra Nevada of southern California, reveal mostly matrix-supported material in which clasts appear suspended in a finer-grained matrix (a diamicton)—a telltale signature of debris flows (see Photograph 5.14). However, there are also lenses of well-sorted sediment moved by flowing water, an indication that more than just debris flows moved sediment onto and across these fans in the past.

Development and urbanization of fans causes numerous and significant geologic hazards. Such development is widespread in the arid southwestern United States but is also occurring in humid regions, such as northeastern North America and parts of Europe. Because of the rapid hydrologic response of steep drainage basins, flash flooding on fans is common in both humid and arid regions. Channels on fans may shift episodically and unpredictably during such events through lateral migration, plugging, and **avulsion** (channel shifting). Large storms and the runoff they generate can cause sediment deposition on fans both from stream flows and debris flows. This sediment and the water that carries it can disrupt infrastructure, destroy homes, and take lives.

Lakes

Lakes are found in many different landscapes and result from a variety of solid-Earth and surficial processes. In humid regions, lakes are usually permanent landscape features on human timescales, although their levels can change in response to climate variability. In arid regions, lakes are commonly ephemeral, reflecting the balance between precipitation delivered mostly to the highlands and evaporation rates that can be extremely high in valleys where the lakes are located. Arid region lakes tend to be shallow and saline, drying to salt and mud flats known as **playas** (see Photograph 7.1).

Lakes are common in glaciated terrain. Glacial erosion can overdeepen valleys, creating bedrock basins that fill with water when the ice melts. Such **tarns** are often arranged in linear strings and are referred to as **pater noster lakes,** after the rosary beads they resemble [**Photograph 7.22**]. Moraines also dam lakes in glacial terrain. These dams can fail catastrophically, triggering massive and hazardous outburst floods, especially during rapid glacial retreat. Advancing or retreating glacial ice can dam rivers and streams, forming ice-marginal lakes that drain when the ice retreats or the lake grows deep enough that the ice-tongue dam floats (see Chapter 13, Digging Deeper). Other ice-marginal lakes formed where isostatic depression, caused by the weight of continental ice sheets, lowered land surfaces around the ice margin, creating depressions that filled with meltwater. These lakes later drained as the ice sheet melted and Earth's surface rebounded.

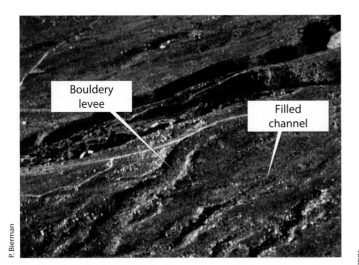

PHOTOGRAPH 7.21 Abandoned Channels on Alluvial Fan. Here, in an aerial view of the Lone Pine Creek fan, there are many abandoned stream channels, filled in places by debris-flow lobes and lined by bouldery debris-flow levees. The channel of Lone Pine Creek is deeply incised and runs diagonally across the photograph. Two-lane road provides scale.

PHOTOGRAPH 7.22 Tarns. A series of tarns, formed by glacial erosion, are visible from the summit of Sunset Peak at the Brighton Ski Resort in Utah.

PHOTOGRAPH 7.23 **Crater Lake.** This photo of Crater Lake, Oregon, shows Wizard Island, a small cone in the center of the lake, which occupies the caldera. The lake formed about 7700 years ago when Mount Mazama exploded, blowing off the upper 1600 meters of the volcano.

Lakes also commonly form in the craters of stratovolcanoes [**Photograph 7.23**] and in the centers of explosive calderas such as Yellowstone. Fed by orographic precipitation, these lakes become prime sources of water to feed subsequent volcanic explosions (when hot lava and lake water mix) and **lahars** (volcanic mudflows). **Maars,** craters formed explosively when magma interacts with groundwater and flashes it to steam, are often filled with water, creating lakes with extremely small drainage basins [**Photograph 7.24**]. Sediment cores from maar lakes have been used to assess geochemical inputs from the atmosphere over time because runoff contributions from the very small drainage basin (only the crater rim) are minimal.

PHOTOGRAPH 7.24 **Maar.** Laguna de Armenia is a maar near the San Miguel Volcano in El Salvador.

Applications

Drainage basins respond on various timescales to external forcings, including climate, tectonics, and human activity. On geologic timescales, tectonic forcing, including isostatic response to erosion, is the dominant control on the size, shape, and slope of drainage basins. On intermediate timescales, climatic forcing is important, and on decadal timescales, the influence of human actions can be dominant.

Land use affects drainage-basin dynamics, changing sediment budgets and altering geomorphological processes and zones of process dominance. Major human impacts on watersheds in Europe began with the advent of deforestation and intensive agriculture in the Bronze Age between 4000 and 5000 years ago. The clearance of forests, for agriculture and for wood, changed sediment budgets as soils began to erode faster than they were being formed—a reversal of the preceding postglacial millennia, during which soils formed more rapidly than they eroded. At a drainage-basin scale, much of the soil and sediment eroded from exposed uplands as the result of human activity never made it to main-stem river channels. Rather, it still resides on the colluvial footslopes, fills upland hollows, or builds terraces in low-order channels. The sediment that did enter the fluvial system often caused aggradation downstream, building bars and raising floodplains. The most glaring effect in the classical world of human-landscape interaction and drainage-basin response was rapid sedimentation in estuarine valleys and siltation of harbors. One famous example is the Roman port of Ostia at the mouth of the Tiber River, which is now several kilometers from the sea.

In eastern North America, New Zealand, China, parts of Canada, and elsewhere, forests have regrown after clear-cutting either naturally, after cleared land was abandoned, or where replanted as part of soil-conservation or reforestation efforts. As a result, sediment loads have diminished. In many places, rivers are incising through decades' to centuries' worth of sediment deposited when the landscape was cleared, the legacy of earlier land use and abuse [**Figure 7.11**]. Such **legacy sediments** are found nearly worldwide, and in many places form a low terrace at or just above the modern floodplain (see Digging Deeper).

Development, both land clearance for agriculture and subsequent urbanization, has specific impacts on drainage basins, changing water and sediment yields and channel behavior over time [**Figure 7.12**]. Removing forests, smoothing the land, and adding impermeable surfaces all affect the volume and speed of runoff (see Chapter 4), changing basin-scale water and sediment budgets. As impervious cover increases, floods become far more common as rainfall cannot infiltrate and greater runoff results. Flood peak magnitude increases and flood duration shortens because less water is stored in the vadose zone and on the landscape.

The removal of wood from streams, either by intentional clearance or indirectly through clearing of streamside forests and the trapping and extirpation of beavers that drop trees into streams to build dams, appears to

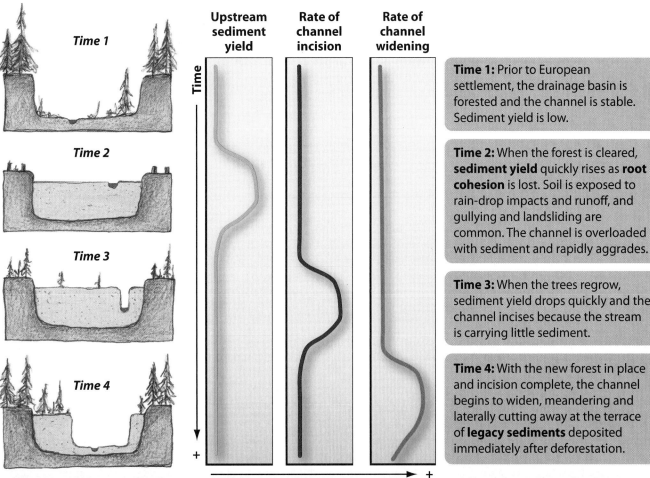

Time 1: Prior to European settlement, the drainage basin is forested and the channel is stable. Sediment yield is low.

Time 2: When the forest is cleared, **sediment yield** quickly rises as **root cohesion** is lost. Soil is exposed to rain-drop impacts and runoff, and gullying and landsliding are common. The channel is overloaded with sediment and rapidly aggrades.

Time 3: When the trees regrow, sediment yield drops quickly and the channel incises because the stream is carrying little sediment.

Time 4: With the new forest in place and incision complete, the channel begins to widen, meandering and laterally cutting away at the terrace of **legacy sediments** deposited immediately after deforestation.

The channel evolution that resulted from European landscape disturbance in North America is a prime example of **complex response.** An initial perturbation, deforestation and other land-use changes such as agriculture, changed hillslope erosion rates and sediment supply to channels. Crossing a **threshold,** channels aggraded. When forests returned, another threshold was crossed and channels incised before starting to widen. The effects of land clearance several centuries ago are still reflected in a complex and interrelated set of landscape scale process and landform changes.

FIGURE 7.11 Channel Change. Channel change resulting from disturbance often follows a trajectory that includes aggradation, then incision followed by widening. [Adapted from Schumm and Rea (1995).]

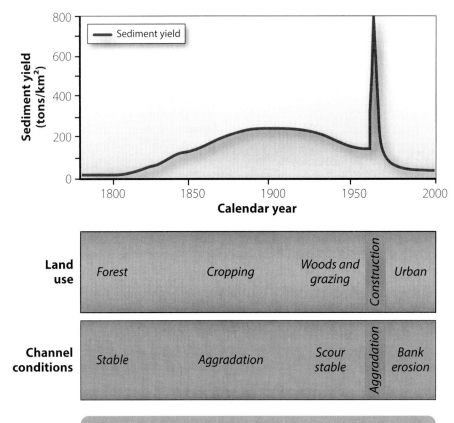

FIGURE 7.12 Sediment Yield and Land Use. Sediment yield is affected by land use and by land development patterns and intensity. Sediment yield peaks during active construction and the accompanying land-surface disturbance. [Adapted from Wolman (1967).]

have changed the fundamental morphology of many forest streams. What were once shallow, anabranching, wood-clogged channels have now become incised, meandering systems. Channel incision isolates floodplains from streams, reducing connectivity, preventing sediment deposition, and reducing the potential of floodplain water storage to attenuate floods. In cases where streams have incised deeply into legacy sediments and are isolated from their floodplains, one stream restoration strategy is to remove the legacy sediment to reconnect the stream and its floodplain. Nature does this over time, but the result of letting nature take its course may be decades to centuries of high-sediment loads downstream and failing streambanks (see Figure 7.11).

Forestry practices have significant, long-lasting, and widespread impacts on drainage-basin geomorphology. Timber harvesting affects watersheds primarily in two ways: (1) Removal of trees from basin hillslopes reduces the apparent cohesion from roots (see Chapter 5, Digging Deeper), encouraging mass movements, decreasing evapotranspiration, and raising water tables (see Chapter 4), thereby decreasing slope stability and increasing runoff. (2) Roads built to remove logs from mountain drainage basins run along or are cut into steep slopes and bridge stream channels, creating constrictions that can act as dams during floods. These roads compact the soil, increase runoff, and channel it onto potentially unstable hillsides, where it can cause gully erosion and mass wasting. Road fill failure is also common. In areas as far apart as Oregon and New Zealand, forestry impacts on drainage basins have produced similar hillslope erosion, high sediment loads, and downstream channel aggradation. These effects are not unique to modern industrial forestry. Similar drainage basin–scale effects were observed in response to deforestation in the French Alps in the late eighteenth and early nineteenth centuries.

Dam construction and removal both have major impacts on drainage basins and their geomorphology. Where dams block river channels, they raise local base levels, triggering aggradation upstream. Over time, reservoirs fill with sediment that would otherwise have been transported downstream [**Photograph 7.25**]. Water released from dams typically carries little sediment and thus incision results downstream. Recently, efforts to remove outdated dams have increased as regulators of fisheries, hydropower, and water-dependent industries reconsider the economic and environmental impacts of dams, as well as safety concerns reflecting the age and location of older dams. Understanding and predicting the impact of dam removal is a relatively new but very important area of research for geomorphologists.

Large dams that block many of the world's major rivers, such as Glen Canyon Dam in Arizona or the Three Rivers Dam in China, clearly and dramatically affect the geomorphology of the downstream river channels. Yet geomorphologists have documented that even small dams, such as those used to power colonial-era mills, can have major effects at a basin scale [**Figure 7.13**] (see Digging Deeper).

Throughout eastern and central North America, small streams and rivers were harnessed for power to cut wood, mill grain, and eventually to generate electricity. In Pennsylvania alone, fieldwork and historical documents suggest there were at least 16,000 mill dams trapping sediment and altering stream morphology at a density of about one dam every 6 or 7 km^2. Now, as many of these dams are failing or being breached, this sediment, left from years of hillslope land clearance and stored for decades to centuries, is being remobilized and is once again entering river channels.

Of particular concern is the release and subsequent transport of this sediment to Chesapeake Bay and other rich but sensitive estuaries in eastern North America. A sustained increase in fine-grained sediment delivery is injurious to estuaries because increased water turbidity decreases light transmission, smothers benthic organisms (like oysters and crabs), and carries sediment-associated nutrients like phosphorus, which cause eutrophication and unwanted algal blooms.

Channel restoration, rehabilitation, and drainage-basin management have become widespread over the past several decades as societal awareness regarding the importance of rivers as both ecological and hydrological systems has increased. Two questions come up when channel or watershed restoration or rehabilitation work is proposed; both require specific understanding of drainage-basin behavior and history.

First, is the postdevelopment hydrology of the drainage basin today similar to that of the basin at the time of the desired restoration objective, known as the **reference condition**? If runoff volumes, timing, and intensity have all changed from development, climate change, or a mix of causes, the historical characteristics of the channel before development (including its slope, width, depth, sinuosity, and pattern) may no longer be stable under contemporary conditions. Thus, it may not be possible to reconstruct the historical or predisturbance river and have such a reconstruction persist.

The second question is pertinent when the watershed has changed enough to preclude reestablishing or achieving the reference condition: How can the channel be rehabilitated so as to meet the objectives of the community in terms of aesthetics and river behavior? Managing drainage basins effectively requires a deep understanding of the complex interactions between water and sediment

PHOTOGRAPH 7.25 Sediment Stored Behind Dam. The 30-meter-tall Rindge Dam is located in Malibu, California. The main arch of the dam was completed in 1924 but now the reservoir is completely filled with sediment from the steep, rapidly eroding watershed. Fish advocates have called for the dam's removal because it is blocking steelhead trout from accessing the upper reaches of Malibu Creek.

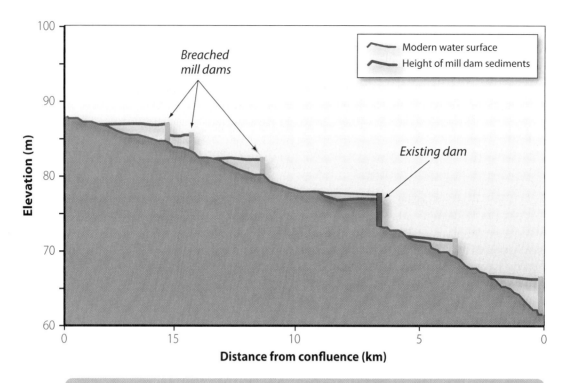

FIGURE 7.13 Mill Dam Sediments. In hilly eastern North America, mill dams and the head they created on small streams were the primary power source during and after colonial settlement. These dams, most long since abandoned, trapped sediments shed from denuded hillsides. [Adapted from Walter and Merrits (2008).]

In many humid-temperate environments, the legacy of mill dams is geomophically important. Once used to impound streams so the potential energy of the falling water could power mills to saw wood, grind grain, and cut shingles, the dams trapped large amounts of sediment shed from deforested landscapes. Today, many of the dams are gone but the **legacy sediments** deposited behind the dams remain in terraces and valley fills.

over time. Geomorphologists, with their understanding of process, form, and landscape history, are particularly well suited to tackling this increasingly important task.

Selected References and Further Reading

Abrahams, A. D. Channel networks: A geomorphological perspective. *Water Resources Research* 20 (1984): 161–188.

Ambers, R. K. R., and B. C. Wemple. Reservoir sedimentation dynamics: Interplay and implications of human and geologic processes. *Northeastern Geology & Environmental Sciences* 30 (2008): 49–60.

Beach, T. The fate of eroded soil: Sediment sinks and sediment budgets of agrarian landscapes in southern Minnesota, 1851–1988. *Annals of the Association of American Geographers* 84 (1994): 5–28.

Bishop, P., T. B. Hoey, J. D. Jansen, and I. L. Artza. Knickpoint recession rate and catchment area: The case of uplifted rivers in Eastern Scotland. *Earth Surface Processes and Landforms* 30 (2005): 767–778.

Brummer, C. J., and D. R. Montgomery. Downstream coarsening in headwater channels. *Water Resources Research* 39 (2003): 1294, doi: 10.1029/2003 WR001981.

Burbank, D. W., J. Leland, E. Fielding, et al. Bedrock incision, rock uplift and threshold hillslopes in the northwestern Himalayas. *Nature* 379 (1996): 505–510.

Church, M., and J. M. Ryder. Paraglacial sedimentation: A consideration of fluvial processes conditioned by glaciation. *Geological Society of America Bulletin* 83 (1972): 3059–3072.

Costa, J. E. Effects of agriculture on erosion and sedimentation in the Piedmont province, Maryland. *Geological Society of America Bulletin* 86 (1975): 1281–1286.

Dietrich, W. E., and T. Dunne. Sediment budget for a small catchment in mountainous terrain. *Zeitschrift für Geomorphologie, Supplementbände* 29 (1978): 191–206.

Gardner T. W. Experimental study of knickpoint and longitudinal profile evolution in cohesive, homogenous material. *Geological Society of America Bulletin* 94 (1983): 664–667.

Hack, J. T. *Studies of Longitudinal Profiles in Virginia and Maryland.* U.S. Geological Survey Professional Paper 294-B. Washington, DC: Government Printing Office, 1957.

Jennings, K. V., P. R. Bierman, and J. Southon. Timing and style of deposition on humid-temperate fans, Vermont, United States. *Geological Society of America Bulletin* 115 (2003): 182–199.

Kelsey, H. M., R. Lamberson, and M. A. Madej. Stochastic model for the long-term transport of stored sediment in a river channel. *Water Resources Research* 23 (1987): 1738–1750.

Leopold, L. B. Downstream change of velocity in rivers. *American Journal of Science* 251 (1953): 606–624.

Montgomery, D. R. Process domains and the river continuum. *Journal of the American Water Resources Association* 35 (1999): 397–410.

Montgomery, D. R. Slope distributions, threshold hillslopes, and steady-state topography. *American Journal of Science* 310 (2001): 432–454.

Montgomery, D. R. Observations on the role of lithology in strath terrace formation and bedrock channel width. *American Journal of Science* 304 (2004): 454–476.

Montgomery, D. R., G. E. Grant, and K. Sullivan. Watershed analysis as a framework for implementing ecosystem management. *Journal of the American Water Resources Association* 31 (1995): 369–386.

Nichols, K. K., P. R. Bierman, M. Caffee, et al. Cosmogenically enabled sediment budgeting. *Geology* 33 (2005): 133–136.

Pazzaglia, F. J., and M. T. Brandon. Macrogeomorphic evolution of the post-Triassic Appalachian mountains determined by deconvolution of the offshore basin sedimentary record. *Basin Research* 8 (1996): 255–278.

Pratt, B., D. W. Burbank, A. M. Heimsath, and T. Ojha. Impulsive alluviation during early Holocene strengthened monsoons, central Nepal Himalaya. *Geology* 30 (2002): 911–914.

Reid, L. M., and T. Dunne. *Rapid Evaluation of Sediment Budgets*. Reiskirchen, Germany: Catena Verlag, 1996.

Reusser, L., P. Bierman, M. Pavich, et al. An episode of rapid bedrock channel incision during the last glacial cycle, measured with ^{10}Be. *American Journal of Science* 306 (2006): 69–102.

Schumm, S. A. *The Fluvial System*. New York: Wiley-Interscience, 1977.

Schumm, S. A., and D. K. Rea. Sediment yield from disturbed earth systems. *Geology* 23 (1995): 391–394.

Shreve, R. L. Statistical law of stream numbers. *Journal of Geology* 74 (1966): 17–37.

Strahler, A. N. Quantitative analysis of watershed geomorphology. *Transactions, American Geophysical Union* 38 (1957): 913–920.

Trimble, S. W. Decreased rates of alluvial sediment storage in the Coon Creek Basin, Wisconsin, 1975–93. *Science* 285 (1999): 1244–1246.

Walling, D. E., and B. W. Webb. *Erosion and Sediment Yield: Global and Regional Perspectives*, Publication no. 236. Wallingford, UK: International Association of Hydrological Sciences, 1996.

Walter, R. C., and D. J. Merritts. Natural streams and the legacy of water-powered mills. *Science* 319 (2008): 299–304.

Whipple, K. X. Bedrock rivers and the geomorphology of active orogens, *Annual Review of Earth and Planetary Science* 32 (2004):151–185.

Whipple, K. X., and T. Dunne. The influence of debris-flow rheology on fan morphology, Owens Valley, California. *Geological Society of America Bulletin* 104 (1992): 887–900.

Wolman, M. G. A cycle of sedimentation and erosion in urban river channels. *Geografiska Annaler* 49A (1967): 385–395.

Wolman, M. G., and A. P. Schick. Effects of construction on fluvial sediment, urban and suburban areas of Maryland. *Water Resources Research* 3 (1967): 451–464.

DIGGING DEEPER When Erosion Happens, Where Does the Sediment Go?

Geomorphologists commonly analyze the connection between erosion, sediment transport, and sediment export to understand the history and dynamics of particular regions or drainage basins. Early workers (e.g., Dole and Stabler, 1909) generally assumed that the volume of sediment issuing from a catchment was equal to that eroded from the catchment slopes. This style of thinking continued until scientists and engineers discovered that much of the sediment eroded by human disturbance of basin hillslopes was not making it into or through river systems. This insight was gained through creating sediment budgets, the construction of which challenged those involved to figure out where the sediment came from and at what rates it was stored within and exported from catchments.

Dietrich and Dunne (1978) formalized such budgeting using a small catchment in the Oregon Coast Range. Their work pointed out the need to determine rates of processes and volumes of sediment in motion and in storage.

The **sediment delivery ratio** quantifies the ratio between the mass delivered at the drainage basin outlet and the mass eroded from basin hillslopes [**Figure DD7.1**]. For very small basins (<km^2) considered over human timescales, one-quarter to perhaps one-half of the eroded sediment makes it to the outlet. For large basins, the sediment delivery ratio on multidecadal timescales is often much lower, invalidating the assumption of equality between erosion and sediment export (Trimble, 1977). For example, by analyzing flow records, using soil-erosion models, and

DIGGING DEEPER When Erosion Happens, Where Does the Sediment Go? *(continued)*

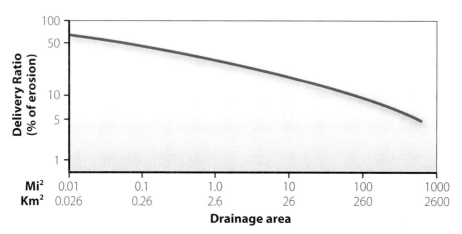

FIGURE DD7.1 The sediment delivery ratio (sediment yield/eroded mass) falls dramatically as basin area increases, suggesting the importance of sediment storage in the lower-gradient reaches of large drainage basins. Note logarithmic axes. [From Trimble (1977).]

mapping deposits in the Coon Creek Basin of Wisconsin, Trimble (1981) created sediment budgets for two time intervals since western settlement of the basin and learned that only 7 percent of the sediment generated by human impact to the basin had been exported [**Figure DD7.2**]. Where was all this recently eroded sediment hiding?

The first clues came from careful examination of soil profiles (Haggett, 1961; Trimble, 1974). Working in the piedmont of eastern North America, an area intensively cultivated for tobacco and other crops between the late 1600s and the mid-1900s, Costa (1975) noted that the upper horizons of hillslope and ridgetop soil profiles were truncated by erosion. Much of the A and O horizons was gone. On the basis of fieldwork, he and others estimated that about 15 cm of soil (on average) had been eroded from the upland during the several-hundred-years-long agricultural period. From this depth estimate, he could calculate a total volume of eroded soil. Then, working on the foot slopes and along river channels, Costa identified accumulations of colluvium and alluvium and estimated their volume. Of the upland soil eroded in his 155 km² study basin by colonial and postcolonial agriculture, about half was still sitting at the base of hillslopes and about 15 percent was retained in river terraces. Streams transported the rest (about 35 percent) out of the basin. The low sediment delivery ratio was now explained. Over human time frames, years to decades, a large proportion of the sediment eroded from slopes was trapped and stored in the basin and thus largely inaccessible to the river system for centuries to millennia.

Starting in the 1960s, extensive fieldwork in east-central North America led to the development of a qualitative model for river and floodplain behavior [**Figure DD7.3**] (Jacobson and Coleman, 1986). Geomorphologists suggested that after European colonists cleared the trees from much of eastern North America, rivers and their floodplains responded to the high loads of sediment moving down channels by aggrading. As agricultural practices changed and degraded farmland was abandoned and reforested, sediment loads dropped and rivers carried less sediment (see Figure 7.12). The response to lowered sediment loads was incision at first, then channel widening (see Figure 7.11). The idea was that the disturbed channels were working to reestablish the pre-settlement equilibrium conditions that included meandering planforms, extensive floodplains, and pool-riffle sequences with coarse-grained gravel bars.

Recent geologic work in east-central North America has provided an alternative explanation for the field observations. Walter and Merritts (2008) suggest that the morphology of today's streams bears little resemblance to the streams in precolonial times. They argue that the tens of thousands of mill dams constructed along the East Coast to harness water power, before the widespread use of fossil fuels began in the late 1800s [**Figure DD7.4**], changed how sediment moved and was stored in eastern North America. Much of the sediment eroded from slopes during the colonial period remains along stream channels and in river valleys where it was trapped behind dams. Although some sediment is stored on floodplains that aggraded when sediment loads were high during the peak of deforestation and land disturbance, much of this sediment was trapped behind the mill dams, creating extensive deposits of fine-grained sediment (over coarser stream gravel) similar to the stratigraphy one would expect to result from overbank sedimentation. If the new interpretation is right, much of the human-induced overbank sedimentation of Jacobson and Coleman (1986) is now the dam-trapped legacy sediment of Walter and Merritts (2008).

Now that upland erosion has been sharply reduced by improved soil conservation practices and natural afforestation resulting from land-use change, streams are carrying less sediment, incising recent sediments, and causing accelerated bank erosion as they adapt to new boundary

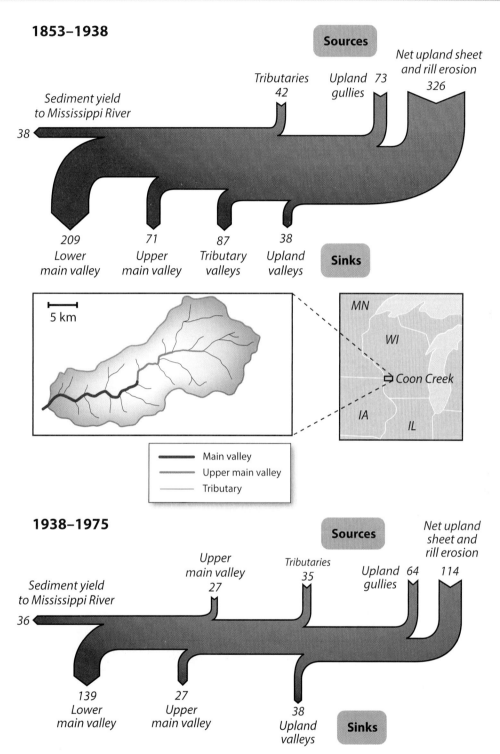

FIGURE DD7.2 Sediment budgets for the Coon Creek drainage basin in Wisconsin for periods before and after 1938. The thickness of the arrows is proportional to the sediment flux and the numbers are annual sediment volumes (10^3 m^3/year). Most of the soil that eroded during and after European settlement never made it to the Mississippi River. Sediment generated by upland sheetwash and rill erosion from cleared lands did not move far; it remains as colluvium on nearby footslopes and as alluvium in the valleys. Soil conservation practices implemented in the twentieth century have reduced the flux of sediment off hillslopes and caused a greater percentage of what was eroded to be retained as colluvium nearby. The flux of sediment to the Mississippi River has remained constant despite large changes in the Coon Creek tributary drainage basin. [From Trimble (1981 and 1999).]

DIGGING DEEPER When Erosion Happens, Where Does the Sediment Go? *(continued)*

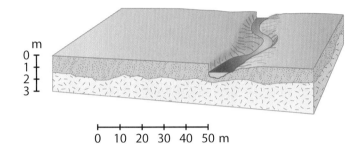
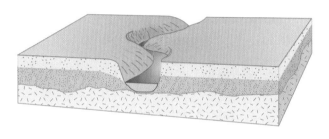
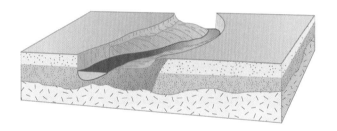

FIGURE DD7.3 Schematic model based on fieldwork in the Maryland Piedmont, showing a narrow meandering stream presettlement with extensive fine-grain, overbank deposits (upper panel). During the period of extensive farming, erosion of the uplands and high sediment loads led to floodplain aggradation (middle panel). Then during very recent time, lowered sediment loads led to incision and the establishment of the floodplain and gravel bars at a new, lower elevation (lower panel). [From Jacobson and Coleman (1986).]

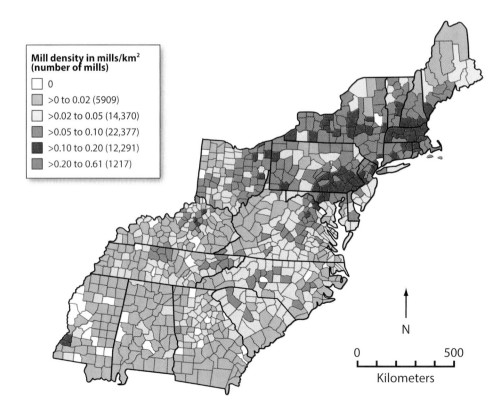

FIGURE DD7.4 This map, made from county-scale U.S. Census data, shows the density of mill dams in the eastern United States, where more than 60,000 mills and mill dams were built in the 1800s. Many areas in the Piedmont and Valley and Ridge provinces, where the density of mill dams was greatest, had one mill dam for every 5–10 km² of land area. [From Walter and Merritts (2008).]

conditions by widening and reestablishing floodplains at lower elevations (see Figure 7.11). Such channel responses to changes in drainage-basin land use often provide the rationale and social catalyst for stream channel restoration and rehabilitation efforts.

This change in thinking has implications for the design of projects that aim to return a stream to a reference condition that reflects fluvial process and form that existed prior to disturbance. If the Walter and Merritts (2008) model is correct [**Figure DD7.5**], then many low-order drainages in eastern North America were swampy, low-gradient, anabranching channels carrying little sediment load. This stands in sharp contrast to the gravely headwater streams with overbank sands previously thought to be representative of the pre-European settlement landscape. It also raises the question of whether society considers the characteristics of precolonial streams to be desirable from an aesthetic and environmental perspective.

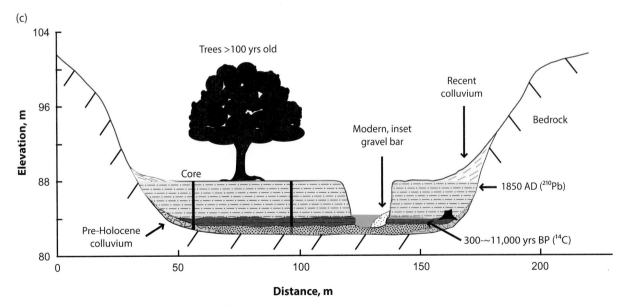

FIGURE DD7.5 Streams along eastern North America often have steep banks of tan fine-grained material overlying dark, organic-rich material, which in turn overlies gravel and bedrock. Photo (a) is from Western Run, Maryland, and photo (b) is from Big Spring Run, Pennsylvania. The scale bars have 0.5-m divisions. Walter and Merritts (2008) propose that fine-grain terraces were deposited in still water caused by mill-pond damming of streams and thus reflect historical land use. They conclude that presettlement streams were not incised into floodplains but rather that the streams carried very little sediment as they moved through wet, marshy valley bottoms (c) Idealized cross section. [From Walter and Merritts (2008).]

DIGGING DEEPER When Erosion Happens, Where Does the Sediment Go? *(continued)*

Understanding the roles of legacy sediment and the present-day response of channels to drainage-basin disturbance over the past centuries has substantial—and usually underappreciated—implications for channel restoration and river-corridor management (Renwick and Rakovan, 2010). In particular, local stream-restoration and rehabilitation projects that do not consider historical changes in the drainage-basin hydrology, sediment supply, and sediment delivery are likely to fail rapidly because the channels they create are not appropriate for contemporary drainage basin conditions.

Costa, J. E. Effects of agriculture on erosion and sedimentation in the Piedmont province, Maryland. *Geological Society of America Bulletin* 86 (1975): 1281–1286.

Dietrich, W. E., and T. Dunne. Sediment budget for a small catchment in mountainous terrain. *Zeitschrift für Geomorphologie Supplementbände* 29 (1978): 191–206.

Dole, R. B., and H. Stabler. Denudation. *United States Geological Survey Water Supply Paper* 234 (1909): 78–93.

Haggett, P. Land use and sediment yield in an old plantation tract of the Serra Do Mar, Brazil. *The Geographical Journal* 127 (1961): 50–59.

Jacobson, R. B., and D. J. Coleman. Stratigraphy and recent evolution of Maryland Piedmont flood plains. *American Journal of Science* 286 (1986): 617–637.

Renwick, W. H., and M. T. Rakovan. Sediment supply limitation and stream restoration. *Journal of Soil and Water Conservation* 65 (2010): 67a.

Trimble, S. W. *Man-induced Soil Erosion on the Southern Piedmont, 1700–1970.* Ankeny, Iowa: Soil Conservation Society of America, 1974.

Trimble, S. W. The fallacy of stream equilibrium in contemporary denudation studies. *American Journal of Science* 277 (1977): 876–887.

Trimble, S. W. Changes in sediment storage in the Coon Creek Basin, Driftless Area, Wisconsin, 1853 to 1975. *Science* 214 (1981): 181–183.

Trimble, S. W. Decreased rates of alluvial sediment storage in the Coon Creek Basin, Wisconsin, 1975–93. *Science* 285 (1999): 1244–1246.

Walter, R. C., and D. J. Merritts. Natural streams and the legacy of water-powered mills. *Science* 319 (2008): 299–304.

WORKED PROBLEM

Question: Describe a conceptual sediment budget (using a flow diagram like that shown in Figure DD7.2) to begin the process of quantification for (1) a nonglaciated 100 km^2 drainage basin in a humid-temperate region, and (2) the same basin in an arid region. First, consider sediment sources and the processes by which sediment is created and delivered to the river network. Then consider areas in the basin where sediment is stored. Finally, consider the rate at which sediment is exported and the processes by which sediment is removed from the basin. Ensure that your sediment budget considers both long and short timescales (thousands of years and decades).

Answer: Sediment budgets for arid and humid regions are distinct because the active geomorphic processes in these regions differ.

In humid regions, weathering converts rock to regolith that is readily eroded to become sediment. Most slopes in humid regions are transport-limited and covered with a mantle of soil. The source terms in a humid-region sediment budget include rock weathering rates and the diffusive transport rates of sediment downslope and advective rates of sediment transport in gullies, rills, and in mass movements. Some sediment is stored in colluvial deposits at the base of hillslopes, some is stored in fans, and other sediment is stored along river and stream valleys as terraces and in floodplains. Sediment is exported from the basin by streams primarily as suspended load during large flow events, such as storms and snowmelt floods. Over thousands of years, the rate at which sediment is supplied from hillslopes is similar to the rate at which it is exported from the watershed. At decadal timescales, these rates are often quite different, with sediment being retained at the base of slopes or along channels and exported by rare but high-magnitude floods.

In arid regions, weathering and sediment production occur more slowly and most slopes are weathering limited and have little if any soil cover. Streams are ephemeral and usually dry. Rare flows can move large amounts of sediment in short times. Source terms in arid regions include advective transport of material off rock slopes by overland flow from gentle slopes, rock fall from steep slopes, and movement of sediment from other parts of the basin (and from other basins) by wind.

In areas where the rocks are weak and there is tectonic activity raising ranges above the valley floors—

for example, around Death Valley, California—large amounts of sediment are stored in fans at the base of steep slopes. In contrast, tectonically stable areas underlain by hard rocks produce little sediment and fans are mostly absent. Depending on the tectonic setting, little if any sediment may leave arid-region watersheds; rather, the sediment may accumulate in closed basins slowly filling them, or some material may be carried out of the basin by wind. Because arid-region processes are so episodic, rates of surface processes over short timescales may bear little if any resemblance to rates over longer timescales.

KNOWLEDGE ASSESSMENT Chapter 7

1. Define a drainage basin.
2. Where in drainage basins does sediment tend to be deposited?
3. Where in drainage basins does sediment most likely originate?
4. Describe the difference between open and closed drainage basins.
5. In what climate and tectonic setting are you most likely to find closed drainage basins?
6. Describe three specific challenges geomorphologists face in creating a sediment budget.
7. Define a sediment rating curve and describe how sediment rating curves are created.
8. Explain why data in sediment rating curves are so variable.
9. Sketch four common types of drainage patterns and suggest a location where each type might be found.
10. Explain the factors leading to each of the four common drainage patterns.
11. Draw a series of sketches to illustrate the difference between superposed and antecedent drainages.
12. Describe the issues related to defining a "first-order" channel.
13. Compare and contrast the Shreve and the Strahler stream ordering classifications.
14. Predict how channel width, depth, and cross-sectional area change downstream in a humid-temperate river network.
15. Why and how does discharge change downstream in an arid-region river network?
16. Describe how both average basin slope and channel slope change as a function of basin area.
17. Where is stream velocity greater—in a mountain cascade or a large lowland river? Explain your answer.
18. Discuss the residence time of sediment in a drainage basin.
19. Explain why fine-grained and coarse-grained sediment have different average residence times in humid-temperate drainage basins.
20. List the four valley segment types and describe the dominant processes in each.
21. Sketch a river longitudinal profile and explain (giving three reasons) why the slope of the longitudinal profile changes in the downstream direction.
22. List three different base levels and describe what processes might lead them to change over time.
23. Define both a knickpoint and a knickzone; explain how they are different and what can cause them to occur.
24. Explain how and why floodplain morphology changes downstream.
25. Predict how and explain why the grain size of sediment carried by a river changes downstream.
26. Explain the difference between depositional and strath terraces.
27. List three reasons why rivers leave behind terraces.
28. Predict where alluvial and debris fans are most likely to be found.
29. List several characteristics that would help you differentiate between alluvial and debris fans.
30. Explain the feedback mechanism that encourages gorge formation and deepening.
31. At what timescales do tectonic, climatic, and human forcings affect drainage basins?
32. Define pater noster lakes and explain how they form.
33. Give two examples of climate changes that can affect drainage basins.
34. Draw a diagram showing how sediment yield might change over time as forests are cleared and land is developed.
35. Define "legacy sediment."
36. Explain the effects of forestry practices on drainage basins.
37. Define a reference condition and argue whether or not it is a valid concept.

Coastal and Submarine Geomorphology

Introduction

Oceans cover 70 percent of Earth's surface, and more than half of Earth's human population lives within 40 kilometers of the marine shoreline. Coastal environments provide resource-rich estuaries, sheltered ports, and beaches for recreation. **Continental margins** are economically important because many oil and gas deposits form and are found in modern and ancient marine environments. Today, however, coastal areas are particularly vulnerable to change in a warming, stormier world in which sea levels are rising.

Much of Earth history is recorded in marine sediments. Unlike upland terrestrial environments that erode over time, depositional coastal and deep marine environments accumulate sediment and thus have the potential to hold a sedimentary record of terrestrial surface processes over time. Only a portion of this record gets preserved over geologic time because subduction recycles most oceanic crust and the overlying sedimentary cover.

Only recently have the principles of geomorphology been applied to understanding submarine landforms and processes. Seafloor topography, the most widespread type of topography on the planet, is hidden from direct observation by water. Thus, the study of seafloor topography, **bathymetry**, uses geophysical images, depth soundings, and seafloor sampling techniques. Geomorphologists interpret seafloor topography using the same general principles of regional context, conservation of mass and energy, and definition of boundary conditions that guide their interpretation of subaerial landforms.

Incoming wave trains shoaling, breaking, and causing sea cliffs to erode between Santa Cruz and Moss Beach, California.

IN THIS CHAPTER

Introduction
Coastal Settings and Drivers
   Tectonic Setting
   Sea-Level Change
   Salinity
   Substrate and Sediment Supply
   Tides
   Waves
Coastal Landforms and Processes
   Rocky Coasts
   Beaches and Bars
   Spits, Tidal Deltas, and Barrier Islands
   Lagoons, Tidal Flats, and Marshes
   Estuaries
   Deltas
   Coastal Rivers
Marine Settings and Drivers
   Currents
   Marine Sedimentation
   Dissolved Load
Marine Landforms and Processes
   Continental Margins
   Abyssal Basins
   Mid-Ocean Ridges
   Trenches
   Coral Reefs
Applications
Selected References and Further Reading
Digging Deeper: What Is Happening to the World's Deltas?
Worked Problem
Knowledge Assessment

Stream systems deliver sediment eroded from continental landscapes to coastal environments, the transitional zone between terrestrial and marine processes. A much smaller but potentially hazardous supply of sediment comes from wave erosion of the coastal margin. Unlike rivers, where water flows only one way (downhill), coastal and estuarine environments are subject to multidirectional flows, the result of rising and falling tides, storms, and wind-driven waves washing back and forth. The interaction of waves, currents, and tides results in coastal erosion, sediment transport along shorelines, and local deposition, giving rise to coastal landforms. Sediment deposition and wave action over glacial–interglacial cycles of rising and falling sea level have shaped the **continental shelves** across which sediment is delivered to the ultimate sink—the deep marine environment.

Sea-level changes greatly affect coastal sediment dynamics. **Estuaries,** partly enclosed coastal water bodies fed by one or more rivers or streams, may form and enlarge when sea level rises, or they may diminish as when, for example, barrier islands erode away. Human impacts are important in coastal zones around the world where land use, river modifications, and coastal engineering structures have substantially altered coastal sediment supply. For example, dams have reduced global sediment delivery to the oceans by roughly half.

This chapter explains the fundamental processes and major landforms of coastal and offshore environments. We consider a variety of coastal settings, their characteristic landforms, and the drivers of geomorphic change in these settings. We provide an overview of offshore geomorphology, using process-based understanding to explain marine geomorphology and the development of submarine landscapes.

Coastal Settings and Drivers

Coastal morphology is broadly determined by tectonic setting, although over time changing sea level is an important control on continental margin morphology and process. Salinity, river inputs of water and sediment, tidal range, and wave action all influence the dynamics of coastal systems and the resulting landforms.

Tectonic Setting

The large-scale geomorphology of a continental margin depends on whether it occurs in the middle or on the edge of a tectonic plate [**Figure 8.1**]. Convergent and transform margins are **active margins,** where the edge of the continent coincides with the boundary between two tectonic plates. Subduction zones (where denser oceanic lithosphere sinks beneath less dense continental lithosphere) typically have a coastal mountain range hosting volcanoes and affected by earthquakes, or a volcanic island arc (where subduction involves oceanic lithosphere). Both types of subduction zone typically have a narrow continental shelf that leads out to a continental slope that drops steeply to a deep submarine trench. Subduction zones surround much of the Pacific Ocean. These collisional (convergent) margins tend to have rocky coastlines that are dominated by erosional landforms [**Photograph 8.1a**], though deltas and other coastal depositional landforms develop near significant sediment sources, such as large rivers.

In contrast, a **passive margin** (or **trailing-edge margin**) is one where the continental margin does not coincide with a plate margin. Consequently, the oceanic and continental crust move in the same direction and speed. Most trailing-edge margins have an exposed coastal plain, part of which was continental shelf during periods of higher sea level. Similarly, the modern continental shelf was the coastal plain during lower sea-level stands, such as during the glaciations that repeatedly punctuated the past 2.7 million years. The eastern continental margins of North America, South America, Australia, and the margins of Africa and the Gulf of Mexico coast are examples of trailing-edge margins. Trailing-edge margins typically have extensive depositional landforms that are built by sediment transport and storage along the coast [**Photograph 8.1b**].

Sea-Level Change

The volume of water in Earth's interconnected oceans plays a major role in coastal dynamics and evolution. Over the past 35 million years, sea level has varied through glacial–interglacial cycles and coastal processes have acted across a wide zone that ranges from meters to tens of meters above today's sea level to the now-submerged edge of the continental shelf.

Shoreline position reflects changes in sea level. When sea level rises, the shoreline advances landward. Falling sea level causes the shoreline to retreat seaward. Depending on the slope of near-shore land, glacial–interglacial sea-level changes can shift coastlines up to tens or even hundreds of kilometers. **Emergent coastlines** are stretches of the coast that have been exposed by relative sea-level fall. **Submergent coastlines** are those that have been inundated by the sea due to a relative rise in sea level.

The cause of changes in the relative elevation of land and sea can be local or global. Local changes in relative sea level are most often caused by uplift or subsidence, driven either by tectonics or by isostatic response to loading or unloading. Sea-level rise and fall can be driven by changes in the volume and size of ice sheets on a millennial timescale and by basin sedimentation and continental erosion on longer timescales. Worldwide sea-level changes driven by changes in ocean volume are referred to as **eustatic.** Such changes are caused by fluctuations in global ice volume as well as **steric** changes (changes in volume) caused by fluctuations in ocean temperature or salinity.

Over the past few million years, the timescale most relevant to geomorphology, eustatic sea level has been closely tied to the amount of glacial ice on land and to the thermal expansion and contraction of ocean water that accompanies climate changes. Sea level falls when a glaciation begins, ice volume at the poles grows, and the average

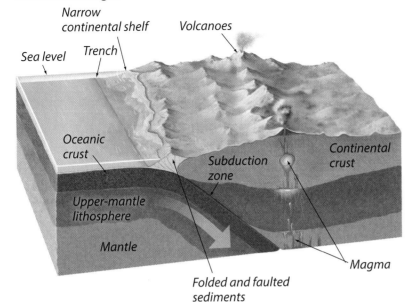

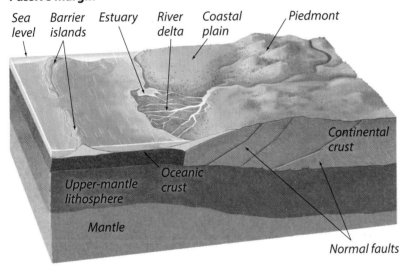

FIGURE 8.1 Collisional Versus Passive Margin Coastlines. The geomorphology of active or collisional margins is very different from that of passive or trailing-edge margins.

global ocean temperature cools, reducing ocean volume. The storage of ocean water as glacial ice and the cooling and contraction of the water left in the oceans leads to marine **regression**; coastlines migrate seaward and fluvial systems extend to reach the sea while cutting down into the coastal plain to adjust to falling base level. Conversely, when glaciers melt and transfer water back into the oceans at the end of a glaciation, sea level rises and causes marine **transgression**, during which shorelines move landward and seawater inundates formerly coastal areas. At the height of the last major glaciation, about 21,000 years ago, global average sea level stood about 130 m below its present level. It has been rising since then. During the most recent deglaciation, sea level rose about 1 cm/yr until about 7000 years ago, then the pace slowed to less than 2 mm/yr about 5000 years ago. Global average sea level then stabilized about 3000 years ago. Now, with climate warming and the ocean warming as well as glaciers melting, sea level is rising several millimeters per year.

The impact of postglacial sea-level rise on coastal geomorphology varies with time since deglaciation and with sediment supply. During the period of rapid sea-level rise from 18,000 to 7000 years ago, coastal valleys flooded and most sediment transported by rivers to the sea was trapped close to the coast. After sea-level rise slowed enough that coastline locations stabilized, those estuaries and fjords that receive little sediment remain unfilled, even though they efficiently trap the sediment supplied to them. In contrast, estuaries that received a large amount of sediment like that of the Columbia River in the Pacific Northwest, have filled to the point that some sediment is carried seaward of the estuary to feed along-coast transport and accumulate

PHOTOGRAPH 8.1 Coastline Morphologies. Coastlines exhibit a great variety of morphologies. (a) The rocky shoreline of a collisional margin at Heceta Head, Oregon. The lighthouse sits on an uplifted marine terrace remnant. (b) This sandy summer beach in Old Lyme, Connecticut, has a small beach face, a wide, flat berm, and a low coastal dune.

on the continental shelf. In areas with a very high sediment supply, like the Nile River delta before the construction of modern dams, the estuary filled to the point where sediment extended the shoreline seaward.

Salinity

The salinity and temperature of seawater are fundamental drivers of coastal and marine processes. Ocean water has approximately 35 parts per thousand (ppt) of salt dissolved in it. For comparison, fresh water in lakes and rivers typically contains less than 0.5 ppt of dissolved mineral material. Brackish water in estuarine environments can have variable salinity but often is between 20 and 30 ppt. Lower salinity translates to lower density, so plumes of river water float on denser salt water where rivers enter coastal and marine environments. In some cases, however, high concentrations of suspended sediment at the mouths of rivers draining steep, rapidly eroding drainage basins increase the density of the water and create bottom-hugging density currents that cause erosion, displacement, and deposition of sediments in deep marine environments well below the depths affected by wave action. Gradients in density, which reflect differences in water-mass temperature and salinity, drive ocean circulation and thus some ocean-bottom sediment transport.

Substrate and Sediment Supply

Like terrestrial landscapes, coastal and marine landscapes are composed of either consolidated bedrock or unconsolidated materials. Many emergent coastlines and headlands are composed of rock outcrops with far greater erosional resistance than submergent coastlines that are mostly made of unconsolidated sediment. This variability in erosional resistance and sediment supply creates strikingly different coastal features and causes different dynamics on rocky, erosional coasts than on depositional shorelines.

Various mechanisms bring sediment to coastal and marine environments, and the particular mix of processes that deliver, store, and remove sediment sets the sediment budget for specific coastal and marine environments [Figure 8.2]. At any site, sediment comes from rivers, coastal erosion, sediment movement along the coast, in situ productivity of biogenic carbonate, aeolian inputs and, at high latitudes, glaciation. In most coastal settings, streams that discharge into coastal and estuarine environments deliver the majority of sediment, although there are exceptions. For example, on volcanic islands such as Hawaii, wave erosion of lava flows provides coastal sediment. Tropical beaches in areas where the uplands are eroding slowly may be made of biogenic coral sand, a product of coral eroded and broken up by wave action.

Once sediment is introduced to a coastal setting, it is generally transported parallel to the coast by the alongshore component of currents generated by waves approaching the shore at an oblique angle. In areas with limited sediment input from rivers and streams, coastal sediment is supplied mainly by cliff erosion that delivers sediment to coastal settings, or by carbonate-producing organisms like corals and shellfish that thrive in clear water away from fluvial inputs of clastic sediment. In modern, high-latitude environments, glaciers and icebergs that calve from tidewater glaciers deliver sediment directly to coastal and marine environments, and have done so in many settings during past periods of cooler climate. Coastal sediment sinks include depositional environments like **dunes** and **lagoons,** as well as deltas, marshes, and carbonate platforms. The dominant sediment source in the deep ocean basins is a steady "rain" of the bodies of single-celled marine plants and animals that sink from the near-surface **photic zone,** and the slow accumulation of wind-delivered dust. Coarse-grained, clastic sediment from the continents occasionally makes its way down the continental slope and into the **abyssal basins** of the deep sea, usually transported by density-driven flows.

FIGURE 8.2 Coastal Sources and Sinks. Sediment moves onto and away from coasts through a variety of different pathways.

Tides

The daily rhythm of the **tides** produces changes in sea level that can be important geomorphically because they control the location at which wave energy is dissipated and sediment deposited. Tidal changes are driven by astronomical factors, specifically, the gravitational interaction of the Earth–moon system, which creates a bulge of ocean water between Earth and the moon and a second antipodal bulge on the other side of the planet [**Figure 8.3**]. The Earth rotates through these bulges, causing two lunar tides per day; in actuality, tidal cycles are about 12.5 hours, the time required for one Earth rotation relative to the moon. The sun also exerts a much weaker gravitational force on the ocean water, producing one tidal cycle per day of lesser amplitude. The geometry of the continents, the bathymetry of the ocean basins and continental margins, the latitude of the coastline, ocean currents, and weather systems all interact with the astronomical factors to determine the water level over time on any specific coastline.

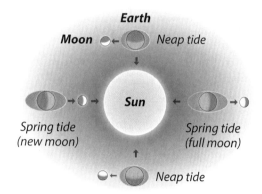

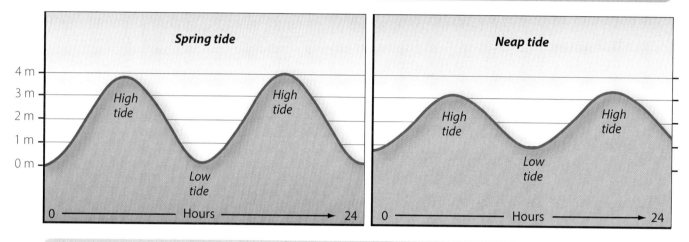

FIGURE 8.3 Tides and Tidal Range. Tides are an important driver of coastal geomorphic processes. Tidal range changes cyclically through time.

Tidal range, the difference between the elevation of high and low tides, is geomorphically important. In the open ocean, tidal ranges are low, usually less than a meter. Along coastlines, tidal ranges can be much greater; thus, the geomorphic effects of the tides as ocean levels rise and fall can be significant. Tidal range is controlled by ocean basin geometry and tends to be lower in basins with restricted connection to the global ocean, such as the Mediterranean Sea, and high where forced by local convergence of flow, such as in the narrow inlet of eastern Canada's Bay of Fundy that has an extreme tidal range of 15 m. **Macrotidal** coastlines experience tides greater than 4 m, **mesotidal** coasts experience intermediate tidal ranges between 2 and 4 m (as is common on the West Coast of the United States), and **microtidal** coasts like those around the Caribbean Sea experience fluctuations of less than 2 m.

Tidal range is driven in part by astronomical forcing (sun and moon position). Maximum tidal range occurs when the gravitational fields of the sun and moon reinforce each other to create a higher-than-average **spring tide**. When the moon and sun are oriented at right angles, a **neap tide** occurs and produces a low tidal range. Consequently, tidal ranges vary over a 28-day period as the orientation of the moon and sun change during the moon's orbital cycle. Spring tidal ranges are typically about 20 percent greater than average, and neap tidal ranges are usually about 20 percent less than average. High spring tides accompany the new and full moon each month, and low neap tides accompany the waxing and waning quarter moons.

Spring tides are important geomorphologically because they allow waves and wave energy to reach farther inland and erode shoreline features. Combinations of high spring tides and high storm surges from storm-generated waves can produce unusually destructive high water and large geomorphic effects. Neap tides are important in the coastal zone because they allow fluvial processes to dominate; for example, at neap tides, flow near deltas may become unidirectional, moving water and sediment offshore.

Tide-generated currents generally flow perpendicular to coastlines, in and out of bays and lagoons. Tidal flows have enough energy to prevent sediment accumulation, maintain tidal inlets, and force seawater into tidal channels in estuaries. The geometry of the seabed along coastlines acts to increase or decrease local tidal ranges, so the effect of tidal flows varies from place to place. Incoming and outgoing tides are often asymmetrical and uneven in strength, and they may roll around a bay rather than moving directly in and out. Rapidly advancing tidal fronts known as **tidal bores** push breaking waves into estuaries and rivers, forcing salt water inland. In the Amazon River, for example, the tide influences river flow as much as 480 km inland.

Waves

Waves are the dominant driving mechanism for coastal processes; they govern erosion and sediment transport in coastal environments. Blowing winds impart energy to the sea surface as the two fluids, air and water, drag on one another sufficiently that kinetic energy is transferred to the water from the air. Over time, that energy becomes focused into discrete water waves. The factors that affect the formation of waves are (1) the duration of the wind event, (2) the velocity of winds, and (3) the **fetch**, the distance that the wind blows across the water surface. Long-duration, high-velocity wind blowing across long distances produces large, energetic waves that travel out from the area where they were generated. Fetch is particularly important in determining wave characteristics. Waves from large storms often cause great changes in coastal environments, but the smaller waves that arrive day and night are the primary shapers of coastal landforms. Fair-weather waves and the longshore currents they generate gradually reshape coastlines after storms.

Waves are described by several basic characteristics [**Figure 8.4**]. Wavelength (λ) is the distance between successive wave crests. Wave period (T) is the time it takes for two wave crests to pass the same point. Wave velocity (V) is given by the ratio of wavelength to wave period:

$$V = \lambda/T \qquad \text{eq. 8.1}$$

Wave height (H), or amplitude, is the vertical distance between a wave crest and its low point or trough and thus equals the amplitude of the water surface rise and fall as a wave moves past a point on the sea surface.

Variable winds blowing across an area of the deep-ocean surface create waves with a wide range of wavelengths and different travel directions. In deep water, waves with longer periods and wavelengths travel more

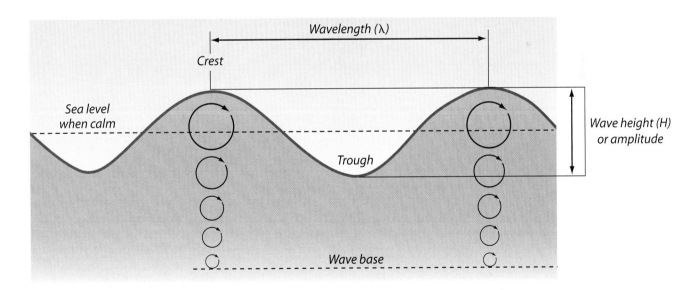

The **wave height (H)**, or amplitude is the elevation difference between the **trough** (low point) and **crest** (high point) of a wave. The **wavelength (λ)** is the distance between the two successive wave crests, and the wave **period** is the time it takes for the wave crest to move through the wavelength. As waves move across the surface of the sea, individual water molecules follow a roughly circular orbit that results in little net motion, advancing with the wave crest and retreating with the wave trough. **Wave base**, below which wave passage has no influence on the sea bed, is generally one-half the wavelength.

FIGURE 8.4 Wave Definition. Waves are described in terms of wavelength and amplitude.

quickly than smaller waves, so waves sort out by size as they travel. This process, known as **wave dispersion**, creates a pattern of well-defined waves that travel outward from the area where they were generated. Wave velocity can be calculated since it is related to the wave period,

$$V = (gT)/2\pi \qquad \text{eq. 8.2}$$

where g is gravitational acceleration. The longest wavelengths create **swell**, regularly spaced waves with low, gently rounded crests that can travel thousands of kilometers across the open ocean with little loss of their original energy, until they approach land. As water depth shallows, pointed wave crests develop and then break to produce **surf**.

The water in a wave moves back and forth, but has little net motion in the deep ocean and does not move with the wave. The propagation of energy imparted to the waveform causes the rhythmic rise and fall of the sea surface in which the wave moves forward, but the water does not. In deep water, the water molecules near the sea surface oscillate and move in circular orbits as a wave passes their location [**Figure 8.5**]. The diameter of the orbit decreases with depth below the water surface and drops off to produce

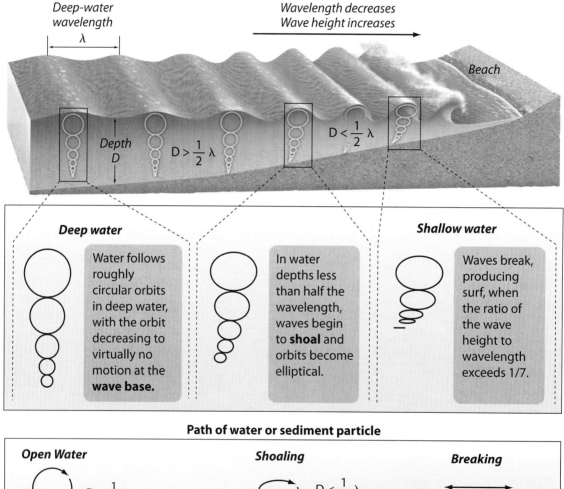

In deep water, waves cause little sediment transport because the water depth, D, exceeds the wave base and is above the ocean floor. In shallower water, the wave base intersects the bed. As waves break in shallow water, bed friction becomes important and wave-induced sediment transport becomes asymmetrical.

FIGURE 8.5 Sediment-Wave Asymmetry. Waves behave differently in shallow and deep water. In shallow water, waves break and move sediment shoreward.

no motion at a water depth greater than one-half the wavelength, a depth referred to as the **wave base**. Below this depth, water molecules are undisturbed as waves pass over the surface. This is why the calmest place to ride out a storm at sea is in a submarine and why waves have little geomorphic effect outside the coastal zone.

As a wave approaches the shore and enters water shallower than half its wavelength, the wave **shoals** as it begins to interact with the seabed. As a wave shoals, the orbital paths of water molecules in the wave flatten and become elliptical, with the water and sediment at the seabed moving back and forth. This interaction exerts drag on the water near the bed and the water exerts a force on the bed that can move sediment. Deeper shoreward flow under the wave crest moves more sediment than the shallower seaward flow under the wave trough, producing an asymmetry that causes net sediment transport toward shore. This is the process by which waves build up beaches.

When a wave enters shallow water and begins to shoal, wave speed decreases due to bottom friction, wave height increases, and wavelength decreases. This causes the wave crest to progressively steepen, and increases the ratio of wave height to wavelength (H/λ). When H/λ exceeds 1/7 (i.e., when wave height exceeds 0.14 times the wavelength), the wave crest loses support, becomes unstable, and breaks, producing surf that dissipates potential energy as kinetic energy in the near-shore environment. The velocity of these shallow-water or translational waves, where D is the depth of water, can be expressed as:

$$V = (gD)^{0.5} \qquad \text{eq. 8.3}$$

Differences in the water velocity, specifically the asymmetry beneath steep and gentle waves, create different types of breaking waves. Beach steepness, which is largely determined by beach sediment grain sizes, affects the type of breakers that develop on a particular beach; those breakers, in turn, affect the grain size of material left on the beach. Gently sloping, fine-sand beaches typically produce **spilling breakers**, through which one can casually wade out into the sea [**Photograph 8.2a**]. The crest of a spilling breaker becomes unstable first and gradually cascades down the advancing wave front as irregular foam. Spilling breakers characterize beaches along the southeastern Atlantic and Gulf of Mexico coasts of the United States.

In contrast, plunging and surging breakers that are more attractive to surfers are generally found along steeper beaches composed of coarse sand and gravel. **Plunging breakers** curl over the front of the advancing wave and fall onto the base of the wave over a short distance, producing a turbulent mass of water that churns up and suspends bottom sediment. Surfers call the tunnels of air under the curling crests of plunging breakers pipelines [**Photograph 8.2b**]. **Surging breakers** maintain unbroken wave crests as they run up the shore [**Photograph 8.2c**]. They typically develop where waves approach steeply sloping beaches and shoal very close to the shoreline.

PHOTOGRAPH 8.2 Types of Breakers. (a) Spilling breakers move gently ashore at Charleston Beach, Rhode Island. (b) Plunging breaker, Cape Ann, Massachusetts. (c) Surging breakers wash over a marine platform at Cape St. Francis, South Africa.

The **surf zone** extends from the seaward limit of breakers to the landward extent of waves that run up the beach face. The **swash zone** is the area covered by **swash** that runs up the beach face. It is exposed after **backwash** flows back down the beach face. This oscillatory action sorts sediment grains by size, shape, and density and creates beaches that are composed of bedload material—the size of which reflects wave energy and sediment source characteristics. Alternation between vigorous swash and sluggish backwash often produces fine lamination in beach sands that is highlighted by contrasting hydraulically sorted layers of light quartz and dark, denser mafic minerals.

As waves shoal and break, they erode material from the seabed. Fine-grained silt and clay are transported seaward in suspension; waves move coarser sand (and gravel on high-energy beaches) back and forth on the seabed, producing net shoreward transport of the coarsest clasts. Swash moves up the beach in the direction the wave was traveling. In contrast, backwash is pulled back into the sea by gravity and moves directly down the slope of the foreshore, perpendicular to the coastline. Because wind-driven waves typically approach the shore at an angle, a zigzag pattern of sediment transport during swash and backwash produces **littoral** or **longshore currents**, called **longshore drift**, that move sediment along the shoreline [**Figure 8.6**]. The direction of longshore currents, and thus that of longshore drift, is governed by the prevailing direction at which waves strike a coastline. Longshore currents produce net transport along a coastline, so material delivered to coastal environments moves parallel to shore, nourishing beaches until it is transported offshore (in submarine canyons) to deeper marine environments below the wave base.

Where waves approach land at an angle to the shoreline, the shoreward ends of the waves reach wave base and begin to shoal before the seaward ends of the waves. This produces an effect known as **wave refraction** that bends the wave crests in response to the near-bottom portion of a wave traveling progressively slower upon moving into

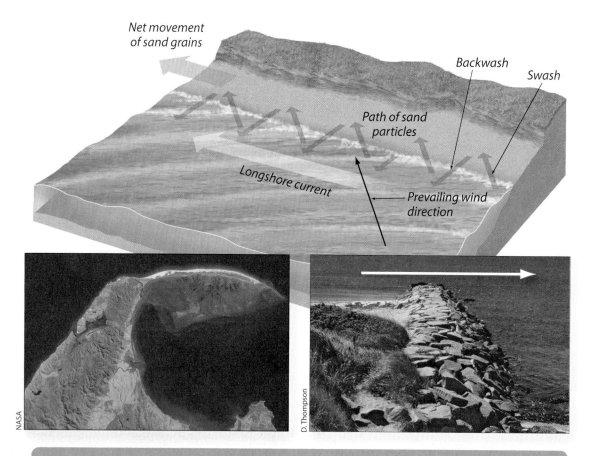

When waves approach the shoreline at an angle, sediment transport on the beach has a fundamental asymmetry that results in net **longshore transport** of sediment. Sediment is transported onto shore by **swash** in the direction the wave was traveling when it struck the coast. **Backwash** travels downslope back into the water under the influence of gravity. The net result is that wave action moves material along the coast. If the sediment moves out into open water, the result is the growth and extension of a **spit**. If the sediment is moving along a beach and encounters an obstruction, such as a groin that projects out from a beach, longshore transport will be disrupted and deposition of material will occur on the up-current side of the obstruction.

FIGURE 8.6 Longshore Drift. Longshore drift moves large amounts of sediment along beaches, due to a fundamental asymmetry of sediment transport by wind-driven waves and gravity.

PHOTOGRAPH 8.3 Wave Refraction. (a) Wave refraction into Lulworth Cove, on the southern coast of England. The dissipation of wave energy in the cove favors beach development. (b) Wave refraction around a headland in northern California favors erosion.

shallower water [**Photograph 8.3**]. Wave refraction along irregular coastlines concentrates erosive energy on headlands that project out into the sea and promotes deposition in protected embayments. Promontories become the focus of erosion, and protected coves become the focus of sedimentation. Thus, wave energy acts to straighten coastlines over time.

Engineered structures built in coastal environments to protect harbors and structures often have profound effects on longshore sediment transport by waves. **Groins** project from a shoreline into the sea [**Photograph 8.4a**]. Because they intercept sediment moving along the coast, sediment accumulates on the upcurrent sides of groins while longshore transport continues to move sediment away from their downcurrent sides. **Jetties** (barriers that project farther out into the ocean than groins) produce a similar effect, leading to a zone of sediment accretion on their upcurrent side and a zone of erosion on their sediment-starved, downcurrent side [**Photograph 8.4b**]. Breakwaters that armor a shoreline or deflect longshore currents also change local patterns of erosion and deposition. Progressive beach erosion because of sediment starvation that results from shutting off or intercepting longshore transport is a major problem in many coastal areas.

Very large, impulse-generated waves known as **tsunamis** wreak havoc on coastlines and can episodically but significantly change coastal morphology. Tsunamis are extremely long-wavelength (typically 100 to 200 km), low-amplitude (< several m) waves that are generated by sudden shifting of the seafloor due to impulses such as earthquakes, submarine landslides, or undersea volcanic eruptions. A tsunami can travel across an ocean and produce geomorphic effects on coastlines far from its source. With wavelengths that exceed the depth of the ocean, tsunamis behave as shallow water or translational waves. Even in water depths of kilometers, they can travel at remarkable velocities (>500 km/hr) and catastrophically inundate

PHOTOGRAPH 8.4 Coastal Engineering Structures. (a) Series of groins interrupt the natural transport of sand by longshore currents at Norfolk, Virginia. Can you determine which way the sand is moving? (b) Channel Islands Harbor breakwater and jetties, southern California.

coastal areas with little to no warning. Because tsunamis are often less than a meter high in open water, they pass unnoticed at sea. Arriving at a coastline, they **shoal** and can grow to produce waves more than 20 meters tall. Some tsunamis arrive as a large wave trough that produces what appears to be an anomalously low tide before the wave crest sweeps inland at devastating speeds (>10 m/s). Coastal configuration and the direction from which a tsunami strikes the coast affect the geometry of the waves and the pattern of coastal inundation at particular locations along a coastline. Like other types of waves, tsunamis interact with the coastal geometry as they approach land and produce wave-refraction and shoaling effects similar to those that result from normal wave action, only at a much larger scale.

Another source of anomalously high waves is **storm surge**, which is produced by strong storm winds that push water shoreward into the coast, in combination with a bulge in the sea surface under the area of lowest atmospheric pressure. Storm surges can raise local sea level 2 to 5 m during hurricanes and push water far inland in low-relief terrain. The additional temporary rise in local sea level because of storm surge amplifies the geomorphic impacts of coastal storms. For example, a storm surge of up to ~8 m during Hurricane Katrina was a key factor in the catastrophic levee breaks that flooded the city of New Orleans in 2005. Along the east coast of North America, storm surges cause major coastal erosion during long-lived nor'easters (very large extratropical storms) because the unusually high water continues over several tidal cycles.

Coastal Processes and Landforms

Coastal landforms reflect erosional and depositional processes as well as the grain size and volume of sediment available for transport in different environments. High-relief, rocky erosional shorelines typically develop on tectonically active continental margins, where erosion of bedrock by wave action creates most coastal landforms, and beaches are small and isolated. Low-relief and drowned, depositional shorelines, in contrast, typically develop on passive margins adjacent to extensive coastal plains composed of unconsolidated, clastic sediment derived from the continental interior. Coastal landforms in such settings typically consist of depositional features produced by longshore transport and deposition of sand by wave action. Beaches are the primary coastal landforms most people know and love, but there are many other distinctive landforms along sandy and rocky coastlines, including tidal flats, estuaries, and deltas [**Figure 8.7**].

There are large differences in rates of coastal retreat and advance between sandy and rocky coasts due to differences in sediment supply, wave action, and resistance to erosion. Sandy coasts can advance if supplied with enough sediment. A falling sea level will also result in coastal advance (forming emergent coastlines), whereas a rising sea level pushes coastal processes and landforms inland (forming submergent coastlines). On eroding shorelines, coastal retreat may be very episodic; a coast retreating at an average rate of a meter per century may retreat 10 meters during a single large storm. Consequently, estimating long-term average rates of coastal retreat is best done using techniques that integrate over decades to millennia, such as air photo analysis and the dating of shoreline platforms.

Rocky Coasts

Sea cliffs that are produced by landward retreat of bedrock slopes undercut by wave action are the most common and striking feature of rocky coasts [**Photograph 8.5**] as well as coasts where well-consolidated glacial deposits dominate the shoreline, such as those of eastern England and parts of Alaska. Active sea cliffs rise steeply from the shoreline at a sharp angle, the steepness of which is maintained by wave erosion at their bases. Inactive sea cliffs that have been raised above the zone of wave attack by tectonic uplift or isostatic rebound typically have a smoothly curving cliff base that has been shaped by subaerial weathering, erosion, and mass wasting. In places where sea-cliff bedrock is resistant, a **wave-cut notch** may develop at the base of the cliff [**Photograph 8.6**]. Relict wave-cut notches at the bases of inactive sea cliffs record former shoreline positions on tectonically active, uplifting coasts.

Landward sea-cliff retreat occurs where wave erosion, undercutting the cliff base, leads to mass failure of the cliff face. If wave action and longshore currents are sufficient to remove the debris, the process repeats and drives the cliff landward. The erosional resistance of the rock or sediment forming the sea cliff, the weathering processes acting to weaken the cliff face, and the energy and height of the waves striking the cliff base together determine the pace of cliff retreat. Sea-cliff erosion rates of several meters per year can occur in unconsolidated material, but rates as low as a millimeter per year are more typical in erosion-resistant rocks like granite.

Modern sea-cliff retreat is driven, in general, by an ongoing response to postglacial rise in global sea level. An additional 0.5 to 1.5 m of sea-level rise that is projected to result from melting of polar ice over the twenty-first century will cause a general acceleration of sea-cliff retreat, but the effect will vary locally depending on lithology, shoreline relief, sediment supply, beach slope, wave energy, and tidal range.

Cliff retreat on rocky coasts also produces a **wave-cut platform** that is beveled off just below the high-tide level (see Photograph 8.6). Wave-cut platforms typically slope seaward at a gentle angle of no more than several degrees and serve as breakwaters that dissipate wave energy and slow sea-cliff erosion. Wave-cut platforms are maintained by a number of processes, including erosion from wave action, the abrasive effect of suspended and bed sediment on the bedrock, and waves sweeping away the weathering products that result from mechanical and chemical disintegration of coastal bedrock that has been exposed

Coastal Processes and Landforms   265

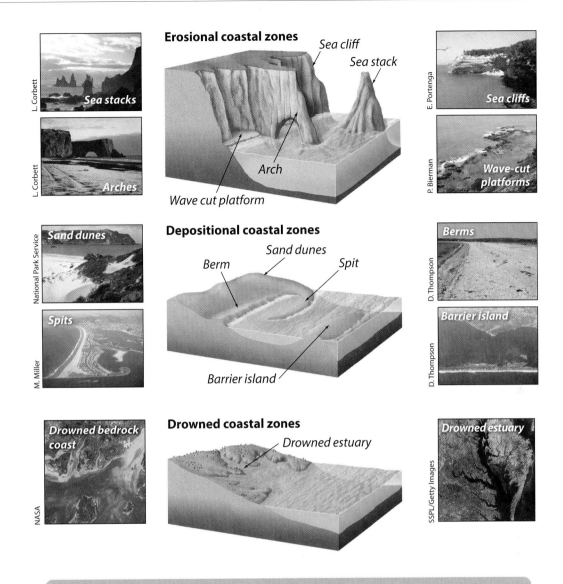

FIGURE 8.7 Erosional and Depositional Coastal Zones. Rocky shorelines and shorelines composed of unconsolidated materials have very different characteristic landforms. Rocky shorelines are dominated by bedrock and as a whole are resistant to erosion, changing slowly over time. Unconsolidated shorelines can change rapidly.

Rocky landforms are common along the erosional coastal zones that typify active, collisional continental margins. Landforms composed of unconsolidated sediment are common along depositional coastal zones typical of trailing-edge, or passive continental margins. Where sea level has risen, relict landscapes are drowned and may not yet have been altered beyond recognition by contemporary coastal processes. Examples of such drowned coastal zones include the coast of Maine and Chesapeake Bay in the United States and Marlborough Sound in New Zealand.

to repeated wetting and drying in the surf zone. On actively uplifting coasts, wave-cut platforms that become elevated above the surf zone provide long-term records of coastal uplift in the form of **marine terraces,** flat-lying features that indicate previous sea level stands.

Sea stacks, caves, and arches are prominent erosional features of irregular rocky coastlines. **Sea stacks** form where wave refraction concentrates wave attack around a headland and erodes through a narrow promontory to produce a small island or pillar that is isolated from the mainland at high tide [Photograph 8.7]. **Sea caves** form through preferential erosion of fractured or less resistant zones of rock exposed in sea cliffs. When a sea cave grows to extend completely through a rock promontory, it forms a **sea arch. Pocket beaches** are those restricted to small inlets and embayments nestled along rocky coasts between sea cliffs and such striking features as arches and sea stacks [Photograph 8.8].

PHOTOGRAPH 8.5 Sea Cliff. This limestone sea cliff (70 m high) is on the coast of the north Bulgarian Black Sea, at Cape Kaliakra.

Beaches and Bars

Beaches are deposits of sand, gravel, or cobbles that form along shorelines directly affected by wave action. Local variations in the grain size and volume of sediment impart variability to beach morphology. Some beaches, such as those on Puget Sound in Washington State or along Cape Cod in Massachusetts, have adjacent stretches of sand, gravel, and cobble beach that result from local differences in the grain size and erosional resistance of glacial sediments exposed in active sea cliffs. Beaches reflect the interplay of sediment transport and sediment supply over a variety of timescales that range from individual storm events to millennia.

Beaches consist of several distinct, shore-parallel zones between mean lower low water (the average elevation of the lower of the two low-water heights of each tidal day) and coastal dunes, a sea cliff, or permanent vegetation [Figure 8.8]. The **offshore zone** extends seaward from the breakers, and a beach can be divided into the **inshore**

PHOTOGRAPH 8.7 Sea Stack. Sea stack at Myers Beach on the Oregon coast.

zone, which consists of the breaker (surf) zone, extending from the base of the swash zone on the beach face to the beach trough and longshore bar; and the **foreshore**, the seaward-sloping beach face exposed at low tide in which **swash** rushes up and **backwash** runs down the beach face. The relatively flat **backshore** is often separated from the beach face by a distinct ridge, or **berm**. On a broad coast lacking a sea cliff, winds may deposit **dunes** inland of the berm (see Chapter 10).

PHOTOGRAPH 8.6 Wave-cut Notch and Platform. This wave-cut notch and wave-cut platform are in Gros Morne National Park, Newfoundland.

PHOTOGRAPH 8.8 Pocket Beach. Sandy pocket beach between rocky outcrops in Noosa, northeastern Australia.

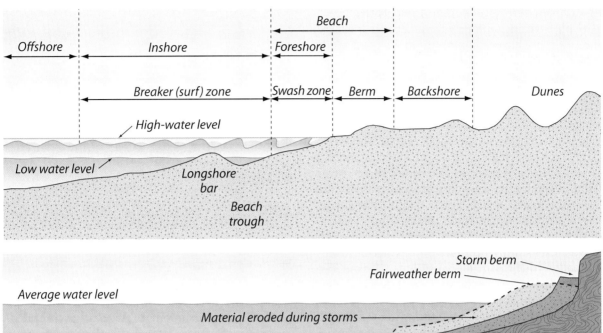

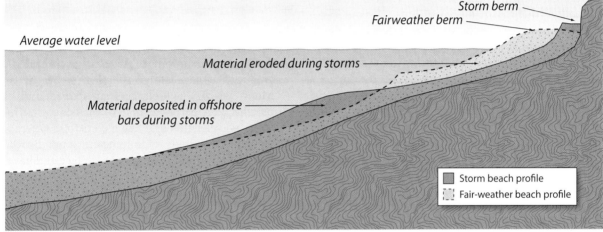

High-energy storms build storm **berms** above the mean high-water mark, while eroding fair-weather berms and steepening the beach profile. Material eroded during storms from the fair-weather berm is stored in offshore bars.

Lower wave energy typically builds a large gently sloping fair-weather berm by slowly transporting material from offshore bars back onto shore, resulting in a seasonal cycle of beach profile change.

FIGURE 8.8 Beach in Cross Section (Storm Versus Fair-Weather Profiles). Beaches change their cross-sectional shape with the seasons as fair-weather waves move sediment shoreward onto berms and then higher-energy storm waves move sediment offshore into bars.

Beaches are dynamic. They are made up of readily transportable granular debris—sand, gravel, or cobbles—subject to movement by regular wave attack, longshore currents, and tidal flows. The fundamental asymmetry in sediment transport during swash and backwash in fair-weather conditions builds up beach berms over time. Waves rush onshore and come to a stop before flow reverses and the water runs back offshore. Because some seawater infiltrates into the permeable beach surface when swash flows up the **foreshore**, backwash has less transport capacity than swash. Less sediment flows back down the beach surface than moves up the beach. During rising and falling cycles of the tides, arriving waves build the beach up to create the berm. The elevation of the berm surface depends on how far swash moves up the beach face at high tide, a distance that scales with the energy (height) of incoming waves.

The slope of the beach face depends on the wave energy striking the coast as well as on the grain size and supply of sediment. Specifically, it is the interaction of swash and backwash processes that governs beach slope. The greater the difference between the ability of swash and backwash to transport sediment, the steeper the beach profile that develops. Backwash rapidly drains down into beaches composed of coarse pebbles or cobbles such that little of the material carried up the beach face is transported back down, building a steep beach face. In contrast, fine-grained beaches stay saturated in the short interval between waves due to their low permeability, limiting backwash infiltration into the beach face and resulting in greater seaward transport and flatter beach slopes. Cobble beaches commonly have slopes that exceed 20°, pebbly beaches have slopes of 10° to 15°, and sandy beaches have slopes of 3° or less.

Wave energy in part controls the flux of sediment on and off beaches and thus beach steepness. High-energy wave attack tends to increase the slope of a beach by eroding the beach face and moving sediment offshore. Thus, the beach slope changes seasonally in regions subject to erosion by large breaking waves accompanying seasonal storminess. Lower-energy waves that strike the coast as swell result in greater net transport of sediment onto the beach face. During calm periods between storms, low-energy waves move beach-forming material shoreward.

Waves build up a **fair-weather berm** during periods between significant storms. Because storms erode the lower-elevation, fair-weather berm and redistribute the sand seaward, they alter beach profiles by steepening the beach face and moving substantial amounts of sediment to offshore bars. Wave attack during very large storms can even erode a notch in the berm, creating temporary sand cliffs along a beach profile. Larger waves from higher-energy storms also may build a higher **storm berm** at the upper limit of wave action, just seaward of dunes. Between storms, beach faces gradually recover and flatten as fair-weather waves move sediment landward again and wave-cut smooth notches.

Storm patterns affect the degree to which beach morphology exhibits regular seasonal change. Years without major storms produce little seasonal change in beach profiles, and anomalous storm events dramatically alter beach profiles regardless of the seasonal pattern along a coastline. In the southeastern United States and eastern Australia, tropical depressions including hurricanes and typhoons are summer and fall events that periodically reform beach profiles. In New England, the most geomorphically effective storms are long-lasting nor'easters, which are most common in the late fall and early spring. In the northwestern United States, winters are stormy and summers are calm. At far northern latitudes, sea ice precludes near-shore winter waves, and there is little seasonal change in beach morphology.

Submerged, near-shore bars commonly form beneath breakers, and the interplay of wave action and sediment transport creates a feedback loop in which breaking waves build bars that, in turn, promote shoaling and thus more breakers. Consequently, bars commonly form at the seaward limit of the surf zone, where waves begin to break, and an associated trough grows between the bar and the beach. The complexity of the relationship between incoming waves and near-shore sediment transport is expressed in the variety of bar types that result from specific coastal currents and configurations. Multiple, shore-parallel bars may reflect the positions of breaking waves of different sizes, breaker positions during high and low tides, or development of breakers that form and break up again and again before reaching shore on low-gradient beaches. Deep bars that form beneath unusually large storm waves sometimes remain stable for long periods because they lie

PHOTOGRAPH 8.9 Beach Cusps. These beach cusps are on a gravel beach with steep backing slope in Baja, Mexico.

beneath the wave base for fair weather and even average storm waves. Offshore bars serve as natural breakwaters that reduce the wave energy affecting the beach.

Many beaches exhibit crescent-shaped and cuspate forms with axes perpendicular to the coast that range in size from wavelengths of a few meters to major capes and embayments hundreds of kilometers long. **Beach cusps** commonly develop on the upper beach face and have wavelengths of less than 30 m [**Photograph 8.9**]. Cusps form in any size of beach sediment, from fine sand to large cobbles. Sediment in the seaward projections, or horns, of cusps is typically coarser than in the intervening embayments. Although their origin remains debated, there is a consensus that cusp spacing depends on wave height and direction and that higher waves generally produce greater cusp wavelengths. Some believe cusps result from standing waves that set up offshore; others believe they are a self-organized feature similar to river meanders. Cusps are typically destroyed by large storms and can reform within a day after storms. In some regions, like the Atlantic coast of the southeastern United States, the periodic spacing of very large shoreline crenulations, called **capes**, may reflect the spacing of rivers or variations in ocean currents.

Spits, Tidal Deltas, and Barrier Islands

Constructional coastal features are formed by sediment deposition in coastal environments. **Spits** are shore-parallel sediment deposits that are connected to the mainland at one end. Spits are found on many coastlines, including on large inland lakes. Longshore transport builds spits by delivering sediment to the end, or tip, of the spit where the sand is deposited, allowing the spit to build out into deeper water. A spit may grow across an existing embayment or extend from a headland to enclose a lagoon that fills with fine sediment once it is protected from wave action. Where tidal flux cuts through spits, it creates **inlets** through which tidal flow moves in and out of the embayment. Tidal currents often reshape the ends of spits into curved

PHOTOGRAPH 8.10 Spit. Dungeness Spit, Strait of Juan de Fuca, Washington; the Olympic Mountains are in the distance.

PHOTOGRAPH 8.12 Barrier Island. Ocracoke Island is a barrier island, southwest of Cape Hatteras. This photograph, taken after Hurricane Isabel (2003) passed within 30 km, shows sand washed over the island into the lagoon.

forms, or **hooks** [Photograph 8.10]. Storm surges may fill in existing tidal inlets with sediment and cut new passages through spits.

Flood-tide deltas form where sand carried by longshore currents reaches a tidal inlet and the incoming tide pushes it landward through the inlet. The flooding (rising) tide carries sand into the lagoon, where it rapidly settles out in the sheltered waters where the waves are weak. Sand transported out of a lagoon by an ebbing (falling) tide sometimes forms an **ebb-tide delta** on the seaward side of a barrier island [Photograph 8.11]. Ebb-tide deltas are generally small compared to flood-tide deltas because they are reworked by waves in the open marine environment and because the outgoing tide usually has a lower sediment-transport capacity than the incoming tide.

Barrier islands are similar to spits but are disconnected from the mainland by tidal inlets at both ends [Photograph 8.12]. They are typically shore-parallel islands located a short distance from the mainland across a lagoon or bay.

Barrier islands are typically several hundred meters to several kilometers across and from 10 to up to 100 km long. Windblown sand dunes on the landward side often form the spine of a barrier island, and tidal flats and marshes fringe the lagoon. The elevation of a barrier island depends on the available sand supply and the strength of dune-forming winds. Most barrier islands, like those that line the coast of South Carolina, are less than 10 m in height. Low barrier islands experience **overwash** by storm surges, which can carve new inlets and deliver sand to interior and back-barrier locations. Inlets migrate in the direction of longshore transport by accumulation of sediment on their upcurrent side and erosion of the downcurrent side of the inlet. Conditions on low-energy to moderate-energy coasts with limited tidal ranges favor development and maintenance of barrier islands.

Barrier islands form by three primary processes: spit elongation, bar submergence, and bar emergence. When a spit becomes long enough to slow transport of water between the ocean and lagoon, a new inlet forms as water levels rise over a spit during storms and splits off the end of the spit into an island. Bar submergence creates barrier islands when an old dune or other topographic coastal high point becomes isolated as sea level rises. Bar emergence occurs either during falling sea level or when a strong storm creates a large offshore bar that becomes exposed as a low island after the storm surge subsides and winds build dunes that raise the bar farther above sea level. Wave refraction sometimes focuses deposition and builds **tombolos,** sand or gravel bars that are inundated at high tide but connect an island with the mainland or another island at low tide.

Coastal barriers migrate seaward or landward in response to changes in sea level, sediment supply, or coastal erosion. Landward migration of coastal barriers is common because of enhanced coastal erosion (due to wave action) driven by recent global sea-level rise. The main processes that move coastal barriers landward are wind transport to and through coastal dunes, washover during large storms,

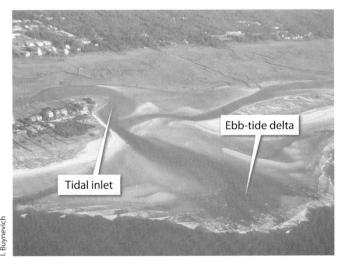

PHOTOGRAPH 8.11 Ebb-tide Delta. The Midway Inlet in South Carolina is a tidal inlet with an ebb-tide delta.

and tidal transport through inlets to lagoons and flood-tide deltas. Seaward growth of coastal barriers does occur close to large sediment sources. Sets of distinctive, prograding **beach ridges** typically accompany seaward movement of the shoreline in regions where rivers deliver abundant sediment to coastal environments or where uplift is occurring.

Offshore barrier islands are common on passive margins, with little tectonic uplift or subsidence—for example, the mid-Atlantic and Gulf of Mexico coasts of the United States. Barrier islands are much less common on active margins, but they can occur near substantial sediment inputs such as river mouths.

Lagoons, Tidal Flats, and Marshes

Lagoons are partially enclosed coastal water bodies on the landward side of spits, barrier islands, and reefs [Photograph 8.13]. The quiet waters of lagoons accumulate sediment that gradually fills in the lagoon embayment unless it is subsiding tectonically. Fill material includes sediment delivered from fluvial sources, marine sediment washed over spits and barrier islands, and biogenic material from high biologic productivity due to quiet water and high nutrient loads.

Tidal flats [Photograph 8.14] are depositional areas within the intertidal zone, the landscape between mean low-tide and mean high-tide levels. In coastal areas that are protected from direct wave attack, fine-grained mud (clay) transported as suspended load in the incoming tidal flux accumulates on a flat surface just below high-tide level. Coarser sand, transported as bedload, accumulates lower on tidal flats, closer to the mean low-tide level, or in intertidal creeks. If sea level is steady, then the rate of sediment supply controls the speed at which tidal flats aggrade or erode. If sea level falls, then tidal flats are more likely to transition from intertidal to terrestrial areas. If sea level rises, tidal flats require a substantial supply of sediment in order to aggrade rapidly enough to keep up with the rising water.

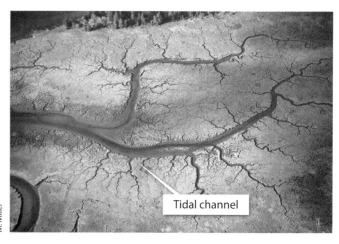

PHOTOGRAPH 8.14 Tidal Flat. Aerial view of branching tidal channels on a tidal flat in Coos Bay, Oregon. Trees in the distance provide scale.

Tidal flats are best developed in macrotidal environments with large tidal ranges because the sediments that would otherwise form beaches, spits, and barriers are instead distributed throughout the coastal zone by waves attacking at a variety of different elevations. Waves die out as they cross broad tidal flats, and tidal currents come and go slowly at their outer, landward reaches, creating a low-energy zone where mud accumulates during slack water at high tides. Salt marshes form in the intertidal zone where salt-tolerant vegetation promotes sediment deposition by introducing roughness that dissipates wave energy and reduces tidal current velocity. Marshes are made up of 5 to 25 percent organic matter; thus, biologic productivity is important as a marsh sediment source, including below-ground productivity. In tropical regions, extensive mangrove vegetation protects coastal environments from erosion and promotes marsh growth by dissipating wave energy [Photograph 8.15].

Recent research suggests that many coastal marshes are recent additions to the landscape, the result of large

PHOTOGRAPH 8.13 Lagoon. The body of water in the right side of the image is a lagoon sheltered by a well-vegetated barrier island on the eastern shore of Virginia.

PHOTOGRAPH 8.15 Mangrove Coast. Mangroves line a beach at the base of a steep coastal slope in Queensland, Australia, near Cape Tribulation.

volumes of sediment filling shallow areas of estuaries when upland deforestation increased sediment yields. Examples of sediment infilled tidal flats and emergent salt marshes that expanded greatly during the past several hundred years in estuaries include San Francisco Bay and the coast north of Boston, Massachusetts. Although afforestation and better land management practices have reduced sediment yields over the past century, the marshes remain because the vegetation changes water-flow dynamics, efficiently traps sediment, and produces organic matter that accumulates in the marsh. Even with sea level rising at >1 mm/yr, these relatively young marshes are so efficient at accreting sediment that they can keep from being submerged.

Estuaries

Estuaries form where rivers discharge into the ocean in topographically confined settings that facilitate mixing of fresh and salt water and promote sedimentation [Figure 8.9].

Coastal-plain estuary

Coastal-plain estuaries occupy drowned river valleys, are V-shaped in cross section, and horn-shaped (triangular) in map view. Coastal-plain estuaries are common on passive margins.

Fjord

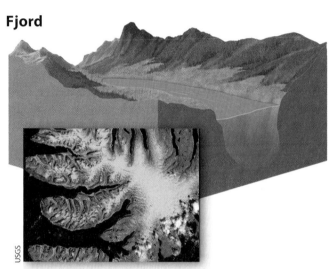

Fjords are drowned glacial valleys, have steep U-shaped cross sections resulting from glacial erosion, and have a shallow sill of glacial sediment at their mouth. Fjords are common at high latitudes where glaciation was pervasive.

Bar-built estuary

Bar-built estuaries, or **lagoons,** are created where a sand **spit** or **barrier island** encloses a shallow embayment. They occur where longshore drift rates are high.

Tectonic estuary

Tectonic estuaries occupy down-dropped basins that have been inundated by seawater. They occur along tectonically active coastlines.

FIGURE 8.9 Estuary Types. Estuaries vary greatly in their plan form morphology, cross-sectional shape, and origin.

Sand traveling as fluvial bedload constitutes less than 10 percent of the riverine sediment that reaches estuaries, and most of that sand is deposited at sea level near the head (upriver end) of estuaries. The suspended load that a river delivers to an estuary, mainly silt and clay, is typically dispersed throughout the estuary and is either deposited or carried out into the ocean. Sediment that is not trapped in estuaries forms the riverine portion of the sediment supply for coastal beaches.

The dynamics of saltwater–freshwater mixing promote estuarine sedimentation. In estuaries, sediment-laden river water flows out over denser salt water and creates a zone of turbid surface water where mixing occurs. The boundary between the fresh water that is moving seaward at the surface and the underlying salt water is called the **halocline.** When silt and clay particles enter this mixing zone, they aggregate to form larger particles. In fresh water, the negative electrical charges on mineral surfaces tend to repel each other. But in brackish water, negatively charged clay mineral flakes and organic matter cluster around cations such as Na^+ and form loose clumps called flocs, a process called **flocculation.** As flocs sink and settle, they become trapped near the river mouth at the head of an estuary by the landward bottom flow of denser salt water. This process makes estuaries excellent sediment traps and leads to the formation of extensive tidal marshes and deltas in coastal environments.

A floc is typically composed of platy clay particles joined end to face in a "house of cards" structure that contains a lot of water. Consequently, estuarine sediments tend to consolidate and dewater as the weight of additional sediment squeezes water out of flocculated clays, leading to gradual subsidence over time. If uplift of the sediments and leaching of the cations occurs before consolidation (and the resultant strengthening) is complete, these estuarine sediments can liquefy quickly if shaken, for example, by an earthquake or an excavation. Such liquefaction caused substantial damage in the 1964 Alaska earthquake.

Different types of estuaries have different cross-sectional geometries and map patterns (see Figure 8.9). **Bar-built estuaries** are shallow features enclosed behind spits or barrier islands. **Tectonic estuaries** are fault-bounded coastal basins that have become flooded by seawater, such as San Francisco Bay. **Coastal-plain estuaries** are extensive embayments that formed as postglacial sea-level rise inundated valleys that were eroded by rivers when sea level was lower. Chesapeake Bay and Delaware Bay are coastal-plain estuaries. Because they were originally cut by rivers, coastal-plain estuaries are usually V-shaped in cross section and have a triangular map pattern that points upstream.

A **fjord** (from Norwegian for "passage") is an estuary in a drowned glacial valley where ice once eroded the valley floor [**Photograph 8.16**]. Fjords are generally deep and U-shaped in cross section and near their mouths have

PHOTOGRAPH 8.16 Fjord. Fjord near Upernavik, central western Greenland, with steep walls and low-relief highlands. Cosmogenic exposure dating indicates that ice last occupied the fjord about 11,000 years ago.

shallow sills that separate them from the marine environment. Most sills are submerged deposits of glacial sediment. Fjords occur along high-latitude coasts, and especially on the windward coasts of continents, where mountains catch snow and build glaciers from the ocean-derived moisture. They are thought to be the result of channelized ice flow excavating areas of weaker rock or preexisting lowland topography.

Deltas

Deltas develop where rivers entering bodies of water supply more sediment than longshore drift and/or tidal currents remove. The result is a wedge of primarily terrestrial material built out into the marine (or lacustrine) environment. Deltas are best developed where rivers with high sediment loads enter sheltered coastal settings like Puget Sound, the Gulf of Mexico, or the Black Sea, and in calm areas behind barrier islands or reefs with small tidal ranges.

Deltaic deposits result when sediment conveyed by rivers enters standing water and river flow diverges because it is no longer constrained by channel banks. The Mississippi River, for example, approaches sea level at the landward side of its delta, but both built and natural levees contain the flow across the delta. Thus, the Mississippi River cannot deposit its sediment load until it finally enters the sea and flow diverges at the delta margin.

In many delta systems, deposition of the river's sediment load leads to development of **distributary channels** that bifurcate downstream. As channels diverge and thus get smaller downstream, sediment-transport capacity drops. This results in more sediment deposition and the formation of in-channel and channel-mouth bars. Such bars increase flow divergence, which reduces transport capacity and triggers further sedimentation. The locus of deltaic sedimentation shifts as coastal depressions become

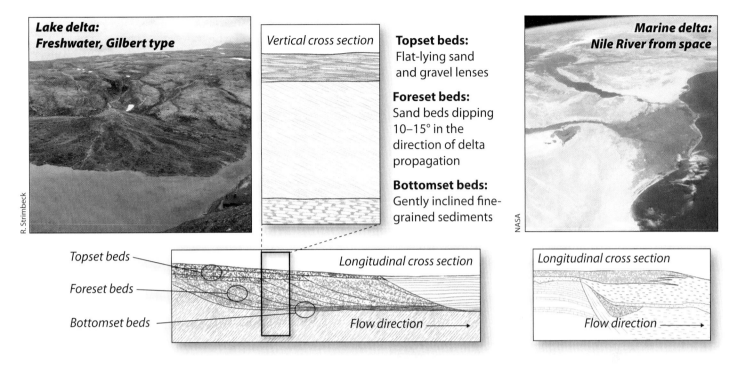

FIGURE 8.10 Delta Anatomy. Deltas, formed as rivers deposit sediment into standing water, have a predictable internal architecture.

filled and streams switch to other, lower locations forming new depositional lobes. For example, the active portion of the Mississippi River delta has switched location many times over the past several thousand years.

The simplest deltas (known as **Gilbert-type deltas**) grow or **prograde** by deposition of thick sediment layers at river mouths, where fluvial sediments settle out and form distinctive topset, foreset, and bottomset deposits [**Figure 8.10**]. These Gilbert-type deltas are most common in smaller, freshwater lakes where the complexities introduced by tides, significant longshore drift, fast-moving currents, and the flocculation is minimal. **Topset beds** are made of sediment that is deposited near the surface of a delta at just above the average water level. **Foreset beds** are inclined lakeward (or seaward), and they form as sediment spills over the edge of the delta. Lateral accretion of foreset beds is the primary process by which a delta advances out into a body of standing water. **Bottomset beds** are horizontal or subhorizontal beds deposited in deep water. They are composed of fine-grained material and are the leading edge of the delta. In studies of now-vanished lakes, including those formed during glaciations and wet periods, the elevation of the topset-foreset contact is used to define the paleolake level.

More complex, marine deltas result when rivers empty into the ocean and deliver sediment at a rate greater than it can be transported away by coastal processes; they have different sedimentary architecture than lake deltas. The overall shape of a marine delta depends on the relative influences of fluvial sediment inputs and stream current strength, wave energy, and tidal energy. Some marine deltas, like Egypt's famous Nile River delta, form an upstream-pointing triangle that looks like the Greek letter delta (Δ) [**Photograph 8.17**]. Other deltas have very different morphologies.

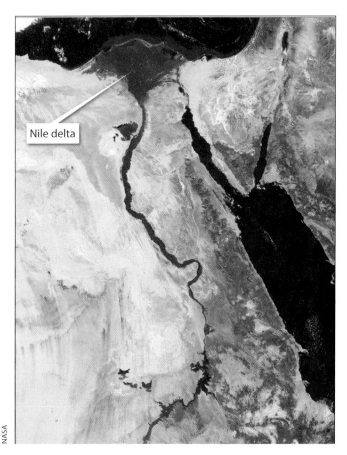

PHOTOGRAPH 8.17 **Nile River Delta.** The Nile delta, bright green from vegetation, shows clearly in this remotely sensed image.

Marine deltas may be river-dominated, wave-dominated, or tide-dominated [**Figure 8.11**]. **River-dominated deltas** typically have a "bird's foot" morphology consisting of an elongated distributary mouth that protrudes at right angles from the coast (like the Mississippi River delta), where river delivery of sediment dominates delta form. River-dominated deltas typically form in areas with low wave energy, low tidal range, and high sediment supply that together allow fluvial processes to dominate. **Wave-dominated deltas** form on coasts where wave attack straightens the coastline, building extensive sandy spits and barrier islands. Wave-dominated deltas form in areas with high wave energy, unidirectional longshore transport, and steep offshore slopes. Longshore drift along wave-dominated deltas reworks the delta front into smooth, cuspate forms. **Tide-dominated deltas** have broad, seaward-flaring, fingerlike sand ridges and islands separating large distributary waterways (like the Ganges–Brahmaputra delta in Bangladesh or the Fly River delta in Papua New Guinea). They tend to form in areas with low wave energy, high tidal range, and little longshore transport. Strong tidal currents enlarge the seaward mouths of distributary channels and create large islands that are oriented orthogonally to the shoreline.

Deltas, marshes, estuaries, and lagoons are particularly sensitive to **subsidence,** which over time causes land levels to decrease unless more sediment is deposited. Much subsidence is the result of **compaction,** driven at least in part by the mass of sediment deposited later than and above the compacting layers. Marsh sediment is rich in organic matter, which both compresses under the weight of overlying sediments and decays over time, losing mass and volume. Parts of major deltas, such as the Mississippi River delta, are subsiding rapidly (6–8 mm/yr) because compaction is ongoing and no longer offset by sediment deposition. This sediment starvation results both from sediment bypassing the delta because the river channel is contained within flood-control levees and also because dams have reduced the sediment load of the Mississippi River by about 50 percent.

Coastal Rivers

Coastal rivers are subject to bidirectional flow and to tidally induced, water-level changes that influence channel morphology and dynamics. Deposition, particularly of suspended sediment, is enhanced in the zone of tidal influence in the lower reaches of coastal rivers. Tidal effects may extend many kilometers upstream in low-gradient river systems typical of passive margin settings. However, the tidal range changes over the year and through each month. The backwater effect of high tides and coincident storm surges can increase river flooding in coastal rivers, such as happened in South Carolina's coastal rivers during Hurricane Hugo in 1989, when the storm surge locally reached an extraordinary 6.1 m. Backwater effects of high tides may cause a rapid decrease in transport capacity near the upper limit of tidal influence. For example, historical maps and accounts indicate that enormous logjams occurred at the head of tidally influenced reaches on rivers around Puget Sound in the Pacific Northwest.

Marine Settings and Drivers

Seafloor processes and morphology typically reflect proximity to active geologic structures at divergent, convergent, and transform plate boundaries as well as proximity to continental and marine sources of sediment. In contrast to the continents, where topography commonly reflects ancient tectonic setting and erosional history, seafloor features are more likely to be the youthful products of recent and active tectonic processes and sedimentation. Portions of some continents are billions of years old, but new seafloor is continuously created at mid-ocean ridges and ancient oceanic lithosphere is consumed at subduction zones. The age of most of the world's seafloor can be measured in millions to tens of millions of years. In addition to the direct effect of plate tectonics

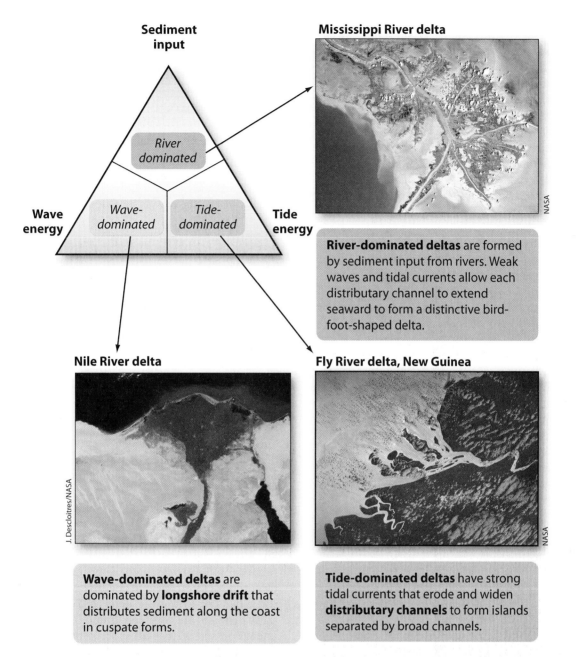

FIGURE 8.11 Delta Process Triangle. The shape of deltas depends strongly on the interaction between coastal and fluvial processes, specifically, mass fluxes as well as wave and tidal energy that affect sediment delivery and redistribution.

on the seafloor morphology, the movement of ocean currents, the delivery of sediment to ocean basins, and the precipitation of dissolved elements are geomorphically important processes.

Currents

Three primary types of currents influence marine topography and depositional environments in deep waters, areas below the wave base and thus protected from wave action. First, surface currents that are generated by friction between the atmosphere and sea surface under persistent wind belts govern the movement of water, heat, and sediment in the upper ~400 m of the ocean. Second, density and temperature gradients drive circulation of the deep ocean waters in what is known as the global conveyor belt (see Figure 1.5). Third, sediment-laden, gravity-driven density currents, called **turbidity currents,** are denser than average ocean water and thus flow down seafloor slopes in much the same manner as subaerial debris flows move down hillslopes. Due to their higher density and thus greater tractive force than clean-water flows, density currents can be highly erosive and can dramatically influence submarine topography.

Marine Sedimentation

Marine sediment consists of material derived from three different sources. **Lithogenic** or **terrigenous** sediments are produced by weathering and disintegration of rock on the continents and are delivered to marine environments by aeolian, fluvial, and glacial processes. **Authigenic** sediments are produced by inorganic precipitation of dissolved components out of supersaturated seawater. **Biogenic** sediments are generated by organic precipitation of dissolved components by organisms.

Terrigenous sediment dominates continental margins where rivers, the main agents of erosion and sediment transport on land, discharge their loads. Streamborne sediments that make it to estuarine and marine environments contain, on average, 90 percent silt and clay (mud) and 10 percent sand. Turbidity currents can carry terrestrial sediment hundreds of kilometers offshore and deposit this material onto the abyssal plains. Such deposits have been cored and used to understand the timing and thus recurrence intervals of such different geomorphically important events as great subduction zone earthquakes and the Lake Missoula glacial outburst floods that carved the Channeled Scablands of eastern Washington State (see Digging Deeper, Chapter 13).

Terrigenous particles are also introduced to marine environments by wind and glaciers. Windblown (aeolian) dust picked up from the continents settles out onto the sea surface far from land and can also be delivered to the oceans by mixing with rain water. Wind-deposited dust is a primary source of lithogenic sediment in the deepest parts of the ocean basins, far from continental margins, and can accumulate to form red clays that blanket the deep-sea floor. Glacial sediments account for a substantial portion of the marine sediment supply in many coastal environments at high latitudes, where flowing ice delivers material ranging in size from fine-grained glacial flour (silt and clay) to large boulders that raft out to sea frozen in icebergs. The cyclic occurrence of coarse, glacially derived sediment in marine cores allows one to infer the onset of glaciation as well as cyclicity in ice sheet behavior.

In tropical regions, where coral reefs are common, calcium carbonate sands derived from reef erosion can accumulate offshore. Such biogenic deposits are typically limited to areas with low terrestrial sediment input and relatively shallow water depths, as the carbonate minerals are not stable in deep, cold water and will dissolve.

Farther out to sea, sediment becomes less important in shaping marine landscapes. The morphology of the abyssal plain primarily reflects the effects of tectonism and volcanism during the creation of crust at the mid-ocean ridges. Over time, as the plate moves away from the ridge, these rocky, solid Earth landforms are draped by a thin blanket of authigenic sediment, marine sediment precipitated directly from seawater. Authigenic marine sediments dominate deep water deposition far from continents where biological productivity is low. In these locations, biologically produced calcium carbonate dissolves because it is soluble in deep, cold seawater. In addition, there are no large inputs of lithogenic sediment because the abyssal plains are far from the continents. Mixed with the authigenic sediment is a small but easily measurable extraterrestrial contribution to ocean sedimentation of interplanetary dust particles. On average, about 40,000 tons (0.04 megatons, or MT) of extraterrestrial dust fall to Earth every year. That sounds like a large amount until one considers that on average, every year, 12,600 MT of suspended sediment is delivered to the oceans by rivers draining the continents.

Dissolved Load

The dissolved load in seawater comes from rivers that carry the ionic products of chemical weathering on land, from volcanic eruptions at mid-ocean ridges, and from air fall and precipitation scavenging of the atmosphere. Some of the dissolved load remains in solution, producing and maintaining the saltiness of ocean water, which is on average about 35 parts per thousand (ppt) dissolved salts. Some of the dissolved load precipitates inorganically from seawater as authigenic chemical sediments. Other dissolved material is precipitated biologically when marine organisms produce mineral material like silica dioxide (SiO_2) and calcite ($CaCO_3$), which settle out as the biogenic oozes that cover vast expanses of the seafloor. Formation of evaporites from concentration and precipitation of the dissolved load can be an important sedimentation process in shallow marine and coastal settings. Tectonic and sea-level changes can isolate ocean basins, as happened when what is now the Mediterranean Sea was briefly isolated from the Atlantic Ocean during the late Miocene (about 6 million years ago). During this period of isolation, the Mediterranean dried up and evaporite deposits, thick beds of salt, were deposited in the basin. Today, these deposits are routinely found in Mediterranean sediment cores as well as on shore, where they are the basis for halite and gypsum extraction industries in Sicily. By the early Pliocene (about 5 million years ago), ocean water again flooded the Mediterranean.

Marine Landforms and Processes

Hidden beneath the waters of the world's oceans are a variety of submarine landscapes that rival in their variety and topography the terrestrial landscapes of the continents. A quick glance at a bathymetric map, on which contour lines represent depth below sea level, reveals an exciting and, until recently, little-known realm of sedimentation, mountain building, and volcanic activity. A variety of remote-sensing techniques—including side-scan sonar, depth profiling, ocean floor drilling, magnetic anomaly analysis, and deep-sea dives—has brought geomorphology to the ocean depths. Plate tectonics sets the location and character of submarine landscapes on the largest scale. Sedimentation and the erosive forces exerted by moving sediment and water continuously reshape the morphology of the ocean bottom.

Continental Margins

Continental margins have aprons of sediment draped over bedrock and extend out from the continents into the ocean basins. The crystalline rocks of continental interiors erode over time, and streams, wind, and ice transport the sediment to the edges of the continents and build them outward. Continental margins cover about 10 percent of Earth's surface and are composed of three primary subenvironments: the continental shelf, slope, and rise (see Figure 1.2).

The **continental shelf** is a smooth surface that slopes an average of less than one-tenth of a degree from the near-shore zone out to the shelf break at the top of the steeper continental slope. The seaward edge of modern continental shelves lies at an average water depth of 130 m, but continental shelves worldwide were subaerially exposed during major glaciations when more water was locked up in polar ice caps, reducing the volume of water in the oceans and lowering sea level. Continental shelves build seaward (prograde) over cycles of deposition and erosion as the shoreline migrates in response to changes in global sea level caused by changing global ice volume. Passive margins usually have wider continental shelves than active margins, the result of higher uplift rates on active margins. In tropical regions, the continental shelves can be covered with extensive submarine platforms built by coral reefs. Such reefs are common in the Caribbean and along the northern coast of Australia.

Offshore sandbanks and linear **sand ridges** occur on continental shelves where sand is abundant and ocean currents and waves are strong enough to transport sand-sized sediment. These features are generally 5 to 120 km long, 0.5 to 8 km wide, and can be up to 40 m tall, with heights that can equal 20 percent of the water depth. Although solitary sand ridges do exist, ridges typically occur in groups called sand-ridge fields. Individual ridges within ridge fields are typically spaced at distances of about 250 times the water depth. Sandbanks are sand ridges that develop in water shallower than 20 m. Most sandbanks are composed of fine to coarse sand, but gravelbanks do form in some areas with swift currents and abundant gravel. Sandbanks form in a variety of locations on continental margins, from the mouths of estuaries (like the Columbia River) to the edge of the continental shelf. Wave dissipation and refraction by offshore sandbanks helps protect beaches from erosion, and sandbanks also provide important nursery and feeding grounds for fish. Offshore sandbanks can present a significant hazard to shipping.

The **continental slope** drops off from the edge of the continental shelf at a depth of about 130 m and extends down to water depths of between 1.5 and 3.5 km. The continental slope is a rugged place. On average, it slopes more than 4 degrees and marks the buried edge of continental crust. Great **submarine canyons,** the largest canyons on Earth, extend down the continental slope [**Figure 8.12**]. Monterey Canyon, a modest submarine canyon on the continental margin south of Santa Cruz on the northern California coast, is about the size of Arizona's Grand Canyon. Even though submarine canyons sometimes exhibit sinuous meanders similar to those that develop in terrestrial rivers, we know that submarine canyons were not carved by rivers because they were submerged even during the lowest sea-level stands.

Submarine canyons are carved by turbidity currents, which transport clastic sediment down the continental slope and deliver it to the long-term geologic sediment sinks of the continental rise and abyssal basins. These currents are particularly erosive due to the high shear stress the dense flows exert on the continental slope. Turbidity currents sort sediments by grain size during transport and produce **turbidites**, distinctive deposits with upward-fining sequences that have coarse sand at the base and grade upward to fine silt and clay because the larger and heavier particles settle out before the lighter, smaller particles.

The **continental rise** is the ramp between the base of the continental slope and the deep abyssal seafloor at depths averaging around 4 km. The continental rise consists of oceanic crust buried beneath submarine fan sediment shed from land and transported across the continental shelf and slope. Accumulation at the head of the continental rise of sedimentary deposits carried by turbidites builds abyssal, or deep-sea, fans—the marine version of terrestrial fans extending down toward the deep seafloor.

During sea-level low stands, rivers and glaciers transport material out across the continental shelf to the shelf break 130 m below today's sea level. Delivery of a substantial volume of clastic sediment directly to the top of the continental slope creates large accumulations of unstable sediment. Large storms or earthquakes can trigger failure of these weak, generally poorly consolidated materials. Gigantic submarine landslides produce great slurries of sediment that move down the continental slope and may trigger tsunami waves.

During sea-level high stands, when base level is higher, continental-margin sediment is typically trapped in river-mouth estuaries and fjords because former fluvial and glacial valleys are inundated with seawater and streams drop their sediment loads farther inland. Deltas form either at the heads of estuaries or at the ocean in locations where fluvial sediment supply has been sufficient to fill estuaries and fjords. The sediment that does escape coastal entrapment is winnowed by waves, so sand is deposited on the near-shore continental shelf and mud is transported

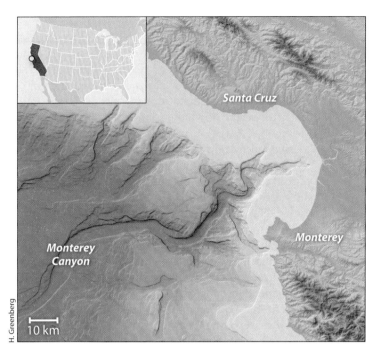

Carved by submarine **turbidity currents,** bottom-hugging flows of sediment denser than the water through which they move, the largest, deepest canyons on Earth are submarine canyons that cross the continental shelf. Monterey Canyon, a modest submarine canyon off the coast of central California, is comparable in size to Arizona's Grand Canyon.

FIGURE 8.12 Submarine Canyons. Submarine canyons dissect the continental shelf and slope and deliver large amounts of terrestrial sediment to the deep ocean.

farther out to the middle shelf. On collisional margins with narrow continental shelves, sediment still makes it out to the continental slope even when sea level is relatively high, as is the case at present. Sedimentary deposits on the shelf grade outward from coarser modern beach and dune sands and gravels to finer-grained deposits that overlie coarse relict fluvial deposits that were laid down during periods of lower sea level. The morphology and sediments on continental margins reflect both the geometry of the continental shelf and the effects of repeated changes in sea level.

Abyssal Basins

The abyssal basins that constitute the deep seafloor between the continental rise and oceanic ridges are composed of linear hills, transform fault traces, normal fault scarps, and sediment-filled grabens in water depths of 3 to 6 km. The vast, low-relief abyssal basins cover more than half of Earth's surface and are divided into erosional bedrock uplands, **abyssal hills,** and depositional sedimentary lowlands that form the broad, smooth **abyssal plains.** Abyssal hills are elongate bedrock ridges that are parallel to the spreading segments of the associated mid-ocean ridge. They cover three-quarters of the floor of the Pacific Ocean and most of the Atlantic Ocean seafloor. They are a dominantly bedrock landscape and include relief up to 900 m that was generated by volcanic action during formation at a spreading center. Abyssal hills are typically blanketed by at least a thin layer of pelagic ooze. The abyssal basins are sedimentary environments that consist of smooth surfaces formed when sediment buried the basement rocks of oceanic crust. They generally characterize the seafloor near continental margins at the base of the continental rise.

Mid-Ocean Ridges

Mid-ocean ridges are underwater mountain ranges that circle the globe like the stitching on a baseball (see Figures 1.1 and 1.4). Also called **seafloor-spreading centers,** they form the network of fractures in Earth's lithosphere where plates diverge and spread apart above zones of convective upwelling in the underlying mantle. Mid-ocean ridges are about 1500 km wide and rise to 3 km above the surrounding ocean floor. Water depths are typically about 2 km along the crests of spreading ridges, but islands can form along ridges where volcanic output is high enough, as is the case in Iceland. Thermal expansion of the lithosphere associated with emplacement of hot basalt elevates the crust along spreading ridges, forming complex patterns of extensional grabens. Ridges typically have a major axial valley along the crest where volcanic eruptions create new oceanic crust. The lithosphere on opposite sides of a mid-ocean ridge cools and becomes more dense as the plates move away from each other; thus, the seafloor sinks to greater depths as it moves out from a spreading ridge.

Local mantle upwellings called **hot spots** produce volcanic eruptions that can build up to form island chains and basaltic plateaus in areas far from mid-ocean ridges. The 6000 km-long chain of the Hawaiian Islands and Emperor seamounts extends west from the "Big Island" of Hawaii to Midway Island, where the chain starts to bend northward to the Aleutian Trench off Alaska, recording changes in the direction of movement of the Pacific plate over a long-lived mantle hot spot.

Trenches

Marine trenches at convergent plate boundaries, where oceanic lithosphere is subducted into the mantle, are the deepest places in the ocean; they average about 8 km in depth. The Mariana Trench in the western Pacific Ocean, where terrestrial sediment loading is low, is the deepest point on Earth's surface at almost 11 km below sea level—several kilometers deeper than Mount Everest is high. In contrast to deep trenches with little sediment load, trenches along convergent margins with high sediment fluxes can be quite shallow because of sediment infilling. The Peru–Chile trench along the west coast of South America is shallow offshore of the northern Andes because steep slopes and intense tropical rainfall drive high rates of erosion that result in substantial sediment delivery. The trench is very deep off the hyperarid coast of Chile, where extremely dry conditions in the southern Andes deliver very little sediment to the marine environment. Farther south, offshore of the formerly glaciated regions of Chile, the trench is shallow due to the high delivery rate of sediment to the marine environment by glaciers.

Coral Reefs

Coral reefs are important features of tropical and subtropical continental margins where coral thrives in the warm waters. Reef-forming coral grows in clear, sediment-free water that is shallow enough to allow light to penetrate (i.e., less than about 60 m deep). Built up on fragments of coral broken by storm waves and the bodies of other marine organisms, coral reefs flourish where vigorous wave action supplies abundant nutrients and oxygen. Large freshwater inputs and turbid water with high sediment concentration inhibit coral growth, so reefs typically develop away from deltas and river mouths.

Charles Darwin first recognized the origin of coral reefs in the 1830s when he sailed through the South Pacific aboard H. M. S. *Beagle*. Based on the islands he visited on this epic voyage, Darwin proposed that there were three basic types of reefs. **Fringing reefs** grow near the shoreline or around islands as a narrow fringe usually a few hundred meters to a kilometer across [**Photograph 8.18**]. Breaks in fringing reefs often occur where streams deliver fresh water and sediment to coastal environments, impeding coral growth. **Barrier reefs** grow offshore and, like barrier islands, are separated from the mainland by lagoons. Barrier reefs, however, are generally broader than barrier islands and can be up to 15 km wide. The growth rate and pattern of a coral reef reflects the balance between coral growth and wave erosion, both of which are concentrated on the seaward side of the reef. The most famous example of a barrier reef, the Great Barrier Reef, extends for more than 2000 km along the eastern coast of Australia.

An **atoll**, Darwin's third variety of reef, is a circular reef that surrounds a lagoon. Atolls rise just a few meters above sea level and lack a central island. Darwin recognized that atolls form when a fringing reef around an island is able to keep pace with subsidence and builds up fast enough to maintain shallow water conditions even as the island sinks beneath sea level [**Figure 8.13**]. Borehole samples from some atolls have found more than 1000 m of coral reef rock above oceanic basement rocks. More recently, atolls were strategically important in World War II, when they were used as positioning sites for troops, as strategic air bases, and as nuclear test sites.

PHOTOGRAPH 8.18 Fringing Reef. Fringing reef surrounds a coral island and shelters the lagoon from wave energy on the southern Great Barrier Reef, Australia, near Heron Island.

Applications

Coastal and marine geomorphology has immediate practical applications. Sand and gravel from coastal environments and from relict estuaries and deltas provide important sources of construction materials in many regions of the world. Most of the world's largest cities are near the ocean, and 75 percent of Americans live in coastal states. Many coastal cities, such as Venice, Italy, and New Orleans, Louisiana, rely upon groundwater pumped from the coastal marine and fluvial sediments beneath their buildings.

Atolls start as **fringing coral reefs** typically growing on the margins of volcanic islands in warm tropical waters.

When gradual growth of a coral reef can keep up with the ongoing **thermal subsidence** of the island (resulting from the cooling of the oceanic crust as it ages), the reef can remain close to sea level.

Eventually, continued subsidence of the volcanic island leads to development of an enclosed **lagoon** surrounded by the originally fringing reef.

FIGURE 8.13 Atoll Evolution. Atolls evolve if reef-building organisms can produce reef material at a rate equal to the rate of volcanic island erosion and subsidence.

These areas are vulnerable to settling and land subsidence, a problem that is compounded by levees and dams that restrict delivery of sediment to and deposition in coastal areas.

On average, North American coastlines are eroding at a rate of just under a meter a year, although rates vary greatly among different regions and depend on local conditions including wave heights, material properties, and land subsidence/uplift rates. The U.S. Army Corps of Engineers has classified about one-quarter of U.S. coastlines as seriously eroding. Severe coastal erosion has profound implications for risk to coastal properties and for geologic hazards during storms. Reduction in the supply of sediment to coastal environments, from both inland dam construction and coastal armoring, affects the sediment supply to beaches and barrier islands.

Historically, efforts to combat beach erosion have focused on engineered solutions. For instance, construction of groins to trap and retain sediment typically causes downcurrent erosion because of longshore sediment transport. Consequently, the construction of some coastal engineering structures leads to the need to build more structures along the shoreline. For example, construction of bulkheads by property owners along the shores of Puget Sound in Washington State and groin fields in New Jersey resulted in extensive beach engineering that greatly altered near-shore environments.

Understanding coastal sediment budgets that consider sediment carried in and out by longshore drift and other processes can help geomorphologists address such problems as harbor siltation and beach erosion. Longshore drift builds spits that block harbors and shipping channels and removes sand from prized beaches. The problems of delivering sand to eroding areas and removing it from aggrading areas have generated ongoing controversies over the role of shoreline stabilization versus planning to accommodate shoreline mobility. Some places have instituted expensive programs of beach nourishment, where sand is routinely pumped from offshore bars onto the beach in order to provide areas for recreation and to protect buildings and roads from erosion. In some of these areas, there is no expectation that the sand will stay in place; rather, the nourishment is regarded simply as beach maintenance.

Development on barrier islands can greatly affect their morphology through changes in sediment delivery and storage. Extensive development of a popular East Coast resort (Ocean City, Maryland) in the 1920s through 1970s motivated extensive efforts to stabilize the nearby barrier inlet by building a jetty at the southern end of the city. With the longshore current running north to south, the jetty stabilized the inlet but reduced the sediment supply to the barrier islands to the south, across the inlet. This caused extensive erosion of Assateague Island off southeastern Maryland, unintentionally shifting the shoreline there almost a kilometer inland.

Inundation from storm surges, tsunamis, and sea-level change are uniquely coastal issues. The ways in which waves interact and refract upon approaching land leads to profound differences in local tsunami and storm surge hazards along vulnerable coastlines. Lagoons and marshes dissipate wave energy and slow wave speed because they are rough; thus, they provide natural coastal defenses from hurricane-driven storm surges and tsunamis. The loss of such buffers can prove disastrous and was one factor contributing to Hurricane Katrina's devastation of New Orleans. The destruction of mangrove swamps that lessen wave energy and provide rich wetlands supporting high biodiversity is an ongoing problem in tropical regions around the world.

Coastal environments stand on the front lines of potential impacts from the anticipated sea-level rise projected to accompany global warming over the next century. Understanding how coastlines behaved during past sea-level rises can help predict future changes. The forecasted rise in sea level over the next century will have its first impacts on low-lying island nations and regions, and coastal geomorphologists expect large changes from even small amounts of sea-level rise in flat-lying coastal areas. Deltas and estuaries that will experience the greatest effects from sea-level rise also happen to be the environments where a majority of Earth's human population lives, where much of the world's seafood is nurtured, and where many of our cities are located.

Understanding the nature of the continental shelf as an environment exposed as land during glaciations and submerged during warmer interglacial periods has a major bearing on regional archaeology and biogeography. Because extensive portions of the continental shelves were land during glaciations, many archaeological sites are now submerged or are buried below sediments on the shelves. For example, ancient coastal environments in Siberia and Alaska and between New Guinea and Australia, now below sea level, are suspected of having been corridors for human migration. Similarly, during glacial sea-level low stands, rivers flowing across the modern continental shelves likely provided refugia for freshwater species, like salmon, during glacial advances that filled drainage basins with ice.

Selected References and Further Reading

Adams, P. N., R. S. Anderson, and J. Revenaugh. Microseismic measurement of wave-energy delivery to a rocky coast. *Geology* 30 (2002): 895–898.

Allen, J. C., and P. D. Komar. Climate controls on U.S. West Coast erosion processes. *Journal of Coastal Research* 22 (2006): 511–529.

Ashton, A., A. B. Murray, and O. Arnoult. Formation of coastline features by large-scale instabilities induced by high-angle waves. *Nature* 414 (2001): 296–300.

Bagnold, R. A. Beach formation by waves: Some model experiments in a wave tank. *Journal of the Institution of Civil Engineers* 15 (1940): 27–52.

Bird, E. *Coastal Geomorphology: An Introduction*, 2nd ed. Chichester, UK: John Wiley & Sons, 2008.

Bradley, W. C., and G. B. Griggs. Form, genesis, and deformation of central California wave-cut platforms. *Geological Society of America Bulletin* 87 (1976): 433–449.

Carter, R. W. G., and C. D. Woodroffe, eds. *Coastal Evolution: Late Quaternary Shoreline Morphodynamics*, New York: Cambridge University Press, 1994.

Darwin, C. *The Structure and Distribution of Coral Reefs*. London: Smith, Elder, 1842.

Davis, W. M. *The Coral Reef Problem*. Special Publication No. 9. New York: American Geographical Society, 1928.

Dixon, T. H., F. Amelung, A. Ferretti, et al. Space odyssey: Subsidence and flooding in New Orleans after Hurricane Katrina as measured by space geodesy. *Nature* 441 (2006): 587–588.

Dyer, K. R., and D. A. Huntley. The origin, classification and modeling of sand banks and ridges. *Continental Shelf Research* 19 (1999): 1285–1330.

Emery, K. O. Continental Margins—Classification and Petroleum Prospects. *American Association of Petroleum Geologists Bulletin* 64 (1980): 297–315.

Ericson, J. P., C. J. Vörösmarty, S. L. Dingman, et al. Effective sea-level rise and deltas: Causes of change and human dimension implications. *Global and Planetary Change* 50 (2006): 63–82.

Fagherazzi, S., and I. Overeem, Models of deltaic and inner continental shelf landform evolution. *Annual Review of Earth and Planetary Sciences* 35 (2007): 685–715.

Galloway, W. E. "Process framework for describing the morphological and stratigraphic evolution of deltaic depositional systems." In M. L. Broussard, ed., *Deltas: Models for Exploration*. Houston, TX: Houston Geological Society, 1975.

Galy, V., C. France-Lanord, and B. Lartiges. Loading and fate of particulate organic carbon from the Himalaya to the Ganga-Brahmaputra delta. *Geochimica et Cosmochimica Acta* 72 (2008): 1767–1787.

Godfrey, P. J. Barrier beaches of the east coast. *Oceanus* 19 (1976): 27–40.

Hayes, M. O. "Barrier island morphology as a function of wave and tide regime." In S. P. Leatherman, ed., *Barrier Islands from the Gulf of St. Lawrence to the Gulf of Mexico*. New York: Academic Press, 1979.

Holland, K. T., and R. A. Holman. The statistical distribution of swash maxima on natural beaches. *Journal of Geophysical Research* 98 (1993): 10,271–10,278.

Inman, D. L., and C. E. Nordstrom. On the tectonic and morphologic classification of coasts. *Journal of Geology* 79 (1971): 1–21.

Jay, D. A., W. R. Geyer, and D. R. Montgomery. "An ecological perspective on estuarine classification." In J. Hobbie, ed., *Estuarine Science: A Synthetic Approach to Research and Practice*. Washington, DC: Island Press, 2000.

Johnson, D. W. The nature and origin of fjords. *Science* 41 (1915): 537–543.

Johnson, D. W. *The Origin of Submarine Canyons: A Critical Review of Hypotheses*. New York: Columbia University Press, 1939.

Kaufman, W., and O. H. Pilkey. *The Beaches Are Moving: The Drowning of America's Shoreline*. Durham, NC: Duke University Press, 1983.

King, C. A. M. *Introduction to Marine Geology and Geomorphology*. London: E. Arnold, 1975.

Kirwan, M. L., A. B. Murray, J. P. Donnelly, and D. R. Corbett. Rapid wetland expansion during European settlement and its implication for marsh survival under modern sediment delivery rates. *Geology* 39 (2011): 507–510.

Komar, P. D. The mechanics of sand transport on beaches. *Journal of Geophysical Research* 76 (1971): 713–721.

Komar, P. D. *Beach Processes and Sedimentation*, 2nd ed. Englewood Cliffs, NJ: Prentice-Hall, 1998.

Komar, P. D., and D. L. Inman. Longshore sand transport on beaches. *Journal of Geophysical Research* 75 (1970): 5914–5927.

Krijgsman, W., F. J. Hilgen, I. Raffi, et al. Chronology, causes, and progression of the Messinian salinity crisis. *Nature* 400 (1999): 652–655.

Moore, J. G., D. A. Clague, R. T. Holcomb, et al. Prodigious submarine landslides on the Hawaiian Ridge. *Journal of Geophysical Research* 94 (1989): 17,465–17,484.

Moore, L. J., and G. B. Griggs. Long-term cliff retreat and erosion hot spots along the central shores of the Monterey Bay National Marine Sanctuary. *Marine Geology* 181 (2002): 265–283.

Mudd, S. M. The life and death of salt marshes in response to anthropogenic disturbance of sediment supply. *Geology* 39 (2011): 511–512.

Parsons, B., and J. G. Sclater. An analysis of the variation of ocean floor bathymetry and heat flow with age. *Journal of Geophysical Research* 82 (1977): 803–827.

Steers, J. A., ed. *Applied Coastal Geomorphology*. Cambridge, MA: MIT Press, 1971.

Syvitski, J. P. M., and Y. Saito. Morphodynamics of deltas under the influence of humans. *Global and Planetary Change* 57 (2007): 261–282.

Trenhaile, A. S. *The Geomorphology of Rock Coast*. New York: Oxford University Press, 1987.

Werner, B. T., and T. M. Fink. Beach cusps as self-organized patterns. *Science* 260 (1993): 968–971.

DIGGING DEEPER What Is Happening to the World's Deltas?

Many of the world's major river deltas are slipping below sea level (Syvitski, 2008; Blum and Roberts, 2009; Syvitski et al., 2009). Why is this happening and why are scientists sounding the alarm in prominent scientific journals?

Deltas are vitally important to the billions of people who inhabit Earth. Their nutrient-rich alluvial soils and extremely flat topography provide wide expanses of excellent agricultural land on which to grow a variety of crops. The rivers that flow onto deltas provide a ready source of irrigation water, and the levees built to contain many of these rivers provide a sense of security from flooding that has encouraged people to settle on productive deltaic lowlands. Deltas are home to half a billion people, almost 10 percent of the world's population. Major cities, including Shanghai, New Orleans, and Bangkok are located on deltas. River deltas can be very large; the Amazon River delta covers almost half a million km^2.

Because deltas form where rivers empty into the ocean, and thus are fine-tuned to sea level, they are vulnerable to flooding and coastal erosion. Their flatness, with gradients as low as 0.00001 (1-mm rise per 100 m run), makes large areas vulnerable to flooding from overflowing rivers and rising seas; a 10-cm rise in sea level could move a delta shoreline landward by 10 km. Such low gradients make flooding on deltas deadly. Syvitski et al. (2009) report that in 2007–2008, delta floods killed more than 100,000 people and drove more than a million others from their homes.

The size and extent of deltas reflect a precarious three-way balancing act between the mass of sediment brought in by rivers, the volume of water in the world's oceans, and the sinking or subsidence of the delta. For example, deltas will shrink if ice sheets melt and/or the ocean warms, causing global ocean volume to increase and eustatic sea level to rise. Conversely, deltas expand if sediment yields increase and rivers supply more sediment to coastal environments. Deltas naturally subside because of the solid-Earth isostatic response to massive sediment loading and because delta sediments compact as they dewater and the organic matter within them decays and oxidizes.

Ocean volume is increasing and eustatic sea level is rising at an accelerating rate due to global warming [**Figure DD8.1**] (IPCC, 2007). After eustatic sea-level stabilized about 7000 years ago, long-term rates of eustatic sea-level rise have been <1 mm/yr. Since about 1940, the rate of global sea-level rise has increased to more than 2 mm/yr. Predictions of future rise are uncertain but suggest that the rate of sea-level rise will continue to increase as both the climate and the oceans warm, melting ice sheets and thermally expanding ocean waters.

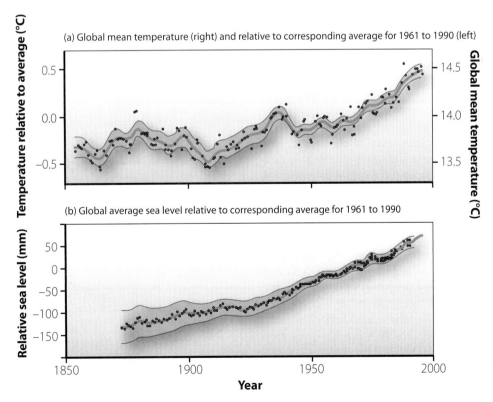

FIGURE DD8.1 Global warming began in earnest after 1900 (top), about the same time that the global average rate of sea-level rise increased (bottom). Global, eustatic sea level has risen about 200 mm in the past 150 years. [From IPCC (2007).]

DIGGING DEEPER What Is Happening to the World's Deltas? *(continued)*

For individual deltas, what matters is the sum of eustatic sea-level change and subsidence. Natural deltaic subsidence rates of a few millimeters per year have been accelerated by human activities. Examples include centimeters per year of subsidence on the Po River delta, from which methane was pumped, and the Chao Phraya delta on which Bangkok is located, where groundwater was withdrawn (Syvitski et al., 2009). Building levees along rivers contributes to subsidence by preventing sediment delivery to large parts of the delta surface. In some places, like the mouth of the Mississippi River, sediment that once spread over the delta surface is now delivered by levee-constrained channels right to the continental shelf. The result of this sediment starvation is that some delta surfaces inhabited by people have subsided enough that they lie below sea level and are kept safe and dry only by levees. Alternately, sediment accumulation in the channel can cause the river bed elevation to rise above the level of the surrounding floodplain.

Syvitski et al. (2009) investigated the critical balance of supply, subsidence, and ocean volume for 33 of the world's deltas. They used SRTM data (see Chapter 2) to determine delta surface elevation, historical maps to determine the number and location of distributary channels in the past, and satellite images to establish the extent of recent flooding. They supplemented these observations with existing data about river sediment loads and eustatic sea-level rise. Their conclusions are sobering. It appears that reduced sediment delivery is the biggest culprit in delta shrinkage. Most deltas are now sinking much faster than sea level is rising because so much sediment is trapped upstream by dams or impounded by levees that delta aggradation has slowed or ceased. Although agriculture initially increased sediment yield from drainage basins, the construction of large dams subsequently starved streams and deltas of sediment [**Figure DD8.2**]. Some deltas, such as that of the Yellow River, today receive no sediment at all.

The Nile delta, home to ~50 million people, is a prime example of a delta in trouble. The Aswan high dam captures about 98 percent of the sediment load that the Nile River carries north from central Africa (Syvitski, 2008). Irrigation channels on the delta now trap what little sediment makes it through the dam. The distributary channel network on the delta has shrunk so much that most of the delta is now disconnected from the river. Where there were once dozens of distributaries spreading river sediment over the delta surface, today there are only two (Syvitski, 2008) [**Figure DD8.3**]. This pattern is not unique to the Nile delta. Syvitski et al. (2009) report that the distributary network shrank through human actions in nearly half of the 33 deltas they studied.

Predicting the future of deltas requires an understanding of the mass balance of sediment in the system as well as predicting how fast the land is going down and the sea

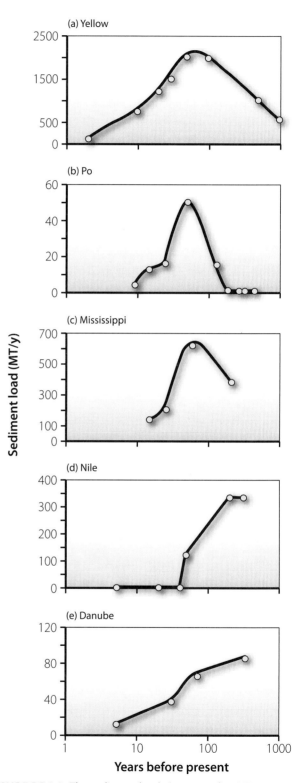

FIGURE DD8.2 The sediment loads (expressed as MT, megatons, millions of tons) of many rivers have changed over recent centuries. Many rivers experienced a peak in sediment loading during periods of intensive agriculture and then a drop in loads as dams were built and sediment moving downstream was trapped. Note the logarithmic scale for time. [From Syvitski (2008).]

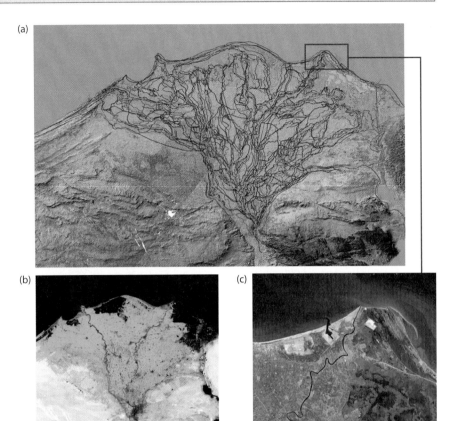

FIGURE DD8.3 The number and pattern of distributary channels on the Nile River delta has changed since the Aswan high dam was built. (a) All the distributary channels indicated on maps made of the delta in 1813, 1831, and 1897 superimposed on SRTM topographic data. (b) A remotely sensed image showing the current situation—only two main distributary channels are still carrying Nile River waters to the ocean. Look closely at part (c) to note the coastal erosion moving sand offshore to the northwest. Compare the older shorelines for this area with the inset in part (a) to see how much loss of coastal land has occurred. [From Syvitski, (2008).]

is going up. Using estimates of sea-level rise, historical rates of subsidence, and long-term average rates of sediment storage, Blum and Roberts (2009) conclude that by 2100 more than 10,000 km² of the Mississippi River delta will be inundated by seawater. The culprit: lowered sediment loading. More than half the sediment that used to flow down the Mississippi River to the delta is now trapped by the numerous dams in the watershed. Syvitski and Milliman (2007) report that the Mississippi River watershed contains more than 50,000 dams. Rates of sediment transfer from the Mississippi River to its delta are so low that sedimentation cannot keep up with even modest rates of eustatic sea-level rise and subsidence. Relative sea-level rise near the delta is anticipated to be on the order of a meter over the next century, flooding much of the low-lying wetlands and coastal marshes.

Can the trend of sinking and shrinking deltas be reversed? Probably not, but the process can be slowed. Kim et al. (2009) created a model of the Mississippi delta and used what we know about sediment loads, sea-level rise, and sediment compaction to suggest that breaching Mississippi River levees below New Orleans could build large amounts of new land [**Figure DD8.4**]. They conclude that if about half of the river flow were spilled through levees onto the subsiding delta, enough flow would remain for navigation to continue in the main channel while 700 to 1200 km² of new land would be created over the next century. This would offset about half the expected land loss. Of course, doing so would require abandoning the century-old policy of containing the Mississippi River in levees, at least for the reach below New Orleans.

Blum, M. D., and H. H. Roberts. Drowning of the Mississippi Delta due to insufficient sediment supply and global sea-level rise. *Nature Geoscience* 2 (2009): 488–491.

Intergovernmental Panel on Climate Change (IPCC). "Summary for Policymakers." In S. Solomon, D. Qin, M. Manning, et al., eds., *Climate Change 2007—The Physical Science Basis: Contribution of Working Group I to the Fourth Assessment Report of the IPCC Change*. New York: Cambridge University Press, 2007.

Kim, W., D. Mohrig, R. Twilley, et al. Is it feasible to build new land in the Mississippi River Delta? *EOS* 90 (2009): 373–374.

Syvitski, J. P. M. Deltas at risk. *Sustainability Science* 3 (2008): 23–32.

Syvitski, J. P. M., A. J. Kettner, I. Overeem, et al. Sinking deltas due to human activity. *Nature Geoscience* 2 (2009): 681–686.

Syvitski, J. P. M., and J. D. Milliman. Geology, geography and humans battle for dominance over the delivery of fluvial sediment to the coastal ocean. *Journal of Geology* 115 (2007): 1–19.

DIGGING DEEPER What Is Happening to the World's Deltas? *(continued)*

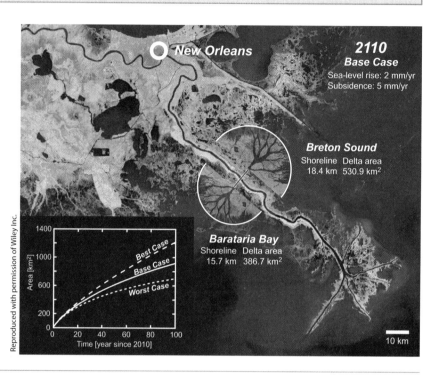

FIGURE DD8.4 This synthetic view imagines what the Mississippi River delta might look like in 2110 if the levees downstream of New Orleans had been breached in 2010. Two new delta lobes form at the breaches. The inset graph shows the rate at which new land will be built once the levees are breached, considering a best-case scenario in which sea level holds steady and subsidence is 1 mm/yr, a base case where sea level rises 2 mm/yr and subsidence is 5 mm/yr, and a worst case where sea level rises at 4 mm/yr while the land subsides at 10 mm/yr. [From Kim et al., (2009).]

WORKED PROBLEM

Question: Consider deep-water waves, created out at sea by a strong storm. Do waves with shorter or longer periods travel more quickly? How does the speed of these deep-water wind waves compare to the speed of a tsunami wave moving over the open ocean? Illustrate your answer graphically by calculating and plotting wave speeds for deep-water wind waves with periods between 1 and 20 s and for tsunami waves with a period of 10 min traveling over the ocean toward shore where water depths vary between 100 and 4000 m.

Answer: For the deep-water wind waves created by a storm, use eq. 8.2 to calculate wave speed. Your graph, as in the example below, should show that wave speed increases linearly as wave period increases. For the tsunami, which behaves as a shallow-water wave, use eq. 8.3. You should find that tsunami waves move much more rapidly than wind-driven waves and that tsunami waves slow when they approach land and the water depth shallows. In general, tsunami waves move much more quickly than wind waves.

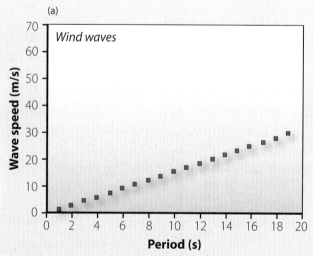
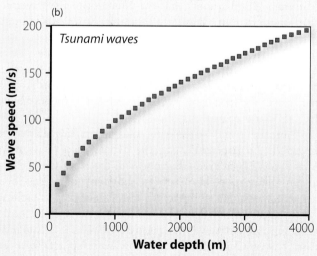

Graphs indicating how (a) the speed of wind waves is linearly and positively related to the wave period, and (b) how tsunami wave speed increases nonlinearly with water depth.

KNOWLEDGE ASSESSMENT Chapter 8

1. How do geomorphologists study Earth's surface under the ocean waters?
2. Contrast shoreline geomorphology at active and passive continental margins.
3. Compare emergent and submergent coastlines and explain the processes that lead to these two different types of coasts.
4. How much salt does ocean water contain and how does this affect its mixing with fresh water from rivers?
5. List and explain the two major controls on local sea level and shoreline position.
6. Identify and explain which part of the tidal cycle is most likely to cause geomorphic change along the coast.
7. Describe the history of sea level globally over the past 21,000 years.
8. Describe the mechanisms that supply sediment to the coastal zone.
9. Explain how tides are geomorphically important.
10. Predict how fetch and wind speed will affect the ability of waves to do geomorphic work.
11. Explain wave base and how it is geomorphically important.
12. Sketch the difference between plunging, spilling, and surging breakers.
13. Explain wave refraction and how it affects the evolution of coastlines.
14. Define longshore drift and explain why it is important for coastal geomorphology.
15. Explain why shoaling occurs and why it is geomorphically important.
16. Predict the geomorphic effects of installing a jetty or a groin in the coastal zone.
17. Define storm surge and explain how it catalyzes coastal geomorphic change.
18. Explain how tsunami waves differ from wind waves.
19. What can a geomorphologist learn from the elevation of wave-cut notches and wave-cut platforms?
20. Sketch a sea stack and a sea arch and explain how they form.
21. Define a beach berm and explain how it is created and changes with the seasons.
22. Explain how spits and barrier islands are genetically related.
23. Predict where a flood-tide delta will form and explain how and why its size differs from an ebb-tide delta.
24. Predict how lagoons evolve over time, specifying the relevant processes.
25. Explain why clay and salt water are important in the behavior of estuaries.
26. List the four major types of estuaries and explain how each forms.
27. Sketch the internal architecture of a delta deposited into a freshwater lake by a stream or river.
28. Explain, from a process perspective, the reason that river-dominated, wave-dominated, and tide-dominated deltas have different shapes.
29. Define lithogenic, authigenic, and biogenic sediments and predict where in the marine realm you would expect to find each kind of sediment.
30. Sketch the large-scale geomorphology of an active and a passive continental margin.
31. Describe how continental margin sedimentation differs between glacial and interglacial periods.
32. Describe a submarine canyon and explain how it forms.
33. Explain, with sketches, Darwin's theory of atoll formation.
34. What type of sediment settles out and covers the abyssal plains of the deep ocean?
35. Explain beach nourishment, how it works, and why it is done.

PART III

Ice, Wind, and Fire

(Chapters 9 to 11)

Glaciers, wind, and volcanoes shape some of Earth's most spectacular landscapes. In Chapter 9, we consider landscapes where geomorphic processes are dominated by the presence of ice. Wind, the landforms it creates, and the sediments it transports are the subject of Chapter 10. In Chapter 11, we consider the geomorphic impact of volcanism on Earth's surface, both the constructional forms that result from eruptions and those that remain after erosion of volcanic landforms.

In this view, a narrow strait passes between Ellesmere Island (at top) and the northwestern Greenland coast. The Petermann Glacier to the right flows north in a narrow fiord from Greenland's central icecap and had been the largest floating glacier in the northern hemisphere until August 2010, when a large portion of it broke off. Landsat ETM + image of the Arctic is at midsummer, near the melt maximum for July 1999. At this season, the snowfields have generally melted away from shore areas, but icecaps and glaciers persist near shore.

Glacial and Periglacial Geomorphology

Introduction

Glaciation is a geomorphically powerful process. The comings and goings of glaciers at high latitudes and high altitudes not only affect the geomorphology of glaciated regions, but the growth and decay of ice sheets and glaciers also affect the rate and distribution of surface processes around the world by changing sea level, affecting global average weathering rates, and shaping topography.

Glaciation is geomorphically powerful enough to influence the hypsometry (elevation/area relationship) of mountain ranges by preferentially eroding high-elevation terrain and limiting the elevation of high peaks. Glaciers excavated the Great Lakes and deposited the sediment that forms marine features including the Grand Banks, Cape Cod, and Long Island. Even when glaciers have melted away, their legacy, in the form of relict or fossil landscapes, controls the pace and distribution of surface processes such as rockfall from steep, glacially carved valley walls.

Glaciers, rivers of ice, and **permafrost,** the frozen ground found at high latitudes and altitudes in areas that lack glaciers, create some of the most spectacular and dynamic geomorphic environments on our planet. Much of their dynamism results from several physical properties of ice. One, on Earth, most ice exists near its melting point, so that **phase transitions** between liquid and solid water [Figure 9.1] occur commonly and account for much of the geomorphic activity of glaciers and permafrost. Two, ice melts when the pressure on it increases. Three, because ice is a weak material near its freezing point, it readily deforms

Margin of the Greenland Ice Sheet about 40 km east of Kangerlussuaq in western Greenland. In the foreground, a granitic erratic boulder sits on glacially sculpted gneissic bedrock. In the distance, Russell Glacier issues from the southwestern margin of the ice sheet. Rocky, light-colored lateral moraines, probably deposited several hundred years ago during the Little Ice Age cold period, fringe the ice on the right along with small meltwater streams.

IN THIS CHAPTER

Introduction
Glaciers
  Glacier Mass Balance
  Glacier Energy Balance
  Accumulation and Ablation of Glacial Ice
  Glacier Movement
  Thermal Character of Glaciers
  Glacial Hydrology
Subglacial Processes and Glacial Erosion
Glacial Sediment Transport and Deposition
  Subglacial Sediments and Landforms
  Ice-Marginal Sediments and Landforms
  Glacially Related Sediments and Landforms
Glacial Landscapes, Landforms, and Deposits
  Landforms of Alpine Glaciers
  Landforms of Ice Sheets
  Geomorphic Effects of Glaciation and Paraglacial Processes
Periglacial Environments and Landforms
  Permafrost
  Characteristic Periglacial Landforms and Processes
Applications
Selected References and Further Reading
Digging Deeper: How Much and Where Do Glaciers Erode?
Worked Problem
Knowledge Assessment

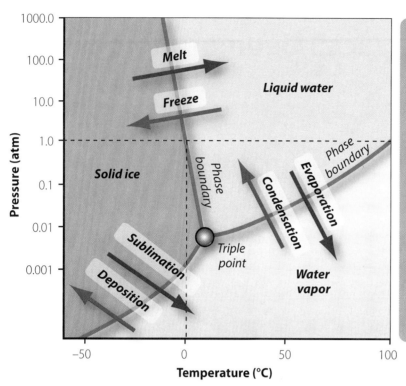

FIGURE 9.1 Phase Diagram for Water. The diagram shows that all three phases of water (vapor, solid, liquid) exist at Earth surface temperatures and pressures.

under gravitational loads. This deformation allows glaciers and frozen soil to flow and transport large amounts of mass, both ice and sediment. This transport of mass shapes landscapes as does glacial **abrasion**, the grinding away of bedrock below the ice, and **quarrying**, the removal of blocks of rock aided by changes in water and overburden pressure.

The erosive power of **alpine glaciers** creates steep cliffs and deep basins characteristic of once-glaciated mountain landscapes. **Continental ice sheets** have scoured whole regions, leaving behind landscapes of smoothed bedrock covered in some places with **till** (unsorted glacial sediment) and covered in other places with sand and gravel sorted and transported by water derived from melting glacial ice. The dramatic and deep **fjords** (deep linear troughs) of coastal Norway, Alaska, Greenland, southwestern New Zealand, and elsewhere owe much of their grandeur to glacial erosion focused in valleys.

Because some landforms and sediment assemblages can be unambiguously attributed to glacial processes, geomorphic mapping indicates the extent of now-vanished ice sheets and glaciers. For example, the distribution of glacial deposits and glacial landforms demarcates the margins of ice sheets that once covered much of the high latitudes and mountain ranges of the world [**Figure 9.2**]. Careful examination of the rock record indicates multiple episodes of glaciation stretching back at least into the Precambrian era, 2.9 billion years ago. Tilllite, or glacial sediment turned to rock (lithified), is preserved in the rock record and in some cases covers striated and polished rock surfaces [**Photograph 9.1**].

Times when continental glaciation was widespread are thought to reflect both favorable atmospheric conditions (low levels of CO_2 and, consequently, less effective greenhouse processes), lowered inputs of solar radiation, and continental arrangements with large land areas near the poles (providing land at cold, high latitudes on which ice sheets can grow from snow that survives the summer melt season). The pacing of glaciation over hundreds of millions of years is controlled by plate tectonics, while solar radiation input, as dictated by cyclic changes in the shape of Earth's orbit, controls the advance and retreat of glaciers over shorter, 100,000-year timescales.

In the last 2.7 million years (the late **Pliocene** and **Pleistocene** epochs), much of Earth's surface, including northern Europe and North America, has been directly and indirectly affected by multiple episodes of continental and alpine glaciation. Significant Quaternary climate changes, which brought Earth in and out of dozens of glaciations and interglaciations in the past several million years, are recorded in deep-sea sediment cores (see Chapter 13), but the evidence for multiple glaciations is much more difficult to find and interpret unambiguously on land. Only the largest and latest glaciations have left consistent and easily recognizable geomorphic signatures on the landscape.

Geomorphic evidence for the most recent glaciation, which reached its maximum extent about 21,000 years ago, dominates the surficial geology of glaciated regions. By mapping moraines and other glacial sediments, we know the maximum extent of the Laurentide Ice Sheet that covered large parts of North America. In some places

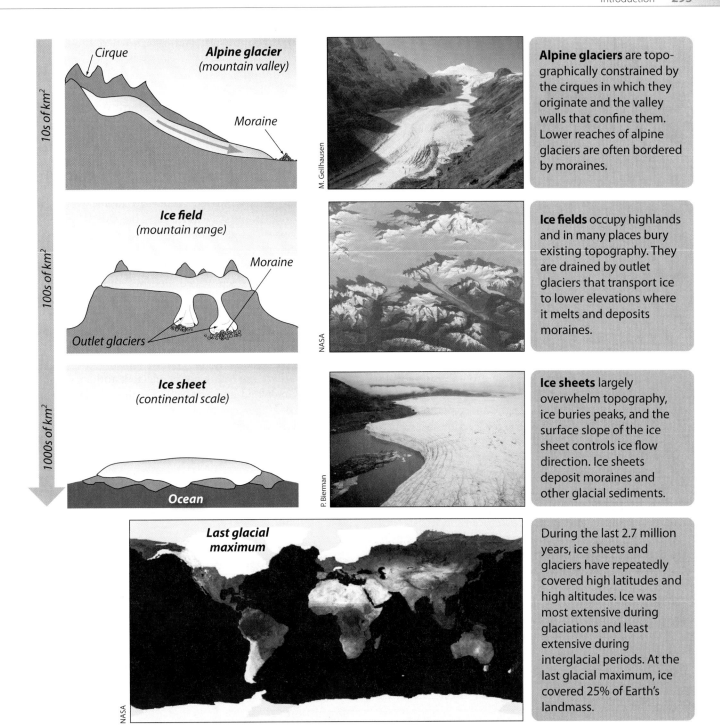

FIGURE 9.2 Bodies of Ice. Ice sheets, ice fields, and alpine glaciers differ in size, location, and their interaction with topography.

outside the most recent ice margin, one finds discontinuous and eroded remnants of material deposited and rock eroded during previous glaciations. This terrestrial geomorphologic record of older glaciations and, by inference a changed climate, is fragmentary. The extent of ice sheets before the last glaciation is poorly known in large part because glaciers usually destroy the evidence of prior glaciations by overrunning, eroding, and incorporating older glacial sediment as they advance.

Glacial and **periglacial** (low temperature but not glaciated) landscapes are distinct from landscapes shaped by fluvial processes. Periglacial environments are defined by both the geomorphic importance of the ice/water phase transition and the presence of seasonally snow-free ground. Much of the geomorphic activity in periglacial environments is induced by the freezing of water and thus climate, and in particular, temperature, is an important driver of periglacial surface processes. Expanses of **patterned ground** that exhibit striking geometric regularity typify periglacial zones and are the result of thermal contraction, the growth of ice lenses, and stirring of soils in periglacial terrain. Periglacial processes are active in many cold regions that were not covered

PHOTOGRAPH 9.1 Ancient Glacial Grooves. Glacial grooves of Permian age eroded into south Australian bedrock are exposed as overlying till (now solidified into a rock known as tillite) is removed by erosion (person shown for scale).

by glacial ice as well as in areas fringing many glacial margins. **Permafrost,** permanently frozen ground, is important in many periglacial environments. Glacial and periglacial environments are similar in a climatic and geographic sense, although flowing glaciers can extend to elevations below periglacial environments. Glacial and periglacial processes involve different percentages of ice and sediment but they share the importance of linked energy and mass balances.

Ice and glaciers are not limited to our planet. Ice covers oceans on Europa (a moon of Jupiter), as well as other moons of Jupiter, Saturn, and Neptune. Water ice is present in regolith covering shadowed craters on the Moon. Remote sensing data indicate that Mars has polar ice caps, which shrink and expand with the seasons. New data, gathered by ground-penetrating radar, have been interpreted to suggest that icy Martian glaciers, extending from mountains at lower latitudes, are preserved under blankets of rocky debris. It is likely that the debris blanket protects underlying ice from melting or subliming, not unlike many debris-covered glaciers on Earth.

Glaciers

Glaciers are persistent, flowing bodies of ice on the land surface that originate as accumulations of snow. Glaciers come in a wide variety of sizes and shapes (Figure 9.2). Largest are **ice sheets** [**Photograph 9.2a**], also known as **continental glaciers,** which in the past have covered extensive areas of the high latitudes including much of North America, northern Europe, and Antarctica. Ice sheets are large enough to influence climate by diverting storms, altering wind directions, and orographically enhancing precipitation. Once ice overtops topography, the slope of the ice surface itself drives glacial flow, allowing ice sheets to move over landscapes in their path. Such overtopping of topography was prevalent in the Northern Hemisphere where the Laurentide Ice Sheet formed over subdued shield topography. In contrast, in Antarctica, where the ice sheet formed over mountains of greater relief, there are many exposed mountain peaks, which influence the direction of ice flow. The Pleistocene-age Cordilleran Ice Sheet, which covered the mountains of western North America, had a similar topographic control of ice flow.

Ice caps ($<50,000$ km^2) are found on the high portions of mountain ranges or on elevated plateaus [**Photograph 9.2b**]. They are smaller than ice sheets and flow largely independent of topography, although the tongues of ice that drain ice caps are often confined to narrow alpine valleys. **Ice fields** are on the same scale as ice caps, but subglacial topography still exerts a significant control on the shape of the ice field and thus on the flow of ice. Ice masses that are confined to valleys are referred to as **valley glaciers** and typically originate in **cirques,** steep-walled hollows the glaciers have eroded into mountain sides [**Photograph 9.2c**]. **Alpine glaciers** may flow out of the mountains and onto lower gradient piedmonts where they are no longer constrained by valley walls [**Photograph 9.2d**]. Cirque glaciers are restricted to basin-shaped cirques high in the mountains.

Today, there are many alpine glaciers and ice caps but only the Greenland and Antarctic ice sheets remain; these are only 10–20 percent smaller in area than they were during glacial maxima but they contain much less ice. Since the last glacial maximum, parts of the Greenland Ice Sheet have lowered hundreds of meters, contributing enough meltwater to raise global sea level about 3 m. Thinning of the Antarctic Ice Sheet since the last glacial maximum has released enough water by itself to raise the world sea level by at least 10 m.

Glacier Mass Balance

Glaciers are best understood by considering their mass balance, which results from the **accumulation** of ice (mostly by snowfall) and its loss by **ablation** (mostly by melting or by calving of large chunks into standing water). You can think about glacier mass balance in the same way that the relationship between income and expenditures influences your bank balance and determines your financial stability.

Ice is added to glaciers by precipitation, both falling directly onto glacier surfaces and added by avalanches and wind carrying snow from above or from adjacent highlands. In high and mid-latitudes, away from monsoonal areas, most glacial ice originates as snow falling during the winter months. In monsoon-affected areas and on mountain peaks in tropics, glaciers accumulate snow during the wet season, which does not need to be winter. As the snow ages over the winter and loses its intricate crystal forms, it changes to larger, more rounded, and compact grains known to skiers as corn snow. Over the melt season, that corn snow further compacts, becoming **firn.** With time, and under the pressure of the overlying snowpack, the firn consolidates to glacial ice, gradually becoming denser and less porous [**Figure 9.3**]. In areas near freezing with high annual precipitation rates, this can happen in just a few

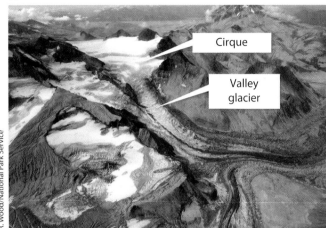
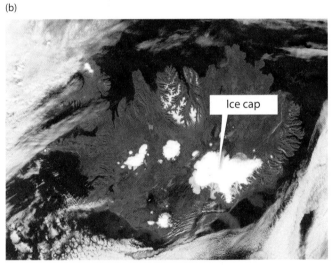
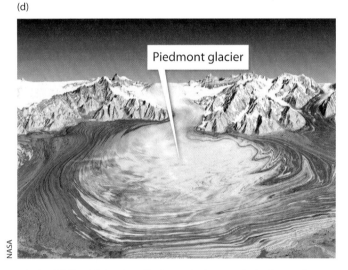

PHOTOGRAPH 9.2 Glacier Sizes and Shapes. Glaciers come in a wide variety of shapes and sizes. (a) The western margin of the Greenland Ice Sheet showing the ice margin and an ice-marginal lake it impounded. (b) Satellite view of Iceland showing several bright white ice caps. (c) Cirque and valley glaciers, Mount Katmai, Alaska. (d) The Malaspina Glacier, a piedmont glacier in southeastern Alaska, spreads into a wide lobe where it is no longer confined by valley walls. View created from a Landsat satellite image and SRTM digital elevation model.

years. In very cold areas with low precipitation rates, the transition to glacial ice can take several tens to a few hundred years. Crystal dimensions increase in glacial ice; older ice crystals can be more than 10 cm long.

Ice is lost from glaciers (ablates) in several ways. Mass leaves the glacier system as ice and snow melt and water drains away in **meltwater streams** (also known as outwash streams) [**Photograph 9.3**]. Some ice is also lost by sublimation, the direct transfer of water from the solid to vapor phase. Sublimation is driven by water vapor pressure gradients when wind moves dry (undersaturated) air over ice surfaces; thus, sublimation is of particular importance for cold glaciers in semiarid and arid regions. Ice can also be lost by **calving** of ice margins into bodies of water, including lakes and the ocean as well as by toppling failures at steep ice margins [**Photograph 9.4**].

Mass balance determines the fate of any particular glacier. If mass losses exceed mass gains, a glacier will thin and the position of the glacial margin will retreat upvalley or toward the center of the ice sheet. It is important to realize that when the position of the ice margin changes, the ice itself continues to flow downgradient; ice margin retreat only happens when the rate of ice ablation at the margin exceeds the rate of ice flow to the margin. If mass gains exceed mass losses, a glacier will grow and the position of its margin will advance.

Alpine glaciers respond rapidly to changes in climate that affect mass balances—they advance and retreat on yearly to decadal timescales. In contrast, large ice sheets and ice caps have greater inertia. Although ice sheets can respond quickly to climate change at their margins (by advancing or retreating) or in their accumulation zones

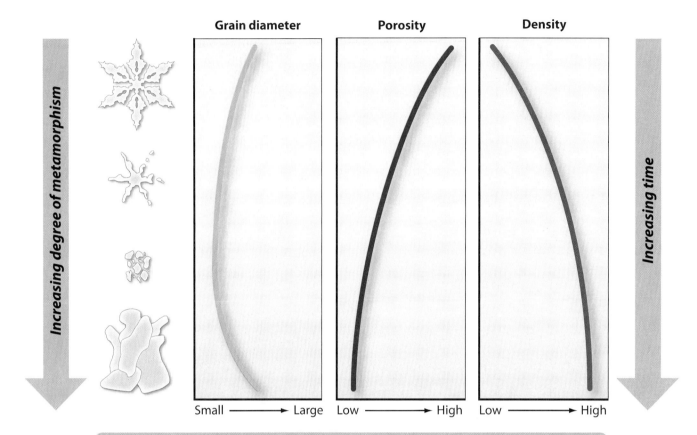

FIGURE 9.3 Snow Metamorphism. Over time, snow metamorphoses, transitioning from intricate forms to rounded grains with lower surface area. Grain diameter, surface area, and snowpack porosity and density change as these transitions continue.

(by thickening or thinning), ice sheets take centuries to respond fully to a new climate state and millennia to disappear in response to changing climate because there is so much ice to melt. For example, the Laurentide Ice Sheet that covered much of North America took more than 12,000 years to largely disappear after the last glacial maximum about 21,000 years ago. The Barnes and maybe the Penny Ice Cap on Baffin Island are thought by some to be the only surviving remnants of the Laurentide Ice Sheet.

Glacier Energy Balance

The energy balance of a glacier is complex but critical to determining whether it will advance or retreat. Energy is lost and gained from glaciers in numerous ways. Sunlight provides energy to the glacier surface, causing mass loss by sublimation (even as the ice remains frozen) and by warming snow and ice until they melt. When the glacier surface is covered by fresh, white, highly reflective snow, its **albedo** (reflectivity) is high and little energy from the Sun is absorbed directly. As snow gets covered in dust, or starts to melt, its albedo lowers, less light is reflected, and the snow absorbs more of the incoming solar energy, increasing the potential for melting. Long-wavelength, blackbody radiation from the ice surface removes thermal energy from the ice, and radiation from the atmosphere and clouds adds energy to the ice. Sensible heat is transferred to the ice by warm air masses and supplies some of the energy to drive the phase transition from ice to water. Large amounts of energy, in the form of latent heat, can be transferred to the glacier surface by the condensation of moisture. A small amount of geothermal heat is transferred from the rock below the glacier. For glaciers, energy

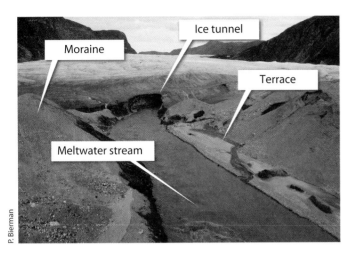

PHOTOGRAPH 9.3 Ice Sheet Margin. Ice tunnel from which large meltwater stream is emerging. Large accumulation of glacial sediment (a moraine) in the foreground was deposited by this outlet glacier draining the Greenland Ice Sheet margin east of Upernavik, west central Greenland. The flat river terrace on the right side of the channel is underlain by ice. The meltwater stream channel is between 50 and 70 m wide.

and mass balances are tightly linked because energy inputs control the melting, sublimation, and calving rates of ice [**Figure 9.4**].

Accumulation and Ablation of Glacial Ice

For mid- and high-latitude glaciers, spatial and temporal changes in glacier **accumulation** and **ablation** are seasonal and predictable. During the winter months, glaciers gain mass as temperature and incoming solar radiation are low and snowfall rates are high. During the summer months, increased solar radiation, minimal snowfall, and higher air temperatures increase ablation, decrease accumulation, and glaciers lose mass.

PHOTOGRAPH 9.4 Glacier Calving. Calving at Lamplugh Glacier, Glacier Bay National Park and Preserve, southeast Alaska. The ice cliff shown is ~30 m high.

Rates of accumulation and ablation are also tied to elevation. Because air masses cool as they are forced to rise over higher elevation terrain, the average air temperature falls between 0.6 and 1°C for every 100 m of elevation gain. This change in temperature with elevation is termed the **lapse rate**. Thus, in their upper reaches, glaciers will be cooler than in their lower reaches, more precipitation will fall as snow and less snow will melt.

Not only does temperature decrease with elevation, but precipitation tends to increase as air masses gain elevation, cool, and the water in them condenses. This **orographic effect** can lead to extremely steep gradients in precipitation. For example, Seattle, Washington, which sits at sea level, gets less than a meter of precipitation annually; in contrast, the flanks of nearby Mount Rainier, at about 2 km elevation, receive almost 3 m. The upper reaches of glaciers usually receive more total precipitation than their lowest reaches and a greater proportion of the precipitation at higher elevations falls as snow. However, in many places, the winter mass balance is close to constant with altitude; there may actually be a decrease in snow accumulation with altitude on the upper portion of glaciers due to wind-driven snow transport downslope. The summer mass balance is typically much more strongly controlled by elevation, causing strong altitudinal control on most net (annual) mass-balance curves (Figure 9.4). This seasonal disparity in mass balance occurs because in the summer, the lower parts of glaciers tend to be above freezing and thus subjected to melt much more often than the upper parts.

Elevation-dependent energy inputs and mass-balance differences result in distinct zonation of glacial process and glacier morphology. One can demarcate areas of any glacier where net mass loss occurs (the **ablation zone**) and where net mass gain occurs (the **accumulation zone**). These zones are separated by the **equilibrium line** at an altitude where, over the average year, mass losses equal gains.

The **equilibrium line altitude** (ELA) reflects the boundary between the area of net ablation and net accumulation on the glacier surface and is controlled by climate. As climate warms or snowfall decreases, the ELA rises. As climate cools, or snowfall increases, the ELA falls. The ELA can be approximated in the field by measuring the elevation of the **firn line** (the boundary between melting glacial ice in the ablation zone and firn, transformed snow, in the accumulation zone) on a glacier at the end of the melt season in early fall.

There are significant positive and negative feedbacks and interactions that change the chance of glacier survival in a warming climate. Consider the feedbacks that promote the stability of alpine glaciers as climate changes. As the ELA rises, the ablation zone extends to higher altitudes, where the average temperature is lower, limiting melt. In alpine settings, the glacier itself is likely to become more topographically confined and thus better shaded as it shrinks within the valley it occupies. Ice sheets behave very differently. Decreasing accumulation or increasing ablation lessens the ice sheet thickness and lowers the elevation

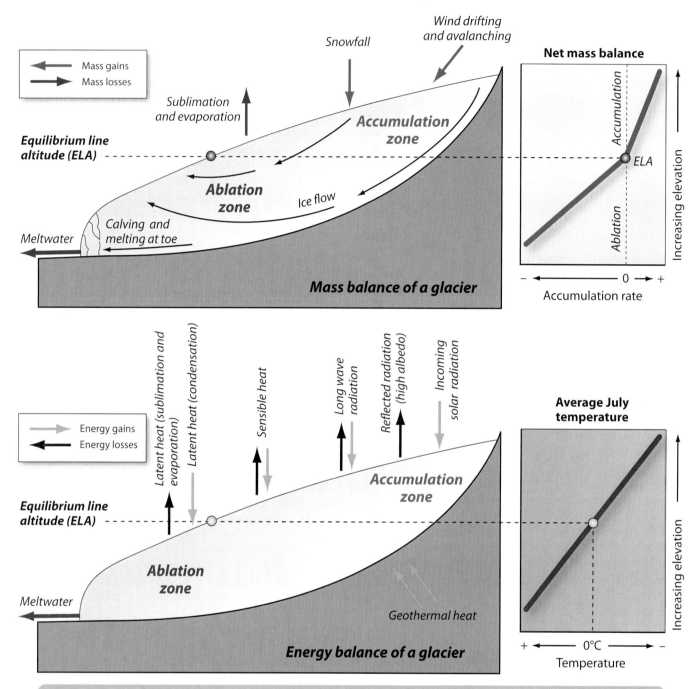

Mass balance is critical to glacier persistence and size. In the **accumulation zone**, more frozen water is deposited (by snowfall, wind drift, and avalanching) than melts away. This excess mass flows as glacial ice into the **ablation zone**, where it leaves the glacial system by melting, sublimation, and calving. In the ablation zone, there is net mass loss. The **equilibrium line** separates the accumulation and ablation zones. Its elevation is similar to the end of summer snow line and the elevation where, in the northern hemisphere, the average July temperature is about 0°C. If the ELA does not change over time, the glacier is in steady state and its mass balance is stable. However, if the ELA rises, the glacier will lose mass and retreat; if the ELA lowers, the glacier will gain mass and advance.

FIGURE 9.4 Glacier Mass/Energy Balance. A glacier's mass balance is determined both by gains of mass from snowfall and loss of mass by ablation. Energy losses and gains by glaciers are largely a function of elevation because of the correlation between elevation and average annual temperature. Arrows in the glacier show the direction of ice flow.

of the ice sheet surface, raising the temperature in the accumulation zone, thereby melting more ice. As an ice sheet begins to shrink and as the average elevation in its accumulation zone lowers, orographically driven precipitation also diminishes and the rate of ice accumulation declines, a feedback that serves to further shrink the ice sheet.

Glacier Movement

Glacial ice moves in several ways as it responds to the shear stress induced by gravity (see eq. 1.3). All glaciers move as a result of internal ice deformation, a process referred to as **ice creep** [Figure 9.5]. Some glaciers also slide at the bed/ice interface, a process referred to as **basal sliding.** Other glaciers move on slowly deforming fluidized sediment (till) between bedrock and the ice. The flow of glacier ice is laminar (like most lava) rather than turbulent (like most water).

The rate of ice deformation or the **strain rate** ($\dot{\varepsilon}$) varies with shear stress ($\tau = \rho g h \sin\theta$) and can be approximated using an experimentally derived equation known as Glen's flow law with the form of

$$\dot{\varepsilon} = A\tau^n \qquad \text{eq. 9.1}$$

where A is a rate parameter that increases with temperature, indicating that warm ice deforms more rapidly than cold ice. Laboratory and field evidence indicate that the exponent, n, has a value of about 3 for ice. Because n is greater than unity, a small change in shear stress causes large changes in strain rate. The model, derived from deforming single ice crystals in the laboratory is consistent with the observation that ice behaves similarly to a plastic material, deforming with increasing rapidity in response to increasing stress [Figure 9.6]. Evidence for ice deformation is preserved both in ice cores and at ice margins, including folded and overturned layers of sediment in glacial ice.

Indeed, it is the rapidly increasing rate of deformation with τ that limits the height and surface slope of continental

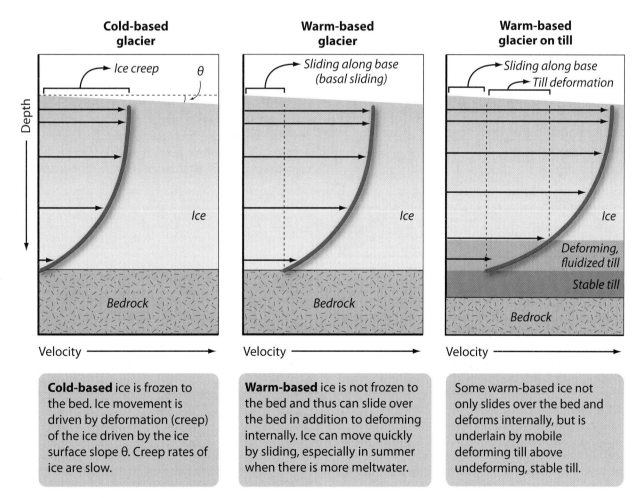

FIGURE 9.5 Glacier Velocity Profiles. Glaciers move as a result of different processes including ice deformation (creep), basal sliding, and deformation of a mobile bed. The relative contribution of each process results in velocity profiles that differ among glacier types.

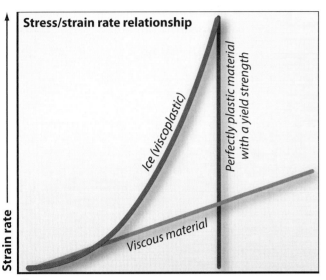

Glen's flow law describes the deformation of ice. At low shear stress, ice strains (deforms) very slowly. Because the strain rate of ice increases rapidly as the shear stress increases, ice subjected to high shear stresses approximates a plastic material. In other words, small changes in shear stress cause large changes in strain rate. Once shear stress reaches a sufficient level, the ice will strain rapidly and significant ice flow will begin. This happens when a glacier reaches a critical ice thickness and surface slope.

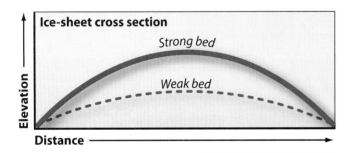

The height and surface slope of an ice sheet reflect basal shear strength. A strong bed (one that requires high shear stress to deform) will generate steep ice margins, whereas a weak bed will result in less steep ice margins. Cold-based ice, frozen to the bed, would also have steep ice margins.

FIGURE 9.6 Glen's Flow Law and the Behavior of Ice. Ice deforms very slowly under low shear stresses. Ice deforms more rapidly at higher shear stresses, behaving more like a plastic material.

ice sheets as well as the thickness and surface slope of alpine glaciers—geomorphically important characteristics because they determine the area of the landscape covered and eroded by ice. To understand why ice sheets can get only so thick, recall eq. 1.3 and consider that increasing slope and increasing thickness both increase shear stress.

As shear stress increases, ice, approximating a plastic material, deforms more quickly, moving more ice and limiting further increases in ice thickness.

Not only does the shear strength of ice matter to glacial flow speeds, the shear strength of the underlying bed, if it is deformable, also matters. In fact, the shear strength of the bed material underlying the ice largely controls the shape of the ice sheet. Weak beds, which have a low critical shear stress (τ) for deformation, result in gently sloping ice sheets. Such beds are common where ice overruns weathered material high in clay and silt or where subglacial water is prevalent. Strong beds that only deform under higher τ are found in areas of scoured rock with little basal water or deformable sediment and result in steep ice margins (Figure 9.6).

When glacial ice terminates in water, such as where ice sheets enter the ocean or alpine glaciers enter ice marginal lakes, ice thickness and water depth determine whether the ice will be grounded or whether it will float. Grounded ice maintains contact with the underlying lakebed or seabed, restricting its flow because resistance to movement (basal shear stresses) is relatively high. Floating ice is buoyed by the water below and can both flow and calve very rapidly because there is no resistance offered from the bed under the floating ice; thus, a floating ice margin can catalyze a rapid loss of ice, in some cases significantly drawing down the volume and elevation of the source glacier or ice sheet. If the floating ice does not melt quickly or if it gets pinned on islands offshore or in an embayment, it can become an **ice shelf,** a large, floating mass of glacial ice still connected to its source glacier. Most ice shelves are found around Antarctica [**Photograph 9.5**]. Ice shelves are connected to their source glaciers and lose mass by calving, sometimes in large, dramatic events. Grounded ice and ice shelves buttress ice upstream, reducing the flow rate. Most coherent floating ice is sourced from cold glaciers, since temperate ice is too weak (and usually too fractured) to hold together.

PHOTOGRAPH 9.5 Ice Shelves. Ice shelves, such as the Ross Ice Shelf, are common in Antarctica. Thin seasonal sea ice is breaking up the in foreground. The thicker ice shelf is in the distance. The ice cliff is ~50 m tall.

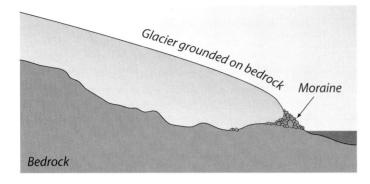

Glacier margins can be **grounded** on land or floating on water. Grounded ice forms moraines. The marginal position is often marked by a moraine and the extent of the glacier changes only slowly (meters to tens of meters per year). Grounding of ice increases resistance to flow, slowing delivery of ice to the margin.

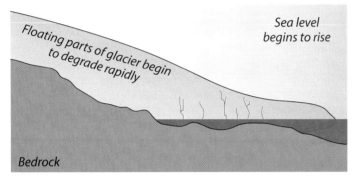

Glaciers terminating in marine environments can be affected dramatically by rising sea level or changes in mass balance. Ocean water begins to float ice that was once grounded; the ice margin degrades rapidly by calving and sheds mass in the form of massive icebergs and sediment, leaving substantial submarine deposits in temperate environments. With less grounded ice, flow speed increases, increasing the rate at which ice is delivered to the margin.

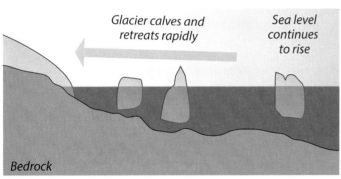

Floating ice provides no resistance to flow and ice from upstream drains rapidly, drawing down inland ice. The glacier continues calving icebergs until it again grounds on bedrock, usually at the head of the **fjord.** Such rapid calving produces a positive feedback as formerly terrestrial ice is delivered to the ocean, further raising sea level.

FIGURE 9.7 Grounded Ice Margin. Grounded ice margins tend to be stable and to retreat slowly. Sea-level rise can float grounded ice and result in rapid calving of once-stable ice masses—an important feedback that can hasten deglaciation.

Rising sea levels that floated glacier termini, increased glacier flow rates, and drew down ice sheets at the end of each glaciation contribute to the positive feedback loop that helps end glaciations. As sea level rises, many formerly grounded ice margins begin to float and calve rapidly, adding more mass to the ocean, further raising sea level and thus accelerating deglaciation [**Figure 9.7**]. The calving of the floating ice itself does not cause sea level to rise; however, calving can lead to drawdown of the grounded and land-based portions of the ice mass, which, as ice is transferred from land to sea and melts, increases the volume of ocean water, raising sea level.

The amount and distribution of subglacial water is a critical determinant of the rate of glacial sliding and subglacial erosion as well as the magnitude of **englacial** (within ice) and subglacial sediment transport. High subglacial water pressures reduce the effective normal stress on the bed, thus reducing the shear stress needed to initiate and maintain movement. Glaciers with abundant subglacial water and substantial subglacial sediment can slide rapidly on fluidized beds of saturated sediment, enhancing both the rate of sliding and of sediment deformation. Sediment is transported both along the bed and in water-filled tunnels within the ice (englacial drainages). Large amounts of subglacial water likely increase erosion rates by speeding flow, thus catalyzing rock abrasion and speeding the quarrying process (also known as **plucking**).

Glacial ice flows from the accumulation zone to the ablation zone (Figure 9.4). In the accumulation zone, the net vertical motion of ice is downward relative to the ice surface. In the ablation zone, the net motion of the ice within a glacier is upward, toward the surface. At the ELA, ice moves parallel to the glacier surface and perpendicular to the ELA. Everywhere, basal glacial ice moves parallel to the glacier bed.

A small percentage of the world's glaciers have been observed to **surge,** a phenomenon where ice flow is very rapid (meters to many tens of meters per day) for a short period of time followed by a longer, quiescent period of much slower flow. Because surges can only rarely be related to other events such as changes in climate or the hydrologic system, they are thought to represent instability inherent to the glacial system. During a surge, large amounts of deforming ice move toward the margin, which commonly advances and then stagnates after the surge. Copious amounts of meltwater are typically released during a glacial surge, consistent with the idea that surges are mediated by changes in bed hydrology, specifically hydraulic head increases that reduce resistance to basal slip. Surging glaciers are found around the world but are particularly common in southeastern Alaska and in areas underlain by soft, sedimentary rock.

In a somewhat morbid twist on glaciology and glacial flow paths, rangers in places like North Cascades National Park in Washington State, in the European Alps, and on Mount Rainier sometimes have the grim job of recovering bodies from the ablation zones of glaciers. By keeping careful records of who fell into crevasses, when they fell in, and where the crevasses were located, they can use a body's glacial journey to determine glacier flow lines and flow rates. Conversely, known glacial flow rates can help them figure out who is melting out of the ice.

Thermal Character of Glaciers

Glaciers can be characterized based on their thermal regime both at the bed and at the ice surface. **Cold-based glaciers** are largely frozen to the substrate and thus have little if any liquid water at their beds. **Warm-based glaciers** are not frozen to their bed and are characterized by moist bed conditions where liquid water is abundant, at least during the melt season.

Glacier surfaces can be characterized as polar, subpolar, and temperate. **Polar glaciers** remain below freezing and have no surface melt at any time of year. **Subpolar glaciers** have surface melt during the summer but most of the glacier remains at temperatures colder than the **pressure melting point** of ice, the temperature at which the ice/water phase transition occurs. Because the freezing point of water drops with increasing pressure, water under thick ice can remain liquid at temperatures a few degrees below 0°C (Figure 9.1). **Temperate glaciers**, with the exception of their near-surface zones during winter, have ice at the pressure melting point throughout.

Large ice sheets, such as in Antarctica, are warm-based at the center. There, ice is very thick, the pressure of the overlying ice has depressed the melting point of water, and the thick ice both insulates the base from the cold air above the ice sheet and traps geothermal heat. In Antarctica, radar remote sensing confirms the presence of liquid water at the bed of the ice sheet, revealing at least 150 lakes trapped beneath the ice. These lakes formed in sub-ice bedrock basins and can be quite large, hundreds of kilometers long with volumes up to several thousand cubic kilometers.

Water in the lakes comes from local pressure-melting at the base of the ice sheet. Similar lakes were likely present below the Laurentide Ice Sheet during the Pleistocene.

The situation in Greenland is different. The Greenland Ice Sheet is warm-based along much of its margin where mean-annual air temperatures are at or above freezing [Figure 9.8]. However, large areas of the Greenland Ice Sheet are frozen to the bed; the basal temperature is $-9°C$ at the thickest spot, the summit, where several ice cores have been drilled.

The critical factor accounting for the difference between the Antarctic and Greenland ice sheets is the effect of vertical ice motion on the basal ice temperature. Ice in Greenland flows rapidly due to high snow accumulation rates, so cold ice near the surface is advected rapidly toward the bed, cooling it significantly. Central Antarctica, by contrast, receives little precipitation and thus accumulation rates are low and the flow of ice is very slow and vertical transport of ice does little to cool the bed. Hence, the bed is melting in most places under Antarctica but not under much of Greenland.

As ice thickness and temperature change over time, so do glacier bed conditions. Some areas of Greenland that were cold-based during glacial times are warm-based today and vice versa, reflecting not only changes in mean annual air temperature, but in ice thickness and the advection of colder or warmer ice toward the bed in the accumulation zone.

The presence or absence of significant liquid water at the bed of a glacier (cold-based versus warm-based ice) has implications for the speed of glacial flow, the efficiency of glacial erosion, and thus the geomorphology of glaciated landscapes. Cold-based ice, where water is not present at the bed, flows only due to internal deformation of the ice. At low temperatures, the rate of this internal ice deformation is low (eq. 9.1).

Regions where the ice is at the pressure melting point (warm-based) are subjected to **scouring**, the removal, by ice sheets, of both rock weathered during interglaciations and fresh rock from below the weathered zone. Much of the bare rock landscape of the Canadian Shield was scoured by warm-based ice. Exposed bedrock in these areas displays **p-forms,** features such as grooves and other streamlined landforms eroded into rock by flowing ice. In contrast, landscapes once occupied by cold-based ice retain intensively weathered rock as shown by the presence of deep saprolite, bedrock tors, and weathering pits in outcrop surfaces. The best clue that ice once covered these surfaces is the presence of isolated erratics left behind when the ice melted away.

Both field observations and very old cosmogenic nuclide exposure ages (Chapter 2) suggest that many now-deglaciated Arctic uplands in Baffin Island, Scandinavia, and Greenland were once covered by non-erosive, probably cold-based ice that was frozen to the bed and slowly moving relative to the fast-moving, wet, erosive, warm-based ice in the surrounding lowlands. Over time, such a contrast in thermal regime, and thus erosion efficiency, can erode large amounts of rock in valleys, deepening fjords while leaving upland surfaces relatively unaltered.

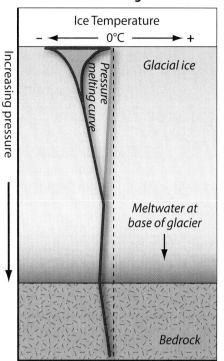
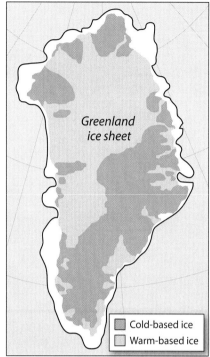

FIGURE 9.8 Cold-Based and Warm-Based Glaciers. Glaciers can be cold- or warm-based. Cold-based glaciers are frozen to their bed with ice temperatures below the pressure-melting point. Warm-based glaciers have basal temperatures at the pressure-melting point. Large ice sheets and ice caps with outlet glaciers have basal thermal regimes that differ over time and space.

Glacial Hydrology

In many ways, glacial hydrologic systems are similar to those found in karst terrain (Chapter 4), with the distinction, of course, that ice melts and carbonate rocks dissolve. Glacial ice, like dense limestone, has low intrinsic permeability, and water moves only slowly through pores in the ice. Both glacial and karst hydrology are dominated by pipe flow; most water and sediment moves through macropores such as fractures that become enlarged into tunnels. Clear evidence of pipe flow is provided by dye-tracer experiments that indicate subglacial water speeds up to 0.5 m/s, far faster than glacier flow (centimeters to meters per day). Radar and bore-hole observations of temperate glaciers further reveal large crevasses extending to within a few meters above the bed. When the summer melt season is in full swing, there is a well-connected plumbing system within most glaciers, connecting the top and sides with the bed and points in between and thus accounting for the mobility of meltwater.

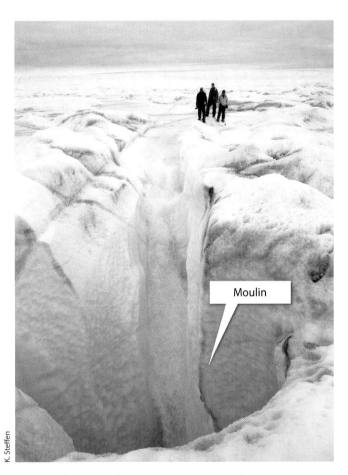

PHOTOGRAPH 9.6 **Moulin.** Large moulin on the western margin of the Greenland ice sheet brings meltwater from the surface to the interior and probably to the bed of the glacier.

Much like the disappearing rivers of karst terrain, glacial streams accumulate water from surface melt in the summer, flow over the ice, and plunge down **moulins**, vertical conduits, to become part of the englacial or subglacial drainage system [**Photograph 9.6**]. Tunnels in the ice, which would otherwise close as ice deforms around them, are kept open by advected heat as well as by frictional heating from turbulent water flow driven by head gradients within the ice-bound hydrologic system. Because water flows through glaciers in response to the pressure gradient (differences in hydraulic head), sediment and water can be transported "uphill" (with respect to the underlying terrain) and some sub-ice channels climb valley sides.

Both theory and field observations suggest a close connection between the character of the bed and the character of the sub-ice drainage network. Under most glaciers, water is present in a variety of places, including pore space in rock and sediment, films between rock and ice, cavities in the ice and rock, and large channels cut into the bed and in the ice. Where the substrate is hard and not deformable, such as over much of the Canadian Shield, drainage will organize into a few large, long-lived channels as flow concentrates. In contrast, where the substrate is soft, deformable, and saturated, there are many small, broadly distributed sub-ice channels that are ephemeral. Thus, it is no accident that water-sorted sediment deposits from former sub-ice channels are found in areas dominated by hard bedrock exposures rather than deformable sediments.

Subglacial drainage influences glacier dynamics, erosion, and sedimentation because glacial hydrology controls in large part the speed of glacier flow (consider also surging glaciers) and the propensity for glaciers to modify their beds. Warm-based ice moves more rapidly than cold-based ice because warm-based ice slides over the bed and over the deformable, wet sediment layer beneath. Subglacial water pressure reduces the effective normal force and thus the frictional resistance to shear at the bed (see Chapter 5). This increases the speed at which the bed deforms and the ice flows, at least in regions that previously were moving slowly or had limited basal water.

Water pressure in tunnels and sub-ice cavities fluctuates over time and space in response to daily and seasonal changes in meltwater production. For example, during the melt season, the speed at which ice moves toward the western margin of the Greenland Ice Sheet increases as more meltwater reaches the bed. It is likely that daily fluctuations in water pressure increase the rate of subglacial erosion by altering the stress regime on the bed, fracturing rock and entraining sediment, and perhaps by inducing **cavitation**, the formation and violent collapse of small gas bubbles that can fracture even the hardest rocks because of the high pressures this stress exerts on the rock surface. This process is similar to that contributing to erosion in bedrock river channels (Chapter 6).

Discharge from glaciers peaks in late summer, when the ice has warmed from long summer days and above-freezing air temperatures [**Photograph 9.7**]. Discharge from glaciers is usually lowest in the morning, reflecting less melt overnight when the inputs of solar energy and sensible heat are low, even during midsummer in polar regions. Discharge in streams sourced from glaciers typically peaks in the late afternoon, reflecting the effectiveness of midday melting and some lag time for meltwater to move toward the glacier margin. Early European explorers in central Asia, far from the glaciated peaks, learned the hard way about meltwater pulses that reached them out of phase with midday heating, often after they had camped near what seemed to be nearly dry streambeds. If you need to cross a glacial outwash stream during the midsummer melt season, it is best to do so in the morning.

The amount of meltwater reaching the bed of ice sheets and glaciers is a critical control on both glacial erosion and ice flow and may well be one of the most important ways by which climate change influences the distribution and stability of glaciers. For example, increasing summer season melt along the margin of the Greenland Ice Sheet has expanded the number and size of meltwater lakes on the surface of the ablation zone [**Photograph 9.8**]. Some of these lakes drain catastrophically through fractures in the ice and in no more than an hour or two deliver large amounts of water directly to the bed of the ice sheet.

PHOTOGRAPH 9.7 **Seasonal Changes.** Compare these two images of the same location on the western margin of the Greenland Ice Sheet where ice is flowing from left to right. In the left image, taken May 2011, the ice is still snow-covered and the outwash stream is so small that one could step over it. In July 2008, at the height of melt season, it was very different. The outwash stream was roaring and all of the snow had melted, exposing sediment-rich basal ice at the margin of the ice sheet. Field of view is about 1 km wide.

Presumably, drainage occurs when a threshold is reached, allowing the water pressure to wedge open fractures until the permeability of the ice suddenly increases. Added water at the bed increases the hydraulic head, decreases the effective normal force, and speeds glacier flow by lowering shear strength of the bed/ice interface.

Occasionally large floods, termed **jökulhlaups**, from the Icelandic words jökull (glacier) and hlaup (burst), issue from beneath ice sheets, glaciers, and glacial deposits that dam lakes. Large jökulhlaups can have exceptional discharges, on the order of 50,000 m³/s, within the range of flow of the Amazon and Mississippi rivers. Such floods transport massive amounts of sediment and have buried outwash stream valleys beneath tens of meters of sand and gravel. In Iceland, jökulhlaups are frequent due to the locally high heat fluxes from the volcanic terrain that melt ice at the base of ice sheets and glaciers. Sub-ice eruptions that melt glacial ice can trigger catastrophic releases of water. Many of Iceland's outwash plains, termed **sandurs**, are formed in large part by jökulhlaups. Jökulhlaups also occur through the rapid release of water stored at the bed of the glacier or failure of glacial dams. Failures of lakes impounded by glaciers or glacial deposits are common in high mountain areas such as the Himalaya and the Andes of South America. Dangerous outburst floods may increase in frequency as global warming causes glaciers to retreat and generates more meltwater. The great Missoula floods that scoured eastern Washington State at the end of the Pleistocene (see Digging Deeper, Chapter 13) and even larger glacial floods in central Asia were jökulhlaups caused by the failure of massive ice dams.

Subglacial Processes and Glacial Erosion

There are several processes by which glaciers erode rock and incorporate unconsolidated material. All are related through the distribution of pressure, water, and connected cavities at the base of the glacier, the interface between the ice and the bed.

The bed of a glacier is rough on a variety of scales generally due to bedrock topography. As warm-based glaciers slide over their beds, cavities can form on the down-ice side of bedrock bumps. **Regelation** can be an important means by which glacial ice moves around small-scale (less

PHOTOGRAPH 9.8 **Meltwater Lake.** Meltwater lake on the surface of the Greenland Ice Sheet near Ilulissat. Such lakes form each summer and can add large amounts of meltwater directly to the bed of the glacier when they drain catastrophically through fractures. Lake is several kilometers wide.

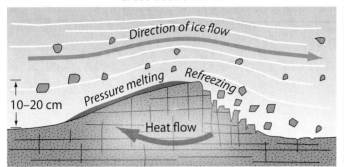
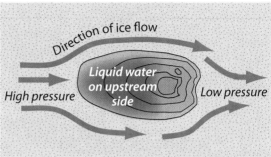

Regelation occurs when ice flows over rock obstacles on the glacier bed. Pressure-melting of ice up-glacier of an obstacle releases water, which then travels down ice to areas of lower pressure and consequently refreezes. This refreezing may incorporate debris and enrich the remaining solution in elements such as calcium, which precipitates and forms calcium carbonate coatings on subglacial rock surfaces. Regelation facilitates the movement of warm-based glacial ice over rough bedrock beds.

FIGURE 9.9 Regelation. Regelation is an effective means of ice flow over rock at the base of glaciers as meltwater moves around small basal roughness elements, on the scale of centimeters to tens of centimeters, and refreezes on the down-ice side.

than tens of centimeters) bumps in the bed. Regelation describes the pressure-melting of ice upstream of an obstruction on the bed where pressures are higher, the movement of that water downgradient, the transfer of heat through the bedrock bump, and the subsequent refreezing of water (and attendant release of latent heat) downstream of the bump in the lower-pressure zone [**Figure 9.9**]. The refreezing process can incorporate debris into the ice as well concentrate dissolved elements in the remaining solution. As a result, some dissolved elements precipitate, such as calcium carbonate that forms rinds on subglacial bedrock outcrops [**Photograph 9.9**].

At a larger scale than regelation, roughness or bumps on the bed below the glacier produce cavities in the ice, changing the distribution of pressure on those bumps. The presence of large sub-ice cavities shifts the load of the ice to bedrock outcrops, especially during times when water pressure below the glacier is falling. Increased pressure on the outcrops can lead to fracturing of the rock. Removal of the fractured rock by entrainment in ice is accomplished as ice flows around and drags the loosened material. Some debris may also freeze onto the base of the glacier. Field evidence for bedrock **quarrying** is clear—large blocks of rock are conspicuously removed along joint planes from outcrops [**Photograph 9.10**]. Most quarrying occurs in low-pressure regions like the down-ice side of hills because large cavities form most readily in such places.

The effectiveness of glaciers as agents of erosion varies spatially, reflecting differences in the intensity and character of erosion processes such as abrasion, quarrying, and glaciofluvial activity as well as the character of the rock below the ice, particularly the spacing of joints. Subglacial and ice-marginal fluvial erosion can be significant in areas of copious meltwater, as evidenced by deep potholes and incised marginal channels once covered by ice but now revealed by deglaciation. Glacially mediated incision can cause erosion near the glacial margin; for example, valley deepening by glaciers may oversteepen adjacent valley walls, promoting mass movement by rockfall onto the ice surface—the source of much **supraglacial** (on top of the ice) debris (see Digging Deeper).

Much of what we know quantitatively about rates of glacial erosion has been inferred from measuring the volume

PHOTOGRAPH 9.9 Subglacial Carbonate Precipitation. Subglacial calcium carbonate precipitates that grew under the ice of Tsanfleuron Glacier, Switzerland. Pocket knife for scale; blade points in direction of ice flow.

PHOTOGRAPH 9.10 **Glacial Bedrock Quarrying.** Jointed and glacially scoured bedrock on western Greenland is ready to be quarried by the next ice advance. This surface was covered by the Greenland Ice Sheet during the Little Ice Age.

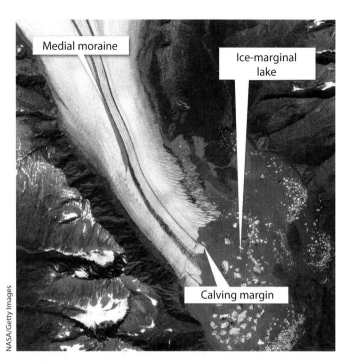

PHOTOGRAPH 9.11 **Ice-Marginal Lake and Rock Flour.** This ice-marginal lake, bordering the calving margin of Bear Glacier on the Kenai Peninsula of Alaska, is blue-green in this natural-color image because very fine rock flour, material abraded by glaciers off outcrops at the base of the ice, is suspended in the lake water. The color of the water depends on the size and abundance of suspended particles. Note the medial moraine.

of sediment carried by glaciers and deposited outside glacial margins and from sediment flux measurements made downstream of glacial margins. Recent work has used tunnels, bore holes, and other means of both observing active processes and placing instruments within and under active ice to better understand glacial erosion processes. For example, time-lapse cameras have been placed in sub-ice cavities, and well-characterized blocks of rock have been placed under flowing glacial ice to measure erosion rates directly over short periods of time.

Glaciers can be effective agents of erosion although rates of glacial erosion vary over time and space. Warm-based glaciers in areas with high precipitation rates (such as southeastern Alaska) erode rock at rates between 1 and 1000 mm/yr with the higher rates over short time frames of years to decades and the lower rates averaged over the last million years. Glaciers in areas with less precipitation erode more slowly, as little as 0.1 mm/yr. Cold-based ice, which is frozen to the bed, does little to erode and entrain subglacial material. Cold-based glaciers preserve the landscapes they cover.

Glacial abrasion produces some of the most distinctive and characteristic evidence of glaciation including **striations, grooves, glacial polish, loess,** and **rock flour.** Rocks of all sizes form tools in the ice that abrade the rock below as they move across the bed while the warm-based ice or mobile sediment layer above is slowly deforming. The deforming ice above moves these erosive tools as it shears. One result of this abrasion is **rock flour,** the finely ground, silt-sized, rock fragments that colors streams issuing from glacial margins a distinctive milky blue-green [**Photograph 9.11**]. Rock flour is important geomorphically and to society. It is the source of much of the fine sediment that, after being carried away from active ice margins by wind and later deposited, forms regionally extensive blankets of **loess** (fine, wind-blown sediment, derived from the German word for "loose"). Loess underlies much of the world's most productive farmland such as the Palouse of eastern Washington State, the midwestern United States, the Russian Steppe, and the Chinese Loess Plateau. Once deposited, loess can erode easily if vegetation removal exposes it to the direct erosive action of water or wind.

Other evidence for abrasion includes smooth **glacial polish** [**Photograph 9.12a**] that typifies fresh outcrops of rock emerging from under glacial ice. **Glacial striations** [**Photograph 9.12b**] are the fine grooves and scratches left on polished rock surfaces by the movement of debris-rich basal glacial ice or deforming basal sediment. **Crescentic gouges** (large semicircular cracks on the rock surface) or chattermarks also form from the pressure of large abrading clasts on the bedrock below the ice [**Photograph 9.12c**]. **Glacial grooves** of varying scales can be found on many outcrops aligned with the paleo ice-flow direction [**Photograph 9.12d**]. Grooves, striations, and gouges are routinely used as flow direction indicators for now-vanished ice sheets. These erosive forms are distinctive; some have survived burial over hundreds of millions of years to document ancient glaciations (such as the Permian glaciation of Australia, Photograph 9.1). Together, quarrying and abrasion can yield streamlined forms indicative of glacial erosion known as **roches moutonnées** (French for "rock sheep"),

(a)

(c)

(b)

(d)

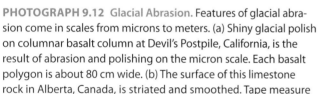

PHOTOGRAPH 9.12 Glacial Abrasion. Features of glacial abrasion come in scales from microns to meters. (a) Shiny glacial polish on columnar basalt column at Devil's Postpile, California, is the result of abrasion and polishing on the micron scale. Each basalt polygon is about 80 cm wide. (b) The surface of this limestone rock in Alberta, Canada, is striated and smoothed. Tape measure provides scale. (c) Crescentic gouges on polished bedrock covered by the Greenland Ice Sheet during the Little Ice Age. Ice flow was in the direction that the shovel handle is pointing, away from the photographer. (d) Glacial grooves in Manhattan Schist, Central Park, New York City, are a few tens of centimeters deep. Flow was parallel to the direction of the grooves.

with gently sloping, smooth, abraded up-ice sides and steeper, plucked down-ice sides [**Photograph 9.13**].

Glacial erosion modifies valleys into distinctive shapes (see Digging Deeper). Ice flowing out of cirques moves down V-shaped stream valleys, preferentially eroding and widening lower valley walls, thereby leaving behind U-shaped valleys [**Photograph 9.14**]. In drainage basins where small tributary glaciers flow into larger trunk glaciers in major valleys, these tributary glaciers are significantly less efficient at eroding rock than the larger trunk glaciers. When the ice melts away, the tributary valleys are left hanging—sometimes hundreds of meters above the floor of the main valley (Photograph 7.12b).

The most dramatic example of glacial erosion is the excavation of deep troughs, known as glacial valleys inland and **fjords** along coastlines where they are at least partially filled by seawater. The location of many fjords appears coincident with areas of increased bedrock fracture density (the result of tectonic forces in the past). Fjords occupy valleys first deepened by fluvial then glacial erosion. They are most common along wet, windward, maritime coastlines where large amounts of ice move through glacier systems, the result of high rates of precipitation and glacier activity. The physical properties of ice makes glaciers an efficient eroder of fractured rock, creating deep fjords and valleys by plucking jointed rock in tectonically inactive cratonic settings such as Norway and Greenland.

Fjords provide a good example of positive feedback. Ice concentrated in valleys is thicker, warmer, and thus flows faster than ice on adjacent highlands, increasing the efficiency of valley-bottom abrasion and quarrying processes. The thicker ice increases the pressure on the bed, lowering the pressure melting point and making it more likely that ice

PHOTOGRAPH 9.13 Roche Moutonnée. Roche moutonnée shows ice movement from right (smooth slope) to left (plucked face) near Lac d'Emosson in the western Alps, Switzerland (person to the left for scale). The roche moutonnée is several meters tall.

elevates it, has been informally termed the **glacial buzzsaw.** While DEM analyses suggest that the glacial buzzsaw mechanism appears relevant in mountain ranges around the world, including the Washington Cascades, the South American Andes, and the Sierra Nevada of California, the buzzsaw is not perfect; the elevation of some mountain peaks extends well above the snowline.

Glacial Sediment Transport and Deposition

The sediment carried by glaciers is transported within (**englacial**), on (**supraglacial**), and under (**subglacial**) the ice before it is eventually deposited on the landscape. The proportion of material transported in, on, and under ice differs among glaciers and depends at least in part upon the underlying landscapes they are eroding. Some glaciers, particularly alpine glaciers incised deeply into weak rocks, carry large amounts of rockfall material supraglacially because rockfall brings in large volumes of debris from above the glacier; indeed, so much material may cover the glacier from rockfall that ice is barely if at all visible [**Photograph 9.15**]. Conversely, large ice sheets that overwhelm topography carry little if any material on their surface. Rather, ice sheets carry most sediment either in the lowermost basal ice or below the glacier in mobile beds of sediment. During transport at or near the bed of a glacier, sediment is broken down or **comminuted** by interaction with nearby grains or the rocky bed below the ice.

Subglacial Sediments and Landforms

The deforming bed and abundant liquid water below warm-based glaciers create diagnostic sediments and landforms that can be preserved when glaciers and ice

in the fjord is warm-based, further enhancing the efficiency of erosion. As erosion is concentrated in the troughs, ice there becomes increasingly efficient at erosion, further deepening the troughs, making a positive feedback loop. Some fjord troughs are so deep they extend well below even the lowest sea levels of the Quaternary, those coincident with past glacial maxima, when ice sheets were at their largest.

The limited relief above numerous glacial cirques high in many mountain ranges, the parallel along-range and cross-range trends in ELA and summit elevations, and the steep topography that such glacial activity creates led to the theory that glacial activity limits the elevation of mountain ranges by chewing away at cirque headwalls. The idea that glacial processes efficiently remove rock elevated some distance above the ELA, and can thus remove material at about the rate that tectonics and/or isostasy

PHOTOGRAPH 9.14 Glacially Eroded (U-shaped) and Fluvially Eroded (V-shaped) Valleys. (a) U-shaped glacial valley near Pangnirtung village on Baffin Island, Canada. (b) V-shaped fluvial valley in Cantwell, Alaska.

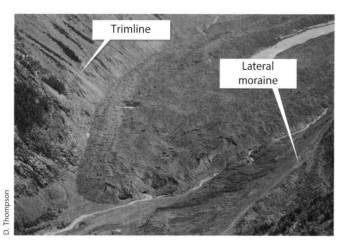

PHOTOGRAPH 9.15 Glacier Ablation Zone. The ablation zone of the debris-covered Emmons Glacier in Mount Rainier National Park, Washington State. High debris load results from rapid weathering of weak volcanic rocks. Most of the debris cover is the result of a single rockfall event in 1963 from nearby Little Tahoma Peak. Recent retreat of this glacier has isolated a lateral moraine on the right, leaving a trimline (area cleared of weathered material by glacial abrasion, often a zone of little or no vegetation) on the cliff to the left. Field of view is several kilometers wide.

sheets vanish. The most ubiquitous sediment left by glaciers is **till** [Photograph 9.16], a specific type of **diamicton** (any unsorted, unstratified deposit). Till is deposited both from below the ice (**lodgment** or **basal till**) and from above the ice as ice melts away (**ablation till** or **melt-out till**). Lodgment till tends to be highly compacted (i.e., low pore space, high density) and may show preferred clast orientations or **fabric**, as well as some fissility, a tendency to split along preferred directions of breaking or weakness, the result of emplacement under the effective pressure of overlying ice (the weight of the ice minus the subglacial water pressure). Ablation till tends to be more heterogeneous and less dense because it was not compacted by the weight of overlying ice. All tills may contain pockets of better-sorted, water-washed material.

Till plains cover large areas that were occupied by continental ice sheets during the Pleistocene, including much of northern North America, Europe, and Asia. Formerly glaciated terrain usually contains erratics, boulders transported from afar on or in glacial ice that were stranded when the ice melted away (see chapter-opening photo). Erratics are often easily recognized because they are different from local rock types or have been weathered very differently than the bedrock on which they sit.

Ice-Marginal Sediments and Landforms

Ice margins are geologically complicated places where numerous different geomorphic and sedimentologic processes occur simultaneously. Thus, the texture and sorting of sediment deposited at and near ice margins varies greatly over short distances and ice-marginal sediment is often

PHOTOGRAPH 9.16 Till. Unsorted, consolidated lodgment till in northern Vermont holds a near-vertical slope. The till is exposed by a landslide where the Huntington River undercuts the bank. Width of view is about 10 m.

disturbed by faulting and folding related to the melting of buried ice blocks, overrunning by advancing ice, and/or by collapse after the ice melts away. Much of the heterogeneity in the ice-marginal environment is due to the interplay of sediment deposition directly from the ice, from meltwater, and from mass movements of sediment and sediment-laden ice. A variety of **ice-contact landforms** are created at and near ice margins. These are known by a variety of different names and are created by several different processes.

Adjacent to former ice margins, one often finds small depressions that are nearly circular in plan view. These **kettles** [Photograph 9.17] result from subsidence of glacial sediments where ice blocks, buried during rapid, ice-marginal sedimentation, melt out, leaving large depressions in the land surface that may fill with water if the water table intersects the depression.

If sediment fills voids in the ice and that ice later melts away, then the sediment will form a high point on the landscape. If the deposit is circular in plan form, it is referred to as a **kame** [Photograph 9.18]. Kames contain both sorted sediment, deposited by flowing water and unsorted material, either till deposited by ice or debris flow sediment originating from the ice surface. **Kame terraces** are benches of fluvial sediment deposited by ice-marginal streams; these streams once flowed between the ice and the valley walls. Where ice-contact deposits are elongated perpendicular

Glacial Sediment Transport and Deposition

PHOTOGRAPH 9.17 Kettle Pond. This kettle pond (~20 m wide) is forming as an ice block melts below a depositional terrace, downstream of the Hugh Miller Glacier, Alaska.

PHOTOGRAPH 9.19 Esker. A small, sinuous esker is all that is now left of a glacial tunnel once filled by meltwater. The esker was deposited beneath the western Greenland Ice Sheet during the Little Ice Age.

to the direction of ice flow, they have been interpreted as **crevasse fills,** the result of sediment-carrying meltwater filling open crevasses. Crevasse fills tend to have more complex patterns in plan view than kame terraces. In all of these deposits, the presence of ice is indicated by faulting and partial collapse caused by the ice beneath and alongside the deposited sediment melting away.

Eskers are elongate, sinuous ridges composed of water-laid, sorted sediment capped and underlain in many cases by till [**Photograph 9.19**]. Eskers are thought to be remnants of subglacial drainage systems, the traces of sand and gravel-filled tunnels that once drained meltwater and sediment near now-vanished ice margins. Eskers often terminate in sediment fans or deltas. Because eskers would be destroyed by significant ice movement, their length and presence implies an immobile or slowly moving zone of ice at the glacial margin, most likely an area where the ice was too thin to deform rapidly. Eskers may also occur where ice flow and subglacial water flow were parallel.

Eskers can disregard surface topography, which makes sense when one considers subglacial water flow does not respond to bed topography but rather hydraulic head gradients in a now-vanished ice sheet.

Glacially Related Sediments and Landforms

Glacially related sedimentation and landforms are not restricted to areas under or near the ice; the influence of glaciers, particularly continental ice sheets, can extend many kilometers from their margins. Glacial sediments are carried from the ice margin by water, wind, and floating ice.

Flowing meltwater transports sediment away from glacial margins generating **outwash plains.** These gently sloping surfaces, underlain by sand and gravel, head at former ice margins [**Photograph 9.20**]. Outwash-laden streams may occupy preexisting drainages, in which case they will deposit fill terraces that will usually be incised after glacial retreat when the sediment load declines. Outwash plains are often pitted with kettles, closed depressions reflecting the presence of now-melted ice blocks. Much of Cape Cod, in eastern Massachusetts, is a pitted outwash plain extending away from a moraine and indicating the former margin of the Laurentide Ice Sheet.

Glacially dammed lakes are common in hilly and mountainous terrain where tongues of ice disrupt drainages and impound both meltwater issuing from the glacier and rainwater. The water, unable to leave the drainage basin by the normal route, backs up until it is high enough to flow though a higher **spillway** (gap) and exit from the watershed. For example, a tongue of the Laurentide Ice Sheet blocked the northern Champlain Valley in Vermont, forcing water that would have flowed north into the St. Lawrence Seaway south into the Hudson River drainage. Some glacial lakes drain catastrophically over or under the ice itself. For example, the Channeled Scablands, deep and intricate drainage ways cut through loess and basalt

PHOTOGRAPH 9.18 Kame. Eroding till and ice-contact sediment exposed in kame near North Adams, Massachusetts.

PHOTOGRAPH 9.20 Outwash Plain. This outwash plain is fed by a meltwater stream (flowing from left to right) from the Greenland Ice Sheet margin east of Upernavik. Gray glacial moraines border the white ice on the left side of the image. A debris fan (paraglacial sedimentation) has formed below the cliffs, which expose flat-lying basalt over deformed gneiss. Glacially streamlined rock is visible in the foreground. Valley is about 2 km wide.

PHOTOGRAPH 9.21 Ice-Contact Delta. An ancient ice-contact delta (at the right end of the lake) now stands isolated in the Holger Danskses Briller Valley, Greenland. During the last glaciation, ice filled the valley (today occupied by the lake) below the cliffs on the left of the image. The field of view in the foreground is several kilometers wide.

in eastern Washington State, are the result of failure of the ice dam that impounded glacial Lake Missoula (see Digging Deeper, Chapter 13). Evidence for similarly massive glacial floods has been found in Siberia, Tibet, the U.S. Midwest, Alaska, and in the English Channel.

Glacial lakes are also held behind dams formed by terminal and recessional moraines deposited during glacier retreat. In the Connecticut River Basin, 300-km-long glacial Lake Hitchcock was impounded by a dam of glacially derived sediment in southern Connecticut. The lake lasted for more than 5000 years. Water flowing into this and other ice-marginal lakes deposited deltas graded to paleolake levels. With the lakes long gone, these glacial-age deltas and their well-sorted sand and gravel are left high and dry above the present-day landscape [**Photograph 9.21**]. By measuring the elevation of abandoned delta topset/foreset contacts, and by mapping the distribution of fine-grained sediment deposited in the deeper sections of these glacially dammed lakes, one can reconstruct the extent and elevation of now-vanished water bodies. Glacial lake deltas are mined intensively for the gravel they contain. Such deposits of readily accessible sand and gravel are economically important in formerly glaciated regions.

Distinctively layered fine-grained deposits, indicative of quiet water with abundant sediment, are commonly laid down in glacier-fed and glacier-dammed lakes. Such deposits commonly display rhythmically bedded couplets of finer and coarser layers, termed **rhythmites** (Photograph 2.1). These couplets may represent annual layers known as **varves,** an assumption that can, in some places, be proven by other dating methods. During varve deposition, fine material settles out when the lake is iced over during the winter and sediment loads are diminished. The coarse material is deposited by currents during the summer when sediment loads are higher and fine material is kept in suspension by turbulence and higher-energy flows. Before the advent of radiocarbon dating, varved glacial sediments were meticulously counted, their thicknesses measured, and the results correlated between outcrops (much like tree rings, Chapter 2) in order to build composite sections and thus estimate the timing of glacial lake formation and the rate of ice-sheet retreat. Rhythmite and varve sequences deposited in proglacial lakes provide important paleoclimatic records. The fine sediment deposited at the bottom of such glacial lakes provided raw material for valley-bottom brickyards.

Glacial lake sediment commonly overlies till, indicating that the retreating ice margin directly bordered a lake (Photograph 2.1). In that case, sediment issuing from tunnels in the ice would be deposited below the water as a subaqueous fan. Indeed, some of the most productive aquifers in New England are in deep mountain valleys where glacial gravels are capped by glacial lake silt and clay. **Dropstones,** anomalously large rocks found isolated in the deposits of fine-grained glacial lake sediment, provide direct evidence for the delivery of sediment-laden, floating ice to water bordering ice margins. In ocean sediments, such dropstones are referred to as **ice-rafted debris** (IRD) and are used as an indication that glaciers were discharging icebergs directly to the ocean. The presence of isolated IRD deposits in marine sediments has been used to determine that limited glaciation began in the Northern Hemisphere almost 40 million years ago. Much more extensive deposits of IRD indicate that continental ice sheets, including the Laurentide, began forming and discharging icebergs to the ocean about 2.6 million years ago [**Photograph 9.22**].

Glacial Landscapes, Landforms, and Deposits

Glaciers leave behind a characteristic, and in many cases diagnostic, set of depositional and erosional landforms, some of which are unique to glacial settings. Glacial landforms were first recognized in alpine settings and later this knowledge was applied to understanding the landforms of continental ice sheets.

Glacial landforms and sediments are formed and deposited under ice, at the ice margin, and distal from the ice [Figure 9.10]. It is important to note that most glacial sediment is deposited in the ablation zone where the ice is melting and releasing material entrained up-glacier. Landform complexes in alpine and continental glacial settings differ in type and scale; thus, we discuss them separately even though there are clear parallels in ice mechanics and process that cause alpine and continental ice to leave behind many similar landforms.

Landforms of Alpine Glaciers

Three closely related erosional landforms—the **cirque**, **arête**, and **tarn**—typify valleys in alpine glacial landscapes

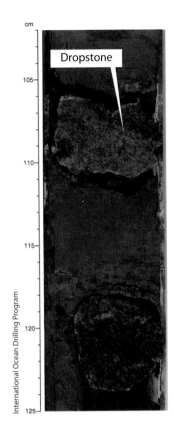

PHOTOGRAPH 9.22 Dropstones. Dropstones (ice-rafted debris) in the core of sandy marine sediment collected by drilling off the coast of southern Greenland. The scale on the left side is in centimeters.

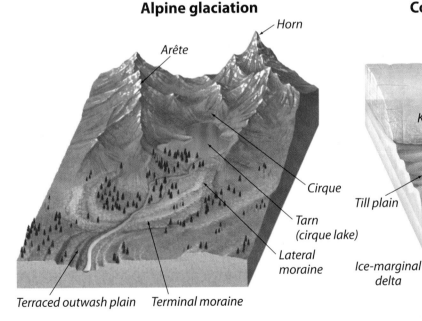

Alpine glaciation leaves a distinctive character to the landscape, including erosional landforms such as **cirques, arêtes, tarns,** and **horns,** as well as depositional features including both lateral and terminal **moraines** and **outwash terraces.** A complex series of moraines can record episodes of glacial advance and retreat.

Continental ice sheets leave distinctive landforms, including moraines, ice-marginal lake deposits such as **deltas, outwash terraces, kames, eskers,** and **kettle ponds. Till plains** cover much of the area once occupied by continental ice; in other areas, streamlined forms such as **drumlins** dominate. Outwash plains and terraces form outside the ice margin.

FIGURE 9.10 Alpine and Continental Glacial Landforms. Characteristic sets of landforms occur at and near retreating continental and alpine ice margins depending on regional topography and the abundance of sediment.

[Photograph 9.23]. Cirques are half bowl-shaped forms cut into mountainsides and backed by a steep **headwall,** the near-vertical face of rock defining the upper limit of the cirque. The processes creating the cirque forms we see today are complex. Basal erosion by ice-driven or meltwater-driven processes must keep up with (and originally exceed) the cirque retreat rate due to freeze thaw and mass wasting. Cirques are deepened by the flow of alpine glaciers but freeze-thaw loosening of blocks on the headwall and subsequent rockfall are also important erosion processes. Most cirques are multigenerational, having been occupied and modified by ice during many glaciations and weathered during the intervening interglacial periods. Many cirques, deepened into closed basins by flowing ice, now hold small lakes, known as tarns.

Arêtes are knife-edge ridges formed as two cirques or parallel U-shaped valleys erode toward one another, leaving only a steep-sided, narrow ridge of rock between. If headward erosion of cirques continues, the upland surrounded by headward-retreating cirques with intervening arêtes may be eroded into an isolated peak of rock known as a **horn.** Alpine valleys with roches moutonnées, rock steps, and irregular profiles, some partially filled with sediments, lie between cirques and the lower reaches of alpine valleys that are lined with lateral moraines.

Terminal moraines or **end moraines** are piles of debris that accumulate at the distal end of glaciers as sediment melts out in the ablation zone and as basal ice deforms and in some cases is thrust up from the bed. The size of a terminal moraine depends on the length of time the terminus remained at one location, the amount of debris in the glacier, and the flux of debris-bearing ice to the glacial margin. Terminal moraines are breached by outwash streams that carry away meltwater from the glacier. **Lateral moraines** form alongside the ablation zones of alpine glaciers as debris both rolls off and melts out of the ablating ice. **Medial moraines** are concentrations of debris where two tributary glaciers merge into one larger glacier (Photograph 9.11).

PHOTOGRAPH 9.23 Common Alpine Glacial Landforms. Cirque, tarn, and arête along the cirque headwall, Longs Peak, Rocky Mountain National Park, Colorado. Glacial features here are cut into metamorphic bedrock and granitic intrusions. Field of view is several kilometers wide.

PHOTOGRAPH 9.24 Glacial Moraines. Little Ice Age moraine complex deposited during the last 1000 years in Vilcabamba, Peru, damming small lakes. The moraine may still be ice-cored. Ice flowed from right to left.

Medial moraines can be preserved as low-standing linear ridges after the ice melts away, although often, they are not preserved at all.

Alpine moraines are built largely of till delivered to the glacier margin, both in basal ice and supraglacially [Photograph 9.24]. The presence of a moraine indicates stability of the ice margin for a period of time sufficient to deliver a morphologically significant pile of debris—rapidly melting ice does not leave well-defined moraines but rather a featureless **till plain.**

Geomorphologists use the presence and location of moraines to understand the extent of prior glaciations and thus to infer past changes in climate (both precipitation and temperature) over time. The outermost moraine reflects conditions most favorable to the advance of ice, high ice accumulation rates and/or cold temperatures. Progressively younger moraines, usually found inside the outermost moraine, provide evidence for later climates less favorable for glaciers. Rarely, alpine moraines may crosscut one another; in that case, crosscutting relationships can be used to determine the relative ages of the moraines. The alpine moraine record is incompletely preserved because younger advances of greater extent than older advances destroy older moraines and thus the geomorphic evidence of those previously expanded glaciers.

Various techniques, including relative weathering, soil development, carbon-14 dating of organic material, and cosmogenic nuclides are used to estimate the age of moraines. Because many alpine moraines are narrow, steep-sided features deposited at least in part on ice, they can erode rapidly, losing mass from their crests and confounding accurate dating by techniques, like cosmogenic nuclide measurement in boulders, that date surface exposure.

Landforms of Ice Sheets

Ice sheets create many of the same landforms as alpine glaciers, such as moraines and till plains, but they also

PHOTOGRAPH 9.25 **Drumlin.** Drumlin covered by a housing development in Friedrichshafen Raderach, Germany. The steep, up-ice side is upper right and the less-steep tail is lower left. Arrow shows ice flow direction.

create forms that are distinctive and not found in alpine settings. Mapping these landforms delineates the extent of now-vanished ice sheets and provides clues about their basal conditions.

Drumlins are streamlined, distinctively glacial landforms. These elongate hills are found in many low-relief areas formerly overridden by continental ice sheets [**Photograph 9.25**]. Famous drumlin fields include those in Wisconsin, upstate New York, and Ireland. Some drumlins are rock-cored, whereas others are made entirely of sediment including both unsorted till (common) and water-washed, well-sorted sand and gravel (less common). Drumlins range from hundreds of meters to as much as several kilometers in length, and are usually hundreds of meters in width and tens of meters in height. Drumlins are oriented in the direction of ice flow and are usually found clustered in drumlin fields where hundreds to thousands of individual drumlins may be preserved. They can be steep on the up-ice side and taper more gently in the down-ice side. The preferred orientation with glacial flow suggests that drumlins are sculpted as the deforming bed of the glacier interacts with materials below it—both rock and preexisting sediments.

When ice sheets cover large continental areas, they change the arrangement of drainages, both by blocking river outlets and by changing the slope of the land via transient isostatic changes in land-surface elevation. The topography of the ice sheet affects head gradients within the ice and therefore the flow direction of meltwater. Ice sheet control of surface drainage leads to a variety of landforms related to surface water hydrology. These include marginal lakes dammed by glacial ice and river channels cut into rock by ice-marginal drainages pinned against valley walls by ice. Such channels are often located above present-day valley bottoms and thus contain little or no water today. Streams in these large channels are far smaller than one would expect given the size of the valley that they occupy and are termed **underfit**.

In some areas, ice sheets and ice fields were not thick enough to cover the landscape completely. Mountain peaks and high plateaus standing above the ice as islands are known as **nunataks**. Ice-free nunataks provided **refugia**, places where plants and animals continued to live during glacial times. When glaciation waned, these refugia provided a source of seeds and animals to repopulate and revegetate the adjacent landscape.

Ice sheets, ice caps, and ice fields left moraines indicative of the their former extent; for example, Long Island and parts of Cape Cod are moraines of the Laurentide Ice Sheet. Such moraines are composed of a mixture of till and outwash sediment and tend to be broader and less steep than those of alpine glaciers. Similar to alpine glacial moraines, older moraines are often overrun and destroyed by later ice advances. Early workers, who studied glacial deposits on land, did not recognize that older moraines and till had been removed and thus believed that there were only four major terrestrial glaciations in the Pleistocene—based on the relative ages of continental ice-sheet deposits. Analysis of deep-sea sediment cores has shown convincingly that there were several dozen major glaciations—the terrestrial deposits of which have largely been erased by a few of the most extensive ice advances.

Geomorphic Effects of Glaciation and Paraglacial Processes

During each glaciation, global climate cooled, ice sheets expanded, and sea level fell as water once in the oceans was taken up and stored in the ice on land. The expansion of ice sheets and the general global cooling altered the operation of the climate and ocean system throughout the whole planet, not just on and around the glaciated regions. The rate of geomorphic processes and the distribution of landforms around the world responded to glacial/interglacial changes in temperature, precipitation, and the resulting distribution of plants and animals.

The legacy of glaciers on our planet is far reaching. For example, alpine glacial erosion likely controls not only the elevation of mountain ranges but, at least in part, the rate of rock uplift. In heavily glaciated mountain ranges, where erosion rates are high, isostatic compensation (Chapters 1 and 12) brings rock toward the surface, partially making up for mass loss by erosion. In glaciated regions, such as large portions of the Northern Hemisphere, pre-Quaternary regolith and deeply weathered soil profiles have been eroded and the landscape is a matrix of bare rock outcrops and till-covered slopes. Large moraine complexes, the distribution of glacial lake sediments, and the orientation of landforms shaped by flowing ice all indicate the large-scale geomorphic impact of continental-scale glaciation.

Changes in the mass of glaciers over time not only alter the land level under and near ice sheets by loading and then unloading the crust (**glacial isostasy**), but they change sea level by storing water on land as ice. When ice sheets expand, sea level drops as much as 130 m and thus so does the base level to which rivers flow around the planet. Even river systems and hillslopes far away from the ice sheets respond geomorphically to this change in a fundamental boundary condition. Falling sea level provides the impetus for river incision and the transport of eroded sediment away from the continents. For example, during times of lower sea level, Chesapeake Bay was a river valley and sediment from the Susquehanna River was deposited far away from the continents. The coming and going of glacial ice thus drives geomorphology and surface processes on a global scale.

The lasting geomorphic impact of glaciers on the postglacial landscape controls the distribution and rate of surface processes. Such **paraglacial** effects (consider them as glacial hangovers) result from direct landscape conditioning by glaciation and deglaciation. For example, recently deglaciated landscapes are typically unstable because they lack vegetation, have glacially oversteepened slopes, and are covered by large amounts of unconsolidated sediment deposited as the ice melted. Immediately after deglaciation, erosion and sediment transport rates are high and then fall off exponentially over many thousands of years. It appears that most landscapes take thousands to tens of thousands of years to readjust to interglacial conditions (longer than many interglacial episodes). Typical paraglacial features of formerly glaciated areas include sediment storage in rockfall talus, debris and alluvial fans, and valley-bottom terraces of outwash. Not only do these features contribute sediment during interglacials, but they also serve as major and easily reworked sources of sediment for incorporation during the next glacial advance. Thus, the long-lasting, landscape-scale memory of past glaciations influences postglacial landscape dynamics in most previously glaciated terrain.

Periglacial Environments and Landforms

About a quarter of Earth's terrestrial surface can be classified as **periglacial**—environments that are not permanently ice-covered but experience below freezing temperatures for much of the year and in which geomorphic processes are largely driven by the seasonal freezing and thawing of water.

Of particular importance in periglacial environments is the downslope movement of water-saturated regolith in thin sheets during the summer, when the upper layers of frozen ground thaw, lose much of their strength, and move downhill over the still-frozen substrate. Most deformation is concentrated at the boundary between frozen and thawed material. This process is called **solifluction**, a term first associated with any downslope movement of saturated regolith but now applied primarily to periglacial processes. Closely related is **gelifluction**, a type of solifluction that describes the downslope movement of regolith by the freezing process en masse, rather than movement along a specific plane in the subsurface (see Figure 5.3, second panel).

Many periglacial landscapes have never been glaciated. Relict permafrost features, left over from colder climates, have been found in places far from the most recent

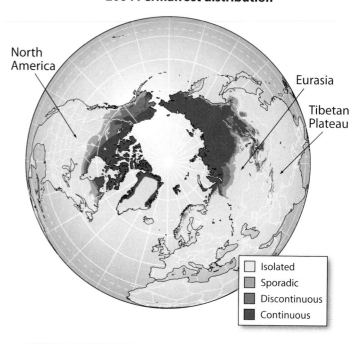

Permafrost distribution is controlled by climate, most importantly mean annual air temperature. In continental areas of Eurasia, far from warm oceans, permafrost is found farther south than in areas near the relative warmth of the ocean. Permafrost also occurs at high altitudes (such as the Tibetan Plateau) because mean annual air temperature decreases with elevation.

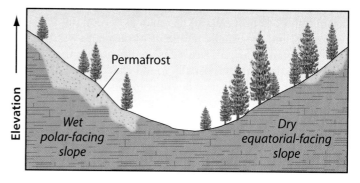

Aspect affects the amount of solar radiation a slope receives and influences the distribution of permafrost. Slopes facing the equator receive more sunlight, have less permafrost, and the permafrost they do have is found at higher elevations.

FIGURE 9.11 Permafrost Distribution. Permafrost occurrence is controlled by the combination of low temperature and low precipitation. Thus, permafrost occurs both at high latitudes and high altitudes.

Pleistocene glaciation including the highlands of South Africa and the southern Appalachian Mountains. As our global climate warms, both natural resource and environmental interests will increasingly focus on periglacial processes and landforms because thawing will disturb now-stable frozen soils and landscapes.

Permafrost

Integral to much of the periglacial environment is **permafrost**, ground that has remained frozen for at least two years and in many cases much longer. Because permafrost is the product of seasonally cold temperatures, usually mean annual air temperatures below −2 °C, it is found at both high elevation (due to cooling of the atmosphere with elevation) and high latitude (the result of reduced annual solar input). For example, elevation-related permafrost is found at mid- and low latitudes on the summits of Colorado's Rocky Mountains at 3800 m above sea level, on the peak of tropical Mauna Kea in Hawaii at 4200 m above sea level, and even at the top of Tanzania's equatorial Mount Kilimanjaro (almost 6000 m above sea level). Latitude-related permafrost begins at about 50° N latitude in North America and continues northward. However, in Asia, the Tibetan Plateau is high enough that permafrost there is found below 30° N latitude [Figure 9.11]. In the Southern Hemisphere, there is scant permafrost outside of Antarctica because there is little landmass in the appropriate latitude range.

The upper decimeters to meters of most permafrost thaw every summer due to increased air temperatures and net solar radiation delivery to the ground surface [Figure 9.12]. The seasonally thawed thickness above permafrost is referred to as the **active layer**. It is often water-saturated as the frozen ground and ice lenses below are less permeable and act as an aquitard. The **permafrost table** defines the bottom of the active layer and indicates the uppermost extent of ground that does not thaw during the summer. **Taliks** (from the Russian *tayat*, meaning "melt") are areas that remain unfrozen—for example, the thawed region below a lake. Permafrost usually contains **ground ice**, both dispersed as pore fillings and concentrated in lenses. Ground ice is most common in the upper tens of meters of permafrost.

Microclimates and the annual distribution of precipitation control the establishment and survival of permafrost. Permafrost is most common where winters are dry and cold (resulting in thin snowpacks and more effective radiational cooling of the soil) and where summer temperatures are not much above freezing. Thick snowpacks reduce conductive

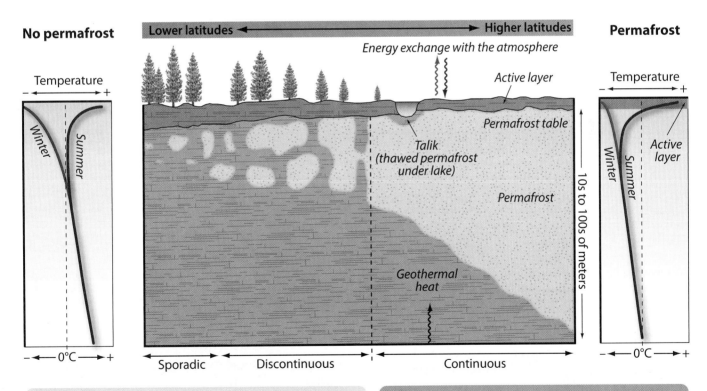

Permafrost can be continuous or discontinuous. Above the permafrost, the **active layer**, which is decimeters to meters thick, thaws every summer. Unfrozen material under the permafrost or within permafrost is known as a **talik**. Talik depths increase under lakes and rivers.

The temperature profile in the upper active layer of permafrost varies seasonally, freezing in the winter and thawing in the summer. Temperature in the permafrost generally increases with depth because of geothermal heat.

FIGURE 9.12 Permafrost Cross-Section and Temperature Profiles. Longitudinal cross-section through the arctic and subsurface temperature profiles showing the distribution of permafrost and seasonal change in temperature profile.

heat loss from the soil, and thereby prevent sufficient chilling to establish or maintain permafrost.

Permafrost is described by its continuity and thickness (some Siberian permafrost is over 1500 m thick, but the world average is closer to 100 m). **Continuous permafrost** implies that all of the ground is frozen, except for the seasonally thawed, active surface layer; any warmer microenvironments do not change the energy balance enough to keep parts of the surface thawed. **Discontinuous permafrost** is indicative of a landscape where some areas remain perennially thawed. **Sporadic permafrost** indicates that unfrozen ground is the rule—not the exception. In general, permafrost becomes more continuous at higher latitudes where mean annual air temperatures are lower.

Permafrost thickness is controlled by the loss of heat to the overlying atmosphere and the gain of heat geothermally from below. Although permafrost thickness is correlated with current mean annual air temperature, it is likely controlled by long-term average climate. Very thick permafrost is long-lived and implies long periods of cold temperatures over multiple glacial-interglacial cycles. It is likely that some permafrost, especially sporadic permafrost at lower latitudes, may be inherited from colder climates of the past. The presence of relict permafrost landforms, in areas such as midwestern North America and the southern British Isles, indicates that at some time in past, the climate was much colder with mean annual air temperature well below 0°C (see Chapter 13).

Characteristic Periglacial Landforms and Processes

Many periglacial landforms and processes are distinctive and differ from their thawed counterparts because the volumetric expansion of water as it goes from liquid to solid phase has the potential to generate significant stresses within rock and soil if the water is confined and cannot expand either into open pores or into fractures. One such situation is top-down freezing that seals water in cracks and pores. Another situation is **segregation ice growth,** during which water and water vapor migration to growing ice lenses increases internal stresses in earth materials. Rapid freezing, cold temperatures, and frequent freeze-thaw cycles all appear conducive to shattering rock. Such **frost shattering** is common in periglacial environments (both arctic and alpine) and often leaves bedrock outcrops hidden under a mantle of angular, shattered rock fragments termed **felsenmeer** (German for sea of rocks) [Photograph 9.26]. Movement of subsurface water is also essential for frost heave (the predominant movement of material by ice growth).

Extensive areas of felsenmeer outside present-day periglacial limits reflect harsher climates during glacial times. Felsenmeer can be an integral part of many different permafrost landforms, including rubble streams oriented downslope, large low-gradient boulder fields, and **cryoplanation terraces** (flat surfaces cut into bedrock hillslopes).

PHOTOGRAPH 9.26 Felsenmeer. Felsenmeer surface of broken rock and sheeted tors on Nelson Crag, northeast shoulder of Mount Washington in the White Mountains of New Hampshire.

These terraces have abrupt backing edges, and the terraces themselves are covered with broken rock. It is not possible to observe the processes that cut these terraces because they appear to form slowly over millennia; however, those studying cryoplanation terraces believe that they are cut by scarp retreat, the result of decades to centuries of **nivation,** the accumulation of snow in erosional depressions termed **nivation hollows** [Photograph 9.27]. In these hollows, both physical and chemical weathering increase because of repeated freeze-thaw cycles and the presence of moisture. Nivation hollows deepen as weathered rock is removed from the slope by mass movement processes.

If the upper, seasonally unfrozen, active layer of regolith and soil becomes saturated with water, mass movements will move material downslope. For example, solifluction,

PHOTOGRAPH 9.27 Nivation Hollow. Nivation hollow (snowy area) in the Bald Mountains, near Jasper, Alberta, Canada, is about 10 m wide. Lighter colored rock (without lichen cover) suggests that a larger area around snow patch was snow-covered earlier in the summer or during previous summer seasons. Photo was taken in early August with snow cover at a minimum.

PHOTOGRAPH 9.28 Solifluction. Solifluction, shown in the Tien Shan Mountains, Kyrgyzstan, moves regolith downslope in lobes with downslope edges about a meter high.

PHOTOGRAPH 9.29 Thermokarst Features. (a) Thermokarst thaw lakes on Alaska's North Slope. Lakes are 1 to 10 km long. (b) Thaw slump located on the Alaskan North Slope in the foothills of the Brooks Range.

the downslope movement of thawed soil and sediment in coherent masses, becomes most active in late spring and summer as the active layer deepens; the soil saturates from snowmelt, rainfall, and melting interstitial ice; the shear strength drops; and the soil weakens. The saturated mass of soil behaves like a viscous fluid moving slowly downslope and forming distinctive, steep-sided, lobate features [Photograph 9.28].

Solifluction is most common on slopes between 5 and 20 degrees and was critical in the development of permafrost-rich, valley-bottom muck deposits that contain Pleistocene megafaunal remains (extinct animals such as mammoths and giant sloths) in Alaska and Siberia. Many steeper slopes (>20°) are too well drained and thus the soils are too strong to flow, whereas the driving force on lower slopes (<5°) is insufficient to overcome the strength of even saturated soil masses. Shallow landslides are also common during the summer months in periglacial terrain. Many of these slides have slip surfaces at the base of the active layer, reflecting both a strength and hydrologic contrast with the underlying frozen material.

Karst-like topography can develop in landscapes where melting of permafrost ice masses produces solution-like features including closed depressions, subsurface streams, and blind valleys. Such **thermokarst** results when ground-ice melts and surficial materials collapse as supporting ice is removed. Thermokarst often initiates through disturbance of the permafrost surface. Some of these disturbances are natural, such as vegetation change, river-channel migration, or mass movements; others are human-induced, such as road building or off-road vehicle use. Disturbance changes the surface energy balance and thus can initiate thawing. If thawing then changes albedo or surface hydrology, melting can become a positive feedback process and the thermokarst can expand rapidly.

The most distinctive and common thermokarst features are thaw lakes [Photograph 9.29a] and thaw slumps [Photograph 9.29b]. **Thaw lakes** occupy shallow depressions that may have once been underlain by higher-than-average concentrations of ground ice. Once thawing begins, the lower albedo of water in comparison to ice allows it to absorb more solar energy. Warm water thermally erodes the shorelines while waves generated on the lake by wind cause physical erosion. Over time, thaw lakes fill with organic and inorganic sediment. **Thaw slumps** are mass movements (Chapter 5) initiated by the melting of ground ice and are often the result of bank undercutting by streams and rivers. Weak, often cohesionless, material mobilized by thawing rapidly flows from thaw slumps, allowing retrogressive failures to continue. Thaw slump scarps can move rapidly back into the landscape and supply large amounts of sediment to river channels.

Patterned ground comes in many different forms (sorted circles, polygons, and stripes) and is common in periglacial terrain [Photograph 9.30]. Some patterned ground (circles and polygons) has coarse margins and finer-grained centers; other patterned ground is unsorted.

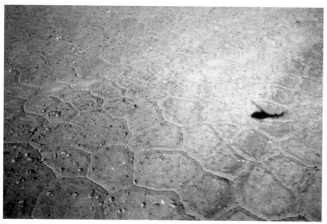

PHOTOGRAPH 9.30 Periglacial Patterned Ground. (a) Sorted stripes near the summit of Mauna Kea, Hawaii, at an elevation of about 4300 m. The stripes are about 15 cm across. (b) Sorted circles, Spitsbergen, Norway. The sorted circles are about 2 m in diameter. This area is an uplifted gravel beach; freeze-thaw cycling in permafrost sorted these circles. (c) Polygons, in the Dry Valleys of Antarctica. These are large features as shown by the shadow of the helicopter for scale.

There are high-centered and low-centered polygons; high-centered polygons have water occupying their margins in the summer, whereas observations suggest that low-centered polygons are actively growing. Stone stripes running down hillslopes are defined by contrasting grain sizes.

A variety of processes have been suggested to explain the development of these features, in particular their sorting. Earlier hypotheses emphasized thrusting and heaving of particles by frost action and the ability of fine grains to fill voids more readily than large grains. Recent explanations center on slow convection in the active layer and preferential movement (and thus sorting) of different grain sizes [Figure 9.13]. Polygon formation is attributed to ice filling ground cracks resulting in ice wedges. When the wedges melt, the resulting voids are often filled by sand, leaving ice wedge casts as indicators of periglacial conditions in the past.

Rock glaciers, present in polar and alpine areas, share many similarities with alpine glaciers and resemble them in form. Unlike alpine glaciers, rock glaciers are completely

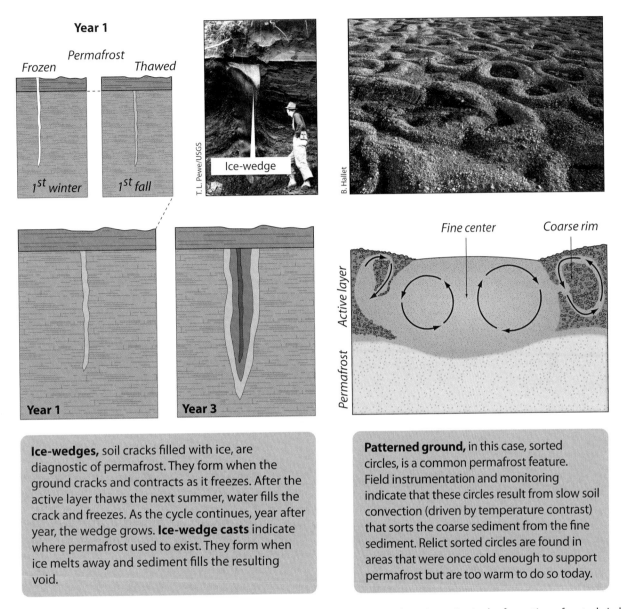

Ice-wedges, soil cracks filled with ice, are diagnostic of permafrost. They form when the ground cracks and contracts as it freezes. After the active layer thaws the next summer, water fills the crack and freezes. As the cycle continues, year after year, the wedge grows. **Ice-wedge casts** indicate where permafrost used to exist. They form when ice melts away and sediment fills the resulting void.

Patterned ground, in this case, sorted circles, is a common permafrost feature. Field instrumentation and monitoring indicate that these circles result from slow soil convection (driven by temperature contrast) that sorts the coarse sediment from the fine sediment. Relict sorted circles are found in areas that were once cold enough to support permafrost but are too warm to do so today.

FIGURE 9.13 Patterned Ground Features. Cross sections showing processes thought to be active in the formation of sorted circles and ice wedges, common forms of patterned ground.

covered with rubble or regolith below their accumulation area and they move at slow rates of centimeters to a few meters per year. Both rock glaciers and alpine glaciers issue from cirques or cliffs, are elongate in plan form, have lobate margins with steep snouts, and move downslope deforming under the force of gravity. Rock glaciers contain ice, which facilitates their flow. This ice can be massive and layered (as has been found in the few rock glaciers cored so far) or it can be interstitial, filling the voids between rubble. The rock cover provides significant insulation, tempering thermal changes experienced by the ice below. Rock glacier genesis is controversial with some arguing that interstitial ice forms in place from percolation of surface water and others suggesting that rock glaciers are simply debris-covered glaciers with ice originating from snowfall in the accumulation zone. It is important to realize that while rock glaciers are a distinct landform, there may be more than one way in which they form (the problem of equifinality, in which the landform is not diagnostic of the process by which it formed).

Pingos (from Inuit, meaning "small hill") are one of the more unusual and distinctive periglacial landforms [**Photograph 9.31**]. Dome-shaped landforms tens to hundreds of meters wide and meters to tens of meters high, pingos are cored with massive ice that may be exposed at their tops. The surface layer of many pingos is extended (in tension) from the growing ice body below. The pingo-forming process illustrates well the distinctive nature of permafrost hydrology. Closed-system pingos (where water does not migrate in from the outside) are typically found in what used to be lake basins or boggy areas. After the lake drains, the surface freezes and the permafrost table

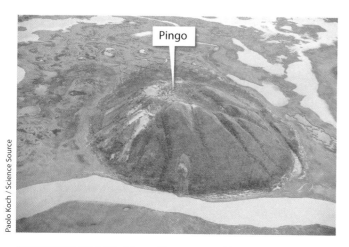

PHOTOGRAPH 9.31 Pingo. Pingo covered by green vegetation and surrounded by river and thermokarst lakes in the Mackenzie River Delta, Canada.

rises, trapping pore water from below and forcing it to expand upward upon freezing. Open-system pingos result from artesian conditions that supply groundwater to the growing ice mass.

Rivers and soils in periglacial regions behave somewhat differently than their temperate-region cousins. Flows in periglacial rivers are highly seasonal—open water may exist for only a few weeks to a few months each year. Extreme runoff events, generated by snow melt, commonly recur every spring and create ice jams that force overbank flow. In addition to physical erosion of riverbanks, flowing water can thermally erode permafrost with which it comes in contact. Permafrost soils are typically very disturbed, or **turbated**, as they are stirred by the melting and freezing of ice and possibly by convection. In places, this cryoturbation is evidenced by folds and discontinuous soil horizons.

Applications

Many of our planet's great cities, much of the population of northern Europe, Asia, and North America, and many of the world's most productive agricultural regions (the loess belts) are located in or just beyond areas once covered by ice sheets. In these regions, understanding the distribution and character of glacially deposited sediments is important for many reasons. The distribution of glacial sediments can control slope stability; glacial lake and glacial marine clays are notorious for landsliding. Oversteepened slopes, the result of glacially induced valley deepening, are prime terrain for rockslides and rockfalls. Aquifers hosted in glacial outwash gravels provide reliable but easily polluted water supplies. Many till-covered uplands in New England, especially in areas where the till is clay- and silt-rich and highly compacted, have low permeability and poorly drained, hydric soils, making septic systems difficult to site. Glacially deposited sand and gravel are important resources for construction.

Mapping the distribution of glacial sediments can be important for solving significant environmental problems. In previously glaciated regions, the surface and subsurface distribution of earth materials, both permeable sand and gravel and less permeable silt and clay, are in large part determined by the geometry of now-vanished glaciers and their associated outwash plains and ice-marginal lakes. Understanding the resulting distribution of glacial sediments can be critical for determining the fate and transport of hazardous materials and the provision of clean drinking water. For example, contaminants released in a valley underlain by a thick sequence of impermeable glacial lake sediments are unlikely to affect an artesian drinking-water aquifer hosted in deeper, outwash-derived sand and gravel.

Understanding the behavior and particularly the mass balance of glaciers over time and into the future is critical as our planet warms. In the past century, most alpine glaciers have retreated, some dramatically. The Greenland Ice Sheet appears to be losing mass and contributing to sea-level rise at a rate greater than anticipated even a decade ago. The most direct impact of melting glaciers is the rise in global sea level as water is transferred from land-based glaciers to the ocean. Indeed, predicting sea-level rise has become the dominant glaciological issue of our time, in part because of the fear that rising sea level could lead to catastrophic positive feedbacks. For example, as sea level rises, the rate at which floating ice margins calve increases. Calving of floating ice does not itself increase sea level, but calving, by reducing buttressing, can increase the velocity of ice streams and thus can increase the flux of inland, terrestrial ice to the ocean. There is also concern that rising sea level could float and destabilize the massive Antarctic ice shelves, allowing more terrestrially based ice there to flow into the sea.

The geologic record clearly indicates that large ice sheets have come and gone over time, and, as a result, sea level has gone up and down. During some of the warmest and longest interglaciations of the past 2 million years, sea level was at least several meters, and perhaps as much as 10 m, higher than it is today. It is sobering to consider that the Greenland and Antarctic ice sheets together hold enough ice that, if melted, sea level would rise about 80 m (7 m from Greenland, 73 m from Antarctica). Most of the world's major cities, and much of its population, are within 80 m of sea level.

Glaciers function as natural water reservoirs, particularly alpine glaciers, in areas as diverse as South America, the Himalaya, and western North America. These glaciers store winter or wet season precipitation as ice and snow and release it slowly as runoff over the summer/dry season months. As the climate warms, many retreating glaciers are forecast to disappear entirely. With them will go reliable summer water supplies in many arid areas because water stored as ice in glaciers provides a buffer against both seasonal and longer-term drought. The storage function that glaciers provide can be replaced by artificial reservoirs, but only at great monetary and environmental costs. Many communities around the world will be severely affected by the loss of glacier water storage, which will reduce summer drinking water supplies and hydroelectric power generation. In the developing world, where funds for investment in water storage infrastructure are scarce, the impact will likely be most severe.

Areas underlain by permafrost have proven to be extremely sensitive to human impacts. Locally, such impacts include excavation, road building, and the presence of structures such as houses, which change the energy balance of the landscape, leading to melting and damaging ground movements.

Globally, our planet's warming climate has already begun to affect periglacial environments. Recent observational data of significant polar warming (at least several °C over the past decades) support the predictions of climate models, which indicate that future warming will be most severe at high latitudes. Warming is already melting permafrost in discontinuous permafrost areas, causing problems for roads, airstrips, and buildings. As arctic warming intensifies, geoengineering problems will become more common and widespread. Not only will the foundations of many structures become unstable, but thermokarst-catalyzed erosion, and the disruption it causes, will likely increase in both extent and severity. The active layer, which is responsible for much periglacial geomorphic activity, will become thicker and stay active for longer each year. Solifluction will become more common in permafrost regions during the summer months. Melting permafrost will release methane, an effective greenhouse gas, causing further warming.

Although we are now in an interglacial period during which the climate is relatively warm and the distribution of ice is restricted, geomorphic and societal effects of past glaciations remain important. In the future, understanding the response of periglacial landscapes to global warming will become increasingly relevant.

Selected References and Further Reading

Agassiz, L. *Études sur les glaciers*. Neuchâtel: Jent et Gassmann, 1840.

Andrews, J. T. Glacier power, mass balances, velocities and erosion potential. *Zeitschrift für Geomorphologie* 13 (1972): 1–17.

Antevs, E. *The Recession of the Last Ice Sheet in New England*. Research Series Number 11. New York: American Geographical Society, 1922.

Ballantine C. K. Paraglacial geomorphology. *Quaternary Science Reviews* 21 (2002): 1935–2017.

Bartholomaus, T. C., R. S. Anderson, and S. P. Anderson. Response of glacier basal motion to transient water storage. *Nature Geoscience* 1 (2008): 33–37.

Benn, D. I., and D. J. A. Evans. *Glaciers and Glaciation*. London: Hodder Arnold, 2010.

Bennett, M. R., and N. F. Glasser. *Glacial Geology: Ice Sheets and Landforms*. Oxford: Wiley-Blackwell, 2009.

Booth, D. B. Glaciofluvial infilling and scour of the Puget Lowland, Washington, during ice-sheet glaciation. *Geology* 22 (1994): 695–698.

Boulton, G. "Processes and patterns of subglacial erosion." In D. R. Coates, ed., *Glacial Geomorphology*, Proceedings, Geomorphology Symposium (1974). Binghamton: State University of New York, 1974.

Brozovic, N., D. W. Burbank, and A. J. Meigs. Climatic limits on landscape development in the northwestern Himalaya. *Science* 276 (1997): 571–574.

Church, M., and J. M. Ryder. Paraglacial sedimentation: Consideration of fluvial processes conditioned by glaciation. *Geological Society of America Bulletin* 83 (1972): 3059–3072.

Clark, P. U., and J. S. Walder. Subglacial drainage, eskers, and deforming beds beneath the Laurentide and Eurasian ice sheets. *Geological Society of America Bulletin* 106 (1993): 304–314.

Cuffey, K. M., and W. S. B. Paterson. *Physics of Glaciers*. Burlington, MA: Butterworth-Heinemann, 2010.

Egholm, D. L., S. B. Nielsen, V. K. Pedersen, and J.-E. Lesemann. Glacial effects limiting mountain height. *Nature* 460 (2009): 884–887.

Evans, D. J. A., ed. *Glacial Landsystems*. London: Arnold, 2003.

French, H. M. Periglacial geomorphology and permafrost. *Progress in Physical Geography* 4 (1980): 254–261.

French, H. M. *The Periglacial Environment*. Chichester: John Wiley, 2007.

Hales, T. C., and J. J. Roering. A frost "buzz-saw" mechanism for erosion of the eastern Southern Alps, New Zealand. *Geomorphology* 107 (2009): 241–253.

Hallet, B. A theoretical model of glacial abrasion. *Journal of Glaciology* 23 (1979): 29–50.

Hallet, B., L. Hunter, and J. Bogen. Rates of erosion and sediment evacuation by glaciers: A review of field data and their implications. *Global and Planetary Change* 12 (1996): 213–235.

Hambrey, M. J. *Glacial Environments*. Vancouver: University of British Columbia Press, 1984.

Hambrey, M. J., and J. Alean. *Glaciers*. Cambridge, UK: Cambridge University Press, 2004.

Harbor, J. M., B. Hallet, and C. F. Raymond. A numerical model of landform development by glacial erosion. *Nature* 333 (1988): 347–349.

Imbrie, J., and K. P. Imbrie. *Ice Ages: Solving the Mystery*. Short Hills, NJ: Enslow, 1979.

Knight P. G., ed. *Glacier Science and Environmental Change*. Oxford: Blackwell Publishing, 2006.

Koppes, M. N., and D. R. Montgomery. The relative efficacy of fluvial and glacial erosion over modern to orogenic timescales. *Nature Geoscience* 2 (2009): 644–647.

Koteff, C., and S. Pessl. *Systematic Ice Retreat in New England*. U.S. Geological Survey Professional Paper 1179. Washington, DC: Government Printing Office, 1981.

MacGregor, K. R., R. S. Anderson, S. P. Anderson, and E. D. Waddington. Numerical simulations of glacial-valley longitudinal profile evolution. *Geology* 28 (2000): 1031–1034.

Meierding, T. C. Late Pleistocene glacial equilibrium line altitudes in the Colorado Front Range: A comparison of methods. *Quaternary Research* 18 (1982): 289–310.

Menzies, J. Drumlins—Products of controlled or uncontrolled glaciodynamic response? *Quaternary Science Reviews* 8 (1989): 151–158.

Menzies, J. *Modern and Past Glacial Environments.* Oxford: Butterworth, 2002.

Mitchell, S. G., and D. R. Montgomery. Influence of a glacial buzzsaw on the height and morphology of the Cascade Range in central Washington State, USA. *Quaternary Research* 65 (2006): 96–107.

Montgomery, D. R. Valley formation by fluvial and glacial erosion. *Geology* 30 (2002): 1047–1050.

Nye, J. F. A method of calculating the thicknesses of the ice-sheets. *Nature* 169 (1952): 529–530.

Owen, L. A., G. Thackray, R. S. Anderson, et al. Integrated research on mountain glaciers: Current status, priorities and future prospects. *Geomorphology* 103 (2009): 158–171.

Porter, S. C. Equilibrium-line altitudes of Late Quaternary glaciers in the Southern Alps, New Zealand. *Quaternary Research* 5 (1975): 27–47.

Reger, R. D., and T. L. Pewe. Cryoplanation terraces: Indicators of a permafrost environment. *Quaternary Research* 6 (1976): 99–109.

Ridge, J. C., and F. D. Larsen. Re-evaluation of Antevs' New England varve chronology and new radiocarbon dates of sediments from Glacial Lake Hitchcock. *Geological Society of America Bulletin* 102 (1990): 889–899.

Sharp, R. P. *Living Ice: Understanding Glaciers and Glaciation.* Cambridge: Cambridge University Press, 1991.

Shaw, J. Drumlins, subglacial meltwater floods, and ocean responses. *Geology* 17 (1989): 853–856.

Sugden, D. E. Reconstruction of the morphology, dynamics, and thermal characteristics of the Laurentide Ice Sheet at its maximum. *Arctic and Alpine Research* 9 (1977): 21–47.

Sugden, D. E. Glacial erosion by the Laurentide Ice Sheet. *Journal of Glaciology* 20 (1978): 367–392.

Tomkin, J. H. Numerically simulating alpine landscapes: The geomorphologic consequences of incorporating glacial erosion in surface process models. *Geomorphology* 103 (2009): 180–188.

Wahrhaftig, C., and A. Cox Rock glaciers in the Alaska Range. *Geological Society of America Bulletin* 70 (1959): 383–436.

Walder, J. S., and B. Hallet. A theoretical model of the fracture of rock due to freezing. *Geological Society of America Bulletin* 96 (1985): 336–346.

Ward, D. J., R. S. Anderson, Z. S. Guido, and J. P. Briner. Numerical modeling of cosmogenic deglaciation records, Front Range and San Juan mountains, Colorado. *Journal of Geophysical Research* 114 F01026 (2009): doi: 10.1029/2008JF001057.

Washburn, A. L. *Periglacial Processes and Environments.* London: Edward Arnold, 1973.

Werner, B. T., and B. Hallet. Sorted stripes: A numerical study of textural self-organization. *Nature* 361 (1993): 142–145.

Young, G. M., V. von Brunn, D. J. C. Gold, and W. E. L. Minter. Earth's oldest reported glaciation: Physical and chemical evidence from the Archean Mozaan Group (~2.9 Ga) of South Africa. *Journal of Geology* 106 (1998): 523–538.

DIGGING DEEPER How Much and Where Do Glaciers Erode?

How much do glaciers erode, modify, and shape the land? This deceptively simple question continues to intrigue geomorphologists and has fueled arguments running back to the nineteenth-century roots of the discipline when glaciers were initially recognized as very effective agents for transporting material.

> *For the moving of large masses of rock, the most powerful engines without doubt which nature employs are the glaciers. . . .*
> —Playfair (1802, p. 388)

In the 1840s, when Agassiz subsequently developed his idea of ancient ice ages based on recognition of the formerly greater extent of European glaciers, the issue of glacial erosion became central to a century-long controversy over whether glaciers could erode valleys into rock or simply entrained and transported the loose, weathered material above rock.

Many early geologists could not accept that ice, a soft material, could erode into much harder bedrock. How could something with low enough density to float in water carve deep valleys into solid rock? Although some argued that permanent snow protected landscapes from erosion (Bonney, 1902; Fairchild, 1905), others such as the German geologist Penck argued that glaciers did not protect but vigorously attacked the rock over which they flowed (Penck, 1905). Penck was not alone. Davis, an American geologist, argued that glacial action sculpted alpine terrain based on how hanging valleys, rock basins, and cirques only occurred in glaciated and formerly glaciated mountains (Davis, 1900).

FIGURE DD9.1 Yosemite Valley before and after glaciation, as depicted by Matthes in his conceptual illustration of the transformation of the incised preglacial valley of the Merced River into the glacially carved terrain of the modern Yosemite Valley, California. [From Matthes (1930).]

Although early arguments over glacial erosion centered on the European Alps, the spectacular topography of California's Yosemite Valley became a focus of such arguments when the first state geologist of California, Whitney, proposed that the steep-walled valley was a down-dropped, fault-bounded graben (Whitney, 1865). Decades later, U.S. Geological Survey physiographer Matthes compellingly argued that glaciers gouged out the previously incised valley of the Merced River to form Yosemite's modern topography (Matthes, 1930) [**Figure DD9.1**]. Numerous field studies went on to verify the association of U-shaped valleys with formerly glaciated landscapes and support the notion that glacial erosion transformed V-shaped river valleys into distinctive U-shaped cross-sectional profiles.

But how does this valley transformation happen? In the 1970s and 1980s, advances in understanding the mechanics of glacial erosion (Hallet, 1979 and 1981) and the development of ice-flow models that included basal sliding and realistic ice rheology paved the way for process-based modeling of glacial valley excavation. Building on these advances, Harbor et al. (1988) developed a numerical model of the development of U-shaped valleys by glacial erosion [**Figure DD9.2**]. The model assumes that glacial erosion is controlled by the basal sliding velocity of the ice, which they modeled as a function of the basal shear stress (Harbor et al., 1988; Harbor, 1992). Simulating the cross-section of glacial flow through an initially V-shaped valley, they found that the greatest erosion initially occurred on the lower valley walls until establishment of an equilibrium U-shaped profile that persisted as long as the valley continued to entrench and excavate a deeper trough [**Figure DD9.3**]. By calibrating the model to observations from modern glaciers, Harbor (1992) was able to show that the transformation from a V-shaped to U-shaped valley cross section could begin in a single glacial cycle.

But by how much did glaciers widen their valleys? It was difficult to tell in most cases because glaciers eroded evidence of the former, preglacial valley profile. Montgomery (2002) studied the unusual case of the Olympic Mountains, which are ringed by some glaciated and some unglaciated valleys. There, valley width, relief, and cross-sectional areas were similar for valleys with drainage areas <10 km^2, but glaciated valleys draining > 50 km^2 had 2 to 4 times the cross-sectional area and up to 500 m greater relief than unglaciated fluvial valleys [**Figure DD9.4**]. With increasing drainage basin area, glaciers excavate far more rock volume than do rivers flowing through fluvial valleys. Unlike river valleys, in which tributaries tend to seamlessly join larger channels at the same elevation, large glacial valleys erode far faster than small valleys, resulting in hanging valleys draining over steep valley walls in formerly glaciated terrain.

Excavation of deep glacial valleys is not spatially uniform, and depends on both the ice thickness and thermal regime of the glacier—whether warm-based or cold-based. Glaciers that are frozen to their bed are static at their base and no motion at the bed means little to no erosion. Glacial erosion is most aggressive where flow is fastest, which is facilitated by thick ice, a warm base where melting and meltwater promote basal sliding, and a steep ice-surface gradient to provide the driving stress. High debris content embedded in basal ice increases the rate of glacial abrasion. Alpine glaciers typically exhibit strong longitudinal patterns of glacial erosion in which the deepest glacial troughs are excavated near and below the glacial equilibrium

DIGGING DEEPER How Much and Where Do Glaciers Erode? *(continued)*

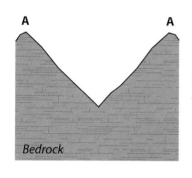

Unglaciated Valley
A–A Unglaciated Valley

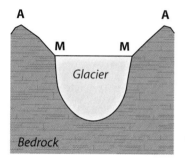

Maximum Vertical Ice Extent
A–A Glacial Valley
M–M Active Glacial Channel and Zone of Glacial Influence

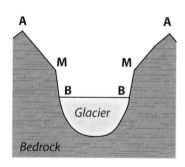

Ice Extent Less Than Maximum
A–A Glacial Valley
M–M Zone of Glacial Influence
B–B Active Glacial Channel

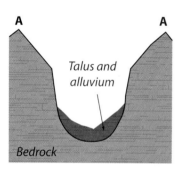

After Deglaciation
A–A Glaciated Valley

FIGURE DD9.2 Schematic illustration of the evolution of an unglaciated V-shaped valley into a U-shaped valley as a result of glacial erosion. [From Harbor (1992).]

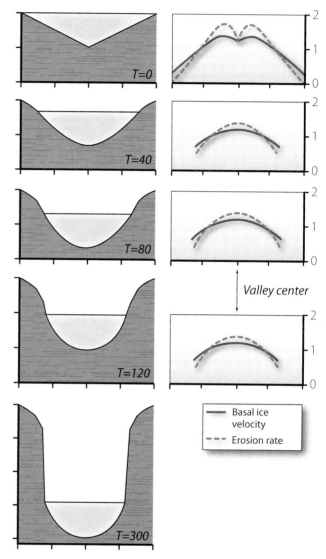

FIGURE DD9.3 Model results for simulation of valley development by glacial scour from an initially V-shaped valley cross section (T = 0) to a steady-state U-shaped cross section (T = 120). Plots are dimensionless and have no vertical exaggeration. Graphs to the right show the corresponding cross-valley profiles of basal ice velocity and erosion rate. [From Harbor (1992).]

line altitude (ELA). Ice sheets have uplands that are cold-based, and deep troughs on their margins where warm-based ice streams are much more effective in excavating terrain. The combination of concentrated erosion in deep ice streams and subglacial fluvial erosion also leads to deeply eroded valleys in the form of fjords and great lakes near ice-sheet margins.

Bonney, T. G. Alpine valleys in relation to glaciers. *Quarterly Journal of the Geological Society* 58 (1902): 690–702.

Davis, W. M. Glacial erosion in France, Switzerland and Norway. *Proceedings of the Boston Society of Natural History* 29 (1900): 273–322.

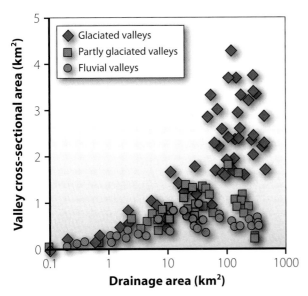

FIGURE DD9.4 Comparison of cross-sectional areas for valleys in the Olympic Peninsula, Washington State, showing the greater amount of rock excavated from valleys with major alpine glaciers (blue diamonds) than from partly glaciated valleys (green squares) and unglaciated fluvial valleys (red circles). [From Montgomery (2002).]

Fairchild, H. L. Ice erosion theory a fallacy. *Bulletin of the Geological Society of America* 16 (1905): 13–74.

Hallet, B. A theoretical model of glacial abrasion. *Journal of Glaciology* 17 (1979): 209–221.

Hallet, B. Glacial abrasion and sliding: Their dependence on the debris concentration in basal ice. *Annals of Glaciology* 2 (1981): 23–28.

Harbor, J. M. Numerical modeling of the development of U-shaped valleys by glacial erosion. *Geological Society of America Bulletin* 104 (1992): 1364–1375.

Harbor, J. M., B. Hallet, and C. F. Raymond. A numerical model of landform development by glacial erosion. *Nature* 333 (1988): 347–349.

Matthes, F. E. *Geologic History of the Yosemite Valley.* Professional Paper 160, United States Geological Survey. Washington, DC: Government Printing Office, 1930.

Montgomery, D. R. Valley formation by fluvial and glacial erosion. *Geology* 30 (2002): 1047–1050.

Penck, A. C. Glacial features in the surface of the Alps. *Journal of Geology* 13 (1905): 1–19.

Playfair, J. *Illustrations of the Huttonian Theory of the Earth.* London: Cadell and Davies; Edinburgh: William Creech, 1802.

Whitney, J. D. *Geology, Volume 1, Report of Progress and Synopsis of the Field-work from 1860–1864.* Geological Survey of California. Philadelphia: Caxton Press, Sherman and Co., 1865.

WORKED PROBLEM

Question: The existence of glaciers implies a balance in mass and energy fluxes into and out of the glacial system. Using what you know about glaciers, create a qualitative model of a glacier considering both mass and energy balance; that is, the glacier is in dynamic steady state. The model, which can be in the form of an equation or a discussion, should show the ways and locations in which mass and energy are gained and lost from the glacier.

Answer: If the glacier is in a dynamic steady state, then both the energy and mass inputs and outputs must be in balance. In other words, the amount of energy gained by the glacier must equal that lost, and, similarly, mass gains and losses from the glacier must be in balance.

Glaciers gain mass from precipitation, blow-in of snow, and condensation of water vapor onto the ice surface. Mass can be added in both the accumulation and ablation zones but is lost primarily in the ablation zone. Glaciers lose mass by calving at their terminus, melting at the surface and on the bed, and sublimation from the ice surface.

Energy is lost and gained from glaciers through a number of different pathways. Sunlight adds energy to glacial surfaces and geothermal heat is added from the bed below. Latent and sensible heat can carry energy both to and from the glacier surface depending on the temperature contrast between the ice and atmosphere. Energy is lost from the ice surface by blackbody radiation.

KNOWLEDGE ASSESSMENT Chapter 9

1. Explain the three physical properties of ice that make it so dynamic near Earth's surface.
2. Compare the size, shape, and location of ice sheets, ice caps, ice fields, and alpine glaciers.
3. By what processes do glaciers lose mass?
4. Predict how a newly deposited snow crystal will change over time before eventually becoming glacier ice.
5. Make a sketch showing the energy balance of a glacier including all the major terms.
6. Explain glacial advance and retreat in terms of mass balance.
7. Predict the effect of elevation on glacial ice accumulation and ablation.
8. Give three reasons why the upper reaches of glaciers function as net accumulation zones.
9. Define the equilibrium line and suggest how you might approximate its location in the field.
10. Describe two ways in which glaciers move.
11. Write Glen's flow law, and describe how it is used.
12. Sketch the movement of ice in an alpine glacier using flow lines in a cross section.
13. Explain why floating and grounded ice margins behave differently.
14. Define cold-based and warm-based glaciers.
15. Explain how regelation works and how it can help glaciers erode rock.
16. List three observations that indicate glaciers are capable of abrading bedrock outcrops.
17. Explain how cirques, arêtes, and tarns are interrelated.
18. Explain the glacial buzzsaw hypothesis.
19. Describe glacial till.
20. What does an esker look like and how does one form?
21. Describe a drumlin and suggest where these landforms are likely to be found.
22. Explain why the nature and texture of ice-marginal sediment is so variable over short distances?
23. How are kettle ponds thought to form?
24. Define outwash and outwash plains and explain how they form.
25. What landform is often used to determine the elevation of now-vanished ice-marginal lakes?
26. Describe the appearance of varved sediment and explain how it was deposited.
27. Define IRD and explain why it is important.
28. Define periglacial.
29. What characteristics best define periglacial environments?
30. Explain the global distribution of permafrost.
31. What do relict permafrost features tell us about paleoclimate?
32. Consider the relationship between energy balance and the thickness, extent, and continuity of permafrost.
33. Define talik and active layer.
34. What is felsenmeer, how is it related to frost shattering, and where might you find it?
35. Define solifluction and discuss when and where it is most active.
36. Where is thermokarst found and what are some of its characteristic features?
37. Sketch the process by which ice wedges and ice wedge casts form.
38. Compare the similarities and differences between glaciers and rock glaciers.
39. What is a pingo and where would you go to find one?
40. How are glaciers, water supply, and climate change related?

Wind as a Geomorphic Agent

10

Arm of a star dune in the Namibian Sand Sea (erg) near Sossusvlei with oryx, a type of antelope, grazing in the foreground. The star dunes here can be up to several hundred meters high.

Introduction

Wind, the coherent movement of the fluid we call air, affects surface processes everywhere on our planet. **Aeolian**, or wind-driven, sediment transport connects different sedimentary environments, some of which are adjacent, others of which are separated by long distances. Along coasts, wind moves sand into and out of storage in coastal dunes and is effectively a local extension of the longshore drift system. On mid-ocean islands, such as Hawaii, wind delivers quartz-bearing dust from Asia thousands of kilometers away. Sediment mantling the abyssal plains of the deep ocean is largely wind-derived, and aeolian dust plays a key role in delivering iron, an important marine fertilizer, to the oceans. Soils developed on wind-transported fine-grained sediment are very fertile (for example, the Palouse of eastern Washington State, famous for wheat farming) and important agriculturally around the world. Desert pavements characteristic of arid regions owe their existence largely to wind.

The geomorphic influence of wind on landscapes is not uniform across the surface of Earth. Wind-driven surface processes are most easily detected where wind-transportable sediment is abundant and where Earth's other geomorphically active fluid, water, and therefore vegetation, is relatively scarce. Such locations include both arid deserts and other areas with significant water deficits, such as coastal zones with highly permeable soils, sandy glacial margins, and exposed sandbars in broad river corridors. In areas where water is abundant, fluvial and hillslope processes are typically quite active, vegetation is dense,

---

IN THIS CHAPTER

Introduction
Air as a Fluid
   Wind Patterns and Speeds
   Vertical Distribution of Wind Speed
   Settling Speed of Particles in Air
Spatial Distribution of Wind-Driven Geomorphic Processes
Aeolian Processes
   Disturbance
   Erosion

   Sediment Transport
   Deposition
Aeolian Features, Landforms, and Deposits
   Aeolian Erosional Features and Landforms
   Aeolian Transport Features and Landforms
   Aeolian Dust Deposits and Loess

Applications
Selected References and Further Reading
Digging Deeper: Desert Pavements—The Wind Connection
Worked Problem
Knowledge Assessment

and the role of aeolian activity is harder to decipher. But aeolian geomorphic features are by no means restricted to deserts; the influence of wind and wind-transported sediment is felt across all of Earth's surface. Wind dominates contemporary surface processes on Mars and perhaps on other terrestrial planets with atmospheres.

Most aeolian sediment is rich in quartz (SiO_2) because the mineral is hard, weathering-resistant, and a major constituent of many rocks exposed at Earth's surface. However, there are dunes made of different minerals. For example, the dunes at White Sands, New Mexico, are made of gypsum (calcium sulfate) derived from a nearby playa, a dry lakebed [**Photograph 10.1**].

The stratigraphic record of wind-deposited sediment can be deciphered to understand the history of our planet. For example, **aeolianite**, indurated coastal dune material cemented by $CaCO_3$ derived from the dissolution of shell fragments within the sand [**Photograph 10.2**], overlies many elevated marine terraces. Aeolianite-capped terraces, dated using a variety of luminescence techniques, provide critical evidence for sea-level heights in the past, allowing us to understand the timing and magnitude of sea-level change over the past half million years.

Biologic activity greatly reduces the relative geomorphic effectiveness of aeolian processes. The roots and decomposition products of vegetation bind soil particles together, providing apparent cohesion. Plants also change airflow patterns and speed, reducing wind's importance as a sediment entrainment and transport mechanism, as well as catalyzing aeolian sediment deposition as rough surfaces slow air movement. Wind is an effective geomorphic agent in deserts primarily because arid regions lack widespread and continuous vegetation cover. Biologic crusts, such as those common on dry-land surfaces, limit wind erosion in deserts. They do this by making the soil surface cohesive and thus

PHOTOGRAPH 10.2 Aeolianite. Several-meter-high Holocene aeolianite (calcium-carbonate-cemented dune sand) outcrop in the Bahamas.

allowing otherwise erodible soils to remain stable even under high winds [**Photograph 10.3**]. Such cohesion allows granular soil material (such as sand and silt) to resist entrainment by wind.

The geomorphic effects of wind can be significant. Strong persistent wind may remove the vegetation-anchoring earth materials including soil, weathered rock, and dune sands. Moving air can be a major agent of vegetation disturbance and thus a catalyst for surface change. Wind-driven waves batter beaches and cliffs and break up arctic sea ice important to the climate system. Prevailing winds carry moisture upslope where orographic precipitation, induced as air masses are forced to rise over mountains, initiates fluvial and hillslope processes. In dune fields, wind transports sand many kilometers. Fine-grained particles, such as silt and volcanic

PHOTOGRAPH 10.1 Sand Dune. Ripple-covered gypsum sand dune at White Sands, New Mexico. Shrubs and small trees provide scale.

PHOTOGRAPH 10.3 Biological Soil Crust. Biological soil crust in Canyonlands National Park in southwest Utah holds sediment in place over rock and resists erosion by wind and water.

ash, are carried around the globe, suspended by wind and deposited far from their sources. Wind-transported dust is a critical component of many soils including those that underlie the breadbaskets of the world.

Air as a Fluid

Air is a fluid moving across Earth's surface. Air behaves similarly to water, except that air is less viscous, and because wind is a reflection of the atmospheric pressure gradient, air flows both up and down topographic gradients. Both flowing air and water are steered by and interact with the land surface; however, the distribution of flowing air and water are different. Because the atmosphere surrounds the planet, all parts of Earth's surface can be affected by wind. In contrast, much of the liquid water on Earth is either in the oceans, where motion and thus kinetic energy are concentrated in currents and waves that do most of their geomorphic work on shorelines, or in stream channels, where kinetic energy is dissipated by flow around roughness elements such as trees, boulders, or rough banks and used to entrain sediment along channel margins.

The material properties of water and air are strikingly different. Air is much less dense than water. At 20 °C, water is more than 800 times denser than air and the dynamic **viscosity** of water (its resistance to deformation) is more than 50 times greater than that of air. These differences mean that transfer of momentum will be more efficient between water and sediment than between air and sediment. Solid material, such as silt and sand, will settle much more rapidly through less dense and less viscous air than through denser and more viscous water [**Figure 10.1**]. In practice, the low viscosity and density of air means that wind can only transport smaller sediment grain sizes such as clay, silt, and sand. For the most part, transport of gravel and boulders must be left to hillslope and fluvial processes. Only in exceptionally windy places (such as the Dry Valleys of Antarctica), where there is scant if any surface vegetation to provide roughness and anchor sediment, is gravel moved by wind.

The movement of air, just as that of water, can be laminar or turbulent. Turbulent flow is chaotic and difficult to predict; the velocity (speed and direction) of turbulent airflow varies rapidly and turbulent flows are effective at mixing the atmosphere. For example, on a calm morning, smoke from a chimney may rise slowly and steadily into the sky. Flow in a stable atmosphere is mostly laminar; there is little turbulence. The physics of laminar flow is determined primarily by the viscosity of air. As the day goes on and Earth's surface is heated by the Sun, packets of warm, less-dense air rise into the atmosphere, and airflow becomes more chaotic or turbulent; the same column of smoke is rapidly mixed and cannot be tracked far from the chimney. Turbulence, and the eddies and wind gusts it spawns, are geomorphically important because bursts of higher-than-average wind speed can initiate and maintain sediment movement.

Unlike fluvial sediment, the deposition of which is constrained by the location of channels and the elevation and location of base level, aeolian sediment, once entrained and in motion, can be transported just about anywhere the wind blows. Windblown sand can move up topographic gradients and around obstacles such as outcrops or hills.

Both water and air are fluids, but water is a much more viscous fluid than air. As a result, particles settle more slowly through water than they do through air. For example, the settling velocity of a 1 mm particle in water is 100 times slower than the same particle settling through air.

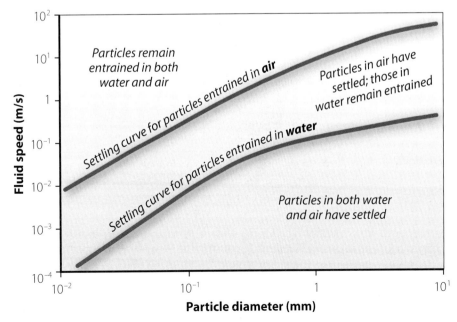

FIGURE 10.1 Comparison of Water and Air as Fluids. The difference in viscosity and density causes particles to settle more rapidly in air than in water.

Globally, aeolian transport moves about 10 percent of the mass of sediment moved by rivers.

Wind Patterns and Speeds

Wind speed matters geomorphically because it, along with vegetation, controls whether a particle of a given size is stable or if it will be moved by rolling, saltation, or suspension. At the global scale, wind speeds are high where the pressure gradient is steep, such as between the adjacent high-pressure and low-pressure systems that drive the trade winds. On a regional scale, wind speeds are often high on mountain ridges, where air is forced up and over topography, and at mountain passes, where air is funneled by topography into a restricted area. At the human scale, wind speeds are lower near the ground as air moves around obstacles such as trees, bushes, and rocks. Such **terrain roughness** creates turbulence and thereby serves to dissipate some of the kinetic energy of the air, lowering the wind speed.

Large atmospheric circulation cells set global patterns of wind speed and direction as well as moisture distribution and thus provide a first-order control over the geomorphic effectiveness of wind. There are substantial differences in average wind speeds and direction around the world [Figure 10.2]. Near the equator, wind speeds tend to be low. These are the **doldrums,** the equatorial intersection of the northern and southern Hadley cells where net air motion is upward (see Figure 1.5). The geomorphic effect of wind in many equatorial regions is limited (other than in coastal areas) because near the equator moisture is commonly sufficient to support dense vegetation, preventing wind erosion and transport of sediment.

Farther north, in the belt of **trade winds,** average wind speeds are higher. In general, between the belts of trade winds and westerly winds (25 to 35 degrees north and south latitude) are dry areas where aeolian activity is more important than elsewhere on Earth. These low latitudes are dry because air masses are generally descending there, the result of convergence between the downflowing limbs of the Ferrel and Hadley atmospheric circulation cells (see Figure 1.5). As the air descends, it warms from the increase in pressure and dries as the **relative humidity** (the percent saturation of the air mass with water vapor) decreases. In general, continental areas where air motion is downward include some of the world's great deserts such as the Sahara (15 to 30 degrees north latitude), and Namib, Kalahari, and Australian deserts (which all lie between 20 and 30 degrees south latitude).

In the Southern Hemisphere, the highest wind speeds are found in the stormy latitudes between the tip of South America and Antarctica. Named by sailors the roaring

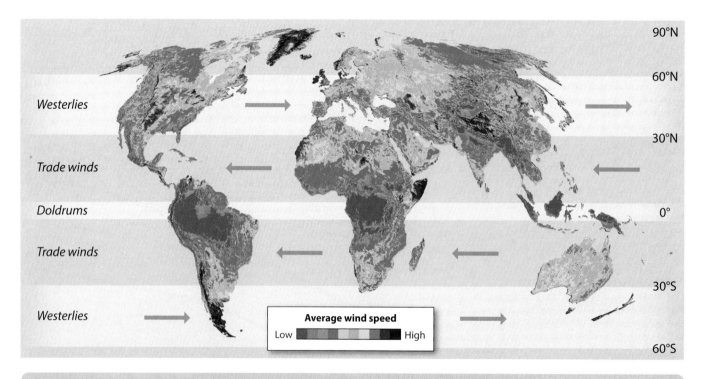

Wind speeds are not uniform on Earth's surface. Winds speeds are low in the **doldrums** near the equator and high in the storm tracks, known as the roaring 40s, at 40° to 50° north and south latitude. Winds are high in mountainous areas such as the Himalaya and the Rocky Mountains. Wind direction at a global scale is controlled by variations in the distribution of incoming solar radiation and the effects of Earth's rotation. These large-scale effects lead to predominant wind directions that vary with latitude.

FIGURE 10.2 Average Wind Speeds on Earth. Average wind speeds are controlled in large part by the position of global circulation cells and major topographic features.

40s, furious 50s, and screaming 60s, wind speeds across the Southern Ocean are on average quite high because there is no land to block the wind and dissipate some of the wind energy through roughness-induced turbulence.

Vertical Distribution of Wind Speed

Wind speed decreases near the boundary between Earth's surface and the atmosphere due to the drag caused by surface roughness. This means that large boulders, tall trees, and high standing outcrops are more likely to experience high winds than material sitting directly on the ground. However, the concentration of wind-suspended sediment is higher closer to the ground; thus, abrasion, one of the geomorphic effects of wind, is more likely to occur at elevations below where the highest wind speeds are found.

The speed of air above the ground surface is usually described by the Karman/Prandtl model, which is based on observations of increasing wind speed with the logarithm of height above the ground [**Figure 10.3**]. At some small height above the ground, mean wind speed goes to zero, the z intercept of the curve, which is known as the **roughness length** (z_0). Physically, z_0 is related to the roughness of the bed and implies there is a boundary layer in which the net wind speed is zero. The Karman/Prandtl model is a simplification that neglects the effects of turbulence and mixing, which are caused both by daily surface heating and cooling and by the evaporation of water. Both effects disturb the logarithmic wind speed profile and change z_0 over short space and time scales.

Settling Speed of Particles in Air

Deposition of wind-transported sediment occurs when lift forces can no longer keep the sediment airborne, that is, when wind speed and the speed of air in turbulent atmospheric eddies fall below those needed to suspend particles. Particle fall speeds depend on grain diameter, a relationship described by Stokes' Law, named after George Gabriel Stokes, who proposed it in 1851:

$$S_s = 2r^2(\rho_p - \rho_f)g/9\mu_f \qquad \text{eq. 10.1}$$

Stokes' Law relates the settling speed (S_s) of a particle of radius r in a fluid to the density (ρ_f in general, ρ_a for air and ρ_w for water, specifically) and dynamic viscosity (μ_f) of the fluid, the acceleration of gravity (g), and the density of the particle (ρ_p). Stokes' Law is applicable to settling in laminar flow conditions, not the turbulent flow conditions

Air, a moving fluid, interacts with boundaries such as the ground surface. This frictional resistance means that near the ground, the wind velocity decreases to zero. Above the ground, wind speed increases. Thus, if grains can be lofted into the flow by impacts of other grains or by turbulent eddies, they will move more rapidly. The y-axis intercept of the height/speed curve, **z_0**, reflects surface roughness and is known as the **roughness length** in the Karman/Prandtl model for air speed. Rough surfaces have a higher **z_0** than smooth surfaces.

FIGURE 10.3 Relationship of Wind Speed and Height Above Earth's Surface. Air speed increases as the logarithm of the height above Earth's surface, with roughness of the underlying surface controlling the value of z_0.

often encountered in nature, particularly near rough surfaces. Nevertheless, Stokes' Law provides a useful way to consider the behavior of very fine particles in a fluid such as air (see the Worked Problem).

Spatial Distribution of Wind-Driven Geomorphic Processes

Wind and wind-carried sediments are present around the globe, but wind is a major or dominant force in landscape modification and sediment transport only in certain geomorphic environments, such as deserts, shorelines, the margins of ice sheets, and areas cleared of vegetation by agriculture or military operations. On other planets, such as Mars, the absence of vegetation and of liquid water allows wind to be a dominant driver of geomorphic change planetwide.

Conceptually, one can consider what has been termed the **sediment state** of a geomorphic system. In the case of aeolian geomorphic systems, defining the sediment state requires (1) identifying the source of wind-transportable sediment, (2) determining its availability for transport, and (3) considering the wind energy available to move the material. In most cases, sufficient wind energy is available to move sediment; the limiting factors are a source of sediment containing wind-transportable grain sizes and the availability of that sediment to move. Sediment availability is modulated by vegetation, moisture content, and the physical nature of the material at Earth's surface. Vegetated, cohesive, moist soils are too resistant to be moved at common wind speeds. In this context, it is easy to see why aeolian processes dominate in areas with loose, dry soils, scarce vegetation, and high wind speeds.

Wind can also be a major process in shaping coastal landscapes. Wind speeds tend to be higher at the coast than inland because the water surface is less rough than the land and because there are thermal contrasts between land and water. The heating and cooling of the land in daily cycles drives regular patterns of onshore and offshore winds. The abundant supply of fine sediment (sand and silt) from beaches and the longshore drift system (see Chapter 8) provide material for wind transport. Salt spray and disturbance from storms tend to limit the continuity and longevity of near-shore vegetation, episodically opening areas of bare soil to erosion by wind.

Landscapes at the margins of ice sheets are frequently affected by wind. Abundant unconsolidated sediment, delivered to the ice margin by streams running over and under the ice, is easily moved by wind once the sediment dries because newly deglaciated areas are unvegetated. Prolonged exposure of such sediment often leaves behind a lag of gravel and cobbles too large for prevailing wind speeds to transport. On glacial outwash plains, meltwater pulses supply fresh sediment each year, sediment that, when reworked, forms the dunes and deposits of wind-transported sediment sand and silt that are common near glacial margins and outwash streams [Photograph 10.4]. The contrast in

PHOTOGRAPH 10.4 Wind-Transported Sediment at Glacial Margins. Aeolian activity is frequent and widespread near glacial margins. Here, a dune of white rippled sand lies on the ablating margin of Nepal's Ngozumpa glacier in the Himalaya.

temperature between large ice sheets and the surrounding terrain can generate **katabatic winds** as cold, dense air descends off the ice. Such winds can kick up large plumes of dust from sparsely vegetated outwash plains [Photograph 10.5].

Areas altered by human activities are frequently modified by wind because disturbance removes anchoring vegetation and disturbs soil structure. Agriculture, construction, and military activities all disturb soil and provide a ready source of material for wind to transport [Photograph 10.6]. Wind can strip fertile topsoil. This fine sediment, lofted into the air from bare ground, fouls internal combustion engines, reduces visibility, and can have significant health effects.

PHOTOGRAPH 10.5 Wind Erosion Along Glacial Outwash Stream. Dust rises in the wind from Myrdalssandur outwash plain around a stream that drains a small ice cap on the southeastern coast of Iceland.

(a)

(b)

PHOTOGRAPH 10.6 Human Activity and Wind. Human activities disturb the ground surface, allowing wind to erode and transport sediment. (a) U.S. Army soldiers shield themselves from airborne dust kicked up from bare ground by a Medevac helicopter taking off outside of Kandahār, Afghanistan. (b) Farmer plowing a bare field raises a cloud of dust that is carried off by the wind.

Aeolian Processes

The geomorphic effects of wind can be considered in terms of disturbance, erosion, transport, and deposition—all related to the sediment state described above. The ability of wind to disturb vegetation communities and erode material depends on both the strength of the wind and substrate resistance (which is related to climate and vegetation, as well as soil properties and moisture content). Transport and deposition of sediment by wind depend on wind speed and its variability over time and space.

Disturbance

Strong winds, primarily during and just after storms, commonly disturb Earth's surface and thus catalyze other surface processes, particularly on hillslopes, the effects of which we consider explicitly elsewhere in the book. Of particular importance is the effect of wind on trees. High winds topple both single trees and whole sections of forests, especially on and near steep ridgelines where wind speeds are generally highest. When trees fall, entire root wads are ripped from the ground, and there is net downslope transport of regolith. This **wind throw** mixes (bioturbates) soils, triggers mass movements, and facilitates bedrock weathering (Photograph 5.6). Storm-induced floods also provide a source of

PHOTOGRAPH 10.7 Dry Lakebed. The white dusty playa is all that remains of Owens Lake in southern California after a century of water withdrawals from the watershed by Los Angeles Water and Power. Here, dust blows off the dry lakebed in a spring dust storm.

unconsolidated sediment that is easily eroded and transported by wind. Deposits of fine-grained overbank flood sediment are particularly susceptible to wind erosion before they revegetate.

Humans are major agents of disturbance, cutting forests, tilling fields, disturbing fragile desert soil crusts, and siphoning off water supplies of lakes in internally drained basins, which then shrink, facilitating wind erosion of the dry lakebeds [**Photograph 10.7**]. The result of anthropogenic disturbance is more dust transport in the atmosphere and, in some places, significant wind erosion of barren areas. Human-induced climate change could greatly decrease Earth's vegetation cover through increased droughts, more frequent forest fires, and the die-back of trees from diseases and insects spreading poleward in a warming climate. Warming and drying of the North American Midwest in response to global warming (as predicted by climate models) could cause reactivation and erosion of currently stable aeolian deposits if the vegetation dies back.

Erosion

Wind exerts forces on earth materials as it moves over them. Such forces include both lifting and shearing components. Critical to understanding the mobilization of granular, noncohesive sediment by wind is the difference between the **fluid threshold** and the **impact threshold**. The fluid threshold is the wind speed needed to erode material when no sediment is in motion. If the force exerted by the wind exceeds the resisting force of the substrate material (frictional strength, particle mass, or cohesion), sediment will be entrained and erosion will begin [**Figure 10.4**]. Fine-grained sediment, with a grain size between a few tens and a few hundred micrometers (thousandths of a millimeter), is the optimal size for erosion and transport by wind as indicated by the wind-speed minimum in the fluid threshold curve in Figure 10.4. The wind-speed minimum arises because grains smaller than about 100 μm

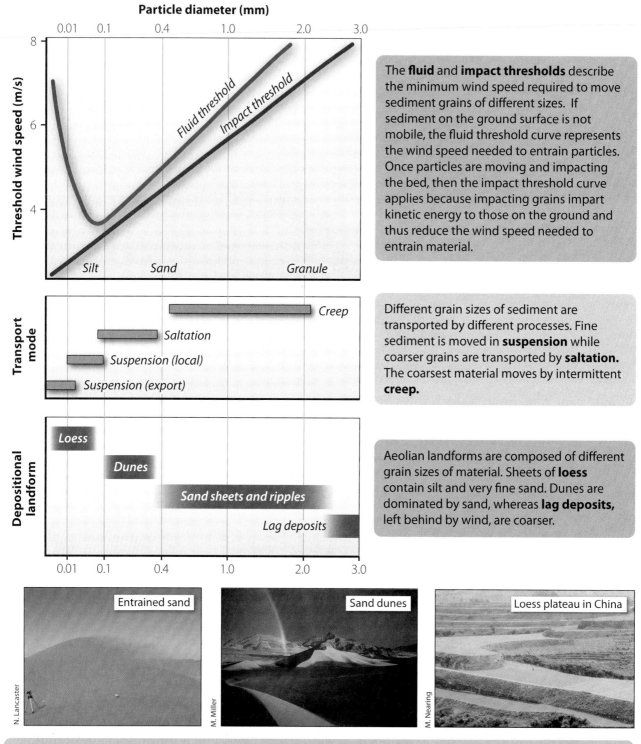

FIGURE 10.4 Wind Speeds Required to Move Sediment. Different wind speeds are required to initiate movement of materials of different grain sizes by wind. Different aeolian landforms are made of grains of different sizes.

(0.1 mm) are resistant to erosion because they are cohesive and because larger grain sizes have greater mass per grain and thus require higher wind speed to move.

The impact threshold is the wind speed required to erode sediment once some sediment is already in motion. Fluid-threshold and impact-threshold wind speeds are different; the impact threshold is always lower. This is easy to remember if one considers the physics of the process. Moving sediment impacts the ground surface, transferring some kinetic energy to surface grains; thus, the kinetic energy from wind necessary to mobilize sediment is reduced by the amount of energy already transferred to the particles by impact. For example, silt (wind-blown sediment of size <63 μm, known as loess) rarely gets mobilized without prior mobilization of fine sand, grains >125 μm. Once sand impacts and abrades a silt-rich ground surface, the critical wind speed for mobilizing the silt is lowered from the fluid-threshold value to the impact-threshold value, which, in the case of small grains, is a manyfold difference in speed (Figure 10.4).

Erosion of granular, noncohesive materials requires a sufficient lift force to remove grains from the bed; thus, in most wind regimes, transport is usually limited to sand-sized and smaller grains. Under the range of wind speeds typical above Earth's surface, coarse-grained granular material does not move because air, being a low-viscosity, low-density fluid, generates insufficient lift forces and momentum transfer.

Cohesive materials are generally resistant to direct wind scour because shear and lift forces generated at typical wind speeds are insufficient to overcome the cohesive strength and detach grains. For soils, apparent cohesion (see Chapter 5) can be provided by vegetation and root structures as well as by biologic crusts and the binding of soil particles by clay and soil moisture (see Photograph 10.3). Damp or water-saturated granular material is less easily eroded by wind. The **surface tension** provided by interstitial water (a type of apparent cohesion) resists aeolian lift forces.

In the southwestern United States, the development of Bt (clay-rich) and Bk (carbonate-rich) horizons during periods of several thousand years between times of active aeolian sediment transport appears to strengthen the soil. The additional cohesion provided by interstitial soil clay and soil carbonate plays a key role in limiting erosion of the previously deposited aeolian sediment.

Moving air often carries sediment, and the impacts of this sediment on cohesive materials can cause erosion by **abrasion**. Such abrasion creates wind polish on surfaces, frosts glass in sandstorms, and creates ventifacts, or wind-eroded rocks [**Photograph 10.8**]. It appears that maximum rates of aeolian abrasion occur within tens of centimeters of the ground surface, reflecting the net influence of the vertical distributions of both airborne-sediment concentration and wind speed.

Rates of aeolian erosion of bedrock are difficult to quantify and as such are poorly constrained in general. Reported global average rates of aeolian erosion range from 0.25 to >10 m/My, depending on assumptions used in the calculation. Rates for specific regions are higher, some in excess of 20 m/My. Point measurements suggest that short-term erosion rates can be higher yet, up to 1 mm in 15 years, which is about 70 m/My.

PHOTOGRAPH 10.8 Basaltic Ventifacts. This large boulder in Death Valley, California, has been eroded by wind and is about a meter wide.

Sediment Transport

Once detached, clay-, silt- and sand-sized material can be moved by wind. Similar to sediment transport in water, sediment moving in air can take different paths, depending on its grain size. These paths reflect the physical balance between the **settling speed** (eq. 10.1) and the lift and shear forces. Wind transport is typically partitioned into three processes: **suspension, saltation,** and **creep** [**Figure 10.5**].

Wind can transport fine-grained sediment in suspension long distances, hundreds to thousands of kilometers [**Photograph 10.9**]. Such sediment settles out downwind of the source, which is generally an arid or semi-arid region with minimal vegetative cover [**Figure 10.6**]. Dust storms, so famous from the Dust Bowl years of the 1930s in North America and today across North Africa and central Asia, are the most dramatic example of wind-suspended sediment. During the Dust Bowl era, dust from agricultural fields on the North American Great Plains rained out on East Coast cities, bringing an eerie darkness to midday. Such dust storms are not a thing of the past. In 2009, red dust from the center of Australia turned the Sydney sky orange at midday [**Photograph 10.10**]. Dust storms, which transport large amounts of fine-grained sediment long distances, are most common in semi-arid climates where sufficient water is available to weather rock to fine grain sizes but vegetation cover, which stabilizes soil surfaces and prevents wind erosion, is scarce. A number of studies show that dust is delivered to Earth's surface by rain associated with dust storms. One can easily verify this phenomenon in arid and semi-arid regions. On those days when only enough rain falls to leave a few raindrops on your

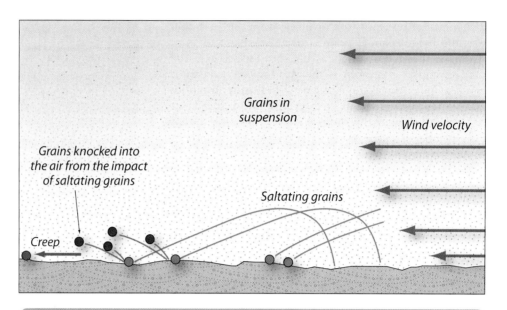

FIGURE 10.5 Processes of Aeolian Sediment Transport. Wind transport of sediment occurs by suspension, saltation, and creep.

Wind transports grains by three different processes. Small grains are kept in **suspension** by turbulence. Larger grains are dislodged from the surface by impacts and **saltate,** moving downwind in ballistic trajectories. The largest grains move by **creep,** pushed forward by impacting grains and rolled forward by the viscous drag of moving air.

windshield but not enough to completely wet and wash the window, let the drops dry and the dust they carried becomes evident.

Saltation is the episodic or hopping movement of particles up from the ground surface and back down over short distances. Most particles are lifted by wind only centimeters to decimeters above the ground, move forward with the wind, and then quickly fall back under the influence of gravity. Because saltating grains cannot rise far above the ground surface, the effective erosion zone for wind is very close to the ground surface as shown by the scalloped bases of telephone poles and fence posts eroded by abrasion and by deposition of wind-carried sediment [**Photograph 10.11**].

Wind creep is the movement of particles that are too large to be lifted by the wind. Some particle movement may be induced by the pressure difference between the

(a)

(b)

PHOTOGRAPH 10.9 Dust Storms. (a) In the 1940s, dust blows through the War Relocation Authority Center, Manzanar, California, to which U.S. citizens of Japanese ancestry were deported during World War II. This camp was in the shadow of the Sierra Nevada north of Lone Pine, California. The large clouds of dust are the result of humans disturbing the desert surface, although dust storms can occur in the absence of human disturbance. (b) Large cloud of windblown dust rolls into Yuma, Arizona.

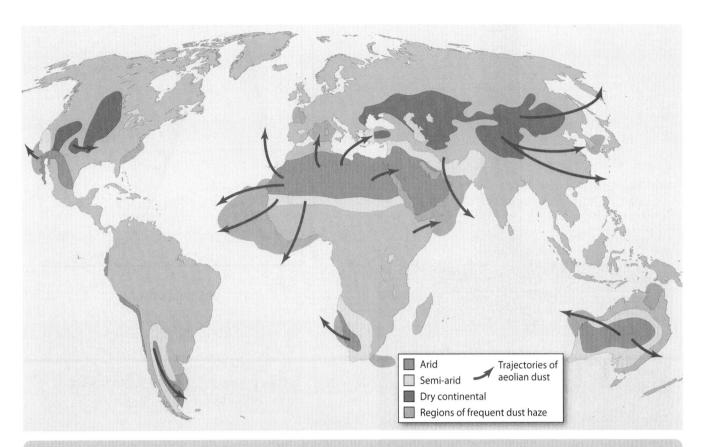

There are many sources of dust in the world—all of which are at least seasonally dry lands. This dust is transported by the prevailing winds and deposited as **loess** on land and contributes to the accumulation of fine sediment on the abyssal plains of the oceans.

FIGURE 10.6 Airborne Dust Sources and Trajectories. The most important sources of airborne dust are drylands. Dust is carried by prevailing winds, the direction of which is determined by overall atmospheric circulation.

PHOTOGRAPH 10.10 Dust Storm. The North Sydney Olympic Swimming Pool and the Harbor Bridge against a backdrop of dust in September 2009, Sydney, Australia. More than 1000 km of the eastern coastline of Australia was affected by the dust, which was the result of gale-force winds blowing southeast from the drought-stricken Northern Territory.

upwind and downwind sides of the particle. In general, creeping particles, although too big to be ejected from the ground surface, are rolled and pushed along the surface by the energy imparted to them by other impacting grains.

The collision between falling particles and those on the ground is important because it transfers energy and thus helps drive creep and eject different particles into the flow. Turbulence in the air results in vortices and gusts of wind that impact the ground surface, loosening grains and initiating aeolian sediment transport. Active saltation lowers the entrainment speed such that slower winds become more effective agents of erosion. Remember the difference between fluid threshold and impact threshold (Figure 10.4). Once saltation begins (when wind speed exceeds the fluid threshold), it is easier to sustain even if wind speed drops a bit (as long as the wind speed remains above the impact threshold).

Sediment transport by wind is a threshold phenomenon with significant feedbacks. At low wind speeds, nothing moves. When the fluid threshold is crossed and motion begins, saltating grains both dislodge other grains and abrade

PHOTOGRAPH 10.11 Wind Transport Profile. This telephone pole shows differential wind transport of material in the San Joaquin Valley after a wind storm in 1977 that stripped the region of much fertile soil in only ~25 hours. More mud is caked on the lower portion of the pole, nearer to the ground where transport of material is greatest. The dust was turned to mud in this area because it was blown across an irrigation canal and picked up moisture.

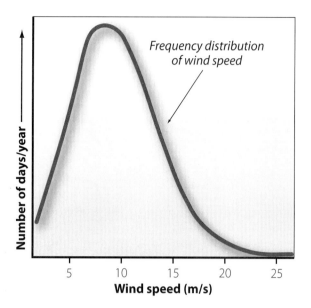

In most areas, the distribution of wind speed is not uniform. There are very few calm days and very few days with extremely high winds. Moderate wind speeds, around 10 m/s, are most common.

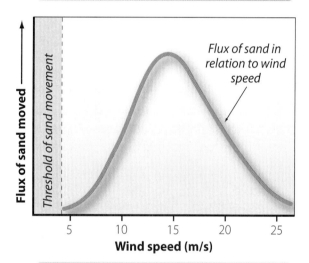

The flux of sand in a dune field is controlled both by the frequency of wind and by the transport efficiency of that wind. At low wind speeds, sand does not move. At high wind speeds, sand transport is very efficient but such winds are rare. The result is that most sand moves at higher-than-average wind speeds, about 15 m/s.

FIGURE 10.7 Wind Speed and Aeolian Transport. Most aeolian sediment is transported by moderate-strength winds. Low-speed wind transports no sediment until a distinct mobility threshold is crossed.

cohesive material—moving grains keep the system in action once action begins. Thus, a small increase in wind speed can translate into a very large increase in aeolian sediment transport. In other words, the rate of sand transport increases nonlinearly with wind speed. If one multiplies the annual record of wind speed with the sediment transport rate as a function of wind speed, it becomes clear that most sand is transported by moderate winds, those moving about 15 m/s (54 km/h) [**Figure 10.7**]. This transport maximum reflects a balance between the increasing sediment transport rate as wind speed increases and the relative rarity of very strong winds.

From the volume of fine-grained sediment delivered to the oceans, it appears that rates of aeolian deposition increased significantly at the onset of major Northern Hemisphere glaciation, about 2.7 million years ago. This increase likely reflects both the large amount of fine-grained sediment delivered from glaciers as well as increased

From Stokes' Law (eq. 10.1), we see that the settling speed is related to the square of the particle radius; thus, large particles will fall much more quickly through fluids than small ones. For aeolian transport, this relationship implies that fine-grained material can stay aloft longer than coarse material. Field data support this theory-based assertion. For example, the grain size of wind-deposited sediment reflects downwind sorting as the coarser particles settle out closer to the source than the finer particles [**Figure 10.8**].

Surface roughness is also important for determining where aeolian sediment will be deposited. Rough surfaces will slow moving air, reducing lift forces and causing sediment to be deposited. This effect is well demonstrated by coppice deposits of sand and silt that form adjacent to shrubs (roughness elements) in the desert [**Photograph 10.12**]; the shrubs disturb the air flow and sediment settles out in the low-wind-speed region downwind of the plant. Bouldery alluvial and debris-flow deposits as well as lava flows can be very rough, locally decreasing near-surface wind velocities and particle lifting forces; thus, these surfaces can function as effective aeolian sediment traps, leading over time to surface burial and, in some cases, the development of desert pavements (see Digging Deeper). Wind-formed features, like dunes, also contribute to surface roughness.

There are several reasons why wind-deposited sediment, on the whole, is better sorted than sediment deposited by moving water. Most important, the size distribution of material transportable by wind is limited: silt, sand, and in extreme cases very fine gravel that slowly creeps or rolls in the highest winds. For sand-sized material, the settling speed is high enough that transport is limited to saltation, an inefficient means by which to move sediment long distances. Silt, because of its low Stokes settling speed, is carried in suspension. With the gravel left behind and transport speeds of suspended silt being much higher than those of saltating sand, it is easy to see how sand dunes and dust deposits derived from the same source can be both very well-sorted and separated by kilometers or more.

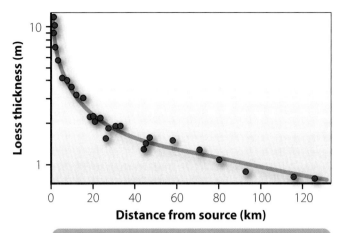

Deposits of **loess** are thickest near their source, in this case the Mississippi River during glacial times when it carried water and silt away from the Laurentide Ice Sheet. Away from the river, loess thickness decreases from more than 10 m near its source to less than a meter 100 km away.

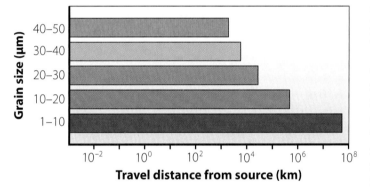

The average transport distance for loess goes up as loess grain size diminishes. Loess size here is given in micrometers, millionths of a meter. Large grains, those 40–50 μm in diameter, might travel 1000 km while grains only a few micrometers in diameter might travel several times around the planet before settling to Earth.

FIGURE 10.8 Wind Transport of Loess. Loess thickness and grain size changes downwind from the source. Grain size and thickness of loess decreases away from the source as larger-diameter grains settle more quickly than smaller grains. [Adapted from Livingston and Warren (1996).]

sediment availability because the climate on average grew colder and drier, and Earth's surface became less well vegetated. Faster wind speeds (responding to increased atmospheric pressure gradients) may have also contributed to increasing sediment transport by wind during glacial periods.

Deposition

Aeolian sediment is deposited when the settling speed of particles exceeds the lift forces provided by turbulent flow.

PHOTOGRAPH 10.12 Coppice Dune. This coppice dune behind a shrub on a dry lakebed (pan) at Sossusvlei in the Namibian Sand Sea (erg) is about 3 m long. Star dunes are in the distance.

Aeolian Features, Landforms, and Deposits

The distribution of both depositional and erosional aeolian landforms is nonuniform and reflects the importance of numerous covarying factors including wind speed, sediment moisture content, substrate erodibility, the area available to accommodate aeolian sediment, surface roughness, and vegetation density. Many landforms of aeolian deposition are also landforms of aeolian sediment movement or translation, meaning that they are temporary storage reservoirs of sediment that are constantly being remolded as sediment is transported through them.

Aeolian Erosional Features and Landforms

Ventifacts are wind-eroded rocks on the scale of decimeters to meters found in a variety of locations. They are prevalent in relatively arid regions or in regions that were arid in the past. Common in mid- and low-latitude deserts, ventifacts are also found in polar regions, particularly along valley bottoms occupied by sediment-loaded outwash streams. Presumably, katabatic winds from nearby glaciers suspended outwash sand and silt that abraded the ventifacts.

Ventifacts are often polished, faceted, and may have pits or **flutes** (elongated pits) on their surfaces (Photograph 10.8). These distinctive surface features suggest that ventifacts formed through abrasion of the immobile surface by saltating or suspended material (sand and silt). Most ventifacts are low to the ground, consistent with high sediment concentrations found in air near the ground surface. Faceting is more difficult to explain and may reflect the influence of different wind directions or the movement of the clasts over time. Such movement could result from frost action in cold regions or shrink-swell behavior in desert soils. The long axis of flutes eroded into rock is parallel to the predominant wind direction; thus, if the clast has not moved, the orientation of flutes can be used to estimate the direction of the strongest winds in the past, those capable of eroding rock.

PHOTOGRAPH 10.13 Yardang. Yardang eroded into soft, fine grain sedimentary rock in Tunisia. Wind blew from left to right.

(a)

(b)

PHOTOGRAPH 10.14 Wind Erosion on Earth and Mars. (a) Field of yardangs cut into soft, sedimentary rock in the Farafra Depression, Egypt's Western Desert. Wind blew from right to left. (b) Grooves sculpted by wind-blown sand near Olympus Mons on Mars. The image shows an area that is about 20 km wide.

Yardangs (from Turkish for "steep bank") are streamlined, positive-relief abrasional forms cut into bedrock or other cohesive earth materials by wind-driven sediment; they are much larger than ventifacts, ranging in size from meters to hundreds of meters in length and have a blunt upwind side and taper downwind like an inverted ship's hull [Photograph 10.13]. Yardangs are most likely formed where the prevailing wind is unidirectional. The long axis of a yardang is oriented in the same direction as the wind that eroded the feature. Fields of yardangs are found in many mid- and low-latitude deserts such as the Sahara [Photograph 10.14a]. Yardangs and unusual elongate erosional grooves (negative-relief features) have also been identified on Mars using remote sensing [Photograph 10.14b], confirming that wind is or was an active geomorphic process eroding the surface of Mars.

Blowouts and **deflation hollows** are areas where sediment has been removed by wind, forming a shallow pit or

Aeolian Features, Landforms, and Deposits **343**

depression. They usually extend no deeper than the water table, where apparent cohesion of the sediment prevents further erosion. Blowouts are most common in areas where sediment is weak, granular, or poorly cemented. For example, blowouts are common in dry lakebeds and in coastal dune fields between vegetated areas. Blowouts are also common in vegetation-stabilized sand sheets and dunes. Sometimes, small dunes composed of fine-grained sediment (termed **lunettes**) form immediately downwind of blowouts [**Photograph 10.15**].

Pavements, concentrations of interlocking clasts on desert surfaces, were once thought to form by deflation, or the selective winnowing and removal of fine sediment by wind [**Photograph 10.16**]. They are common on gently sloping surfaces in arid regions, and are especially well-developed on the low-gradient distal sections of alluvial fans. Although deflation can leave a residual lag of clasts, evidence collected since the 1980s suggests that many pavements are born at the surface and probably result from the gradual incorporation of wind-deposited silt over time (see Digging Deeper).

Aeolian Transport Features and Landforms

Landscapes dominated by aeolian processes contain a variety of landforms and features indicative of sediment transport by wind. At all scales, these features are composed predominately of granular, noncohesive material, primarily sand.

Regional accumulations of sand, known as sand seas or **ergs** (from the Arabic for "dune field") are the largest aeolian features [**Photograph 10.17**]. Ergs, which can cover

PHOTOGRAPH 10.15 Lunette and Blowout. Small lunette in Great Sand Dunes, Colorado, with larger blowout in the distance. The blowout is the source of sand in the lunette. Shrubs provide scale.

(a)

(b)

PHOTOGRAPH 10.16 Desert Pavements. Well-developed desert pavements have a single layer of interlocking, heavily rock-varnished clasts underlain by a stone-free, columnar, fine-grained Av horizon and a reddened B horizon. (a) View of pavement surface in Panamint Valley, California. Rock hammer is included for scale. (b) Cross-section view of a desert pavement in the Negev Desert, Israel, underlain by a 15-cm-thick Av horizon.

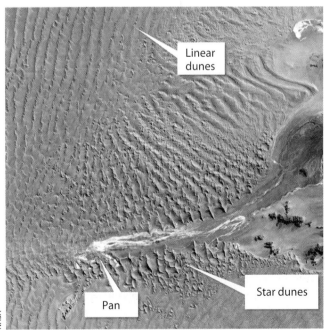

PHOTOGRAPH 10.17 Sand Sea. Namibian Sand Sea (erg) in the central Namibian desert is bordered by rocky mountains to the east and the coast to the west. The ephemeral Tsauchab River ends in the sand sea, its waters evaporating in the pan, or dry lakebed. There are both linear and star dunes. Width of photograph is about 75 km.

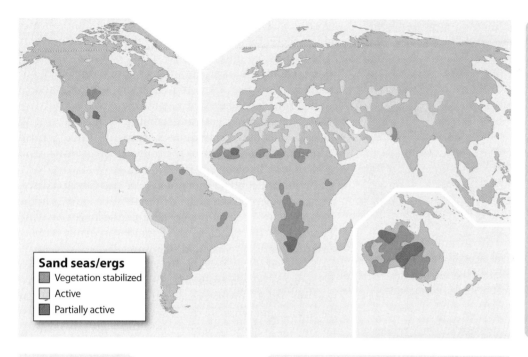

Ergs (sand seas) are found scattered throughout the mid- and low latitudes. Their presence reflects both the presence of sand and a dry climate with little vegetation to stabilize the soil. Today, some ergs are active with moving sand throughout, others have no activity because the sand is stabilized by vegetation. Ergs with limited activity have areas where vegetation cover is insufficient to completely stabilize the sand.

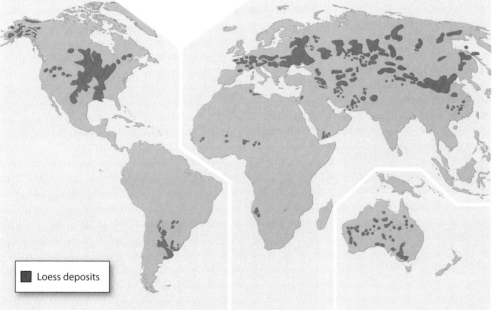

Loess deposits are widespread on Earth's surface and have varied origins. Some loess is derived from deserts, such as that in the Chinese loess plateau; other loess is sourced from glacial outwash, such as that along the Mississippi River or in other high latitude regions where glaciers or ice sheets were common.

FIGURE 10.9 Global Distribution of Ergs and Loess. Ergs are most common in dry, low- and mid-latitude locations with little vegetation. Accumulations of loess are found downwind of dust sources including deserts and the locations of former ice sheets. [Adapted from Thomas (1989) and Hugget (2003).]

thousands to hundreds of thousands of square kilometers, contain sand derived from either longshore drift along the coast or from direct deposition by rivers. Most active ergs are in arid and semi-arid regions and many are at or near the 30° latitude bands [**Figure 10.9**].

Relict ergs are common in areas that were either drier (less vegetation) or had greater sediment supplies in the past when climate conditions were different. For example, the Sand Hills of Nebraska in central North America have sufficient moisture in today's climate regime to retain a cover of stabilizing vegetation, but were an active erg during drier times in the Pleistocene and Holocene. Erg reactivation does occur, either from disturbance, such as the removal of vegetation, or from climate change when drying causes vegetation to die off. Ergs form over multiple cycles of activity. Optical luminescence dating of cores collected from ergs shows that sand has moved during different periods in the past.

At a smaller scale, and often located within ergs, are **dunes**. Dunes have many different forms, which can be interpreted to reflect different sand-transport processes and boundary conditions. Lee-slope (or **slip-face**) avalanching is common to all dune types [**Figure 10.10** and **Photograph**

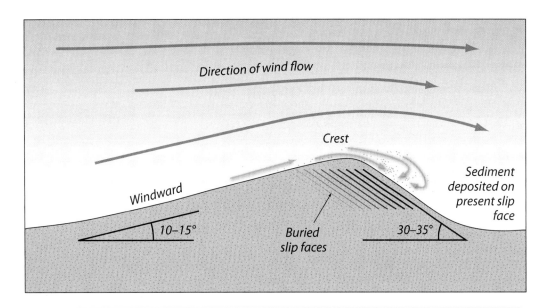

Wind carries sediment up and over dunes. Grains **saltate** up the backslope of dunes and avalanche down the steeper slip face. The dune moves or progrades downwind, burying previous slip faces and producing cross-bedded sand deposits.

FIGURE 10.10 Dune Migration. Free dunes migrate as sand is blown from the windward (upwind) to leeward (downwind) side of the dune. [Adapted from Summerfield (1991).]

10.18]. A fundamental categorization considers whether the dune is associated with (anchored to) a topographic feature or whether it is free to move across the landscape. Examples of **anchored dunes** include those downwind of topographic obstructions and vegetation (Photograph 10.12) as well as **climbing dunes** or sand ramps attached to cliffs and steep slopes. The orientation of active dunes reflects today's wind direction(s); by inference, the orientation of now-stabilized dunes reflects paleowind direction(s).

The creation and maintenance of both free and anchored dunes can be considered through the physics of sand transport. Creating a dune requires depositing sand that is intermittently in transport. Consider a patch of sand on an otherwise hard (elastic) substrate. If a saltating grain lands on the sand patch, more of its kinetic energy will be dissipated (by displacing other grains and imparting momentum to them) than if the grain bounced off the hard substrate. Over time, this contrast in behavior with impact causes small sand patches to grow as impacting sand grains lose energy and are trapped. Once a dune begins to form, sand accelerates over the low-gradient, upwind side, cascades over and flows down the downwind side before being dropped by the divergent, decelerating flow. Rates of dune movement downwind vary greatly; typical values appear to be tens of meters per year.

Free dunes develop independent of topography [Photograph 10.19 and Figure 10.11]. The number and

PHOTOGRAPH 10.18 Slip-face Avalanching. Gypsum sand avalanching down the slip face of a dune at White Sands, New Mexico.

PHOTOGRAPH 10.19 Star Dune. Large star dune that developed and moved independent of topography in Sossusvlei region, Namib-Naukluft National Park, Namibia.

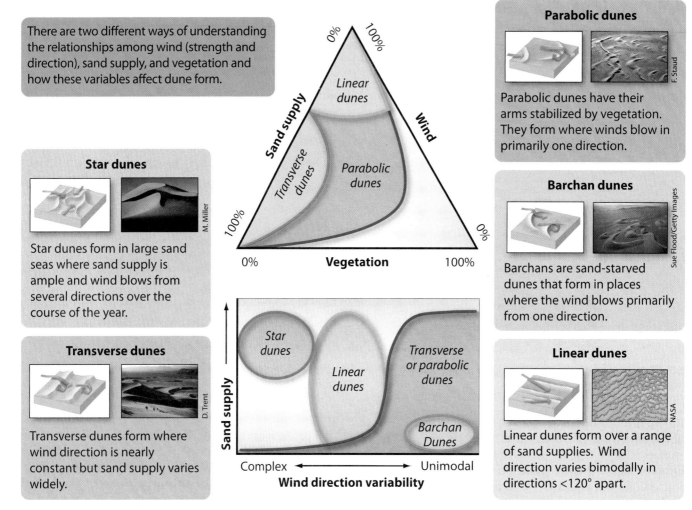

FIGURE 10.11 Dune Types. The morphology of sand dunes and the processes that control their shape and mobility reflect both sand supply and the variability of wind direction. [Adapted from Hack (1941) and Livingston and Warren (1996).]

orientation of slip faces are key to interpreting the process of dune formation and inferring the wind regime. Free dunes are categorized as transverse or linear. Sand transport is perpendicular to the crest of **transverse dunes** and parallel to the crest of **linear dunes** that have slip faces on both sides of the crest—active alternately as wind direction changes seasonally. **Star dunes** have slip faces oriented in several different directions; they are interpreted as reflecting dune development in an environment where wind direction changes seasonally, resulting in little net sand transport. Note that the key factor in the formation of star dunes is not the frequency of wind-direction changes, but that winds come from several different directions during the year.

There are two types of transverse dunes. Both form **crescentic ridges** in areas where there is little available sand and the wind blows predominately from one direction. The tips of **barchan dunes** move more rapidly downwind than the slower main body of the dune, presumably because there is less energy dissipation by loose sand at the tips on the firm desert surface than over the main part of the dune. In contrast, **parabolic dunes** have their upwind arms anchored by vegetation. Many parabolic dunes form in areas where the underlying sand is vegetated and unavailable for transport until the stabilizing vegetative cover is disturbed. When the vegetation is disturbed, parabolic dunes originate from **blowouts,** bowl-shaped, erosion pits (Photograph 10.15) formed where wind erodes a formerly stabilized sand surface.

Superimposed on these larger dunes are smaller **ripples,** ubiquitous asymmetric bedforms that cover dune surfaces [Photograph 10.20]. The upwind sides of ripples are gently sloping and the downwind slopes are generally steeper, like dunes. Ripples are ubiquitous in sandy areas without vegetation. Individually, ripples tend to be short-lived, their orientation adapting rapidly to changing wind direction. Sediment grain size appears to control the wavelength and height of ripples. Most ripples have wavelengths of centimeters to meters and heights of centimeters, and they are oriented perpendicular to the

PHOTOGRAPH 10.20 Ripples on Dunes. Ripples cover this large sand dune at the Great Sand Dunes National Monument, Colorado.

wind flow [**Photograph 10.21**]. Low slope angles for ripples, 2 to 7 degrees for stoss slopes and 2 to 10 degrees for lee slopes suggest that suspension rather than saltation and avalanching are the dominant sand-transport processes on these fine-scale features.

Repetitive patterns are common characteristics of aeolian features. Such replication leads to **dune fields** (extensive areas where dunes are of similar size and shape), and on the small scale it leads to fields of similarly shaped ripples on the slopes of dunes.

Aeolian Dust Deposits and Loess

A globally important effect of aeolian sediment transport is the deposition of silt-size, wind-blown sediment known as **loess**. Loess is the most geomorphically important aeolian sediment because of its tremendous geographic spread, its significance as parent material for agricultural soils, its erodibility, and the ability of continuous loess sections to preserve a long record of Quaternary climate and paleoenvironmental change.

The source of loess varies. Isotopic and geochemical tracing shows that some loess is derived from glacial outwash, other loess is derived from broad, poorly vegetated dryland basins, and some loess is derived from the erosion of soil developed on weak, fine-grained rock such as shale. Today, the fine-grained material that is deposited as loess originates primarily from the world's deserts. In glacial times, much loess was derived from glacial outwash plains—an inference supported by the thinning and fining of loess deposits away from these sources (Figure 10.8).

Distinct beds of loess cover an estimated 5–10 percent of Earth's surface and up to 30 percent of the United States [**Photograph 10.22**]. Blankets of thick loess cover much of midwestern North America, the Pampas of Argentina, and central Europe. In Alaska, extensive deposits of loess preserve important archaeological and fossil sites. Parts of China retain an astonishing thickness of loess (locally more than 100 m) in the huge Chinese loess plateau [**Photograph 10.23**] that records more-or-less continuous deposition (at varying rates) for more than 3 million years. The Chinese loess preserves one of highest-resolution, long-term terrestrial records of changing climate and landscape response on the planet.

Loess deposition tends to be ongoing, forming **cumulic** (or accumulating) soils (see Chapter 3); however, the rate of loess deposition changes over time as conditions in both the source and deposition areas change. In many places, loess deposition rates increased during glacial times when the climate was drier, colder, and windier. During interglacial times, loess deposition slowed, allowing soil development to proceed at a rate sufficient to create distinct soil horizons

PHOTOGRAPH 10.21 Sand Dunes and Loess. Sand blowing across ripples in the dune fields of the Gobi Desert. Saltating sand grains collide and fracture if wind speeds are high enough, forming silt that is deposited as desert loess.

PHOTOGRAPH 10.22 Fertile Loess Deposits. Thick, fine-grained, tan loess deposits in the Palouse region of southeastern Washington State are fertile areas in which to grow wheat. Here, the loess is exposed in a road cut several meters high, below the wheat. The rounded hills are typical of areas with thick loess, which blankets and buries pre-existing topography.

PHOTOGRAPH 10.23 China Loess Deposits. In the loess plateau of northern China, the hillslopes are extensively terraced to reduce erosion of the loess from runoff.

[Photograph 10.24]. If distinct layers of loess are seen in soil profiles, accumulation rates must have been high, otherwise loess deposited at low rates would have been stirred into and mixed with the soil by biological and physical turbation processes.

Even in areas where distinct blankets of loess cannot be mapped, the addition of wind-deposited dust may also significantly alter soil properties (see Digging Deeper). Some surfaces, such as desert pavements, are excellent natural dust traps. Incorporation of dust infiltrating down into a permeable soil may eventually clog pore spaces and decrease permeability enough to alter the intensity and pace of chemical weathering as well as the stability of surficial materials. If water cannot infiltrate because subsurface pores are clogged by fine sediment, it will run off, creating rills and eventually gullies. Near the ocean and on dry inland salt lakes or playas, wind can erode and carry sediment rich in salts (Photograph 10.7). When these salts are deposited on the surface of rocks they can increase rates of physical weathering as they dissolve and recrystallize (see Chapter 3).

Fine-grained soils (the parent material of which is often loess) are particularly vulnerable to both wind and water erosion if stripped of their vegetation cover. For example, during the Dust Bowl era of the 1930s, wind eroded loess-derived soils from the North American Great Plains and sent clouds of soil, in the form of dust, eastward. These soils had been stable since the last glaciation, held tightly under the thick mollisol A horizons and the root mats of prairie grasses until sod-busting plows removed the vegetative cover. When nearly a decade of drought hit the Great Plains in the 1930s, there was nothing to hold the dry, fine-grained soil in place, and it blew away [Photograph 10.25].

Recent research has shown that many soils contain large amounts of dust delivered from far-off sources by wind. For example, the presence of elements such as chromium, thorium, and zirconium in soils on the calcium carbonate-dominated Caribbean island of Barbados indicates significant

PHOTOGRAPH 10.24 Paleosols and Loess. Red paleosols alternate with loess in a section exposed on the Chinese loess plateau. The exposure is 150 m high and includes loess deposited over the past 2 million years.

PHOTOGRAPH 10.25 Dust Bowl Soil Erosion. Wind stripped soil from fields during the Dust Bowl years of the 1930s, burying farms in sand and dust, as in this photo from Dodge City, Kansas, 1935.

aeolian contributions including volcanic ash from the island of St. Vincent, dust eroded from Africa, and Mississippi River Valley loess. U.S. Geological Survey data indicate that African dust is an important component of soils developed on many western Atlantic Ocean islands and that aeolian dust, both because of the nutrients associated with it and because it contains clay with high cation-exchange capacity that increases nutrient retention, is important for sustaining regional vegetation. Similarly, soils in the Hawaiian Islands contain significant amounts of continentally derived dust from Asia, identified by its quartz content. We know this dust is derived from afar because the basaltic rocks of Hawaii contain no quartz.

The global flux of dust changes significantly over several different timescales. Climate controls dust flux on geologic timescales; most important are linked changes in water balance and global hydrology [**Figure 10.12**]. Over glacial-interglacial cycles, dustiness ebbs and flows with the coming and going of ice sheets. Glacial periods tended to be dustier because extensive outwash plains provided a source for dust, the climate was drier, vegetation near the ice was reduced, and the atmosphere was likely windier. On millennial timescales, warmer and drier climates, such as those prevalent in some parts of the world during the middle Holocene, likely led to increased dustiness. In the last millennia, humans have greatly increased the global dust flux over its natural

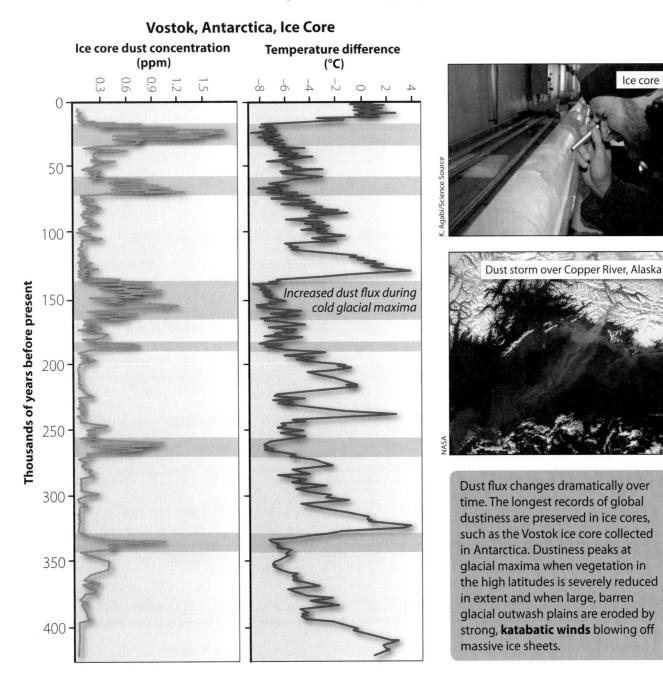

FIGURE 10.12 Dust Flux Changes Over Time. The ice core from Vostok in Antarctica records changes in dust flux over the past 400,000 years. More dust is delivered to the ice during glacial periods when temperatures are cold, vegetation is reduced, and the climate is, in general, drier.

background levels, primarily by removing vegetation, plowing fields, grazing animals, and using unpaved roads for vehicles.

Applications

Understanding wind's direct and indirect effects on Earth surface processes and the distribution of earth materials is critical for a variety of land management applications. Wind's direct geomorphic effects are most easily noticed where water is scarce and fluvial and hillslope processes are slow or limited in extent. The indirect effects of aeolian processes are far more widespread but less apparent. They include biological and physical disturbances caused by wind and the pervasive aeolian addition of moisture-holding, fertile, fine-grained parent material to soils, including rich soils used for agriculture that are critical for human survival in much of the world. Wind transport can move fine-grained material long distances in suspension before depositing it in diverse environments including silt-rich soil A horizons that support desert pavements (see Chapter 3; see also Digging Deeper) and the sediments that coat the deep abyssal plains of the world's oceans (see Chapter 8).

Wind both erodes and deposits materials at Earth's surface. Wind-induced erosion can rapidly strip valuable topsoil from agricultural areas, lofting fine-grained material in suspension as great dust storms and depositing it downwind tens, hundreds, even thousands of kilometers away. The massive 2009 dust storm in Australia that blanketed Sydney in an orange haze and the 2012 storm that shut down Arizona's interstate highways indicate the widespread and continuing nature of aeolian sediment transport.

Dry, terminal lake basins are important sources of silica- and metal-bearing dusts at a global scale. Some of these dusts can contain toxic or hazardous air pollutants. For example, desiccation of several large, internally drained lakes including Owens Lake in southern California (Photograph 10.7) and the Aral Sea in central Asia have allowed strong winds to erode a mix of salts that precipitated as these lakes dried. People and livestock living downwind are exposed to toxic dust that includes arsenic, chromium, copper, molybdenum, nickel, lead, silica, and uranium at levels equal to or higher than those found in industrialized areas.

Loess, wind-deposited silt, is the parent material for some of Earth's most fertile soils. Because loess soils are fine-grained, they are susceptible not only to wind erosion but to erosion by flowing water. Rates of soil loss can be astounding. In parts of the Palouse of eastern Washington State (Photograph 10.22), where wheat yields are extraordinarily high on native soils, more than a meter of this rich, wind-derived soil has been lost in the past century. Such rates of erosion are unsustainable and if they continue, will result in severely diminished crop yields. Various conservation techniques have been developed and applied to reduce wind erosion of agricultural soils including planting wind breaks and orienting fields perpendicular to predominant winds. Conversely, encroachment of wind-transported sediment can be a problem in coastal zones and around oases where dunes and sand are migrating [Photograph 10.26].

Human actions can destabilize regions underlain by fine-grained, wind-deposited sediment. The removal of trees and shrubs for agriculture in marginal, semi-arid landscapes, such as portions of west Africa, triggers desertification (drying of the land surface) by changing the soil's water-holding capacity and the regional water balance. Both the change in water balance and the removal of effective cohesion provided by the plants can expose soils to greatly accelerated wind erosion.

Deposition of wind-transported sediment can have unexpected consequences. For example, human disturbance of stable soil profiles in the southwestern United States has increased dust loads to Rocky Mountain snowpacks. The darkening of the snow has reduced the reflection of sunlight, increased spring melting, and reduced summer streamflow, leading to water-management concerns.

There are numerous examples of dunes being reactivated when the vegetation is removed. For example, once-stable coastal dunes on Cape Cod, Massachusetts, became extremely active when European settlers removed the trees that held the sand in place. Revegetation, instituted by the U.S. National Park Service over the past several decades, has slowed rates of dune migration and kept most of the sand off the main highway. Post-glacial dunes in interior New England also were reactivated by colonial deforestation. There, revegetation programs in the 1930s planted hundreds of thousands of trees to stabilize bare, sandy slopes.

Military activities in the world's deserts are inexorably linked to and often constrained by aeolian processes (Photograph 10.6a). Some dust storms are natural, the result of strong winds, often generated by severe convective storm cells. However, human impacts on fragile desert soils often exacerbate the problem. For example, the disturbance of desert pavements by wheeled and tracked military vehicles exposes the fine-grained soil below. Once

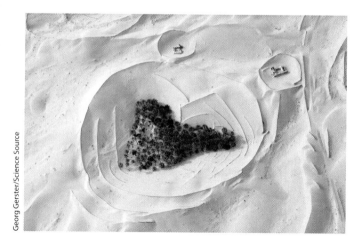

PHOTOGRAPH 10.26 Wind and Oases. This aerial view of Souf Oasis in Algeria shows the series of wind fences placed to protect the palm trees from being overrun by sand moved by wind in the Sahara Desert.

a pavement has been disturbed, it can take decades to reestablish; meanwhile, rare but intense rainfalls and strong desert winds scour the silt and lift billowing clouds of dust into the air. The resulting high concentrations of airborne particles clog air filters and damage engines and military equipment.

Stabilized or buried dunes can serve a variety of purposes that benefit society. For example, aquifers such as the Ogallala, which supports much of the irrigated agriculture of the western United States, are hosted in part within permeable, ancient dune sands. Areas of western North America covered by sand sheets with high infiltration rates are less likely to be affected by erosive processes that lead to deep arroyo cutting and massive sediment export because the permeable nature of the sand retards sapping and retrogressive headcut advance. On the Colorado Plateau, a surprisingly large area of the landscape that was once dominated by fluvial erosion has become progressively buried by aeolian materials, perhaps the result of Pleistocene aridity. This influx of aeolian sediment had resulted in profound changes in subsequent landscape evolution mostly related to differences in infiltration capacity and cohesion.

Wind-deposited sediments are an important archive of Earth's history. For example, the orientation of dunes and wind-sculpted features such as yardangs allows us to decipher wind patterns of the past, and older, continuous loess deposits preserve in their geochemistry and physical sedimentology a record of a changing Pleistocene climate. Deciphering geologic records of aeolian activity helps us to understand Earth's response to prior warming, cooling, wetting, and drying. Understanding the response of Earth's aeolian system to past changes in climate is of particular importance as we try to predict our planet's reaction to human-induced climate change.

Selected References and Further Reading

Bagnold, R. A. *The Physics of Blown Sand and Desert Dunes*. Chapman and Hall: London, 1941.

Bauer, B. O. Contemporary research in aeolian geomorphology. *Geomorphology* 105 (2009): 1–5.

Bristow, C. S., G. A. T. Duller, and N. Lancaster. Age and dynamics of linear dunes in the Namib Desert. *Geology* 35 (2007): 555–558.

Ewing, R. C., G. Kocurek, and L. W. Lake. Pattern analysis of dune-field parameters. *Earth Surface Processes and Landforms* 31 (2006): 1176–1191.

Gillies, J. A., W. G. Nickling, and M. Tilson. Ventifacts and wind-abraded rock features in the Taylor Valley, Antarctica. *Geomorphology* 107 (2009): 149–160.

Goudie, A. S., and N. J. Middleton. *Desert Dust in the Global System*. Berlin, New York: Springer, 2006.

Hack, J. T. Dunes of the western Navajo Country. *Geographical Review* 31 (1941): 240–263.

Hugget, R. J. *Fundamentals of Geomorphology*. Abingdon, Canada: Routledge, 2003.

Kocurek, G., and N. Lancaster. Aeolian system sediment state: Theory and Mojave Desert Kelso dune field example. *Sedimentology* 4 (1999): 505–515.

Laity, J. E. "Landforms, landscapes, and processes of aeolian erosion." In A. D. Abrahams and A. J. Parsons, eds., *Geomorphology of Desert Environments*, 2nd ed. New York: Springer, 2009.

Lancaster, N. *Geomorphology of Desert Dunes*. London: Routledge, 1995.

Livingstone, I., and J. E. Bullard. "Dust." In A. D. Abrahams and A. J. Parsons, eds., *Geomorphology of Desert Environments*, 2nd ed. New York: Springer, 2009.

Livingstone, I., and A. Warren. *Aeolian Geomorphology: An Introduction*. New York: Longman, 1996.

Livingstone, I., G. F. S. Wiggs, and C. M. Weaver. Geomorphology of desert sand dunes: A review of recent progress. *Earth Science Reviews* 80 (2007): 239–257.

Macpherson, T., W. G. Nickling, J. A. Gillies, and V. Etyemezian. Dust emissions from undisturbed and disturbed supply-limited desert surfaces. *Journal of Geophysical Research* 113 (2008): F02S04. doi: 10.1029/ 2007JF000800.

McKee, E. D., ed. *A Study of Global Sand Seas*, Professional Paper 1052. United States Geological Survey, 1052.

Montgomery, D. R., J. L. Bandfield, and S. K. Becker. Periodic bedrock ridges on Mars. *Journal of Geophysical Research* 117 (2012): E03005. doi: 10.1029/2011JE003970.

Muhs, D. R. The geologic records of dust in the Quaternary. *Aeolian Research* 9 (2013): 3–48.

Muhs, D. R., J. R. Budahn, J. M. Prospero, and S. N. Carey. Geochemical evidence for African dust inputs to soils of western Atlantic islands: Barbados, the Bahamas, and Florida. *Journal of Geophysical Research* 112 (2007): F02009. doi: 10.1029/2005JF000445.

Parsons, A. J., and A. D. Abrahams, eds. *Geomorphology of Desert Environments*. New York: Springer, 2009.

Summerfield, M. A. *Global Geomorphology*. New York: Prentice Hall, 1991.

Sun, J., and D. R. Muhs. "Mid latitude dune fields." In S. A. Elias, ed., *Encyclopedia of Quaternary Science*. Amsterdam: Elsevier, 2007.

Thomas, D. S. G. *Arid Zone Geomorphology*. Chichester: Wiley-Blackwell, 2011.

Warren, A., A. Chappell, M. Todd, et al. Dust-raising in the dustiest place on Earth. *Geomorphology* 92 (2007): 25–37.

Wells, S. G., L. D. McFadden, and J. D. Schultz. Eolian landscape evolution and soil formation in the Chaco dune field, southern Colorado Plateau, New Mexico. *Geomorphology* 3: 517–546.

Wolfe, S. A. "High latitude dune fields." In S.A. Elias, ed., *Encyclopedia of Quaternary Science*. Amsterdam: Elsevier, 2007.

Yair, A., and Y. Enzel. "The relationship between annual rainfall and sediment yield in arid and semi-arid areas: The case of the northern Negev." In F. Ahnert, ed., *Geomorphological Models—Theoretical and Empirical Aspects*, Catena Supplement No. 10, pp. 121–135. Cremlingen-Destedt, Germany: Catena Verlag, 1987.

DIGGING DEEPER Desert Pavements—The Wind Connection

Stone pavements, flat, closely packed collections of clasts, cover many desert surfaces. These pavements are present on residual weathering mantles, overlie lava flows, and form on alluvial deposits. In many areas, the exposed surfaces of pavement stones are covered by a thick coating of dark, shiny rock varnish, a manganese-iron-rich surface coating found on stable rock surfaces in arid regions.

Pavements are known by different names in different regions of the world. They are gibber plains in Australia and regs in North Africa. In central Asia, gobi is the term used to describe pavements. In the southwestern United States, they are known as desert pavements. Some pavements are exceptionally old, such as those in Israel for which ^{10}Be measurements suggest ages of several million years, making the plains on which the pavements reside some of the oldest large-scale landforms found to date (Matmon et al., 2009).

Most pavements are only a clast or two thick and are underlain by vesicular, fine-grained A horizons centimeters to a few tens of centimeters thick with columnar structure indicative of shrink-swell behavior driven by clays. The A horizon is composed predominantly of quartz-rich, sandy silt and is punctuated by vesicles, small air-filled pockets. It is known as an Av horizon with the "v" indicating that it is vesicular. The A-horizon vesicles result from trapped air bubbles locked in place when the soil dries after rainfalls. Below the Av horizon is a massive, fine-grained B horizon with few clasts [Figure DD10.1].

As a landform, pavements were long enigmatic. Pavement formation has been attributed to a variety of processes including selective fluvial erosion washing away fine material, deflation by wind leaving a lag of coarse clasts, and upward migration of stones carried by a clay-rich argillic B horizon (born-in-place model). Wind figures prominently in the latter two of these three explanations.

Sheet flooding (overland flow) has been suggested as a mechanism by which to move clasts and to remove the silt and thus the matrix between the clasts. There is evidence that sheet flooding can effectively move clasts and even arrange them in a crude mosaic where they lay next to each other on an impermeable substrate, in one case, a recent basalt flow (Williams and Zimbelman, 1994). However, the sheet-flood model of pavement formation does not explain well the presence of the stone-free A horizon underlying the pavement. Similarly, the deflation model, where wind removes the fines and leaves the clasts, cannot explain the underlying Av horizon [Figure DD10.2].

The born-in-place or accretionary model (McFadden et al., 1987) has evolved from observations over the past several decades (starting with Cooke, 1970), is gaining wider acceptance, and has been tested using measurements of a cosmogenic stable nuclide, ^{3}He (Wells et al., 1995).

The accretionary model posits that clasts originally at the surface stay at the surface as fine-grained, wind-transported

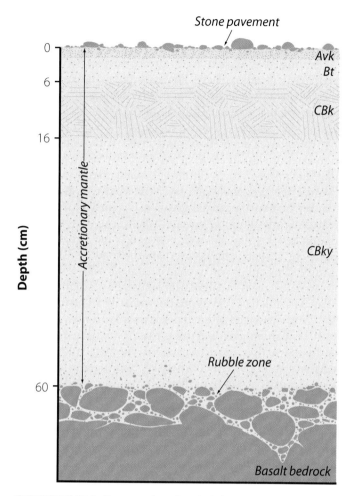

FIGURE DD10.1 Cross section of typical desert pavement shows the one- to two-clast-thick layer at the surface underlain by the thin Av horizon with some calcium carbonate (the letter k designation). Below the Avk horizon, is a Bt horizon (where clay has accumulated, as indicated by the letter t). It is underlain by a two-part, carbonate-rich C horizon where the letter k indicates the presence of carbonate and the letter y indicates the presence of gypsum. The letter B following the C indicates that C-horizon properties dominate but that some B-horizon properties are present. The basalt bedrock from which Wells et al. (1995) believed the surface clasts originated is at the base of the cross section. [From Wells et al. (1995).]

material accumulates beneath them. The process is thought to work like this. The initially rough, rocky surface (such as a fresh deposit on a desert fan or a new basalt flow) functions as a dust trap, slowing the wind and encouraging deposition of aeolian silt and clay. Each year, small amounts of aeolian material are added to the surface. The silt is washed between the clasts and into the developing soil, accumulating in thickness over time.

Clay, brought in on the wind with the dust, is critical to cumulic soil development (the thickening of soil over time). Because clay shrinks and swells as it wets and dries, its

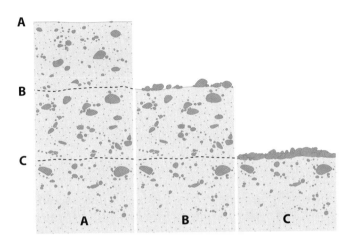

FIGURE DD10.2 Deflation, the removal of fine material from geomorphic surfaces by wind, has long been proposed as a means by which pavements could form. In this view, the pavement is the lag deposit of gravel left behind because wind speeds, although sufficient to remove silt and sand, were insufficient to remove gravel. Over time (indicated by the letters A, B, and C), the amount of gravel on the surface increases as fines from the underlying deposit are removed. The dashed line shows the depth to which the surface will erode in later time steps. [From Cooke (1970).]

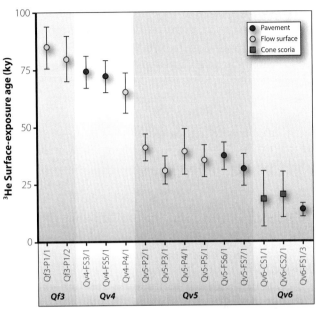

FIGURE DD10.3 Cosmogenic ^{3}He exposure ages measured in desert pavement clasts, basalt flow surfaces, and scoria from basalt flows of four different ages in Cima volcanic field, Mojave Desert. For any one unit, the ages of the clasts in the pavement are indistinguishable for the ages of the flows from which they were derived. This study shows that clasts in the pavements had the same exposure history as the flows, which would be possible only if the clasts had remained at the surface since they were erupted. [From Wells et al. (1995).]

behavior enables aeolian material to get below the pavement. Shrinking produces conduits (fractures) that enable transport of soil water and associated suspended clay, solutes, and fine silt below the soil surface, making possible continued deposition and pedogenic alteration of the fine-grained material. Over time, soil development in the fine-grained dust produces peds defined by columnar structure. After thousands of years, the source of the clasts, be it a lava flow or an alluvial deposit, is buried by accumulated aeolian material while the loose clasts continue to "float" at the surface. If this model is correct, pavements are born at the surface and rise over time as dust accumulates below them.

The accretionary model was verified for pavements in the Mojave Desert using the cosmogenic nuclide, ^{3}He. Wells et al. (1995) collected clasts from several pavements developed on basalt flows of varying age in the Cima volcanic field, an active area of basaltic volcanism in southern California. The concentration of ^{3}He in pavement clasts closely matched that in samples collected from basalt outcrops near the pavements [**Figure DD10.3**]. Such a match could only happen if both the pavement clasts and the basalt-flow surfaces had the same exposure history (exposure age), revealing that the pavement clasts, just like the sampled basalt flow surfaces, have been at the surface since they were incorporated into the pavement.

The accretionary pavement model is attractive because it can explain all of the observed features of desert pavements: the paucity of clasts in the B horizon, similar cosmogenic ages for clasts and associated outcrops, and the thick Av horizon.

Cooke, R. U. Stone pavements in deserts. *Annals of the Association of American Geographers* 60 (1970): 560–577.

Matmon, A., O. Simhai, R. Amit, et al. Desert pavement–coated surfaces in extreme deserts present the longest-lived landforms on Earth. *Geological Society of America Bulletin* 121 (2009): 688–697.

McFadden, L. D., S. G. Wells, and M. J. Jercinovich. Influences of eolian and pedogenic processes on the origin and evolution of desert pavements. *Geology* 15 (1987): 504–508.

Wells, S. G., L. D. McFadden, J. Poths, and C. T. Olinger. Cosmogenic ^{3}He surface-exposure dating of stone pavements: Implications for landscape evolution in deserts. *Geology* 23 (1995): 613–616.

Williams, S. H., and J. R. Zimbelman. Desert pavement evolution: An example of the role of sheetflood. *Journal of Geology* 102 (1994): 243–248.

DIGGING DEEPER Desert Pavements—The Wind Connection (continued)

WORKED PROBLEM

Question: Stokes' Law (eq. 10.1) describes the settling of solid material though a viscous fluid assuming laminar-flow conditions:

$$S_s = 2r^2(\rho_p - \rho_f)g/9\mu_f$$

First, calculate the settling speed (S_s) of coarse silt (63 μm diameter, r = 0.0000315 m) and sand (500 μm diameter, r = 0.000250 m). Assume a rock density (ρ_p) of 2700 kg/m³ and a density of air (ρ_f) of 1.225 kg/m³. The dynamic viscosity of air (μ_f) at 20 °C is 1.983×10^{-5} kg/(m × s).

Then, consider how long the sand and silt grain would each take to settle from the top of a 1-km-high dust cloud that resulted from a major sand storm.

Answer: Once you fill in the variables in eq. 10.1, the only change you need to make is r, the radius of the particle, which for consistency with other variables in eq. 10.1, needs to be expressed in meters. The answer is in meters per second (do the unit analysis):

$$S_s = 2r^2(2700 - 1.225)9.8/(9 \times 1.983 \times 10^{-5})\,\text{m/s}$$

Multiplying and dividing out all the nonvariable terms, this expression reduces to

$$S_s = r^2 \times 2.96 \times 10^8 \text{ m/s}$$

Incorporating grain radius yields a settling speed of about 18.5 m/s for 500 μm grains, medium sand. For 63 μm grains, coarse silt, the settling speed is about 0.29 m/s. Hence, it would take sand grains only (1000 m/18.5 m/s) or 54s to settle from a 1 km elevation, whereas it would take silt grains (1000 m/0.29 m/s) more than 3400s or nearly an hour to settle the same distance.

KNOWLEDGE ASSESSMENT Chapter 10

1. Identify in what settings aeolian processes are the dominant geomorphic actors and explain why.
2. What is the predominant mineral in most aeolian sediment? Explain why.
3. Explain the influence of biologic activity on the intensity of aeolian geomorphic processes.
4. List some of the geomorphic effects of wind.
5. Compare and contrast the physical properties of water and air and explain what differences control the geomorphic effectiveness of these fluids.
6. Define turbulence and explain why it is important for understanding aeolian geomorphic processes.
7. What is a sediment state and why is it useful for understanding aeolian geomorphic systems?
8. Explain the difference between fluid and impact thresholds.
9. Explain how erosion processes differ for cohesive and noncohesive materials.
10. What is ventifaction and how and where does it occur?
11. What is aeolianite, how does it form, and where are you most likely to find it?
12. What was the effect of widespread Quaternary glaciation on rates of aeolian sediment deposition?
13. Make a diagram illustrating the differences in movement and typical particle size in transport for suspension, saltation, and creep.
14. What is Stokes' Law and why is it important for understanding aeolian geomorphic processes?
15. Which grain size is optimal for wind erosion? Explain why.
16. What is loess and what is the primary control on the grain size and thickness of loess deposits?
17. What is a yardang and where might you find one?
18. Draw a cross section of a desert pavement and explain the most commonly accepted mechanism of formation.
19. Explain how soil development strengthens soil over time, reducing the likelihood of wind erosion.
20. Compare the sizes and longevity of ergs, dunes, and ripples.
21. Give two examples of the aeolian geomorphic effects of high surface roughness.
22. Dunes can be separated into two distinct categories. List those categories and explain how they differ.
23. Explain how barchan, parabolic, and star dunes form and move, highlighting similarities and differences.
24. Give three examples of the importance of dust/loess in soil formation.
25. Explain how we know that the flux of dust carried by wind changed over time and why such changes have occurred.
26. Describe how aeolian sediment transport impacts humans and how humans impact aeolian sediment transport.
27. Explain how aeolian sediment transport contributes to the fertility of soils.

Volcanic Geomorphology

11

Introduction

Volcanoes are a critical part of the Earth system—making new crust, changing the shape of Earth's surface, emitting gases that change climate, affecting weather, and loading rivers draining volcanic terrain with sediment. Volcanic processes are dramatic geomorphic agents, creating new and distinctive landforms during eruptions. Volcanoes contain large volumes of relatively weak, often unconsolidated material high above surrounding river valleys and adjacent lowlands. Volcanic landscapes are shaped by disturbance from repeated eruptive episodes that create land at high elevation, erode preexisting landforms, rearrange drainage courses, overload rivers with rocks and mud, and cover landscapes in rock and erupted rock particles. Geomorphic work is done both by active volcanic processes and on the volcanic landforms that volcanic processes create.

The distribution of volcanoes on Earth is not uniform. Volcanic processes and landforms are concentrated along plate margins and hot spot tracks. In some tectonic environments, such as along subduction zones, spreading centers, and hot spots, volcanoes are the dominant geomorphic agent. They generate earthquakes, lava flows, and landslides, changing Earth's surface by erosion and/or deposition of materials. Although much of Earth's surface has no active volcanoes, volcanic activity affects the entire planet through the influence of volcanic emissions, both dust and gases, on climate. Most of the water on Earth and the gases in our atmosphere were emitted through volcanoes.

Volcanoes are landforms created when molten rock (**magma**) reaches Earth's surface. As such, volcanoes often mark zones of lithospheric weakness where magma

P. W. Lipman/USGS

The May 1980 eruption of Mount St. Helens volcano in southwestern Washington State dramatically changed the landscape in the surrounding area, removing the north flank of the mountain, covering thousands of square kilometers in ash, and choking streams that drained the volcano with debris. Geologists in foreground provide scale against ash cloud that erupted in August 1980.

IN THIS CHAPTER

Introduction
Distribution and Styles of Volcanism
   Magma Chemistry and Volcano Morphology
   Tectonic Forcing and Volcanic Provinces
Eruptive Mechanisms and Products
   Lava Flows
   Pyroclastic Flows and Falls
   Volcanic Gases
Eruption Sizes and Types

Volcanic Landscapes
   Landscapes of Basaltic Volcanism
   Landscapes of Silicic Volcanism
Processes of Volcanic Landform Evolution
   Geomorphic Effects of Magma Intrusion
   Biologic Colonization
   Denudation and Aging
   Mass Movements
   Lahars
   Volcano-River Interaction

   Hydrologic Considerations
   Erosional Landforms
Applications
Selected References and Further Reading
Digging Deeper: Geomorphic Effects of Volcano Sector Collapse
Worked Problem
Knowledge Assessment

emission is focused. Magma can be erupted **effusively**, flowing onto Earth's surface as **lava**. Magma can also erupt explosively. This dichotomy forms the basis for eruption classification as well as the explanation for fundamentally different emplacement processes, eruptive products, and landforms.

Large-scale volcanic landforms can be long-lasting. Even after a volcano stops erupting, its landscape-scale geomorphic effects continue. Ongoing geomorphic processes in volcanic terrain include erosion of the volcano, lithologic control on the biologic colonization of new volcanic landscapes, hydrothermally induced weathering of volcanic rocks, and volcano/river interactions that influence both local and regional hydrology and ecology. Another key linkage is that volcanic processes and their products control the development of hydrologic flowpaths (surface and subsurface) that, in turn, localize geomorphic processes.

Volcanoes and thus volcanic geomorphology are not limited to Earth; remote-sensing data show that volcanism has been active on Venus, Mercury, and the Moon. On the moon, extensive fields of volcanic rocks as young as 800 million years cover 16 percent of the visible surface (the maria, from Latin for "sea"). The largest volcano in the solar system, Olympus Mons, stands 27 km above the surrounding plains on Mars, and ongoing eruptions have been photographed on Jupiter's moon Io.

In this chapter, we present volcanoes as geomorphic systems that have distinctive geographies, dynamics, landforms, and evolutionary trajectories. Understanding volcanic geomorphology requires a basic understanding of volcanic rocks and volcanic processes. Most important is understanding how styles of volcanism vary widely depending on the tectonic setting, and that these stylistic differences set the stage for geomorphic processes that follow.

Distribution and Styles of Volcanism

If you could see Earth from space, you would recognize the importance of volcanic activity in shaping the major geomorphic features of our planet [**Figures 11.1** and **11.2**]. Planetary-scale landforms shaped by volcanism include **mid-ocean ridges**, the linear elevated rift zones where new oceanic crust is created, and **volcanic arcs**, linear arrays of volcanoes parallel to subduction zones. Regional-scale volcanic landforms visible from space include large **calderas**, craters left by massive, explosive eruptions, and **flood basalt provinces**, areas where huge volumes of basalt (a dark rock low in Si and high in Fe) covered the landscape, burying most preexisting landforms.

Volcanic systems create new crust and change ocean-bottom topography at spreading centers, the largest and most continuous of which are the mid-ocean ridges. Mid-ocean ridges are large, low-gradient landforms; their morphology reflects the supply of magma and the speed of spreading. Fast spreading produces raised ridges (East Pacific Rise) from thermal uplift, whereas slow spreading centers, such as the Mid-Atlantic Ridge, are dominated by down-dropped grabens (fault-bounded basins) superimposed on broad, thermally uplifted ridges. **Seamounts**, underwater volcanoes, form from magma sources on the flanks of ridges.

Volcanic arcs result from subduction and are defined by a series of volcanoes aligned parallel to, but displaced down-dip from, the trench axis where subduction occurs. Subduction under continental lithosphere leads to strings of volcanoes on land, parallel to the coast—the Cascade volcanic arc in the northwestern United States, for example. Subduction under oceanic lithosphere creates an **ocean island arc**, a string of volcanic islands. The distance from the trench to the line of volcanoes (known as the **arc-trench gap**) is determined by the dip of the subducting slab. Under normal geothermal gradients (the increase of temperature with depth below Earth's surface), the slab must descend to a depth of about 100 km before it is hot enough to melt overlying mantle and the water-rich slabs of downgoing rock and marine sediments that generate magma [**Figure 11.3**]. Shallowly dipping slabs result in long arc-trench gaps. Steeply dipping slabs generate short arc-trench gaps.

Caldera eruptions cause geomorphic change on a massive scale. The evacuation of magma chambers in large catastrophic eruptions removes preexisting topography, leaving in its place fault-bounded depressions. The best-known example of a caldera is Crater Lake in the Cascade Mountains of Oregon, which formed about 7700 years ago when Mount Mazama erupted 50 to 60 km³ of volcanic material [**Photograph 11.1**]. However, the caldera at Crater Lake is tiny when compared to the immense Yellowstone Caldera (90 km long, 50 km wide). Other major calderas in North America include Long Valley, California, and the Valles (Jemez) Caldera in northern New Mexico [**Photograph 11.2**].

The legacy of caldera eruptions also includes the emplacement of large amounts of silica-rich **volcanic ash** (volcanic eruptive particles <2 mm), such as the Bishop Tuff in Long Valley, which smothers and smooths the preexisting

PHOTOGRAPH 11.1 Crater Lake. Occupying the caldera of Mount Mazama, Crater Lake was formed 7700 years ago. Wizard Island is the small volcano emerging from the lake. The caldera is 9 km across.

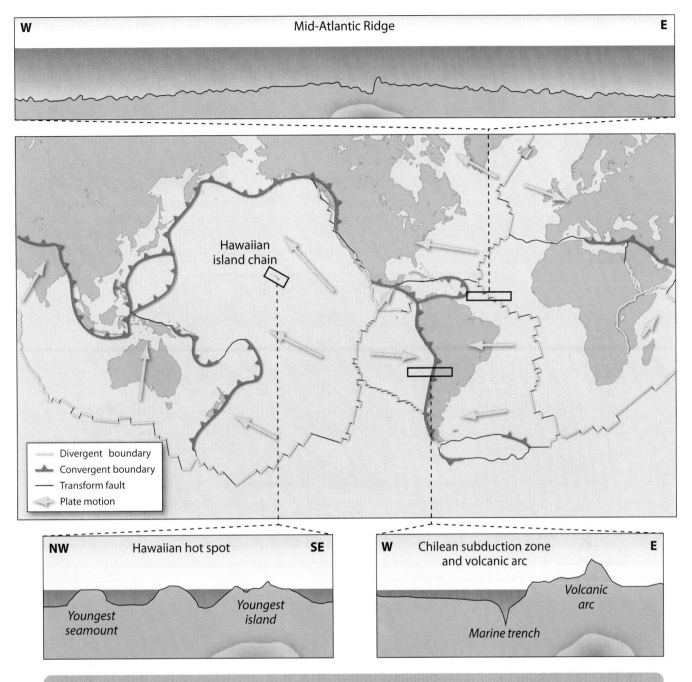

Different **plate boundary** types have different morphologies, different associated types of volcanism, and are found in different parts of the world. Oceanic **hot spot** volcanoes, such as the Hawaiian Islands, are isolated high points in the ocean basins. The Mid-Atlantic Ridge is a more subtle topographic feature, whereas **subduction zones** are characterized by steep and deep trenches and adjacent, high topography of the **volcanic arc**. The cross sections are vertically exaggerated.

FIGURE 11.1 Volcanically Active Plate Margins and Hot Spots. The cross sections of different volcanically active continental margins (including subduction zones and mid-ocean ridges) as well as hot spots reflect the style of volcanism and the physical characteristics of the erupted lava.

landscape [**Photograph 11.3**]. A caldera complex can erupt repeatedly over hundreds of thousands to millions of years, forming nested basins down-dropped on caldera-margin normal faults. Between caldera eruptions, viscous magma produces **resurgent domes** (steep-sided lava extrusions) within the caldera complex [Photograph 11.2 and **Photograph 11.4**].

The largest eruptions, **flood basalts,** can continue intermittently for millions of years and cover thousands to tens of thousands of square kilometers with basaltic lava piling

Volcanic activity is concentrated along plate margins and at hot spots. Different tectonic settings produce different styles of volcanism and thus create different and characteristic types of landscapes.

Oceanic island arc

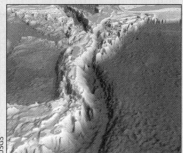

Island arcs are long, linear features parallel to oceanic subduction zones; they are made up of islands built by **stratovolcanoes** over a basaltic platform.

Mid-ocean ridge

Spreading centers at **mid-ocean ridges** generate mountain chains thousands of kilometers long, the thermally buoyant crust of which rises above the adjacent ocean floor.

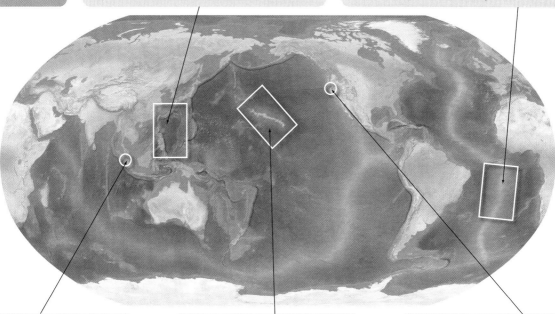

Caldera

Explosive, silica-rich eruptions can create immense volcanic **calderas** like the one that underlies Yellowstone National Park or the one at Toba in Indonesia, as shown here.

Hot spot

Oceanic hot spots generate linear chains of basaltic islands, which, as they thermally subside, become **seamounts.** Older islands are progressively more eroded and slowly sink below the ocean surface.

Continental volcanic arc

Volcanic arcs, created by offshore **subduction zones** and built onshore by the eruptions of stratovolcanoes, run parallel to the coast of continents.

FIGURE 11.2 Large-Scale Volcanic Geomorphology. The large-scale geomorphology of different volcanically active areas on Earth is related to magma composition and thus tectonic setting.

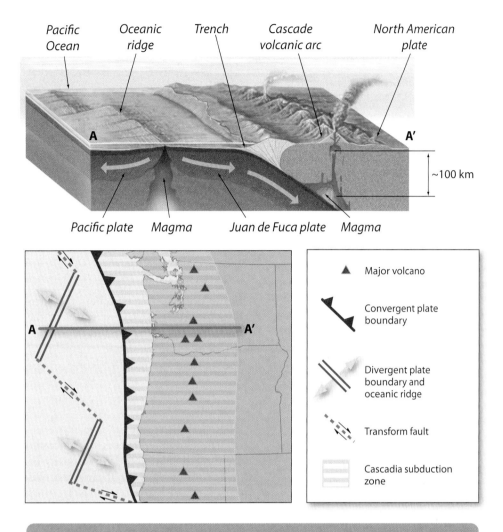

FIGURE 11.3 Geomorphology of Volcanic Arcs. Subduction-related stratovolcanoes often lie in a line parallel to the subduction zone reflecting the geometry of the subducting slab.

Subduction of oceanic lithosphere beneath continental lithosphere creates both distinctive arc-trench morphology and a chain of **stratovolcanoes** inland of the subduction zone. These volcanoes are aligned because partial melting of rock and generation of magma occur when the downgoing slab reaches warm enough temperatures, usually at a depth of about 100 km.

up to depths of a kilometer or more. Major flood basalt provinces include the Deccan Traps (India) and the Columbia Plateau of Washington and Oregon in western North America. The impact of flood basalts on landscape evolution is both immediate (blocking valleys and changing drainage networks) and long-lasting, their flow units eroding in characteristic stair-step topography with shattered flow tops being highly conductive to groundwater flow [**Photograph 11.5**].

The global factors controlling the distribution of volcanoes, the style of volcanism, the type of lava that volcanoes emit, and the overall geomorphology of volcanic terrain lie largely in the domain of two subdisciplines of geology—geochemistry and tectonics.

Magma Chemistry and Volcano Morphology

The intensity and spatial extent of volcanically induced geomorphic change is directly related to the composition of the magma feeding the volcano. Magma gas content and chemistry, particularly the amount of silica and water that magma contains, determine eruptive style, eruptive products, and thus the geomorphic effects of volcanic eruptions. The link between magma composition and geomorphology is the **viscosity**, or resistance to flow, of magma (**Table 11.1**).

Magma becomes less viscous (flows more easily) if it is hotter, contains fewer crystals, and has less silica (basaltic composition). Conversely, highly viscous magma is relatively cool and high in silica (andesitic or rhyolitic composition).

TABLE 11.1

Physical and Chemical Characteristics of Lava Types

Material	Viscosity (Pa s)	Si content (%)
Basaltic lava	50	45–52
Andesitic lava	10^6 to 10^7	57–63
Rhyolitic lava	10^{11} to 10^{12}	69–77
Water	5×10^{-4}	N/A

As magma nears the surface and erupts, its viscosity influences lava flow thickness and the steepness of flow margins.

Eruption explosivity is determined by magma viscosity. Highly viscous, silica-rich magmas prevent gases, which **exsolve** (come out of solution) as the magma moves toward the surface, from escaping rapidly. This allows pressure to build up and increases the potential for explosive eruptions. In contrast, less viscous basaltic lava rarely degases explosively.

The yield strength of lava, its ability to resist flowing, is particularly relevant to volcanic geomorphology. One can use a simple approximation that disregards the control on lava-flow cooling and emplacement caused by crust formation to determine the yield strength at the point in time when the flow stops. Assuming that lava is a perfectly plastic material that has a finite yield strength, deformation

PHOTOGRAPH 11.3 **Bishop Tuff.** Exposure of Bishop Tuff that erupted from Long Valley Caldera and covered the landscape about ~750,000 years ago.

(flow) does not occur unless the shear stress driving the flow exceeds the yield strength of the flow. Thus, if you know the thickness (h) and density (ρ) of a lava flow and the slope (θ) on which it stopped flowing, you can calculate its yield strength (S_y) from:

$$S_y = \rho g h \sin\theta \qquad \text{eq. 11.1}$$

where g is gravitational acceleration.

The character, speed, and morphology of lava flows are directly related to lava viscosity because more viscous lava requires a greater shear stress to deform than less viscous lava. Viscous lava flows are thick, resulting in steep landforms when they cool. The gravitational driving stress specified by $\rho g h \sin\theta$, must be great enough to deform viscous lava. Only steep slopes (high θ) covered by thick flows (high h) generate sufficient shear stress to move this viscous, sticky, Si-rich lava. When a high-viscosity andesite flow reaches the low slope at the base of a volcano, it does not have the shear stress to continue

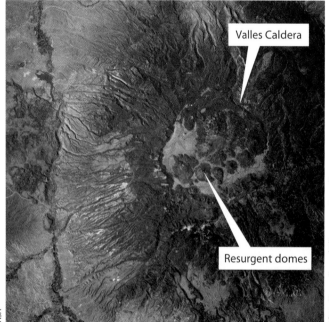

PHOTOGRAPH 11.2 **Valles Caldera.** The 19-km-wide volcanic Valles or Jemez Caldera in New Mexico is circular and there are resurgent domes within the caldera walls.

PHOTOGRAPH 11.4 **Resurgent Dome.** Steep-walled, resurgent dome in the Valles (Jemez) Caldera, New Mexico, sits above the caldera floor.

Tectonic Forcing and Volcanic Provinces

Magma is generated and volcanoes are present at all tectonic boundaries (Figures 11.1 and 11.2) but less commonly within plates. Volcanoes exist where conduits, zones of weakness, form and bring magma to the surface. Such zones of weakness are most often found where rock is warm and soft or where there is extensive tectonic activity that weakens rock, such as hot spots, mid-ocean ridges, faults, and areas of crustal extension.

Eruptive mechanisms and the resulting volcanic geomorphology are related to tectonic setting, which in large part controls magma chemistry and gas content; together, these two variables determine eruptive style. Explosive eruptions are most common at convergent margins and on hot spot tracks in continental provinces, whereas nonexplosive, effusive eruptions are common in basaltic provinces such as spreading centers and oceanic hot spots.

Subduction zone volcanism produces high-relief landforms. Rugged subduction zone topography reflects both the upthrusting of rocks from collision and the resulting compression, as well as steep, largely intermediate (andesitic) stratovolcanoes known for their explosive eruptions. Additional volcanism (usually basaltic) can also occur in extensional subdomains within the otherwise compressive subduction system. Subduction-related volcanic arcs can produce prodigious amounts of lava and ash because magma is continually supplied by subduction and melting of oceanic lithosphere. Consider that several such arcs have each produced thousands of cubic kilometers of volcanic material (the volume of Lake Michigan) over just the past 2 million years. Arc stratovolcanoes are mostly andesitic but many are constructed on a large mafic platform (made up of rocks, such as basalt, dominated by iron and magnesium-rich minerals) that comprise most of the erupted volume.

Mid-ocean ridge volcanism creates large amounts of new crust every year, much more than subduction zones do. Basaltic magma, derived from shallow mantle sources, rises up and erupts at the ridges, which are characterized both by normal faults along their axes and transform faults connecting ridge segments (see Figure 11.3). This lava is rapidly quenched by the cold seawater creating characteristic **pillow basalts**, rounded forms caused by the rapid chilling of molten basalt. At and near mid-ocean ridges, **hydrothermal** circulation through the hot rocks rapidly weathers new basalt and basaltic glass, sometimes forming deposits of metals (sulfide ores). As new rock is created at the ridge and older rock moves away from the ridge over time, the older ocean floor is slowly covered by sediment raining out from the water above; this cover subdues the local relief of the sea floor by burying the volcanic topography. As the new crust cools, it grows steadily denser and subsides. Thus, the ocean floor, on average, gets steadily deeper away from mid-ocean ridges (see Chapter 8).

PHOTOGRAPH 11.5 Flood Basalts. Stacked flood basalts, Palouse River, Washington, erode to form characteristic stair-step topography reflecting flow units.

moving. Thus, the flow stalls, cools, and solidifies. Conversely, less viscous basaltic magma can travel in thinner flows on lower slopes before resisting forces exceed driving forces and the flow stops and solidifies.

The viscosity contrast among basaltic, andesitic, and rhyolitic lava determines the large-scale geomorphic character of volcanic landforms associated with each type of lava [**Figure 11.4; Photograph 11.6**]. Basaltic **shield volcanoes** (edifices many kilometers wide, built over time by numerous flows) are larger and less steep than either andesite-dominated **stratovolcanoes** (edifices kilometers wide and built of layered andesite flows, ash, and mudflow deposits), or rhyolite-dominated **volcanic domes** (steep-sided plugs of rock often less than a kilometer wide). Individual basalt flows can extend tens of kilometers over the landscape; more viscous, silica-rich (silicic) lavas flows (andesite and rhyolite) are typically restricted to areas immediately adjacent to the volcano from which they originated.

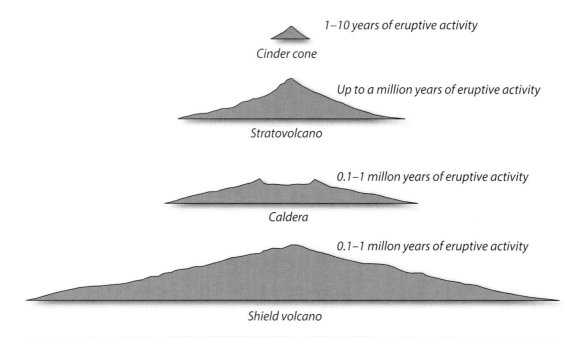

FIGURE 11.4 Volcano Size and Shape. The size, shape, and slope of volcanic edifices vary greatly from cinder cones that are only kilometers wide to shield volcanoes hundreds of kilometers wide at their bases. [Adapted from Decker and Decker (2006).]

When tectonic plates drift over hot spots, the volcanic results are geomorphically diverse. For example, consider the hot spot track in northwestern North America [**Figure 11.5**]. When the North American plate first moved over the hot spot in the Miocene (17 million years ago), copious flows of fluid basaltic lava (some several hundred cubic kilometers in volume) covered eastern Washington and Oregon with hundreds of meters of rock. These flood basalts overwhelmed low-lying areas of the state, flowed through the Columbia River gap in the Cascade Mountains, reached the Pacific Ocean, and thus can be used to infer that the Columbia River Valley is at least Miocene in age. As the North American plate drifted southwestward, volcanism became active farther east in Montana and Wyoming, forming the features for which the Yellowstone area is famous, including the giant rhyolitic caldera eruptions 2.1 million, 1.2 million, and 640,000 years ago that ejected so much volcanic ash that traces can still be found over most of North America.

In oceanic settings, hot spot volcanoes emit large amounts of basaltic lava, creating submarine plateaus and volcanic islands. Oceanic hot spot volcanic edifices (such as the Hawaiian Islands) are the largest landforms on the planet, towering more than 10,000 m from the ocean floor to their peaks over 4000 m above sea level. Because these ocean volcanoes rise thousands of meters from the abyssal plains to the ocean surface, they are much larger than any continental subduction zone volcanoes. As they weather subaerially and move away from the thermally uplifted area nearest the hot spot plume, the mean elevation of each island diminishes progressively until it becomes an atoll covered in coral sand and surrounded by a fringing reef (Figure 8.13). Eventually, the rate of coral growth on the reef can no longer keep up with the thermal subsidence rate and the atoll sinks below the waves, becoming a flat-topped seamount under water.

Volcanism also occurs along some major strike-slip and normal faults, typically those where there is a tensional or stretching component to their motion. For example, along the range-front fault system that defines the eastern margin of the Sierra Nevada in southern California, there are numerous basaltic cinder cones [**Photograph 11.7**]. Strike-slip fault volcanism is common in parts of Mexico. The volcano El Chichón, which erupted in 1982 and killed several thousand people, is located along a transform fault.

PHOTOGRAPH 11.6 Lava Types. Different types of lava create different sizes and shapes of volcanic landforms. (a) Basaltic shield volcano, Fernandina, Galapagos Islands. (b) The Three Sisters in Oregon are andesitic stratovolcanoes. (c) Steep-sided Little Broken Top and Rock Mesa rhyolitic lava flows in the Three Sisters Wilderness, Oregon.

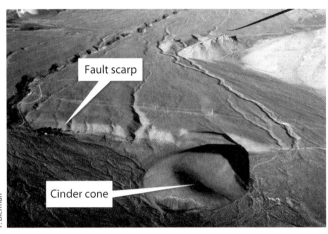

PHOTOGRAPH 11.7 Cinder Cone. Fish Springs cinder cone, cut by a fault scarp of the Owens Valley fault system, is 300,000 years old and mostly on the downthrown block. Variably offset alluvial and debris-fan deposits wrap around the cinder cone. Fan-crossing stream channels have incised the upthrown block in response to base-level fall induced by the normal faulting.

Eruptive Mechanisms and Products

Volcanoes emit a variety of eruptive products—**lava flows, pyroclastic materials,** and **volcanic gases**—each of which is geomorphically important on different spatial and temporal scales.

Lava Flows

Eruption of lava, molten rock, is the most salient characteristic of volcanoes. Lava flows come in all shapes and sizes, their geomorphology in large part determined by the composition of the parent magma. Basaltic magma, being relatively fluid, can move large distances, covering preexisting topography and resurfacing landscapes. On low-relief terrain, fluid basaltic lava creates long, gently sloping flows and can be erupted either from discrete conduits (**vents**) or from long, linear **fissures.** Smaller basaltic flows tend to be longer than they are wide. Such flows typically follow preexisting valleys. If there are no valleys, basalt flows create deposits that are long and sinuous [Photograph 11.8a], reflecting the flow of the lava funneled by the chilled, immobile margins of basalt that restrict lateral movement of the lava.

In contrast, silica-rich lava flows are more geographically restricted in their impact. Stiffer andesitic and dacitic lava generally creates thick, steep, short flows [**Photograph 11.8b**], whereas very stiff rhyolitic lava commonly extrudes in plugs and domes that do not flow significant distances (see Photograph 11.6c). Silicic lavas extrude slowly, creating steep-sided landforms such as rhyolitic domes, which tend to increase rather than diminish local relief. Obsidian forms from noncrystalline lava. Most

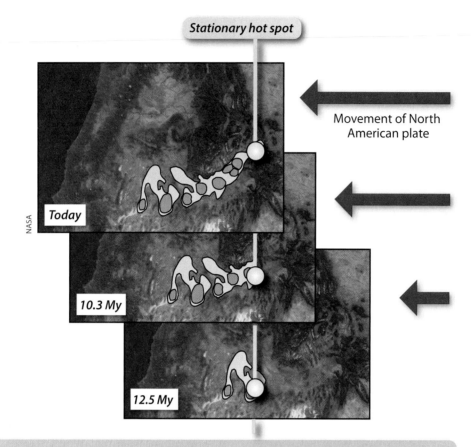

Hot spots represent a source of heat originating in Earth's mantle. As a continental plate moves over the stationary hot spot, the hot spot, a thermal anomaly, leaves its mark in the form of volcanoes and volcanic rocks. Here, a hot spot under the North American plate is responsible for the volcanism in the Snake River Plain and in Yellowstone—volcanism that gets younger to the east.

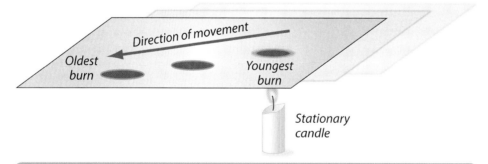

In a now famous demonstration, J. Tuzo Wilson, a scientist whose work helped convince geoscientists that plate tectonics was a valid and useful theory, used a stationary candle and a moving piece of paper to demonstrate how hot spots worked. The candle was a physical model of a hot spot. It was stationary while the paper (the plate) moved over it, leaving a track of burned paper analogous to the strings of volcanoes that mark the passage of tectonic plates over hot spots.

FIGURE 11.5 Hot Spot Movement. Hot spots leave volcanic tracks. The track of the Yellowstone hot spot records changes in the position of the North American plate as it moved westward over the past 12.5 million years.

(a)

(b)

PHOTOGRAPH 11.8 Lava Viscosity. Lava of different viscosity creates flows with different morphology. (a) Basaltic lava of low viscosity flows over the landscape in Hawaii. The glowing breakout is ~60 cm across. (b) Silica-rich dacite lava at Mount St. Helens creates a steep-side dome about a year after the 1980 eruption.

PHOTOGRAPH 11.9 Basalt Flow. In 1973, along the Chain of Craters Road in Volcano National Park, Hawaii, a fluid basalt flow moved over these trees, coating them with lava that was rapidly quenched and thus hardened around the trees. Charcoal is preserved inside the lava coating.

large obsidian flows are rhyolitic. Glassy obsidian was valued by Native Americans for its ability to be worked into sharp projectile points and knives.

Basaltic lava flows can be dated by K/Ar (potassium/argon) if they are at least hundreds of thousands of years old. Much younger felsic flows (a few tens of thousands of years) can be dated using K/Ar because they contain more K and thus more K-derived argon relative to K-poor basaltic lavas of comparable age. Flows of any composition less than 50,000 years old can be dated by ^{14}C if there is any fossil organic material, such as charcoal, associated with the flow [**Photograph 11.9**]. Recently, cosmogenic techniques, specifically ^{3}He in olivine, have been used to date basalt flows, but accurate dates require that uneroded flow surfaces can be found and sampled.

Pyroclastic Flows and Falls

Pyroclastic materials, rock fragments of volcanic origin, are expelled from volcanoes by rapidly expanding gases. These gases exsolve from rising magmas as confining pressures are reduced during eruption. Pyroclastic materials result from the **exsolution** process; the glass shards represent the material between the exploding vesicles in the lava and, once erupted, become the volcanic ash that coats landscapes. Eruptions are driven by the expansion and exsolution of gases dissolved in the magma. Exsolution is a classic example of a positive-feedback cycle. As the magma rises toward the surface, confining pressure decreases and more gas exsolves, sustaining eruptions until pressure in the magma chamber is reduced.

Pyroclastic materials come in a wide variety of sizes, from very fine-grained dust (volcanic ash) that plugs air filters and chokes people, to volcanic bombs (large masses of airborne lava) the size of cars [**Photograph 11.10**]. Pyroclastic materials can either be deposited gently as **pyroclastic falls** or more violently in density-driven **pyroclastic flows** (dense mixtures of hot rock and gas emanating from volcanoes that erupt explosively).

Pyroclastic falls of volcanic ash bury landscapes, smoothing preexisting topography. Such pyroclastic falls are more likely to come from volcanoes erupting silica-rich lava. Near the volcano, pyroclastic fall deposits can be meters thick and quite coarse. Far from the source, fall deposits thin and the grain size decreases; only the smallest particles are carried long distances from the volcano. The distribution and thickness of pyroclastic fall deposits are used to determine wind directions during past eruptions [**Figure 11.6**]. Perhaps the most famous pyroclastic fall deposits are those that buried Pompeii with 2 m of ash in less than 24 hours during the early phases of Mount Vesuvius' CE 79 eruption. These initial, pyroclastic-fall deposits preserved much of the city before multiple pyroclastic flows swept over the buried town, killing its remaining inhabitants and preserving them in ash.

PHOTOGRAPH 11.10 Pyroclastic Materials. Light gray ashfall deposits from the eruption of the Valles (Jemez) Caldera near Los Alamos, New Mexico, with some larger volcanic bombs included. The ash is overlain by an obsidian flow.

debris into a massive, solid rock (**welded tuff**). Such tuff can become very hard and extremely resistant to erosion. For example, the welded Bishop Tuff, which erupted from Long Valley Caldera more than 700,000 years ago, forms extensive tablelands in southeastern California. This erosionally resistant plateau of welded tuff continues to influence landscape evolution three quarters of a million years after the eruption.

The chemical content of pyroclastic debris (**tephra**) can be analyzed and used to correlate distinctive volcanic ash beds—a technique known as **tephrochronology**. If the right minerals, such as sanidine (potassium feldspar), are present and their grain size is large enough, tephra can be dated directly by argon isotopes, providing important time constraints on the eruptive history of the source volcano. If the tephra cannot be dated directly, then the composition of the tephra can be compared to those in regional or global databases and if a match is found, the age and origin of the tephra can often be established by correlation. Because very fine-grained, chemically distinctive tephra from large volcanic eruptions is carried far and wide and deposited in geologic archives including lake sediments, ice sheets, river terraces, and hillslope soils, chemically fingerprinted tephra layers can be used both for correlating deposits and as time markers [Photograph 11.12].

Volcanic Gases

Volcanoes emit a wide variety of different climatically active and therefore geomorphically important gases including carbon dioxide, sulfur dioxide, and water. At depth, these gases are dissolved in magma. As the magma ascends toward Earth's surface, and the confining pressure lowers, so does gas solubility in magma causing gases to exsolve and form bubbles in the magma. The bubbles are preserved as **vesicles** in volcanic rocks; the extreme case

Pyroclastic flows can reshape entire landscapes in just a few moments. Such flows are made up of volcanic rock and ash mixed with hot gases and follow topography, hugging the ground. The largest flows overwhelm topography, creating large, extensive plateaus such as the Yellowstone, Wyoming, and Taupo, New Zealand, volcanic tablelands. Pyroclastic flows originate in different ways including failure of hot lava domes, collapse of eruption columns, and landsliding of whole sectors of volcanic edifices.

Many small pyroclastic flows are triggered by the collapse of newly extruded lava domes near volcanic summits. These flows move down volcano slopes, often following preexisting stream valleys, depositing granular materials on the volcano's flanks and in adjacent lowlands [Photograph 11.11]. Large pyroclastic flows, such as those generated by the collapse of eruption columns or the massive caldera eruption of Yellowstone 2.2 million years ago or of Long Valley 0.76 million years ago, can create **ignimbrite sheets**, which bury preexisting topography for thousands of square kilometers. If hot pyroclastic materials are deposited quickly, the heat can weld the

PHOTOGRAPH 11.11 Pyroclastic Flow. Pyroclastic flows move down Mayon volcano in the Philippines during the 1984 eruption.

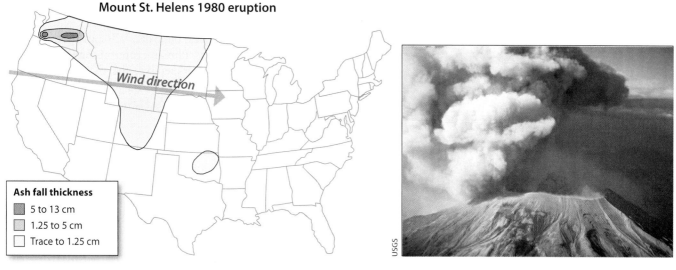

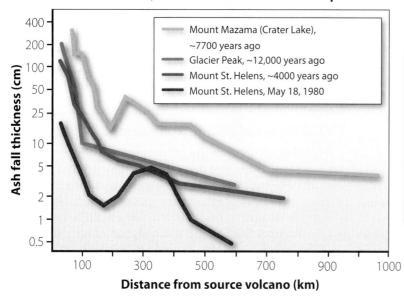

FIGURE 11.6 Ash Fall Distribution. Eruption size, wind speed, and wind direction determine the thickness and distribution of ash falls from explosively erupting volcanoes. [Adapted from Hoblitt et al. (1987).]

being **pumice**, a rock so **vesicular** its density is less than that of water and so it floats and can be carried long distances (hundreds to thousands of kilometers) by ocean currents.

The emission of gases and particulates by volcanic eruptions can alter global climate for months to years. Volcanic emissions, particularly sulfate aerosols, decrease the transmission of sunlight through the atmosphere and thus cool the planet. These sulfate aerosols eventually rain out of the atmosphere as sulfuric acid; for example, the acidity of ice collected from cores in Greenland increased by a factor of more than 4 after the sulfur-rich, explosive eruptions of Greece's Santorini volcano (1645 BCE) and Indonesia's Tambora volcano (1815).

The massive eruption of Tambora is also thought to have caused the year without a summer, 1816, when snow fell every month of the year in the northeastern United States and in much of Europe. Some literary critics have suggested that the cold, dark, and stormy weather of 1816 inspired Mary Shelley to write her famous novel, *Frankenstein*, after spending a very cold summer on Lake Geneva in Switzerland. Even smaller eruptions, such as the 1991 event at Mount Pinatubo in the Philippines, temporarily reduced atmospheric transmission of sunlight by more than 10 percent, likely cooling the northern hemisphere ~0.5°C for a year or two after the eruption.

PHOTOGRAPH 11.12 Volcanic Ash Bed. Light gray Mazama ash deposited 7700 years ago on these hillslopes along the Snake River in Oregon is now exposed in a road cut through an alluvial fan. The ash provides a time marker that allows calculation of fan-sediment deposition rates over the past 7700 years. Columbia River flood basalt flows underlie the hillslope.

Eruption Sizes and Types

Volcanic eruptions come in a wide variety of sizes and intensities. Eruption types can be classified based on their explosivity [Figure 11.7]. The least explosive eruptions are called Icelandic. These effusive eruptions usually involve basaltic lava and build volcanic plateaus and shield volcanoes. Hawaiian-type eruptions are also typically basaltic in composition and are only weakly explosive. They generate lava fountains caused by the expansion of magmatic gases. Because the lava cools as it flies through the air, Hawaiian-type eruptions can create **cinders** or **scoria**, fragments of volcanic rock. The proportion of erupted material that is fragmented into cinders during Hawaiian eruptions is small. Strombolian eruptions are more violent than Hawaiian eruptions with sporadic, explosive gas emissions lofting large blocks of cooling lava out of the crater and onto the volcano's slopes; when and where the slopes exceed the angle of repose for this coarse, granular material, they collapse. Strombolian eruptions often involve basaltic to intermediate composition magma. Plinian eruptions are very explosive and the resulting eruptive material contains a large percentage of new volcanic glass from the erupting lava. Plinian eruptive columns can rise many kilometers into the atmosphere and ejecta is spread tens to hundreds of kilometers away from the vent. Most Plinian eruptions are driven by intermediate or felsic magmas but there are basaltic examples, usually from high-water-content basalts that provide the gas (steam) to fragment the erupting magma.

The **Volcanic Explosivity Index (VEI)** is a frequency/magnitude metric (values range from 0 to >8 on a log arithmetic scale, like earthquake magnitude) used to classify explosive eruptions. The VEI considers primarily the volume of fragmented ejecta (not lava) and thus, with careful field mapping of volcanic deposits, VEI can be calculated for prehistoric eruptions. Plinian eruptions have high VEIs. To place the scale in perspective, the massive Tambora eruption in 1815 had a VEI of 7. The Mount St. Helens 1980 eruption had a VEI of 5. Hawaiian eruptions have low VEIs.

In general, the recurrence interval for volcanic eruptions is inversely related to the VEI. Only a few very large eruptions (VEI > 8) might happen every million years. These would be large, caldera eruptions. Eruptions with a VEI ≥ 4 happen every year or two and globally we expect 15 eruptions with VEI ≥ 2 every year.

Volcanic Landscapes

Volcanic eruptions construct landscapes, create landforms, and leave characteristic deposits. Because the style of eruption and the resulting landforms are controlled in large part by the composition of the erupting magma, we consider basaltic and silicic landscapes separately.

Landscapes of Basaltic Volcanism

The relatively fluid nature of basaltic lava controls the processes, landforms, and deposits of basaltic landscapes such as the Hawaiian Islands, various rift environments, and areas of volcanic arcs where basaltic eruptions occur. Basaltic lava issues from discrete vents building symmetrical volcanoes through the accumulation of lava and pyroclastic debris. A prime example is Paricutín, which was born in the Mexican highlands on February 20, 1943, after two months of increasingly frequent earthquakes in the region [Figure 11.8]. The eruption began when a small vent formed in the middle of a cornfield. The vent opened quickly, starting with small ash emissions and building into a sizable cinder cone in 24 hours. Paricutín is an example of a water-rich arc basalt; the eruption was explosive. Two-thirds of the 1.3 km^3 total volume of material erupted was pyroclastic. The eruption lasted 9 years and when it was over, there was a cone >400 m high with lava flows covering almost 25 km^2.

Most basaltic volcanoes are a mix of pyroclastic material and lava flows. The pyroclastic material, commonly referred to as cinders or scoria, is emitted as molten rock, cools in the air and falls to the ground near the vent forming a symmetrical form known as a **cinder cone** (Photograph 11.7) with slopes generally at or near the angle of repose for granular material (see Chapter 5), in this case, the cinders. Cinder cones typically form in a single eruption and are thus **monogenetic**. Often, lava flows emerge through the same vent as the cone, breaching the cone wall; the flows may issue as fountains. Hardened lava with a smooth, ropey surface texture is known as pahoehoe (Hawaiian word for "paddle"). Blocky flows

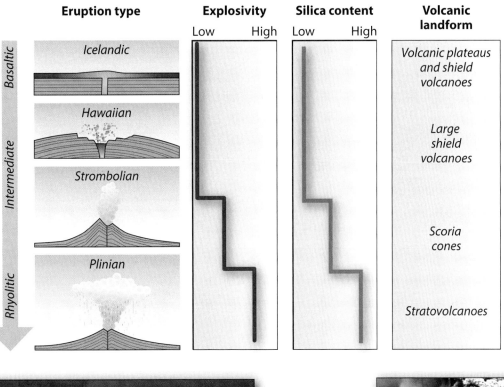

The silica and water content of **magma** control its behavior and the geomorphology of the resulting volcanic landform. Wet, low-silica magmas are less viscous and less explosive than dry, high-silica magmas. Low-silica magmas result in less steep landforms and less explosive eruptions.

Eruption types are related to tectonic setting. **Strombolian** eruptions are common at island arcs and **Hawaiian** eruptions are common in basaltic volcanic provinces, particularly at ocean hot spots. **Icelandic** eruptions often occur at rift zones. **Plinian** eruptions are most common at continental subduction zones.

FIGURE 11.7 Volcanic Form and Process. Eruption type and the resulting volcanic landforms are related to magma composition, which determines lava viscosity and eruption explosivity.

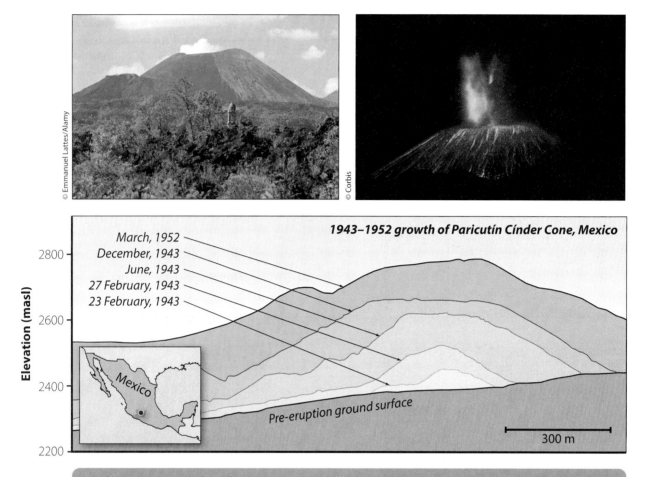

Cinder cones usually form from single eruptions lasting days to years. They grow over time and are often breached by lava flows that issue from them. Paricutín is one of the best known modern examples. It appeared suddenly in a Mexican farmer's field and spewed ash and lava for years. The cone itself is a minor contribution to the erupted volume. Much more important were the tephra deposits (about 2/3 of the total mass), which blanketed a wide area around the volcano and the lava flows from the base of the cone, which covered the towns of Paricutín and San Juan Parangaringatiro.

FIGURE 11.8 Growth of a Cinder Cone. Volcanic edifices, such as cinder cones, change shape over time as eruptive products cover their surfaces. [Adapted from Lockwood and Hazlett (2010).]

are termed *aa* (from Hawaiian for "stony") [**Photograph 11.13**]. The difference in the surface form of blocky aa and ropey pahoehoe lava results from differences in lava temperature, moisture content, and/or lava effusion rate.

Basaltic eruptions can also occur through long fissures, chains of vents from which ash is emitted and lava flows. Such fissures are most common in extensional environments such as Iceland but are also found on hot spot volcanoes including the Hawaiian Islands. Basalt flows can travel long distances when lava is transported through lava tubes, in part because the excellent insulating properties of rock keep the lava warm and allow it to flow at low viscosity [**Photograph 11.14**].

Repeated eruptions of fluid basaltic lava build large shield volcanoes. The surface of a large shield volcano is made up of flows of various ages. Thus, the degree of chemical and physical weathering may differ greatly in nearby areas, depending on when they were last covered by lava. Over time, old flow surfaces degrade and become covered in soil, derived both from weathering of the flow itself and from the accumulation of younger ash and/or aeolian dust. Chronosequences of weathered lava flows provide fertile research grounds for understanding the linked biological, chemical, and physical changes that alter newly formed landscapes over time.

As basalt flows weather and are incised by streams or glaciers, they reveal their internal structure. Most flows have a rubbly basal or interflow zone that formed when the flow moved over cold ground and the molten lava rapidly cooled and solidified, or when rubble that formed on the flow surface fell off the flow front and was overridden by the advancing flow. This basal rubble zone is permeable to groundwater and can become an aquifer. Above the rubble is a zone of massive lava, commonly fractured into distinct hexagonal columns known as the **colonnade.** The columns are formed as the lava in the center of the flow slowly cools and contracts [**Photograph 11.15**]. Where exposed by erosion, they form scenic features such as Devil's Postpile in the Sierra Nevada or Giant's Causeway in northern Ireland. In many cases, the joints fractures are sealed by vein-filling cements. The columns are oriented perpendicular to the cooling front and the colonnade often functions as a low-permeability aquitard. The rubbly top of uneroded flows may also be permeable, which promotes infiltration and reduces runoff.

Landscapes of Silicic Volcanism

The archetypical depiction of a volcano as a tall, snow-capped cone is the form of a **stratovolcano,** the dominant volcanic landform at convergent margins. Stratovolcanoes are composite landforms, generally made up of intermediate composition lava flows, pyroclastic debris, and debris flow deposits. Stratovolcanoes are steep because the relatively high silica content of the andesitic lava they typically erupt makes it quite viscous and resistant to flow. A smooth cone without deeply incised valleys is likely young and probably postglacial in age. Older cones, such as Mount Rainier, predate the last major glaciation and are deeply dissected by glacial valleys [**Photograph 11.16**]. Stratovolcanoes are formed by multiple eruptions occurring over time; thus, they are **polygenetic.**

After a large, explosive eruption, stratovolcanoes are slowly rebuilt by the gradual extrusion of viscous lava as plugs and domes. For example, during the May 1980 eruption, Mount St. Helens lost ~500 m of elevation as the summit

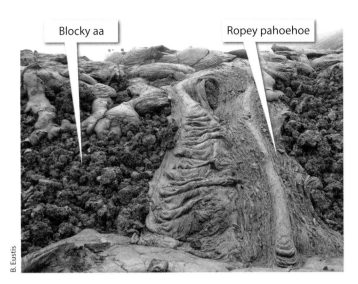

PHOTOGRAPH 11.13 Basaltic Lava Flow Morphology. Two different types of basalt flow morphology. Here, ropey pahoehoe flows over an older, underlying blocky aa flow.

(a)

(b)

PHOTOGRAPH 11.14 Lava Tubes. Fluid basalt can travel kilometers in lava tubes because they effectively insulate the hot lava and prevent it from chilling and quenching. (a) This skylight is an opening into a lava tube flowing from the Pu'u O'o vent, Kilauea volcano, Hawaii. (b) Fossil lava tube, Lava Beds National Monument, California.

PHOTOGRAPH 11.15 Basalt Columns. Look up! Svartifoss, a small waterfall in Skaftafell National Park on the southern coast of Iceland, flows over a sequence of columnar basalts, here viewed from below.

(a)

(b)

PHOTOGRAPH 11.16 Stratovolcanoes. Glaciation erodes stratovolcanoes and destroys their symmetrical form. (a) Arenal stratovolcano in Costa Rica has never been glaciated and retains its smooth form. Note cloud of volcanic gas erupting from the summit. (b) Mount Rainier from the air in late summer showing minimum glacial ice extent and deep troughs carved by glaciers active in the Pleistocene and Holocene.

PHOTOGRAPH 11.17 Lava Dome. "Whaleback" feature, the lava dome forming in the crater of Mount St. Helens volcano, is increasing local relief and rebuilding the summit of the mountain.

and north flank collapsed. In the 30 years since the eruption, intermittent eruptions of thick, pasty lava have built a dome in the crater. The dome, which was over 300 m high before partially collapsing, is slowly rebuilding the mountain. It and the collapsing crater inner walls are filling the crater with volcanic rock and debris [Photograph 11.17].

Processes of Volcanic Landform Evolution

Volcanic landform evolution is dynamic. Large volcanoes, projecting thousands of meters into the atmosphere, create their own climates, orographically increasing precipitation and commonly providing the snow and freezing temperatures necessary to support erosive glaciers that tear into volcanic edifices. High heat fluxes in volcanically active areas speed weathering, weakening rock, and forming clays capable of sustaining debris flows.

The potential energy of rocks high up on a volcanic edifice is quickly transformed to kinetic energy when rocks, landslides, and mudflows tumble down steep volcanic slopes. As volcanoes age and erode, the removal of eruptive products such as lava flows and aprons of pyroclastic material exposes the underlying intrusive igneous rocks. Over time, such erosion first removes volcanoes and then shapes landscapes where the differential erosivity of intrusive igneous rocks controls landform morphology.

Geomorphic Effects of Magma Intrusion

The intrusion of magma below and into a volcanic edifice can be monitored and has a variety of geomorphic effects while a volcano is active. Magmatic intrusions under volcanoes can tilt the land surface without the eruption of lava. Such tilting is monitored as a precursor to possible eruptions. The tilting can be significant. For example, during the 2000 eruption of Usu volcano in Japan, magmatic intrusion caused sufficient deformation that a stream reversed its flow direction. Magma intrusion into Mount St. Helens bulged the volcano's north side nearly 150 m in the month before the cataclysmic 1980 eruption, visibly tilting the upper surface.

Prior to many eruptions, earthquakes occur near and under the volcano, the result of magma inching toward the surface, breaking rock as it moves. These precursor earthquakes, particularly periods of **volcanic tremor,** high-frequency ground vibration that marks magma movement, have become an important means of predicting eruptions [Figure 11.9].

Using earthquakes as a volcanic warning sign is nothing new. Archaeological evidence including broken stone stairways, the lack of fatalities, and the removal of household effects from Akrotiri, the city buried by ash from the 1645 BCE eruption of Santorini, on the island of Thera in Greece, suggests the populace had warning of the impending eruption. Did precursor volcanic

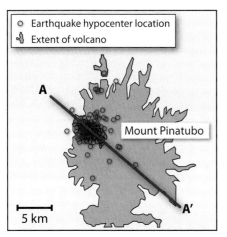
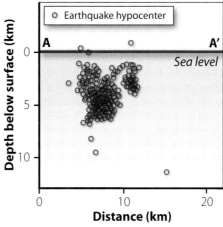

Using seismometers, the spatial location of earthquakes can be mapped in three dimensions. These quakes show the movement of **magma** below the volcanic edifice of Mount Pinatubo in the Phillipines before it erupted catastrophically in 1991. Knowing the location of magma helps with predicton of where volcanic eruptions will occur.

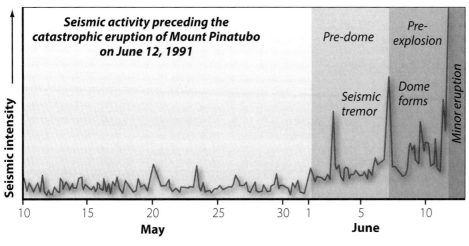

The seismic intensity (energy release) from earthquakes increased in the weeks before Mount Pinatubo's first 1991 eruption. **Seismic tremor,** high frequency ground vibration, indicated that magma was moving into the volcano, magma which formed a **dome** and then led to a minor eruption.

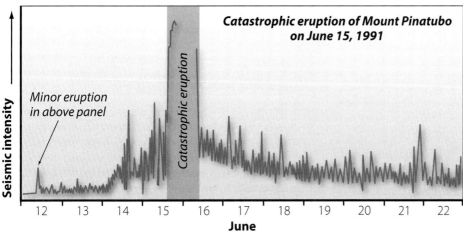

Seismic intensity increased after the first, minor eruption, peaked during the catastrophic eruption of Mount Pinatubo, and then waned slowly as the eruption subsided.

FIGURE 11.9 Volcanically Triggered Earthquakes. As magma moves below and into volcanoes it triggers earthquakes as rocks around the magma conduit crack. These earthquakes allow mapping of magma intrusions and prediction of explosive eruptions. [Adapted from Decker and Decker (2006).]

earthquakes encourage everyone to leave before Santorini exploded in a massive caldera-forming eruption? In contrast, the people of Pompeii in Italy seemed to have been more complacent. Many were trapped and killed in their homes by pyroclastic flows during the 79 CE eruption of Mount Vesuvius.

Biologic Colonization

New rock produced by volcanoes may initially be sterile, but it does not stay that way for long. Even before an eruption ends, biologic activity begins with insects and bacteria arriving on the wind. In basaltic terrain, recolonization is catalyzed by the invasion of plants and animals from "islands" formed where lava flows around a high point in the landscape, leaving patches of intact vegetation known in Hawaii as **kipukas** (from the Hawaiian for "opening" or "hole"). Similar to glacial refugia, kipukas speed the reestablishment of ecosystems and are important factors in plant and insect evolution. In moist climates, lichens colonize volcanic rocks rapidly, initiating the process of weathering. Dust fall is also important, bringing in nutrients from afar. Two recent eruptions have provided useful but contrasting case studies in colonization after volcanic disturbance.

In 1963, the island of Surtsey formed off the southern coast of Iceland in a four-year-long series of pyroclastic and lava eruptions. Within several years after volcanic activity stopped, sea birds began nesting on the island. The organic material in and around their nests promoted the establishment of small plants. Even so, more than 50 years later, this exposed North Atlantic island remains mostly barren, as the basaltic lava weathers slowly in the cold subarctic environment.

In contrast, the 1980 eruption of Mount St. Helens devastated the forested landscape surrounding the mountain, burying the soils under decimeters to meters of debris. Many thought biologic recovery would take decades, but it was far more rapid. Because the eruption happened in May, much of the area near the volcano was still buried under a thick winter snowpack. The snow protected both seeds lying on the soil surface and small animals such as rodents from the heat of the pyroclastic flow deposits. After the eruption was over, these surviving animals brought seeds and soil to the surface, speeding recolonization [**Photograph 11.18**]. Today, many of the areas laid bare by the 1980 eruption are again heavily vegetated—some naturally, some replanted for forestry.

Denudation and Aging

Volcanic deposits weather as they age. Weathering rates depend in part on the mineralogy and in part on the glass content of volcanic rocks. Mafic volcanic rocks are rich in minerals that are not stable at Earth's surface. Thus, in general mafic volcanic rocks weather more rapidly than felsic volcanic rocks (see Chapter 3). However, felsic rocks that are high in glass and low in quartz can weather rapidly, too.

Volcanic glass and minerals weather to clay, changing the permeability of material comprising volcanic landforms. Volcanic landforms with coarse surface textures such as cinder cones and lava flows also function as efficient dust traps. Over time, dust fills otherwise connected pores, reducing the porosity, permeability, and thus the infiltration capacity of the landform. The cinder cone or lava flow, where rainfall once passed easily through porous soils, will then generate overland flow during intense rainfall events. Such flow can carve gullies and erode previously stable landforms [**Photo-**

PHOTOGRAPH 11.18 Mount St. Helens Vegetation. Sequence of images from NASA Landsat showing Mount St. Helens before (1973), a few years after (1983), and 20 years after the 1980 eruption. Vegetation removed by the 1980 eruption quickly regrew and now recovers much of the landscape denuded in 1980. Images are approximately 50 km across.

graph 11.19]. Thus, cinder cones become increasingly incised and gullied with age, providing a means to estimate relative age.

Mass Movements

Mass movements, including debris flows, slides, and slumps, remove material from volcanic edifices before, during, and after eruptions [**Photograph 11.20**]. Volcanoes provide an ideal location for mass movements because they have steep slopes, orographically enhanced rainfall, and an abundant supply of fine-grained, volcanic rocks and pyroclastic material weakened over time by hydrothermally enhanced weathering.

PHOTOGRAPH 11.19 Cinder Cone. Degraded and gullied cinder cone on the wet side of Oahu, Hawaii.

The eruption of Mount St. Helens opened the eyes of geomorphologists to a process they had not appreciated before, the failure of large portions of a volcanic edifice by mass movement [Figure 11.10]. A magnitude-5 earthquake, just a kilometer or two beneath the north flank of the volcano, signaled the beginning of the 1980 eruption. Only 15 seconds after the earthquake, the north side of the volcano gave way in a massive landslide, releasing a tremendous lateral blast of hot ash, gases, and rock. Bulging from the intrusion of magma, and shaken loose by the earthquake, the entire northern sector of the mountain collapsed, mobilizing as a **debris avalanche**—a mixture of ash, ice, and weathered volcanic rock—that moved rapidly downslope, filling the river valley below. The magma chamber was "uncorked" (as confining pressure was instantaneously removed) and the eruption began. The eruption, including the debris avalanche and rock fragments, buried much of the landscape in hot ash, which included dismembered pieces of glaciers and snowbanks. Many hours later, as blocks of ice and snow melted within the hot debris, the debris flows began. Having seen such a massive landslide (termed a **sector collapse**) happen, geomorphologists began to notice evidence indicating that such collapses happened elsewhere in the past—for example, at Mount Shasta in California and in the Canary Islands off West Africa (see Digging Deeper).

Large portions of basaltic volcanoes can also fail and move down slope. During the 1973 Heimey eruption in Iceland, loading of one part of the volcanic edifice from eruptive products exceeded the strength of the underlying rock and triggered a massive slope failure. Volcanic ocean islands, such as Hawaii, are prone to massive gravitationally driven slides as lava flows load upper slopes and erosion along the coast removes mass from the lower slopes. Such megaslides have repeatedly removed large amounts of rock from these islands, causing local earthquakes, generating tsunamis, and leaving behind distinctive evidence on the seafloor in the form of large, lobate deposits. Towering vertical sea cliffs, such as those of the Na Pali Coast on Kauai, and elsewhere in Hawaii, are the head scarps of these megaslides [Photograph 11.21].

Lahars

Volcanic debris flows, known as **lahars**, have significant geomorphic impacts. Many such flows are generated as the heat from eruptions melts snow and ice on or near volcano summits; they can also result from dome collapse, heavy rain on newly erupted ash, and catastrophic crater-lake drainage such as occurred at Mount Ruapehu, New Zealand, in 2007 [Photograph 11.22]. However initiated, a lahar mixes water with both weathered volcanic rock and/or newly erupted material to form highly mobile flows that thunder down valleys leading away from the volcanic edifice at speeds up to 20 m/s. Lahars erode debris from the flanks of volcanoes, deepening channels and growing in volume as they incorporate eroded material. As lahars run out onto lower-gradient slopes surrounding volcanoes, the flows fill and then pour out of river channels, building broad fans and burying adjacent lowlands. In addition to the direct geomorphic effects of erosion and deposition, lahars can dam river valleys, creating lakes that may themselves fail catastrophically.

Lahars also occur on extinct volcanoes and on volcanoes dormant between eruptive periods. Hurricanes delivering large amounts of rain in short periods to the steep, unstable slopes of volcanoes trigger lahars even when the volcano is not active; for example, in three days, Hurricane Mitch (1998) dropped nearly a meter of rain on Casita volcano in Nicaragua, triggering a massive lahar. The lahar killed 2500 people in two towns built on the deposits of previous lahars. Land-use change probably

PHOTOGRAPH 11.20 Mount St. Helens After Eruption. Looking south at Mount St. Helens stratovolcano in southern Washington State. The north side of the mountain is gone, having failed in a giant landslide, the debris of which is being actively incised by a large channel originating just below the crater. The photograph within this photograph shows how symmetrical the cone was prior to eruption.

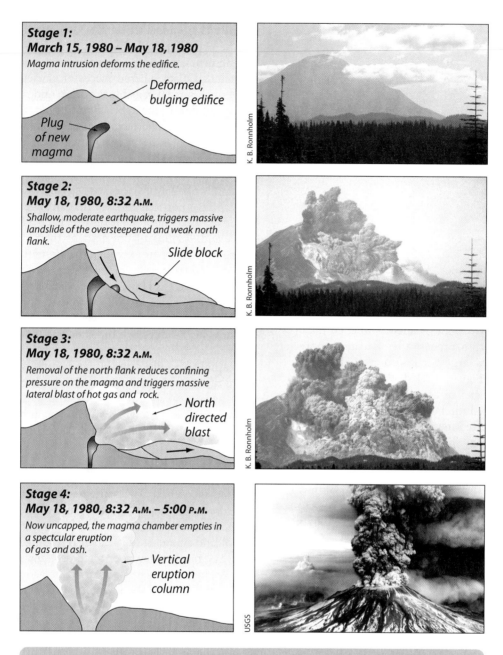

FIGURE 11.10 Time Series: Mount St. Helens Erupting. The eruption of Mount St. Helens involved a sequence of events, beginning with magma intrusion and ending with the earthquake-induced giant landslide, which led to the destructive lateral blast, the debris avalanche, and the Plinian eruption.

exacerbated the problem; 80 percent of the sediment transported by the lahar came from areas that had been deforested.

The most disastrous lahar in recent memory descended Nevado del Ruiz volcano, Colombia, in 1985 after a small eruption. The lahar incorporated pyroclastic flows that melted some of the mountain's 25 km² cap of snow and ice. The lahar grew to more than 4 times its original volume by eroding material from the streambeds as it flowed down the volcano, after which it buried the them with debris. Lahars, being debris flows, can deliver boulders large enough that they cannot be removed by fluvial processes and so may dictate channel pattern and roughness.

Basaltic and andesitic lava have sufficiently low viscosity to follow existing topography and flow down stream channels and canyons. Once such intracanyon lava flows cool and harden, they are often more resistant to erosion than the surrounding terrain. Thus, an ancient river course (now filled with volcanic rock) can become a meandering ridge crest as the surrounding terrain erodes faster than the volcanic rocks that define the old river valley. Many volcanic terrains provide excellent examples of such **topographic inversion** [**Figure 11.11**], where erosion-resistant volcanic rock hardened in topographic lows only to become preserved as topographic highs due to preferential erosion of the surrounding weaker rocks.

The volume of rock eroded by rivers from dated lava flows and volcanic deposits is used to calculate rates of incision and erosion. For example, in Papua New Guinea, lava flows dated with K/Ar have been used to calculate rates of mass loss from volcanic terrain by measuring the volume of valleys eroded into the volcano. In the Valley of Smokes, near Katmai, Alaska, diversion of the Ukak River onto rock by 200 m of ash provided a natural experiment through which the rate of river incision into sandstone and siltstone bedrock could be measured. Basaltic lava has repeatedly poured into the Grand Canyon, blocking the Colorado River. By dating the lava flows, one can both calculate the rate of canyon incision and the time at which the river began to occupy its current position, between 4 million and 6 million years ago.

If the pyroclastic material is not hot enough to weld, it remains unconsolidated and thus easily eroded. Explosive eruptions, such as that which occurred at Mount St. Helens in 1980, cover the land surface in an easily eroded blanket of ash and debris-avalanche sediment and fill stream channels with lahar deposits. After such an eruption, sediment yields can rise by several orders of magnitude in rivers and streams draining the affected area. High rates of sediment export from volcanically disturbed hillslopes diminish rapidly (over several years) as the erodible material from the steepest and most accessible areas is removed but excess sediment may continue to issue from disturbed channel networks for decades, especially during wet seasons and during years of above average precipitation or storminess [**Figure 11.12**]. Most of the material eroded from Mount St. Helens since 1980 is debris-avalanche and lahar sediment, the sediment most accessible to stream erosion.

PHOTOGRAPH 11.21 Head Scarps. Near-vertical cliffs are the head scarps of mega-landslides that tore away the basalt of the Hawaiian Islands, here on Molokai near Kalaupapa.

town of Armero in mud and killed more than 23,000 people.

Volcano-River Interaction

Volcanoes around the world interact with rivers. For example, rivers draining volcanoes are repeatedly affected by lahars, which both erode channels and fill

PHOTOGRAPH 11.22 Lahar. Fluid lahar from Mount Ruapehu in New Zealand fills a river channel. The lahar originated when a tephra dam collapsed and the volcano's summit lake drained.

Hydrologic Considerations

Volcanoes influence the flow of both surface and subsurface water. Created and intruded by lava, volcanic areas have steep subsurface temperature gradients and can provide ready sources of geothermal energy including steam and hot rock.

Stage 1: Young lava flow

Fluid **lava**, basaltic or andesitic, driven by gravity, flows away from its vent and down a river valley. The lava cools and fills the valley.

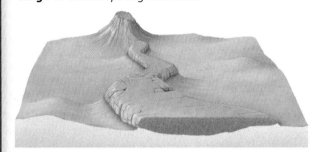

Stage 2: Landscape begins to erode

The cooled and hardened lava is more resistant to erosion than the rock into which the river valley was cut. The valley walls erode and the lava flow remains above the landscape.

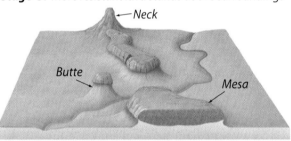

Stage 3: More resistant lava stands above surroundings

Over time, the valley walls are removed by erosion, leaving what is left of the lava flow as high points on the landscape. The topography has inverted with the former lava-filled river valley now occupying the high ground.

The St. George, Utah, airport sits on top of a lava flow that once filled a now-vanished river canyon. The canyon was cut into relatively soft sandstone before it was filled with basalt from a nearby volcano. Over time, the sandstone eroded away, leaving the flat-topped basalt flow—a perfect example of inverted topography.

FIGURE 11.11 Topographic Inversion. Topographic inversions are common in volcanic terrain because lava follows surface drainages, and the volcanic rocks it forms are often more resistant to erosion than the rocks bordering river valleys. [Adapted from Lockwood and Hazlett (2010).]

Layering of lava flows (see Photograph 11.5) creates aquifers and aquitards, contrasting zones of higher and lower permeability, setting the stage for springs commonly found at and near the base of volcanoes. Many of these springs are warm or hot if groundwater circulating in the aquifer contacts still-hot volcanic rock. Geysers are a special case of volcanic springs in which rapid heating causes a dramatic eruption of water and steam from the ground. Just the right geometry of subsurface conduits allows steam pressure to build up under water before lifting that water toward the surface, releasing the hydrostatic pressure and allowing the water to flash

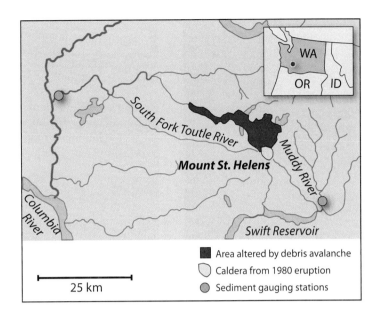

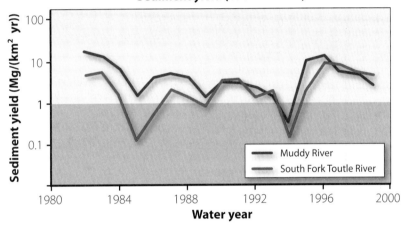

The 1980 eruption of Mount St. Helens dropped a large amount of easily erodible ash and sediment onto the landscape. In the years after the eruption, the rivers draining the mountain carried much of this material away. Sediment yields were initially high, then dropped quickly from the eruption until 1984 but increased after that during years of high runoff, such as 1996. Dark gray shaded region at bottom of plot shows the range in mean sediment yields for several other western Cascade Range rivers. Note logarithmic y-axis scale.

FIGURE 11.12 Sediment Yields After the Eruption of Mount St. Helens. Sediment yield in rivers draining Mount St. Helens peaked after the 1980 eruption then dropped off. After the initial pulse of sediment resulting from the eruption, the annual sediment yield was well correlated to the annual runoff volume. [Adapted from Major et al. (2000).]

into yet more steam in a positive-feedback cycle leading to an eruption.

Magma interaction with groundwater or snow and ice generates **phreatomagmatic** eruptions—eruptions involving both molten rock and water. Where this happens, the steam turns otherwise effusive basaltic eruptions violent and explosive, emitting large amounts of ash and steam in hydrovolcanic activity. A recent example, the eruption of Iceland's Eyjafjallajökull volcano through the Mýrdalsjökull glacier, produced enough ash to disrupt air travel across Europe for weeks in the spring of 2010 [**Photograph 11.23**]. Much of that ash was likely generated by the interaction of hot lava with glacial ice in a phreatomagmatic eruption—at least during the first few days of the eruption.

PHOTOGRAPH 11.23 Phreatomagmatic Eruption. Phreatomagmatic eruption (2010) at Eyjafjallajökull volcano, Iceland, showing white steam cloud, dark ash cloud, and fountain of basaltic lava over snow.

PHOTOGRAPH 11.25 Table Mountains. Icelandic tuyas or flat-topped table mountains formed when lava erupted through a now-vanished ice sheet. The upper surface of the tuya indicates the minimum elevation of the ice at the time of the eruption. Bláfjall (1200 m elevation) is on the left, Sellandafjall on the right. Many of these tuyas date to the last deglaciation, between 10,000 and 14,000 years ago suggesting that unloading when ice sheets melt may trigger increased eruptive activity.

Phreatomagmatic eruptions can leave characteristic, steep-sided craters in the landscape referred to as **maars**, from the German name applied to crater lakes in the Eifel district of Germany [**Photograph 11.24**]. For eruptions in the marine realm, steam eruptions and the emission of ash dominate volcanic emissions until the vent edifice grows sufficiently tall that lava does not flow into the sea but rather pours onto dry land. From then on, lava flows, rather than the emission of steam and ash, dominate the eruption. A recent example is the birth of the Icelandic island of Surtsey. **Phreatic** eruptions are steam explosions (with no magma ejected). Phreatic eruptions can be quite violent and produce large amounts of ash, but they do not involve molten rock (magma).

Subglacial eruptions occur on high-altitude volcanoes and in near-polar latitudes. Such eruptions are common in Iceland and leave distinctive landforms. **Table mountains** (or **tuyas**) are the remnants of volcanoes that erupted through larger, Pleistocene glaciers that have now retreated. Their top elevations provide a minimum constraint on the height of now-vanished ice because the steep sided, flat-top volcanoes were once ringed by ice [**Photograph 11.25**]. Hot lava erupting through or onto glaciers can also trigger jökulhlaups, glacial floods produced as glacial ice rapidly melts.

Many active and dormant stratovolcanoes have deep summit craters, the result of explosive eruptions. Some summit craters contain lakes. Volcanic activity, earthquakes, or breaching or overtopping of the crater wall all may cause crater lakes to drain catastrophically. As water spills from the lake down the steep slopes and drainages of the volcano, the flow incorporates debris and can transform from a raging torrent of water into a devastating lahar, such as the typhoon-generated lahars that wreaked havoc in the lowlands surrounding Mount Pinatubo in the Philippines after its 1991 eruption (see Photograph 11.22).

Erosional Landforms

Once volcanoes become inactive, they, and the landscape around them, begin to erode. As erosion proceeds, it reveals the internal magmatic plumbing system of the volcano because the rock that cooled below the surface is often harder, less jointed, and thus more resistant to erosion than either rock that cooled quickly or pyroclastic debris that makes up much of the near-surface material in volcanic terrain. Specific landforms of volcanic erosion include **necks**, the solidified remains of the conduits that fed magma to the vent at the center of the volcano [**Photograph 11.26**]. Shallow intrusions of magma into fractures often produce resistant vertical **dikes** and horizontal **sills** in the subsurface, particularly if the country rock into which they intrude is weak [**Photograph 11.27**]. Dikes can be both

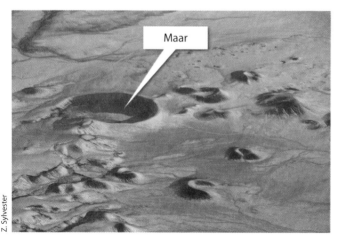

PHOTOGRAPH 11.24 Maar. Lunar Crater, in Nevada, is a kilometer-wide maar, a crater formed when hot magma explosively flash-boiled ground water, resulting in a steam explosion.

Applications

Understanding volcanic geomorphology has significant practical applications across a wide range of fields including geologic hazard reduction, paleoclimatology, and economic development. Because the influence of volcanoes on the landscape is both short-term (explosive eruptions) and longer-term (increased sediment loads after eruptions), a geomorphologist's appreciation of deep time and landscape evolution are critical for understanding volcanic systems. Geomorphic mapping and detailed geochronology can delineate and date prior volcanic events, providing information about the scale and recurrence intervals of volcanic hazards including eruptions and lahars. In many cases, the main volcanic hazard is not the direct effect of the eruption but the subsequent redistribution of material downslope.

Volcanic process and form control both the immediate hazards during an eruption and posteruptive hazards. Consider, for example, the contrast in the relative hazards of hot spot basaltic volcanism in Hawaii with the explosive nature of subduction zone–driven eruptions in northwestern North America. During nonexplosive basaltic eruptions, fluid lava flows down topographic gradients allowing specific hazard zones to be delineated rapidly and reliably. In Hawaii, volcanologists use the concept of "lava sheds" (analogous to watersheds) to delineate potential hazard zones from different possible vent locations. In contrast, the explosive andesitic volcanism of subduction zones is much less predictable. Explosive eruptions, sector collapse, pyroclastic flows, and lahars can happen with few if any precursors and can overwhelm or even destroy preexisting landforms and stream channels.

Posteruptive volcanic hazards can be long-lasting and significant, reflecting the downstream effects of volcanic debris erosion, lahars, and sediment deposition following the initial eruption. For example, volcanic edifices such as Mount Rainier weather rapidly as large amounts of orographically induced precipitation interact with hot rock. As volcanic glass and feldspars turn to clay, the andesite weakens, making these steep mountains susceptible to landslides that evolve into massive mud and debris flows. In the past several thousand years, two major mudflows (the Electron and Osceola flows) originated on the flanks of Mount Rainier and thundered toward Tacoma, Washington, covering many square kilometers of river-bottom land beneath tens of meters of debris and even filling an arm of Puget Sound. Similar events are recorded in well-documented Native American oral traditions. Today the same river valleys are filled with shopping malls, residences, and car dealerships, leading the state of Washington to establish "lahar early-warning systems" upstream.

Volcanic emissions are critical for society as well as for scientists. More than 10 percent of the world's population lives on volcanically derived soils. In volcanic terrain, rich soils, developed on weathered tephra, support diverse

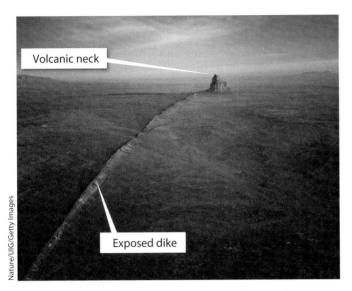

PHOTOGRAPH 11.26 Volcanic Neck. Shiprock in south central Colorado, a volcanic neck with steep cliffs, is exposed because it was more resistant to erosion than the rock the magma intruded. A dike extends away from the volcanic neck. Over 1000 m of erosion have revealed the volcanic plumbing system.

linear, filling joints, and radial, leading away from a magma intrusion. In areas that are eroded so deeply that subsurface magma chambers are exposed, granitic batholiths, representing the roots of stratovolcanoes, host their own distinctive landforms (see Chapter 3).

PHOTOGRAPH 11.27 Volcanic Dike. Vertical volcanic dike near Ward Spring, Big Bend National Park, Texas, is more resistant than the surrounding rock and thus stands out above the landscape.

agricultural activities; some lava flows, such as the flood basalts of Washington and Oregon, host large aquifers useful for irrigation.

Volcanic geomorphology has benefitted many other scientific disciplines. For tectonicists, the edifice age and orientation of oceanic hot spot tracks, such as the Hawaiian Island–Emperor Seamount chain, were critical in the development of plate tectonic theory. Potassium/argon ages on lavas making up this chain of subaerial and subaqueous volcanic landforms provide a means by which to calculate rates and orientations of plate movement over geologic timescales, in this case 70 million years. For archaeologists, tephra deposits in places such as Pompeii and Santorini have preserved the remains of entire cultures intact, providing a "gold mine" for understanding long-vanished cultures.

Selected References and Further Reading

Ahnert, F. Functional relationships between denudation, relief, and uplift in large mid-latitude drainage basins. *American Journal of Science* 268 (1970): 243–263.

Crandell, D. R. *Postglacial lahars from Mount Rainier volcano, Washington.* U.S. Geological Survey Professional Paper 677. Washington, DC: 1971.

Crandell, D. R., and D. R. Mullineaux. Potential hazards from future eruptions of Mount St. Helens volcano, Washington. *U.S. Geological Survey Bulletin* (1978): 1383-C.

Crandell, D. R., and H. H. Waldron. A recent volcanic mudflow of exceptional dimensions from Mount Rainier, Washington. *American Journal of Science* 254 (1956): 349–362.

Decker, R., and B. Decker. *Volcanoes.* New York: Freeman, 2006.

Freidrich, W. *Fire in the Sea: The Santorini Volcano.* Cambridge: Cambridge University Press, 1999.

Glicken, H. Rockslide-debris avalanche of May 18, 1980, Mount St. Helens Volcano, Washington. *U.S. Geological Survey Open-File Report 96-677.* Washington, DC: 1996.

Gran, K., and D. R. Montgomery. Spatial and temporal patterns in fluvial recovery following volcanic eruptions: Channel response to basin-wide sediment loading at Mount Pinatubo, Philippines. *Geological Society of America Bulletin* 117 (2005): 195–211.

Hoblitt, R. P., C. D. Miller, and W. E. Scott. Volcanic hazards with regard to siting nuclear-power plants in the Pacific Northwest. *United States Geological Survey Open-File Report, 87-297* (1987).

Iverson, R. M., S. P. Schilling, and J. W. Vallance. Objective delineation of lahar inundation hazard zones. *Geological Society of America Bulletin* 110 (1998): 972–984.

Karlstrom, K. E., R. S. Crow, L. Peters, et al. ^{40}Ar/^{39}Ar and field studies of Quaternary basalts in Grand Canyon and model for carving Grand Canyon: Quantifying the interaction of river incision and normal faulting across the western edge of the Colorado Plateau. *Geological Society of America Bulletin* 119 (2007): 1283–1312.

Licciardi, J. M., M. D. Kurz, and J. M. Curtice. Glacial and volcanic history of Icelandic table mountains from cosmogenic ^{3}He exposure ages. *Quaternary Science Reviews* 26 (2007): 1529–1546.

Lockwood, J. P., and R. W. Hazlett. *Volcanoes: Global Perspectives.* Sussex, UK: Wiley-Blackwell, Sussex, 2010.

Major, J. J., and C. G. Newhall. Snow and ice perturbation during historical volcanic eruptions and the formation of lahars and floods: A global review. *Bulletin of Volcanology* 52 (1989): 1–27.

Major, J. J., T. C. Pierson, R. L. Dinehart, and J. E. Costa. Sediment yield following severe volcanic disturbance— A two-decade perspective from Mount St. Helens. *Geology* 28 (2000): 819–822.

Meyer, D. F., and H. A. Martinson. Rates and processes of channel development and recovery following the 1980 eruption of Mount St. Helens, Washington. *Hydrological Sciences Journal* 34 (1989): 115–127.

Oppenheimer, C. Climatic, environmental and human consequences of the largest known historic eruption: Tambora volcano (Indonesia) 1815. *Progress in Physical Geography* 27 (2003): 230–259.

Parfitt, E. A., and L. Wilson. *Fundamentals of Physical Volcanology.* Oxford, UK: Blackwell, 2008.

Pierson, T. C., and R. J. Janda. Volcanic mixed avalanches: A distinct eruption-triggered mass-flow process at snow-clad volcanoes. *Geological Society of America Bulletin* 106 (1994): 1351–1358.

Pierson, T. C., R. J. Janda, J.-C. Thouret, and C. A. Borrero. Perturbation and melting of snow and ice by the 13 November 1985 eruption of Nevado del Ruiz, Colombia, and consequent mobilization, flow, and deposition of lahars. *Journal of Volcanology and Geothermal Research* 41 (1990): 17–66.

Procter, J., S. J. Cronin, I. C. Fuller, et al. Quantifying the geomorphic impacts of a lake-breakout lahar, Mount Ruapehu, New Zealand. *Geology* 38 (2010): 67–70.

Punongbayan, R. S., C. G. Newhall, and R. P. Hoblitt. "Photographic record of rapid geomorphic change at Mount Pinatubo, 1991–94." In C. G. Newhall and R. S. Punongbayan, eds., *Fire and Mud—Eruptions and Lahars of Mount Pinatubo, Philippines,* Seattle: University of Washington Press, 1996.

Scott, K. M. *Origins, behavior, and sedimentology of lahars and lahar runout flows in the Toutle-Cowlitz River system.* U.S. Geological Survey Professional Paper 1447-A. Washington, DC: 1988.

Siebert, L. Large volcanic debris avalanches: Characteristics of source areas, deposits, and associated eruptions. *Journal of Volcanology and Geothermal Research* 22 (1984): 163–197.

Thouret, J.-C. Volcanic geomorphology: An overview. *Earth-Science Reviews* 47 (1999): 95–131.

Valentine, G. A., D. J. Krier, F. Perry, and G. Heiken. Eruptive and geomorphic processes at the Lathrop Wells scoria cone volcano. *Journal of Volcanology and Geothermal Research* (2007). doi: 10.1016/j.jvolgeores.2006.11.003.

Vallance, J. W., and K. M. Scott. The Osceola mudflow from Mount Rainier: Sedimentology and hazard implications of a huge clay-rich debris flow. *Geological Society of America Bulletin* 109 (1997): 143–163.

Van Rose, S., and I. F. Mercer. *Volcanoes*. London: British Museum (Natural History), 1991.

Vitousek, P. M., G. H. Aplet, J. W. Raich, and J. P. Lockwood. "Biological perspectives on Mauna Loa Volcano: A model system for ecological research." In J. M. Rhodes and J. P. Lockwood, eds., *Mauna Loa Revealed: Structure, Composition, History, and Hazards*. American Geophysical Union Monograph 92. Washington, DC: 1995.

Wells, S. G. Geomorphic assessment of late Quaternary volcanism in the Yucca Mountain area, southern Nevada: Implications for the proposed high-level radioactive waste repository. *Geology* 18 (1990): 549–553.

Whipple, K. X., N. P. Snyder, and K. Dollenmayer. Rates and processes of bedrock incision by the Upper Ukak River since the 1912 Novarupta ash flow in the Valley of Ten Thousand Smokes, Alaska. *Geology* 28 (2000): 835–838.

DIGGING DEEPER Geomorphic Effects of Volcano Sector Collapse

For volcanic geomorphology, the eruption of Mount St. Helens in 1980 was a dramatic affirmation that the "present is the key to the past." The collapse of the volcano's north flank was captured on film in a now-famous sequence of images (see Figure 11.10) and the several cubic kilometers of debris that the collapse left behind were studied extensively. The results solved a long-standing mystery and spawned a whole new way of thinking about volcanic landform evolution and volcanic hazards (Brantley and Glicken, 1986; Tilling, 2000).

Extensive hummocky deposits at the base of many terrestrial volcanoes had puzzled geomorphologists for decades. The deposits were variously attributed to piedmont glaciers, lahars, phreatic explosions, and landslides (Seibert, 1984). Some, such as the deposit extending north from Mount Shasta, a stratovolcano in northern California, were immense. That deposit, which slopes gently away from the volcano, has been dated to between 300,000 and 360,000 years ago, covers nearly 700 km², includes over 45 km³ of debris, and incorporates masses of andesite lava tens to hundreds of meters wide (**Figure DD11.1**; Crandall, 1989; Tilling, 2000).

After the Mount St. Helens 1980 eruption, it did not take long for geologists to put the pieces of the puzzle together. The hummocky landscapes surrounding many volcanoes were reinterpreted as debris avalanche deposits from the largest landslides on Earth. More than 150 volcanic examples were quickly identified. Measurements showed that the volume of debris avalanches was well correlated with the volume of material missing from the volcano (**Figure DD11.2**; Siebert, 1984).

Debris avalanches from sector collapses that removed large parts of a volcanic edifice were geologically common events; field mapping and geochronology suggested that they occurred somewhere on the planet on average every 25 years (Siebert, 1984). The location of these massive

FIGURE DD11.1 After geologists observed the 1980 eruption of Mount St. Helens, including the sector collapse and debris avalanche that began the eruption, they reinterpreted the hummocky terrain to the north of Mount Shasta (shown here) as evidence for a massive debris avalanche (>45 km³) that occurred between 300,000 and 360,000 years ago. [From Tilling (2000).]

failures appears to be related to the stress field around the volcano as indicated by the orientation of dikes (e.g., Gee et al., 2001). Triggering mechanisms included injection of magma into volcanoes during eruptive cycles and hydrothermally induced weathering of volcanic rocks that reduced the strength of the volcanic edifice (Reid, 2004).

About the same time that geomorphologists working on land were solving the debris-avalanche puzzle, marine geologists were using new remote-sensing techniques involving side-scan sonar to map the ocean floor (Moore, Normark, and Holcomb, 1994a, 1994b). They too were finding extensive hummocky terrains, some of which

DIGGING DEEPER Geomorphic Effects of Volcano Sector Collapse *(continued)*

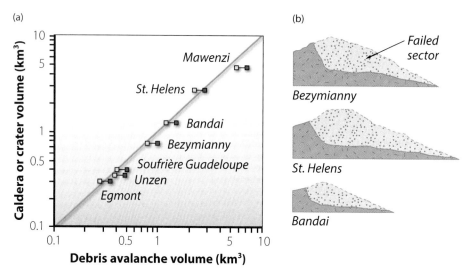

FIGURE DD11.2 Fieldwork and measurements showed that the source of volcanic debris avalanches was the adjacent volcanic edifice. (a) The volume of material missing from seven volcanoes matched well the volume of the debris avalanches near those volcanoes. The blue squares represent the measured volume of debris; the yellow squares represent the volume of debris corrected for a 25 percent expansion during transport and deposition. (b) Cross sections for three of these volcanoes show the shape of the landslide scars left after sector collapses. Shading shows the area removed by the failures. [From Siebert (1984).]

dwarfed those found on land [**Figure DD11.3**]. Such failures were first found around the Hawaiian Islands but are now being identified on volcanic islands around the world including areas around Stromboli volcano in the Mediterranean Sea [**Figure DD11.4**], Réunion volcano in the Indian Ocean, and around several volcanoes that make up the Canary Islands in the Atlantic Ocean (McGuire, 2003).

The submarine topographic data also solved the puzzle of marine gravels deposited high on the slopes of the Hawaiian Islands. For years, geologists had debated how these marine deposits were emplaced. Some argued for shoreline processes and uplift while others believed that giant waves had emplaced these gravels at high elevation (reviewed in Moore, Bryan, and Ludwig, 1994). The arguments were hard to resolve because the elevation history

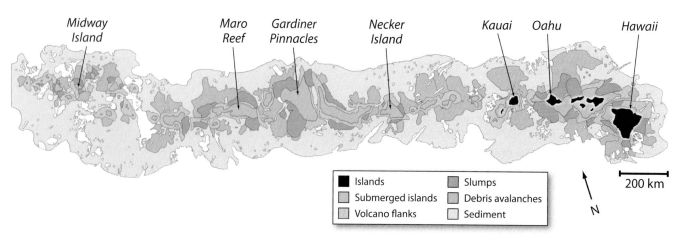

FIGURE DD11.3 Interpretive map showing large-scale geomorphology of the Hawaiian Islands and the adjacent seafloor made from remote sensing data collected by the Gloria side-scan sonar system. The map shows large areas where debris from both slower-moving slumps and rapidly moving debris avalanches covers the ocean floor around both the islands that are emergent today and those that were once emergent but have now subsided and are submerged. [From Moore, Normark, and Holcomb (1994b).]

of deposits on the Hawaiian Islands is complicated by the interplay of thermal uplift, sea-level change, isostatic responses to loading, and eventually thermal subsidence. Now, with the understanding provided by the Mount St. Helens eruption and the sea floor data, the origin of the gravels became clear; they were the result of mega-tsunamis generated by massive landslides both into the ocean and below its surface (McMurtry et al., 2004).

The geomorphic impacts and geologic hazards caused by sector collapses and the resulting debris avalanches are significant. On land, such landslides spawn lahars that can be far-traveled, impacting communities tens of kilometers from the volcano. Sector collapses associated with the intrusion of magma can release lateral blasts, horizontally directed pyroclastic emissions. On volcanic islands, sector collapses can trigger large tsunamis as huge masses of material enter the water. The effect of these tsunamis is greatest near the landslide with run-ups measured in hundreds of meters but they can also be far-traveled, affecting areas thousands of kilometers away (McMurtry et al., 2004; Ward and Day, 2001; **Figure DD11.5**).

Brantley and Glicken (1986) succinctly capture the impact of one eruption on geomorphic thinking: "As Mount St. Helens illustrated so dramatically on May 18, 1980, debris avalanches from volcanoes pose significant hazards to people and property. Debris avalanches may occur without warning, move great distances at high speed, cover large areas, initiate lateral blasts, and, if they

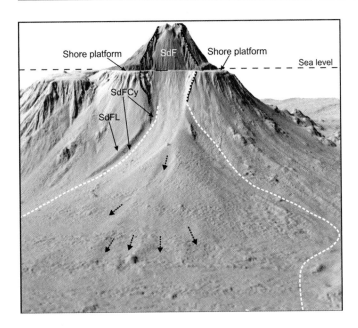

FIGURE DD11.4 Stromboli is a large stratovolcano rising more than 3000 m above the floor of the Mediterranean Sea. This view, derived from digital elevation models, shows the scar from a major edifice collapse (labeled SdF) and the submarine fan that resulted. There is a twofold vertical exaggeration. The dashed line (white) shows the limit of the submarine fan; black, hatched lines show the limit of failure and short black arrows indicate troughs on the fan surface. A prominent levee on the side of fan is labeled SdFL and the submarine canyon bordered by the levee is labeled SdFCy. [From Romagnoli et al. (2009).]

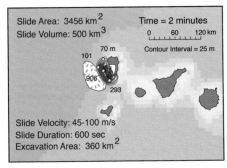 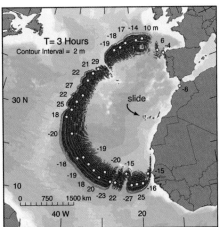 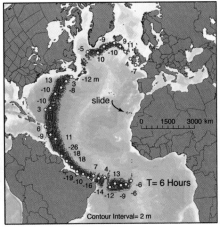

FIGURE DD11.5 A tsunami created by a sector collapse on the Cumbre Vieja volcano in the Canary Islands, east of North Africa, could cause a tsunami that would sweep across the Atlantic Ocean (Ward and Day, 2001). Here, a model of that tsunami predicts both the timing of the wave (indicated on each panel) and the wave height (in meters, indicated next to the wave fronts in each panel). This initial modeling suggested that waves many meters high could impact the east coast of North America. Later models (e.g., Lee, 2009) by different authors suggested the waves would be much smaller. [From Ward and Day (2001).]

DIGGING DEEPER Geomorphic Effects of Volcano Sector Collapse *(continued)*

enter the sea, cause tsunamis. The May 18 eruption was the first time eyewitness accounts and photographs documented the emplacement of a large volcanic debris avalanche. The debris-avalanche deposit at Mount St. Helens has provided a basis for interpretation of similar deposits elsewhere and has led to the realization that large-scale gravitational slope failures of volcanoes are more common than previously thought. Since 1980, volcanic hazard assessments have included consideration of hazards posed by debris avalanches in addition to other, more common products of eruptions, such as pyroclastic flows, lahars, lava flows, and tephra."

Brantley, S. R., and H. Glicken. Volcanic debris avalanches. *Earthquakes & Volcanoes* 18 (1986): 195–206.

Crandell, D. R. Gigantic debris avalanche of Pleistocene age from ancestral Mount Shasta Volcano, California, and debris-avalanche hazard zonation. *United States Geological Survey Bulletin* 1861 (1989): 14–18.

Elsworth, D., and S. J. Day. Flank collapse triggered by intrusion: The Canarian and Cape Verde Archipelagoes. *Journal of Volcanology and Geothermal Research* 94 (1999): 3340.

Gee, M. J. R., A. B. Watts, D. G. Masson, and N. C. Mitchell. Landslides and the evolution of El Hierro in the Canary Islands. *Marine Geology* 177 (2001): 271–293.

Lee, H. J. Timing of occurrence of large submarine landslides on the Atlantic Ocean margin. *Marine Geology* 264 (2009): 53–64.

McGuire, W. J. Volcano instability and lateral collapse. *Revista* 1 (2003): 33–45.

McMurtry, G. M., P. Watts, G. J. Fryer, et al. Giant landslides, mega-tsunamis, and paleo-sea level in the Hawaiian Islands. *Marine Geology* 203 (2004): 219–233.

Moore, J. G., W. B. Bryan, and K. R. Ludwig. Chaotic deposition by a giant wave, Molokai, Hawaii. *Geological Society of America Bulletin* 106 (1994): 962–967.

Moore, J. G., W. R. Normark, and R. T. Holcomb. Giant Hawaiian landslides. *Annual Review of Earth and Planetary Sciences* 22 (1994a): 119–144.

Moore, J. G., W. R. Normark, and R. T. Holcomb. Giant Hawaiian landslides. *Science* 264 (1994b): 46–47.

Reid, M. E. Massive collapse of volcano edifices triggered by hydrothermal pressurization. *Geology* 32 (2004): 373–376.

Romagnoli, C., P. Kokelaar, D. Casalbore, and F. L. Chiocci. Lateral collapses and active sedimentary processes on the northwestern flank of Stromboli volcano, Italy. *Marine Geology* 265 (2009): 101–119.

Siebert, L. Large volcanic debris avalanches: Characteristics of source areas, deposits, and associated eruptions. *Journal of Volcanology and Geothermal Research* 22 (1984): 163–197.

Tilling, R. I. Mount St. Helens 20 years later: What we've learned. *Geotimes* 45 (2000): 14–19.

Ward, S. N., and S. Day. Cumbre Vieja Volcano—Potential collapse and tsunami at La Palma, Canary Islands. *Geophysical Research Letters* 28 (2001): 3397–3400.

WORKED PROBLEM

Question: Calculate the yield strength (S_y) of two different lava flows assuming they behave as plastic materials given the information in the table below. Discuss your results and their implications for the evolution of volcanic landforms.

Location	Thickness (h, meters)	Slope (θ, degrees)	Rock type
Mauna Loa, Hawaii	5	2	basalt
Mount St. Helens	80	15	andesite

Answer: When the lava flows stopped, their yield strengths must have been equal to the shear stress (see eq. 11.1). The shear stress is $\rho g h \sin\theta$, where $\rho = 3200$ kg/m³ for basalt, 3000 kg/m³ for andesite. The acceleration of gravity, g, is 9.8 m/s² and h and θ are listed in the table. The yield strength is 5470 pascals kg/(m s²) for basalt in this example and 380,000 pascals kg/(m s²) for andesite. The short, steep nature of most andesite flows reflects the higher yield strength for andesite (by a factor of ~70). Because andesite cannot flow down low-gradient slopes, it creates the steep volcanic cones we know as stratovolcanoes.

KNOWLEDGE ASSESSMENT Chapter 11

1. What is the difference between magma and lava?
2. Explain how plate-tectonic setting influences the predominant style of volcanism.
3. Predict where one is likely to find effusive eruptions and where one is likely to find explosive eruptions.
4. Where do you find the largest volcanoes on Earth?
5. Define and explain the arc-trench gap.
6. Predict the large-scale geomorphic impacts of a hot spot track under oceanic lithosphere and under continental lithosphere.
7. List three types of geomorphically important materials produced by volcanic activity.
8. What factors control the viscosity of lava?
9. Explain how the viscosity of lava affects the shape of lava flows and of volcanoes.
10. What are pyroclastic materials and how do they form?
11. Explain the process that forms welded tuff.
12. What approximation can be used to determine the yield strength of lava?
13. Give three examples demonstrating that volcanic processes are geomorphic agents capable of changing landscapes.
14. What causes volcanic earthquakes?
15. What are flood basalts and where are some found?
16. Describe the relationship between pyroclastic flows and ignimbrite sheets.
17. Describe how lahars form and how they affect the geomorphology of volcanic terrains.
18. Describe sector collapse and explain where it was first identified.
19. Provide two examples of caldera eruptions and describe their geomorphic impact.
20. What is a cinder cone and how does it form?
21. Compare and contrast the shape, size, and formation processes of shield and stratovolcanoes.
22. Describe the internal structure of a thick basalt flow after it has cooled.
23. What process rebuilds the peaks of stratovolcanoes after major eruptions?
24. List several factors that result in the rapid weathering of volcanic rocks.
25. Explain how and why cinder cones evolve after the end of the eruption that formed them.
26. List several biologic activities that are critical to the colonization of new volcanic terrain.
27. Explain how topographic inversion occurs in volcanic terrain.
28. Explain the causes of phreatic eruptions and list some of the characteristic landforms and eruptive products.
29. What is a table mountain, how does it form, and what can you learn from its elevation?
30. Which types of volcano are typically monogenetic and which are typically polygenetic?
31. Explain how gases are integral to the volcanic eruption process and how they influence volcanic geomorphology.
32. Explain the volcanic explosivity index (VEI) and how it is determined.
33. What is tephrachronology and how is it useful to geomorphologists?
34. List two landforms that result from the exposure of igneous rocks emplaced below Earth's surface.

PART IV

The Bigger Picture
(Chapters 12 through 14)

Tectonics and climate govern landscape process, evolution, and form by changing important boundary conditions including base level, uplift rate, temperature, and precipitation. Together, these characteristics set the pace of weathering, erosion, and sediment transport. Chapter 12 considers how tectonic processes shape landscapes and how tectonic influences can overwhelm those of erosional processes and earth materials to produce distinctive features. In Chapter 13, we examine the influence of climate on the rate and distribution of surface processes and consider the distribution of landforms indicative of specific climatic conditions. Chapter 14 brings together ideas presented earlier in the book to understand how landscapes change and evolve over time. After reading these three chapters, you should understand how tectonics affects landscapes, how landscapes both respond to and influence climate, and how landscapes evolve over time in response to changes in the processes shaping Earth's dynamic surface. Our goal for these chapters is to integrate the more specific material presented earlier in the book into a broad treatment through the big-picture lenses of tectonics, climate, and landscape evolution.

Structural deformation resulting from India's collision with Asia is apparent in the folded terrain to either side of the Himalaya, which defines the southern edge of the Tibetan Plateau. Deposition of material eroded from the elevated areas formed the extensive alluvial plains along the Indus and Ganges rivers. Image is a composite of Moderate Resolution Imaging Spectroradiometer (MODIS) natural color imagery superimposed on a digital elevation model of India, Tibet, and southeast Asia from the Shuttle Radar Topography Mission data. The daily MODIS data from August 2002 were stacked together and the least cloudy sample was selected to produce a nearly cloud-free image.

Tectonic Geomorphology

12

Introduction

The tectonic processes that move rocks shape the global distribution of continents, mountains, and ocean basins that, in turn, control the distribution and intensity of surface processes and the resulting landforms. Tectonic processes elevate rocks above sea level, where weathering prepares rock materials for sculpting into landscapes through erosion by wind, rain, rivers, and glaciers.

Earth's high-elevation surface features coincide with the boundaries of tectonic plates because rock uplift and deformation are focused at such boundaries. Tectonically active regions give rise to topography with spectacular vertical relief, whereas tectonically quiescent areas host broad lowlands and older, lower-relief uplands. Geomorphically intriguing landscapes include those that are tectonically quiescent at present but topographically significant (such as the southern Appalachian Mountains). Equally impressive are tectonically active areas raised to high elevation but lacking significant local relief (such as the high plateaus of Tibet and the Altiplano in South America). At the edge of these flat-lying highlands, spectacular landscapes develop, the result of erosional processes acting on the steep margins of these uplifted plateaus.

The imprint of tectonics on geomorphology is apparent not only in the size, extent, and location of mountain ranges, but in the localized steepness of river profiles, the character of mountain slopes, and in the form of river networks that develop along regional joint patterns. Tectonics sets the stage for, and sometimes directs, the

Alaska's Denali Fault Zone last ruptured on November 3, 2002, in a 7.9 magnitude earthquake. The fault scarp just west of the Delta River had an offset of about 5 m. Ground cracks, like the one pictured, were up to 3 m deep.

IN THIS CHAPTER

Introduction
Tectonic Processes
    Uplift and Isostasy
    Thermal and Density Contrasts
Tectonic Settings
    Extensional Margins and Landforms
    Compressional Margins and Landforms
    Transform Margins and Landforms
    Continental Interiors
    Structural Landforms

Landscape Response to Tectonics
    Coastal Uplift and Subsidence
    Rivers and Streams
    Hillslopes
    Box 12.1: Drainage Area-Slope Analysis
    Erosional Feedbacks
Applications
Selected References and Further Reading
Digging Deeper: When and Where Did that Fault Last Move?

Worked Problem
Knowledge Assessment

work of erosion. Tectonic processes also form particular small-scale landforms that provide clues as to the style and pace of tectonic deformation.

Tectonics influences geomorphology both actively and passively through the effects of rock uplift and subsidence on relief, spatial patterns of slope, and the juxtaposition of bedrock of different mechanical properties. **Active tectonic controls** involve landscape response to ongoing deformation of the land surface, such as the response of a river's slope to fault offset and regional uplift. **Structural controls** (sometimes referred to as passive tectonic controls) are those that influence landforms and landscape dynamics indirectly through spatial patterns in the erosional resistance of rocks. In tectonically active regions, geologic structure and lithology develop only minor geomorphic expression during active mountain building, as uniformly steep slopes and rapid erosion mask the underlying geologic structure and make structure difficult to infer from hillslope morphology. Once active tectonic forcing ends, lithology and structure emerge as the major controls on landforms, sometimes allowing one to map or infer geologic structure from topography.

This chapter examines landscape response to tectonics, beginning with how different tectonic settings give rise to broad landscape assemblages with characteristic landforms that reflect specific tectonic processes. We discuss landforms indicative of deformation and explore geologic controls on landforms that can persist long after active tectonic forcing ceases. We also consider how high rates of erosion in tectonically active areas influence the growth and development of geologic structures by spatially focusing and sustaining both erosion and **isostatic rebound** (see Chapter 1) over geologic time.

Tectonic Processes

The motion of Earth's tectonic plates drives the lateral and vertical movement that results in extension, compression, uplift, and subsidence—all of which are important in creating characteristic landcapes at the largest scale. Maintaining the mean elevation of a landscape requires that the tectonic processes driving rock uplift counteract the mass removed by erosion above sea level. Landscapes continually erode and, given time, eventually come to reflect the balance between the forces driving rock uplift and the erosional potential arising from the steepness of slopes and rivers draining them—the form of the land itself. Although uplift of the land surface is generally the result of the movement of tectonic plates, buoyancy due to thermal and density contrasts also elevate areas above neighboring regions. In addition, the isostatic response to erosion raises fresh rock to replace eroded material even as the average elevation of the land surface lowers (see Chapter 1).

Each of the primary mechanisms driving tectonic geomorphology—plate tectonics, **dynamically supported topography** (i.e., topography supported by heat and density contrasts in the crust and mantle), and isostatic response—results in different amounts and styles of rock uplift.

Dynamically supported topography is a key component of mean landscape elevation and is not independent of plate tectonics. Both plate tectonics and dynamically supported topography are consequences of a convecting thermally inhomogeneous mantle.

The common element among these tectonic processes is that rock uplift, and spatial gradients in rock uplift, set the stage for erosion to sculpt the land. In general, the relatively small areas of spectacular mountainous topography that fascinate tectonic geomorphologists occur in extensional rift zones, zones of collisional tectonics along continental margins, and areas that were formerly in such tectonic settings (see Chapter 1, Figure 1.4).

The interaction of tectonic plates elevates rock masses above sea level, where they are subject to subaerial erosion. Spreading centers along mid-ocean ridges and continental rifts make new crust and create ocean-bottom topography far from where plates collide, crumple, and deform to make mountains. Where oceanic and continental plates collide, cold and dense oceanic lithosphere (the outer solid part of Earth, including the crust and uppermost mantle) sinks into the mantle beneath less-dense continental material. This creates a subduction zone along which oceanic sediments can be scraped off the downgoing slab to form a thickened wedge of sedimentary rock that forms a coastal mountain range in front of a plate-margin-parallel range of active volcanoes. Examples of such a wedge include the Olympic Mountains and Oregon Coast Range that rise seaward of the volcanic arc of the Cascade Mountains in western North America.

Volcanic arcs like Alaska's Aleutian Islands [**Photograph 12.1**], or the northern and southern Andes in South America, lie above the point where subducted material descends deep enough (about 100 km) to drive volatiles (like water) from

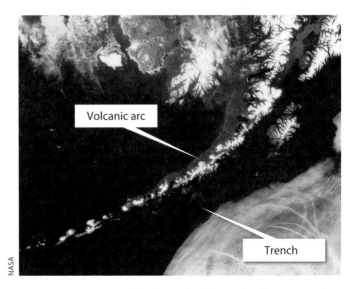

PHOTOGRAPH 12.1 Aleutian Volcanic Arc. Satellite image of the Aleutian Islands, showing the volcanic arc that parallels the adjacent trench. Clouds obscure lower right of image.

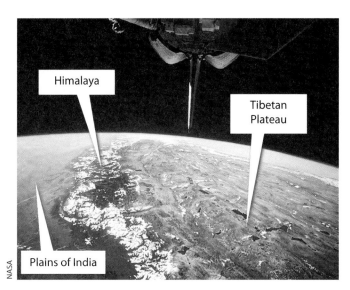

PHOTOGRAPH 12.2 **Himalayan Front.** View from the space shuttle looking west at the Himalayan front, with the foreland basin and plains of India to the left, and the Tibetan Plateau to the right.

PHOTOGRAPH 12.3 **East African Rift System.** Synthetic image based on digital topography of East African Rift Valley, looking north (top of image). Several large lakes occupy the rift. On the uplifted western (left-hand side) rift flank, the Ruwenzori Mountains (Uganda) are about 5000 m high.

the subducting and metamorphosing rock and sediment and up into the overlying mantle. These volatiles depress the melting point of rock enough that it partially melts, sending lower-density magma rising toward the surface. Thus, near oceanic subduction zones, chains of volcanic islands parallel to the subduction zone characterize the topography above the subducting plate. The horizontal distance from the subduction trench to the arc of volcanoes, and thus the width of the depositional forearc seaward of the arc, varies between many tens and a few hundred kilometers, depending on the angle at which the slab subducts. Steep subduction reduces the distance because the slab more quickly reaches a depth (and thus temperature) where partial melting and magma generation occur.

The nature of plate boundaries governs tectonic influences on landscapes. The mean elevation of Earth's surface generally reflects the thickness of continental crust; thicker continental crust leads to higher average elevation because continental crust is less dense than oceanic crust. When two lithospheric plates carrying continental crust collide, neither is dense enough to sink into the mantle. Instead, they crumple together in a collision that creates high mountains, like the Himalaya, through crustal thickening that builds up a low-density root; isostatic compensation of this root elevates the range well above sea level [**Photograph 12.2**].

Where two plates diverge in a rift zone, they create a spreading center associated with crustal thinning and mantle upwelling. Thermal buoyancy of the underlying, rising mantle and flexural uplift of the unloaded rift flank elevate rift-valley flanks above surrounding areas [**Photograph 12.3**]. At rifts, no crustal root is created, and this uplift is thermal and dynamically driven. Without a crustal root to drive a long-term isostatice response, uplift at rifts is relatively short-lived and ends when the lithosphere cools.

Where two plates slide past each other, the result is a series of long, linear landforms [**Photograph 12.4**] with differential uplift only if there are bends in the fault that cause local extension or compression.

Within each of these settings, tectonics influences geomorphology across a wide range of spatial scales, from the regional physiography to local fault interactions that raise some areas and cause others to subside. At each of these scales, understanding the influence of tectonics on geomorphology centers on landform analysis and knowledge of the underlying tectonic and geomorphic processes.

Uplift and Isostasy

Several distinct types of uplift are defined based on their reference frame: surface uplift, uplift of rock material, and

PHOTOGRAPH 12.4 **Fault Trace.** The San Andreas Fault viewed from the air looking along the Carrizo Plain in southern California. Two streams are offset (right laterally) in this 150 m length of the fault.

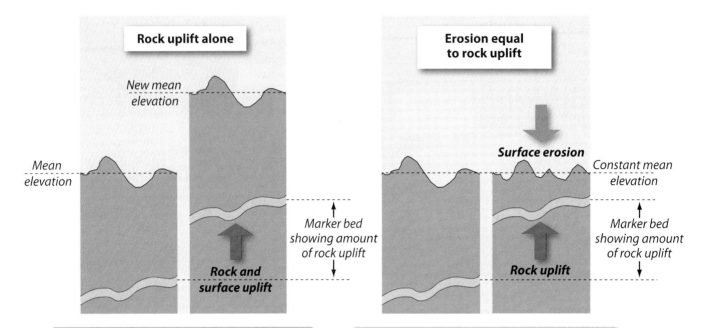

FIGURE 12.1 Surface Uplift Versus Rock Uplift and Exhumation. Relationship between surface uplift and the uplift and exhumation of rocks, illustrated through the end-member cases of no erosion and of erosion equal to rock uplift.

exhumation [**Figure 12.1**]. **Rock uplift** refers to changes in a rock's vertical position relative to a fixed datum such as sea level. **Surface uplift** refers to the change in the elevation of the land surface (again referenced to a fixed datum) rather than the uplift of a specific rock layer. Although the uplift of rocks and surfaces are tied to the same datum (usually sea level), rock uplift equals surface uplift only if no erosion occurs. If there is erosion, the surface uplift must be less than the rock uplift. **Exhumation** of rock describes the uplift of rocks relative to the ground surface through erosion that brings subsurface rock closer to the ground surface by stripping off the overlying material (lowering the ground surface). Unless countered by erosion, tectonically driven rock uplift will increase the land surface elevation. Surface uplift (U_S), uplift of rock (U_R), and erosion (or exhumation, E) are all interrelated:

$$U_S = U_R - E \qquad \text{eq. 12.1}$$

where rock uplift (U_R) is the sum of tectonically driven rock uplift (U_T) and isostatically driven rock uplift (U_I)

$$U_R = U_T + U_I \qquad \text{eq. 12.2}$$

Erosion and rock uplift are not independent because when erosion occurs it triggers an isostatic response that raises rock toward the ground surface. Consequently, the resulting net surface lowering from erosion is much less than the depth of rock removed—due to the uplift of rock by **isostatic compensation** (see Chapter 1). The magnitude of such compensation is given by $U_I = (\rho_c/\rho_m)E$, where ρ_c and ρ_m are the density of the crust and mantle, respectively.

Hence, the resulting net change in surface elevation due to erosion will be equal to:

$$U_S = E\,[(\rho_c/\rho_m) - 1] \qquad \text{eq. 12.3}$$

For each meter of rock stripped off the landscape, isostatic response lifts the underlying rock back up by about 82 cm due to the density contrast between the less dense eroded crust ($\rho_c \approx 2.7$ g/cm^3) and the more-dense underlying mantle ($\rho_m \approx 3.3$ g/cm^3). When tectonic uplift ceases (i.e., $U_T = 0$), the isostatic response to erosion (U_I) becomes the dominant form of rock uplift. Consequently, in tectonically quiescent regions, erosion itself plays a prominent role in rock uplift as the isostatic response to erosion continues to exhume rock, bringing once-deeper rock closer to the ground surface in response to removal of overlying mass.

Tectonic convergence over the lifetime of a mountain range builds up a thick crustal root. This root isostatically sustains elevated topography long after tectonic activity ceases because erosion must remove an amount of rock equivalent to many times the total relief before the root is reduced enough that it no longer supports elevated topography. Because postorogenic erosion rates are relatively slow, it can take surface processes a very long time to erase a mountain range supported by a deep crustal root (see Worked Problem).

Over geologically short timescales, isostatic compensation is influenced by the wavelength (size) of the load and by the **flexural rigidity** (stiffness) of Earth's lithosphere, which distributes and partially supports topographic loads [**Figure 12.2**].

Tectonic Processes 395

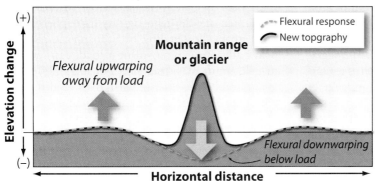

Elements of lithospheric flexure

Flexure, or bending, of the lithosphere in response to the load from a mountain range, glacier, or sedimentation is distributed across a broader area than that of the load itself, causing **flexural downwarping** below the load and **flexural upwarping** away from the load. The resulting topography reflects both the load and the stiffness of the underlying lithosphere.

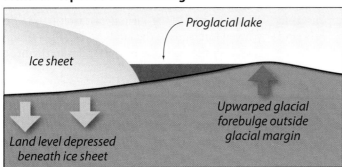

Flexural response to ice loading

The weight of ice depresses the land beneath the ice sheet and results in upwarping of a **forebulge** outside the ice margin. This creates a depression between the ice margin and forebulge, in which proglacial lakes can form.

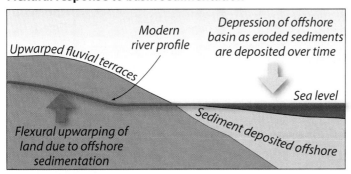

Flexural response to basin sedimentation

Flexural response to offshore sedimentation, or basin sedimentation at the foot of a mountain range, can cause flexural upwarping on land, producing uplifted fluvial terraces inland from the basin.

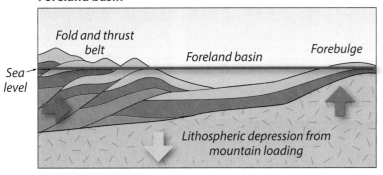

Foreland basin

Flexural depression can produce a foreland basin on the margin of a fold and thrust belt at the toe of a compressional mountain range. Such basins subside isostatically as sediment is deposited in them, in places producing deposits many kilometers thick.

FIGURE 12.2 Plate Flexure. Plates have flexural strength that distributes the loading or unloading of material, resulting in far-field depressions, bulges, and tilting.

The flexural rigidity of the lithosphere means that isostatic rebound or subsidence are distributed across a wide area, in contrast to how local erosion can vary greatly across the landscape. The flexural rigidity of the crust varies regionally and controls the wavelength and amplitude of isostatic response to either the addition of a topographic load (such as from a rising mountain range produced by lateral convergence of rock material) or the removal of a load (whether from erosion of rock or melting of glacial ice).

The wavelength of isostatic response is on the order of tens to hundreds of kilometers, exceeding the width of individual valleys. Thus, it is possible that mountain peaks can rise in response to the excavation of valleys if mass is preferentially removed from valley bottoms [**Figure 12.3**], triggering isostatic uplift across a broader region. As much as a quarter of a mountain's height can result from the incision of nearby valleys. It is no coincidence that many high peaks in the Himalaya rise above adjacent, deeply incised river valleys.

The magnitude of isostatic compensation for erosion or glacial melting and the spatial scale over which it occurs are related to the flexural rigidity. Hot, thin lithosphere is weak and has low flexural rigidity, resulting in greater isostatic compensation. The greater strength of cold, thick lithosphere produces higher flexural rigidity that acts to reduce the amount of isostatic compensation and spreads it across a broader area. The ongoing rebound of Scandinavia following deglaciation is an example of how the removal of a load (the ice sheet) triggered rock uplift that, in turn, produced slow surface uplift over a broad area of cold, high-rigidity lithosphere.

Flexure, the bending of lithospheric plates from the local addition or removal of a load, can indirectly cause far-field uplift or subsidence. For example, flexure causes upland source areas to rise when mass eroded from a mountain range fills a neighboring sedimentary basin. Just such a thing happened when material from the Sierra Nevada was deposited in California's Great Valley and to a lesser degree when material eroded from the Appalachian Mountains was deposited on the Atlantic shelf.

Foreland basins are depressions that develop adjacent and parallel to a mountain belt from flexural isostatic subsidence due to deposition of sediment shed from the uplands. For example, long-term deposition dominates the great alluvial forelands of the Indus and Ganges-Brahmaputra rivers, and on the east side of the Andes in the headwaters of the

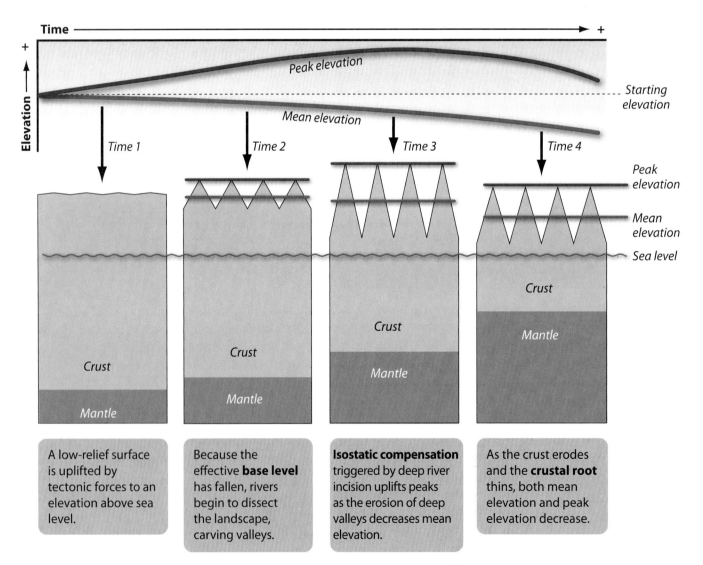

FIGURE 12.3 Isostatic Uplift of Mountain Peaks. Incision of deep valleys removes mass from mountains, resulting in isostatic compensation at broader length scales. Hence, even as mean elevation of the range lowers, peak elevation can be driven higher by rock uplift. [Adapted from Burbank and Anderson (2001).]

Amazon. Depositional foreland basins drive their own subsidence as the weight of deposited sediment (with a typical bulk density of about 1.6 to 1.7 g cm^{-3}) compresses previously deposited material and isostatic response causes roughly 50 cm of subsidence for each meter of material deposited. Piles of sediment >10-km-thick have accumulated in broad forelands adjacent to steep mountain fronts of the Himalaya and the Andes due to the isostatic depression of the lithosphere during millions of years of sediment deposition. Not all of the sediment shed off the Himalaya and the Andes makes it to the sea. Much is trapped in extensive, low-relief depositional zones (subsiding basins) along the Ganges and the Amazon rivers.

Thermal and Density Contrasts

Tectonic uplift and the resulting landscape response are not limited to collisional settings along plate margins. Thermal contrasts related to tectonic forcing can elevate topography wherever heating from subsurface sources produces lower-density, and thus more buoyant, lithosphere. Lithospheric thinning that accompanyies upwelling of hot rock in extensional settings, such as rift zones, can lead to **thermal uplift** of landscapes as buoyancy contrasts raise warm, less-dense lithosphere above adjacent cooler, denser lithosphere. The resulting physiographic upwarping decays over time as rocks cool with distance from the thermal anomaly. Thus, the elevation of deep-sea ridges decreases away from the spreading center at the ridge crest because the newly extruded rock cools and becomes denser as it moves laterally away from the rift over time (see Chapter 8). Large areas of thermal buoyancy, associated with rising mantle plumes, can result in dynamically supported topography, high-standing regions that owe their elevation to thermal forcing. In such settings, the lack of a thick crustal root means that topography rapidly becomes subdued as the lithosphere begins to cool. Dynamic support from mantle upwelling, which can result in higher elevations than would otherwise be supported by the crustal root (as is argued by some for southern Africa), further illustrates the connections between surface topography and deep-Earth processes. Low-standing topography, such as deep ocean trenches near subduction zones, can also be dynamically maintained; in this case, Earth's surface is depressed by sinking lithosphere.

Not only are thermal contrasts central to igneous processes that lead to volcanoes and volcanic landforms, but thermally driven uplift associated with mantle plumes also influences continental drainage patterns and volcanism in continental interiors (e.g., Yellowstone, Montana) or on plate margins (e.g., Iceland). The hot spot responsible for forming the Hawaiian Islands left a trail of progressively older (and now lower) islands to the west from the big island of Hawaii, where thermally driven uplift is buoying up the oceanic lithosphere. Along the island chain, older islands have subsided more because the older lithosphere below them has progressively cooled; their age is also reflected in their landscapes. The degree of dissection, rock weathering, and soil development all increase as the islands age (see Chapter 11).

PHOTOGRAPH 12.5 Salt Glaciers. Dark-colored salt flowing from breached anticlines in the Zagros Mountains in southern Iran, a zone of collision and uplift. The salt is less dense than the surrounding rock and moves upward as diapirs. When these diapirs reach the surface, the salt flows ductilely under gravity-induced stress.

In addition to tectonically induced density contrasts, there are stratigraphic situations in which low-density materials, like salt, underlie higher density rock. This creates gravitational instabilities through which buoyancy drives uplift and structural deformation. When a body of salt, such as a deposit of marine evaporites, is buried beneath denser overburden, the salt may rise up through the overlying material to form a **diapir**, a bulbous intrusion that can take odd shapes and form a salt dome where it displaces and deforms overlying strata. Rising salt bodies that reach the surface can flow out over the ground as salt glaciers [Photograph 12.5], which move much like conventional glaciers, only much slower, at rates on the order of meters per year. Although they are rare on Earth, being concentrated in extremely arid regions such as the Zagros Mountains in Iran, salt glaciers also appear to occur on Mars (although there they might be made of sulfates and other salts rather than sodium chloride). Layers of salts interstratified with other types of rock provide zones of weakness along which substantial deformation may become focused.

Tectonic Settings

Tectonic setting is the primary control on the global pattern of regional physiography and landscape character. The regional tectonic settings of active plate margins, passive continental margins, and continental interiors strongly influence landforms through styles of tectonic deformation and uplift, differences in dominant lithologies, and changes in the degree of fracturing (which affects erosion resistance). Distinct topography characterizes plate margins with different types of lithosphere (oceanic or continental). Because tectonic plates generally consist of both oceanic and continental lithosphere, it is useful to consider separately the dominant controls on the topography of extensional zones (including passive continental margins), compressional margins, transform margins, and continental interiors.

Tectonic setting and structural geology influence landforms through the direct action of faulting and isostasy and through the indirect influences of spatial variability in erosion resistance generated by folding, faulting, and offset of rocks of differing lithology. **Faults** are discontinuities in Earth's lithosphere that record deformation and movement and that are now boundaries between rocks that did not originally form next to one another. Springs preferentially occur along fault traces due to interception and blockage of subsurface water flow across the fault and/or enhanced discharge through a permeable zone of crushed rock along the fault. Faults can influence topography either through surface uplift and offset or through fault-influenced patterns of differential erosion, which can create erosional relief where highly erodible rocks are juxtaposed against erosion-resistant rocks. **Fault scarps** are steep linear slopes that reflect the direct topographic expression of fault offset [**Photograph 12.6**]. Offset across fault scarps in seismic events is highly variable but can be as great as 10 m for a single earthquake. The height of a fault scarp often reflects multiple discrete episodes of offset.

Below, we consider the types of landforms that typically develop in association with fault systems along **normal faults** with extensional offset, **reverse faults** with compressional offset, and **transverse faults** with lateral (strike-slip) offset. The topographic influences of tectonics and structural geology depend on the rate and type of crustal deformation, the three-dimensional geometry of the resulting geological structures, and differential erosion of rock involved in the deformation.

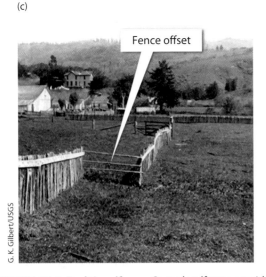

PHOTOGRAPH 12.6 Fault Landforms. Some landforms provide surface evidence of faulting. (a) In the distance, a thrust fault scarp in Tibet offsets a fan and a river terrace on which the vehicle is parked. (b) Normal fault scarp in limestone in the northern Galilee of Israel strikes directly through a village. (c) Fence offset by the 1906 San Francisco earthquake along the strike-slip San Andreas Fault in Marin County, California. (d) Normal fault in alluvium generated by the 1959 earthquake at Hebgen Lake, Montana.

Extensional Margins and Landforms

Zones of divergence (extension) are places where continents split apart, producing geomorphologically distinctive landscapes that include escarpments, rift valleys, volcanoes, and alluvial and debris fans. At extensional plate boundaries, tectonic forces pull plates apart and form rift zones that can, over time, mature into a spreading center and mid-ocean ridge. Such tension (extension) produces fracturing and normal faulting that facilitate the eruption of lava and the erosion of rock masses. Although mid-ocean ridges constitute most of the world's extensional plate boundaries, divergence within continents forms major active **rift valley systems** like those of the East African Rift system and the Rio Grande Rift in New Mexico and Texas [Figure 12.4]. Rifting is the first step in the formation of passive margins like the east coast of South America and the west coast of southern Africa.

Zones of continental extension range in scale from individual **pull-apart basins** within transform fault systems like the Dead Sea Rift of the Jordan Valley to regional extension along high-angle normal faults that extend through the shallow crust in the Basin and Range Province of western North America. The East African Rift system, long thought to be the cradle of human evolution, is still in the early stages of breaking up the heart of Africa. The separation of the Arabian Peninsula from Africa across the Red Sea, to which the East African Rift is connected, represents a more advanced stage of rifting and the birth

Where tectonic spreading centers occur on land, they form extensional **rift zones.** Where they occur in ocean basins, they form **mid-ocean ridges.** Both are characterized by an axial valley and upwarped rift margins. As a continental rift zone develops and widens, it can progress from upland topography like the East African Rift Zone to form an incipient ocean like the Red Sea. With time, the Red Sea may continue to widen and eventually become an ocean like the Atlantic.

Extensional **rift zones** often have a down-dropped **axial valley** filled with accumulations of lake sediments and alluvium shed from steep rift flanks that drain into the rift. Gentler slopes drain away from the rift flanks outboard of the upwarped rift margin. Hot mantle material upwelling beneath the thinned crust of the axial valley drives **thermal uplift** and **dynamic support** of the overlying topography. Lakes can develop in down-dropped grabens between high-standing horsts.

Extensional zones contain diagnostic **normal faults** that delineate range fronts, can offset alluvial fans, and form triangular bedrock facets that represent the exhumed fault surface. Hourglass-shaped valleys extend into the range front between the facets.

FIGURE 12.4 Rift Zones—Extensional Settings. Rift zones characterize extensional tectonic settings both on land and in the ocean, and they produce landforms characteristic of normal fault zones and extension.

of a new ocean basin. Spreading at the Mid-Atlantic Ridge created the Atlantic Ocean basin and progressively separated Africa and South America while creating the largest and longest mountain chain in the world. Ancient rifting created continental margins and the world's great escarpments (see Chapter 14).

In most rift zones, the central axial valley is part of a system of interconnected local depressions atop a great ridge. Typically, several strands of rift valleys together define an integrated rift system. The topography within rift zones can be broken up into extensive blocks separated by substantial escarpments. Large normal fault scarps define the steep inner shoulder of a rift leading down into closed, internally drained depressions that form deep sedimentary basins and elongated lakes parallel to the rift axis [**Photograph 12.7**].

Rift zones are geomorphically distinct. Rockslides typically contribute to high erosion rates on the steep slopes facing into the rift zone where deep canyons feeding numerous fans are found. Short, steep channels drain down the rift-margin escarpment and into a down-dropped chain of basins along the bottom of the rift, which can host either lakes (such as Lake Tanganyika in East Africa) or longitudinal rivers (such as the Rio Grande in New Mexico).

Regional drainage typically flows away from the uplifted rifted margins. Rivers and streams draining away from rift shoulders are characterized by relatively low gradients and minimal sediment supply. Warping due to thermal uplift of rift shoulders can lead to drainage reversals or captures along an actively developing rift system, with smaller channels more likely to be diverted due to their lower stream power and limited ability to incise bedrock. For example, small rivers originating near the Atlantic coast of South America first flow west down the upwarped rift flank away from the coast before turning to the east and emptying into the Atlantic Ocean—the result of capture by trunk streams breaching the uplifted rift shoulders.

Large rivers can incise rapidly enough to keep pace with the rising rift shoulders. For example, the major rivers of Africa and South America (with the exception of the Nile) flow toward the passive (trailing-edge) Atlantic margin over what was once the rift shoulder. Many of these rivers flow over great knickzones along the middle of their courses. While some of these knickzones are lithologically controlled, others reflect knickzone retreat from the original rift margin since the breakup of the supercontinent Pangaea more than 100 million years ago.

Passive continental margins unaffected by active tectonic processes develop after rifting has advanced sufficiently that oceanic crust is created at the spreading center and the two continents involved are drifting apart. Together, thermochronologic and cosmogenic nuclide data suggest that rift shoulders rapidly eroded back and then stabilized to form steep, long-lived escarpments characteristic of passive margins (such as those along the southern African or eastern Australia coasts). A slowly eroding, low-relief coastal plain typically lies on the ocean side of the escarpment. Above the escarpment are low-relief highlands that also erode slowly. The escarpment itself, however, is eroding somewhat faster and thus slowly retreats. Over time, through-going rivers dissect escarpments creating large topographic embayments and isolated plateau surfaces.

Lithology and structure greatly influence the geomorphology of passive margins that evolve from rifts. For example, in southern Africa and along the east coast of North America the passive margin runs parallel to the location of ancient mountain ranges. Today, rocks that were pervasively deformed and metamorphosed at depths of many kilometers are exposed on the surface, their coarse-scale structures (such as anticlines and synclines) clearly exposed by differential erosion.

Regional extension in continental settings results in distinctive large-scale topography even if it does not lead to rifting. For example, the linear north-south trending mountain ranges that collectively define the Basin and Range province of western North America are due to normal faulting resulting from east-west tectonic extension of the region. These fault-block mountains are thought to have formed from the stretching of a regionally extensive plateau [**Figure 12.5**]. In the Basin and Range, the valleys are generally tectonically down-dropped along normal faults (**grabens**), and the mountain ranges are uplifted as **horsts**, representing the toppled ends of gigantic tilted and back-rotated crustal blocks. Fault-bounded grabens typically form broad, flat-floored valleys that may create closed, internally drained depressions. Some, such as Death Valley in California, are below sea level.

Normal faults often form distinct fault-bounded range fronts. Where rivers or streams incise V-shaped valleys across such a range front, the intervening slopes are characterized by distinctive triangular facets [**Photograph 12.8**]. Along many extensional range fronts, active faulting visibly offsets

PHOTOGRAPH 12.7 Dead Sea Rift. The internally drained Dead Sea and the adjacent steep, faulted rift flank in Israel. This is the steep, western (inboard) side of the rift. The view is to the north.

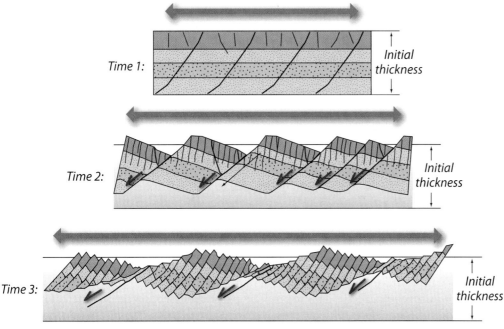

FIGURE 12.5 Tectonic Extension—Basin and Range Province. The Basin and Range physiographic province of western North America is the product of east-west extension that produced numerous north-south trending ranges and valleys.

moraines or fans developed at the topographic break in slope that separates bedrock uplands from alluvial lowlands [**Photograph 12.9**]. Dating such offset features provides a means of calculating rates of fault movement and patterns of fault offset.

Compressional Margins and Landforms

Compressional **orogens**, mountain ranges composed of linear belts of highly deformed rock, form along collisional plate boundaries where tectonic convergence leads to thickening of the crust that, in turn, elevates mountains. Where lithospheric plates converge, tectonically active mountain ranges are generally arrayed as linear belts of high elevation and significant topographic relief. Examples include the western margins of North and South America and continental collision zones like the Himalaya. Mountains at active compressional orogens generally consist of deformed sedimentary rock intruded or interlayered with igneous rock.

Compressional orogens typically have high rates of deformation and uplift. This uplift drives river incision that, in turn, leads to steep slopes and frequent landsliding. These mass movements deliver large sediment loads to steep, energetic rivers with high flows capable of rapidly

PHOTOGRAPH 12.8 Triangular Facets. Triangular facets on a range front from normal faulting at the base of the Tendoy Mountains in southwest Montana.

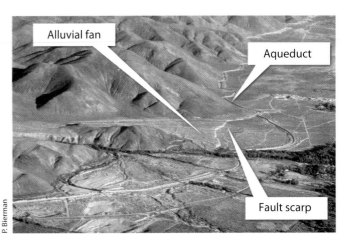

PHOTOGRAPH 12.9 Faulted Fan. Lone Pine strand of the Owens Valley fault zone offsets late Pleistocene alluvial fan. The white fault scarp (center, at the foot of the hills) cuts across the Los Angeles aqueduct, which runs around the toe of the fan.

transporting all the sediment delivered to them. In active orogens, deeply incised rivers commonly flow through narrow bedrock valleys or deep gorges cut into rock. Different types of convergent orogens have different topographic characteristics that we describe below [**Figure 12.6**].

When two plates carrying continental crust converge, neither is as dense as the underlying mantle and thus neither will readily sink. Consequently, where continents collide, material piles up and the crust thickens, leading to dramatic high-standing mountain ranges like the Himalaya. Continental collisions lack significant volcanic activity and have a zone of thrust faulting that typically separates the upland drainage system of bedrock channels from the alluvial channels flowing across the sedimentary basin that develops in the depositional foreland where rivers leave the uplands. Longitudinal drainage paralleling the strike of a mountain system can develop because of orogen-parallel flexural subsidence of the foreland; in contrast, short, steep channels typically drain the mountain front. Large rivers that cross major mountain ranges (like the Indus and Tsangpo/Brahmaputra that begin on the Tibetan Plateau and flow across the Himalaya) are generally thought to be **antecedent rivers** older than the mountains (see Chapter 7). These rivers were able to maintain their courses by vigorous erosion across the rising mountain range. Landforms typical of compressional orogens include upland bedrock channels and hillslopes with thin soils and little if any saprolite (weathered rock) due to rapid erosion and therefore short residence times of materials on slopes.

In contrast, volcanic arcs form along subduction zones where an oceanic plate subducts beneath either a continental or oceanic plate. Volcanic arcs may be either continental, in which an oceanic plate subducts beneath a continental plate, or an **island arc**, in which an older, colder, and denser oceanic plate subducts beneath another plate that is younger, warmer, and thus more buoyant. In either case, a wedge of sedimentary rock scraped off the downgoing, subducting plate can rise seaward of the mountains of the volcanic arc that parallels and is fed by the subduction zone.

The island of Taiwan is an example of a **sedimentary wedge** formed in an ocean-continent collision. Mountain building and erosion in Taiwan are particularly rapid because of the combination of a wet tropical climate, weak rocks, and a huge tectonic influx of material into the orogenic wedge due to the great thickness of incoming continental sediments on the Eurasian plate. In regions where the incoming plate has a lot of sediment on top, thick wedges of deformed, highly erodible, fractured sedimentary rock characterize compressional orogens.

The rocks exposed along the spines of active volcanic arcs consist of massive lava flows in oceanic island arcs, and more erodible pyroclastic deposits in continental arcs (see Chapter 11). Some deeply exhumed mountain ranges, such as the British Columbia Coast Range of Canada or the Sierra Nevada of California, expose hard, relatively massive undeformed rocks (like granite) that represent the exhumed roots of ancient volcanic arcs. The Sierra Nevada, for example, consists of the eroded roots of an ancient continental volcanic arc (like the modern Cascade Mountains of Oregon and Washington State) that was shut off from its magma source more than 65 million years ago.

Reverse (or thrust) faulting is the dominant form of tectonic displacement in compressional orogens, and typically has indirect topographic expression. **Thrust faults** are shallow reverse faults that place older rocks over younger rocks [**Photograph 12.10**]. Thrust faults can extend over hundreds of kilometers, but typically do not produce distinct fault scarps. Many reverse faults either terminate in **blind thrusts** (that do not produce scarps) before reaching the surface or splay out into multiple fault traces near the surface.

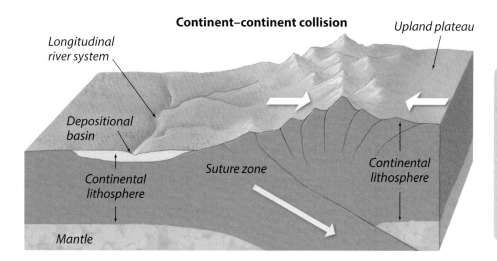
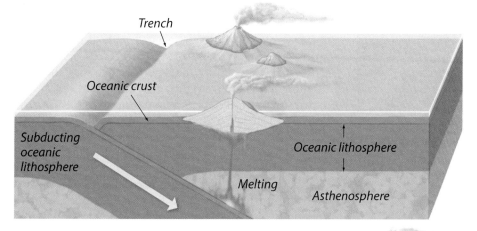
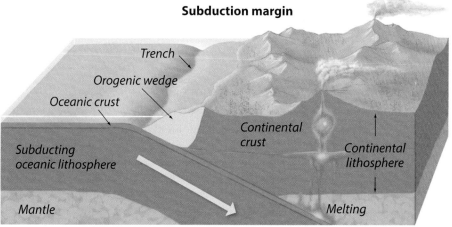

FIGURE 12.6 Collisional Tectonic Settings. Collisional tectonic settings include continent–continent collisions with a major mountain range bordered by a foreland sedimentary basin, ocean–ocean collisions forming oceanic island arcs, and continental-oceanic subduction zones.

Reverse faults can have substantial topographic expression at large scales, such as the topographic front of the Himalaya, which is an impressive escarpment generated by long-term offset along thrust faults. In regions where rates of deformation are higher than rates of erosion, compressional tectonic deformation and folding may directly affect topography. In such cases, structural synclines (troughs) form valleys, and the crest of structural anticlines (arches) form topographic ridges, the best known of which are in the Zagros Mountains in Iran and the Ventura anticline in

PHOTOGRAPH 12.10 Thrust Fault. Reverse (thrust) fault in the northern Gulf of Corinth between Athens and Delphi in Greece. The unit below is clastic sedimentary rock. The unit above is a deformed fault mélange of originally sedimentary material. The fault plane is reddish brown.

southern California. Where a river breaches an anticlinal ridge, surveying and dating sets of river terraces that have been bowed upward and deformed across the rising fold (anticline) reveal the pattern of uplift centered on the crest or hinge of the anticline and allow calculation of the rate of surface uplift [**Figure 12.7**].

Transform Margins and Landforms

Plates slide past one another at transform margins, where strike-slip faults dip steeply creating a distinctive set of landforms [**Figure 12.8**]. Transform margins can be simple with one distinct fault trace or complex with activity distributed along a series of faults that define a broad shear zone. Such shear zones produce distinctive landforms and topography resulting from lateral movement.

A regional component of extension or compression across a transform margin produces distinct extensional or compressional features. **Transtensional margins,** such as the Gulf of California, have a string of down-dropped basins, or grabens, along the shear zone. In contrast, compressional transform margins (**transpressional margins**) typically have fault-zone parallel mountains due to compression distributed across the shear zone. A good example of transpressional transform margin topography are the central California Coast Ranges that extend to either side of the main San Andreas Fault, reflecting compression across the margin and complex interactions between fault strands.

On a more local scale, as material moves laterally through bends along a strike-slip fault, the geometry of the fault system creates zones of local uplift or subsidence where compression or extension occurs. This results in the development of compressional pop-up structures in **restraining bends** and extensional pull-apart basins in **releasing bends** [**Figure 12.9**]. The Santa Cruz Mountains, south of San Francisco, California, are an example of a mountain block uplifted due to lateral compression through a restraining bend along the San Andreas Fault. Conversely, San Francisco Bay lies in a subsiding area (transtensional) between the San Andreas and Hayward faults along the same fault system.

Strike-slip faults are associated with distinctive landforms that can be used to map fault traces. Streams that cross strike-slip faults may be offset across the fault, resulting in distinctive right-angle bends that can be used to determine offset directions and magnitudes. Fault offset may even behead streams, separating channels from their sources. Trenching and dating of truncated or offset alluvial deposits along such **offset streams** can reveal the amount and timing of movement across a fault. Distinctive **shutter ridges** form where lateral fault offset moves a ridge in front of a stream, deflecting its course. Similarly, **sag ponds** develop where drainage courses are impounded or small depressions form along the trace of strike-slip faults [**Photograph 12.11**]. Transform faults also occur in marine settings where they cut and laterally offset mid-ocean ridges.

Continental Interiors

Landforms associated with continental interiors include deeply weathered low-relief terrain, areas of extensive loess deposition, and large alluvial rivers flowing across broad floodplains (where there is an extensive orogenic sediment source). Some continental interiors have both low slopes and low overall relief, leading to development of internal drainage where rivers end in closed depressions.

Continental interiors in humid, tropical regions develop thick weathering profiles, whereas those in arid regions generally have bare rock slopes. Consequently, overland flow in arid landscapes and diffusive processes in humid landscapes (see Chapter 5) dominate hillslope transport as there typically is only limited, localized landsliding due to the predominance of gentle slopes. Weathering processes greatly influence landforms in such low-gradient environments, and isolated high-standing rock outcrops, known as **inselbergs** (see Chapter 3), tend to be best developed in continental interiors [**Photograph 12.12**].

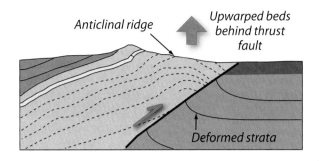

Anticlinal ridges develop above **thrust faults** due to compression. Motion along the fault may deform strata near the fault due to drag.

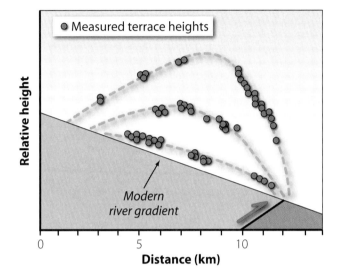

Patterns of **rock uplift** due to fault offset can be preserved in upwarped alluvial (depositional) terraces or strath (erosional) terraces, as illustrated by this example from Tien Shan, Tibet. Plotting the relative height of terrace surfaces can reveal the pattern of deformation associated with thrust fault movement.

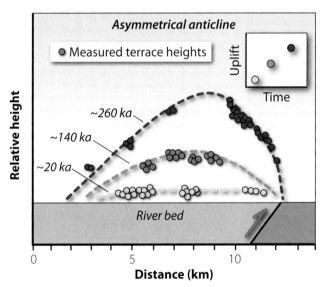

In this diagram, the modern river profile has been made horizontal to emphasize the warping of the terraces over time. Anticlinal upwarping may either be symmetrical (not shown) or asymmetrical (as shown). Dating of upwarped sediments can be used to determine rates of uplift along the underlying fault.

FIGURE 12.7 Anticlinal Ridges. Uplift of anticlinal ridges associated with thrust faults may deform stream terraces. The age and pattern of deformation of these terraces may be used to investigate the history of uplift along the fault. [Adapted from Molnar et al. (1994) and Burbank and Anderson (2001).]

The vast interiors of continents primarily consist of relatively flat areas known as **cratons**—tectonically stable regions of relatively low relief, typically rising no more than a few hundred meters above sea level. Cratons are underlain by ancient continental crust with the low-elevation and low-relief terrain characteristic of regions of prolonged tectonic stability. Cratons consist of **shields,** areas composed of complexly deformed crystalline rocks, and **platforms,** areas where younger, undeformed sedimentary rocks lay on top of far older crystalline basement rock.

Plateaus are high-elevation, low-relief surfaces separated from surrounding land by steeper slopes. The world's high elevation plateaus—Asia's Tibetan Plateau and South America's Altiplano—are tectonically constructed landforms

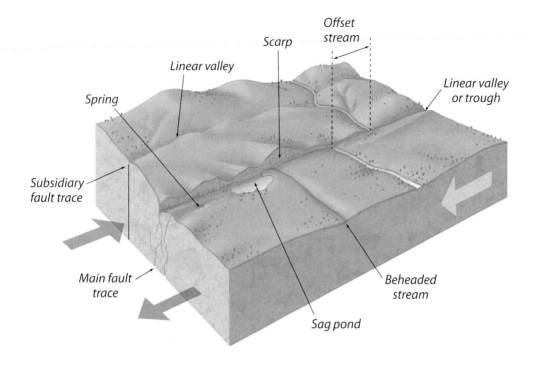

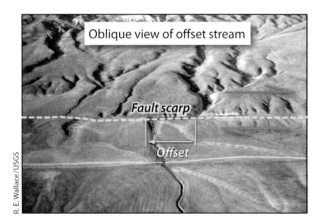
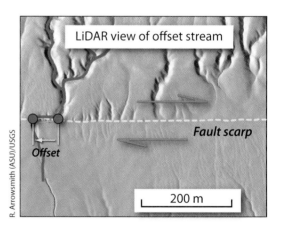

Strike-slip fault zones produce a characteristic suite of landforms that reflect both linear offset (**beheaded streams, offset channels**) and weakening of rock or tectonic extension near the fault (**linear valleys, sag ponds**). LiDAR and aerial photographs can reveal the location and amount of recent offset on strike-slip faults.

FIGURE 12.8 Strike-Slip Fault Zones. Characteristic landforms of strike-slip fault zones include sag ponds and offset stream channels, as well as evidence of changes in hydrology such as springs.

that formed when the lithosphere thickened enough that its strength limits further increases in elevation [**Figure 12.10**]. When continental crust reaches a height of 4 to 5 km, geothermal heating weakens the base of the lithosphere enough to make it susceptible to lateral extrusion upon further thickening. This effectively limits the mean height of topography, as further thickening leads to lateral flow at depth that controls plateau elevation. Once a plateau reaches this critical height, continued tectonic convergence leads to widening of the plateau.

Some continental interiors have substantial topography reflecting in large part the location of former orogens, now tectonically inactive. The combination of isostasy and a thick root ensures that tectonically inactive mountain ranges persist, lasting far longer than one would calculate by considering erosion rates alone (thanks to isostatic rock uplift offsetting ~82 percent of erosion).

The interiors of continents are for the most part tectonically quiescent. Earthquakes and fault scarps are few and far between and many are associated with old structures

Releasing bend

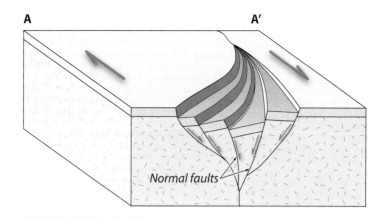

Normal faults
(ticks on lower side)

Releasing bends result in local development of **pull-apart basins** in which extension is accommodated on a nested set of normal faults. Pull-apart basins often have a rhomboidal shape and may contain sag ponds.

Restraining bend

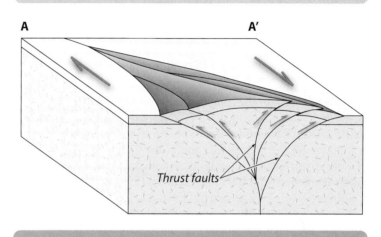

Thrust faults
(ticks on upper side)

As material is squeezed through a **restraining bend,** by movement along a strike-slip fault, compression results in surface uplift distributed along a nested set of thrust faults.

FIGURE 12.9 Restraining and Releasing Bends. Interaction of strike-slip fault geometry and topography produces zones of structural uplift that result from compression at restraining bends whereas subsidence occurs in extensional pull-apart (releasing) bends.

PHOTOGRAPH 12.11 Sag Ponds. The Denali Fault (Alaska) runs obliquely across this image, and a line of sag ponds mark its trace (image is several kilometers across).

PHOTOGRAPH 12.12 Inselberg. An inselberg rising above the low-relief coastal plain of Namibia near the dunes of Soussevlei. Sport utility vehicle is shown for scale.

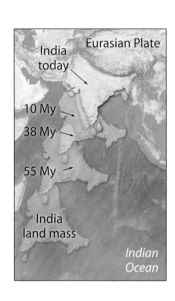 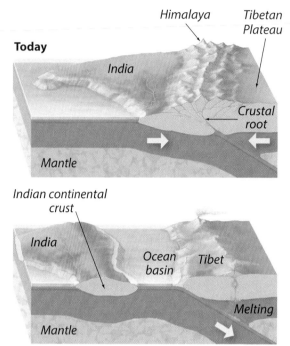 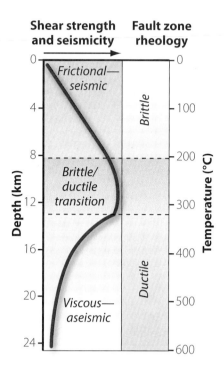

As India moved into Asia, the continental lithosphere of both continents was not dense enough to subduct and ongoing collision resulted in **crustal thickening**. Over time, the **crustal root** grew thicker and the Tibetan Plateau rose and, as the collision continues, it grows wider. The thickening crust, rising mountains, and high rates of erosion affect the shape and behavior of the Himalaya. Rapid erosion advects warm rock toward the surface, and the thickened crust causes rock at depth to cross the brittle/ductile transition. This change in the process and rate by which the rock deforms allows rock to more easily and rapidly extrude out the sides of the range, removing some of the mass brought in by continental collision and limiting the height of the Tibetan plateau.

The shear strength and rheology (flow characteristics) of the lithosphere vary with depth due to the effects of increasing pressure and temperature. In the uppermost brittle crust, frictional strength increases to a depth of around 8 to 10 km. At higher pressures and temperatures below about 13 km depth, shear strength decreases and the crust becomes ductile, producing aseismic viscous flow. Tectonic convergence that produces exceptionally thick crust can lead to lateral flow of ductile material at depth.

FIGURE 12.10 Formation of the Tibetan Plateau. The Himalaya and Tibet formed as a result of crustal thickening sufficient to change solid-Earth behavior at depth, limiting the crustal thickness supportable by the underlying mantle and thereby forming the high-elevation, low-relief Tibetan Plateau.

such as failed rift margins. Earthquakes that do occur can be strong and affect large areas because of the relative strength of cratonic, continental crust. Such mid-continental quakes have caused liquefaction of weak, wet sediments (for example, in the New Madrid Fault Zone in south-central North America) and left distinctive linear fault scarps (as at Tennent Creek in northern Australia).

Structural Landforms

After tectonically driven rock uplift declines or ceases, structure and lithology exert a pronounced influence on topography as rates of weathering and erosion outpace rates of rock uplift, folding, and deformation. Different lithologies or structures determine slope form as preferential erosion of weaker rocks leaves more resistant rocks standing out in relief. In regions with large variations in erosion resistance, steeper slopes tend to form on more erosion-resistant rocks, whereas gentler slopes characterize more erodible formations.

The topographic expression of deformation structures, such as folds, results from differential erosion of rocks. For example, geologic structure is readily apparent in the Appalachian Mountains of eastern North America, where large differences in the erodibility of the underlying geology govern ridge and valley patterns [**Figure 12.11**]. Resistant

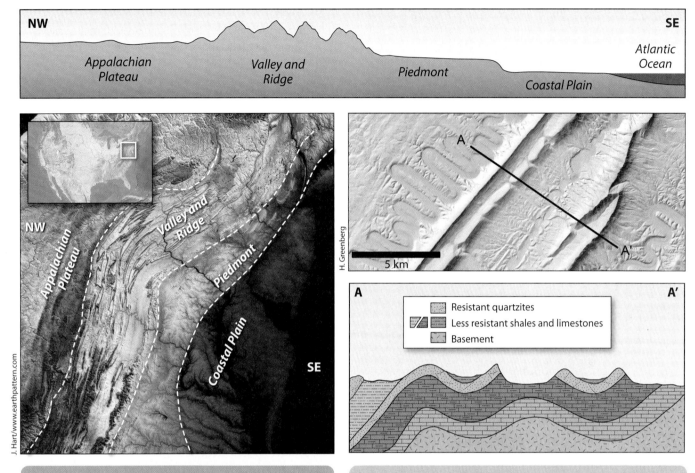

FIGURE 12.11 Tectonically Inactive Mountains—the Appalachians. The valley and ridge morphology of the Appalachian Mountains reflects the strong influence of geological structure and lithology.

units (sandstone and quartzite) form ridges and weaker units (shale and limestone) form valleys [**Photograph 12.13**]. Over time, stream patterns become adjusted to the underlying geologic structure, and streams preferentially flow along the weakest rocks or in fracture zones. In general, there is greater structural control on landforms in ancient, no-longer-active orogens than in active orogens because in rapidly uplifting regions like the Himalaya and Taiwan uniformly steep (threshold) slopes obscure the underlying geologic structure and patterns of rock uplift.

As erosion cuts through hard, ridge-forming beds, it can expose weaker underlying rocks, which can then erode quickly. This can lead to a topographic inversion, in which geologic structure and topography are out of alignment. Where erosion breaches the crest of a structural anticline, exposure of more erodible rocks along the crest of the anticline can produce a topographic low along the structural high through excavation of an **anticlinal valley.** Similarly, a **synclinal ridge** reflects the preservation of structurally low but erosion-resistant rocks as a topographic high in the structural trough of a fold.

The inclination or dip of underlying beds can greatly influence landform development [**Figure 12.12**]. **Monoclines** occur where the dip of strata increases locally, but the beds do not turn over, producing a structural and topographic step. The relative inclination of stratified (layered) rocks

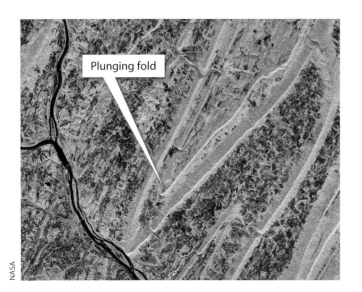

PHOTOGRAPH 12.13 Appalachian Topography. This false-color radar image shows the large, plunging folds of the Valley and Ridge physiographic province near Sunbury, Pennsylvania. The ridges are held up by resistant sandstone and the valleys are dominated by shale and limestone. Black trace at left of image is the Susquehanna River cutting across structure. The view is about 30 km wide and north is to the upper lefthand corner.

relative to the hillslope angle influences slope forms. Erosion of gently dipping beds leads to the development of pronounced dip slopes where the hillslope angle parallels rock bedding. **Cuestas** are asymmetric slopes that are elongated in the down-dip direction. More symmetric slopes where the underlying beds dip approximately 45 degrees are known as **hogbacks**. Distinctive **flatirons**, named due to their resemblance to the now antiquated household appliance, develop from differential erosion of a resistant rock layer

---

The inclination of stratified rocks relative to the hillslope angle influences slope forms. Erosion of gently dipping beds leads to the development of pronounced dip slopes where the hillslope angle parallels the angle of rock bedding.

Monocline

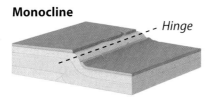

Monoclines form where strata are folded but do not turn over—their shape resembles a carpet draped over a stair step. The dip of the strata increases over the hinge line along the axis of the monocline.

Cuestas

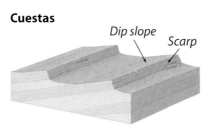

Cuestas are asymmetric slopes that are elongated in the down-dip direction. More symmetric slopes, where the underlying beds dip approximately 45°, are known as **hogbacks.** Slopes developed from erosion-resistant rock layers that dip more steeply than the hillslope angle form **flatirons.**

Flat-lying, erosion-resistant beds that protect underlying rock lead to the development of flat-topped, steep-walled **mesas.** Continued erosion of a mesa can lead to an isolated summit area known as a **butte,** like the famous outcrops of Monument Valley in Utah and Arizona.

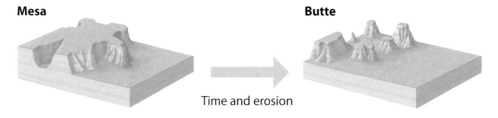

FIGURE 12.12 Structural Landforms. The orientation of dipping beds of rock can control the shape of landforms, especially in areas where soil cover is minimal.

PHOTOGRAPH 12.14 Joint-influenced weathering patterns and landforms. (a) Checkerboard weathering pattern caused by chemical weathering along intersecting joint sets in the El Capitan Granite, Yosemite, California. Trees provide scale. (b) Fins left by preferential erosion of joints in sandstone in the Maze District, Canyonlands National Park. The fins are 20 to 40 m high.

that dips more steeply than the hillslope angle. In flat-lying strata, erosion-resistant cover beds that protect underlying rock lead to the development of **mesas,** flat-topped plateaus like those of northern Arizona. Mesas are often protected by an erosion-resistant cap rock that retards erosion on the mesa top, leading to steep-walled sides. Continued erosion of a mesa can lead to an isolated summit area known as a **butte,** like the famous outcrops of Monument Valley in Utah and Arizona. In arid landscapes, where soil mantles are thin and slope form reflects bedrock properties (see Chapter 5), the influence of differential erosion resistance in flat-lying layered sedimentary rocks can be expressed as **cliff-and-bench topography,** in which erosion-resistant rocks (like well-cemented sandstone and lava flows) form cliffs, whereas weaker interstratified rocks (like shale) form benches.

Joints, fractures in rocks along which little or no motion has occurred, can result from regional tectonic stresses or shrinkage upon cooling, and can have significant topographic expression. Joint-related landforms are particularly visible in arid environments with bedrock hillslopes. Regional tensile stresses may produce joints in even relatively strong rocks because the tensile strength of rock is far less than its compressive or shear strength. On a smaller scale, igneous rocks often have well-developed joint sets due to shrinkage upon cooling, for example, columnar joints in thick, slowly cooled, basalt flows (see Photograph 11.15). Joints can greatly influence differential weathering and erosion because they provide conduits along which water and plant roots can preferentially penetrate, break up, and erode rock masses (see Chapter 3). In rocks with well-developed vertical jointing, continual erosion along joint surfaces can produce checkerboard weathering patterns [**Photograph 12.14a**] or leave the unjointed rock standing in relief as narrow fins [**Photograph 12.14b**], the jointed (more fractured) rock having been removed more rapidly by erosion.

Landscape Response to Tectonics

Tectonics affects landscapes in a variety of different ways. Tectonically induced base-level fall and uplift can increase topographic relief and steepen river profiles and hillslopes (see Chapter 7). Uplift of the land surface relative to local or global base level (surface uplift) will trigger an increase in local slope that sweeps up through a river system, driving a wave of incision toward the headwaters of a drainage basin as the increased topographic gradient (slope) influences the rate and type of erosional processes.

Tectonic forcing can reorganize drainages and change drainage-basin boundaries. For example, continental rifting opens new basins and splits existing drainages. Tectonic tilting can reverse a river's flow direction or cause one stream to erode headward more rapidly than another, eventually beheading the captured stream and creating a new and expanded drainage network. Such truncated basins are common in tectonically active areas and are often identified by fluvial gravels stranded at high elevation, far above any modern-day stream channels. A good example can be found in the northern Galilee in Israel, where basaltic gravel from the Syrian highlands sits atop the local limestone bedrock, even though today there is no fluvial connection between the basalt outcrops and the gravels. More subtle responses to tectonic forcing are common in drainage basins where a rise in base level leads to aggradation and basin filling, whereas a falling base level causes incision and the creation and preservation of terraces.

Topographic slope drives erosional processes more directly than elevation. The effect of elevation on erosion occurs indirectly through its control of climate. Flat surfaces at high altitude erode slowly, whereas steep slopes at low elevation erode relatively quickly.

Coastal Uplift and Subsidence

Changes in relative sea level, whether from global sea-level change, tectonic rock uplift, or subsidence, leave their mark on coastal landscapes. The coastlines of many tectonically active continental margins consist of rocky coasts where actively eroding sea cliffs define the shoreline angle (see Chapter 8). During times when sea-level rise matches the rate of coastal rock uplift, wave action can abrade a broad, gently sloping (~1 degree) wave-cut platform that extends seaward from the base of the sea cliff to below the level of tidal influence. If sea-level rise is less than the rate of coastal rock uplift, then the sea cliff and wave-cut platform are abandoned, which results in the formation and exposure of a low-relief marine terrace with a fossil sea cliff separating it from the next higher terrace.

Over multiple glacial/interglacial cycles of global sea-level rise and fall (due largely to variations in the amount of glacial ice), sequences of marine terraces can form on actively rising coastlines, with younger terraces closest to sea level and older, increasingly eroded terraces at progressively higher elevations. The series of marine terraces formed by uplifted coral reefs on the coast of New Guinea provides a classic example of landforms resulting from the interaction of changing sea level with tectonically driven surface uplift [Photograph 12.15]. On this and many other rising, tectonically active coastlines, such as those of northern California and Japan, the heights of individual marine terraces can be correlated with sea-level high stands; such correlation allows one to estimate coastal uplift rates back through time from the relative elevations of different marine terraces [Figure 12.13].

PHOTOGRAPH 12.15 Marine Terraces. Flights of marine terraces along the rapidly uplifting Huon Peninsula on the northeastern coast of New Guinea. The highest terrace is now several hundred meters above sea level and hidden under the low clouds.

Coastal subsidence can occur for a variety of reasons. Subsidence can accompany large subduction-related earthquakes along active margins. Drowned forests and marsh deposits covered by layers of tsunami-deposited sand along the coasts of northern California, Oregon, and Washington testify to long periods of slow coastal uplift separated by episodes of rapid subsidence during large earthquakes

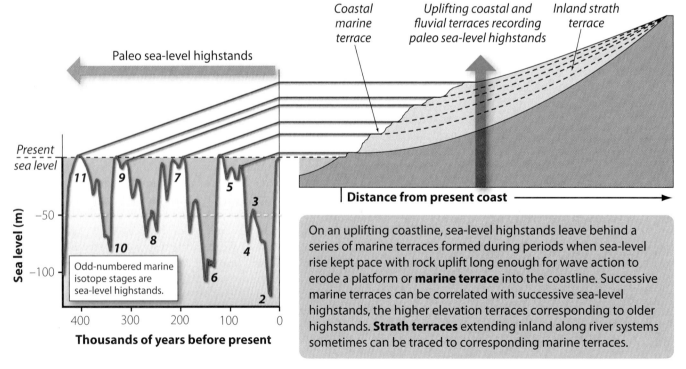

FIGURE 12.13 Marine Terraces Along Uplifting Coastlines. On an uplifting coastline, the elevation of marine terraces and inland bedrock strath terraces may be correlated with past sea-level highstands to estimate their age and the local uplift rate.

respond to a rising base level through deposition in their lower reaches, building a depositional wedge or ramp of sediment [Figure 12.14]. Above this depositional zone, the channel network remains uninfluenced by the change other than now being graded to a higher base level.

Along rivers discharging to an uplifting coastline, flights of strath (bedrock) terraces abandoned above the modern river floodplain record progressive incision of the river. Such terraces can be correlated by height, just like marine terraces, and often can be traced up through a river network. By dating terraces at different elevations, one can determine river down-cutting rates and, by inference, surface and rock uplift rates.

River-channel morphology can serve as a sensitive indicator of tectonic deformation because channel patterns are closely related to both channel slope and sediment supply (see Chapter 6). For example, in response to localized steepening of their longitudinal profile, braided channels tend to incise and convert to a single-thread channel, whereas meandering channels may increase in sinuosity or braid before incising in response to tectonic steepening. More extreme changes in channel gradient can convert an alluvial channel reach to a bedrock channel reach. Greater increases in slope lead to development of distinct knickpoints or knickzones where channels narrow, steepen, and incise gorges or canyons.

The effect of vertical fault offset along a river profile depends on the magnitude and sense of offset. Sufficient upward displacement on the downstream side of a reverse fault can create a low-gradient reach upslope or even impound a lake, whereas upward displacement on the upstream side of a normal fault will create a knickpoint, or waterfall. The relationship between a river's course and the underlying bedrock structure provides clues as to whether the course of the river predates or postdates the topography (see Chapter 7). Lateral deformation of drainage patterns, via strike-slip fault offset, can produce asymmetry in drainage-basin form if sustained over geologic time.

Hillslopes

The steepness of hillslopes can respond to changes in tectonic forcing—up to a point. Soil-mantled slopes can only become so steep before landsliding begins to limit further steepening. In low- to moderate-gradient landscapes, slope steepness generally increases with increasing rock uplift rates, a characteristic of **sub-threshold slopes.** However, the angle of soil-mantled and well-fractured bedrock slopes can only increase to an upper limiting or threshold mean hillslope angle between 30 and 40 degrees, depending on soil and rock strength. Once hillslopes steepen to **threshold slopes,** erosion rates will respond to further increases in rock uplift through more frequent landsliding as hillslopes cannot steepen further to keep pace with ongoing river incision. The contrasting behavior of sub-threshold and threshold slopes (see Chapter 5) means that hillslope angles in low-gradient, sub-threshold terrain can

PHOTOGRAPH 12.16 Ghost Forest. Ghost forest along the Copalis River, southwestern Washington. The dead trees (snags) were victims of tidal submergence caused by tectonic subsidence during the great CE 1700 Cascadia subduction-zone earthquake. The trees died when their roots were immersed in salt water as the land subsided during the earthquake.

[Photograph 12.16]. Such subsidence can result in brief periods of marine inundation even on an actively uplifting coast. Long-term subsidence due to sedimentary loading and compaction from ongoing deposition characterizes large estuaries such as the Mississippi River delta. Subsidence along the mid-Atlantic coast of North America reflects the slow decay of the glacial forebulge, formed as a result of the displacement of mantle material from beneath the Laurentide Ice Sheet in the Late Pleistocene.

Rivers and Streams

Rivers and streams respond to tectonic forcing through adjustments in slope. Spatial variations in rock uplift along a river profile can lead to local reaches steeper, or flatter, than expected along a river's longitudinal profile. A plot of the downstream values in the stream gradient index (see Chapter 7) can identify such anomalous reaches where faults or lithology might be influencing channel slopes. Methods for analyzing such deviations also include DEM-based drainage area-slope analyses [Box 12.1]. Steep sections of a river profile tend to erode faster than gentler reaches, causing knickpoints (see Chapter 7) to migrate upstream and in extreme cases can lead to one river capturing and diverting the flow from another.

The relationship between a drainage basin and its base level can change through tectonic subsidence, uplift of the land, or a rise or fall in sea level. Base level changes greatly affect the locus of sedimentation in coastal zones and alluvial rivers. In contrast to how the effects of a base-level drop propagate up through the channel network, base level rises primarily affect reaches near river mouths, drowning deltas and turning coastal valleys into estuaries and bays. Rivers

BOX 12.1 Drainage Area-Slope Analyses

A formal way to assess the adjustment of channel slopes, and therefore river profiles, to tectonic activity comes through positing a balance between rates of rock uplift (U_R) and river incision (E) to predict the form of steady-state river profiles. Generally, the erosive potential of a river may be expressed as a function of its drainage area (A) and its local slope (S), as a steeper river with a larger drainage area, and thus greater discharge, and will have greater power to cut down into rock:

$$E = K \cdot A^m \cdot S^n \qquad \text{eq. 12.A}$$

where the scaling exponents m and n are in many cases found to have values of 1.0 and 0.5, respectively, and K is a constant to characterize bedrock erodibility and the role of climatic factors and basin geometry in scaling the river's discharge. For the idealized case of a steady-state river profile eroding everywhere along its length at the rock uplift rate, $U_R = E$ and thus

$$U_R = K \cdot A^m \cdot S^n \qquad \text{eq. 12.B}$$

which may be rearranged to yield an expression for how channel slope would be expected to vary as a function of drainage area

$$S = (U_R/K)^{(1/n)} \cdot A^{-(m/n)} = K_s \cdot A^{-\theta_c} \qquad \text{eq. 12.C}$$

where $K_s = (U_R/K)^{(1/n)}$ and $\theta_c = m/n$. Equation 12.C predicts that the steady-state profile of an incising bedrock river will plot as a straight line on a logarithmic graph, with a slope equal to $-m/n$ and a coefficient (K_s) directly related to the ratio of the uplift rate to the bedrock erodibility, climate, and basin geometry (U_R/K). The coefficient K_s is termed the steepness index and the exponent θ_c is the profile concavity.

Subject to the assumptions of constant erodibility and rock uplift, area-slope plots can be used to identify locations along a bedrock river profile where uplift rates change; such changes will show up as breaks in slope on such plots [Figure 12.B1]. Portions of an upland river system with higher rock-uplift rate (U_R) will plot at higher slopes for the same drainage area than will reaches with lower rock-uplift rates. Values of K_s determined for different reaches of a river system can be used to estimate relative differences in the value of rock-uplift rates in different portions of a river system, if, of course, one has accounted for any differences in lithology, discharge, and bedrock erodibility, which may influence the denominator of the ratio that makes up K_s. Interpreting drainage area-slope analyses requires consideration of potential lithologic or tectonic factors that may influence the analysis, such as the presence of knickpoints propagating up through a channel network or large differences in bedrock erodibility.

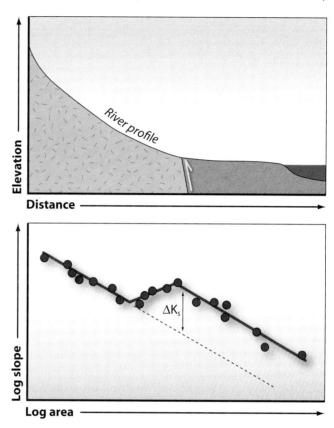

Deviations from the expected equilibrium profile of a bedrock river can be used to quantify differences in the erodibility or uplift/erosion rate through area-slope analysis of a river profile. With river slope (S) related to drainage area (A) through $S = K_s A^{-\theta}$, the relationship between A and S will plot as a straight line on logarithmic axes, with a slope equal to $-\theta$ and a y-intercept of K_s. Hence, deviations from a log-linear trend may be interpreted as due to changes in either the rock erodibility or uplift rate.

FIGURE 12.B1 Drainage area and slope are inversely related down river systems. Locations where the slope/area trend along a river diverges from the general trend can be used to identify locations where rock uplift rate or erodibility may have changed.

respond to an increased uplift rate or base-level fall through steepening. In contrast, the morphological response of landscapes with steep, threshold slopes would be most pronounced in stream profiles or in the stripping of soil to expose bare bedrock slopes, and the proliferation of deep-seated bedrock landslides.

The development of **inner gorges**, zones of steeper slopes low on valley walls that create a distinct valley-within-a-valley morphology [**Photograph 12.17**]. Inner gorges represent a response to either falling base level or an increase in uplift rate. Gorge walls steepen where rivers incise into bedrock faster than hillslopes erode. Where river incision

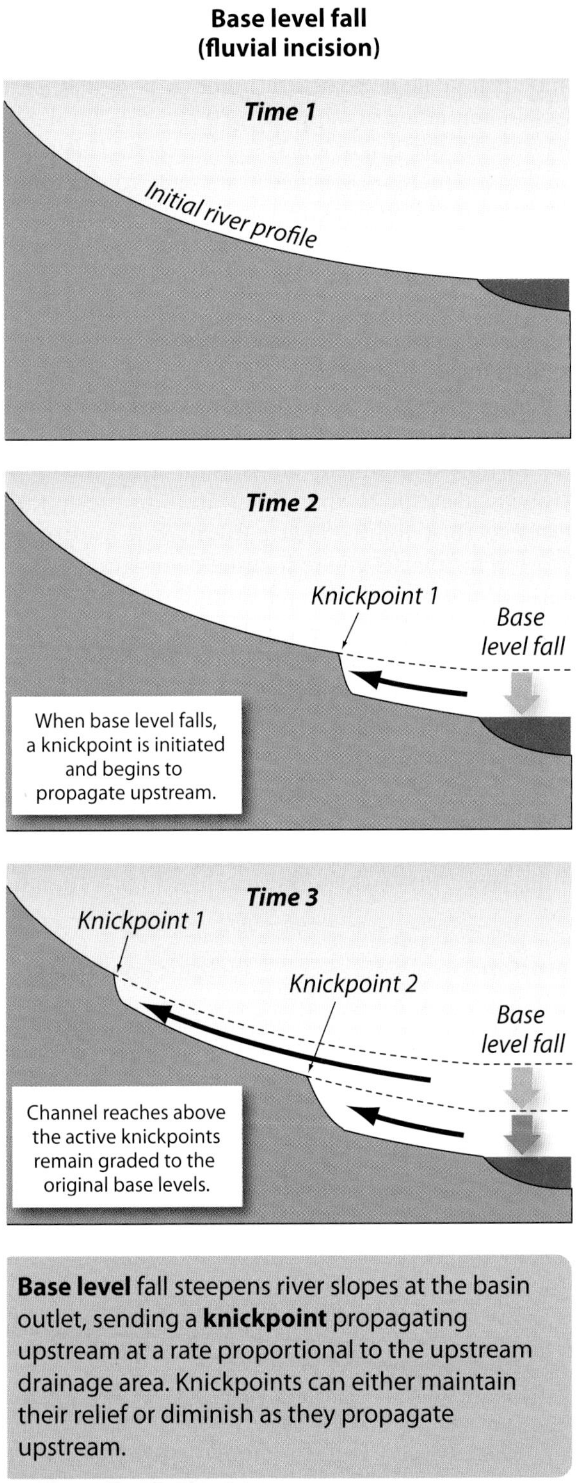

Base level fall (fluvial incision)

Time 1
Initial river profile

Time 2
Knickpoint 1
Base level fall

When base level falls, a knickpoint is initiated and begins to propagate upstream.

Time 3
Knickpoint 1
Knickpoint 2
Base level fall

Channel reaches above the active knickpoints remain graded to the original base levels.

Base level fall steepens river slopes at the basin outlet, sending a **knickpoint** propagating upstream at a rate proportional to the upstream drainage area. Knickpoints can either maintain their relief or diminish as they propagate upstream.

Base level rise (aggradation)

Time 1
Initial river profile

Time 2
Base level rise
Delta

As base level rises, the river deposits its load of sediment.

Time 3
Continued base level rise
Foreset beds
Topset beds
Buried delta
Bottomset beds

Subsequent rises in base level shift the locus of deposition farther inland.

The influence of a rising base level shifts the locus of sedimentation inland, predominantly affecting estuarine and lowland river systems. The direct influence of a base level rise is restricted to aggradation in the downstream end of a river system as the system adjusts.

FIGURE 12.14 Fluvial Responses to Base Level Change. The rise and fall of base level result in very different river, estuarine, and sedimentary basin responses.

rates increase in response to increased tectonic uplift, the lower portions of valley walls steepen first, creating the distinct break in slope that defines an inner gorge.

The sediment supply of rivers can be affected by earthquakes or large storms that deliver large loads of hillslope-derived sediment to river systems in the form of seismically induced or precipitation-induced landslides. The resulting increase in downstream sediment loading may cause channel aggradation and braiding, and temporarily convert bedrock channel reaches into alluvial reaches.

Erosional Feedbacks

Coupling between erosion and tectonics produces a range of effects from continental-scale topography down to the

PHOTOGRAPH 12.17 Inner Gorge. Inner gorge in the Himalaya between Tenboche and Namche, Nepal. Note the large (~10 m) boulder stuck between the inner gorge walls.

scale of geological structures along individual river valleys [**Figure 12.15**]. For instance, erosion of deep valleys can focus exhumation, resulting in preferential rock uplift (through isostatic rebound) along major rivers. This can produce a **river anticline** defined by the crest of a structural fold running parallel to the river, with the greatest uplift centered along the course of the river.

This odd situation of a river flowing along the structural high, with the topographic low perched atop the spine of the structural anticline, characterizes major trans-Himalayan rivers, like the Arun River just east of Mount Everest. The anticlinal geologic structure running along the Arun River has an amplitude of more than 10 km, structural relief comparable to the height of Mount Everest. This and similar geologic structures oriented transverse to the trend of the compressional mountain range are the youngest deformational structures in the Himalaya. Their young ages indicate that they developed in response to incision along the rivers and thus that the rivers did not simply take advantage of preexisting structure to establish their courses across the rising range. In other words, the focused erosion along the course of major, range-crossing rivers unroofs rocks at speeds greater than in areas away from the river courses.

Sustained gradients in erosion also occur at much larger scales due to spatial variation in precipitation and tectonic rock uplift. Such patterns can lead to tectonic-erosion feedbacks where focused denudation is coupled to active deformation. Greater rainfall on the windward side of mountain ranges can result in differential exhumation that strongly influences structural development on either side of the drainage divide. This asymmetry deeply exhumes rocks on the wetter, windward side of ranges, such as on the rain-drenched western slope of the southern Alps of New Zealand. In the dramatic cases of where the Indus and Tsangpo rivers—the most powerful and erosive in Asia—slice through deep gorges at either end of the Himalaya, extremely rapid river incision (>10 mm/yr) has

River anticline

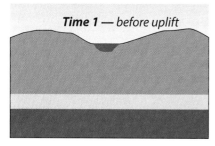

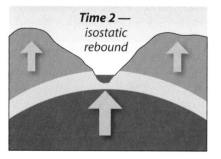

River anticlines form where mountain rivers, rapidly eroding into flexurally weak rock, remove enough mass that they catalyze focused exhumation and isostatic rebound. This focused rock uplift, driven by differential erosion, deforms rock, creating anticlines that run beneath some major rivers.

Mountain front erosion

Heavy precipitation can drive rapid erosion on steep mountain fronts. This focused erosion can remove so much mass so quickly that warm rock from below the range flows (advects) to the mountain front in response to erosion. The end result is that erosion drives the movement of rock and can elevate rock of high metamorphic grade above lower grade rock.

Tectonic aneurysm

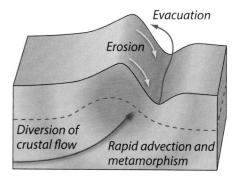

Tectonic aneurysms are thought to form where large Himalayan rivers carve deep canyons through rapidly uplifting rock. Rapid erosion drives crustal flow, advecting hot rock toward a self-sustaining zone of locally weak crust. The anomalously hot and weak rock flows toward the aneurysm, supplying more mass for erosion and leading to extremely high rock uplift rates.

FIGURE 12.15 Tectonics and Geomorphic Feedbacks. Erosion influences solid-Earth processes by removing mass from the surface, thereby inducing crustal flow, deformation, and uplift. [Adapted in part from Zeitler et al. (2001).]

resulted in the development of deeply exhumed geologic structures expressed as a bull's-eye pattern of young, high-grade metamorphic rocks in the region surrounding the rapidly eroding gorges.

Applications

Understanding the role of tectonics in setting the rate and type of geomorphic processes and the distribution of landforms is key to understanding the geomorphology of many regions around the world. Explicitly considering and understanding the linkage between solid-Earth and surface processes has important applications to landscape management, sustainability, and natural-hazard reduction. The tools and approaches of tectonic geomorphology are central to understanding why certain types of rocks and landforms are often found together in different parts of the world. Structural patterns inherited from ancient tectonic deformation can greatly influence modern landforms and surface processes through sustained variability in erosion resistance. In rapidly uplifting areas of the world, such as the Himalaya, tectonic uplift and the limited strength of rocks keeps slopes poised to fail at threshold steepness. Frequent landslides make road maintenance in tectonically active mountain belts a challenge.

Understanding the dynamic nature of Earth's surface, as driven by tectonics, is the foundation for understanding the nature and magnitude of many geologic hazards. Tectonic geomorphology has direct applications to understanding seismic hazards. For example, building and development is limited, and more highly regulated, along active fault zones in California than it is elsewhere in the state. Mapping the distribution and alignment of fault-related landforms is a standard technique employed in seismic-hazard assessments. The physiographic expression of faults and fault zones can be used to map the traces of faults, evaluate seismic hazards, and identify locations at risk of direct fault offset. Tectonic geomorphology is useful for assessing earthquake recurrence intervals for faults along active mountain fronts in both densely populated areas, such as along California's San Andreas Fault and in relatively inaccessible regions, such as the High Himalaya. In particular, identifying the character, extent, and history of motion across fault zones allows geologists to estimate the size and characteristics of past and thereby future earthquakes. Geomorphological techniques are commonly used in site investigations for seismically vulnerable infrastructure, such as nuclear power plants and dams.

The broad patterns of rock uplift, governed by tectonic processes, provide the raw material and template upon which erosion works to sculpt landscapes. In this manner, tectonic setting influences both the tempo and spatial patterns of landscape-forming processes, specifically, the rate at which rock weathers and soil erodes. Tectonics controls differential rates of erosion through slope steepness and the extent and rate of soil formation. Thus, regions where rapid uplift outpaces weathering have thin soils that erode rapidly.

In contrast, wet, tectonically inactive regions, where weathering outpaces exhumation, develop deeply weathered, thoroughly leached, nutrient-poor soils and thick zones of saprolite. The resulting differences in soil type and fertility greatly affect both ecosystem characteristics and dynamics and the ability of a landscape to sustain intensive land uses, such as farming. The terracing of mountainsides, a practice common to many different cultures around the world, reflects a human adaptation to accelerated erosion of the soil needed to farm on mountainsides steepened by tectonics.

Selected References and Further Reading

Ahnert, F. Functional relationships between denudation, relief, and uplift in large mid-latitude basins. *American Journal of Science* 268 (1970): 243–263.

Anderson, R. S. Evolution of the Northern Santa Cruz Mountains by advection of crust past a San Andreas Fault bend. *Science* 249 (1990): 397–401.

Anderson, R. S., A. L. Densmore, and M. A. Ellis. The generation and degradation of marine terraces. *Basin Research* 11 (1999): 7–20.

Andrews, D. J., and T. C. Hanks. Scarp degraded by linear diffusion: Inverse solution for age. *Journal of Geophysical Research* 90 (1985): 10,193–10,208.

Atwater, B. F. Evidence for great Holocene earthquakes along the outer coast of Washington state. *Science* 236 (1987): 942–944.

Baldwin, J. A., K. X. Whipple, and G. E. Tucker. Implications of the shear-stress river incision model for the timescale of post-orogenic decay of topography. *Journal of Geophysical Research* 108 (2003): 2158. doi: 10.1029/2001JB000550.

Bilham, R., and G. King. The morphology of strike slip faults: Examples from the San Andreas Fault, California. *Journal of Geophysical Research* 94 (1989): 10,204–10,226.

Burbank, D. W., and R. S. Anderson. *Tectonic Geomorphology*. Malden, MA: Blackwell Science, 2001.

Burbank, D. W., J. Leland, E. Fielding, et al. Bedrock incision, rock uplift, and threshold hillslopes in the northwestern Himalaya. *Nature* 379 (1996): 505–510.

Colman, S. M., and K. Watson. Age estimated from a diffusion equation model for scarp degradation. *Science* 221 (1983): 263–265.

Crosby, B. T., and K. X. Whipple. Knickpoint initiation and distribution within fluvial networks: 236 waterfalls in the Waipaoa River, North Island, New Zealand. *Geomorphology* 82 (2006): 16–38.

Densmore, A. L., M. A. Ellis, and R. S. Anderson. Landsliding and the evolution of normal-fault-bounded mountains. *Journal of Geophysical Research* 103 (1998): 15,203–15,219.

England, P., and P. Molnar. Surface uplift, uplift of rocks, and exhumation of rocks. *Geology* 18 (1990): 1173–1177.

Finlayson, D., D. R. Montgomery, and B. H. Hallet. Spatial coincidence of rapid inferred erosion with young metamorphic massifs in the Himalayas. *Geology* 30 (2002): 219–222.

Flint, J. J. Stream gradient as a function of order, magnitude, and discharge. *Water Resources Research* 10 (1974): 969–973.

Gardner, T. W. Experimental study of knickpoint migration and longitudinal profile evolution in cohesive homogeneous material. *Geological Society of America Bulletin* 94 (1983): 664–672.

Gilchrist, A. R., and M. A. Summerfield. Differential denudation and flexural isostasy in formation of rifted-margin upwarps. *Nature* 346 (1990): 739–742.

Gilchrist, A. R., M. A. Summerfield, and H. A. P. Cockburn. Landscape dissection, isostatic uplift, and the morphologic development of orogens. *Geology* 22 (1994): 963–966.

Hack, J. T. Stream-profile analysis and stream-gradient index. *Journal of Research of the United States Geological Survey* 1 (1973): 421–429.

Kirby, E., and K. Whipple. Quantifying differential rock-uplift rates via stream profile analysis. *Geology* 29 (2001): 415–418.

Koons, P. O. The topographic evolution of collisional mountain belts: A numerical look at the Southern Alps of New Zealand. *American Journal of Sciences* 289 (1989): 1041–1069.

Lavé, J., and J. P. Avouac. Active folding of fluvial terraces across the Siwalik Hills Himalaya of central Nepal. *Journal of Geophysical Research* 105 (2000): 5735–5770.

Martel, S. J., T. M. Harrison, and A. R. Gillespie. Late Quaternary displacement rate on the Owens Valley Fault Zone at Fish Springs, California. *Quaternary Research* 27 (1987): 113–129.

Matmon, A., P. Bierman, and Y. Enzel. Pattern and tempo of great escarpment erosion. *Geology* 30 (2002): 1135–1138.

Merritts, D., and W. B. Bull. Interpreting Quaternary uplift rates at the Mendocino triple junction, Northern California, from uplifted marine terraces. *Geology* 17 (1989): 1020–1024.

Merritts, D. J., K. R. Vincent, and E. E. Wohl. Long river profiles, tectonism, and eustasy: A guide to interpreting fluvial terraces. *Journal of Geophysical Research* 99 (1994): 14,031–14,050.

Molnar, P., R. S. Anderson, and S. P. Anderson. Tectonics, fracturing of rock, and erosion. *Journal of Geophysical Research* 112 (2007): F03014. doi:10.1029/2005JF000433.

Molnar, P., E. T. Brown, B. Clark, et al. Quaternary climate change and the formation of river terraces across growing anticlines on the north flank of the Tien Shan, China. *Journal of Geology* 102 (1994): 583–602.

Montgomery, D. R. Valley incision and the uplift of mountain peaks. *Journal of Geophysical Research* 99 (1994): 13,913–13,921.

Montgomery, D. R., and D. Stolar. Revisiting Himalayan river anticlines, *Geomorphology* 82 (2006): 4–15.

Morisawa, M., and J. T. Hack, eds. *Tectonic Geomorphology*, Boston: Allen & Unwin, 1985.

Ouimet, W. B., K. X. Whipple, and D. E. Granger. Beyond threshold hillslopes: Channel adjustment to base-level fall in tectonically active mountain ranges. *Geology* 37 (2009): 579–582.

Ollier, C. D. *Tectonics and Landforms*. London and New York: Longman, 1981.

Pazzaglia, F. J. and T. W. Gardner. Late Cenozoic flexural deformation of the middle U.S. Atlantic passive margin. *Journal of Geophysical Research* 99 (1994): 12,143–12,157.

Pritchard, D., G. G. Roberts, N. J. White, and C. N. Richardson. Uplift histories from river profiles. *Geophysical Research Letters* 36 (2009): L24301.

Roe, G. H., K. X. Whipple, and J. K. Fletcher. Feedbacks among climate, erosion, and tectonics in a critical wedge orogen. *American Journal of Science* 308 (2008): 815–842.

Schumm, S. A., J. F. Dumont, and J. M. Holbrook. *Active Tectonics and Alluvial Rivers*. Cambridge: Cambridge University Press, 2000.

Seeber, L., and V. Gornitz. River profiles along the Himalayan arc as indicators of active tectonics. *Tectonophysics* 92 (1983): 335–367.

Sieh, K. E., and R. H. Jahns. Holocene activity of the San Andreas fault at Wallace Creek, California. *Geological Society of America Bulletin* 95 (1984): 883–896.

Strahler, A. N. Hypsometric (area-altitude) analysis of erosional topography. *Geological Society of America Bulletin* 63 (1952): 1117–1141.

Trudgill, B. D. Structural controls on drainage development in the Canyonlands grabens of southeast Utah. *American Association of Petroleum Geologists Bulletin* 86 (2002): 1095–1112.

Valensise, G., and S. N. Ward. Long-term uplift of the Santa Cruz coastline in response to repeated earthquakes along the San Andreas fault. *Bulletin of the Seismological Society of America* 81 (1991): 1694–1704.

Wager, L. R. The Arun River drainage pattern and the rise of the Himalaya. *Geographical Journal* 89 (1937): 239–250.

Whipple, K. X. The influence of climate on the tectonic evolution of mountain belts. *Nature Geoscience* 2 (2009): 97–104.

Whipple, K. X., and G. E. Tucker. Dynamics of the stream-power river incision model: Implications for height limits of mountain ranges, landscape response timescales, and research needs. *Journal of Geophysical Research* 104 (1999): 17,661–17,674.

Whittaker, A. C., P. A. Cowie, M. Attal, et al. Contrasting transient and steady-state rivers crossing active normal faults: New field observations from the Central Apennines, Italy. *Basin Research* 19 (2007): 529–556.

Willett, S. D. Orogeny and orography: The effects of erosion on the structure of mountain belts. *Journal of Geophysical Research* 104 (1999): 28,957–28,982.

Wobus, C. W., K. V. Hodges, and K. X. Whipple. Has focused denudation sustained active thrusting at the Himalayan front? *Geology* 31 (2003): 861–864.

Wobus, C. W., K. X. Whipple, E. Kirby, et al. "Tectonics from topography: Procedures, promise, and pitfalls." In S. D. Willett, N. Hovius, M. T. Brandon, and D. M.

Fisher, eds., *Tectonics, Climate, and Landscape Evolution*. Geological Society of America Special Paper 398. Boulder, CO: Geological Society of America, 2006.

Zeitler, P. K., A. S. Meltzer, P. O. Koons, et al. Erosion, Himalayan geodynamics, and the geomorphology of metamorphism. *GSA Today* 11 (2001): 4–9.

DIGGING DEEPER When and Where Did that Fault Last Move?

Understanding when, where, and how much a fault last moved is important both to scientists and society. Earthquakes associated with active fault offset have damaged infrastructure in the past and present an ongoing hazard. Characterizing the location, timing, and amount of fault movement is critical for determining the rate of solid-Earth deformation, an important geomorphic control on landscape evolution.

In lightly vegetated and undeveloped terrain, the surface expression of faults—their scarps—can be mapped in the field and by using aerial photographs. Fault scarps are one of the most geomorphically distinct surface expressions of plate tectonics and a direct way to document deformation and infer seismic shaking in the past. However, in forested terrain or in urban areas, fault traces were difficult to map until the advent of LiDAR. Now, high resolution LiDAR-based digital elevation models (DEMs) are being used to map scarps beneath dense vegetation or in areas that have been heavily disturbed by development. For example, in Washington State, LiDAR clearly revealed the trace of the Tacoma Fault Zone (Sherrod et al., 2004; **Figure DD12.1**), which carbon-14

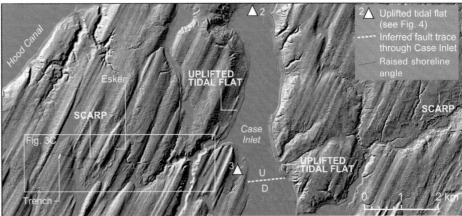

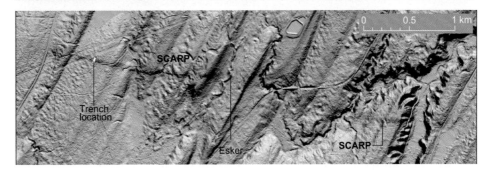

FIGURE DD12.1 (a) Look closely at this orthophotograph of the landscape near Tacoma, Washington; it is unlikely you can find the fault scarp. (b) In this LiDAR DEM, the fault scarp is visible (and labeled). The elongated ridges, trending NE/SW, are glacial drumlins. (c) The LiDAR image is greatly enlarged, showing the linear fault scarp even more clearly. [From Sherrod et al. (2004).]

DIGGING DEEPER When and Where Did that Fault Last Move? (continued)

dating of associated organic material indicates has been active in the past 1000 years.

Geomorphologists can estimate the age of fault scarps by studying how they erode and change shape predictably over time. Scarps evolve from steep and angular to gentler and more rounded [**Figure DD12.2**]. In scarps formed in unconsolidated materials such as sand, gravel, and glacial till, material moves from the steep fault scarp face and from the sharp upper inflection to form a depositional wedge of colluvium at the base of the scarp. Most of this mass appears to move by diffusive processes similar to creep (see Chapter 5); thus, the rate of mass movement off the scarp face is proportional to the gradient or slope of the scarp. Researchers have calibrated the change in fault scarp shape over time by determining or assuming a diffusion coefficient so that scarps of unknown age can be dated (Hanks et al., 1984). Diffusion dating of fault scarps is complicated by the observation that diffusion coefficients differ as a function of scarp height and orientation (Pierce and Colman, 1986).

Clues to the timing and magnitude of past earthquakes and the offset they caused are preserved both on the fault scarp itself and in the adjacent colluvial wedge (material shed off the scarp and deposited at its base). Trenching colluvial wedges provides an indication of the number of events and their displacement as each event will often bury or displace identifiable soil horizons. Radiocarbon analysis of buried organic material, as well as luminescence dating techniques when organic matter is absent, are used along with detailed trench-wall mapping to infer the history and timing of sediment deposition onto the wedge. Since most wedge sedimentation occurs during and just after each earthquake steepens the fault scarp, understanding the history of wedge sedimentation and soil development allows geomorphologists to infer faulting history [**Figure DD12.3**].

In the absence of a colluvial wedge, or if the wedge contains no datable material, then geomorphologists measure the offset of distinct landforms. Offset moraines, pluvial lake shorelines, and river terraces are common datums (Wallace, 1977) that can be dated directly, or their age can be assumed based on climatic inferences [**Figure DD12.4**]. In some cases, offset historical structures, such as a Crusader castle wall in Israel (Ellenblum et al., 1998), provide chronologic control and act as strain gauges.

Bedrock scarps, because they generally lack colluvial wedges, remained a dating challenge until the development of cosmogenic nuclide techniques. In 2001, Gran et al. made ^{36}Cl isotopic measurements in samples collected at regular intervals down the limestone exposed on a 9-m-high normal fault scarp in northern Israel. The scarp bisects a village, and homes are built at the top of the upthrown block to take in the view. Because cosmic rays penetrate below the ground surface and because a variety of different combinations of displacement and earthquake timing could generate similar isotope concentration profiles down the scarp, Gran et al. (2001) fit many different models to their data in a process termed optimization modeling [**Figure DD12.5**]. This allowed them to conclude that the fault had moved several times during the Holocene, most recently in the past few thousand years. Motion was greatest in the mid-Holocene, a time when archaeologists identified fatalities in a nearby cave, the collapse of which was likely triggered by shaking on the fault they dated.

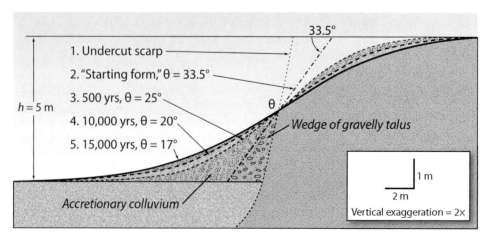

FIGURE DD12.2 Schematic diagram of a one-event scarp in unconsolidated material and its adjacent colluvial wedge (labeled "accretionary colluvium"). Scarp shapes, as predicted by the diffusion equation over time, are shown along with average slopes (in degrees). The assumed "starting form" of a 33.5-degree slope is the estimated angle of repose for colluvium. Over time, the scarp gets less steep and the wedge gets larger and thicker. [From Pierce and Coleman (1986).]

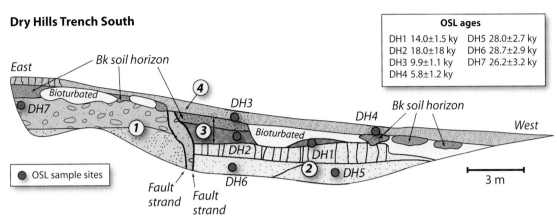

FIGURE DD12.3 Trench log depicting the colluvial wedge formed by erosion of a normal fault in northern Great Basin of western North America. The upthrown block is on the left; the downthrown block on the right. The two strands of the fault are the solid black, nearly vertical lines. The trench reveals three faulting events dated by optically stimulated luminescence (OSL). Unit 1 is not exposed in the hanging wall (graben) but only on the upthrown block; it and unit 2 are the same age. The earliest event caused the deposition of unit 2 about 28,000 years ago. Note that unit 2 is capped by a soil as indicated by the short vertical lines. The next event shed material off the scarp, forming what is now unit 3 (14,000–18,000 years ago). The latest event shed unit 4 between 5,000 and 10,000 years ago. [From Wesnousky et al. (2005).]

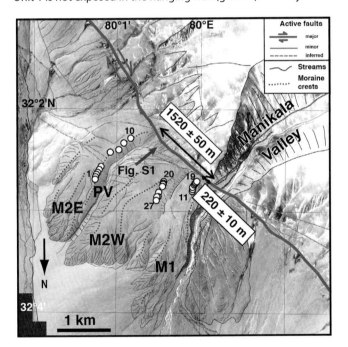

FIGURE DD12.4 In Tibet, the Karakorum Fault is a major, 1200 km-long feature north of the Himalaya. The fault laterally and vertically offsets many moraines deposited during cooler or wetter times. Here, ice flowing down the Manikala Valley formed two sets of moraines at different times. The fault has since offset the moraine complexes laterally from the source valley. Cosmogenic exposure ages (sample sites shown by white circles) show the earlier moraines, labeled M2, are ~140,000 years old and the younger moraines (M1) are ~20,000 years old. Using the measured offsets of 1520 ± 50 m (M2) and 220 ± 10 m (M1) for the two sets of moraines and their ages, Chevalier et al. (2005) suggest a long-term slip rate of about 1 cm/yr. This is many times larger than the deformation rate measured by InSAR (see Chapter 2) over a recent 8-year period, implying that slip on the fault varies over time. [From Chevalier et al. (2005).]

Chevalier, M. L., F. J. Ryerson, P. Tapponnier, et al. Slip-rate measurements on the Karakorum Fault may imply secular variations in fault motion. *Science* 307 (2005): 411–414.

Ellenblum, R., S. Marco, A. Agnon, et al. Crusader castle torn apart by earthquake at dawn, 20 May 1202. *Geology* 26 (1998): 303–306.

Gran, S., A. S. Matmon, P. R. Bierman, et al. Displacement history of a limestone normal fault scarp northern Israel from cosmogenic ^{36}Cl. *Journal of Geophysical Research* 106 (2001): 4247–4265.

Hanks, T., R. C. Buckham, K. R. Lajoie, and R. E. Wallace. Modification of wave-cut and faulting controlled landforms. *Journal of Geophysical Research* 89 (1984): 5771–5790.

Pierce, K. L., and S. M. Colman. Effect of height and orientation (microclimate) on geomorphic degradation rates and processes, late-glacial terrace scarps in central Idaho. *Geological Society of America Bulletin* 97 (1986): 869–885.

Sherrod, B. L., T. M. Brocher, C. S. Weaver, et al. Holocene fault scarps near Tacoma, Washington, USA. *Geology* 32 (2004): 9–12.

Wallace, R. E. Profiles and ages of young fault scarps, north-central Nevada. *Geological Society of America Bulletin* 88 (1977): 1267–1281.

Wesnousky, S. G., A. D. Barron, R. W. Briggs, et al. Paleoseismic transect across the northern Great Basin. *Journal of Geophysical Research* 110 (2005): B05408 doi: 10.1029/2004JB003283.

DIGGING DEEPER When and Where Did that Fault Last Move? *(continued)*

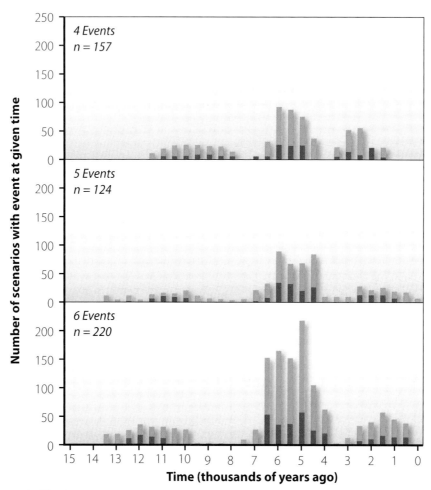

FIGURE DD12.5 Measuring the ^{36}Cl distribution on a fault scarp does not yield a unique age estimate for prior motion; rather, many different models (timing, size, and number of fault offsets) can be fit to the data. By running multiple models and looking for consistency between results, one can have higher confidence in the estimated timing of paleoearthquakes. These histograms are a summary of many different models; the best fit models are shown in blue. The shapes and modes of the three histograms are similar, indicating that no matter if one assumes that 4, 5, or 6 events created the present-day fault scarp, the mid-Holocene (4000 to 7000 years ago) was the most active time for earthquakes. [From Gran et al. (2001).]

WORKED PROBLEM

Question: The average elevation of the High Himalaya is about 5 km, with peaks rising substantially higher. The average long-term erosion rate in the Himalaya is about 1–2 mm/yr and the average erosion rate of inactive orogens, such as the Appalachian Mountains, is only about 0.02 mm/yr. How long would it take to erode the world's highest mountain range down to sea level once tectonic rock uplift ceased, and how much rock would be eroded off in the process?

Answer: A simple estimate obtained by dividing the height of the range (5 km) by the average erosion rate (1–2 mm/yr) yields just 2.5 to 5 million years. However, this estimate is far too short due to neglect of isostatic rebound. Recall that surface uplift equals the uplift of rock less the amount of rock eroded (eq. 12.1). The effect of isostatic compensation on net surface elevation may be estimated using equation 12.3: $U_S = E\,[(\rho_c/\rho m) - 1]$. For the case of $\rho_c \approx 2.7$ g/cm^3 and $\rho_m \approx 3.3$ g/cm^3, the ratio $\rho_c/\rho_m = 0.82$ indicates that once tectonically driven rock uplift ceases, erosionally driven rock uplift will compensate for 82 percent of the mass removed from the surface by erosion. Hence, equation 12.3 may be recast as $U_s = -0.18$ E. Rearranging and solving for E, yields $E = -5.6\, U_s$, indicating that 5.6 km of rock needs to be eroded in order to result in 1 km of net surface lowering. Hence, 28 km of rock needs to be eroded to reduce a 5 km high mountain range to sea level (5 km · 5.6). At a pace of 1–2 mm/yr, this would take 14–28 million years. But this itself represents an underestimate of the total time involved because the erosion rate of the range would decline to well below 1 mm/yr as the topographic relief of the range diminished and hillslopes relaxed to gentler gradients once tectonic forcing ceased. For example, the current erosion rate of the southern Appalachian Mountain range is about 0.02 mm/yr. At this pace it would take almost 420 million years to reduce the 1500 m relief of the southern Appalachians to sea level [1500 m/(0.18 · 0.00002 m/yr)].

KNOWLEDGE ASSESSMENT Chapter 12

1. Explain why many of our planet's major surface features correspond to current or ancient boundaries between tectonic plates.
2. Explain the difference between active and passive tectonic control on landscapes.
3. In what tectonic settings do lithology and structure exert strong control on geomorphology? Explain why this is the case.
4. How do the drivers of rock uplift change over time?
5. Explain the role of erosion in driving both rock and surface uplift.
6. Explain how changes in base level can be inferred from the landscape.
7. Use a sketch to explain and define the different types of uplift.
8. List and explain several examples of how density contrasts drive geomorphic change.
9. Provide several examples illustrating the geomorphic expression of active faulting.
10. What is an anticlinal valley and how does it form?
11. Why is it difficult to read underlying structure from landscape features in areas like the Himalaya and Taiwan?
12. What are the geomorphic expressions of joints and in what climate zones are such expressions easiest to detect?
13. Give two examples of different tectonically-controlled drainage patterns and explain what controls their geometry.
14. When and where does coastal subsidence happen?
15. Make a sketch showing the three most important geomorphic features of an uplifting coast.
16. How can fluvial terraces be used to estimate uplift rates?
17. How do hillslopes respond to increasing uplift rates?
18. What is an inner gorge and what might inner gorges tell us about uplift rates?
19. What can be learned from the longitudinal profile of a river?
20. Explain how the stream gradient index works and why it is useful for studying the effect of tectonics on landscapes.
21. Sketch an area/slope plot and explain why it is geomorphically useful.
22. Provide two examples of how tectonics creates knickpoints and knickzones.
23. What is a river anticline and how does it form?
24. Compare and contrast the large-scale geomorphology (characteristic landscape features) of compressional orogens, extensional rift zones, and continental interiors.
25. Explain sedimentary wedges, noting where they are found.
26. Explain how rift zones control the orientation and character of drainage networks.
27. Explain colluvial wedges, noting where they are found.

Geomorphology and Climate

13

Introduction

Climate, the long-term average of day-to-day weather, varies dramatically over the surface of our planet and can be described in terms of the temporal distribution and variability in the amount of precipitation, the speed of the wind, the range of temperature, and the relative humidity. Climate strongly influences geomorphology through its effects on the rate and character of atmospheric and surface processes, such as the volume, duration, and type of precipitation (rainfall versus snowfall), runoff, and flood flows, as well as the distribution of vegetation. Together, these climate characteristics set the dimensions of channels, the nature of soils, and the pace and temporal variability of fluvial and hillslope processes, such as sediment transport by mass movements.

Earth's climate is not steady but changes over a wide variety of timescales. On the timescale of plate tectonics, millions of years, large-scale climate zonation reflects the arrangement of continents and resulting oceanic and atmospheric circulation as well as changing rates of volcanism, carbon dioxide (CO_2) emission, carbon sequestration in sediment, and thus atmospheric CO_2 content. On intermediate timescales, thousands of years, changes in Earth's orbit alter the seasonal distribution of incoming solar radiation. Over years to decades, the ash and aerosols that volcanic eruptions spew into the atmosphere cool the planet.

Racetrack Playa, an ephemeral lake in Death Valley National Park, California, is known for its enigmatic moving rocks and the tracks they leave behind. One theory suggests that wind moves the rocks over mud slickened by rain, but calculations indicate that only the most extreme winds could overcome the friction of the playa mud. Other theories suggest that wind could move rocks if they are locked in a sheet of ice that buoys them and provides more surface area for wind to act upon.

IN THIS CHAPTER

Introduction
Records of a Changing Climate
   Landform Records of Climate
      Change
   Lake and Marine Sediment
   Ice Cores
   Windblown Terrestrial Sediment
Climate Cycles
   Glacial Cycles
   Orbital Forcing
   Local Events—Global Effects
   Climate Variability Within a
      Climate State
   Short-Term Climate Changes

Geomorphic Boundary Conditions
   Precipitation and Temperature
   Vegetation, Fire, and Geomorphic
      Response
   Base Level
Climatic Geomorphology
   Köppen Climate Classification
   Climate-Related Landforms and
      Processes
   Relict Landforms
Landscape Response to Climate
   Glacial–Interglacial Changes
   Isostatic Responses
   Climatic Control of Mountain Topography
   Climate Change Effects

Landscape Controls on Climate
   Regional Climate
   Earth's Energy Balance
   Hydrologic Cycling
   The Atmosphere
Applications
Selected References and Further Reading
Digging Deeper: Do Climate-Driven Giant Floods Do Significant Geomorphic Work?
Worked Problem
Knowledge Assessment

Climate zones have been mapped worldwide, and some landforms, such as moraines and arid-region playa lakes, are directly related to specific climatic conditions. Other landforms, such as sand dunes (see Chapter 10), occur in a wide variety of climatic settings and thus are not climatically diagnostic [**Photograph 13.1**]. For example, sand dunes are common in hyperarid regions, such as the Skeleton Coast of Namibia, because there is little vegetation to anchor noncohesive sand. But, dunes are also found along many humid-temperate beaches and along glacial margins (see Photograph 10.4). The link here is not climate but the ready availability of sand, unsecured by the roots of vegetation.

Geologists use a variety of continuous climate archives, including sediment and ice cores, to decipher the timing and magnitude of past climate changes. Studies of such environmental records have revealed that over the past 2.7 million years, glacial-interglacial cycles have become pronounced, shifting global climate dramatically and repeatedly on 10^4 to 10^5 year timescales. **Relict landforms**, such as glacial cirques, permafrost ice wedge casts, shorelines above modern lake levels, and vegetated dune fields, provide geomorphic evidence for such climatic change. The **last glacial maximum** (**LGM**) was the peak of the most recent glacial expansion, about 22,000 years ago. At the LGM, Earth's surface was, on average, about 5°C colder than during the warmest time of the **Holocene** (the last 11,700 years). In general, high latitudes cooled the most; the tropics cooled the least.

The ocean and its currents influence climate dynamics and thus the pace, type, and distribution of geomorphically important Earth surface processes. Usually the oceans buffer the planet from rapid change, but in some cases, oceans can amplify small changes when modes of ocean circulation and heat transport change suddenly. For example, large amounts of glacial meltwater, released as late Pleistocene ice sheets melted away, appear to have capped the North Atlantic with buoyant fresh water and temporarily altered (perhaps for centuries) heat transport patterns in the ocean.

Climate, through its control on the growth and shrinkage of ice sheets over the past several million years, has repeatedly taken sea level up and down by more than 100 m. Thus, rivers flowing to the sea have experienced numerous base-level rises and falls over time (see Figure 7.9). The history of these base-level changes is written in terrace sequences and knickzones throughout the world but can be difficult to decipher because climate, tectonics, and other effects on sediment supply can also catalyze terrace formation (see Figure 7.10). The continental shelves owe their planar form in part to beveling by waves, as the seas repeatedly transgressed and then regressed, and in part to the deposition of sediments during periods of high sea level (see Chapter 8).

There are feedbacks and interactions between solid-Earth processes and climate that affect geomorphology. Surface uplift changes airflow patterns and induces orographic precipitation. Heavy rainfall on the windward slopes of mountain ranges speeds erosion there, causing an isostatic response that moves rock more quickly toward the surface on the windward slopes than on the leeward slopes. Loading of the crust and mantle by ice sheets and greatly expanded pluvial lakes during glacial times results in isostatic compensation. Earth's surface sags under the load only to rebound after the ice retreats or the lakes dry. Glacial erosion is capable of limiting the height of mountain ranges.

Not only does climate affect surface processes, but the rate and distribution of surface processes affect climate. Geomorphic influences on climate include the consumption of CO_2 through weathering of fresh minerals exposed by erosion (see Chapter 3); changes in the global energy balance driven by the extent and **albedo** (reflectivity) of large glaciers, sea ice, and snow-covered land (see Chapter 9); and the recycling of water on a massive scale by plant transpiration (see Chapter 1).

(a)

(b)

PHOTOGRAPH 13.1 Sand Dunes. Sand dunes can form in a variety of climates because the major control on their presence is a source of mobile sand. (a) Partially vegetated coastal sand dunes along the beach at Channel Islands National Park, Santa Barbara and Ventura counties, California. (b) Sand dunes in Tadrart Acacus, a hyperarid desert area in western Libya, part of the Sahara Desert. Rock outcrops are heavily coated in dark, shiny rock varnish.

In this chapter, we examine the relationships between geomorphology and climate. First, we consider various geologic and instrumental records documenting changes in climatic variables that drive geomorphic response over a wide variety of temporal and spatial scales. Then we examine the variability and influence of climate on geomorphic processes before discussing the distribution of resulting climate-sensitive landforms over time and space. Finally, we consider both how landforms respond to changes in climate and how the landscape itself can drive climate change.

Records of a Changing Climate

Geomorphologists have explored and interpreted a variety of natural archives to understand and quantify geomorphic changes and the changes in climate that drive them over a variety of timescales. Such archives include lake and marine sediments, glacial ice, and terrestrial sediments such as loess. Some archives, such as lake cores, preserve more or less continuous sedimentation while other archives, such as paleoflood deposits, record individual events. Archives differ in their degree of time-averaging; for example, soils integrate the record of climatically driven pedogenic processes over thousands to tens of thousands of years while a debris flow deposit may result from a single, exceptional rainstorm. As new data reveal the magnitude and effects of human-induced climate change, it becomes increasingly important to understand both the natural range of climate variability as well as past geomorphic responses to climate change.

Landform Records of Climate Change

Some landforms are direct or indirect indicators of a changed climate. The challenge lies in dating the landforms and, in some cases, showing that the landforms result from changing climate and not changes in other factors such as tectonics. In much of the world, glacial and periglacial features are the landforms most indicative of a changed climate because they directly reflect changes in temperature and/or precipitation. Periglacial features require mean annual temperatures below freezing. Glacial landforms are less diagnostic climate indicators, because the glaciers that form such landforms respond to changes in both precipitation and temperature.

In alpine glacial systems, moraines far down valley from the limit of present-day ice indicate very different glacial mass balances in the past, either suggesting increased snow and ice accumulation and/or decreased **ablation** (ice loss) as the result of changing climate. The presence of now ice-free cirques, such as those on Mount Washington in New Hampshire, indicate that in the past, conditions were once sufficient to support alpine glaciers. If alpine moraines can be mapped, then equilibrium line altitudes can be estimated using a variety of methods, including the empirical **accumulation area ratio** technique (which relies on the observation that on average two-thirds of an alpine glacier's area lies in the accumulation zone [**Figure 13.1**]). Such data indicate that equilibrium line altitudes lowered during the LGM on average about 1000 m but lowering amounts varied greatly around the world because of localized conditions affecting ice accumulation and ablation rates. Continental glacial deposits (such as till and outwash) are broadly indicative of a cooler climate. Direct interpretation of the continental glacial record in terms of temperature and precipitation at a specific location is not possible because many different variables (e.g., bed materials, bed thermal status, and climate patterns) affect the size of ice sheets (see Chapter 9).

Climate-induced changes in sea level have left a significant mark on the landscape. Drowned valleys and deep estuaries along many coastlines (such as eastern North America) are evidence for postglacial sea-level rise. Similarly, many barrier islands (see Chapter 8) were born when sea level was lower and are today marching shoreward as sea level continues to rise. During periods of lower sea level, large portions of the continental shelves were exposed. This exposure is particularly important in terms of human and animal migration and archaeology. Lowered sea level exposed land and narrowed open water crossings, such as the Torres Strait and Bering Strait, during glacial times, facilitating migrations between continents and the peopling of both Australia and the Americas.

The most extreme interglacial periods, occurring about 130,000, 300,000, and 400,000 years ago were warm enough and/or long enough that large volumes of the Greenland and most likely the Antarctic ice sheets melted. This, along with expansion of the ocean water from warming (which for the warming of the past century accounts for about 40 percent of sea-level rise), increased the volume of the oceans so that that global sea level was as much as 12 m higher than it is today. During such sea-level high stands, waves cut shoreline platforms that today are preserved above sea level as marine terraces [**Photograph 13.2**].

Fluvial features, including river terraces and knickzones (where river gradient steepens abruptly), are more difficult to interpret as climatically significant landforms. Some terraces, such as those made up of glacial outwash and traceable upstream to moraines, are clearly climatic in origin, for example those originating from ice cap outlet glaciers [**Photograph 13.3**]. However, terrace formation can also result from base-level changes and changes in river discharge and sediment loading. Although these changes can be driven by climate, they can also be related to tectonics and drainage basin adjustments such as stream capture, thus complicating any climatic interpretation of river terraces (see Chapters 7 and 12). Knickzones can form in response to climate change, expressed as lowered sea level when ice sheets grow, but they can also reflect active faulting as well as structural and lithologic discontinuities. Understanding landform setting and context are

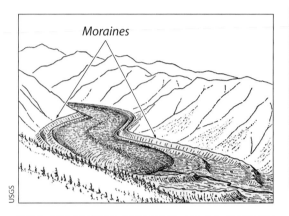

Alpine **moraines** can be mapped in the field and used to define the down-valley extent of now-vanished glaciers. Moraines, which are built by deposition of material from melting ice, are found only in the ablation zone. In the accumulation zone, ice extent is defined by the extent of glacially polished rock and trimlines where weathered rock has been removed by glacial erosion.

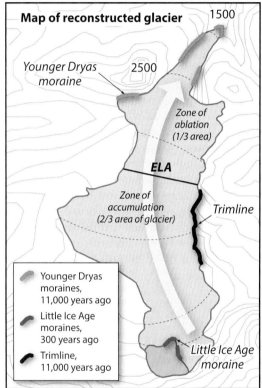

Using a topographic map and the location of mapped moraine segments and trimlines, one can sketch the outline of the glacier that once filled the valley. Empirical studies have shown that, on average, about 2/3 of an alpine glacier's surface area lies in the **accumulation zone** and 1/3 lies in the **ablation zone.** Using this **accumulation area ratio** (AAR) of 2/3, one can define the former equilibrium line altitude or **ELA,** the boundary between the accumulation and ablation zones. The surface of the reconstructed glacier can be contoured. In the accumulation area, the contours are convex up glacier because ice flow is convergent. In the ablation zone, the contours are convex down glacier because the flow is divergent. You can create a cross section of the vanished glacier using the contour map-based glacier reconstruction.

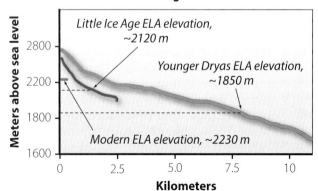

During the Younger Dryas cold period between about 12,800 and 11,500 years ago, the ELA in the Alps fell almost 400 m below its elevation today. During the Little Ice Age (between 1300 and 1850 CE), the ELA dropped 110 m from today's elevation of 2230 m. Assuming that ELA changes reflect only cooling, and considering a lapse rate of 1° C per 100 m elevation, the Little Ice Age was about a degree cooler than today. Younger Dryas times were about 4° C cooler.

FIGURE 13.1 Reconstructing Vanished Glaciers. Using the accumulation area ratio (AAR) method, the equilibrium line altitude (ELA) of now-vanished glaciers can be reconstructed from a topographic map and field mapping of lateral and terminal moraines as well as trimlines. [Adapted from Sailer et al. (1999).]

PHOTOGRAPH 13.2 Dissected Marine Terrace. Determining paleo–sea levels in tectonically stable continental regions is a challenge because uplift rates are very low and thus older marine terraces remain close to present-day sea level. Here, along the southeastern coast of South Africa, is a dissected marine terrace that was likely cut about 400,000 years ago during MIS (marine isotope stage) 11, when global sea level was higher than today.

structures made of opal (amorphous silica) and known as **phytoliths**. Phytoliths can be preserved in soil A horizons and separated for analysis. Because the shape of phytoliths differs depending on plant type, they can be useful for documenting the type of vegetation that once covered a site; for example, phytoliths can be used to distinguish between grass and forest cover, a difference that can be used to infer climate change.

Lake and Marine Sediment

The stratigraphy and composition of marine and lake sediment cores can be used to decipher changes in climate and differences in geomorphic processes and their rates over time.

Some sediment shows a clear cyclicity in the color or grain size of the material it contains. For example, cores from postglacial, humid-temperate zone ponds are typically dominated by gyttja (from the Swedish, for "mud" or "ooze"), the dark, fine-grained, organic-rich muck that squeezes between your toes when you walk in for a swim. Cores from such ponds reveal thin bands of gray or tan sand and silt [**Photograph 13.4**], interpreted as storm deposits resulting from flooding-induced erosion in the pond's watershed. In marine cores from the North Atlantic,

critical to any climatic interpretation of river terrace sequences and knickzones.

Soils developed on landforms preserve evidence of changes in climate. For example, relict calcic (Bk) horizons (accumulations of calcium carbonate in soil B horizons found in areas where no calcium carbonate is being deposited in soils today) directly indicate a change in the ratio of precipitation to evaporation (see Chapter 3). Other indicators of climate change are less direct. Many plants create microscopic

PHOTOGRAPH 13.3 Origin of River Terraces. Some flights of river terraces are climatic in origin. Here, along the Snake River, outwash terraces at Deadman's Bar are composed of material that came from the termini of outlet glaciers of the Southern Yellowstone–Absaroka Mountains ice cap. The upper terraces are of last glacial maximum age. The lowest terraces are postglacial.

PHOTOGRAPH 13.4 Paleoflood Layer. Core section with centimeter scale on the right side showing gray, sandy, paleoflood layer sandwiched between organic-rich pond sediment. The sand was deposited when a large storm hit the basin, causing runoff and sediment transport. Core was taken from Chapel Pond in northern Vermont.

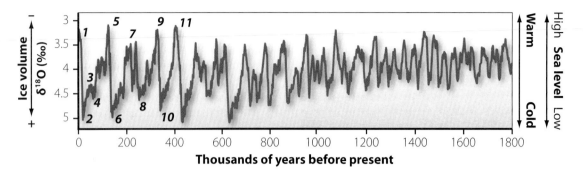

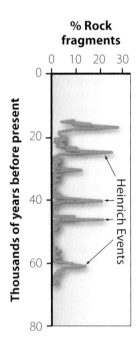
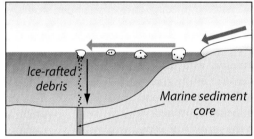

FIGURE 13.2 Marine Sediment Record of Changing Quaternary Climate. Stable oxygen isotope ratios of marine foraminifera (CaCO₃ shells) reflect the amount of water stored in the planet's glaciers and the temperature of the oceans. Ocean sediment cores also reveal periods when ice-rafted debris dumped from icebergs was abundant. [Data from Lisiecki and Ramyo (2005).]

grain size and stratigraphic analyses reveal isolated accumulations of large stones embedded in a matrix of poorly sorted debris [**Figure 13.2**]. Known as **ice-rafted debris (IRD)**, such material is the smoking gun of glaciation in marine cores and indicates where and when icebergs, calved from tidewater glaciers, dropped loads of previously ice-bound sediment. In glacial lake sediments, changes over time in the thickness and grain size of annual silt, clay, and sand layers (varves) can be interpreted both in terms of sediment delivery from changing summer climate (seasonal runoff) and ice retreat (distance from the paleo–ice margin). Warm summers and nearby ice result in thick varves composed of coarse sediment. During cold winters and when the ice margin is farther away, varve layers are thinner and composed of finer-grained sediment.

Lake and marine sediments can be analyzed using a variety of techniques; chemical, physical, and isotopic data are used to infer changes in climate from established

Packrat

Midden

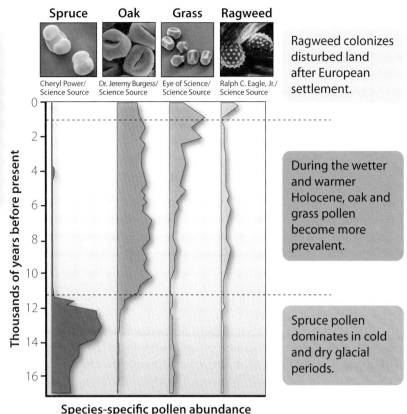

In arid regions, packrats scavenge vegetation, return it to their nests, and urinate on it, creating a resinous material called **amber-rat** that preserves characteristic macrofossils for identification by paleoclimatologists. These accumulations are termed **middens.**

Pollen grains are preserved in geologic archives such as lake mud that can be cored. The cores can be dated and the pollen can be identified to the species level by its characteristic shapes. Counting pollen grains of different types allows reconstruction of the vegetation that used to grow near the sample site and illustrates how that vegetation assemblage changed over time. This example is from Nelson Lake, Illinois.

FIGURE 13.3 Vegetation Responds to Climate Change. Animals, such as packrats, collect vegetation and store it in their middens. Pollen, preserved in lake sediments, can be identified to determine ancient vegetation assemblages.

relationships between climate and the measured parameter. For example, the inorganic chemical composition and stable oxygen isotope ratios of preserved marine organisms can be used to estimate ocean temperature. In other cases, **pigments** (organic molecules extracted from sediment) are analyzed because their composition relates to water temperature. Many sediments preserve biologic debris, a rich source of information about climate change. In some cases, analysis of **pollen** and **macrofossils** reveals the species of plants in local or regional vegetation communities; such fossils are usually well-preserved in lake sediments because oxygen levels are low enough to prevent substantial decay.

Pollen analysis is one of the most widely applied tools in paleoclimate research. The morphology or shape of pollen grains released by plants is unique for each genus [Figure 13.3] and pollen is durable and thus well-preserved. Qualitative pollen analysis is straightforward. If the pollen of cold-tolerant species such as spruce is found in sediments and spruce trees are not found near the pond or lake today, one can infer climate was likely colder in the past. Finding the pollen of species like oak and pine indicates a warm climate, and ragweed pollen increases after western settlement and land clearance. Determining how much temperature change is represented by a change in the species composition of fossil pollen requires comparison with modern analogs in different climatic and geographic zones. Because some pollen, like pine, can travel tens to hundreds of kilometers, pollen records of climate are more regional than local.

Macrofossils (primarily plant parts such as cones, leaves, and needles) are often recovered from the same lake cores used for pollen analyses and are useful because they provide information about only the biologic communities within the watershed of the lake. Insect parts preserved in sediment also tell a climate story. For example, distinctive pieces of chitinous beetles' exoskeletons preserve well in lake sediment. Because different beetle species live in different climate zones, and because each species of beetle has distinctive skeletal parts, the distribution of beetle pieces found in prehistoric sediment can be used to reconstruct climates of the past.

In arid regions, where lakes are scarce and ephemeral, the content of packrat **middens** (nesting sites in caves and rock crevices) is used to infer the paleodistribution of vegetation (Figure 13.3). Packrats (small desert rodents) gather bits of vegetation from the area around their middens and then urinate on the debris. Over decades and centuries, the fossilized plant debris and urine build up into a solid mass of well-preserved organic material known as **amber-rat**. Former plant communities and thus paleoclimate can be determined by analyzing the species distribution of the preserved plant parts—which can be dated by ^{14}C.

In marine sediment cores, the chemical and stable isotopic character of various biota, in particular **foraminifera**, single-celled, calcium carbonate secreting animals [**Photograph 13.5**], have been used to determine both the paleotemperature of the ocean as well as to estimate global ice volume (Figure 13.2). Most species of foraminifera are small, the size of fine sand. The ratio of Ca/Mg in calcite, the mineral that makes up the shells of these tiny marine creatures, changes with water temperature, providing a paleothermometer for ocean water.

Both ocean water temperature and the amount of water locked up in ice caps, ice sheets, and glaciers determine the oxygen isotope ratio of the shells secreted by marine organisms. $^{18}O/^{16}O$ works as a paleotemperature indicator because lower $^{18}O/^{16}O$ ratios are correlated with lower ocean water temperatures. Global ice volume estimates work because $H_2^{16}O$ has a slightly higher vapor pressure than $H_2^{18}O$, and thus $H_2^{16}O$ is preferentially evaporated from the ocean. As glaciers grow on land and store the isotopically lighter water from the ocean in their ice, the $^{18}O/^{16}O$ ratio in seawater increases. Foraminifera incorporate some of this oxygen into their $CaCO_3$ shells, die, drop to the ocean floor, and are preserved in sediments. The ratio of stable oxygen isotopes in the shells, ^{18}O versus ^{16}O, can be interpreted as a paleothermometer (if calculations are made to account for the loss of ocean water to ice-sheet storage) and as a measure of global ice volume (if other temperature records are used to correct the measured isotope ratio for temperature dependence).

The cyclical nature of marine oxygen isotope changes is clear, and the major peaks and troughs in the curve (Figure 13.2) are identified as numbered **marine oxygen isotope stages** (**MIS**). Even-numbered stages are times when climate was cooler than average, ice sheets were larger, and sea level was lower. Odd-numbered stages are times when climate was warmer than average, ice volume was less, and sea level was higher. Dating of these stages was first done by tuning (stretching) marine records to match predictions of warm and cool times deduced from calculations of Earth's changing orbit and thus the latitudinal distribution of solar radiation. More recently, the MIS timescale has been updated by other, more direct dating approaches; this updating has largely validated the original orbitally tuned timescale.

Ice Cores

Ice, which can be cored and collected from ice sheets as well as smaller glaciers, preserves a detailed record of climate in cold regions, both in the Arctic and Antarctic as well as in high mountains [**Figure 13.4**]. Ice cores contain the water that fell as snow on the glacier, along with dust and fragments of volcanic ash as well as chemical aerosols and gases (such as CO_2), which become trapped as the snow consolidates into impermeable ice containing discrete bubbles (see Figure 9.3), closing off exchange with the atmosphere. The concentration of CO_2, CH_4, and other gases trapped in bubbles in the ice tells us the past composition of Earth's atmosphere. From analysis of glacial ice cores, we now know conclusively that the concentration of these greenhouse gasses has varied closely with temperature over at least the past 800,000 years. Layering in ice cores allows counting of annual bands much like tree rings and so the upper portion of ice cores can be dated precisely. Such layering can persist for thousands to tens of thousands of years in ice where accumulation rates are high and ice deformation rates are low.

Much of what we know about paleoclimate has come from ice cores, in particular measuring the stable oxygen isotope composition of the ice. For example, the $^{18}O/^{16}O$ ratio of glacial ice can be used to infer the air temperature (which in polar regions was as much as 15°C colder than today at the last glacial maximum). As air cools, the $^{18}O/^{16}O$ ratio of precipitation becomes more depleted; there is less ^{18}O.

Ice cores preserve the chemistry of the snow and they are archives of dust fall. Changing acidity levels in the ice

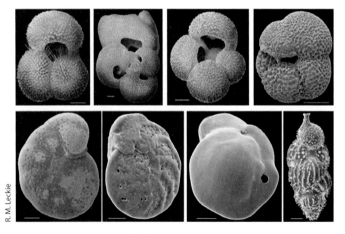

PHOTOGRAPH 13.5 Planktonic and Benthic Foraminifera Used in Paleoclimatic Studies. The upper row illustrates planktonic, near-surface dwelling species. The lower row illustrates benthic, deep water foraminifera. Grey bar under each foraminifera is 100 μm long.

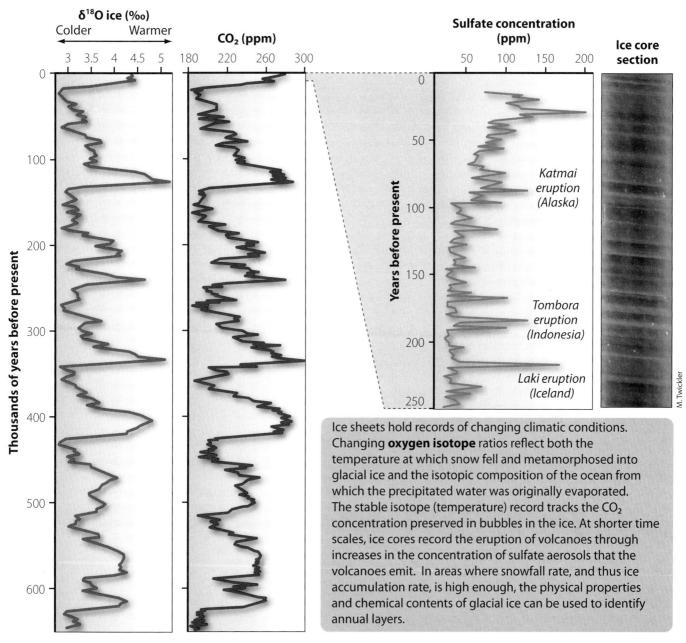

FIGURE 13.4 Climate Records Preserved in Glacial Ice. The stable oxygen isotope composition, CO_2 concentration, and sulfate concentration are used to interpret paleotemperature, atmospheric composition, and the history of volcanic eruptions. [Data from Siegenthaler et al. (2005).]

are related to volcanic eruptions because such eruptions emit gases, such as SO_2, that react with water to form acid (Figure 13.4). Sometimes, volcanic ash is found in an ice core. If tephrochronology can be used to identify the ash and if the ash has been dated isotopically, then the age of the ice layer in which the ash was found can be determined. Such ash dates are used to cross-check layer counting and to extend ice-core chronologies beyond the age (below the depth) at which discrete layers are preserved.

Windblown Terrestrial Sediment

Loess, silt-sized wind-blown dust, preserves a record of both sediment availability and windiness, both of which are related to climate (see Chapter 10). Loess deposits are found downwind of significant sediment sources; for example, much fine sediment was sourced from the broad outwash plains that bordered glaciers and ice sheets [Photograph 13.6]. The middle of North America has deep and fertile loess deposits left behind when strong winds scoured fine sediment from massive outwash plains

PHOTOGRAPH 13.6 Outwash Plain. Outwash plain (braided stream) from which wind is transporting dust (which could be deposited as loess) in Kagbeni, Nepal.

PHOTOGRAPH 13.8 Paleosol. Scientists sample a dark red paleosol (upper section) developed on tan loess (lower section) in the Loess Plateau of China.

of the Laurentide Ice Sheet. Rich prairie soils, which are today intensively farmed, formed in this loess. The most famous and best-studied loess sheets cover the Loess Plateau of China [Photograph 13.7]. Here, loess that originated from deserts to the west, rather than directly from glaciers, has been deposited for at least several million years, and is >100 m thick in places. The Chinese loess sequence contains many **paleosols** or buried soils [Photograph 13.8]. Each paleosol indicates a period of soil formation and stability when the climate was relatively warm and wet in the source and deposition areas; therefore, dustiness was reduced, loess deposition rates were low, and biologic activity on the plateau was high. Each loess layer indicates a time when climate was colder and dustier in the source area and therefore loess deposition rates increased, exceeding rates of soil development.

Loess deposits are a valuable climate and geomorphic archive but they are difficult to date. Young loess (<40,000 years old) can be dated using radiocarbon analysis of associated organic matter. Loess with ages up to a few hundred thousand years can be dated using luminescence methods (see Chapter 2). Older loess is dated primarily using paleomagnetic methods (if it has normal polarity, the loess was deposited <780,000 years ago; reversed polarity indicates that the loess is older). With such imprecise dating, older loess provides a less well-constrained record of changing regional climate over time. In Alaska and other volcanic regions, tephrochronology of ash found in loess beds can be used to date loess deposits.

Climate Cycles

A variety of records clearly indicates that Earth's climate changes over time and some of those changes are cyclical (Figures 13.2 and 13.4). Over the timescale relevant for most geomorphology, the last few million years of Earth history, climatic changes have periodically plunged our world into glaciations that spread ice over much of the Northern Hemisphere and some of the Southern Hemisphere. On the timescale of decades, climate variability associated with a variety of large-scale atmospheric and ocean processes has had significant geomorphic effects.

Earth's climate largely reflects an energy balance between incoming solar short-wave radiation and outgoing long-wave radiation. Thus, the dominant explanation for longer-term climate variability considers changes in the seasonal distribution of incoming solar radiation on Earth's surface due to predictable changes in Earth's orbit. Variations in atmospheric and oceanic heat and water transport can influence climate on shorter timescales. The brightness of the Sun may also change over decadal to millennial timescales, possibly affecting climate.

PHOTOGRAPH 13.7 Loess Plateau. The Loess Plateau of China is covered by tens to hundreds of meters of loess deposited over the Quaternary period. Deforestation and intensive land use have caused significant soil erosion. Here, trees are planted in an attempt to stabilize the easily eroded loess.

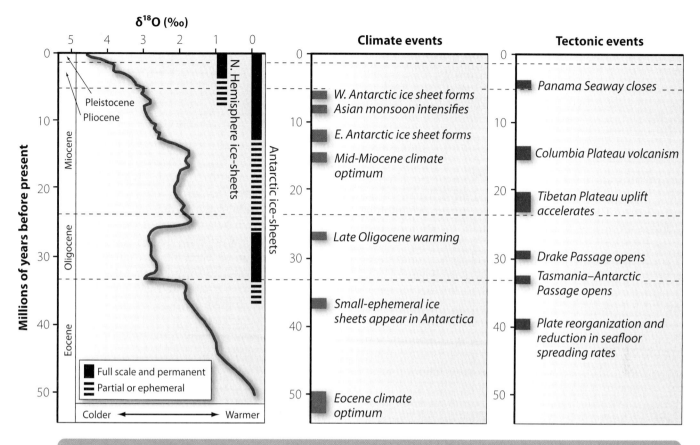

FIGURE 13.5 Climate and Tectonics. Over the past 50 million years, Earth's climate has changed because of tectonic events, including uplift, opening of ocean passages, and large volcanic eruptions. [Adapted from Zachos et al. (2010).]

Glacial Cycles

The most recent period of Earth history during which glaciation was common began in earnest about 2.7 million years ago, although large amounts of ice began to cover parts of Antarctica several tens of millions of years earlier [Figure 13.5]. Most of Greenland remained ice free until the early Pleistocene (~2.7 million years ago), although it appears some ice began to accumulate in Greenland during the Pliocene (7 million years ago), and there are hints of ice in Greenland (evidence from marine sediment records of ice-rafted dropstones) as early as 38 million years ago.

Early glaciations, identified by isotopic excursions in deep-sea sediment records, were fast paced and symmetrical with cooling and warming occurring in a regular, smooth pattern on a ~40,000-year cycle (Figure 13.2). About a million years ago, something changed abruptly. The cycles became much longer (~100,000 years) and asymmetric, with slow cooling and ice expansion, and then very rapid warming and quick demise of extensive ice sheets; these rapid warmings are referred to as **terminations.**

There is no consensus regarding why the cyclicity changed. Some suggest that the change in cyclicity was the result of decreasing atmospheric CO_2 and the concurrent cooling of the planet, the result of which was larger and less responsive ice sheets. Another explanation considers evolving ice sheet interactions with previously weathered regolith. The regolith idea suggests that after the soft, extensively weathered, pre-Quaternary regolith was stripped away, Northern Hemisphere ice sheets began flowing over hard, crystalline bedrock, which increased basal shear stresses ($\tau = \rho g h \sin\theta$, see Chapter 9); thus, ice sheets thickened and thereafter responded more slowly to orbital forcing (hence the jump from 40,000-year to 100,000-year cycles). This inference is consistent with data showing that after the first million years of extensive glaciation, the ocean oxygen isotope record suggests that global ice volume increased, but the terrestrial record (glacial till) shows that ice extent did not change. It appears that about a million years ago, ice sheets went from being thin and flat to thicker and more voluminous.

Orbital Forcing

Over thousands to tens of thousands of years, the shape of Earth's orbit around the Sun and the tilt of Earth on its axis change. These orbital variations control the amount and distribution of solar radiation incident upon Earth's surface by latitude and by season. Accounting for such changes can reliably predict the distribution of solar radiation on Earth's surface over time and space [**Figure 13.6**].

Three characteristics of Earth's orbit interact to determine the spatial distribution of incoming **solar radiation;** each characteristic has a different period over which it changes. The **eccentricity** of Earth's orbit around the Sun (the change from a more circular to a more elliptical orbit) varies with a ~100,000-year period. A more elliptical orbit causes greater seasonality. The **obliquity** (the variation in the angle of Earth's tilt as it rotates) varies on a ~41,000-year cycle and changes seasonality. When the tilt is greater, there is more contrast between summer and winter temperatures. Cool summers better preserve snow from the previous winter and thus initiate ice sheet growth. The **precession of the equinoxes,** or the variation in the direction of Earth's rotation axis as it orbits the Sun, also changes seasonality but with ~22,000-year cyclicity. Together, these three orbital characteristics define **Milankovitch cycles,** named in honor of Milutin Milankovitch (1879–1958), a Serbian engineer who made the first detailed calculations indicating that such cycles could control Earth's climate. Although orbital variations change the seasonal distribution of insolation for a specific time and place on Earth's surface up to 20 percent, they do little to change total annual insolation for Earth as a whole, which varies over time by no more than 0.3 percent.

Understanding exactly how changes in solar radiation (specifically at high northern latitudes where ice sheets are born) translate into planet-wide glaciations, interglaciations, and rapid changes in paleotemperature has not come easily and has revealed numerous interactions, feedbacks, and resulting amplifications between the solid Earth, the atmosphere, and Earth-surface processes. Glaciations begin when cool summers allow some snow at high Northern Hemisphere latitudes (nominally 65°) to survive the melt season and thus begin to build up into glacial ice. Once this buildup begins, the ice and snow have greater albedo (reflectivity) than the forest or tundra they covered, and thus reflect more of the incoming solar radiation. This positive feedback further cools the planet, encouraging expansion of the nascent ice sheet. The effect of greenhouse forcing (changes in the atmospheric concentration of CO_2, CH_4, and water vapor resulting from glaciation and its effects) is responsible for the general synchroneity of climate change in the Southern and Northern hemispheres.

Local Events—Global Effects

Abrupt and globally synchronous paleotemperature changes likely reflect rapid reorganization of ocean circulation patterns and the resulting redistribution of heat around the planet. An important component of global oceanic circulation is the formation of cold, salty, dense water in the North Atlantic, part of the ocean's **thermohaline circulation,** flow driven by density contrasts that depend both on temperature (thermo) and salinity (haline). The sinking of this water to form **North Atlantic Deep Water** (**NADW**) maintains a current that moves 20 times the average flow of all the world's rivers.

Changes in the volume of NADW flow can alter the delivery of tropical warmth to the North Atlantic through small changes in the flow of both intermediate depth and surface currents, perhaps amplified by changes in sea ice distribution. For example, diminished northward heat transport by ocean currents (likely because of a sudden outburst of cold, fresh water into the North Atlantic, see Digging Deeper) contributed to the rapid cooling of the North Atlantic, Europe, and eastern North America known as the **Younger Dryas** cold episode 12,800 to 11,500 years ago. The name Dryas refers to an alpine plant that appeared at lower altitudes during this chilly time period. Sea ice cover expanded significantly at this time, likely driving a large temperature change around the North Atlantic because ocean-atmosphere heat exchange was suppressed. Evidence from Greenland ice cores suggests that average temperatures there plummeted about 15°C within a few decades and that the millennium-long cold spell caused glacial margins to advance and changed plant communities around the North Atlantic.

Similar flooding of the North Atlantic by fresh, less-dense glacial meltwater appears to have altered ocean heat flow on at least six occasions during the last glacial period, rapidly changing climate at least regionally. These episodes of flooding, called **Heinrich events** (Figure 13.2), were caused by Northern Hemisphere ice sheets episodically discharging large volumes of fresh water and icebergs to the North Atlantic. Evidence for Heinrich events comes from the abrupt appearance of coarse, terrestrial sediment in ocean sediment cores collected far from land; such terrestrial material could only have been brought far out into the ocean by icebergs. Heinrich events seem to have been particularly frequent when the Laurentide Ice Sheet was partially extended. In this configuration, small changes in glacier extent forced meltwater discharges to alternate between the Mississippi River drainage (greater

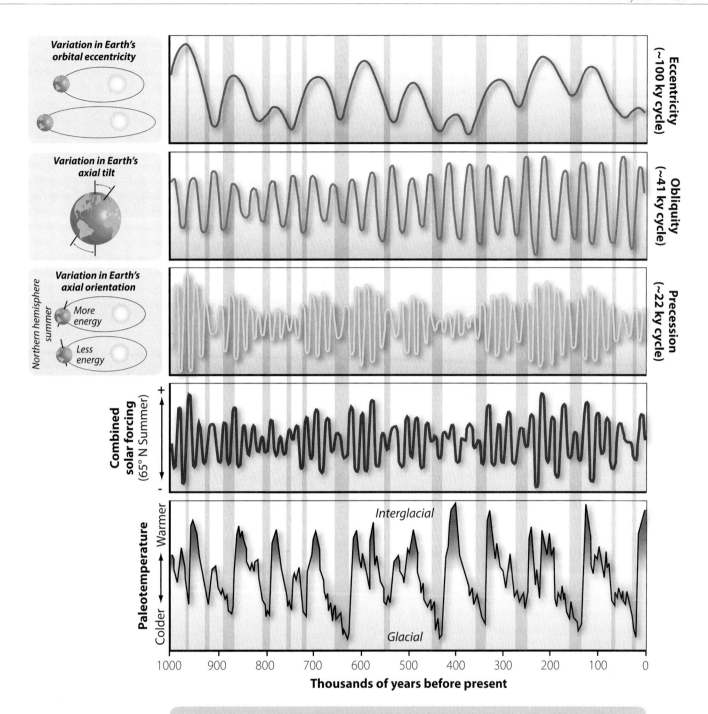

The geometry of Earth's orbit around the sun varies in a regular and predictable fashion with three main orbital parameters (**eccentricity, obliquity,** and **precession**) changing on 100,000-, 41,000-, and 22,000-year cycles, respectively. Together, these orbital parameters act to change the seasonal distribution of solar radiation (energy) incident upon Earth. When little solar energy is incident on high northern latitudes, some of the snowpack survives the summer melt season and begins to grow an ice sheet. The solar forcing is smooth and symmetrical, but the paleotemperature record has many fewer variations of greater amplitude. This reflects ocean/atmosphere interactions and amplification of the solar forcing.

FIGURE 13.6 Orbital Forcing of Climate Change. The geometry of Earth's orbit changes the seasonal distribution of radiation incident on Earth's surface over time.

ice extent) and the Saint Lawrence River drainage (lesser ice extent).

Significant changes in North Atlantic climate, recorded as changes in isotope ratios in ice cores, are associated with some Heinrich events. The causes of these rapid climate changes, known as **Dansgaard–Oeschger (DO)** events, remain elusive, since there are many more of them (about 20 during the last glacial period) than there are massive iceberg discharges (Heinrich events).

Climate Variability Within a Climate State

Within a glacial or an interglacial period, there can be variability sufficient to drive significant geomorphic change. Such variability can occur on the scale of decades, centuries, and millennia.

During the Holocene interglacial period (the last 12,000 years), climate has been in general much more stable than it was during the previous glacial interval. The Holocene was in general warmest during the **Altithermal** or **Holocene climatic optimum** (several millennia in the mid-Holocene, between 8000 and 5000 years ago) and coolest during the **Neoglacial** (the last several thousand years). The distribution of moisture during the Holocene changed as wind belts and storm tracks shifted latitude in response to changing Northern Hemisphere summer insolation.

Despite the relative climatic stability of the Holocene, there were significant, climatically driven landscape changes. For example, between 8000 and about 5000 years ago, in response to a slightly warmer mid-Holocene climate, various lines of evidence indicate the Greenland Ice Sheet retreated inland, perhaps many kilometers, behind its present margin. The most compelling evidence for retreat comes from radiocarbon dates of marine mollusk shells reworked by the ice and included in the till of historic moraines. These mid-Holocene age clams indicate that the glacier margin must have been located up fjord, allowing the mollusks to grow in marine waters in locations now far inland. During the last several thousand years, the Neoglacial, the ice tongues then readvanced, scooping up the mollusks and delivering them to the glacial margin mixed with glacial sediment. Such Neoglacial advances are also typical of mountain glaciers worldwide.

Different changes occurred away from the glacial margin. For example, the woodland-prairie ecosystem boundary in North America moved eastward, responding to mid-Holocene warming and drying and changing the nature and intensity of soil-forming processes. In New England, spruce and fir trees, which thrive in a cool, wet climate, replaced warmth-loving pine trees over the last few millennia. It is likely that fire frequency changed along with the vegetation, and if the last 100 years of warming is a clue to past biotic-landscape interactions, the tree line was probably higher, stabilizing steep mountain slopes in the warmer mid-Holocene.

More recently, and on a shorter timescale (centuries), the North Atlantic climate warmed sufficiently (perhaps a

(a)

Ice Sports, c. 1610 (panel), Avercamp, Hendrick (1585–1634)/Mauritshuis, The Hague, The Netherlands/The Bridgeman Art Library

(b)

P. Bierman

PHOTOGRAPH 13.9 Climate Change Over the Past Millennium. Climate change has affected society's interaction with the landscape. (a) The painting *Ice Skating Near a Village*, by Hendrick Avercamp, shows a Dutch scene during the Little Ice Age, about 1610, depicting colder winters. Several hundred years of cold temperatures between 1300 and 1850 allowed ice to form regularly on water bodies, such as rivers and lakes, that today rarely if ever freeze. (b) Brattahlid, Erik the Red's Viking farmstead in southern Greenland, is today nothing more than foundations around which sheep wander. During the Medieval Optimum, Viking society thrived here but the cooling that led into the Little Ice Age ended Viking settlement on Greenland about 1300 CE.

degree Celsius on average) to allow the Vikings to settle Greenland and Newfoundland during the **Medieval Warm Period** (900–1300 CE) but soon after, the climate cooled, slipping into what has been termed the **Little Ice Age**, several hundred years (1300–1850 CE) of temperatures cool enough that the canals of Holland froze regularly in winter, the Norse settlements were abandoned, glaciers advanced in the Alps destroying buildings that were in their way, and crops failed repeatedly in the cool, wet weather setting the stage for extensive famine [**Photograph 13.9**]. Other societally important climate changes include drying of the Sahara Desert and the advance of sand dunes over streambeds and settlements about 5000 years ago [**Photograph 13.10**].

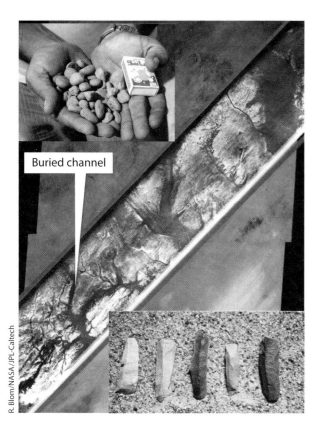

PHOTOGRAPH 13.10 Sahara Desert. Today the Sahara Desert is dry and sandy but in the early to mid-Holocene, before 5000 years ago, it was much moister. In 1982, imaging radar flown on the space shuttle Columbia detected a large network of river channels buried beneath just a few meters of sand. The radar swath in gray shows the channels. The background image in orange shows the sand-sheet surface. Fieldwork confirmed the presence of stream-rounded pebbles (upper left) and Neolithic artifacts (lower right) left by people attracted to the water resources near the channels.

Short-Term Climate Changes

Repeated and persistent oscillations in patterns of rainfall and runoff on yearly to decadal timescales affect the tempo, distribution, and intensity of surface processes. On the basis of barometric pressure comparisons and sea-surface temperature distributions, a number of climate patterns characterized by short-term variability have been identified. The best-known climate patterns are the **North Atlantic Oscillation (NAO)**, the **Pacific Decadal Oscillation (PDO)**, and the **El Niño–Southern Oscillation (ENSO)**. Despite the name "oscillation," variability of the PDO and NAO are indistinguishable from random noise with the time between different phases lasting from weeks to decades. In contrast, ENSO is a true oscillation, varying with a period of 2–7 years.

ENSO events (defined by sea-surface temperatures and atmospheric pressure changes in the equatorial Pacific) can be geomorphically significant. The phase during which sea-surface temperatures are high in the eastern Pacific is defined as an **El Niño** event. El Niño events are associated with heavy rains, floods, and landslides along the western coast of South America and the southwestern coast of North America—precipitation that greens the Mojave Desert as long-dormant seeds sprout after soaking rains. On the west coast of the United States, in California, El Niño brings storminess that causes large ocean waves, high river flows, and warm average temperatures. Such changes increase beach erosion, the frequency of debris flows and landslides, and the number of rain-on-snow flooding events. In contrast, during El Niño years, hurricanes and the landslides, floods, and coastal erosion they cause are suppressed along the Gulf Coast and eastern North America. Coral records from the tropics show that ENSO variability existed throughout the Holocene and the last glacial period. Analysis of varves from glacial Lake Hitchcock, which occupied the Connecticut River Valley for >4000 years in the Late Pleistocene, shows distinct changes in the amount of runoff and thus possibly rainfall at the 3–5 year timescale, consistent with ENSO climate oscillations at the end of the last glacial period.

The NAO, defined as the pressure gradient between the Icelandic Low and the Azores High, is related to the intensity of storms affecting the North Atlantic, including northern North America and Europe. Evidence for long-term NAO variations was identified by tracking the frequency of storm-induced inorganic sediment layers in New England lake cores dated by ^{14}C measurements on macrofossils (leaves, cones, and twigs; Photograph 13.4). Geomorphically effective changes in runoff and sediment transport (storm sediment layers) were attributed to the NAO because they matched in phase, frequency, and timing layers of ice containing high levels of sea salt recovered by ice coring from the center of the Greenland Ice Sheet. This increase in sea salt in the glacial ice was attributed to strong, salt-laden ocean winds blowing inland to central Greenland. The cause of NAO changes is not understood but may relate to small changes in solar output, which in climate models tend to produce an NAO-like response. Such changes may also explain the pattern of temperature and precipitation changes during the Little Ice Age cooling.

Geomorphic Boundary Conditions

Changing climate directly affects almost all geomorphically important variables including the frequency, intensity, and duration of precipitation, temperature, and both local and global base levels to which rivers and streams are graded. These changes are linked through the intensity of the global hydrologic cycle, the activity of which increases with temperature and with the temperature gradient between the poles and the equator.

Precipitation and Temperature

Linked changes in precipitation and temperature over time control the availability of water on the landscape and thus the frequency, magnitude, and rate of many surface processes. For example, increases in temperature at the end of the Pleistocene in North America's desert southwest reduced effective moisture and vegetation cover leading to the generation of less runoff. In some cases, this drying appears to have decreased sedimentation on alluvial and debris fans. In other cases, increasing aridity and the concomitant loss of soil-anchoring vegetation appears to have increased erosion and sediment transport from slopes. Unfortunately, at most arid-region study sites, dating of erosion and deposition episodes is imprecise enough (due to the lack of preserved organic carbon) that it is not possible to conclusively link climate changes and geomorphic response.

Climate also controls the intensity and spatial extent of large-scale weather patterns and storms. For example, **monsoons**, the seasonal reversal of winds and the resulting torrential summer rainstorms, are caused by a seasonal change in the location of the subtropical semipermanent high-pressure cells. Monsoon-associated precipitation tends to be best developed where elevated land masses heat up during the summer months, causing the air above them to warm, become less dense, rise, and pull moist air from a nearby ocean up steep topography where it cools, the moisture condenses, and torrential rains result. Such monsoons, including those in India, Australia, and southwestern North America, are geomorphically important, shaping channels, moving sediment off slopes, and carrying large sediment loads. During the last glacial maximum, monsoonal precipitation appears to have been less than current values, but monsoon strength increased in the early to middle Holocene. In the Himalaya, this heavy early Holocene precipitation caused channels to fill with alluvium, presumably because increased landslide frequency stripped soil off slopes more rapidly than the rivers below could remove the sediment.

A variety of information sources tell us that the frequency and intensity of hurricanes and coastal storms change over time. Instrumental and written records, including ships' logbooks, provide detailed information about storms that occurred during the past few hundred years. On a longer timescale, geologic records can be used to determine the timing and location of paleostorms that flood coastal regions and wash over barrier islands forcing salt water, sand, and marine organisms into coastal freshwater ponds and lagoons. In order to detect coastal paleostorms, geomorphologists analyze sediment from cores and outcrops. Coarse, sandy layers are indicative of storm surges, and the presence of saltwater organisms reveals flooding of coastal marshes and ponds by seawater. Such geologic detective work indicates that hurricanes were more frequent than today between 1000 and about 3400 years ago along the U.S. Gulf Coast and that storm intensity has increased over the past several hundred years along the coast of Maine.

Vegetation, Fire, and Geomorphic Response

Climate, vegetation, fire, and the rates and distribution of landscape-scale processes are tightly linked because climate, in particular moisture availability, temperature, and wind speed, determines the type and density of vegetation covering the landscape. Vegetation accelerates weathering and soil development through mechanical disruption, decomposition of leaf litter, the production of organic acids, and the acidification of regolith by CO_2 emitted during respiration (see Chapter 3). Roots also provide apparent cohesion, strengthening soil and holding it on hillslopes (see Chapter 5).

Fire is a geomorphic agent that can quickly remove vegetation, making slopes susceptible to erosion until vegetation regrows [**Photograph 13.11**]. Removing vegetation from the landscape reduces surface roughness and increases the speed and erosivity of overland flow. Fire kills trees and over time the roots from fire-killed trees rot, reducing effective soil strength. In some ecosystems, such as the chaparral of southern California, fire releases hydrophobic (water-repellent) compounds from the plants, coating the soil. This coating diminishes infiltration, increasing the risk of erosive runoff

PHOTOGRAPH 13.11 Post-fire Erosion. After the great Yellowstone forest fires of 1988, a single thunderstorm that dumped rain on burned slopes caused rilling and gullying in Madison Canyon, Yellowstone National Park, Wyoming.

and the chance of generating debris flows when heavy rain strikes.

Fire can act as a direct geomorphic agent. The heat of fire rapidly expands the surface of rocks; yet the interior of rocks, especially those exposed to rapidly moving fires, remains cool due to rock's low thermal conductivity. This differential heating and expansion results in significant stresses within fire-heated boulders and outcrops. These stresses crack and erode exposed rock over time (see Photograph 3.6).

Climate determines the type and density of vegetation that can grow on a landscape and thus the fuel load; it also determines the moisture content of that fuel and its susceptibility to ignition during dry seasons. Climate strongly influences the frequency of fire and the resulting geomorphic effects, particularly the generation of debris flows and the deposition of debris flow material in fans at the base of slopes and its subsequent transport through river systems. Periods of aggradation in the past, dated with charcoal buried in debris flow sediment, correlate strongly to times when climate was warmer and drier, such as the Medieval Warm Period (900–1300 CE). Today's warming and drying climate in western North America, as well as heavy fuel loads due to a century of wildfire suppression, mean that the next century likely will see more fires and increased geomorphic activity, specifically hillslope erosion, debris flows, and fan aggradation in areas affected by fire (Photograph 13.11).

Sediment yield is the amount of sediment transported out of a drainage basin over time. It reflects both the rate at which the basin erodes and the efficiency of sediment transport mechanisms that remove sediment from the basin. Early measurements, such as those plotted in the classic Langbein-Schumm curve [Figure 13.7], showed maximum sediment yield from catchments located in semi-arid climates. This curve was created using data (mainly from the tectonically inactive central United States where drainage basins are underlain by sedimentary rocks) collected from

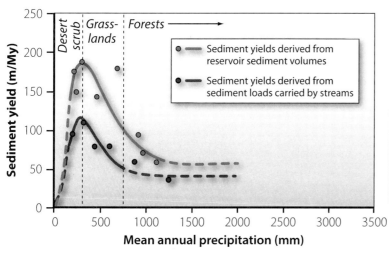

The Langbein-Schumm curve relates **sediment yield** (the amount of sediment leaving a watershed as determined by either suspended sediment load or reservoir sedimentation over a known time period) to mean annual precipitation. Langbein and Schumm found that semi-arid drainage basins had the highest sediment yields. The data for this compilation were collected from central North America and represent sediment transport during a period when people were altering land use through agriculture.

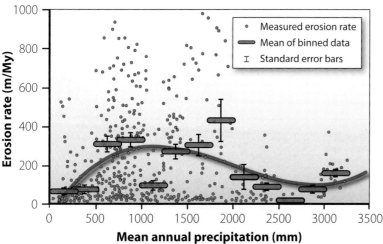

A compilation of over 1000 measurements of ^{10}Be made in river sediments around the world shows great variability in long-term erosion rates for similar amounts of precipitation (green dots are the data). On average, arid regions (<500 mm/yr of rainfall) had low rates of erosion, humid regions eroded more quickly, and erosion rates were lower in areas where mean annual precipitation was very high, perhaps reflecting the importance of vegetation stabilizing slopes.

FIGURE 13.7 Sediment Yield, Erosion, and Climate. Sediment yield data and cosmogenic estimates of long-term, drainage-basin scale erosion rates offer different views on how mean annual precipitation affects erosion and sediment yield. [Data from Langbein and Schumm (1958) and Portenga and Bierman (2011).]

170 gauging stations as well as reservoir sedimentation data; thus, the sediment yield data reflect both human impact and a short (years to decades) integration time. One plausible explanation for the humped shape of the resulting curve is that arid climates are so dry that sediment yield was limited by rainfall and runoff (transport-limited) even though there was little if any vegetation to hold regolith in place. In humid climates, vegetation anchored sediment making the slopes supply-limited. In semi-arid climates, the sparse vegetation provided little erosion resistance and intense thunderstorms produced erosive runoff.

In contrast, a 2011 compilation of more than 1000 measurements of the cosmogenic nuclide ^{10}Be in fluvial sediment from around the world shows that long-term, basin-scale erosion rates vary up to two orders of magnitude even if the amount of precipitation is similar. The difference between the global isotopic estimates of erosion rate and the Langbein-Schumm compilation of predominantly central United States sediment yield data likely reflect a combination of the major influence of global variability in tectonics and lithology, as well as differing integration time and the influence of human impacts on short-term rates of sediment yield. Such a comparison points out the importance of understanding all of the variables that influence both the rates of sediment generation and the rates of sediment delivery from a watershed.

Base Level

Sea level, the base level for the world's major rivers, repeatedly fell more than 100 m between maximum interglacial and glacial stages, reflecting changes in the volume of ocean water as ice sheets advanced and retreated. Volumetric control on sea level is termed **eustatic** and includes not only sea-level fluctuations driven by the expansion and contraction of ice sheets and the amount of water they hold on the 1000-year to 10,000-year time scale but also the thermal expansion and contraction of ocean water driven by changes in global temperature. Streams respond to lower sea level by incising. Incision propagates upstream, rapidly removing unconsolidated alluvium and triggering the formation and migration of knickzones in bedrock channels. As sea level increases, base level rises, triggering aggradation, initially in areas nearest the sea, but then spreading upstream a limited distance like a wave.

Incision and aggradation can occur simultaneously on different parts of the landscape (see Figure 12.14) because the effects of base-level fall and the incision it causes take time to translate up a drainage network. Given enough time, the effects of a major base-level fall will eventually propagate up to the head of the channel network (and onto the hillslopes) through the processes of knickzone retreat, channel incision, and hillslope lowering by erosion.

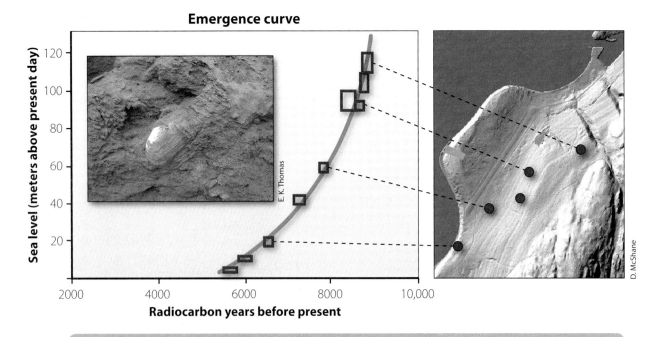

As an ice sheet shrinks, the land begins to rebound as the load on Earth's mantle is reduced. At the land-sea interface, **wave-cut terraces** can form from large storms and beach sediments are deposited. If the material in terraces of known elevation can be dated (usually using ^{14}C dating of marine shells or driftwood), an **emergence curve** may be constructed that represents local sea level over time. Uplift can result in stacked flights of terraces rising above the modern shoreline.

FIGURE 13.8 Emergence Curves. Local emergence curves describe the change in sea level at one place on Earth's surface reflecting both eustatic sea-level change and the isostatic response of the solid Earth. [Adapted from Ten Brink (1974).]

The passage of knickzones and the lowering of local base level will increase local gradients and erosion rates, changing the amount of sediment delivered to the channel network. In contrast, the effects of a base-level rise are limited to estuarine and lowland river systems where channels will flood and sediment will aggrade. Prime examples of such aggradation include the drowned landscape of the Maine coast and the distal end of lakes in previously glaciated areas where postglacial, isostatically driven tilting raises land closest to the ice sheet, drowning stream and river valleys farther away from the ice.

Paleo–sea levels are a challenge to determine because it is difficult to find a stable frame of reference as both sea level and land level change over time. At many sites around the world where the land is rising relative to the sea, geomorphologists have created **emergence curves** (graphs of land level over time) by dating shoreline features and measuring their elevation [Figure 13.8]. Dating usually relies on ^{14}C analysis of either driftwood found on raised beaches or fossils found in marine sediment underlying wave-cut terraces. Differences in the amount and timing of region-specific isostatic responses cause local sea-level histories to differ, sometimes markedly, from global **eustatic sea-level curves,** which reflect primarily the volume of water in the world's oceans [Figure 13.9].

Consider the northeast coast of North America, which for the most part was overrun by ice and isostatically depressed at the last glacial maximum. As the ice melted, and the volume and area of the Laurentide Ice Sheet diminished, the load on the crust decreased and the land rebounded. As long as the rate of local rebound exceeded the rate of eustatic sea-level rise from worldwide melting of glacial ice, land emerged from the sea [Figure 13.10]. However, when global sea-level rise (eustatic) later became faster than the local rate of isostatic rebound, then the land that initially emerged isostatically was resubmerged by eustatic sea-level rise. The land reemerged again above sea level as isostatic response, though waning, continued after the eustatic sea level stabilized about 8000 years ago, because all of the Northern Hemisphere ice sheets, except Greenland, had melted.

Such complex emergence scenarios are more likely if the ice readvances during periods of rapid eustatic sea-level rise. Indeed, this is the case in Greenland, where the ice margin in some places retreated perhaps kilometers in the mid-Holocene only to readvance in the last few thousand years. This readvance caused mantle displacement and isostatically driven crustal subsidence of a meter or two. This subsidence has left a clear stratigraphy in near-shore deposits. There, one finds saltwater-tolerant

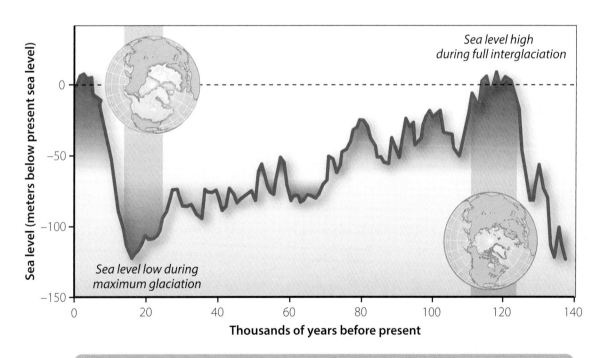

As ice sheets grow, they lock up water that otherwise would fill the ocean basins. During glaciations, sea level falls proportional to the amount of water stored in ice caps and glaciers. During interglaciations, sea level rises as ice caps and glaciers shrink and meltwater returns to the oceans.

FIGURE 13.9 Glacial-Interglacial Eustatic Sea-Level Changes. During the past few million years (the Pleistocene), global, or eustatic, sea-level change reflects the amount of water held in ice sheets. The last glacial–interglacial cycle saw sea level change 130 m over 100,000 years.

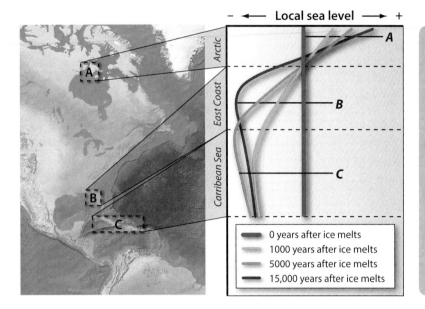

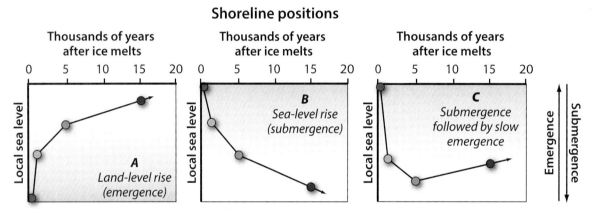

FIGURE 13.10 Variability of Local Sea-Level Histories. Local sea-level curves during and after deglaciation differ between locations. [Adapted from Clark et al. (1978).]

flora and fauna deposited over the remains of freshwater vegetation.

The advance of glaciers in response to small changes in climate can be detected not only from moraines but, if the geometry of the landscape is right, in the sediment of **threshold lakes**. A threshold lake typically receives sediment only from its watershed until a tongue of glacial ice crosses a topographic threshold and begins to deliver glacial sediment directly to the lake [Photograph 13.12]. When this happens, the character of the sediment changes as the concentration of fine, glacial sediment (glacial flour) increases manyfold. Dating the first and last occurrence of this glacial sediment can accurately delineate the timing of glacial advance and retreat.

Climatic Geomorphology

The relationship of some landforms to specific climate zones led to the development of **climatic geomorphology**, a school of thought that suggested climate primarily controlled the distribution and shape of many landforms. Taken at face value, such a suggestion makes sense because temperature, precipitation, and wind regimes

PHOTOGRAPH 13.12 **Threshold Lake.** In Tasiussaq, west Greenland, is a lateral moraine with a threshold lake to the left that collected meltwater and sediment from the ice when the glacier deposited the moraine.

As climate changes, so does the location of climate zones and their boundaries. Paleoclimatological studies using pollen and macrofossils document past shifts in climate zones. Contemporary climatological data allow mapping of climate zone shifts over the past century while computer models, based on estimates of greenhouse gas–driven climate change, predict the location of climate zones well into the future. Shifting climate zones mean changes in the intensity and duration of hydrologic phenomena (such as rainfall and runoff) and thus changes in the rate and distribution of Earth surface processes including landsliding, sheetwash, and the generation of debris flows.

Climate-Related Landforms and Processes

Both characteristic landforms and the distribution and rate of surface processes are often related to climate. In some cases, the geomorphic effects of climate are obvious and significant. For example, glaciers, glaciation, and periglacial processes dominate in many polar and some alpine climates. In dry climate zones, playa lakes and desert pavements are common landforms and aeolian transport of sediment is an important geomorphic process. Much of the geomorphic effect of climate is filtered through vegetative systems. For example, the paucity of vegetation in arid regions does not support the significant root reinforcement of soils common in more humid climates. Arid-region slopes >15° to 20° rarely have significant soil cover—a contrast to the steep, yet soil-mantled slopes of heavily vegetated, humid uplands.

Climate strongly affects the behavior of hillslopes and rivers. For example, rivers in dry climate zones flow only intermittently. Without the roots of vegetation to stabilize their banks, arid-region channels often shift large distances laterally during rare but high-discharge floods. Bankfull flows are much rarer events in arid than in humid regions. Flows often decrease downstream in arid-region streams as water flowing downstream from higher, wetter headwaters infiltrates into the bed of the channel or into alluvial fan surfaces (see Figure 4.12). This infiltration results in very different relationships between channel dimensions and basin area for arid-region versus humid-region streams. Specifically, channel width, depth, and cross-sectional area increase downstream for humid-region channels (gaining streams) and often decrease for arid-region channels (losing streams) in which the volume of flow diminishes with distance from wetter highlands.

Slopes in arid regions are often weathering-limited and thus bare rock outcrops are common with little soil or regolith cover. This leads to rapid runoff and "flashy" hydrologic responses because infiltration rates are low and there is little if any shallow subsurface flow. In the wet tropics, warm temperatures and abundant moisture catalyze deep weathering that produces thick, clay-rich saprolites and lateritic soils—conditions conducive to gullying. In contrast, well-developed soils in temperate climates support large trees, the roots of which bind soil together

(their averages, ranges of variability, and frequency and intensity of variation) define climate and the resulting vegetation cover that shape landforms through the action of surficial processes. In practice, climatic control on landforms is not always clear-cut because some landforms occur in multiple but different climate zones, and the influence of tectonics, especially in terms of base-level control, can make it difficult to isolate the influence of climate.

Köppen Climate Classification

The most widely used classification of climate zones was proposed by climatologist Vladimir Köppen in 1884 and, although it has been revised several times since, still bears his name. Köppen classified the world's land areas into five major climate zones: equatorial, arid, warm temperate, snow, and polar [**Figure 13.11**]. The definitions were based on measured values for precipitation and temperature as well as vegetation assemblages. There are numerous climatic subzones in the Köppen climate classification, reflecting more specific climate phenomena or controls such as summer drought and proximity to the coast and coastal moisture.

The distribution of climate zones roughly follows latitude (see Chapter 1), although there are clearly relationships with elevation and location downwind of major mountain chains. Different climate zones are distinguished by a variety of characteristics. For example, areas in the dry (arid) climate zone have significant moisture deficits—evaporation exceeds precipitation. Cold temperatures characterize polar regions. In temperate regions, precipitation exceeds evaporation, leading to a positive moisture budget so that runoff is routinely generated.

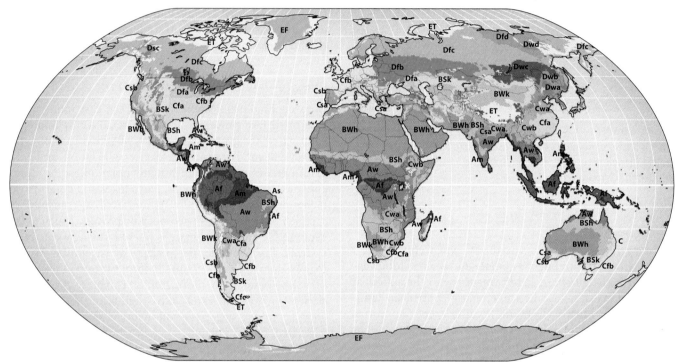

The Köppen system is the most widely applied classification of climate zones, which some geomorphologists believe can be associated with predictable sets of landforms and active surface processes. The system uses a three-letter system to classify zones according to a main climate, precipitation, and temperature. Polar regions are characterized by two letters showing whether they are tundra or ice. Climate zones are based on the annual amount and seasonal distribution of precipitation as well as the range and average of temperatures. The main climate zones reflect global wind patterns and the location of major mountain ranges.

FIGURE 13.11 Köppen Climate Zones. Climate zones reflect the global distribution of precipitation and temperature.

on steep slopes and increase porosity when they die and rot; these roots thus help prevent mass movements until the exceptional storm strikes.

Relict Landforms

Regular and significant changes in Earth's climate over the past several million years have left behind **relict landforms**, those that that are no longer forming or being maintained by active geomorphic processes—for example, moraines on the warm, semi-arid floor of Owens Valley in southern California [Photograph 13.13] or in southern Connecticut, places where ice would never survive a summer today. Many geomorphologists devote much of their careers to interpreting the paleoclimatic significance of relict landforms.

Many relict landforms are either directly or indirectly related to the comings and goings of glaciers and ice sheets. These include moraines that mark the extent of former glaciers, glacial outwash deposits along what were once meltwater streams, and periglacial features that reflect the presence of now-vanished permafrost. In mid-latitudes, dunes stabilized by vegetation are common, the result of increased moisture, lower sediment supply, lighter winds, and more vegetation during the Holocene than the Pleistocene.

PHOTOGRAPH 13.13 Moraines. Here, on the western side of semi-arid, temperate Owens Valley, California, are a set of moraines. The Sierra Nevada, from which the glaciers that deposited the moraines originated, are visible in the distance.

PHOTOGRAPH 13.15 Pluvial Lakes. Series of pluvial (glacial period) wave-cut shorelines and terraces (extensive horizontal surfaces) in the Great Salt Lake Basin, Utah, record falling lake levels after the last glacial maximum.

In the desert southwest of North America, there are clear, relict geomorphic indicators of a much wetter time. For example, in the Mojave Desert, where today only tens of millimeters of rain fall annually, in the Pleistocene there were a series of **pluvial lakes** connected by rivers. The extent of these lakes is demarcated by now-dry beach ridges, fossil waterfalls [**Photograph 13.14**], and erosional shoreline notches cut into rock by pounding waves as well as terraces deposited at the lake margin [**Photograph 13.15**]. Today's Great Salt Lake is just a shadow of its former self. During glacial times, its surface area was >10 times larger than today. That expanded lake, known as Lake Bonneville, left many distinct, relict shoreline features including spits and bars now sitting high and dry above desert valley bottoms. The drying of desert lakes after glacial times reflects a changing water balance across the landscape. In some cases, the change resulted primarily from decreased precipitation. In other cases, lake contraction was driven mostly by increased evaporation, the result of warmer air temperatures and less cloudiness.

Landscape Response to Climate

Landscapes respond to changing climate because the type, rate, and spatial distribution of active geomorphic processes, such as weathering, erosion, and sediment transport, depend to a large degree on climate.

Glacial–Interglacial Changes

Over the past few million years, the most important climate changes have been regular shifts between glacial and interglacial climates. Occurring on regular cycles taking many tens of thousands of years, the climate cooled slowly as ice sheets thickened and covered large areas in the Northern Hemisphere. Then, in comparatively rapid terminations lasting about 10,000 years, these ice sheets collapsed, melting back rapidly as the climate warmed. Large areas of Earth's surface were exposed from under ice, ecosystems were reestablished, and nonglacial surface processes began remolding the glacial landscape.

The effects of glaciation extended far beyond the borders of the ice sheets. Immediately adjacent to the ice, fierce **katabatic winds,** spawned by cold, dense air draining

PHOTOGRAPH 13.14 Fossil Falls. Fossil falls in southeastern California have not always been dry. Waters from Pleistocene Owens Lake overflowed the basalt during the ice ages, polishing and fluting the rock surface.

off the ice, whipped loess into the sky from barren outwash plains along raging meltwater streams. In many areas where today's temperatures are warm (such as the southern Appalachian Mountains and parts of Africa and Australia), there are distinctive clues indicating that the climate was once much colder, with mean annual air temperatures sufficiently cold (below freezing) to form ice wedge casts, patterned ground, rock glaciers, block fields, solifluction lobes, and even thermokarst and relict pingos (see Chapter 9)—all indicating a permafrost-dominated landscape [**Photograph 13.16**].

The advent of glaciation about 2.7 million years ago appears to have been geomorphically significant globally, even in places far away from ice margins and isolated from the effects of glacially induced rises and falls of sea level. Throughout the world, more and coarser sediment appears in the geologic record between 2 and 4 million years ago. Some think this change in sediment character and abundance reflects an increasingly variable climate that destabilized landscapes and removed readily erodible regolith generated by long periods of weathering; others suggest that a cooler, drier climate reduced vegetation cover, increasing erosion.

The effect of glaciation does not end when the ice melts away but continues as nonglacial surface processes modify glaciated landscapes. **Paraglacial sedimentation** refers to the material deposited as a result of the disequilibrium landscapes left after deglaciation. For example, steep cliffs that characterize deeply incised glacial valleys rapidly shed sediment after the buttressing ice melts away, generating large talus slopes at their base. Steep, unvegetated slopes of till erode rapidly after deglaciation, creating fans and colluvial deposits. Permeable glacial deposits, such as outwash gravel, are transmissive to groundwater while less-permeable, glacial lake sediment functions as an aquitard, limiting flow. When exposed on steep slopes, this massive, fine-grained lacustrine sediment often fails in large rotational landslides.

Isostatic Responses

Climate change causes surface processes to redistribute mass across Earth's surface. This redistribution of mass drives isostasy, a solid-Earth process, which then affects surface processes for millennia after glaciation has ceased. For example, the growth of large glaciers and pluvial lakes adds mass to the crust and the upper mantle beneath responds, slowly flowing away from the load. The lithosphere then deforms, downwarping beneath the load and upwarping with much lower amplitude away from the load. This long-wavelength upward movement raises land tens to hundreds of kilometers away from the load forming what is known as the **forebulge**. As an ice sheet melts or a pluvial lake dries, the mantle, although quite viscous, responds to the decreased load. The land beneath the load rises, rapidly at first, and then more slowly over time. Farther away from the load, the forebulge relaxes and the once uplifted land subsides [**Figure 13.12**]. Complete relaxation of isostatically induced, land-level changes can take thousands to tens of thousands of years; indeed, the forebulge formed in response to the LGM Laurentide Ice Sheet is still slowly lowering and causing the gradual inundation of coastal Maryland and Virginia. A similar isostatic response happens near the coast as sea level rises and falls, changing the load on the continental shelf.

The magnitude of postglacial uplift can be deduced in a variety of ways. In recently deglaciated areas, such as the coast of Greenland, continuously recording high-precision GPS measurements document both the deglacial trend (the land is rising over time as the ice sheet shrinks) and the annual oscillation as mass (ice and snow) accumulates in the winter and melts away in the summer. Raised shorelines in Scandinavia and the Hudson Bay region of Canada testify to the land rising out of the ocean after deglaciation (Figure 13.8).

Ancient deltas, formed where rivers and streams entered standing bodies of water, such as large, glacially dammed lakes, can also be used as paleo-level lines. Presuming such deltas were deposited synchronously, their present-day elevation indicates the differential amount of isostatically driven uplift. Where the ice was thicker, deltas rose more after the ice melted away and today have higher elevations than deltas closer to the former ice margin, where ice was thinner and isostatic depression was less.

Shoreline and delta locations can be combined into maps and the uplift amounts contoured. These **isobases**,

PHOTOGRAPH 13.16 Ice Wedge Cast. An ice wedge cast, evidence of permafrost conditions in New England, disturbs well-stratified outwash sand and gravel in Connecticut. The ice wedge likely formed just after deglaciation while the climate was still very cold.

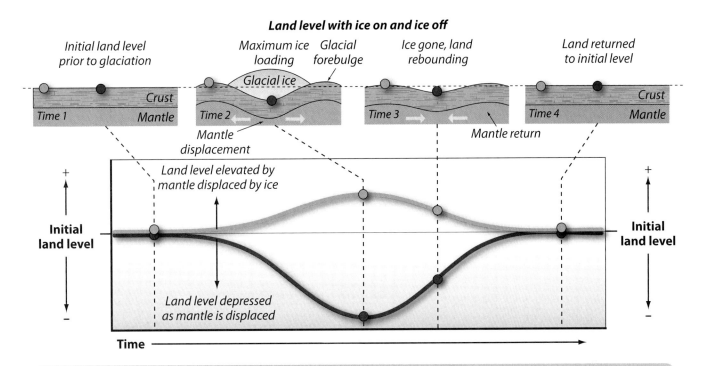

FIGURE 13.12 Land Levels Change in Response to Changing Ice Sheet Volume. During glaciation, land under the ice is isostatically depressed. Outside the ice margin, the land rises as mantle flow creates a forebulge.

or lines of equal uplift, can be used to infer the mass distribution of the now-vanished ice and the magnitude and orientation of the uplift field. Differential postglacial uplift can change the appearance of contemporary landscapes. For example, greater uplift in the northern Champlain Valley of Vermont and New York caused the Lake Champlain Basin to tilt southward after deglaciation, drowning river valleys in the southern part of the basin [**Figure 13.13**].

Climatic Control of Mountain Topography

One of the most striking examples of climate-landscape interactions on the scale of mountain ranges has been called the **glacial buzzsaw** [**Figure 13.14**]. In the Andes, the mountains that run along the west coast of South America, latitudinal trends in the **snowline** (the elevation above which snow survives through the summer, an approximation of the ELA) track variations in the height of the range. Such correspondence in this and other mountain ranges where peak heights rise a limited distance above the ELA has been interpreted as implying that the elevation of mountain ranges is kept in check by glacial erosion—the glacial buzzsaw cutting away at rock as valleys are excavated and cirque headwalls cut back into the range. In other words, glacial erosion is able to match the rate of rock uplift.

The amount of sediment shed from a continent can influence the behavior of a subduction zone and the shape of the resulting trench and mountain range. For example, the trench off the west coast of South America varies in depth depending on the amount of sedimentary fill originating from the uplifted western margin of South America. Where the climate is wetter, sediment yield is higher, more sediment loads the oceanic crust, and the trench is filled and bathymetrically shallow. Where the climate is very arid, less sediment is delivered to the trench, which as a result is less filled and deeper. Not only is the trench morphology affected by sediment yield but so is the morphology of the mountains. The Andes are wider where the climate is more arid. There are likely two reasons that the mountains are larger where it is drier. First, erosion rates are less in the hyperarid Atacama Desert than in wetter parts of the range. Second, the lack of soft, easily deformable sediment on the downgoing slab increases the shear stress between the downgoing plate and the oceanic crust; thus, the plate boundary, because it is stronger, can support the load of more massive mountains.

Climate Change Effects

Glacial-interglacial climate changes are large enough and long enough to cause considerable geomorphic effects

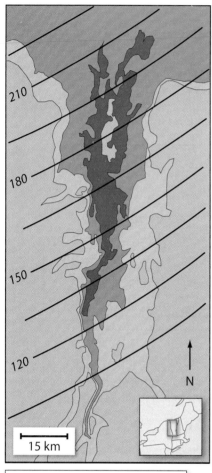

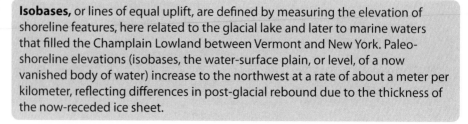

Isobases, or lines of equal uplift, are defined by measuring the elevation of shoreline features, here related to the glacial lake and later to marine waters that filled the Champlain Lowland between Vermont and New York. Paleo-shoreline elevations (isobases, the water-surface plain, or level, of a now vanished body of water) increase to the northwest at a rate of about a meter per kilometer, reflecting differences in post-glacial rebound due to the thickness of the now-receded ice sheet.

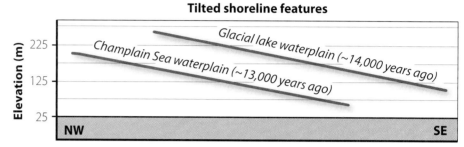

These cross sections show shoreline features, such as beach ridges and deltas, that were once horizontal but are now at different elevations due to differential post-glacial uplift. These features, deposited at the same time in the glacial lake that once filled the Champlain Lowland, are now more than 100 m higher in the north than those farthest south. Not only is the paleoglacial lake water-surface plain tilted to the northwest but so are landforms related to the waterplain of the Champlain Sea that filled the Champlain Basin when eustatic sea level rose enough that ocean water, pouring through the Saint Lawrence River valley, flooded the Champlain Lowland replacing the glacial lake. Several thousand years of continued isostatic uplift then caused local sea-level fall and modern freshwater Lake Champlain was born about 10,000 years ago.

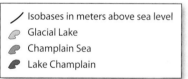

FIGURE 13.13 Uplift Triggered by Deglaciation. Mapping uplifted geomorphic features such as deltas deposited into now-vanished glacial lakes allows definition of isobases, contours of equal isostatic uplift. [Adapted from Chapman (1937).]

over much of Earth. However, geomorphic responses to changes in climate are not everywhere the same. Recent compilations of dozens of local temperature records indicate that temperature changes during and after the last glacial cycle were both broadly synchronous (peak of the last glacial maximum, ~22,000 years ago; peak of the Altithermal or the mid-Holocene warm period, ~7700 years ago) and similar in the Northern Hemisphere and Southern Hemisphere.

Precipitation changes were much more variable. Some parts of the world became wetter and others became drier as the climate changed. Consider the pluvial lake systems of the Sahara and Mojave deserts. In the Mojave Desert, pluvial lakes expanded during glacial times. As glaciation ended, the lakes dried as temperatures increased and cloudiness and precipitation decreased. In the Sahara Desert, glacial times were dry but lakes there filled with water during the early to mid-Holocene (10,000 to 5000 years ago) in response to shifting storm patterns and the location of moisture-laden winds. It was during the wetter mid-Holocene that channels apparent in space shuttle radar images, and verified in the field by archaeologists, carried water across the now-parched Sahara Desert (Photograph 13.10).

Not only does climate change over time but the frequency of extreme events, and their intensity, also changes. Understanding how the distribution and character of extreme events change is important because some geomorphic processes (such as mass movements, gully initiation, and fluvial sediment transport) are driven not by average conditions but by extreme conditions. In many landscapes, only long-lasting or high-intensity storms can provide enough energy (through rainfall, wind, or runoff) to cross stability thresholds and initiate significant geomorphic activity (see Chapter 1).

Consider, for example, how the transport of large cobbles requires a critical shear stress provided by a deep-enough

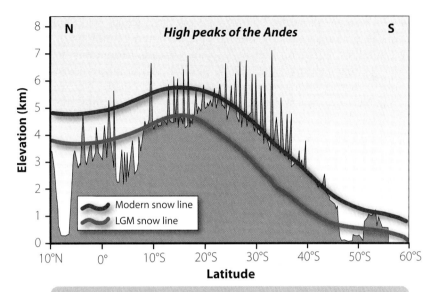

The overall topography of the South American high peaks tracks the last glacial maximum (LGM) **snow line** or ELA toward the equator and the modern snow line at higher latitudes. The correspondence between snow line and topography has been interpreted as erosive mountain glaciers controlling topography. The nearly 1 km displacement between the modern and LGM snow line reflects climate change (warming) between glacial and interglacial times.

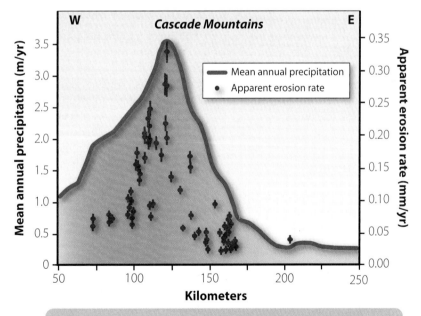

Rock erosion rates, inferred from **U/He dating** of apatite crystals in granitic rocks, are highest on the west slope of the Cascade Mountains of western Washington where the most precipitation (up to 3.5 m/yr) falls. Erosion rates are lowest in the rain shadow on the east slope of the range where less than half a meter of precipitation falls annually.

FIGURE 13.14 Climate and Erosion. Active glaciers remove rock and thus keep the average elevation of mountain ranges similar to the elevation of the glacial equilibrium line altitude. Erosion is focused on the windward side of mountain ranges where precipitation is greatest and the glaciers are largest and most active. [Adapted from Porter (1977) and from Reiners et al. (2003).]

and fast-enough flood flow (see Figures 6.6 and 6.7), and how the triggering of landslides requires critically high pore pressure, provided by rainfall of sufficient intensity and duration (see Figure 4.1). Other examples include landslides in the forested southern Appalachian Mountains or the well-vegetated highlands of New Zealand. Such failures occur in large numbers and region-wide only when very large hurricanes or typhoons strike. Consequently, there is great interest in refining long-term, geologic records of paleo-hurricanes, floods, and monsoons. These events can be so geomorphically effective that they change the rate of erosion and thus sediment yield over time.

The effects of climate change on landscapes depend strongly on the changes in active surface processes that a changing climate causes. Consider the effect of rising and falling sea level driven by warming and then cooling climate and the concomitant contraction and expansion of ice sheets. As ice sheets grow and sea level falls, the effect on a major river system entering the ocean is likely to be incision as potential energy increases; this incision could leave strath terraces along the channel and/or cause a knickzone to move upstream. When sea level begins to rise, knickzone retreat slows and the river aggrades, perhaps burying the strath terraces in alluvium. The process and the resulting landforms are very different when sea level falls and when it rises, an example of hysteresis. Second, consider the advance of a glacier when climate cools or precipitation increases. Such an advance obliterates landforms created by previous glaciations and then deposits a moraine at the glacial maximum. When the climate warms, the moraine is abandoned and a series of recessional moraines may form. These moraines slowly degrade as surface processes redistribute and remove mass. Warming and cooling climate result in very different surface process responses.

Landscape Controls on Climate

Climate-landscape interaction is not a one-way street. Surface processes, and the landforms they generate, influence climate by changing the planet's energy balance, hydrologic cycle, and atmosphere.

Regional Climate

The combined effect of dominant wind direction and topography on local and regional climate, and thus on geomorphology, is often pronounced. Because air masses ascending mountain ranges are forced to rise and thus cool **adiabatically** (by expansion but without heat transfer), they become saturated with water vapor, triggering **orographic** precipitation [Figure 13.15]. As a result, **windward** (upwind) slopes are much wetter than **leeward** (downwind) slopes. Such hydrologic asymmetry increases erosion rates on the windward side of mountain ranges through a combination of increased runoff, higher stream power, and more frequent mass movements. Increased snowfall lowers snow lines, which result in large, active glaciers. For example, long-term (millions of years) erosion rate data from the Cascade Mountains of Washington State are well correlated to the modern distribution of precipitation (Figure 13.14). On the wet west side of the range, long-term erosion rates and thus rock uplift rates are high. Erosion rates and rock uplift rates are lower on the drier, less rapidly eroded east side. Over geologic time, this can result in the development of different geologic structures on either side of a range, as the loss of mass by erosional processes drives an isostatic response that draws rock upward in addition to or in place of the push of tectonics.

On longer timescales, the arrangement of continents has influenced climate and the distribution of climate-sensitive landforms (Figure 13.5). For example, as Antarctica pulled away from South America and Australia, the circum-Antarctic current developed, thermally isolating the continent from the rest of the Southern Ocean, allowing the landmass to cool, and likely helping to trigger the onset of continental-scale glaciation in Antarctica about 30 million years ago. The growth of Antarctic ice sheets lowered sea level as water was sequestered and also increased planetary albedo as the area of highly reflective ice and snow expanded. Thus, the climate-changing effect of establishing the Antarctic ice sheet was felt worldwide.

Earth's Energy Balance

The most significant landscape-climate linkage is related to ice sheets and glaciers. Ice and snow have high albedos, reflecting much of the solar energy delivered to their surfaces. Thus, during glaciations, a larger percentage of incoming solar radiation is returned to space than during interglaciations. As glacial ice begins to melt and saturate with water, its albedo decreases, more solar energy is absorbed (accelerating melting), and less is reflected. Once the ice and snow melt away, revealing rock and soil or allowing the growth of dark green vegetation, albedo is further reduced and even more solar energy is retained on and near Earth's surface. This albedo decrease further warms the planet in a positive **feedback loop.** Climate models suggest that if the Greenland Ice Sheet melts away due to climate warming driven by increased greenhouse gases, it would not form anew even if greenhouse gas concentrations were reduced to preindustrial levels. Similarly, loss of sea-ice cover on the Arctic Ocean, another effect of global warming, will reduce albedo as dark seawater absorbs more sunlight than white snow and ice.

Hydrologic Cycling

Feedbacks link biologic, geomorphic, and hydrologic systems. For example, removing trees from the Amazonian rain forest has dramatically reduced transpiration. The loss of transpiration has reduced atmospheric water content and thus the volume of precipitation in the Amazon Basin. Because Amazonia is a partially closed system, with

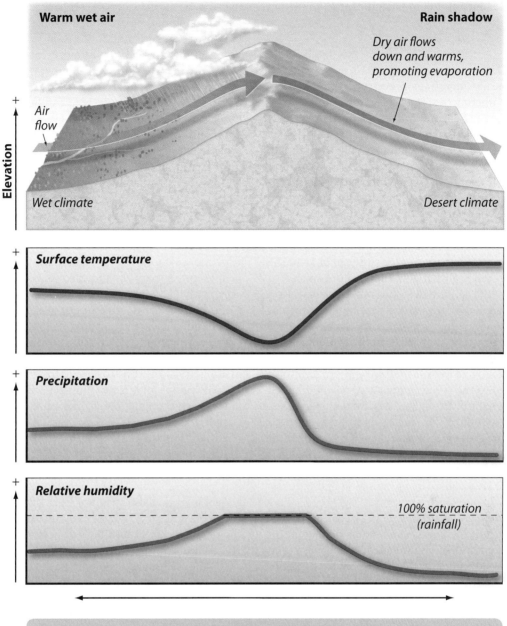

FIGURE 13.15 Orographic Effects on Climate. Mountains strongly affect local climate by forcing masses of air to rise and thus cool, leading to precipitation as water vapor in the rising air mass reaches saturation and condenses.

significant amounts of water recycling within the basin, it is unclear if sufficient water now exists to reestablish the full extent of the original rain forest. In some places, savanna encroachment is establishing a new regime for soil formation, bioturbation, and geomorphic activity.

Tectonically driven surface uplift of large areas, such as the Tibetan Plateau, can accentuate monsoonal circulation due to summer heating of the uplands and the air masses above them. Such rising air unleashes large volumes of precipitation in short periods, filling rivers in extreme runoff events and doing significant geomorphic work.

The Atmosphere

Two geomorphically important processes, volcanic emissions and the weathering of silicate rocks, alter atmospheric physical properties and chemistry. These important forcings affect climate on very different timescales.

Volcanic emissions can dramatically change climate over the short term, i.e., months to years. During large and explosive volcanic eruptions, such as those typical of subduction zone volcanism, particulate matter and sulfate aerosols (Figure 13.4) are injected into the upper atmosphere where they are carried around the world by high-level winds (the **jet stream**). This airborne material changes the energy balance of Earth's surface since small airborne particles reduce the transparency of the atmosphere to solar radiation, and sulfate aerosols provide nucleation sites for water vapor, increasing the rate of cloud formation. Because clouds are light-colored, they increase Earth's albedo, reflecting rather than adsorbing incoming solar radiation. Both effects tend to cool the planet as typified by the climatic effects of the great Tambora eruption (1815), which led to the "year without a summer" in much of the Northern Hemisphere. In northeastern North America, snow fell every month of the year. Such eruption-induced changes in climate are typically short-lived (on the order of months to years) because of the limited atmospheric residence time of both particulate matter and sulfate aerosols. Both are washed from the atmosphere by precipitation and particles settle out under the influence of gravity.

The climatic effects of massive flood basalt eruptions, such as those producing the Deccan Traps at the end of the Cretaceous or the Columbia Plateau flood basalts in the Miocene, last longer than those of short-lived, explosive eruptions of silicic volcanoes (Figure 13.5). Due to the volume of magma erupted, these long-duration, high-volume events can release large amounts of CO_2 gas, changing the chemistry and thus the radiation balance of the atmosphere. Because the processes that remove CO_2 from the atmosphere occur relatively slowly in comparison to precipitation scrubbing of particles, the greenhouse-induced climatic effect of high-volume eruptions is longer lasting.

Weathering of silicate minerals drives climate on much longer timescales than volcanic eruptions by slowly removing CO_2, a greenhouse gas, from the atmosphere (see Chapter 3). Such rock weathering is the source of dissolved loads in rivers, which carry cations and bicarbonate icons to the oceans where organisms make $CaCO_3$ shells from these components of carbonate minerals. Once locked up in mineral form, CO_2 is removed from the atmosphere and the greenhouse effect it causes is reduced.

Over the past 20 years, some scientists have argued that pulses of uplift, such as the growth of the Himalaya, have affected global climate by changing the global rate of rock weathering. They suggest that surface uplift steepens topographic gradients, increasing stream power, and causing incision. River incision leads to steeper slopes, more mass wasting, and more exposure of fresh, unweathered rock. If the global rock-weathering rate increases, this will draw down atmospheric CO_2 as more carbon is sequestered. Such weathering-catalyzed removal of CO_2 from the atmosphere is thought to have reduced the greenhouse effect over the Cenozoic era, cooling climate and thus exacerbating if not catalyzing glaciation in the Pliocene and Pleistocene (Figure 13.5). Removal of CO_2 by weathering reactions is a slow process and will not compensate, on human timescales, for the anthropogenic additions of CO_2 to the atmosphere.

Applications

Climatic geomorphology is of particular importance to people and society. In the face of continued, and likely accelerating, climate warming driven by emission of CO_2 from the burning of fossil fuels, it is important to understand and predict the response of landscapes to changing climate. Warming climate will cause geomorphic changes related to rising sea level in coastal regions as the result of melting ice sheets. Loss of permafrost will destabilize soils in cold regions and loss of alpine glaciers will reduce summer runoff, changing the geomorphology (and hydrology) of mountain channels.

It is informative to place previous climate changes in the context of predictions for the next century. Most climate models predict an increase in global average temperatures between 2–4°C by 2100 (about half the rise that typifies glacial-interglacial transitions). Sea levels are predicted to rise at least 20 to 60 cm by 2100 unless feedbacks increase the rate of ice sheet disintegration, in which case, sea-level rise could be much greater.

Analogs from the geologic past suggest that warming estimates from models are reasonable. Such analogs suggest that over the next hundreds to thousands of years, sea level will rise many meters as the Greenland and Antarctic ice sheets melt. For example, emission of CO_2 during the middle Miocene eruption of the Columbia River flood basalts raised atmospheric CO_2 levels only about half as much as human emissions have over the last century but resulted in a 3°C average warming of Earth. Because little water was locked up in polar ice caps, Miocene sea level was tens of meters higher than today.

Understanding the behavior of ice sheets and the speed at which they react to changing climate is critical, because the transfer of water from melting glaciers and ice sheets to the ocean will be one of the most devastating results of a warming climate. Accelerating sea-level rise will inundate low-lying areas, including many of the world's most important cities. Melting of the Greenland Ice Sheet in its entirety would raise global sea level by 6 to 7 m, enough to submerge many coastal cities, such as New York. Not only will cities be lost but deltas, estuaries, and their marshes, the nurseries for many important marine species and a buffer against storm surges, will disappear as rates of sea-level rise exceed rates of sedimentation and these coastal landforms vanish beneath the waves of a rising sea.

As both alpine glaciers and polar ice sheets melt, sea level will rise as it has before, most recently at the termination of the last glaciation, starting about 22,000 years ago. However, over the past several million years, sea level was rarely higher than it is today. Only during a few particularly warm or long interglaciations, such as marine oxygen isotope stages (MIS) 5e, 9, and 11, did the Greenland and/or the Antarctic ice sheets melt back more than they have during the Holocene. There is evidence, some of it ambiguous and still controversial, that during at least one interglaciation (MIS 11, about 420,000 years ago) sea level was as much as 12 m higher than it is today. Did this extreme sea-level rise reflect the disappearance of the Greenland Ice Sheet? Perhaps. There is evidence in and below the ice, including pollen indicative of boreal forests, for a much-reduced Greenland Ice Sheet during MIS 5e, 9, and/or 11.

Many feedbacks associated with global warming have the potential to amplify initial climate changes. For example, the melting of both sea ice and polar ice sheets reduces planetary albedo, changing Earth's energy balance. Other positive feedbacks include the release of methane (a powerful greenhouse gas) from melting permafrost as arctic soils warm. Large amounts of methane are also stored in sediment on the continental shelf in the form of **clathrates**, complex mixtures of ice and organic molecules unstable at temperatures above 4°C. Warming seawater will catalyze thermal decomposition of these clathrates, causing release of this methane and weakening of the sediment in which it is contained. Not only will such releases accentuate the greenhouse affect and climate warming but they could also trigger giant submarine landslides with the potential to generate devastating tsunamis.

On a warming planet, the distribution of atmospheric water vapor will change. The hydrologic cycle will become more active because at higher temperatures the atmosphere can hold more water (the saturation vapor pressure of water increases with temperature). Speeding the hydrologic cycle is likely to increase both the rate and frequency of precipitation, and thus runoff, in many areas. More water in a warmer atmosphere will likely lead to stronger and longer duration storms—conditions conducive to mass movements and debris flows. If discharge increases, streams are likely to incise and widen, changing their courses.

Models suggest that the extra water will not be evenly distributed. Some parts of the planet may get wetter, such as northeastern North America; other parts, such as Australia and the middle of North America, are predicted to become drier. In areas that become drier, aeolian activity is likely to increase. Once-stable dunes may reactivate as vegetation dies back. In a hotter and at least seasonally drier climate, the increased size and severity of wildfires will be an important driver of hillslope erosion and debris flow generation. Predicting the nature and extent of landscape response to climate change presents a challenge that draws on geomorphological insights and expertise.

Selected References and Further Reading

Alley, R. B. Abrupt climate change. *Scientific American* 291 (2004): 62–69.

Anderson, R. Y., B. D. Allen, and K. M. Menking. Geomorphic expression of abrupt climate change in southwestern North America at the glacial termination. *Quaternary Research* 57 (2002): 371–381.

Ballantyne, C. K. Paraglacial geomorphology. *Quaternary Science Reviews* 21 (2002): 1935–2017.

Bond, G., et al. Persistent solar influence on North Atlantic climate during the Holocene. *Science* 294 (2001): 2130–2136.

Broecker, W. S., and G. H. Denton. The role of ocean-atmosphere reorganizations in glacial cycles. *Quaternary Science Reviews* 9 (1990): 305–341.

Broecker, W. S., and G. H. Denton. What drives glacial cycles? *Scientific American* 262 (January 1990): 48–56.

Brozovic, N., D. W. Burbank, and A. J. Meigs. Climatic limits on landscape development in the northwestern Himalaya. *Science* 276 (1997): 571–574.

Bull, W. B. *Geomorphic Responses to Climatic Change*. New York: Oxford University Press, 1991.

Bull, W. B., and A. P. Schick. Impact of climate change on an arid region watershed: Nahal Yael, southern Israel. *Quaternary Research* 11 (1979): 153–171.

Burnett, B. N., G. A. Meyer, and L. D. McFadden. Aspect-related microclimatic influences on slope forms and processes, northeastern Arizona. *Journal of Geophysical Research* 113 (2008): F03002 doi:10.1029/2007JF000789.

Chapman, D. Late-glacial and postglacial history of the Champlain Valley. *American Journal of Science* 34 (1937): 90–125.

Church, M., and J. M. Ryder. Paraglacial sedimentation: A consideration of fluvial processes conditioned by glaciation. *Geological Society of America Bulletin* 83 (1972): 3059–3072.

Clark, J., W. Farrell, and W. Peltier. Global changes in postglacial sea level: A numerical calculation. *Quaternary Research* 9 (1978): 265–287.

Clark, P. U., S. J. Marshall, G. K. C. Clarke, et al. Freshwater forcing of abrupt climate change during the last glaciation. *Science* 293 (2001): 283–287.

Dansgaard, W., S. J. Johnsen, H. B. Clausen, et al. Evidence for general instability of past climate from a 250-kyr ice-core record. *Nature* 364 (1993): 218–220.

Delcourt, P. A., and H. R. Delcourt. Late Quaternary paleoclimates and biotic responses in eastern North America and the western North Atlantic Ocean. *Paleogeography, Paleoclimatology, Paleoecology* 48 (1984): 263–284.

Dunne, T. Stochastic aspects of the relations between climate, hydrology, and landform evolution. *Transactions of the Japanese Geomorphological Union* 12 (1991): 1–24.

Enghold, D. L., S. B. Nielsen, V. K. Pedersen, and J.-E. Lesemann. Glacial effects limiting mountain height. *Nature* 460 (2009): 884–887.

Fagan, B. *The Long Summer: How Climate Changed Civilization*. New York: Basic Books, 2004.

Funder, S., N. Abrahamsen, O. Bennike, and R. W. Feyling-Hanssen. Forested Arctic: Evidence from North Greenland. *Geology* 13 (1985): 542–546.

Gomez, B., L. Carter, N. A. Trustrum, et al. El Niño-Southern Oscillation signal associated with middle Holocene climate change in intercorrelated terrestrial and marine sediment cores, North Island, New Zealand. *Geology* 32 (2004): 653–656.

Hallet, B., L. Hunter, and J. Bogen. Rates of erosion and sediment evacuation by glaciers: A review of field data and their implications. *Global and Planetary Change* 12 (1996): 213–235.

Harris, S. E., and A. C. Mix. Climate and tectonic influences on continental erosion of tropical South America, 0-13 Ma. *Geology* 30 (2002): 447–450.

Hartshorn, K., N. Hovius, W. B. Dade, and R. L. Slingerland. Climate-driven bedrock incision in an active mountain belt. *Science* 297 (2002): 2036–2038.

Hebbeln, D., F. Lamy, M. Mohtadi, and H. Echtler. Tracing the impact of glacial-interglacial climate variability on erosion of the southern Andes. *Geology* 35 (2007): 131–134.

Huntington, K. W., A. E. Blythe, and K. V. Hodges. Climate change and Late Pliocene acceleration of erosion in the Himalaya. *Earth and Planetary Science Letters* 252 (2006): 107–118.

Intergovernmental Panel on Climate Change (IPCC). *The Physical Science Basis*. Cambridge, UK: Cambridge University Press, 2007.

Kidder, D. L., and T. R. Worsley. A human-induced hothouse climate. *GSA Today* 22 (2012): 4–11 doi: 10.1130/G131A.1.

Lamb, S., and P. Davis. Cenozoic climate change as a possible cause for the rise of the Andes. *Nature* 425 (2003): 792–797.

Langbein, W. B., and S. A. Schumm. Yield of sediment in relation to mean annual precipitation: *EOS, American Geophysical Union Transactions* 39 (1958): 1076–1084.

Lisiecki, L. E., and M. E. Raymo. Plio-Pleistocene climate evolution: Trends and transitions in glacial cycle dynamics. *Quaternary Science Reviews* 26 (2007): 56–69.

Mayewski, P. A. Holocene climate variability. *Quaternary Research* 62 (2004): 243–255.

Meyer, G. A., S. G. Wells, and A. J. T. Jull. Fire and alluvial chronology in Yellowstone National Park: Climatic and intrinsic controls on Holocene geomorphic processes. *Geological Society of America Bulletin* 107 (1995): 1211–1230.

Mitchell, S. G., and D. R. Montgomery. Influence of a glacial buzzsaw on the height and morphology of the Cascade Range in central Washington State, USA. *Quaternary Research* 65 (2006): 96–107.

Molnar, P. Climate change, flooding in arid environments, and erosion rates. *Geology* 29 (2001): 1071–1074.

Molnar, P., and P. England. Late Cenozoic uplift of mountain ranges and global climate change: Chicken or egg? *Nature* 346 (1990): 29–34.

Montgomery, D. R., G. Balco, and S. D. Willett. Climate, tectonics, and the morphology of the Andes. *Geology* 29 (2001): 579–582.

Noren, A. J., P. R. Bierman, E. J. Steig, et al. Millennial-scale storminess variability in the northeastern United States during the Holocene epoch. *Nature* 419 (2002): 821–824.

Orombelli, G. Climatic records in ice cores. *Terra Antarctica* 3 (1996): 63–75.

Peel, M. C., B. L. Finlayson, and T. A. McMahon. Updated world map of the Köppen-Geiger climate classification. *Hydrology and Earth System Sciences* 11 (2007): 1633–1644.

Peizhen, Z., P. Molnar, and W. R. Downs. Increased sedimentation rates and grain sizes 2-4 Myr ago due to the influence of climate change on erosion rates. *Nature* 410 (2001): 891–897.

Peteet, D., D. Rind, and G. Kukla. "Wisconsinan ice-sheet initiation: Milankovitch forcing, paleoclimatic data, and global climate modeling." In P. U. Clark and P. D. Lea, eds., *The Last Interglacial-Glacial Transition in North America*. Boulder, CO: Geological Society of America Special Paper 270 (1992): 53–69.

Pollack, H. N., and S. Huang. Climate reconstruction from subsurface temperatures. *Annual Review of Earth and Planetary Sciences* 28 (2000): 339–365.

Portenga, E., and P. R. Bierman. Understanding Earth's eroding surface with ^{10}Be: *GSA Today* 21 (2011): 4–10.

Porter, S. C. The glacial ages: Late Cenozoic climatic variability on several time scales. In *Man Nature—Makers of Climatic Variation*. College Station: Texas A & M University, 1977.

Pratt, B., D. W. Burbank, A. M. Heimsath, and T. Ojha. Impulsive alluviation during early Holocene strengthened monsoons, central Nepal Himalaya. *Geology* 30 (2002): 911–914.

Raymo, M. E., L. E. Lisiecki, and K. H. Nisancioglu. Plio-Pleistocene ice volume, Antarctic climate, and the global δ^{18}O Record. *Science* 313 (2006): 492–495.

Raymo, M. E., W. F. Ruddiman, and P. N. Froelich. Influence of late Cenozoic mountain building on ocean geochemical cycles. *Geology* 16 (1988): 649–653.

Reiners, P. W., T. A. Ehlers, S. G. Mitchell, and D. R. Montgomery. Coupled spatial variations in precipitation and long-term erosion rates across the Washington Cascades. *Nature* 426 (2003): 645–647.

Sailer, R., H. Kerschner, and A. Heller. Three-dimensional reconstruction of Younger Dryas glaciers with a raster-based GIS. *Glacial Geology and Geomorphology* (1999), rp01.

Shakun, J. D., and A. E. Carlson. A global perspective on Last Glacial Maximum to Holocene climate change. *Quaternary Science Reviews* 29 (2010): 1801–1816.

Siegenthaler, U., T. F. Stocker, E. Monnin, et al. Stable carbon cycle-climate relationship during the Late Pleistocene. *Science* 310 (2005): 1313–1317.

Sonett, C. P., and H. E. Suess. Correlation of bristlecone pine ring widths with atmospheric ^{14}C variations: A climate-sun relation. *Nature* 307 (1984): 141–143.

Steig, E., E. Brook, J. W. C. White, et al. Synchronous climate changes in Antarctica and the North Atlantic. *Science* 282 (1998): 92–95.

Ten Brink, N. W. Glacio-isostasy: New data from west Greenland and geophysical implications. *Geological Society of America Bulletin* 85 (1974): 219–228.

Whipple, K. X., E. Kirby, and S. H. Brocklehurst. Geomorphic limits to climate-induced increases in topographic relief. *Nature* 401 (1999): 39–43.

Wolman, M. G., and R. Gerson. Relative scales of time and effectiveness of climate in watershed geomorphology. *Earth Surface Processes and Landforms* 3 (1978): 189–208.

Zachos, J., M. Pagani, L. Sloan, E. Thomas, and K. Billups. Trends, rhythms, and aberrations in global climate 65 Ma to present. *Science* 292 (2001): 686–693.

DIGGING DEEPER Do Climate-Driven Giant Floods Do Significant Geomorphic Work?

Many people think of climate change as the ups and downs of average global temperature over time and thus the slow comings and goings of glaciers or the gradual wetting and drying of land outside the glacial margin. But there is a long-standing argument that the geomorphic effects of climate change can be much more dynamic, even catastrophic. The first hints of sudden and massive climate-induced geomorphic change came about a century ago, the result of detailed fieldwork by one man in the semi-arid plains of eastern Washington.

Using horses and primitive automobiles, a young geologist, J Harlen Bretz, traversed the rugged, loess-covered basalt terrain of eastern Washington. He was fascinated by the deep canyons and large deposits of gravel and river sediment at locations inconsistent with the size and course of the Columbia River that flows through the area today. Massive dry waterfalls and plunge pools were fed by empty channels where the loess was stripped away and the basalt deeply potholed by long-vanished rivers [**Figure DD13.1**]. Current ripples made of gravel were up to 15 m high and filled valley bottoms. Immense boulders were deposited in broad fan-form features where the flow diverged. All of these features were evidence implying massive flows of water [**Figure DD13.2**].

Bretz thought these features were best explained by a giant flood and coined the term **Channeled Scabland** to describe the landscape they left behind (Bretz, 1923). The flood he envisioned was so large, he termed it a "debacle." But, without being able to identify a source of the floodwaters and because his hypothesis invoked a catastrophic

(a)

(b)

FIGURE DD13.2 Oversize geomorphic evidence shows the magnitude of the floods that swept over the Channeled Scablands. (a) Boulders many times the size of a field vehicle were moved kilometers by floodwaters onto the Ephrata fan in eastern Washington. (b) Giant current ripples cover the top of a massive gravel bar in eastern Washington. Roads and a grain silo provide scale. [From Baker (2009).]

event for which there was no modern analog, Bretz's ideas were ridiculed for decades by most geologists who held to an overly restrictive interpretation of **uniformitarianism**— that processes operating at typical modern rates must be able to explain all past landscape features.

Then, in 1942, Pardee published a paper entitled "Unusual currents in glacial Lake Missoula," describing glacial Lake Missoula as the source of Bretz's flood (Pardee, 1942). There, ~2500 km³ of water had been impounded by

FIGURE DD13.1 Dry Falls is the result of the great Missoula floods pouring across the Channeled Scablands of eastern Washington. Floodwaters moved left to right pouring over the basalt cliffs and eroding the deep plunge pools and sinuous channels, here filled with standing water.

> **DIGGING DEEPER** Do Climate-Driven Giant Floods Do Significant Geomorphic Work? *(continued)*

a tongue of glacial ice that blocked the Clark Fork River. Distinctive megaripples on the ancient lakebed showed that the ice dam failed, rapidly draining the lake and providing a source for Bretz's flood. Analysis of maps and newly available air photographs sealed the case and allowed the routes of the floods to be determined in detail [**Figure DD13.3**]. Careful examination of backwater sediment, deposited in tributary valleys of the Columbia River, showed evidence that the scabland was flooded not once as Bretz had thought, but many times. In the late Pleistocene alone, over 40 massive floods poured down the Columbia as the ice dam repeatedly breached and reformed (Waitt, 1980).

The idea of massive floods as significant agents of geomorphic change has gained traction over the past several decades but there are still controversial flood hypotheses. For example, Shaw and Gilbert's analysis of the morphology and orientation of several North American drumlin fields (Shaw and Gilbert, 1990) suggested to them that some drumlins were shaped by massive subglacial floods. They posited that such floods likely resulted from the release of meltwater stored subglacially and that subglacial floods were a glacial-hydrologic response driven by a warming climate. Calculations suggest that these floods would have raised sea level >20 cm in days to weeks and up to a few meters over several years (Shaw, 1989).

However, many researchers (e.g., Benn and Evans, 2010) question the subglacial flood hypothesis for the origin of drumlins, and the hypothesis remains untested (and may be untestable).

It is indisputable that large episodic floods of glacial meltwater have affected the world's oceans, perturbing circulation, and thus further changing climate (Clark et al., 2001). For example, abrupt changes in ocean water oxygen isotope composition after the last glacial maximum are consistent with large, episodic inputs of isotopically light (^{18}O-depleted) glacial meltwater [**Figure DD13.4**]. Floods of glacial meltwater into the North Atlantic likely caused the region-wide Younger Dryas cooling event from 12,800 to 11,500 years ago (Broecker, 2006). The climatic effects of major glacial-lake floods continued well into the Holocene. About 8400 years ago, as the last section of the Laurentide Ice sheet melted away, it released over 100,000 km^3 of water from proglacial Lake Agassiz into Hudson Strait (Clarke et al., 2004). For nearly 200 years after this discharge, the ice in cores collected from the Greenland Ice Sheet shows a 5-degree drop in average temperature over central Greenland and increased Northern Hemisphere dustiness from cooling and drying of the landscape.

Nearly 100 years after Bretz first suggested that megafloods have major geomorphic impacts, the data now

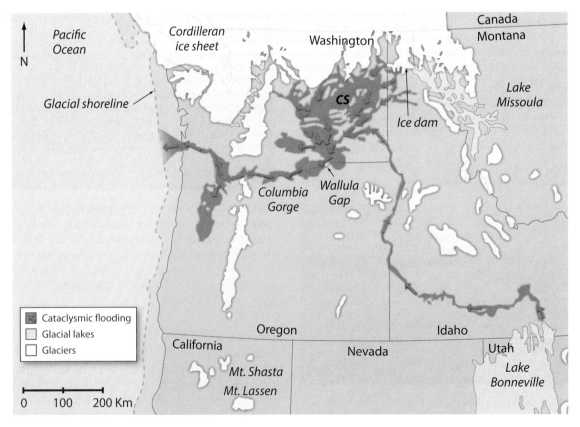

FIGURE DD13.3 The Channeled Scablands (CS) occupy much of eastern Washington State. This map shows ice-dammed Lake Missoula and pluvial Lake Bonneville in light blue, ice sheets and ice caps in white, and the scabland channels in dark blue with flow directions shown by red arrows. [From Baker (2009).]

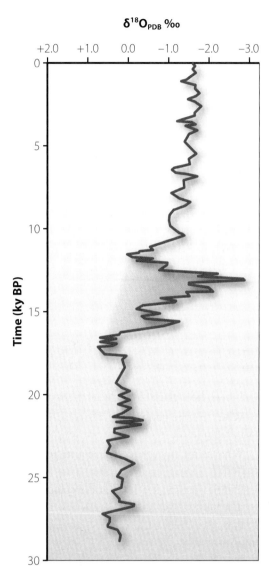

FIGURE DD13.4 Record of the stable oxygen isotopic composition of seawater in the Gulf of Mexico created by analyzing foraminifera, the fossils of small marine organisms isolated from marine sediment cores. The large isotopic excursions (to more depleted values) between 17,000 and 12,000 years ago (shaded area of the graph) are thought to represent pulses of glacial meltwater pouring down the Mississippi River. [From Shaw (1989).]

show that megafloods were widespread around glacial margins, including such places as Siberia and Alaska. Bretz's thinking has also greatly influenced the interpretation of surface features on other planets. For example, using Earth as an analog, there is now consensus that the streamlined islands and channels on Mars formed billions of years ago in giant Martian outburst floods (Baker, 1978).

Baker, V. R. The Channeled Scabland—A retrospective. *Annual Review of Earth and Planetary Sciences* 37 (2009): 6.1–6.19.

Baker, V. R. The Spokane Flood controversy and the Martian outflow channels. *Science* 202 (1978): 1249–1256.

Benn, D. I., and D. J. Evans. *Glaciers and Glaciation*. London: Hodder Education, 2010.

Bretz, J H. The Channeled Scabland of the Columbia Plateau. *Journal of Geology* 31 (1923): 617–649.

Broecker, W. S. Abrupt climate change revisited. *Global and Planetary Change* 54 (2006): 211–215.

Clark, P. U., S. J. Marshall, G. K. C. Clarke, et al. Freshwater forcing of abrupt climate change during the last glaciation. *Science* 293 (2001): 283–287.

Clarke, G. K. C., D. W. Leverington, J. T. Teller, and A. S. Dyke. Paleohydraulics of the last outburst flood from glacial Lake Agassiz and the 8200 BP cold event. *Quaternary Science Reviews* 23 (2004): 389–407.

Pardee, J. T. Unusual currents in glacial Lake Missoula. *Geological Society of America Bulletin* 53 (1942): 1569–1600.

Shaw, J. Drumlins, subglacial meltwater floods, and ocean responses. *Geology* 17 (1989): 853–856.

Shaw, J., and R. Gilbert. Evidence for large-scale subglacial meltwater flood events in southern Ontario and northern New York State. *Geology* 18 (1990): 1169–1172.

Waitt, R. B. About forty last-glacial Lake Missoula jokulhlaups through southern Washington. *Journal of Geology* 88 (1980): 653–679.

WORKED PROBLEM

The Intergovernmental Panel on Climate Change as well as climate modelers have made predictions about how Earth's climate will change over the next 100 years. These predictions are based on assumptions about the amount of carbon dioxide and other greenhouse gases that humans will add to the atmosphere and take into account a variety of landscape, atmospheric, and oceanic feedback mechanisms.

Question: Consider the geomorphic impact of the following climate change predictions in terms of hillslope and river processes including the effects of vegetation.

1. As an environmental planner in the humid, temperate northeastern United States, describe the geomorphic impact of a warming climate in which winter snowpacks and river ice are reduced, the frequency and extent of summer drought is increased, mean annual

precipitation goes up, hurricane frequency and intensity increase, and the number of exceptionally hot days doubles.
2. As a dry-land rancher in central Australia where the climate is semi-arid, you face predictions of less annual rainfall, longer droughts, and higher summer temperatures. What changes are likely to happen to the landscape you farm?

Answer: In the humid northeast, climate change will cause a variety of impacts to both hillslopes and rivers. Thinner snowpacks and warmer winters will shift the timing of spring floods earlier and shorten their duration, and will increase the frequency of rain-on-snow flooding. Summer drought and more heat waves will reduce stream and river baseflow during summer months, impacting aquatic organisms and the ability of farmers to take water for irrigation. Heavier rainfalls and stronger hurricanes may cause more frequent and extensive landslides as precipitation intensity-duration thresholds for slope stability are crossed more often. Higher river flows during storms may move bedload and erode banks that would otherwise be stable, causing channel change. Changes in vegetation amount and character as the result of droughts could also influence hillslope and channel dynamics through the effect of roots on soil strength.

In the dry lands of central Australia, rising temperatures and decreased rainfall will reduce already scant vegetation, limiting the number of animals that can be grazed and potentially destabilizing hillslopes and valley bottoms when heavy rains do come. Baseflow and hyperheic flow in streams will diminish, reducing the number and density of riparian plants including trees that stabilize the banks. In the absence of root reinforcement from vegetation and the bioturbation caused by plants, these rains are likely to cause more surface erosion, creating rills and gullies as precipitation intensity exceeds infiltration rates. Eroded material will move downslope and accumulate on valley bottoms, particularly if flows in main channels are discontinuous and unable to move sediment far from its source.

KNOWLEDGE ASSESSMENT Chapter 13

1. Explain how climate differs from weather.
2. Provide two examples of how geomorphologists create records of paleostorms.
3. Define climatic geomorphology.
4. Give two examples of how ocean circulation affects terrestrial climate and explain how each works.
5. Predict and explain the effect that surface uplift of a mountain range will have on the distribution of precipitation across and downwind of the mountains.
6. Give an example of how lake sediments are used to document climate change over time.
7. Define "ice-rafted debris" and explain its origin.
8. Explain how oxygen isotopes can be used to infer paleoclimatic changes using two different geologic archives.
9. What is the relationship in ice cores between the concentration of CO_2 and paleotemperature?
10. Define loess and explain where you would be likely to find it.
11. Predict the response of climate if rock weathering rates increase.
12. Explain how volcanic eruptions can influence climate.
13. What is the physical explanation for how erosion of preglacial regolith may have changed the behavior of ice sheets and the cyclicity of glaciations?
14. What are the three primary Milankovitch cycles and together how are they thought to control climate change over time?
15. Explain how glacial-interglacial cycles affect sea level.
16. What are the geomorphic effects, both near the coast and inland, of changing sea level over time?
17. What is an emergence curve and how is it constructed?
18. Define eustatic sea level and explain how and why it changes over time.
19. Give examples of how El Niño is important geomorphically.
20. What caused lakes to form in arid southwestern North America during the last glaciation?
21. How did the elevation of Earth's surface change under and just outside of Pleistocene ice sheets when they were fully advanced?
22. Explain how geomorphic evidence can be used to deduce the magnitude of glacial isostatic response.
23. Explain how the location of glacial moraines is used to estimate the change in ELA over time.
24. Give an example of the linkage between biologic, geomorphic, and hydrologic systems.
25. Compare the magnitude of climate changes predicted for the next 100 years due to human-induced global warming with those that occurred between glacial and interglacial times.
26. Describe a positive feedback that takes place when the Earth warms.
27. Who was J Harlen Bretz and how did he revolutionize geologic thinking?

Landscape Evolution

14

Introduction

The geomorphological processes that shape landscapes and drive landscape evolution reflect the interaction over time of tectonic, climatic, topographic, geologic, and biotic characteristics and processes. Earth's landscapes are the result of a continuous long-running experiment in which both the starting point (initial conditions), and the subsequent changes in external drivers (boundary conditions) are at best only partially known. Through the concepts and principles discussed in previous chapters, geomorphologists decipher how a landscape formed, unravel the history it records, and discern the relevant surface processes. Such knowledge can help explain the distribution and state of soil and water resources, identify and predict the effects of human actions and land use, guide efforts to mitigate environmental impacts, and link ecosystem conditions and characteristics to climatic, hydrologic, and geologic drivers.

Landscapes evolve over many spatial and temporal scales, from the carving of shallow rills into the bare soil of a denuded hillside during a single rainstorm to the growth and decay of entire mountain ranges through geologic time. Across all scales, understanding landscape evolution requires measuring topographic change and identifying dominant surface processes before applying geomorphological models to interpret and understand the behavior of specific landscapes. Depending on the problem one is trying to address, landscape evolution needs to be considered over different temporal and spatial scales. An agricultural scientist cares most about sediment transport at plot scales over timescales of hours to days. Geomorphologists studying the evolution of mountain ranges focus on broader

Horseshoe Bend, a meander of the Colorado River deeply incised into bedrock downstream of Glen Canyon Dam. Note the vegetated point bar immediately above river level and the strath terrace partway up the bedrock valley wall. What can you read about the history of this landscape from these features?

IN THIS CHAPTER

Introduction
Factors of Landscape Evolution
   Tectonics
   Climate
   Topography
   Geology
   Biology
Models of Landscape Evolution
   Conceptual Models
   Physical Models
   Mathematical Models

Landscape Types
   Steady-State Landscapes
   Transient Landscapes
   Relict and Ancient Landscapes
   Basin Hypsometry and Landscape Form
Rates of Landscape Processes
   Uplift Rates
   Erosion Rates
   Spatial and Temporal Variability

Applications
Selected References and Further Reading
Digging Deeper: Is This Landscape in Steady State?
Worked Problem
Knowledge Assessment

spatial scales and longer time intervals, measured in thousands to millions of years, as they investigate how landscapes evolve toward forms that dominantly reflect erosional processes—as long as tectonic uplift is sustained. When an event perturbs or changes landscape-forming processes, the form of the land may retain evidence of the change for some time and provide a morphologic record of landscape evolution in the form of relict or inherited landscape features.

The history and evolution of landscapes determine the ability of Earth's surface to host and sustain human activity, and the pervasive influence of human actions on Earth surface processes provides a fundamental motivation for understanding landscape change over time. Deciphering the linkages among landscape processes is essential for properly assessing the effects of human activity on our environment and its capacity to sustain us. In this chapter, we discuss landscape evolution and address the conditions under which landforms reflect particular processes or the history of previous events.

Factors of Landscape Evolution

In 1941, Hans Jenny proposed his now-famous factors of soil formation to systematize thinking about the effects of environmental factors (climate, organisms, relief, parent material, and time) on the rate and style of soil development. Here, we define a similar set of key factors influencing landscape evolution. The importance of geologic, atmospheric, and biotic characteristics and processes in landform development makes it instructive to consider tectonics, climate, topography, geology, and biology as the basic factors governing landscape evolution [**Figure 14.1**]. Although each of these factors is useful in its own right for understanding landscape evolution, the interactions and feedbacks among them are important controls on landscape properties and evolution, shaping a range of characteristic landforms and distinctive landscapes.

Landscapes evolve in response to internal or external forcing over timescales from millennia to many millions of years. The appropriate timescale for landscape analysis varies with the spatial scale. A single hillside can respond over shorter timescales than an entire mountain range. Moreover, the dependent and independent variables influencing landscape evolution change over different timescales. Much like the distinction between (short-term) weather and (long-term) climate, geomorphological processes that appear **stochastic** (random) over short timescales may behave more predictably and even seem steady over long timescales.

Over short timescales, factors such as geologic structure and lithology behave as fixed influences (independent variables) to which geomorphological processes respond. However, over a sufficiently long (geologic) time frame, these factors are dependent variables that respond to changes in tectonic or topographic forcing. Consider, for

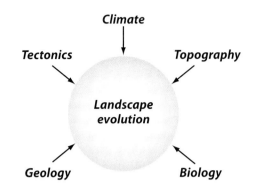

Landscape evolution reflects five factors, including the interaction of tectonics, climate, and topography along with the effects of geology and biology on Earth surface processes.

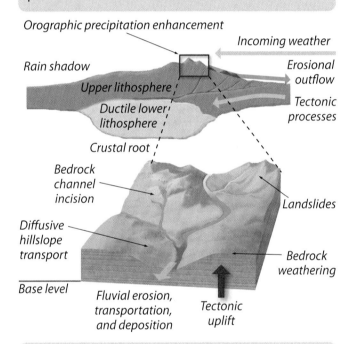

Tectonic uplift supplies mass to geomorphic systems. That mass is altered, transported, and removed by a variety of different surface processes.

FIGURE 14.1 Factors of Landscape Evolution. Tectonics, climate, topography, geology, and biology all contribute to the evolution of landscapes on Earth's surface.

example, the slope of a mountain river. In any given year, the river's slope is fixed and serves to drive flow velocities and rates of sediment transport. But considered over geologic time, the slope of a riverbed changes in response to rock uplift and erosion of the upland that it drains.

Tectonics

Surface uplift, the rate of change of landscape elevation relative to a fixed datum such as sea level, is a fundamental

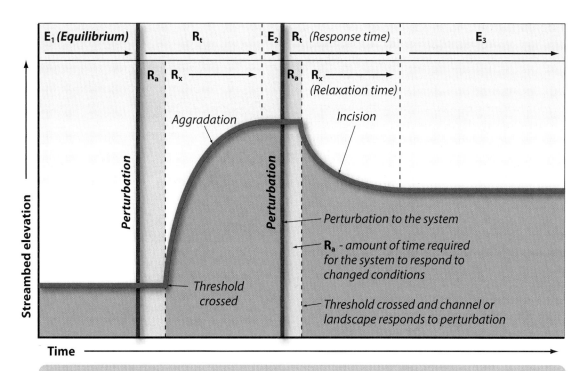

FIGURE 14.2 Landscape Response to Disturbance. Landscapes respond to disturbance with characteristic response and relaxation times. [Adapted from Bull (1991).]

control on landscape evolution. Tectonic uplift governs the size and physiographic character of mountain ranges that develop in convergence or rift zones. Changes in plate motion or tectonic setting influence landforms as tectonically driven increases or decreases in topographic slope propagate through a landscape. Topographic response to tectonic forcing may occur over an extended **relaxation time,** such as the prolonged isostatic response to deglaciation or the time it takes for a river profile to adjust to a change in the pace of tectonically driven rock uplift [**Figure 14.2**].

Rock uplift rates vary both locally and regionally in response to tectonic and erosional processes. On local scales, vertical offset across individual faults can create scarps or mountain fronts where one can readily identify sharp differences in rock-uplift rates and amounts. Tectonically active regions (like the Himalaya) are rugged mountains, whereas nearly featureless plains close to sea level dominate tectonically stable areas (like southern Australia). At the broadest scale, the association of mountains with active and ancient plate boundaries shows how tectonic setting dominates the style and pace of landscape change, even long after the driving plate motion ceases. Tectonic and isostatic processes determine the locations of dominantly depositional environments through the style and rate of crustal subsidence.

The uplift (or subsidence) of rock relative to **base level,** the level to which a landscape would erode given enough time, provides a fundamental control on landscape evolution. Base levels can be either global (sea level) or local. On a continental scale, global sea level acts as a base level for drainages entering the ocean. Over glacial–interglacial timescales, sea level has gone up and down by more than 100 m, alternately driving valley-bottom incision and aggradation in coastal landscapes. On a local scale, base level can be fixed for short times at a particular elevation, such as at the spillway of a landslide dam, the lip of a cirque lake, or the location of a resistant rock outcrop in a riverbed. Changes in base level directly affect the upstream erosional, depositional, and hillslope processes that drive landscape evolution.

Climate

Climate affects landscape evolution and geomorphic processes directly through the amount, intensity, and type of precipitation and indirectly through climate's effects on vegetation type and density. For example, the relative importance of chemical versus mechanical weathering depends on the climate. Chemical weathering processes are most important (given similar tectonic conditions) in hot, wet regions like the humid tropics, where chemical erosion rates (mass removal in solution) can approach 0.1 mm/yr. The development of deep-weathering profiles with thick saprolite greatly influences landscape evolution in warm, wet environments. Mechanical weathering tends to exert a dominant influence in environments subject to seasonal freeze-thaw cycles. Precipitation seasonality and intensity also matter. For example, intense monsoon rainfall leads to mechanical erosion in excess of 1 mm/yr in steep upland environments of Asia and the volcanic islands of the western Pacific (such as Java and the Philippines). The high plateaus of the Altiplano in the Andes and the Tibetan Plateau in the Himalaya lie astride the latitude-controlled global belt of deserts (between 30° and 35°; see Chapters 1 and 10) and reflect the ability of tectonically driven crustal convergence and uplift to outpace climatically limited erosion rates over geologic time.

Climate shapes the character of surface processes active in a landscape and greatly affects their intensity and thus the style and pace of landscape evolution. Changes in climate can leave behind inherited landscapes that can persist for as long as it takes to reshape Earth's surface and bring the topographic signature of landscape-forming processes into alignment with the contemporary climate. In areas with hard rocks, low surface gradients, and low rates of erosion, landscape features or characteristics formed under former climates can last for millions of years. Vegetation type and the presence of bedrock or soil-mantled hillslopes are strongly influenced by climate. Similarly, the amount of runoff from hillslopes, and thus the character of rivers, differ between arid and temperate regions (see Chapter 6).

Although the amount of precipitation falling onto a landscape is a general indicator of the pace and extent of active geomorphological processes driving landscape evolution, the type of precipitation (rain or snow) also matters. For example, it is common for only isolated peaks to rise more than a few hundred meters above the snowline or equilibrium line altitude in glaciated landscapes because rock elevated higher by tectonics is efficiently removed by glacial erosion (see Chapter 13). Fjord development is another example of the importance of global climate patterns on landscape evolution. Globally, fjords are best developed where the windward sides of mountain ranges rise from sea level because the high moisture flux, delivered as snow, feeds large, erosive glaciers [**Photograph 14.1**]. The global distribution of deserts where aeolian processes shape the land is likewise controlled by climate. In the past, intense winds blowing off great ice sheets and broad deserts shaped the global deposition pattern of thick loess that blankets the world's most productive agricultural regions in the U.S. Midwest, eastern Europe, and northern China (see Chapter 10).

Topography

Topography and relief control the erosional processes that, in turn, shape topography and give rise to characteristic forms, such as convex hillslopes and concave river profiles. Topography itself drives both the pace (tempo) and spatial pattern of geomorphological processes because relief, the elevation of rocks above base level, provides the potential

PHOTOGRAPH 14.1 Fjord. This fjord near Upernavik, central western Greenland, has steep walls tumbling away from low-relief highlands. Cosmogenic dating indicates that ice last occupied the ~2-km wide fjord about 11,000 years ago.

energy that fuels gravity-driven erosion. Consequently, the form and state of topography influence the future evolution of topography. Gentle slopes and the low potential-energy gradient of such slopes mean little downslope sediment transport by runoff or hillslope processes. Erosion generally increases as slopes steepen, until the development of threshold slopes limits further relief development (see Chapters 5 and 12).

The absolute elevation of topography influences climate and precipitation in ways that affect landscape evolution. Consider how the rise of a mountain range oriented perpendicular to the predominant wind direction leads to orographic precipitation on the windward side of the range and thus a wet-side, dry-side dichotomy that affects rates of erosion (see Figure 13.15). Persistent spatial gradients in the rate of erosion lead to differences in long-term exhumation rates, thereby influencing the development of geological structures. Topography can fundamentally influence hydrology when a range rises high enough that the mountains capture snow instead of rain, leading to glacial erosion and a different suite of landscape-forming processes than in warmer lowlands. Finer-scale topography also matters because it influences the formation and pattern of jointing and fractures, such as one sees in the exfoliation sheeting of the granite walls of Yosemite Valley [**Photograph 14.2**]. In addition, aspect is important in shaping some landscapes. At mid- to high latitudes, north-facing and south-facing slopes receive vastly different amounts of solar radiation. This asymmetry influences spatial patterns of freeze-thaw action and soil moisture that, in turn, affect plant communities, erosion rates, and slope steepness.

Topography itself helps determine the mix of processes acting to erode and transport sediment. The development of steep topography leads to a dominance of landslide processes and bedrock river systems. In contrast, subdued topography, such as that of ancient continental landscapes (cratons), leads to a slow pace of landscape change. The

PHOTOGRAPH 14.2 Exfoliation. The Royal Arches of Stoneman Meadow in Yosemite Valley, California, result from exfoliation along joints and are more than 300 m tall. As fractures developed in the cliff face, fragments of exfoliated rocks slid down the cliff.

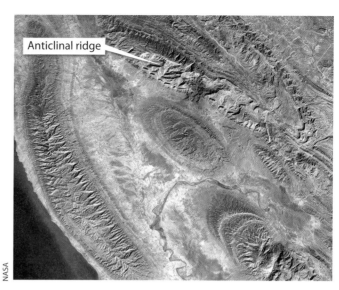

PHOTOGRAPH 14.3 Anticlinal Ridge. Anticlinal ridges, such as this 80-km wide one in the Zagros Mountains of Iran, are a hallmark of tectonic deformation.

ancient surfaces of low-relief cratons are characterized by hillslopes dominated by diffusion-like processes, alluvial rivers, internal drainage, dune fields at low latitudes, and ice sheets at high latitudes.

Geology

Earth materials influence landscape evolution through their physical properties, particularly erosional resistance (including the development of discontinuities and fractures), and how geologic structure imparts spatial patterns of erosion resistance. The cohesive strength and erodibility of different soils and rock types vary tremendously; loose, saturated sand cannot hold high and steep slopes, but solid, unfractured granite is capable of supporting kilometer-high cliffs.

The degree of fracturing influences bedrock erodibility. Highly sheared and fractured sedimentary rocks in tectonically active mountain ranges provide little erosion resistance because of low cohesion and the propensity for clay-rich rocks (e.g., mudstone) to weather mechanically by shrink/swell. In contrast, the relatively undeformed crystalline rock typical of deeply exhumed terrain is highly resistant to erosion due to high cohesion and limited fracturing. Whereas slopes formed on intact, unfractured bedrock tend to be relatively stable, those formed on weak, tectonically sheared rock, such as that found in much of the northern California Coast Ranges, experience widespread earthflows. Lithology and fracture development also influence the ability of rivers and streams to cut into bedrock (see Chapter 6).

Geologic structure imparts landscape-scale spatial patterns of erosion resistance (and thus slope) through the distribution of strong and weak rock units. Slopes in erosion-resistant units generally are steeper than those on highly erodible units. Geologic structures govern the development of topography either indirectly through the effect of differential bedrock erodibility or directly when rock uplift outpaces erosional processes, such as in the rise of anticlinal ridges like the Yakima folds in central Washington State or the Zagros Mountains in Iran [**Photograph 14.3**]. The joint system resulting from regional tectonic stress influences rock-strength properties and thus landscape evolution. Rocks will in general be weak where they are tectonically sheared, such as in and near fault zones. The resulting shear can make the regional rock strength **anisotropic** (unequal in different directions), causing the landscape to have a specific grain or a preferential orientation of ridges and valleys. In tectonically inactive terrain with variable lithology, topography often clearly reveals geological structure.

In high-relief, rapidly uplifting terrain, the topography generally bears little direct relationship to the underlying geological structures, because the development of steep threshold slopes close to the friction angle for fractured rock (see Chapter 5) obscures such distinctions and precludes inferring structure from topography. At the other extreme, extensive **erosion surfaces** or **pediments** are found in tectonically quiescent areas. These gently sloping topographic surfaces also truncate geological structures. At both high and very low rates of erosion, the underlying geologic structure exerts little influence on topographic form.

Biology

Vegetation is a key factor affecting hydrologic processes, erosion potential, and sediment yields. Thus, landscapes can be expected to exhibit significant response to changes in the type and age of vegetation, whether naturally or anthropogenically induced. The geomorphological signature of biota includes effects on a number of different processes that alter the shape of individual landforms, such as channels and hillslopes. Changes in vegetation communities can trigger significant geomorphic responses directly through changes in the binding effect of plant roots or indirectly through changes in evapotranspiration.

Biologically induced changes can affect both valley-bottom landforms and hillslopes, and different species of plants and animals influence landscape evolution across a wide range of scales. For example, beavers building dams turn valley bottoms into wetlands. Trees large enough to jam rivers and streams when they topple can locally block and divert flow, creating anastomosing, island-filled channel patterns. Likewise, the loss of root strength reduces both hillslope and streambank cohesion, and the ensuing erosion can alter river morphology from meandering to braided, allowing sediment to pour off hillslopes. Plants not only help hold soil in place, they help make it. Mechanical disturbance by the growing roots of vegetation accelerates conversion of rock to soil, especially in humid temperate environments with lush vegetation.

Today, the direct and indirect effects of human activity affect primary geomorphological process around the world. On the tropical island of Madagascar, portions of the landscape eroded in deep gullies (lavakas) after forest clearing exposed thick, deeply weathered, and highly erodible soils and saprolite to intense tropical rainfall [**Photograph 14.4**]. Although the rate of gullying increased after land clearance, some gullies predate human settlement, suggesting that gullying is a normal process of landscape evolution in this terrain. Geomorphic responses to vegetation change are not restricted to forested terrain. Overgrazing of grassland valley bottoms accelerated **arroyo** (Spanish for "small stream") entrenchment in California and across the U.S. Southwest [**Photograph 14.5**]; here, too, there is good evidence that some arroyos downcut in cycles for millennia prior to large-scale grazing. Land-use change increased the rate and area affected by incision.

Not all vegetation has the same effects on landscapes. The replacement of deep-rooted native grasses by shallow-rooted introduced species (non-native plants) led to increased landsliding in California and elsewhere. Landscape-scale disturbance from extensive plowing of the organic-rich grassland soils of the U.S. Midwest caused the catastrophic Dust Bowl during a severe drought in the 1930s when there was little root strength left to hold fine-grained soil in place during high winds [**Photograph 14.6**]. In case after case, human-caused changes in landscape-scale vegetation patterns have produced substantial geomorphic changes.

PHOTOGRAPH 14.5 Arroyo. An arroyo (gully) cut several meters deep through fine-grained sediment in El Paso County, Colorado.

PHOTOGRAPH 14.4 Lavakas. Lavakas, or deep amphitheater gullies, are common in some areas of Madagascar. Although many lavakas formed or were reactivated after forest clearance, many others predate deforestation. Much of the sediment shed from lavakas is stored in adjacent fans and has yet to make it to river and stream channels. Field of view is about a kilometer wide.

PHOTOGRAPH 14.6 Dust Bowl. In the 1930s, massive dust storms devastated the North American Great Plains after sod breaking and plowing by settlers was followed by long-lasting drought. "Fleeing a dust storm" by Arthur Rothstein, who was a photojournalist for the Farm Security Administration, shows farmer Arthur Coble and sons walking in the face of a dust storm in Cimmaron County, Oklahoma, April 1936.

Models of Landscape Evolution

Landscape evolution models may be conceptual, physical, or mathematical. Models range in scope from describing the development of individual landforms, such as hillslope profiles or river terraces, to fully three-dimensional computer simulations of whole-landscape evolution in response to tectonic and climatic change. **Conceptual models** qualitatively generalize understanding of landscape evolution and the role of surface processes in different contexts. **Physical models** allow us to experiment with system behavior in a controlled fashion using analog systems at scales where we can measure features or rates of particular interest, relevance, or importance. **Mathematical models** formalize system behavior and interactions using equations and allow computation of relationships between process and form. These three different types of models provide complementary approaches for investigating landscape evolution.

All models are at best representations of nature. The appropriate choice of model for any investigation depends on the goal of the investigation, the factors one wishes to vary, and the desired level of detail and realism. Models allow us to formalize our understanding of geomorphic systems and better predict their behavior.

Conceptual Models

Across scales, conceptual models of landscape evolution embody nonquantitative (intuitive) understanding of system behavior and are useful for organizing thinking and for interpreting context as one attempts to understand particular landscapes. The weaknesses of conceptual models is in their nonpredictive, qualitative, and typically simplified nature.

Geomorphologists tend to view landscape evolution through two complementary conceptual frameworks. One view focuses on the **transient response** of landscapes to changes in either boundary conditions (tectonics, base level, and climate) or to internal system dynamics (such as a major landslide) that trigger time-dependent, landscape-scale responses. The other view focuses on the development of **dynamic equilibrium** in which landscape characteristics vary over time around a central tendency, an average value or shape producing **steady-state** landforms with a relatively time-independent character. Either framework (transient or steady-state) may be applied over different spatial scales, from the evolution of particular landforms to the development of whole landscapes. Equilibrium or steady-state landforms dominantly reflect the processes involved in shaping them. Transient landforms bear a strong imprint of the landscape history.

At the end of the nineteenth century, William Morris Davis formalized the idea that landscapes change systematically in response to external forcing through a geographical cycle of landform development [**Figure 14.3**]. In Davis's view, following an initial episode of rapid uplift, landscapes evolve through a cycle of stages he characterized as youth, maturity, late maturity, and old age. He thought that dissection typified youthful, recently uplifted landscapes,

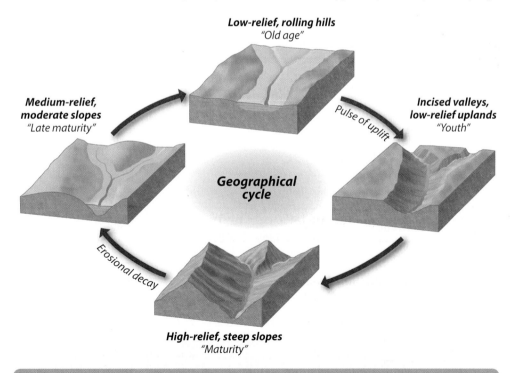

The **geographical cycle** of William Morris Davis describes landscape response to a discrete pulse of uplift that leads to rejuvenation, causing incision, topographic dissection, and relief development followed by erosional decay back to more subdued landforms.

FIGURE 14.3 Davis's Geographical Cycle. The cycle describes a conceptual model of how landscapes evolve in a predictable sequence after an episode of uplift.

that steep, rugged mountains were indicative of mature landscapes, that gentle rolling hills were characteristic of late maturity landscapes, and that broad plains typified old, erosionally degraded landscapes. Davis also introduced the concept of a **peneplain**, an erosional surface with virtually no relief that represented the final state of landscape evolution. Although geomorphologists today address landscape evolution through the context of plate tectonics, Davis's view of transient landscape response to uplift still has conceptual utility because patterns and rates of tectonic forcing change over time.

Steady-state landscapes are those for which overall characteristics, such as average slope and drainage density, do not change significantly over time. Even as the landscape is locally perturbed by erosional events including landslides and floods, the larger landscape maintains a dynamic equilibrium as individual landforms respond to episodic disturbances. A landscape where the surface is literally unchanging (a pointwise perfect steady state) only occurs in the perfectly flat terrain of a Davisian peneplain and in mathematical models; the idea has little currency in the real world where surface processes proceed incrementally and episodically.

Landscapes evolve into a persistent assemblage of landforms if the other factors controlling landscape evolution (tectonics, climate, geology, and vegetation) remain relatively consistent over time and space and thus respond similarly to landscape-shaping events [**Figure 14.4**]. In this way, the concept of a steady-state landscape in dynamic equilibrium provides a framework for understanding how broad variations in tectonic and climatic setting influence landscape characteristics. For landscapes in a dynamic equilibrium in which the vagaries of erosion, transport, and deposition alter individual landforms but maintain overall landscape characteristics, the connection between process and form can be quantified through the development of **characteristic forms** that reflect the physiographic signature of different processes, such as convex hillslopes in regions where soil creep is the dominant downslope mass transport process.

Landscapes change in response to forcing by factors controlling landscape development. These transient effects alter landforms and disrupt the development of characteristic forms. For example, a drop in base level sends a wave of incision progressing upstream, expressed on the landscape as knickzones along main-stem rivers and their tributaries.

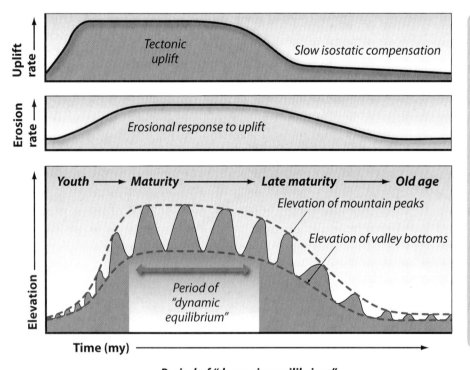

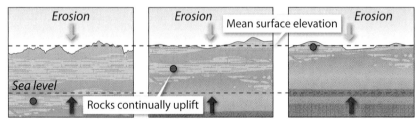

FIGURE 14.4 Conceptual Models of Landscape Evolution. Landscapes are constantly changing but over long timescales (millions of years) they may achieve a steady state in which rates of erosion and uplift remain balanced.

Landscape evolution reflects changing relationships between uplift, erosion, and the development of relief. A period of increased tectonic uplift results in rates of rock uplift exceeding rates of erosion, causing surface uplift and relief development. **Steady-state topography** results when tectonic forcing is sustained for long enough that the landscape reaches dynamic equilibrium in which erosion and rock uplift rates are balanced. Once the rate of tectonic uplift declines, topography and relief gradually decay through erosion and isostatic rock uplift that reduces net surface lowering rate to a fraction of the erosion rate.

During a prolonged period of **dynamic equilibrium,** steady-state topography can develop. At steady state, surface elevation remains nearly constant as rocks continually uplift and erode.

Termination of tectonic rock uplift due to shifting plate motion results in the progressive decay of mountain ranges. If changes in the factors controlling landscape evolution occur faster than the landscape can respond, it can compromise the development of characteristic forms. For example, falling base level can preclude development of a concave river profile. Although it can be challenging to differentiate between the effects of different factors controlling landscape evolution and the transient effects of changes in the history of such influences, the conceptual models of steady-state and transient landscapes provide useful frameworks for understanding physiographic development.

Landscapes vary in their sensitivity to and ability to recover from disturbances or changes in the magnitude and frequency of external forcings. In this sense, **landscape sensitivity** characterizes the propensity of a landscape to evolve in response to changes in external forcings. A landscape in which vegetation is critical for stabilizing hillslopes, such as steep slopes with weak, sandy soil, may be very sensitive to land-use change, and forest clearing in particular. The converse concept of **landscape stability** expresses the resilience of a landscape to externally imposed changes. Upland landscapes graded to a local base level, such as a lake, are buffered from changes in global sea level because they are disconnected from the ocean. Landscape evolution reshapes landscape characteristics over a range of timescales, from the instantaneous damming of a river by a bedrock landslide to the grindingly slow, many million-year breakup of continents.

Physical Models

Physical models of landscape evolution are simplified or scaled-down versions of nature that can be manipulated to investigate geomorphic behavior. Such models provide a way to explore the characteristics and dynamics of geomorphic systems under controlled conditions (Photograph 2.18). The ability to manipulate scaled models allows researchers to isolate the influence of individual processes or system components, observe and measure processes that either occur infrequently or are impossible to measure in the field, and explore the interaction of processes under differing circumstances. Physical models allow direct experimentation and the ability to control or isolate individual factors.

Flume experiments, for example, offer an attractive way to study fluvial processes, the dominant landscape-sculpting agent in many nonglaciated landscapes. Studies of sediment transport by rivers and streams rely on flume experiments because they allow experimental control and replication not available in natural, full-scale channels. Consider the problem of measuring bedload transport rates in a gravel-bed stream. Not only can it be both difficult and dangerous to make such measurements during high-flow events, but bedload transport does not always occur at predictable (let alone convenient) times for measurement. It is difficult to predictably alter a natural stream's sediment supply, but simple to add sediment in a flume. Consequently, flume experiments provide valuable insight in studies of sediment transport, allowing researchers to isolate and examine streambed response to differences in flow depth and velocity, as well as sediment size, shape, and load.

Physical models have proven extremely useful in a wide range of applications. They are particularly useful for illustrating fundamental characteristics of system behavior, such as the relationship between the angle of internal friction of earth materials and the angle of repose of slopes made of granular material (see Chapter 5). Studies of hillslope processes have employed piles of sand and slopes made of various types of beans to investigate hillslope evolution and slope instability. Researchers have even employed loudspeakers blasting amplified music beneath sand piles to study the effects of vibration on the stability of slopes dominated by frictional strength. Although scale models can provide insights into system behavior, they need to be designed and interpreted carefully to ensure their applicability and relevance to the problem or question they are intended to address.

The key weaknesses of physical models are scaling and simplification. Physical models generally involve rescaling the size of modeled features because building a full-size system is impractical for most applications. If the size of the system being modeled requires downscaling, then so do the materials making up the model. One would not, for example, use gravel and cobbles in a tabletop flume model of sediment transport by a mountain stream; rather, the sediment size would be scaled down with sand representing the gravel. Issues of scale not only involve the dimensions of the modeled system and its components, but its physical properties as well, for example fluid density and viscosity. Inevitably, compromises are necessary. For example, in flume experiments, it is impossible to scale hydraulic flow conditions for both the viscosity of the fluid and the density and size of particles (see Chapter 6).

Mathematical Models

Geomorphological applications of mathematical models range from analytical force balances used to characterize the behavior of individual processes to computer simulations that integrate the behavior of multiple processes across entire landscapes. Mathematical modeling is unique because it can provide direct evidence of the linkages between process and form over long time spans, and as such provides an ideal way to investigate geomorphic response to climate and tectonic forcing over time frames too long to be examined experimentally.

The key challenge for mathematical models is to identify and formalize in equations the relevant geomorphic processes. Mathematical models allow exploration of how interactions between and among processes influence the resulting landforms. Indeed, one of the major geomorphic roles of mathematical models is testing whether we can predict the form or attributes of landforms using just a few dominant variables. Although there are many ways to portray geomorphological processes using mathematical models, two general approaches, **process models** and **transport laws,** warrant particular attention in the context of understanding landscape evolution.

Process models include force balances used to represent geomorphic thresholds, such as the failure of a hillslope or the initiation of gravel transport on a mountain streambed. Force balances can illustrate the relative importance of key variables involved in a process (see eq. 5.8 and Figure 1.11). More elaborate process models might simulate the translation of rainfall into runoff and the routing of flood flows from a landscape and through the channel network. Incorporation of simple process models based on Digital Elevation Model (DEM)-derivable attributes (such as drainage area and slope) into Geographic Information System (GIS) routines allows predictions of landscape-scale spatial patterns and variability in geomorphological processes that may be used to identify areas of particular interest, such as the location of failure-prone slopes or flood-prone terrain.

Transport laws are generalized mathematical expressions that formalize relationships between sediment movement and key governing variables. They are informally termed laws because they are thought to characterize the essential system behavior. Assuming steady state, transport laws can be used to solve mathematically for the characteristic topographic forms expected to result from particular processes, such as soil creep as a function of slope or river incision driven at a rate proportional to stream power.

Many geomorphologists refer to slope-dependent transport as **diffusive** because the rate of transport depends on the topographic gradient (slope), analogous to the diffusion of heat driven by temperature gradients, with steeper slopes eroding more rapidly than gentler ones. There is no influence of upslope drainage area on the rate of diffusive processes. Diffusion-like processes tend to reduce relief and fill in local depressions. Diffusive sediment transport may be expressed as

$$q_s = DS^n \qquad \text{eq. 14.1}$$

where q_s is the rate of sediment transport, D is a rate constant termed the hillslope diffusivity, S is the slope, and the exponent n is taken to be 1 in linear diffusion, >1 for nonlinear diffusion, and <1 for some (viscoplastic) creep processes.

For example, debris shed from a fault scarp accumulates at the base of the slope due to the lower gradient (and therefore declining ability of the system to transport material) at the base of the slope. Over time, an initially linear fault scarp will round and become subdued in a manner well described by diffusion. For the case of spatially uniform, steady-state uplift, diffusive sediment transport leads to the development of convex hillslope profiles, on which gradients progressively increase downslope [**Figure 14.5**] (see Box 5.1).

Processes for which the upslope drainage area influences rates of sediment transport are considered **advective** because the entrained material generally moves along with (and thus is advected by) flows that increase downstream. Advective processes tend to incise valleys and create relief. Advective sediment transport may be expressed as

$$q_s = K A^m S^n \qquad \text{eq. 14.2}$$

where q_s is the sediment transport rate, K is the fluvial rate parameter, A is drainage area, S is the slope, and the exponents m and n vary, depending upon the process under consideration. A physically reasonable basis for fluvial transport in humid regions based on shear stress leads to values of 0.5 and 1, respectively, for m and n. The inclusion of the area term (A) on the right-hand side of eq. 14.2 means that downstream areas (large A) will have higher sediment transport rates than upstream areas (small A) with the same slope. Thus, for the case of spatially uniform, steady-state uplift, advective transport leads to progressive decline of slope downstream; the trade-off between the greater area (and thus discharge) of downstream locations and the steeper slope of upstream locations equalizes erosion rates along the profile, producing a characteristic concave form in which stream gradients progressively decrease downslope (see Box 7.1).

Empirical studies have determined D (hillslope diffusivity) and K (fluvial rate parameter) values for a range of environments and found that D varies by roughly a factor of 2, whereas K varies by more than five orders of magnitude. Consequently, the length of diffusive hillslope convexities and the pace of advective sediment transport are widely variable between landscapes in different settings.

Mathematical models that incorporate particular landscape geometries, characteristics, and transport laws can be used to simulate the action and interaction of surface processes that create both individual landforms and entire landscapes. One of the most robust findings of the many mathematical landscape evolution models developed over the past several decades is that the competition between diffusive and advective processes produces "realistic"-looking landscapes composed of rounded hills and an integrated valley network. Diffusive processes dominate small drainage areas, producing convex rounded hillslopes near drainage divides, whereas advective processes dominate large drainage areas, producing integrated, concave valley systems.

Mathematical models of landscape evolution can be used to explore the impact on landform development of the full range of potential variables and tectonic or climatic settings. For example, mathematical models of landscape evolution driven by the competition between advective channel incision and diffusive hillslope processes illustrate how the steady-state development of sharp ridge crests or broad slopes depends on the relative values of the hillslope diffusivity (D) and the combined influences of

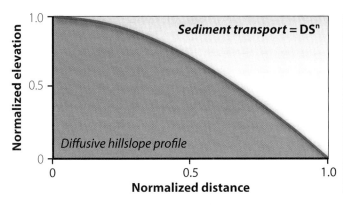

The equilibrium profiles of soil-mantled hillslopes tend to be **convex,** with slope (S) increasing progressively downslope, which accommodates the increasing flux of soil moved diffusively down the slope by processes such as soil creep. D is hillslope diffusivity and n is a scaling exponent, typically considered to be equal to 1.0.

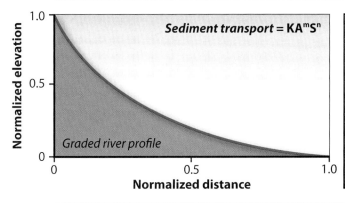

The equilibrium profiles of rivers are **concave,** reflecting the lower slope (S) needed to transport material as drainage area (A), and thus river discharge, increases downstream. K is a constant that incorporates bedrock erodibility and drainage basin geometry, and m and n are scaling exponents, often considered to equal 0.5 and 1.0, respectively, for fluvial transport.

FIGURE 14.5 Equilibrium Hillslope and River Profiles. Characteristic forms for convex hillslopes and concave graded river profiles depend on the relationship between sediment transport, slope, and area.

bedrock erodibility, climate, and drainage basin geometry (K) [**Figure 14.6**].

Mathematical models of landscape processes and evolution vary greatly in their approach and ability to describe accurately the geomorphology of landscapes. Strengths of mathematical models include the ability to investigate and illustrate feedbacks and system behavior and to test hypothesized connections between process and form. Weaknesses of mathematical models include the problem that most field measurements of sediment transport or landscape change sample a far shorter time period than the landscape-forming timescales of numerical models, which complicates calibration of rate parameters (D and K) and makes it difficult to relate them to specific geomorphological processes.

Landscape Types

In the broadest sense, there are four landscape types: steady-state, transient, relict, and ancient. These four landscape types define distinct settings in which the factors of landscape evolution operate on the template set by the history of past tectonic and climatic forcing.

Steady-State Landscapes

Under steady, tectonically driven rock uplift, a landscape may develop and maintain steady-state characteristics for tens of millions of years. During such long periods of steady uplift, individual geomorphic events may disturb system conditions, and modify individual landforms, but

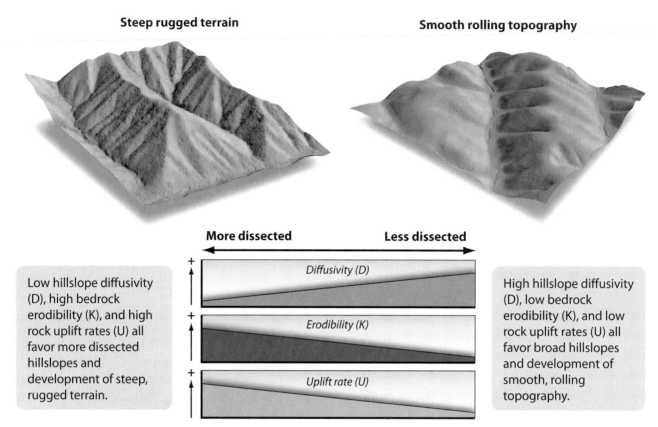

FIGURE 14.6 Degrees of Landscape Dissection. Hillslope diffusivity, rock erodibility, and rock-uplift rate influence the character and degree of landscape dissection. [Adapted from Roering et al. (2007).]

the landscape overall maintains a characteristic condition and form, reflecting a balance between long-term erosion and uplift. For example, rates of rock uplift and erosion are closely matched despite the occurrence of large but transient uplift and erosion events in tectonically active regions like the Oregon Coast Range, Taiwan, and the Southern Alps of New Zealand [**Photograph 14.7**]. Atmospheric scientists find it useful to distinguish short-term weather from long-term climate, and a parallel distinction is useful for considering how steady-state landscapes may experience short episodes of greater-than-average activity (such as in response to extreme rainfall during a hurricane or a major earthquake) or prolonged periods of inactivity (such as during severe droughts).

At collisional plate margins, the size and steepness of a mountain range grow until the erosional flux off the range matches the tectonic flux of material into the range. This is geomorphology at the largest spatial scale. At collisional margins, tectonically bulldozed crustal material piles up to form **orogenic wedges** that grow until the slope of the range front reaches the limiting angle that the lithosphere can support—much like how sand piles up to a critical angle in front of an advancing bulldozer blade [**Figure 14.7**]. Orogenic wedges develop a **taper** to their cross-range profile that reflects large-scale mechanical controls (rock strength and friction on the plate interface) and imposes coupled limits on the width and height of mountain ranges. Faster tectonic convergence leads to a wider, taller mountain range. Once a rising mountain range attains its critical, or limiting, profile, the mass balance between uplift and erosion is maintained even as patterns in rainfall and tectonic uplift continue to influence spatial patterns of erosion and rock exhumation. In such settings, greater rainfall on the windward side of a mountain range results in greater erosion and thus deeper

PHOTOGRAPH 14.7 Steady-State Topography. Straight mountain slopes at Lindis Pass in the Southern Alps, South Island, New Zealand, an area where rock uplift and erosion rates are thought to be closely matched. The two-lane road in center of image provides scale.

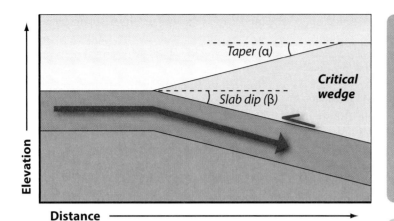

Convergence at a compressional margin can result in the development of a wedge of lithospheric material that grows over time until it reaches an equilibrium form, termed a **critical wedge,** in which the rate of erosion balances the input of mass by convergence and the rangefront taper (α) equals the dip (β) of the downgoing slab.

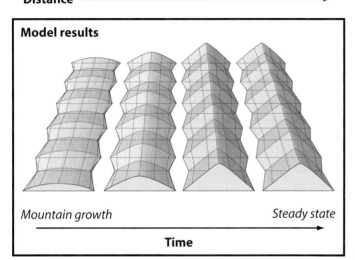

Numerical models that incorporate both diffusive and advective sediment transport can show how initially low-relief surfaces, once uplifted, can become realistic-looking mountain ranges. After sufficient time, they achieve a dynamic steady state in which landscape form remains constant as long as tectonic mass input and erosional output from the range are balanced (forming and maintaining the critical wedge).

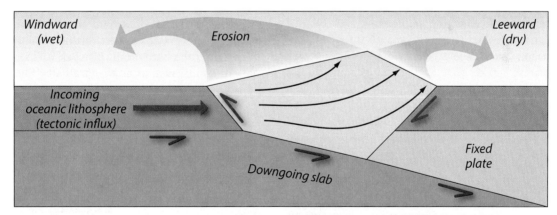

The tectonic influx of material into a critical wedge is balanced by the erosional flux of material out of the wedge from the wet windward and dry leeward sides of the orogen (mountain range). Individual particle paths of material through the orogen (arrows) vary across the range, with material eroded off the windward side of the range following shallow trajectories and material eroded from the leeward side of the range experiencing deeper burial and higher pressures and temperatures prior to erosion. For a given slab dip, and therefore a given wedge taper, a greater rate of tectonic influx will be balanced by development of a wider and therefore taller range.

FIGURE 14.7 Critical Wedge. Considering the mass balance and overall form of collisional boundaries helps geologists predict the large-scale geometry and behavior of mountain ranges. [Adapted from Willet (1999).]

exhumation than in the rain shadow on the leeward side of the range, as is the case in the Southern Alps of New Zealand and the Cascade Range in the Pacific Northwest.

Because of the long time it takes to erode away mountains (due to the thickness of their crustal roots), an assumption of steady state is a reasonable short-term approximation for mountain ranges after tectonic forcing ceases because the isostatic response to erosion ensures that most (~82 percent) of the mass lost to erosion is replaced by rock uplift. For example, cosmogenic nuclide studies have established that erosion rates over the past 10,000 to 100,000 years in the Appalachian Mountains are on the order of 20 m per million years, similar to the rate of rock uplift in the region over the past 60 to 100 million years as determined by thermochronology. Thus, even though the Appalachians are gradually wearing away because the tectonic activity that elevated them ended tens of millions of years ago, the landscape approximates geomorphic steady state over hundreds of thousands of years because of the isostatic uplift of rock in response to erosion.

Transient Landscapes

Transient landscapes are those in the process of responding to a change in base level, uplift rate, or climate [**Photograph 14.8**]. The distinction between steady-state and transient landscapes depends on timescale. Over long spans of geologic time, all landscapes are transient due to changes in climate and tectonic forcing as the motion of continents shifts direction and speed, altering continental configurations, mountain range elevation, uplift rates, and styles of rock deformation.

In transient landscapes, response times vary with location because it takes time to propagate change across the landscape. This lag results in a **complex response**, during which some areas of the landscape respond slowly and others quickly. This differential response time means that some areas are reacting to the most recent forcing, while others are reacting to forcings that occurred long ago (or to the secondary effects of those older forcings).

Consider, for example, how the signal from a drop in sea level moves through a landscape. Initially, the oversteepened reaches of rivers and streams near the coast will erode more rapidly, sending a wave of incision sweeping inland as knickpoints advance up the river system. Locations upstream of the knickpoints will not yet "know" about the downstream change in base level. Similarly, when sea level rises and incision stops downstream, knickpoints upstream will keep retreating headward, out of sync with the base-level control.

Knickpoints generated by base-level fall migrate upstream at rates determined by lithology and the upstream drainage area, which is a crude proxy for runoff, stream power, and erosion capacity. Thus, the pace of headward advance decreases as knickpoints propagate up through river networks. The distance upstream at which knickpoints are found on tributaries will be different for knickpoints of the same age on different branches of the same river system, because knickpoints on larger, more powerful tributaries are capable of retreating faster than those on smaller tributaries. The rate of knickpoint propagation determines the approximate timescale for river basins to respond to uplift or base-level lowering.

In contrast, the retreat of lithologically controlled knickpoints that form where strong differences in rock strength occur is controlled through plunge-pool undermining and the excavation of weak rocks exposed in the base of a falls, such as at Niagara Falls [**Photograph 14.9**;

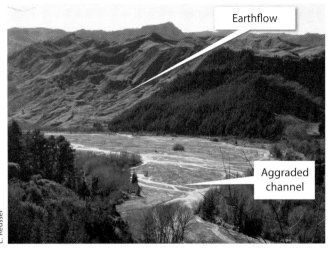

PHOTOGRAPH 14.8 Transient Landscape. The Waipaoa Basin on North Island, New Zealand, has steep slopes and earthflows resulting from incision triggered by uplift and base-level fall. Low-relief uplands are "unaware" of the incision below. Massive aggradation of the channel resulted from deforestation-triggered erosion upstream. The main channel is ~100 m wide.

PHOTOGRAPH 14.9 Niagara Falls. Niagara Falls, on the United States–Canada boundary (Ontario, Canada, and New York State), is a classic knickpoint. Here, hard dolomitic rock overlies soft shale and holds up the falls, which have retreated more than 11 km since they formed at the end of the last glaciation, ~12,500 years ago.

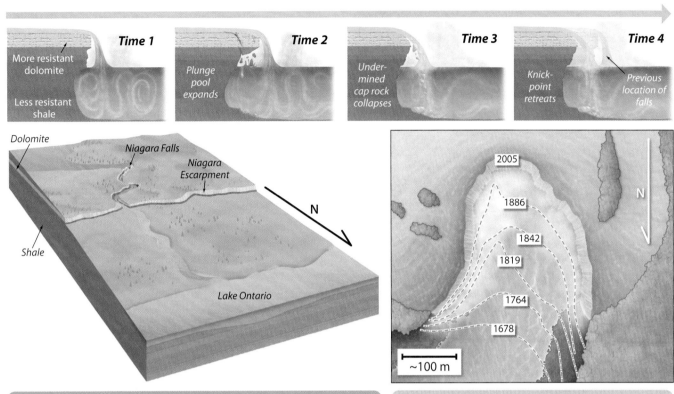

FIGURE 14.8 Knickpoint Retreat. Some knickpoints, like Niagara Falls, retreat by the undermining of soft rock and subsequent collapse of a stronger caprock.

Figure 14.8]. In such cases, the weathering rate of the headcut-forming material through the formation of new joints and fractures sets the pace for knickpoint retreat. Both lithology-controlled and incision-driven knickpoints propagate upstream, but how knickpoints advance upstream depends on whether headward retreat is controlled by stream power or weathering-mediated toppling of material from the headcut (or both).

Faulting can induce transience in landscapes. For example, recently exposed faults are undissected, planar surfaces. Over time, erosion increasingly rounds, subdues, and dissects fault scarps. In addition, differential erosion of rocks with contrasting erodibility can accentuate topographic relief across a fault zone, resulting in erosional **fault-line scarps**. Some fault-line scarps develop significant topographic expression due to differential erosion of the rocks exposed on either side of an inactive fault. Other fault-line scarps represent parallel slope retreat of the upthrown block back from an active fault trace [**Photograph 14.10**].

Transient landscape response may be quite rapid at the scale of individual valley bottoms. Channel **entrenchment** refers to substantial incision of the bed that leads to floodplain abandonment such as occurs when arroyos incise (see Photograph 14.5). Entrenchment can occur in response to changes in valley-bottom vegetation, increased flood flows downstream of urbanized areas, base-level lowering, stream channelization (straightening), and decreased sediment supply such as downstream of dams; all but the last of these changes can initiate a wave of incision that propagates upstream (see Figure 7.11).

Channel entrenchment is usually followed by channel widening as the system readjusts. Continued lateral migration and bank erosion can eventually carve out enough space for a new floodplain to begin forming, entrenched down below the level of the original abandoned floodplain, which then becomes a terrace. Transient upstream entrenchment often causes downstream aggradation by temporarily increasing the sediment supply delivered to channels, another example of a basin-scale complex

PHOTOGRAPH 14.10 Rift Shoulder. Gebel Hammam Faraun rises hundreds of meters above the Gulf of Suez. It is a rift shoulder in the Dead Sea–Red Sea rift zone. The steep slope to the right of the high point is a fault-line scarp eroded back into pre-rift carbonate rocks that are exposed on the uplifted rift shoulder. There are sulfur-rich springs at the actual rift-bounding normal fault near the shoreline. The dark colored soft sediments were deposited after rifting and include shale and sandstone.

response. In extreme cases, agricultural soil erosion can lead to the development of deeply incised gully systems, such as Providence Canyon State Park, the "little grand canyon" of Georgia. Increased runoff from agricultural fields began incising this now almost 50-m deep gully system not long after the area was first plowed in the 1820s [Photograph 14.11].

Changes in the course of drainage systems represent another type of transient landscape response. Drainage rearrangement through **stream capture, drainage diversion,** or **beheading** [Figure 14.9] occurs when a stream or river impinges upon and redirects all or a portion of a neighboring drainage network.

Stream capture is a bottom-up process in which headward retreat of a stream with a steeper gradient (or higher discharge) than a neighboring channel can erode back into and divert flow from a portion of the other stream's headwaters. Although stream capture has been widely invoked to explain anomalies in regional drainage patterns, it is now thought to be a relatively rare event because of the unusual circumstances required for it to occur. Stream capture requires both a significant elevation difference and a low divide between neighboring drainage basins, and faces the problem of how headward surface erosion could breach a drainage divide when stream channels do not extend to the divide. Many unusual bends in river systems may actually be structurally controlled.

Drainage diversion is a top-down process in which the headwaters of a river system are diverted into a neighboring system. Drainage diversion occurs when the course of a river or stream is blocked and flow is directed off in a new direction, such as through fault uplift that blocks and diverts a river, or when a landslide or glacier dams a river, impounding a lake that fills and then spills across a former drainage divide and flows in a new direction. River diversion becomes progressively more difficult as the relief of a valley system increases because commensurately higher impoundments are required to overtop the valley walls.

Beheading is the process through which a river network appropriates part of the adjacent drainage basin without preservation of surficial drainage courses. This can occur where a retreating escarpment consumes the area formerly occupied by plateau streams draining in the opposite direction above the escarpment. Analogous situations can develop when a sea cliff erodes back into coastal terrain.

The evidence for drainage rearrangement may be either geological, in the form of distinctive sedimentary deposits or structures related to the paleo-drainage system, or morphological, in the form of unusual bends in river systems (elbows) or **wind gaps** [Photograph 14.12], streamless valleys that extend to the drainage divide at the site of suspected captures or diversions. In most cases, suspected drainage rearrangement is inferred on the basis of landscape-scale morphological evidence, although tracking of rock types in river terraces has also been used, for example, the presence of basalt clasts in terrace deposits of a river that currently contains no basalt in its watershed. Stream captures may explain sharp changes in flow direction, such as **barbed drainage** in which tributaries flow in the opposite direction from the mainstem channel. In the case of a stream capture, a distinct knickpoint should be found in the captured stream, upstream of the point of capture. Deciphering the history of drainage rearrangement can be important for understanding long-term changes in

PHOTOGRAPH 14.11 Providence Canyon. This canyon in Georgia is a large, young, and deep gully complex eroded into weathered rock and saprolite. The gully complex initiated in a plowed field in the early nineteenth century.

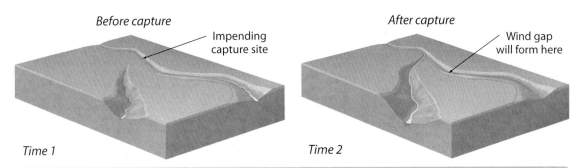

Stream capture occurs when headward erosion of a stream valley captures the drainage of another stream. Such headward erosion may be accomplished by intercepting groundwater flow as the headward-cutting stream approaches the one that is about to be captured. The capturing stream gains drainage area, discharge, and thus stream power, allowing it to incise rapidly.

Drainage diversion occurs when a landslide or glacier blocks a river or stream, impounding a lake. If the outlet to the new lake spills over a drainage divide, it can divert the drainage course into a receiving channel that may rapidly incise. The new stream orientation may persist once the lake drains, if incised deeply enough.

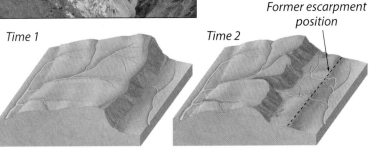

Stream **beheading** occurs when the headwaters of a channel erode into and capture part of an adjacent drainage basin, such as occurs during escarpment retreat or coastal sea cliff erosion.

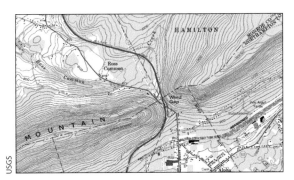

Wind gaps are dry valleys that cut through ridges. They are thought to form when streams, eroding headward, intersect and capture the drainage of other streams. The reaches of the captured stream just below the capture site are left high and dry with no contributing area.

FIGURE 14.9 Stream Drainage Rearrangement. Stream capture, drainage diversion, and stream beheading change the arrangement of surface drainages.

regional sediment supply and in understanding species distributions and rates of species divergence in regional river systems.

Transient landscapes do not necessarily evolve rapidly. The development of rifted continental margins presents an example of slow transient landscape response. The world's great escarpments, such as those in southern Africa, represent places where landscapes rapidly responded to ancient continental breakups and then continued to evolve slowly over tens of millions of years. Low-temperature thermochronometry and cosmogenic nuclide data from high-elevation, passive-margin escarpments in Namibia,

PHOTOGRAPH 14.12 Wind Gap. Dolls Gap, West Virginia, is a classic example of a wind gap because it has no stream flowing through it today. Wind gaps were carved by the streams that once flowed through them when valley elevations on either side stood higher than they do today. A stream eventually captured the flow of the Dolls Gap stream, leaving the nearly kilometer-wide gap itself literally high and dry, an example of stream capture.

southeastern Australia, and southern Africa indicate that these great escarpments reached their present configurations within a few million to tens of millions of years after continental breakup and since then have evolved far more slowly, in great part due to the small drainage areas of the basins on the rift-facing escarpment, produced by rift-flank uplift diverting streams away from the rift.

Relict and Ancient Landscapes

Relict landscapes are those in which topography formed under former climates is preserved under modern conditions [**Photograph 14.13**]. The most common relict landscapes are formerly glaciated regions out of equilibrium with the post-glacial climate. Much of the topography of mountain ranges like the Swiss Alps, the Himalaya, and the once-glaciated upland of the Sierra Nevada formed under conditions and through processes no longer actively shaping the landscape. Similarly, the landforms within and immediately around formerly glaciated lowlands, like around Seattle, Washington, and across much of the North American midcontinent, make little sense if interpreted only through the lens of contemporary processes. The subglacial meltwater channels that carved down into a broad glacial outwash plain to shape the topography around Puget Sound are no longer operating. Many alpine landscapes are still influenced by the transient effects of now-vanished late Pleistocene glaciers, such that reworking, erosion, and transport of glacially related sediments still dominate the sediment system. Cycling back and forth between glaciated and unglaciated conditions can lead to a landscape in perpetual disequilibrium if the landscape

PHOTOGRAPH 14.13 Relict Landscape. The Wasatch Mountains in Utah contain numerous cirques that eroded toward one another to form glacial horns and arêtes during past glaciations. Today, these are relict features because glaciation is no longer active in this area.

relaxation time (Figure 14.2) is longer than the periodicity of climate forcing.

Relict landscapes may also reflect the lasting imprint of catastrophic processes no longer in operation. Portions of northern central Asia and the megaflood-sculpted landscapes of eastern Washington State, including the dry waterfalls and deep potholes along the arid valleys of the Channeled Scabland [**Photograph 14.14**], are relict landscapes shaped by discharges impossible to produce through modern processes (see Digging Deeper, Chapter 13). Likewise, the great outwash channels on Mars, where there is little surface water today, show that deep flows left a persistent mark on the surficial morphology under radically different conditions in the past [**Photograph 14.15**]. Some landscapes are polygenetic in that their present form and characteristics are the integrated result of multiple episodes of landscape-forming processes under varying climates.

Ancient landscapes are those that have proven stable through extended periods of geologic time, and thus may integrate a long series of climates and conditions

PHOTOGRAPH 14.14 Dry Falls. Dry Falls, carved by the great Missoula floods pouring across the Channeled Scablands of eastern Washington State, form a relict landscape shaped in the Pleistocene. The image is 3 km across. When active, the falls were the largest by volume in the world.

PHOTOGRAPH 14.16 Ancient Landscape. This mesa (~100 m high) capped by resistant silcrete (silica-cemented soil) formed under a different climate regime and is now eroding slowly in the semi-arid climate of the Hamersley Ranges, western Australia.

[Photograph 14.16]. The surficial preservation of **paleosols**, fossil soils formed under conditions different than at present, provides a clue that the landscape is ancient. In the humid tropics, where deep weathering profiles can develop, some ancient soils are so thoroughly weathered and leached that they consist of ore-grade concentrations of iron and aluminum. In regions where deep weathering creates an irregular bedrock surface, subsequent erosion of the weathered mantle can leave behind rocky irregular surfaces known as **etchplains** (see Figure 3.13). Ancient landscapes generally occur on low-relief, nonglaciated continental surfaces common in the stable cratons of Africa, North America, South America, and Australia.

PHOTOGRAPH 14.15 Relict Landscape on Mars. Exceptionally deep (8 km) chasma (from Greek, meaning "yawning hollow") on Mars are relict landscapes formed by fluid flow and landsliding that are no longer active. This image of Valles Marineris, was taken from orbit with a ground resolution of approximately 300 m per pixel by NASA's *Mars Odyssey* spacecraft.

Basin Hypsometry and Landscape Form

Hypsometry describes the relationship between area and elevation in a drainage basin or landmass (see Figure 1.2). Hypsometric curves can be created by digitizing topographic maps or by analyzing DEMs and then calculating the proportion of the area at different elevations. The shape of hypsometric curves for different drainage basins can vary greatly, from concave to straight or convex.

Hypsometric curves reflect both landscape history and the signature of active geological and geomorphological processes [**Figure 14.10**]. For example, in basins containing glacially dissected plateaus, such as the fjord landscapes of the North Atlantic margin, hypsometric curves are convex with most land at high elevation (the plateaus between the deep fjords). Analysis of hypsometry along the western margin of South America shows different but regionally consistent curve shapes along the axes of the Andean Mountains, reflecting dominant erosional processes and geography in each region. For example, in the northern Andes, fluvial processes are dominant and the curves are concave-up, reflecting graded river and slope profiles. Hypsometric curves of the central Andes reflect the high elevation of the flat and extensive plateau, the Altiplano. In the southern Andes, glaciers have eroded away the highest elevations and limited the height of the mountaintops. The hypsometric curves for many large African drainage basins, including the Nile, Orange, and Niger rivers, all show very little land at low elevations. The dearth of land at low elevation has been interpreted as a signal of renewed, recent uplift along continental margins bordered by steep escarpments.

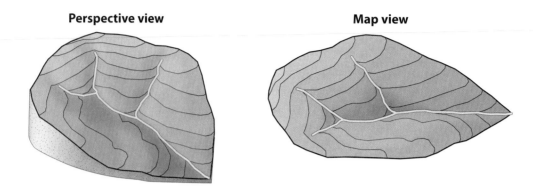

The **hypsometry** of a drainage basin quantifies its form by considering how much of the basin area is above or below a given elevation.

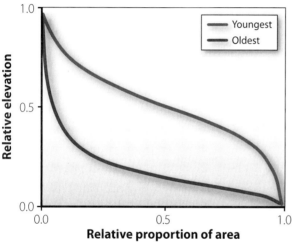

The hypsometric integral is defined as the area under the hypsometric (area-elevation) curve for a particular landscape. The hypsometry of a drainage basin can change as a result of changes in tectonic boundary conditions. Recently uplifted, young terrain will have a greater proportion of its area at higher elevations, and thus a larger hypsometric integral, whereas fluvially dissected topography of older terrain will have less area at higher elevations, and thus a smaller hypsometric integral.

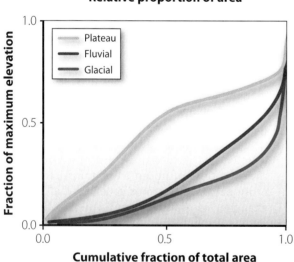

The cumulative area of a landscape increases with elevation but the shape of the curve differs for broad, undissected plateaus, fluvially dissected terrain, and glaciated (or formerly glaciated) landscapes. Arid plateaus have large areas at high elevation, whereas high elevations are typically under-represented in glaciated landscapes due to valley widening by glacial erosion. Fluvially sculpted topography tends to have smoother, concave cumulative elevation curves.

FIGURE 14.10 Basin Hypsometry. The area-elevation relationship of a drainage basin (its hypsometry) changes over time as an uplifted landscape becomes dissected and also reflects the type and distribution of surface processes active in the basin. [Adapted from Strahler (1952) and Montgomery et al. (2001).]

Rates of Landscape Processes

Timescale is a fundamental concept when considering landscape evolution because the timescale for changing individual landforms is not the same as that for changing the whole landscape (Figure 1.8). The pace of geomorphological processes varies greatly, from meters per second in the case of rapid processes like flash floods and debris flows to meters per million years in the case of tectonic processes that play out over geologic time. A landscape in steady state over several million years would not be in steady state over 100 million years as it rises tectonically and decays erosionally.

Considered broadly, the pace of tectonics governs rates of geomorphological processes at active margins, while climate and isostatic response to denudation set the pace of erosion in tectonically passive regions such as ancient cratons. In part because of the wide range in rates of uplift and erosion (micrometers to millimeters per year), measured rates are reported in a variety of units, including millimeters per year and meters per million years (1 mm per year = 1 km per million years). Rates of erosion and sediment transport are also often reported in terms of kilograms per square meter per year (or tons per hectare per year), which may be converted to an equivalent lowering rate (meters per year) through dividing by the density of the transported material (kilograms per cubic meter). Different methods of determining uplift and erosion rates produce results that integrate over different time frames and thus can be compared to detect and evaluate changes in rates over time. The relative importance of frequent versus rare events is important to consider when comparing rates determined by different methods.

Uplift Rates

Rates of uplift can be measured in terms of the upward movement of either the land surface or the underlying column of rock relative to sea level (or any other datum). For the case of no erosion, the two measures of uplift are equivalent. But as this is hardly ever the case, it is worth noting that most conventional measures of uplift except for surveying (see below) measure **rock uplift** rather than **surface uplift**. Thus, for the general case where erosion does indeed occur, local uplift rates determined by these measures will overestimate landscape-scale surface uplift. Rock uplift driven by isostasy will proceed at a substantial fraction of the erosion rate (~82%), reflecting the contrast between crustal and mantle density. Tectonically driven rock uplift rates can reach 10 mm/yr in the most rapidly uplifting environments, such as portions of the Himalaya or the southern Alps of New Zealand. Surface uplift occurs when rates of rock uplift outpace erosion and increase landscape relief (and slopes) until either erosion rates match the rock uplift rate or the development of threshold slopes precludes further increases in relief (see Chapter 5).

A record of rock and surface uplift is prominently archived in flights of marine terraces along the coastline of many active tectonic margins. As erosional features etched onto an actively rising shoreline during periods when sea level rises at the same pace as rock uplift, marine terraces rise along with the rocks into which they are carved (see Photograph 12.15). Consequently, they record both net surface uplift (as long as they do not erode) and also act as markers that record rock uplift referenced to a sea-level datum, which itself moves vertically as ice sheets come and go and sea level moves up and down.

Flights of strath terraces preserved along river systems likewise record rock uplift and surface uplift as long as river gradient does not change and the terraces do not erode. However, the timing and amount of local rock uplift may be very different from the age of abandonment of the strath terrace and the amount of incision below the strath. Consider, for example, the case of a strath terrace formed by base-level fall that propagates upstream as a knickpoint and encounters a resistant lithology. The strath may form over a long time after the base-level fall (and may not be the same age along its entire length).

Uplift or downcutting rates can be determined from terraces at known elevations by establishing their age through either cosmogenic nuclide exposure dating or radiocarbon dating of associated deposits. Rates of uplift recorded by marine and strath terraces reflect the combined influences of tectonic and isostatic uplift.

Rates of surface uplift can be measured using surveying techniques such as precise leveling repeated over the course of several decades. In addition, high-resolution satellite altimetry and global positioning systems (GPS) can be used to measure rates of surface uplift directly. Coastal tide gauge records extending back as much as a century in some places can provide high-resolution records of land surface uplift, as long as one factors in the continuing (and increasing) pace of eustatic sea-level rise. (The average late Holocene eustatic sea-level rise of about 1 mm/yr increased to about 2 mm/yr in the late twentieth century.) In some circumstances, historical or archaeological evidence can inform estimates of surface uplift or subsidence, such as where ancient irrigation canals are offset or where ancient docks are displaced relative to modern sea level.

Erosion Rates

Landscape erosion rates are influenced by both the **erosivity** of the processes acting to entrain and transport material and the **erodibility** of earth materials. Erosivity is influenced by the amount and style of precipitation and runoff, and the energy available to drive geomorphological processes, as controlled by landscape relief and local slopes. The erodibility of the landscape is influenced by the nature and characteristics of both vegetation and bedrock, and the presence, type, thickness, or absence of soil or surficial deposits. The greatest erosion rates occur where both erosivity and erodibility are high.

Erosion rates tend to scale with rates of tectonic forcing. Rapidly uplifting areas tend to erode quickly, and tectonically quiescent areas tend to erode more slowly. Likewise, steeper slopes and higher-relief terrain tend to be correlated with higher erosion rates. The influence of climate and vegetation are more complicated. Although greater precipitation increases the erosivity of geomorphological processes, the resulting denser vegetation reduces the erodibility of the ground surface. Mechanical denudation rates have been found to correlate with precipitation variability (and seasonality), and chemical denudation tracks mean annual precipitation. Solute loads increase with greater precipitation.

Erosion rates vary by at least six orders of magnitude around the world, from less than a micrometer per year on boulders in the flat, extremely arid Atacama Desert of South America to as much as 10 mm/yr in the steep, narrow gorge of Tibet's Tsangpo River [Figure 14.11]. Transient, short-term erosion rates can be even higher as landscapes adjust to changes in boundary conditions or internal disturbances. Erosion rates match rates of rock uplift in steady-state landscapes, and there is a broad pattern to the range of typical erosion rates, with ancient cratonic landscapes eroding at rates measured in meters per million years (<0.01 mm/yr), soil-mantled terrain eroding at rates of tens to hundreds of meters per million years (0.01 to <1 mm/yr), and mountainous environments and alpine terrain eroding at rates of hundreds to thousands of meters per million years (>0.1 mm/yr). The comparable range in global rates of sediment export from glacially and fluvially dominated landscapes indicate that tectonic uplift ultimately determines long-term erosion rates, although climate, topography, geology, and vegetation can influence the magnitude and pattern of erosion over shorter timescales. Transient landscape response can result in spatially and temporally restricted extreme rates of erosion and sediment export.

Each method of determining erosion rates involves assumptions that directly affect the meaning of the calculated rates (see Chapter 2). A common indirect measure of sediment yields, and by inference of erosion rates, relies on equating the volume of sediment removed from a landform or watershed with that deposited in some kind of natural or artificial sediment trap over a known time interval. For example, rates of reservoir infilling can provide information on sediment yield from upstream areas over recent timescales (decades), and the volume of alluvial fans can be used to estimate minimum sediment yields and, by assuming 100% sediment transfer to the fan, erosion rates for areas upstream over longer timescales. The total volume of clastic marine sediment in offshore

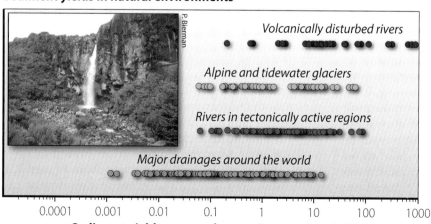

FIGURE 14.11 Sediment Yield and Erosion Rates. Global compilation of short-term sediment yields and longer-term erosion rates span many orders of magnitude and reflect both natural and human forcing. [Adapted from Montgomery (2007).]

basins can be used to estimate long-term sediment yields over millions of years.

The most common method for estimating sediment yields is the integration of suspended sediment concentrations measured in runoff at gauging stations along rivers. Calculating erosion rates from such data relies on the usually invalid assumption that erosion and sediment yield are equivalent. Study after study has shown that over decades to centuries much of the sediment eroded from hillslopes is deposited at the foot of colluvial slopes or stored as alluvium on floodplains. Thus, hillslope erosion rates can be very high while sediment yields in the drainage network are far lower. Indeed, much of the sediment eroded from uplands deforested by colonial and postcolonial settlement of the eastern and central United States has yet to make it through the channel network (see Figures 7.2 and 7.13).

At a drainage basin scale, the ratio of eroded material to the amount of sediment that leaves the drainage basin (its sediment yield) is termed the **sediment delivery ratio** (see Figure DD7.1). This ratio is time-dependent and area-dependent. The sediment delivery ratio increases with the timescale over which it is integrated and decreases with increasing basin area. In large drainage basins over short timescales (years to decades), the sediment delivery ratio is usually much less than 1 as sediment is stored on colluvial footslopes and deposited in broad floodplains; it can be as little as 0.05. In small drainage basins lacking significant areas for long-term sediment storage, the long-term sediment delivery ratio is generally close to 1, although in the short term, large amounts of sediment may leave slopes but fail to enter channels, lowering the sediment delivery ratio.

Indirect measures of erosion rates also include those that rely on determining the concentration of cosmogenically produced nuclides such as ^{10}Be on exposed land surfaces or in river sediments (see Chapter 2). Cosmogenic nuclide concentrations reflect an integrated average erosion rate over the timescale required to move material through the upper several meters of the land surface, which varies with the erosion rate but is generally within the past few thousand to a million years—the Holocene and some of the Pleistocene. Longer-term measurements of exhumation determined using thermochronometry are complicated by the potential for high-relief terrain to perturb local geothermal gradients, by the advection of anomalously hot material from deeper in the crust toward the surface in areas of rapid erosion, and by the difficulty of knowing past geothermal gradients.

Mass export rates determined by analysis of river systems can be separated into chemical and mechanical components determined by magnitudes of the solute (dissolved) and solid (sediment) load. Most estimates of the solid load of a river only include the suspended load, as bedload transport rates are difficult to measure. Suspended load can be readily estimated from samples obtained from the water column, automated turbidity measurements, and by remote sensing of the water surface color. Estimates of sediment load obtained from measuring only the suspended load underestimate the total load of a river, because they neglect the dissolved load and bedload. This is not much of an issue for estimating the total load of large rivers such as the Amazon, for which bedload has been estimated to account for about 1 percent of the total sediment load. Estimating bedload is more of a problem for high-gradient mountain streams. Although it is widely assumed that the fraction of the total load carried as bedload is typically around 10 percent of the total load, the proportion of bedload can be higher in mountain rivers draining rapidly eroding terrain.

The dissolved component of mass export from drainage basins ranges from just a few percent in rivers draining rapidly eroding mountains, such as the Himalaya, to almost 90 percent of the total load of rivers draining limestone terrain and slowly eroding cratonic environments, such as the St. Lawrence River in eastern Canada [**Figure 14.12**]. Generally, rates of mechanical and chemical mass loss from landscapes (denudation) are positively correlated, with both tending to be higher in mountainous regions because high rates of mechanical weathering produce greater surface area for chemical weathering. On a global scale, the solid load of rivers is about 6 times greater than the solute load.

Spatial and Temporal Variability

Geomorphic processes are not uniform in time and space; rather, they occur episodically and at different places at different times. For example, short historical records of sediment transport and erosion rates often do not include the effects of large events; thus, long-term erosion rates may exceed those inferred from direct, short-term measurements. Conversely, contemporary human disturbance of specific areas on the landscape can lead to localized sediment yields far greater than those determined from long-term integrators, such as cosmogenic nuclides or thermochronometry.

Infrequent large events generally produce more dramatic effects on the landscape than do more frequent smaller events, but the integrated effects of more common but smaller events can cause substantial landscape change over time. In many geomorphic systems, such as alluvial rivers in humid/temperate climate zones, it is the intermediate size and frequency events (for example, the mean annual flood) that in total do the most geomorphic work (see Chapter 6).

Studies of sediment accumulation rates recognize that the average sedimentation rate decreases with the time span over which it is measured because of the episodic nature of sedimentation. In formerly glaciated regions, sediment generation rates determined by modern process studies may exhibit little relationship to long-term erosion rates determined by cosmogenic nuclides or thermochronometric approaches. This temporal disconnect results from large changes in sediment storage and massive sediment loading during deglaciation. Understanding the connection between erosion and sediment-transport rates and the timescale over which they are representative is an important consideration for geomorphological studies.

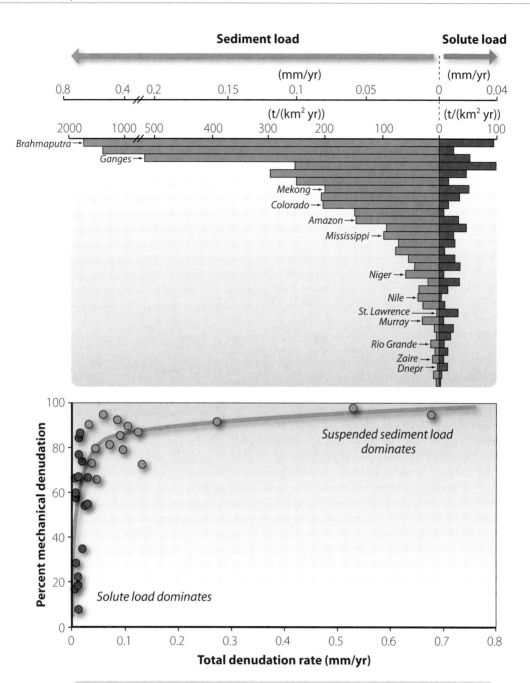

FIGURE 14.12 Dissolved and Mechanical Loads. Both the dissolved and mechanical loads carried by the world's largest rivers vary greatly and are in general related to tectonic setting. [Adapted from Summerfield and Hulton (1994).]

Sediment yield generally decreases with increasing drainage basin area due to increased trapping and storage of sediment downstream (in landforms such as floodplains and terraces) and the general decline in average basin slope as basin area increases [**Figure 14.13**]. Globally, the highest short-term sediment yields are reported from volcanically disturbed river systems where fresh deposits of loose fine-grained material blanket steep, devegetated slopes. Ranges of glacial and fluvial sediment yields are comparable in tectonically active regions where rock uplift is the ultimate pacesetter of landscape-scale erosion. In contrast, basin-scale erosion rates measured

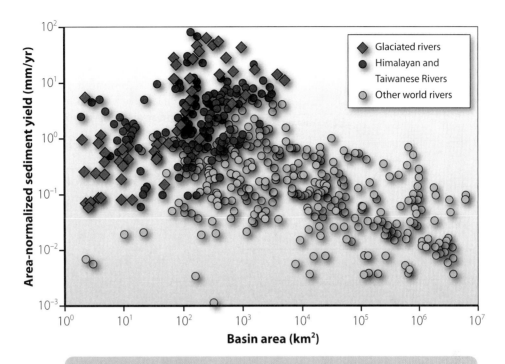

Area-normalized **sediment yields** tend to decrease with increasing drainage area due to localized areas of high erosion rate in upland environments and increasing floodplain storage of material along large rivers. Glaciated rivers (blue diamonds) and high-gradient rivers (e.g., Himalayan and Taiwanese Rivers, red circles) have similar ranges of sediment yields, spanning two to three orders of magnitude.

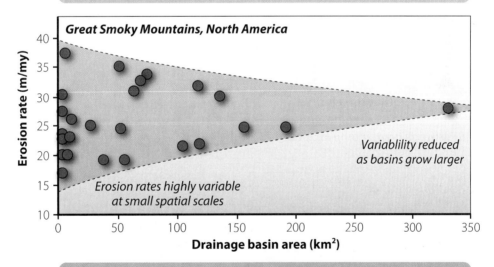

Erosion rates inferred from the concentration of **cosmogenic nuclides** (typically ^{10}Be) in fluvial sediment are more variable at small spatial scales and exhibit less variability at larger spatial scales. The variability at small spatial scales is likely the result of both differences in local erosion rates and the influence of spatially restricted events (such as landslides and rockfalls) on the concentration of cosmogenic nuclides in river sediment. Because rivers draining large basins mix sediment from many small basins, variability in nuclide concentration has a narrower range of values with increasing drainage basin area.

FIGURE 14.13 Erosion, Sediment Yield, and Basin Area. Sediment yields vary with basin area but span similar ranges in glaciated and nonglaciated basins. Erosion rates of small basins tend to be more variable than those for larger basins. [Adapted from Koppes and Montgomery (2009) and Matmon et al. (2003).]

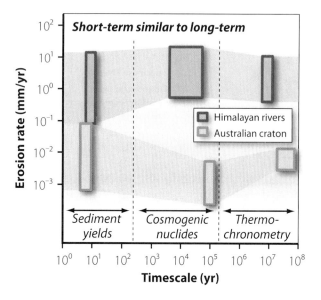

In some landscapes, short-term and long-term erosion rate estimates are comparable. **Sediment yields** from Himalayan rivers are comparable to **erosion rates** derived from cosmogenic nuclides and thermochronometry, as are those from Australian cratonal landscapes. Such landscapes are in **erosional steady-state,** or **dynamic equilibrium.**

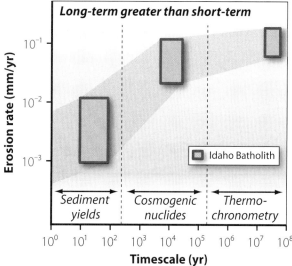

In some areas, long-term erosion rates exceed short-term, modern sediment yields. This discrepancy is likely due to the lack of large erosion events occurring over the time that modern sediment yield data were collected or to greater erosion rates during glacial periods than under the modern climate conditions.

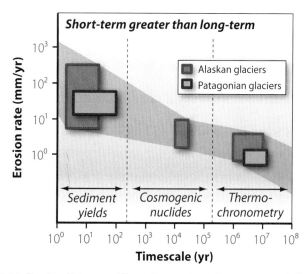

In glaciated terrain, erosion rates derived from short-term sediment yields tend to exceed those from longer-term rates measured by cosmogenic nuclides and thermochronometry because high short-term rates reflect rapid sediment delivery during recent glacial retreat.

FIGURE 14.14 Erosion Rates over Time. Apparent erosion rates may differ when integrated over different timescales. [Adapted from Koppes and Montgomery (2009) and Kirchner et al. (2001).]

cosmogenically appear to have no dependence on basin area because such isotopic measurements primarily reflect the erosion rate of basin hillslopes. Within a landscape, cosmogenically determined erosion rates exhibit higher variability at smaller drainage areas but collapse toward a central tendency in larger basins (Figure 14.13).

Erosion rates measured over different timescales may differ substantially or be comparable [**Figure 14.14**]. The variety of different relationships between short-term and long-term erosion rates illustrates the importance of measuring and comparing erosion rates over a variety of time and space scales. These differing relationships also emphasize the need to test hypotheses about variables controlling erosion rates in a wide variety of landscapes over a substantial range of both time and spatial scales. Landscape evolution is scale-dependent, and integration of surface processes over differing spatial scales can be useful for investigating the outcome of particular processes and evaluating hypotheses based on linkages between process and form.

Applications

Over the past century and a half, the field of geomorphology has matured from initial, isolated attempts to describe and unravel the basic forces shaping the surface of our world to the point where predicting landscape change over time is now possible at a variety of spatial scales. Understanding whole-landscape evolution is important for seeing how human-induced changes in Earth surface processes influence system characteristics. For example, increased sediment supply from denuded hillslopes causes aggradation and flooding downstream; increases in stormwater flows after urbanization cause channel incision.

One way to forecast landscape behavior is to look at the history of landscape change in places where nature and people have conducted similar but unintentional experiments, such as removing trees to clear land for agriculture. Another way is to study contemporary processes to gain insight into system dynamics. The two approaches define complementary avenues of geomorphological inquiry. In a changing world, we need to draw on both to forecast the short-term and long-term impacts of land use, agricultural practices, and a shifting climate. In order to forecast landscape response to global or regional environmental change, one needs to understand what controls the geomorphic resilience of a particular landscape or drainage basin. The answer, of course, depends on the types of processes, settings, and relationships discussed in this book.

Geomorphological processes have long influenced the course of civilizations. At the downstream end of the world's great rivers, the fertile alluvial lowlands of Mesopotamia, Egypt, and China all reflect long-term sediment deposition that built well-watered floodplains where agricultural societies developed and prospered for millennia on fields underlain by sediment eroded from distant uplands. In assessing the environmental challenges facing humanity over the next several centuries, it is hard to overstate the importance of understanding human impacts on the processes determining the trajectory of landscape evolution. In many regions, human actions have changed landscape resilience to disturbance, particularly through the removal or alteration of vegetation.

Recent changes in the pace and distribution of landscape-altering processes are relevant for environmental management, specifically because runoff and the sediment it entrains travel downslope. In other words, the impacts of land use in upland landscapes affect societies and ecosystems in the lowlands. For example, channel widening and entrenchment in response to increased storm runoff following upland urbanization (as when houses are built on steep slopes because of the great views) can rapidly and radically alter channels downstream in the lowlands. As the landscape evolves under human-forcing, the resulting channel incision, bank erosion, and sedimentation all have adverse consequences for property owners who lose their homes, municipalities that maintain infrastructure such as roads, and aquatic ecosystems that are affected by changes in turbidity and streambed material.

The wide range of distinct landscapes around the world reflects different tectonic and climatic settings, as well as the particular geological, geomorphological, and environmental histories that shaped landscape evolution in different regions. As long as landscapes retain slopes and the potential energy gradients they represent, erosional and depositional processes will continue to drive the evolution of landscapes. One can study landscapes for the intellectual challenge and satisfaction, because of their intrinsic beauty, or for purely practical reasons relating to the conservation and management of our planet's changing surface. Whether geomorphology is a casual interest or the intended focus of your career, we encourage you to join us in appreciating the fascinating variety of landscapes that grace our Earth.

Selected References and Further Reading

Anders, A. M., G. H. Roe, D. R. Montgomery, and B. Hallet. The influence of precipitation phase on the form of mountain ranges. *Geology* 36 (2008): 479–482.

Anderson, R. S. Evolution of the Santa Cruz Mountains, California, through tectonic growth and geomorphic decay. *Journal of Geophysical Research* 99 (1994): 20161–20179.

Bishop, P. Drainage rearrangement by river capture, beheading and diversion. *Progress in Physical Geography* 19 (1995): 449–473.

Bookhagen, B., R. C. Thiede, and M. R. Strecker. Late Quaternary intensified monsoon phases control landscape evolution in the northwest Himalaya. *Geology* 33 (2005): 149–152.

Braun, J. Quantitative constraints on the rate of landform evolution derived from low-temperature thermochronology. *Reviews in Mineral Geochemistry* 58 (2005): 351–374.

Brown, R. W., M. A. Summerfield, and A. J. W. Gleadow. Denudational history along a transect across the Drakensberg Escarpment of southern Africa derived from apatite fission track thermochronology. *Journal of Geophysical Research* 107 (2002): 2350. doi:10.1029/2001JB000745.

Brozovic, N., D. W. Burbank, and A. J. Meigs. Climatic limits on landscape development in the northwestern Himalaya. *Science* 276 (1997): 571–574.

Bull, W. B. *Geomorphic Responses to Climate Change*. New York: Oxford University Press, 1991.

Butler, D. R. *Zoogeomorphology: Animals as Geomorphic Agents*. Cambridge, MA: Cambridge University Press, 1995.

Church, M., and O. Slaymaker. Disequilibrium of Holocene sediment yield in glaciated British Columbia. *Nature* 337 (1989): 452–454.

Clark, M. K., and L. H. Royden. Topographic ooze: Building the eastern margin of Tibet by lower crustal flow. *Geology* 28 (2000): 703–706.

Cockburn, H. A. P., R. W. Brown, M. A. Summerfield, and M.A. Seidl. Quantifying passive margin denudation and landscape development using a combined fission-track thermochronology and cosmogenic isotope analysis approach. *Earth and Planetary Science Letters* 179 (2000): 429–435.

Cox, K. G. The role of mantle plumes in the development of continental drainage patterns. *Nature* 342 (1989): 873–877.

Davis, W. M. Base level, grade, and peneplain. *Journal of Geology* 10 (1902): 77–111.

Davis, W. M. The geographical cycle. *Geographical Journal* 14 (1899): 481–504.

Derricourt, R. M. Retrogression rate of the Victoria Falls and the Batoka Gorge. *Nature* 264 (1974): 23–25.

Dietrich, W. E., D. Bellugi, L. Sklar, et al. "Geomorphic transport laws for predicting landscape form and dynamics." In P. Wilcock and R. Iverson, eds., *Prediction in Geomorphology*, Washington, DC: American Geophysical Union, 2003.

Dietrich, W. E., C. J. Wilson, D. R. Montgomery, et al. Erosion thresholds and land surface morphology. *Geology* 20 (1992): 675–679.

Dunne, T. Rates of chemical denudation of silicate rocks in tropical catchments. *Nature* 274 (1978): 244–246.

Finnegan, N. J., B. Hallet, D. R. Montgomery, et al. Coupling of rock uplift and river incision in the Namche Barwa-Gyala Peri massif, Tibet. *Geological Society of America Bulletin* 120 (2008): 142–155.

Gabet, E. J., and T. Dunne. Landslides on coastal sage-scrub and grassland hillslopes in a severe El Niño winter: The effects of vegetation conversion on sediment delivery. *Geological Society of America Bulletin* 114 (2002): 983–990.

Gilbert, G. K. The convexity of hilltops. *Journal of Geology* 17 (1909): 344–350.

Gilbert, G. K. *Report on the Geology of the Henry Mountains*: U.S. Geographical and Geological Survey of the Rocky Mountain Region. Washington, DC: Government Printing Office, 1877.

Goudie, A. *The Changing Earth: Rates of Geomorphic Processes*. Oxford, UK: Blackwell, 1995.

Granger, D. E., D. Fabel, and A. N. Palmer. Pliocene-Pleistocene incision of the Green River, Kentucky, determined from radioactive decay of cosmogenic ^{26}Al and ^{10}Be in Mammoth Cave sediments. *Geological Society of America Bulletin* 113 (2001): 825–836.

Hack, J. T. Interpretation of erosional topography in humid temperate regions. *American Journal of Science* 258-A (1960): 80–97.

Hancock, G. S., and R. S. Anderson. Numerical modeling of fluvial strath-terrace formation in response to oscillating climate. *Geological Society of America Bulletin* 114 (2002): 1131–1142.

Herman, F., E. J. Rhodes, J. Braun, and L. Heiniger. Uniform erosion rates and relief amplitude during glacial cycles in the Southern Alps of New Zealand, as revealed from OSL-thermochronology. *Earth and Planetary Science Letters* 297 (2010): 183–189.

Howard, A. D. A detachment-limited model of drainage basin evolution. *Water Resources Research* 30 (1994): 2261–2285.

Howard, A. D., W. E. Dietrich, and M. A. Seidl. Modeling fluvial erosion on regional to continental scales. *Journal of Geophysical Research* 99 (1994): 13971–13986.

Kelsey, H. M., D. C. Engebretson, C. E. Mitchell, and R. L. Ticknor. Topographic form of the Coast Ranges of the Cascadia Margin in relation to coastal uplift rates and plate subduction. *Journal of Geophysical Research* 99 (1994): 12245–12255.

Kirchner, J. W., R. C. Finkel, C. S. Riebe, et al. Mountain erosion over 10 yr, 10 k.y., and 10 m.y. time scales. *Geology* 29 (2001): 591–594.

Kirkby, M. J. "Hillslope process-response models based on the continuity equation." In D. Brunsden, ed., *Slopes: Form and Process*, London: Institute of British Geographers Special Publication No. 3, 1971.

Knox, J. C. Valley alluviation in southwestern Wisconsin. *Annals of the Association of American Geographers* 62 (1972): 401–410.

Kooi, H., and C. Beaumont. Escarpment evolution on high-elevation rifted margins: Insights derived from a surface processes model that combines diffusion, advection, and reaction. *Journal of Geophysical Research* 99 (1994): 12191–12209.

Kooi, H., and C. Beaumont. Large-scale geomorphology: Classical concepts reconciled and integrated with contemporary ideas via a surface processes model. *Journal of Geophysical Research* 99 (1996): 3361–3386.

Koons, P. O. The topographic evolution of collisional mountain belts: A numerical look at the Southern Alps, New Zealand. *American Journal of Science* 289 (1989): 1041–1069.

Koons, P. O. The two-sided orogen: Collision and erosion from the sand box to the Southern Alps. *Geology* 18 (1990): 679–682.

Koppes, M. N., and D. R. Montgomery. The relative efficacy of fluvial and glacial erosion over modern to orogenic time scales. *Nature Geoscience* 2 (2009): 644–647.

Leopold, L. B., and W. Bull. Base level, aggradation, and grade. *Proceedings of the American Philosophical Society* 123 (1979): 168–202.

Loget, N., and J. van Den Driessche. Wave train model for knickpoint migration. *Geomorphology* 106 (2009): 376–382.

Matmon, A., P. R. Bierman, J. Larsen, et al. Temporally and spatially uniform rates of erosion in the southern Appalachian Great Smoky Mountains. *Geology* 31 (2003): 155–158.

Matmon, A., Y. Enzel, E. Zilberman, and A. Heimann. Late Pliocene to Pleistocene reversal of drainage systems in northern Israel: Tectonic implications. *Geomorphology* 28 (1999): 43–59.

Melhorn, W. N., and R. C. Flemal, eds. *Theories of Landform Development*. London and Boston: Allen and Unwin, 1975.

Milliman, J. D., and J. P. M. Syvitski. Geomorphic/tectonic control of sediment discharge to the ocean: The importance of small mountainous rivers. *Journal of Geology* 100 (1992): 525–544.

Mitchell, S. G., and D. R. Montgomery. Influence of a glacial buzzsaw on the height and morphology of the central Washington Cascade Range, USA. *Quaternary Research* 65 (2006): 96–107.

Montgomery, D. R., G. Balco, and S. Willett. Climatic, tectonics, and the morphology of the Andes. *Geology* 29 (2001): 579–582.

Penck, W. *Morphological Analysis of Landforms*. New York: St. Martin's Press, 1953.

Persano, C., F. M. Stuart, P. Bishop, and T. Dempster. Deciphering continental breakup in eastern Australia using low-temperature thermochronometers. *Journal of Geophysical Research* 110 (2005): B12405. doi:10.1029/2004JB003325.

Renwick, W. Equilibrium, disequilibrium, and non-equilibrium landforms in the landscape. *Geomorphology* 5 (1992): 265–276.

Roering, J. J., J. W. Kirchner, L. S. Sklar, and W. E. Dietrich. Experimental hillslope evolution by nonlinear creep and landsliding. *Geology* 29 (2001): 143–146.

Roering, J., T. Perron, and J. Kirchner. Functional relationships between denudation and hillslope form and relief. *Earth and Planetary Science Letters* 294 (2007): 245–258.

Sadler, P. Sediment accumulation rates and the completeness of stratigraphic sections. *Journal of Geology* 89 (1981): 569–584.

Schumm, S. A. "Geomorphic thresholds and complex response of drainage systems." In M. Morisawa, ed., *Fluvial Geomorphology*, Binghamton: State University of New York, 1973.

Schumm, S. A., and R. W. Lichty. Time, space and causality in geomorphology. *American Journal of Science* 263 (1965): 110–119.

Stock, J. D., and D. R. Montgomery. Geologic constraints on bedrock river incision using the stream power law. *Journal of Geophysical Research* 104 (1999): 4983–4993.

Strahler, A. N. Hypsometric (area-altitude) analysis of erosional topography. *Geological Society of America Bulletin* 63 (1952): 1117–1141.

Summerfield, M. A., ed. *Geomorphology and Global Tectonics*. New York: John Wiley & Sons, 2000.

Summerfield, M. A., and N. J. Hulton. Natural controls of fluvial denudation rates in major world drainage basins. *Journal of Geophysical Research* 99 (1994): 13871–13883.

Syvitski, J. P. M., C. J. Vörösmarty, A. J., Kettner, and P. Green. Impact of humans on the flux of terrestrial sediment to the global coastal ocean. *Science* 308 (2005): 376–380.

Thornes, J. B., and D. Brunsden. *Geomorphology and Time*. London: Methuen, 1977.

Trimble, S. W. Contribution of stream channel erosion to sediment yield from an urbanizing watershed. *Science* 278 (1997): 1442–1444.

Whipple, K. X. The influence of climate on the tectonic evolution of mountain belts. *Nature Geoscience* 2 (2009): 97–104.

Whipple, K. X., and G. E. Tucker. Dynamics of the stream-power river incision model: Implications for height limits of mountain ranges, landscape response timescales, and research needs. *Journal of Geophysical Research* 104 (1999): 17661–17674.

Willett, S. D. Orogeny and orography: The effects of erosion on the structure of mountain belts. *Journal of Geophysical Research* 104 (1999): 28957–28981.

Willett, S. D., C. Beaumont, and P. Fullsack. Mechanical model for the tectonics of doubly vergent compressional orogens. *Geology* 21 (1993): 371–374.

Willett, S. D., and M. T. Brandon. On steady states in mountain belts. *Geology* 30 (2002): 175–178.

Willgoose, G. Mathematical modeling of whole landscape evolution. *Annual Review of Earth and Planetary Sciences* 33 (2005): 443–459.

DIGGING DEEPER Is This Landscape in Steady State?

How do you know whether a landscape is in **steady state** or is changing in some way over time? The answer and the tools used to find the answer depend on the temporal and spatial scale at which you ask the question. To think more about steady state as a concept, we examine two geomorphically important end-members, tectonically active mountain ranges and long-dead orogens.

The southern Appalachian Mountains, a tectonically inactive mountain belt running parallel to the Atlantic passive margin, have catalyzed geomorphic thinking for more than a century. Here, William Morris Davis did fieldwork that led to the conceptual model of a **geographical cycle** of landscape evolution for which he is best known (Davis, 1899; Figure 14.3). Davis interpreted concordant summits that rise to comparable elevations as evidence of former **peneplains** (low-relief erosion surfaces) and invoked renewed uplift after planation to explain incision and valley formation. Davis's model presumed that landscapes were rejuvenated by episodic pulses of uplift and that topography then decayed following a predictable trajectory—a decidedly non–steady-state approach. In contrast, Hack (1960) rejected the Davisian interpretation and made an explicitly steady-state argument. He suggested that Appalachian topography reflected the strength of the underlying bedrock and that the landscape en masse changed little over time. He referred to this overall steady state as **dynamic equilibrium.** How is one to evaluate these two seemingly incompatible views of the same landscape?

In part the answer lies in what one means by steady state. Willet and Brandon (2002) recently described several types of steady state applicable to active mountain ranges, two of which are testable and particularly relevant geomorphologically: (1) **topographic steady state**, where geomorphological metrics such as the average elevation or relief of the range remain similar over time; and (2) **flux steady state,** during which the mass of material brought into mountain belts by the movement of the tectonic plates is equal to that removed by surface erosion. We may also ask the geomorphologically relevant question of whether the rate of sediment generation (erosion) on hillslopes matches the rate of sediment export by rivers?

Critical to any rigorous evaluation of steady state is an understanding of the rate at which landscapes change over time and space. Until recently, such an assessment was not possible because geomorphologists had few tools capable of measuring the pace and spatial distribution of landscape evolution. Most critical to assessing questions of steady state is the ability to compare rates of, and thus the balances between, uplift and denudation over different timescales. In the late 1800s, Davis could hypothesize but he could not test his hypotheses quantitatively, although within a few decades the first relevant data would begin to appear.

The first continental-scale erosion rate estimates were made over a century ago by Dole and Stabler (1909). They used very fragmentary records of the mass of material moving downriver per unit time (particulate **sediment yield** and **dissolved load**), along with the critical (and incorrect) assumption that erosion and sediment yield were balanced (a short-term steady state) to estimate that the continental United States was lowering at about 30 m per million years.

The trouble is, long-term erosion rates calculated from short-term estimates of fluvial sediment and dissolved load transport are often in error because the fundamental assumption, that of steady state, is violated. Trimble (1977) pointed out the fallacy of assuming stream equilibrium over human timescales and the importance of time-varying sediment storage, on the landscape, in channels, and behind dams. Trimble's work showed that in the southeastern United States, only 6 percent of the sediment eroded historically from hillslopes could be accounted for by sediment yield [**Figure DD14.1**]. The rest of the sediment remained stored at the foot of colluvial slopes, in floodplain sediments, as terraces alongside channels, and in reservoirs. Clearly, on the human timescale, the southern Appalachian Piedmont was a non–steady-state system.

Dividing the mass of sediment delivered downstream by the mass of sediment eroded off the landscape results in the **sediment delivery ratio,** which for short-term sediment-yield data decreases as the size of the basin increases. This decrease reflects the change in typical basin geomorphology with basin area. Specifically, as basin size increases and stream gradient lowers, sediment storage on footslopes, in overbank deposits, on terraces, and in broad flood plains, becomes more important. Thus, most compilations of sediment yields expressed as erosion rates decline with increasing basin area (Figures 14.13, upper panel, and DD7.1). Unlike erosion rate estimates based on sediment yields, those calculated from the concentration of cosmogenic nuclides in river sediment (see Chapter 2) show no dependence on basin area and thus are thought to record more faithfully the long-term erosion rate of source basins (Figure 14.13, lower panel).

Using a combination of sediment yields, cosmogenic nuclide data, thermochronometric information, and incision rates, different definitions of landscape steady state are being tested worldwide. In the Coast Range of Oregon, Reneau and Dietrich (1991) found that rates of sediment production on hillslopes (as indicated by dated sediment accumulation in colluvial hollows) matched river sediment yields well, a result confirmed by later cosmogenic nuclide analysis of river sediment. Several hundred kilometers north, Pazzaglia and Brandon (2001) found that incision rates, deduced from dated terraces on the Clearwater River, matched long-term, tectonic uplift rates well for the orogenic wedge of the

DIGGING DEEPER Is This Landscape in Steady State?

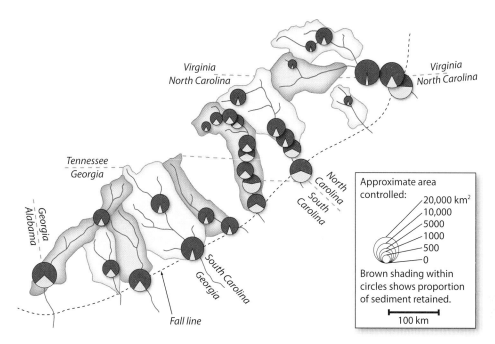

FIGURE DD14.1 Map of the southeastern United States showing rivers (thin blue lines), their drainage basins (shaded areas), and reservoirs (indicated by circles). Circle size increases with drainage basin area upstream of the reservoir (see the key). The larger the dark brown shaded part of the circle, the more recently eroded sediment is retained behind the dam. The data show that most of the sediment eroded from slopes in the past remains trapped in these basins. [From Trimble (1977).]

Olympic Mountains developed above the Cascadia subduction zone in Washington State. Both of these studies confirm flux steady state but on different temporal and spatial scales.

The development of thermochronologic and cosmogenic nuclide techniques provides rate data on a variety of timescales. In the southern Appalachian highlands, rates of erosion averaged over vastly different integration times (10^2 to 10^8 years) are similar (few tens of meters per million years) and match contemporary sediment yields well [**Figure DD14.2**], suggesting relatively steady erosion of the range over time driven by the isostatic response to

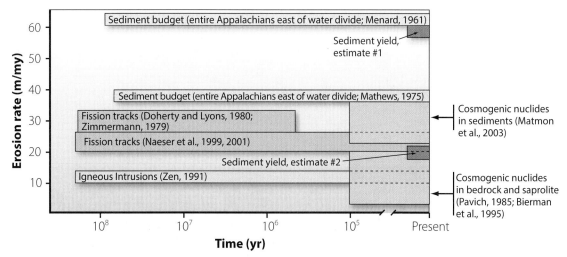

FIGURE DD14.2 Diagram showing average erosion rate over time for the southern Appalachian Mountains. The horizontal length of the bar is the integration time. The vertical position of the bar represents the erosion rate. Citations are given in the original paper (Matmon et al., 2003). Although different methods of estimating erosion rates integrate over different timescales, the results are similar. Appalachian Mountain erosion rates are a few tens of meters per million years. [From Matmon et al. (2003).]

DIGGING DEEPER Is This Landscape in Steady State? (continued)

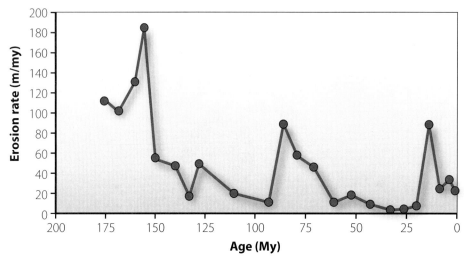

FIGURE DD14.3 Graph showing the erosion rate of the Appalachian Mountains estimated by considering the volume of sediment deposited offshore in the Atlantic Ocean over time. High rates of erosion before 150 million years ago are related to rifting along the now-passive Atlantic margin. Over the past 150 million years, rates of erosion inferred from sediment volumes have ranged from more than 80 to less than 10 m per million years, averaging 20 m per million years, similar to results shown in Figure DD14.2. [From Pazzaglia and Brandon (1996).]

erosion (Matmon et al., 2003). The sedimentary record offshore of eastern North America suggests that rates of sediment export from this passive margin varied, when considered as an erosion rate, by as much as 10-fold over the past 150 million years [**Figure DD14.3**]. We do not yet know whether apparent changes in sediment export rate on the multimillion-year timescale represent changing rates of erosion, changing rates of sediment storage, or inaccuracies in our interpretation of the sedimentary record.

We now have the data to recognize that in some locations and over some timescales, steady state is not a viable hypothesis. For example, in the Idaho batholith (Kirchner et al., 2001) and in southeastern Australia (Tomkins et al., 2007), thermochronologic and cosmogenic erosion rates match well but short-term sediment yields are many times lower [**Figure DD14.4**]. These studies suggest that climate shifts (such as from the Pleistocene to Holocene) and rare, and thus unsampled, storm events move large amounts

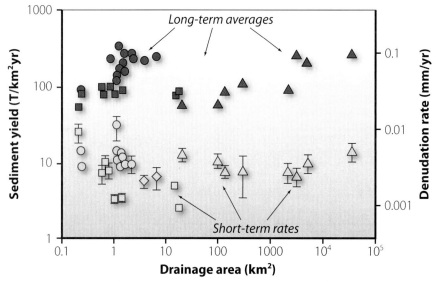

FIGURE DD14.4 Plot comparing short-term sediment yields (yellow symbols) to long-term, cosmogenic ^{10}Be estimates (blue symbols) of erosion rate in the Idaho batholith area, western North America. The discrepancy between short-term rates and long-term averages likely reflects sediment moved during exceptionally large storm events that occur so infrequently that they are not included in the short-term record. [From Kirchner et al. (2001).]

DIGGING DEEPER Is This Landscape in Steady State?

of mass off the landscape and through the river network. While the steady-state assumption may be valid for these landscapes at millennial and greater timescales, it is invalid over decades and centuries—the time frame over which sediment yields are commonly measured.

With new techniques providing quantitative information about rates of landscape change, this is an exciting time for geomorphology. We can now directly test hypotheses concerning landscape evolution and evaluate theories of landform development. For example, if erosion rates and sediment yields are in balance, we know that sediment storage is not changing. If erosion rates integrated over different time intervals, such as those provided by cosmogenic nuclides, a variety of thermochronometers, and interpretations of the sedimentary record all line up, then steady state is likely. If erosion rates are not the same over different time frames, the steady-state assumption is not valid.

Davis, W. M. The geographical cycle. *Geographical Journal* 14 (1899): 481–504.

Dole, R. B., and H. Stabler. Denudation. *United States Geological Survey Water Supply Papers* 234 (1909): 78–93.

Hack, J. T. Interpretation of erosional topography in humid temperate regions. *American Journal of Science* 258A (1960): 80–97.

Kirchner, J. W., R. C. Finkel, C. S. Riebe, et al. Mountain erosion over 10 year, 10 k.y., and 10 m.y. timescales. *Geology* 29 (2001): 591–594.

Matmon, A., P. R. Bierman, J. Larsen, et al. Temporally and spatially uniform rates of erosion in the southern Appalachian Great Smoky Mountains. *Geology* 31 (2003): 155–158.

Pazzaglia, F. J., and M. T. Brandon. A fluvial record of long-term steady state uplift and erosion across the Cascadia Forearc High, western Washington State. *American Journal of Science* 301 (2001): 385–431.

Pazzaglia, F. J., and M. T. Brandon. Macrogeomorphic evolution of the post-Triassic Appalachian Mountains determined by deconvolution of the offshore basin sedimentary record. *Basin Research* 8 (1996): 255–278.

Reneau, S. L., and W. E. Dietrich. Erosion rates in the southern Oregon Coast Range: Evidence for an equilibrium between hillslope erosion and sediment yield. *Earth Surface Processes and Landforms* 16 (1991): 307–322.

Tomkins, K. M., G. S. Humphreys, M. T. Wilkinson, et al. Contemporary versus long-term denudation along a passive plate margin: The role of extreme events. *Earth Surface Processes and Landforms* 32 (2007): 1013.

Trimble, S. W. The fallacy of stream equilibrium in contemporary denudation studies. *American Journal of Science* 277 (1977): 876–887.

Willett, S. D., and M. T. Brandon. On steady states in mountain belts. *Geology* 30 (2002): 175–178.

WORKED PROBLEM

Question: A central geomorphological question is the conditions under which transient landscape evolution occurs versus conditions in which a steady-state, dynamic equilibrium persists. What are the dominant factors affecting whether a landscape is in steady state over geological timescales?

Answer: Central to whether a landscape is in steady state is the stability of boundary conditions over time including the rate of rock uplift, the climate, and base level. Over long-term, or geological, timescales a steady tectonic and climatic regime will eventually lead to development of a steady-state landscape. How long would this take? A minimum estimate can be made by dividing the topographic relief (the range in elevation in a landscape, from the highest to lowest point) by the rock uplift rate. Hence, for the Himalaya, with an average elevation of 5 km and an erosion rate of 2 mm per year, it would take at least 2.5 million years for the landscape to develop steady-state topography (assuming that the erosion rate ≈ the uplift rate). Given that there have been numerous glacial–interglacial oscillations over the same time frame, it is unlikely that the range is in a strict topographic steady state over shorter timescales. However, the average elevation of the range may still be in a steady state as the tectonic forcing of India colliding with Asia has been maintained for well over 10 million years.

The key factor in addressing the conditions under which transient landscape evolution occurs versus conditions in which a steady-state, dynamic equilibrium persists is the time it takes for the landscape to respond to changes in external forcing (i.e., changes in climate or tectonics). Specifically, how long does it take a change in base level to propagate through the landscape (spatial scale) over which one is concerned. This question is in many ways best answered by understanding the rate at which landscapes erode and the rate at which rivers incise—two intimately linked parameters.

KNOWLEDGE ASSESSMENT Chapter 14

1. List the five factors governing landscape evolution.
2. Explain why old mountain ranges are so long-lived.
3. Explain why only half the sediment shed off the Andes makes it to the ocean.
4. List three factors influencing bedrock erodibility.
5. Explain why topography can be used to map the underlying bedrock structure and lithology in some areas but not in others.
6. What limits the height of the Tibetan plateau and the Altiplano?
7. Explain how climate affects the processes and tempo of landscape evolution.
8. List three ways in which glaciers affect the evolution of landscapes.
9. Predict how steep forested hillslopes will respond if trees are removed.
10. Explain the importance of time frames in considering the factors important for landscape evolution.
11. Who was William Morris Davis and what did he propose about landscape evolution?
12. Define and contrast steady-state and transient landscapes.
13. Provide examples of steady-state and transient landscapes.
14. Explain how landscapes in dynamic equilibrium behave.
15. Give an example of how physical models are used to study landscape evolution and point out their limitations.
16. Define a conceptual model and provide an example of a conceptual model important to geomorphology.
17. Explain how and why mathematical models are used in geomorphology.
18. Compare and contract diffusive and advective sediment transport.
19. Contrast the two settings in which knickpoints form.
20. Provide an example of stream capture driving landscape evolution.
21. What is the most common relict landscape?
22. What is the range of erosion rates measured on Earth?
23. Explain how marine and river terraces can be used to estimate rock and surface uplift rates.
24. Define the sediment delivery ratio.
25. Explain how and why the sediment delivery ratio changes as a function of basin scale and integration time.
26. List three ways by which erosion rates can be estimated and compare their integration times and the assumptions underlying each method.
27. Worldwide, how do chemical and mechanical denudation rates compare?
28. Provide examples to explain why erosion rates vary over time in some landscapes and not in others.

Table of Variables

A	drainage area (km²), or constant in Glen's flow law	SS	shear strength (Pa)
A_{cs}	cross-sectional area (m²)	ST	storage
(a, c, k)	hydraulic geometry equation coefficients	S_y	yield strength (Pa)
(b, f, m)	hydraulic geometry equation exponents	t	time
C	cohesion (kPa)	T	wave period (s)
c_1, c_2	constants of integration	V	wave velocity (m/s)
C_L	channel length (km)	V_L	valley length (km)
D	water or flow depth (m), or hillslope diffusivity (m²/yr)	U	flow velocity (m/s), or uplift rate
d_{50}	median diameter of the bed-forming grains (mm)	U_I	isostatically driven rock uplift
DD	drainage density (km/km²)	U_R	uplift of rock
E	erosion rate (mm/yr, or m/My), or state of equilibrium	U_S	surface uplift
ET	evapotranspiration (mm)	U_T	tectonically driven rock uplift
Fr	Froude Number	W	stream width (m)
FS	factor of safety (SS/τ)	W_b	rate of bedrock weathering (mm/yr, or m/My)
g	gravitational acceleration (m/s²)	W_i	weight of slice in method of slices
H	wave height, or amplitude (m)	x	distance (m)
h	hydraulic head, or height of the water table above a slide plane (m)	X_c	distance from ridgecrest to channel head (m)
I	inputs	x_d	distance from drainage divide (km)
i	index variable	z	elevation (m)
k	rate constant	z_0	roughness length
K	hydraulic conductivity (m/s), or fluvial rate parameter	z_i	thickness of a column of rock, soil, water, ice, or air (m)
K_d	landscape diffusivity, k/ρ_s (m²/yr)	z_b	elevation of top of bedrock (m)
K_r	recession constant	z_s	soil thickness (m)
K_s	steepness index	λ	wavelength (m)
l	length (m)	Δ	change in . . .
L	channel length (km)	Σ	summation of . . .
L_c	length of channel (m)	Ω	stream power per unit channel length (watts/m)
L_p	lag-to-peak (hr)	α	natural log of recession constant, $\ln K_r$, or taper slope of critical wedge
L_v	length of valley (m)	B	slab dip in critical wedge
m	ratio of the saturated soil thickness (h) to the total soil depth (z_s) above the slide plane that is saturated ($m = h/z_s$), or exponent in sediment transport equations	β_i	angle of failure plane from vertical at center of mass, method of slices
n	Manning roughness coefficient, or exponent in Glen's flow law, or exponent in sediment transport equations	$\dot{\varepsilon}$	strain rate
N	number of samples or observations	φ	angle of internal friction (degrees)
O	outputs	μ	pore water pressure (Pa)
P	pressure (Pa)	μ_f	viscosity of a fluid
p	probability (of an event)	π	constant, 3.14159
P_w	wetted perimeter (m)	θ	slope angle (degrees)
PE	potential energy	θ_c	concavity index
PmC	¹⁴C activity, percent modern carbon	ρ	density (kg/m³)
Q	discharge; volumetric rate of fluid flow (m³/s)	ρ_c	crustal density (kg/m³)
Q_b	bedload sediment transport rate, or bedload supply (kg/s)	ρ_f	fluid density (kg/m³)
Q_{in}	flux of mass in (kg/s)	ρ_m	mantle density (kg/m³)
Q_{out}	flux of mass out (kg/s)	ρ_p	particle density (kg/m³)
Q_p	peak discharge (m³/s)	ρ_r	density of bedrock (kg/m³)
q_s	mass transport; volumetric flux rate (m²/yr)	ρ_s	density of soil or sediment (kg/m³)
Q_t	discharge at time t (m³/s)	ρ_w	density of water (kg/m³)
R	hydraulic radius (m)	σ	normal stress (Pa)
r	rank (in order of observations), or radius of particle	σ'	effective normal stress (normal force per unit area due to the weight of the material less the buoyant force) (Pa)
R_a	lag time before response		
R_T	response time	σ_t	tensile strength (Pa)
R_x	relaxation time	τ	shear stress (Pa)
RI	recurrence interval (yr)	τ_c	critical shear stress (Pa)
S	channel, water, or land surface slope (degrees)	τ_c^*	Shields parameter (dimensionless critical shear stress)
S_s	particle settling speed (m/s)	τ_{cb}	critical shear stress to incise bedrock (Pa)
SL	stream gradient index	ω	unit stream power (watts/m²)

Index

Note: Page numbers followed by the letter f indicate figures; page numbers followed by the letter t indicate tables; and page numbers in *italics* indicate photographs.

A horizons, 92, 93f, *94, 95, 99,* 101f, 352
aa lava, 368–370
ablation, 297, 427
ablation till, 310
ablation zone, 297, 298f, 301, *310,* 428f
abrasion, 211, 337
   and streamflow, 189–190
   and wind, 333
abyssal basins, 256, 278
abyssal fans, 277
abyssal hills, 278
abyssal plains, 9f, 278, 329
accelerator mass spectrometer, *59*
accelerator mass spectrometry (AMS), 54, 71
accommodation space
   alluvium storage, 202
accretionary wedge, 12f
accumulation
   of glaciers, 297–299
accumulation area ratio (AAR), 427, 428f
accumulation zone, 297, 298f, 301, 428f
acid rain
   pH value, 86t
active plate margins, 11, 254, 255f, 277, 357f
active remote sensing, 45f, 46
active tectonic controls, 392
activity (decay rate), 69
adiabatic cooling, 452
advective fluvial processes, 170f, 470
aeolian abrasion, 337
aeolian activity
   geologic records, 351
aeolian dust deposits, 348–349
aeolian erosion
   rates of, 337
aeolian features and landforms, 336f, 342–343
aeolian processes, 334, 335–341
   and landscape evolution, 464
aeolian sediment, 330, 331–332, 351
   dating of, 56
aeolian sediment deposition, 330, 341
aeolian sediment transport, 329, 337–341, 350
aeolian transport features and landforms, 343–347
aeolianite, 330
aerial photography, 47
African plate, 8f
Agassiz, L., 324

aggradation, 442
   and channel response, 208
   and hillslope erosion, 222
aggradational terraces, 235
air
   as a fluid, 331–334
   viscosity and density of, 331
air turbulence, 339
airborne dust sources and trajectories, 339f
airflow, turbulent, 331
Alaska
   earthquake, *155,* 272
   erosion, 7
   surging glaciers, 302
Alaska Range, *7, 202*
Alaskan North Slope, *319*
albedo
   of glaciers, 296
   and solar energy, 452
   temperature gradients, 14
Aleutian Trench, 279
Aleutian volcanic arch, 392
alfisols, 98, 99
alluvial banks
   bankfull flow, 187
alluvial bed cover, 213f
alluvial channel morphology, 207
alluvial channel reaches, 202, 413
alluvial channels, 182, *183,* 191, 197
alluvial fans, 49, 237, 238–239, 343, *402*
   abandoned channels on, *239*
   at divergent boundaries, 12f
alluvial mid-channel island, *184*
alluvial sediments, 238
alluvial storage, 180
alluvial terraces, 99, *217*
alluvial valley segments, 228, 229f
alluvial valleys, 230
alluvium, 182–183, 220f
   in channels, 184
   in drainage basins, 247f
   on hillslopes, 147
alpine glacial settings, 313
alpine glacial systems
   mass balances of, 427
alpine glaciers, 292, 293f, 294, 295
   landforms of, 313–314
   thickness and slope of, 300
alpine landscapes, *18*
alpine moraines, 314
alternate bars, 200
Altiplano, 11, 391, 405, 464, 479

altithermal climatic optimum, 438
alumina octahedra, 87
aluminosilicates, 87, 91
aluminum oxides, 86
Amazon River, *75,* 197, 227, 397
   drainage basins, 217, 218
   tides, 259
Amazon River Basin, 22, 222, 452
Amazon River delta, 283
Amazonia, 452–453
Amazonian rain forest, 452–453
amber-rat, 431f, 432
amino-acid racemization, 51, 52t
analog experiments, 65–66
analytical force balances, 469
anastomosing channels, 199f, 200, 201–202
anchored dunes, 345
ancient deltas, 448
ancient dune sands, 351
ancient landscapes, 471, 478–479
Andean Mountains
   subduction zone, 11
Andes, *75,* 309, 396, 397, 479
   continental collision zone, 11
   glaciers, 305
   landslides, *154*
andesitic lava, 360–361, 363, 371, 377
andesitic volcanoes, 361
andisols, 97
angle of internal friction, *see* friction angle
angle of repose, 148
angle-of-repose slopes, *77*
anhydrite
   conversion to gypsum, 83, 88
animals
   distribution of and climate zones, 19
   and landscape evolution, 466
   and soil development, 79, 90
anisotropic rock strength, 465
Antarctic Ice Sheet, 294, 302, 322, 427, 452, 454, 455
Antarctica, 435
   ancient glacial ice, 32f
   glaciation of, 452
   ice shelves, 300
antecedent drainage, 225
antecedent rivers, 402
Anthropocene Epoch, 20
anticlinal ridges, 405f, *465*
anticlinal valley, 409
ants, mound-building, 19

Appalachian Mountains, 11, 22, 48, 59, 391, 474, 490
  crustal root of, 9
  drainage density, 169
  erosion rates, 491–492
  landslides, 452
  limestone, 14
  and precipitation, 114
  root network density and strength, 176
  topography, 410
  valley and ridge morphology, 408, 409f
apparent cohesion
  of plant roots, 161
  and vegetation, 163t
applied geomorphology, 28–29
aquifers, 123, 129f, 322, 377
aquitards, 377
arc-trench gap, 356
arches
  of pocket beaches, 265
Arctic landscapes, 18
arêtes, 313, 314
argon/argon ($^{40}Ar/^{39}Ar$) dating, 54
arid channel systems, 187
arid regions
  lakes, 239
  rainfall, 116
  soils, 93f, 98
aridisols, 98
armor layer, 128f
arroyo, 466
artificial levees, 209
aseismic creep, 46
ash fall distribution, 367f
Asia-India collision, 389
asthenosphere, 6
Aswan high dam, 284, 285f
at-a-station hydraulic geometry, 187–189
Atacama Desert, 16, 18, 449, 482
Atlantic Ocean seafloor, 278
Atlantic Ocean basin, 400
atmosphere
  and climate, 454
atmospheric circulation
  and climate zones, 14–15, 15f
  convective cells of, 15–16, 332
atolls, 279, 280f
Australia
  interior of, 11
Australian craton, 13
Austral-Indian plate, 8f
authigenic sediments, 276
Av horizons, 93f, 94, 98, 343, 352
Avk horizon, 352f
avulsion, 187, 193, 206f, 239, 187

B horizons, 92–95, 101f, 337, 343, 352
backscattering
  of radar energy, 47

backshore, 266
backswamps, 207
backwash, 261–262, 266, 267
backwater effects
  of high tides, 274
Baffin Island, 59, 302, 309
bajada, 238
bank erosion, 193
bank failure, 183
bankfull flow, 28, 131–132, 187, 445
bar-built estuary, 271f, 272
bar emergence, 269
bar submergence, 269
barbed drainage, 476
barchan dunes, 346
barrier islands, 265f, 269, 271f, 281
barrier reefs, 279
bars, 198
basal ice velocity, 325, 326f
basal sliding, 299
basal till, 310
basalt columns, 371
basalt flows, 365, 370–371
basaltic cinder cones, 362
basaltic eruptions, 370
basaltic lava, 360t, 362, 365, 377
basaltic lava flows
  dating, 365
basaltic magma, 361, 363
basaltic volcanism, 12f
  landscapes, 368–371
basaltic volcanoes, 368, 375
base level, 442–444
base-level changes, 413, 415f, 426
  and landscape evolution, 463
base-level fall and uplift, 411, 442–443
base saturation, 89, 93f
baseflow, 116, 123, 124f, 126f, 127, 129, 180
Basin and Range Province, 399, 400, 401f
basin morphology
  and channel networks, 223–227
basin relief, 230–231
basin-scale erosion rates, 442
basin-scale precipitation models, 132
basin-scale processes, 219–223
basin sedimentation, 254
batholiths, granitic, 381
bathymetry, 253
beach cusps, 268
beach erosion, 281
beach face, 256, 267–268
beach nourishment, 281
beach profile change, 267
beach ridges, 270
beach slope, 267–268
beaches, 262, 264
  sediment transport and supply, 266
beaches and bars, 266–268
bed and bank material
  of river channels, 181–184
bed and bank roughness, 185

bed-surface grain size, 208
bedding-parallel strike valleys, 224
bedding planes
  orientation of and slope stability, 153, 154
bedforms, 185, 187, 198
bedload, 59–61, 184, 194f, 195f, 196, 197, 208
bedload deposition
  and floodplains, 206
bedload sediment
  colluvial stream reaches, 202
  transport of, 195, 197
bedrock
  erodibility, 211
  frost shattering, 82
  inselbergs and tors, 103, 104f
  landslides, 150, 153, 154, 166
  physical weathering, 80
  and soil formation, 107, 108
  water-sculpted, 112
bedrock channel incision, 212
bedrock channel reaches, 202, 413
bedrock channels, 183, 211
bedrock erosion, 83, 211–212, 213f
  and rock types, 180
  thresholds, 190
bedrock hillslopes, 146, 165
bedrock hollows, 172f
bedrock incision, 189–191, 212
bedrock landscape, 278
bedrock lithology
  and fluvial erosion processes, 211
bedrock meanders, 229
bedrock river incision
  control of rates of, 211–212
bedrock river systems, 464
bedrock scarps, 420
bedrock valley segments, 228
bedrock valleys, 229–230
  sediment storage, 222
beetles, 431–432
benchmarks, 45f, 46
Bering Strait, 427
berms, 256, 265f, 267
beta particle emissions, 54, 55f, 69
bicarbonate ions, 86
biodiversity
  human impact on, 33f
biogenic sediments, 276
biologic colonization
  and volcanoes, 373–374
biologic crusts
  and wind erosion, 330
biologic debris, 431
biological activity
  and soil development, 90
  soil infiltration, 119
  and weathering, 78
biological weathering, 77–80
biology
  and landscape evolution, 462, 464–466

biosphere, 6, 16–20
biotite, 84, 85
bioturbation, 78
    and sheetwash, 151
Bishop Tuff, 356–357, *360*, 366
Black Sea, *266*, 272
Blackhawk landslide, 158
blind thrusts, 402
blind valleys, 135
blowouts, 342–343
    parabolic dunes from, 346
Blue Mountains, Australia, 23
Blum, M. D., 285
bornhardts, 81f, 102
bottomset beds, 273
boulder-bed channels, 232, *233*
boulders, 184
bouldery snout, 157f
boundary conditions
    and sediment flux, 27f
Boundary Creek, *186*
Bowen's Reaction Series, 84
brackish water
    salinity, 256
braid bars, 201
braided channels, 199f, 200, 201, 204, 413
braided rivers, *179*
breakers, 261
breakout, *365*
breakwaters, 263
British Columbia Coast Range, 402
Bronze Age, 32f
buoyant force, 147
buried dunes, 351
buried soils, 100, 101f
burrowing animals, 19, 20, 78, 91
    and soil creep, 151
    and soil infiltration, 119
buttes, 410

C horizons, 95, 352f
calcite, 276
    dissolution of, 86
    in soil, 89
    thermal expansion of, 82
calcium carbonate
    and cohesion, 149
    dating of, 56
    desert soils, 19
    precipitation of, 135
    sands, 276
    soil development, 106f
    subglacial bedrock outcrops, 306
    see also aeolianite
calcrete (caliche), 104
caldera eruptions, 356–357
calderas, 356, 357, 358f
CALIB program, 70
calibrated relative dating, 50–52, 52t
calibrated soil development rates, 99
calibration, 54

caliche (calcrete), 103
calving
    of floating ice, 301, 322
    of ice margins, 295
Canadian Shield, 302, 304
Canyonlands National Park, 155, *330*, *410*
Cape Cod, Massachusetts, *94*, 266, 291, 311, 315, 350
capes, 268
capillary force, 120
    and water, 116
capillary fringe, 120
caprocks, *104*
carbon, 31
carbon atoms, 55f
carbon dioxide
    Earth's atmosphere, 31
    removal from atmosphere, 454
carbon-12 (^{12}C), 54
carbon-14 (^{14}C), 53–54, 55f
carbonate rocks, 86
carbonates
    global cycle of, *31*
    soils, 91
carbonic acid, 86
Carrizo Plain (California), 67f, *393*
cascade channels, 203f
Cascade Mountains, *163*, 356, 309, 392, 402, 452
    continental collision zone, 11
    precipitation, 115
    transverse drainage, 225
cascade reaches, 202–203
Cascade volcanic arc, 356
Cascadia subduction zone, *413*
catchment average value, 63
cation exchange, 88f, 89
cations
    in rock-forming minerals, 85
caves, 135
cavitation
    glacial, 304
centroid lag time, 126f
cesium-137, 61–62
channel avulsions, 207
channel bed
    shear stress, 196f
channel bends
    flow through, 192f
channel change, 241f
channel confinement, 231–232
channel entrenchment, 475
channel erosion, *61*
channel head, *125*, 170–172
channel incision, 189, 211
channel initiation, 170–172
channel meanders, 207
channel migration, 191–193
    bed and bank material, 182
    and distribution of terraces, 237
    and floodplains, 206

    lateral, 207
    and vegetation, *193*
channel-mouth bars, 272
channel network change, 413
channel networks
    and basin morphology, 223–227
    valley segments, 228
channel ordering, 225
channel patterns, 199–212
    sediment erosion and deposition, 191
channel processes
    and regional climate, 180
channel-reach morphology, 202–205
channel reaches
    gorges, 234–235
channel response, 207–208
channel restoration and rehabilitation, 243
channel roughness, 185, *186*, 188f
channel sediment supply, 211
channel slope, 180
channel units, 202
Channeled Scablands, 276, 311–312, 457–458, 478, 479
channels, 22, 179–209, 218
    alluvial valley segments, 230
    bedforms, 198
    discharge variability, 181
    on fans, 239
    root stabilizing effect, 184
    sediment mobilization, 181
    stream grading, 180
characteristic forms, 468
checkerboard weathering
    patterns, 410
chelation, 89
chemical weathering, 77–80, 83–89
    and landscape evolution, 463
chemographs, 128f, 129
Chesapeake Bay, 243, 272, 316
Chilean subduction zone, 357f
China
    loess deposits, 347, *348*
chromium, 348
chronometers, 44
chronosequence, 48
cinder cones, 50, *238*, 362, *363*, 368, 370f, 374, 375
cinders, *see* scoria
circular particle accelerator, 71f
circum-Antarctic current, 452
cirque glaciers, *295*
cirques, 294, 309, 313–314, 427
clast-supported alluvial sediments, 238
clast-supported fluvial deposits, 157
clastic load, 197
clastic sediment, 277
clasts, 353
    pavements, 352
    and sediment movement, 65, 220
    step-pool reaches, 204
clathrates, 455

clay minerals
  base saturation in, 89
  formation of, 87–88
  hydration of, 83
  and hydrolysis, 87
clay soil, 119t
clay weathering sequence, 88f
clays
  cohesion of, 149t
  crystal and sheet structure of, 88f
  formation of, 84
  friction angles, 148, 149t
  and wind, 352–353
clear-cutting
  and landslides, 175, 176f
cliff-and-bench topography, 410
climate, 14–16
  and channel processes, 180
  effects on hydrology and geomorphology, 115–116
  and fire, 441
  geomorphic effects of, 445
  and geomorphology, 425–455
  and glacier advance, 444
  landscape controls on, 452–454
  and landscape evolution, 462, 463
  landscape response to, 447–452
  and soil development, 90
  and solar insolation, 14
  and surface processes, 464
  and tectonics, 435f
  and vegetation, 441
climate archives, 426
climate change, 426–434, *438*
  effects of, 449–452
  and floods, 458
  glacial-interglacial, 447–448
  human impact on, 33f
  and isostatic responses, 448–449
  landform records, 427–429
  and landscapes, 452
  natural archives, 427
  short-term, 439
  and soil, 429
climate cycles, 434–439
  temporal scales, 23
climate geomorphology, 444–447
climate-related landforms and processes, 445–446
climate variability, 438
climate warming, 454
climate zone shifts, 445
climate zones, 14–16, 17f, 426
  distribution of, 445
  and geomorphic processes, *18*
  and Köppen climate classification, 445
  precipitation and temperature ranges, 17f
  and vegetation zones, 17f
climatic forcing
  drainage basins, 240
climbing dunes, 345

closed drainage basins, 218–219
closure temperature range, 63, 64f
clouds
  water budget, 115
coastal barrier migration, 269
coastal dunes, 256, 329
coastal environments, 253, 254
  erosion and sediment transport, 259
coastal erosion, 254, 257f
  and deltas, 283
  United States, 281
coastal geomorphology, 253–275
coastal landforms, 254
coastal landscapes, 412–413
  and wind, 334
coastal marshes, 270–271
Coastal Plain (North America), 22, 23f
coastal-plain estuaries, 271f, 272
coastal processes and landforms, 264–274
coastal retreat and advance, 264
coastal rivers, 274
coastal sediment budgets, 281
coastal sediment dynamics, 254
coastal sediment sinks, 256, 257f
coastal sediment sources, 257f
coastal settings and drivers, 254–264
coastal storms, 440
coastal subsidence, 412–413
coastal uplift, 412
coastline morphologies, *256*
cobble-boulder channels, 199
cobbles
  beaches, 266, 267
  transport of, 450–452
coefficient of thermal expansion, 82
cohesion, 147, 148, 149
  and soil texture, 95
cohesive materials
  and wind scour, 337
cohesive strength, 149, 195f
cold-based glaciers, 302, 303f, 307, 325–326
cold-based ice, 302, 304
collisional margins, 255f, *256*
collisional tectonic settings, 403f
colluvial aprons, 49
colluvial channels, 202, 228
colluvial fan, *222*
colluvial reaches, 202
colluvial soils, 90
colluvial storage, 222
colluvial valley segments, 228
colluvial valleys, 228, 229f
colluvial wedge, 420, 421f
colluvium, 147, 220f
  in drainage basins, 247f
  in hillslopes, 172f
colonnade
  of lava, 371
Colorado Plateau, 22, 351
Colorado River, 139, 140f, 184, 197, 223, 377

Columbia Plateau, 359, 454
Columbia River, *368*, 457
comminuted sediment, 309
compaction, 274
complex responses
  channel evolution, 241f
  and geomorphic thresholds, 26–27
  and transient landscapes, 474
compressional margins and landforms, 398, 401–404
compressional orogens, 401–404
compressive strength, 147
concave hillslopes, 164f
concavity index values, 228
conceptual models
  of landscape evolution, 467–468, 469
confining stress, 148
congruent dissolution, 86
consequent streams, 225
conservation of energy, 24
conservation of mass, 24
constitutive equations, 25
continent arrangements
  and climate, 452
continent-continent collisions, 403f
continental collision zones, 11
continental collisions, 11, 12f, 408f
continental crust, 6, 8–11, 12f
  buoyancy, 12f
  density of, 9, 10f
  differentiation of, 31
  plate convergence, 402
  and subduction zones, 11
  thickness of, 393
continental erosion, 254
continental expansion zones, 399
continental glacial deposits, 427
continental glacial settings, 313
continental glaciation, 292
continental glaciers, 294
continental ice sheets, 292, 303f
continental interiors, 398
  landforms of, 404–408
continental-margin sediment, 277
continental margins, 253, 277–278, 400
  active and passive, 11
  geomorphology of, 254
continental-oceanic subduction zone, 403f
continental plates collisions, 392
continental rifting, 411
continental rifts, 392
continental rise, 277
continental shelves, 11, 254, 277, 281, 426
continental slope, 277
continents
  cratons, 9f
  elevation of, 6, 8–11
  formation of, 31
  interior of, 406
continuous permafrost, 317f, 318
convergent boundaries, 8f, 11

convergent margins, 254, 255f
convergent plate boundaries, 12f
   marine trenches, 279
   subduction zones, 11
convergent slopes, 163, 164f
convex hillslopes, 164f, *166*, 167
Coon Creek Basin, 246, 247f
coppice deposits, 341
coprolite, 146, *147*
coral reefs, 276, 277, 279
   dating rocks of, 56
coral sand, 256
corals, 66
Cordilleran Ice Sheet, 294
core stones, *13*
corn snow, 294
cosmic radiation, 31
cosmic ray flux, 70
cosmic ray protons, 55f
cosmogenic exposure ages, 59
cosmogenic isotopes
   and rates of soil production, 107, 108
cosmogenic nuclide analysis, 52t, 58–59
cosmogenic nuclides, 58–59
   and drainage basin erosion, 63
   erosion rates, 62
   production of, 58f
   sampling, *59*
Coulomb criteria, 148f
Coulomb equation, 147
Cr horizon, 95
Crater Lake, *240*, 356
crater lakes, 380
cratons, 9f, 11, *13*, 405, 465
crescentic gouges, 307, *308*
crescentic ridges, 346
crest (of wave), 259f
crevasse splays, 206
crevasses, 206
critical angle
   of hillslopes, 166
critical flow, 187
critical shear stress, 196f
   and flow velocity, 195
critical wedge, 473f
cross-cutting, 48
crust (Earth), 6–8
   density of, 394
     tectonic plate movement, 11
   see also continental crust; oceanic crust
crustal roots, 8–11, 394, 408f
   mountain ranges of, 10f
crustal thickening, 408f
cryoplanation terraces, 318
cryoturbation, 96–97, 322
cryptogamic crusts, 19
crystalline rocks, 14
cuestas, 410
cumulic soils, 90, 347

currents
   and marine topography, 275
cutbank erosion, 191, 193
cutbanks, 192f, 193, *217*
cyclic time, 27
cyclones, 114

dacitic lava, 363, *365*
dam construction and removal
   and drainage basins, 243
damming
   of rivers, 139
dams, 254
Dansgaard-Oeschger events, 438
Darcy's Law, 121f, 122
Darwin, Charles, 19, 107, 279
dating methods, 69–71
   numerical, 48, 52–59
   relative, 48–52, 52t
daughter isotope, 54
Davis, William Morris, 324, 467–468, 490
Dead Sea–Red Sea rift zone, *476*
Dead Sea Rift, 399, *400*
Dead Sea rift shoulder, *13*
Death Valley, *237*, 337
debris avalanche deposits, 383, 384f
debris avalanches, 156, 375, 384f, 385, 386
debris fans, 157f, *158*, 238, *312*
debris flow behavior
   and flumes, 65
debris flow–delivered sediment, 238
debris flow deposits, *158*, 238, *239*
debris flow sediment
   and channels, *183*
debris flows, 7, 156–157, 374, 375
   channel responses, 208
   off clear-cut slope, 177f
   colluvial channels, 202
   colluvial valleys, 229f
   forest roads, *163*
   Tien Shan mountains, *5*
debris levee, 157f
decay counters, 62f, 69
Deccan Traps (India), 359, 454
Deep Springs Valley, 218
deep-sea dives, 276
deep-sea ridges, 397
deflation hollows, 342–343
deflation model, 352, 353f
deglaciation
   land rising from, 448
   and sea-level curves, 444f
   and uplift, 450f
dehydration
   of minerals, 83
   as weathering mechanism, 83
delta floods, 283
delta process triangle, 275f
delta shrinkage, 284
deltaic deposits, 272
deltaic sedimentation, 272–273
deltaic subsidence, 284

deltas, 272–274, 281, 283–285, 286f, 313f
   erosion, 24
   future of, 284
   sediment supply, 277
dendritic patterns
   of channel networks, 223, 224f
dendrochronology, 52t, 52–53, 53f
deposition
   aeolian sediment, 341
   and meander migration, 192f
   sediment, 195f
deposition zone
   of debris flow, 157
depositional areas
   floodplains and estuarine areas, 25f
depositional coastal zones, 265f
depositional sinks, 24–25
depositional terraces, 235, 236f
depth profiling, 276
desert lakes
   drying of, 447
desert loess, 347
desert pavement, *see* pavements
desert rock
   fracture from thermal stresses, 82
desert varnish, *see* rock varnish
desertification, 350
deserts
   global belt of, 15f, 16
   and landscape evolution, 464
   plant communities in, 19
   vegetation zones, 18
   winds in, 330
detention storage, 132
diamicton, 239, 310
diapir, 397
differential erosion, 398
differential weathering, 102
differentiation
   of continental crust, 31
diffusion dating, 420
diffusion-dominated hillslopes, 167
diffusive hillslope processes, 170f
diffusive processes, 150–152, 470
   hillslope profiles, 168f
Digital Elevation Models (DEMs), 47, 470
digital total stations, 44
dikes, vertical, 380–382
dilative materials, 157
dip slope rockslide, *154*
dip slopes, *147*, 410
disaggregation (of rocks), 80
disappearing streams, 136f
discharge, 126, 128f, 181, 182f
   in channels, l85–186
   rating curve, 187
discharge regime, 208
discharge variability
   in channels, 187–189
discontinuous permafrost, 317f, 318
disintegration (of rocks), 80

dissections, 467–472f
dissolution
    in channel-bed rocks, 189, 191
    and soil profiles, 89
    and weathering, 85
dissolved ions, 77, 128f
dissolved load, 107, 194f, 197, 276, 490
distance determining
    laser ranging, 44–46
distributary channels, 272, 275f
disturbances, geomorphic, 28
    landscape response to, 463f
    spatial scales of, 28
divergence zones, 399
divergent plate boundaries, 8f, 11, 12f
divergent slopes, 163, 164f
doldrums, 332
dolines, 135
domes, 102
downslope creep rates, 152
downstream trends, 225–227
    of channel networks, 232
drag force, 26f
drainage, 476, 478
drainage area
    slope analyses, 414
drainage area–slope relations, 228
drainage basin area
    and stream channel characteristics, 225
drainage basin hypsometry, 479, 480f
drainage basin land use, 249
drainage basin landforms, 233–240
drainage basin outlets
    sediment delivery, 245
drainage basin shape, 225
drainage basins, 22, 25f, 217–244,
        484–487
    basin-scale processes, 219–223
    channel bed material, 232, 233
    elevation continuum, 227
    hydrograph shapes, 127–128
    laboratory experiments, 65
    and mass export, 483
    process domains, 228–230
    sediment budgets, 219–222
    sediment flux, 60f
    tectonic subsidence, 413
    uplands to lowlands, 227–233
    valley segment types, 228–230
drainage density, 166–170
    badlands, 169
    distance of channel head from
        drainage divide, 170f, 171f
    terrestrial, 168, 169
drainage diversion, 476, 477f
drainage divides, 22, 166, 217, 228
    and drainage density, 170f
drainage patterns, 223–225
drainage rearrangement, 476–477
drainage superposition, 225
drainage systems
    of transient landscapes, 476

driftwood
    carbon-14 analysis of, 443
dripstone, 134
driving forces, 25, 26f
    and response time, 28
driving shear stress, 161
driving stresses, 159–160, 161
dropstones, 312, 314
drowned bedrock crust, 265f
drowned coastal zones, 265f
drowned estuary, 265f
drumlins, 313f, 315, 458
dry falls, 479
dune fields, 330, 347
dune-ripple channels, 203f, 204
dune types, 346f
dunes, 198, 256, 257f, 266, 330, 334,
        344–347
duration
    of precipitation events, 113
duricrusts, 102, 103
Dust Bowl era, 337, 338, 348, 466
dust fall, 374
    ice core archives of, 432
dust flux, 349–350
dust loads
    Rocky Mountain snowpacks, 350
dust storms, 337, 339, 350, 466
dust traps, 374
dustfall
    and soil development, 98
dynamic equilibrium, 27, 467, 468f,
        486f, 490
    landscapes in, 468
dynamic steady state, 27
dynamically supported topography,
        392

E horizon, 92, 93f, 94
Earth
    energy balance, 452
    habitability of, 31–33
    oceans, 253
    surface characterizing techniques, 45f
Earth-moon system
    and tides, 257, 258f
earthflows, 158, 159
earthquake recurrence intervals, 417
earthquake shaking
    and landslides, 161
earthquakes, 263, 408
    hazard levels and predicting, 67–68
    plate boundaries and, 8f
    rapid subsidence, 412–413
    and volcanoes, 355, 372, 373f
earthworms, 119
East African Rift, 399
    divergent plate boundary, 11
East African Rift Valley, 393
East Pacific Rise, 356
    volcanoes and, 8f
ebb-tide delta, 269

eccentricity
    of Earth's orbit, 436, 437f
ecological restoration, 29
ecosystems, 16–18
    global distribution of, 18–20
effective discharge, 187
effective normal stress, 149
effusive eruptions, 361
El Niño–Southern Oscillation, 439
Electron and Osceola flows, 381
electron exchange, 83
elemental mass fluxes, 107
elevation terracing, 47
elevation
    and climate zones, 16
    of Earth's surface, 393
eluviation, 89
Elwha sampler, 197
emergence coastlines, 254, 264
emergent curves, 443, 442f
Emperor Seamount, 13, 279
end moraines, 314
endogenic processes, 6
energy gradients
    atmospheric and oceanic circulation
        and, 14
englacial sediment, 301, 309
entisols, 95, 99
entrainment, 195, 196f
entrenchment, 208
entropy
    and life, 31
ephemeral channels, 180, 199
ephemeral streams, 130, 183, 184, 217
episodic floods, 458
equatorial easterlies, see trade winds
equatorial regions
    wind effects, 332
equatorial zone
    temperatures, 15–16
equilibrium
    between landforms and processes, 27
equilibrium line, 297, 298f
equilibrium line altitude (ELA), 297,
        298f, 325–326, 428f
    last glacial maximum, 427
ergs, 343–344
erodibility
    of earth materials, 481
erosion
    of continental shelves, 11
    estimating rates of, 59
    and isostatic compensation, 9, 396
    and landscape evolution, 464
    material routing, 24–25
    and meander migration, 192f
    measuring at outcrop and hillslope
        scales, 61–62
    of mountain belts, 11
    and rock types, 14
    and rock uplift, 394
    and sediment, 245–250

and soil profiles, 89
of streambeds, 195f
and structural landforms, 408–411
source-to-sink framework, 25
erosion control
    and slope processes, 173
erosion pins, 61
erosion rates, 63, 452, 481, 482f, 490, 493
    hillslopes, 166
    and landscape evolution, 481–483
    temporal scales, 23
    and thermochronometry, 64f
    and time, 486
erosion resistance
    and landscape evolution, 465
    rock types and structures, 13–14
erosion surfaces
    and landscape evolution, 465
erosion zones
    steep headwaters, 25
    for wind, 338
erosional coastal zones, 265f
erosional feedbacks, 415–416
erosional landforms, 380–381
erosional shoreline notches, 447
erosional steady state, 486f
erosional terraces, 235, 236f
erosivity, 481
erratics, 310
eruptive products, 363–367
eskers, 311, 313f
estuaries, 22, 230, 254, 271–272, 281
    drainage basins, 25f
    sediment and post-glacial sea level, 255–256
estuarine valley segments, 228, 229f
etchplains, 103, 104f, 479
eustatic changes
    sea level and ocean volume, 254
eustatic sea level, 254, 283, 284
eustatic sea level curves, 443
eustatic sea-level rise, 481
eustatic volume control on sea level, 442
evaporation, 116
    hydrologic cycle, 19f
evaporites, 86, 276
evapotranspiration, 116–118
event likelihood
    precipitation, 114
event-driven transport
    of sediment, 21f
exfoliation (of rocks), 80–82
exfoliation joints, 81f
exfoliation sheets, 80, 81f
exhumation (of rock), 394
exogenic processes, 6
expandable clays, 88f
expansive clays, 91
expansive soils
    shrink-swell behavior, 83
experiments
    geomorphic processes, 63–67

explosive eruptions, 361, 381
exsolution process, 365
extensional margins and landforms, 399–401
extensional plate boundaries, 399
extensional tectonic settings, 399f
extensional zones, 398

factor of safety, 161, 162f, 163
failure surface, 153
fair-weather berm, 268
falling limb, 127, 129
falls, 153, 158
fan head, 238
fan toe, 238
fans, 237–239
    see also alluvial fans
fault-block mountains, 400
fault landforms, 398
fault-line scarps, 475, 476
fault movement, 419–420, 421f, 422f
fault-related landforms, 417
fault scarps, 49, 363, 398, 400, 402, 419–420
    degradation, 49f
    locating, 68
    Owens Valley Fault Zone, 7
fault zones, 417
faulting
    and transience in landscapes, 475
faults, 398
feedback loop, 452
feldspar
    dating of, 56
    hydrolysis of, 101
    weathering, 84f, 85
felsenmeer, 82, 318
felsic flows, 365
fence offset, 398
Fernandina volcano, 363
Ferrel convective cells, 15f, 16, 332
ferricrete, 103
fetch, 259
field experiments, 63–65
field surveys, 44–46
fill terraces, 235, 236f, 237
fins, 410
fire spalling, 82
fire
    as geomorphic agent, 440–441
    and mass loss from rock surfaces, 82
firn line, 297
first-order streams, 225
first-order topography, 6
fission tracks, 63, 64f
fissures, 363
fixed rate constant, 51
fjord troughs, 309
fjords, 271f, 272, 292, 308–309, 326
    and landscape evolution, 464
flash flooding, 180
    on fans, 239

flashy drainage basins, 127
flatirons, 410–411
flexural downwarping, 395f
flexural rigidity, 394–395, 396
flexural upwarping, 395f
flexure, 395f, 396
floating ice, 300
    calving of, 301, 322
flocculation, 272
flocs, 272
flood basalt eruptions, 454
flood basalt provinces, 356
flood basalts, 357–359, 361
flood flows, 181
flood forecasting, 132
flood frequency, 130–132
flood recurrent intervals, 131–132
flood stage, 131
flood-tide deltas, 269
floodplain aggradation, 248f
floodplain connectivity, 231–232
floodplain development, 206f
floodplain sediment residence times, 222
floodplains, 22, 25f, 192f, 205–207
    alluvial valleys, 230
    restoration and widening, 249
    sediment storage, 222
    and sinuosity, 199f
    terraces, 235
floods
    and channel roughness, 185
    defined, 130
floodwaters, 111
flow depth
    at channel cross sections, 189
    and sediment transport, 194
flow recession, 128
flow roughness, 185
flow stage, 187
flow stone, see dripstone
flow tubes, 186
flow velocity, 187
    in channels, 185–187, 188f
    downstream, 227
    and sediment, 195
flowpaths, 122–126
    and evapotranspiration, 117f
flows, 153, 156–158
    and discharge variability, 187–189
fluid threshold, 335, 336f, 339
    and wind speeds, 337
fluid threshold curve, 335, 336f
flumed experiments, 469
flumes, experimental, 65, 66
flutes, 342
fluvial channels, 228
fluvial erosion, 211
    valley segments, 230
fluvial processes, 179, 185–193
    external controls on, 180–184
fluvial relief, 230

fluvial sediment, 232
    dating, 56
fluvially eroded valleys, 309
flux steady state, 490, 491
foraminifera, 432, 459f
force balances, 25–26
    of slides, 162f
forebulge, 395f, 448
foreland basins, 396–397
foreset beds, 273
foreshore, 266, 267
forest behavior, 487
forest clearing, 20
forest soils, 93f, 94, 98
forest stream channels, 184, 204
forested floodplains, 207
forestry practices
    and drainage basin morphology, 242
forests
    vegetation zones, 18–19
fossil waterfalls, 447
fracturing
    of rock mass, 80–82
fragipan, 98
free dunes, 345–346
free oxygen, 85
freeze-thaw cycles, 82, 464
friction angle, l47–149
frictional forces, 26f
frictional resistance to flow, 185
fringing reefs, 279, 280f
frost-shattering, 82, 318
Froude number, 187, 198
funnel ants, 78

gabbro, 91
gaining streams, 129–130
gamma particles, 62f
Ganges-Brahmaputra delta, 274
Ganges–Brahmaputra rivers, 218, 389, 396, 397
gases
    exsolving, 360
gelifluction, 316
gelisols, 96–97
Geographic Information Systems (GIS), 47–48, 225, 470
geologic structures, 13–14
    and landscape evolution, 462
geology
    and landscape evolution, 462, 465
geomorphic boundary conditions, 439–444
geomorphic hydrology, 111–137
geomorphic processes
    applications of, 28–29
    measuring rates of, 59–63
    spatial and temporal scales, 21f
geomorphic provinces, 218
geomorphology
    and climate effects, 115–116

defined, 3, 5–6
process and form, 22
geosphere, 6–14
geothermal gradient, 63
geysers, 377
ghost forest, 413
gibber plains, 352
gibbsite, 87, 88, 91
Gilbert-type deltas, 273
glacial abrasion, 292, 307–308
glacial and periglacial geomorphology, 291–323
    and climate change, 427
glacial bedrock quarrying 306, 307
glacial buzzsaw, 309, 449, 451f
glacial cycles, 435–436
glacial debris, 78
glacial deposits, 315–316, 448
glacial episodes, 69
glacial erosion, 292, 305–309, 324–326, 327f, 449, 451f
    and landscape evolution, 464
    valley segments, 230
    valleys, 308, 309
glacial erratics, 78, 82
glacial floods, 305
glacial flow rates, 302
glacial grooves, 294, 307, 308
glacial hydrology, 303–305
glacial ice, 294–295
    accumulation and ablation, 297–299
    ancient, 32f
    hillslopes, 147
    melting of, 443
    volcanoes, 372
glacial-interglacial climate changes, 447–448, 449–450
glacial-interglacial cycle, 443f
glacial–interglacial eustatic sea level changes, 443f
glacial isostasy, 315
glacial lake deltas, 312
glacial lake floods, 458
glacial lake sediments, 312, 322, 430, 448
glacial lake systems, 450
glacial lakes, 311–312
    sediment and till, 48
glacial landforms, 313–315
glacial landscapes, 293, 313
glacial margins
    wind-transported sediment, 334
glacial melting
    isostatic compensation, 396
glacial meltwater, 458, 459f
glacial moraines, 48–49, 51, 312
glacial periods
    and dust, 349
    Younger Dryas, 33
glacial polish, 307, 308
glacial sediment transport and deposition, 309–312, 313

glacial sediments, 276
    distribution of, 322
glacial sliding, 301
glacial striations, 307
glacial till, 436
glacial valley excavation, 325–326
glacial valleys, 308, 309
glacial water storage, 323
glacially eroded valleys, 309
glacially mediated incision, 306
glacially related sediments and landforms, 311–312
glaciated terrain
    lakes in, 239
glaciation, 291, 292–293, 448
    ancient, 307
    Austrian Alps, 18
    early, 435–436
    effects of, 447, 448
    end of, 301
    geomorphic effects of, 315–316, 445
    onset of, 276
    and sea level, 254–255
glacier bed, 302, 305–306
glacier calving, 297
glacier energy balance, 296–297
glacier mass balance, 294–296, 322
glacier mass/energy balance, 298f
glacier movement, 299–302
glacier velocity profiles, 299f
glaciers, 289f, 291, 294–305, 452
    as agents of erosion, 306
    discharge from, 304
    geomorphic effects of, 445
    and landscapes, 16
    as natural water reservoirs, 322
    and record of climate, 432
    sizes and shapes, 295
    thermal character of, 302, 303f
    upper reaches of, 297
    vanished, 482f
Glen Canyon Dam, 139, 140f, 243
Glen's flow law, 299, 300f
gley soils, 125
global climate
    and human activity, 140
    and volcanic eruptions, 367
global conveyor belt, 275
Global Positioning System (GPS), 43, 45f, 46
global warming, 283, 323
    and climate change, 455
    coastal environments, 281
Gloria side-scan sonar system, 384f
Goldich's Weathering Series, 84
gorge incision, 235
gorges, 234
grabens, 155, 278, 356, 400
graded steam, 181
grain sizes
    analysis, 65
    drainage basins, 232

Grand Canyon, 165, 184, 377
   limestone, 14
   rock types, 13
   suspended silt and sand, *223*
granite, 14
   and grus, 83
   hydration of, 83
   strength of, 147
granite pebble
   weathering rind, *50*
grassland soils, 93f, 98
   and plant roots, 161
grassland stream channels, 184
grasslands
   vegetation zones, 18
gravel
   beaches, 266, 267
   transport of, 197
   water flow, 121
gravel armor layer, 199
gravel bar, *223*
gravel-bed channels, 198, 232, *233*
gravel point bars, *217*
gravelbanks, 277
gravity force, 26f
Great Falls of the Potomac River, *111, 234*
Great Plains, 22
   Dust Bowl era, 348
Great Salt Lake, 219, 447
greenhouse effect, 454
greenhouse forcing, 436
greenhouse gases
   and climates, 14
   and life, 31
Greenland, 435, 443
   ancient glacial ice, 32f
   cold-based ice, 302
   drained lake, *18*
   ice mass of, 46
   Viking settlement, 438
Greenland Ice Sheet, *291, 294, 295,* 302, 304, *305, 307, 308, 311, 312,* 322, 427, 438, 439, 452, 454–455, 458
groin fields, 281
groins, 263, 281
ground ice, 317
grounded ice, 300
groundwater, 112, 117f
   and Darcy's Law, 122
   flow direction, 121
   and hydrologic cycle, 19f
   interaction with surface flow, 129–130
   losses, 133
   recharging of, 129f
   residence times, 133
groundwater flow
   and rock erosion, 150
groundwater flux, 121f
groundwater hydrology, 118–126

growth curves
   and lichenometry, 51f
grus, 83, *101*
Gulf of Mexico, 272, 459f
   atmospheric moisture, 16
Gulf Stream, 15f
gullies, 150, 374
gully development
   and animals, 19
gullying, 466
gypsum, 134, 330
   hydration of, 83
gypsum extraction, 276
gyttja, 429

habitable zone
   around stars, 31
Hadley convective cells, 15–16, 15f, 332
Half Dome, 80
half-life
   of carbon-14, 54, 55f
halite extraction, 276
halite
   hydration of, 83
halocline, 272
hand levels, 44
Hawaii, *133*
   basaltic lava flow, *365*
   cinder cone, *50*
Hawaiian hot spot, 357f
Hawaiian Island-Emperor Seamount chain, 382
Hawaiian Islands, 279
   formation of, 13
   hot spot, 397
   hot spot volcanoes, 362
   mapping, 384f, 385
   soil and dust, 349
   volcanism landscape, 368
Hawaiian-type eruptions, 368, 369f
Hayward faults, 404
head scarps, 375, 377
head
   of landslides, 156f
headlands, *230*, 263
headwall, 314
headwater areas, 25
heave, 151–152
Heinrich events, 436–438
helium-3 cosmogenic nuclide, 353
hematite, 85
hillslope-channel continuum, 229f
hillslope curvature, 167
hillslope erosion, 150, 222
   and slope angle, 166
hillslope erosion rates, 168f, 483
hillslope evolution, 166, 169f
hillslope geometries, 146f, 164f
hillslope geomorphology, 145–173
hillslope hollows, *230*
hillslope processes
   soil transport, 107

hillslope profile, 470, 474
hillslope sediment transport
   and rainsplash, 150
   sheetwash, 151
hillslope topography, 145–173
hillslopes, 22, 413–415
   angle of and erosion, 166
   catenas, 101f
   and climate, 445–446
   field experiments, 63
   laboratory experiments, 65
   landslide recurrence intervals, 28
   measuring landscape change, 61–62
hilltops, 145
Himalaya, 197, *389, 393, 397,* 402, 403, 408f, 409, 416, 440
   continental collision zone, 11
   crustal roots, 9
   erosion of, 24
   glaciers, 305
   river gorges, 235
   transverse drainage, 225
Himalayan Mountains
   terrace remnants, 235
historical geomorphology, 22
histosols, 95–96
hogbacks, 410
hollows, 22, 124, *125,* 171–172, 228
   hillslopes, 162–163, *166*
Holocene climatic optimum, 438
Holocene Epoch, 31, 32f, 55f, 63, 420, 426
   dust, 349
   precipitation, 440
   sea levels, 230
honeycomb weathering, *83*
hooks (of spits), 269
horns (of cusps), 268
horns (landform), 313f, 314
horsts, *155,* 400
Horton Overland Flow, *120,* 123, 124f, 128
Horton's laws, 226f
hot spot basaltic volcanism, 381
hot spot movement, 364f
hot spot tracks, *355*
hot spot volcanoes, 362
hot spots, 11–13, 279, 357f
human activities
   as geomorphic agent, 138–141
   and geomorphic processes, 20
   and global systems, 32–33
   and landscape evolution, 466
   and sediment generation, 246
   and soil infiltration rate, 119
   and wind, *335*
humid-region drainage basins, 226–227
humid regions
   precipitation events, 116
hummocky terrain, 383–384
humus, 95
Hurricane Hugo (1989), 274

Hurricane Isabel (2003), *111*, 269
Hurricane Katrina (2005), *115*, 264, 281
Hurricane Mitch (1998), 375
hurricanes, 114, *115*, 440
 and Gulf of Mexico, 16
 temporal scales, 23
 and volcanoes, 375
hydration, 88–89
 of minerals, 83
 as weathering mechanism, 83
hydraulic conductivity 119f, 121, 122t
hydraulic geometry variables
 basin area, 226–227
hydraulic gradients, 122
hydraulic head, 121–122
hydraulic radius, 185
hydraulic wedging, 191
hydrofracturing, 82
hydrographs, 126–129
hydrologic cycle, 16, 19f, 111, 439, 452–454, 455
 and climate change, 140
 and precipitation, 112
hydrologic landforms, 133–136
hydrology
 and climate, 115–116
 and humans and landscape change, 138–141
hydrolysis, 84, 86–87
hydrosphere, 6, 14–16, 19f
hydrothermal circulation, 361
hydrous oxides, 84
hydrovolcanic activity, 379
hyporheic flow, 124
hypotheses testing, 44
hypsometric curves, 479, 480f
hypsometric integral, 480f
hypsometry, 9f, 48, 479
hysteresis, 128f, 129, 220, 452

ice, 294
 and climate zones, 16
 forces on, 26f
 melting rate, 140f
 physical properties of, 291
 regelation, 306
 shear strength, 300
ice caps, 289f, 294, 295, 315
ice-contact delta, *312*
ice-contact landforms, 310
ice-core chronologies, 433
ice cores
 and climate, 432–433
 and sediment deposit, 67
ice creep, 299
ice deformation rate, 299, 300f
ice fields, 293f, 294, 315
ice-flow models, 325
ice jams, 180
ice loss
 from glaciers, 295
ice-marginal fluvial erosion, 306

ice-marginal lakes, 239, *307*, 313f
ice-marginal sediments and landforms, 310–311
ice margins, 310
ice motion
 vertical, 302
ice-rafted debris (IRD), 312, 430
ice rivers, 291
ice sheet behavior, 276
ice sheet margin, 297
ice sheets, 291, 293f, 294, 295–296, 297–299, 315, 322, 326, 452
 and climate, 426
 and climate change, 452
 and climate record, 432
 collapse of, 447
 and delta shrinking, 283
 disintegration of, 454
 and ice-rafted debris, 312
 and land level changes, 449, 450f
 landforms of, 314–315
 landscapes at, 334
 melting of, 443, 448
 and sea-level change, 67
 sea-level rise and fall, 254
 sediment transport, 309
 slope of, 299–300
 warm-based, 302
ice shelf, 300
ice wedge casts, 448
ice wedges, 320, 321f
icebergs, 312
 density of, 10f
 and isostatic compensation, 8, 9
Iceland, 278
 ice caps, *295*
 jokulhlaups, 305
Icelandic-type eruptions, 368, 369f
igneous rocks, 6, 82
ignimbrite sheets, 366
illite, 87
illuviation, 89
impact threshold, 335, 336f, 337, 339
in-channel bars, 272
inceptisols, 95, 99
incision, 442
incongruent dissolution, 86, 87
increment borers, 52
India
 collision with Asia, *389*
 plains of, *393*
Indus River, *389*, 396, 416
infiltration, 124f
infiltration rates, 117f, 119–120
 and plants, 14
infinite-slope model, 159, 160–161, 162
infinite-slope stability equation, 174
initiation zones
 translational slides, 155
inlets, 268, 269
inner gorges, 235, 414–415, *416*

inselbergs, 102–103, 404, *407*
inshore zone, 266
insolation, 14, 80
integration time, 59
intensity
 of precipitation events, 113
intensity-duration threshold
 of precipitation events, 113
interferometry data, 46
interflow, 123
interglacial periods
 and climate change, 427
intermittent stream flow, *130*
intermittent streams, 130
Interoferometric Synthetic Aperture Radar (InSAR), 45f, 46
interstitial cements
 and cohesion, 149
ion exchange, 83, 84
iron
 and chelating agents, 89
 insolubility of, 86
 oxidation of, 85
 rock varnish, 50
iron oxide
 in soil, 89
island arcs, 12f, 402
island chains
 formation of, 279
isobases, 448–449, 450f
isostasy, 6–11, 448
isostatic compensation, 6, 9–11, 394–396, 426, 450
 erosion and glacial melting, 396
isostatic responses, 392, 396, 448–449
isostatic uplift, *see* isostatic compensation
Israel
 Dead Sea rift, *13*
 pavements in, 352

Jenny, Hans, 462
jet stream, 454
jetties, 263
joint exfoliation, *465*
joint patterns, 80–82
joint systems, 82
joints, 410
jokulhlaups, 305, 380
junction angles
 dendritic channel networks, 223

K horizon, *94*, 95
K/Ar dating method, *see* potassium/argon (K/Ar) dating
Kalahari Desert, 16, 332
kame terraces, 310–311
kames, 310, *311*, 313f
kaolinite, 87–88, 91
Karman/Prandtl model, 333
karst hydrology, 303
karst landform, 133–136

karst landscapes, 136f
karst terrain
    water flow, 121
karst topography, 102, 134
katabatic winds, 334, 447–448
kettle ponds, *311*, 313f
kettles, 310, 311
key members (obstructions to flow), 198
kinetic energy, 24
    hydrologic cycle, 19f
kipukas, 374
knickpoint retreat, 475f
knickpoints, 233–234, 413, 474–475
knickzone retreat, 442–443
knickzones, 233, 400
    and climate change, 427–429
Köppen climate classification, 445
Köppen climate zones, 446f

laboratory experiments, 65–66
lagoons, 256, 257f, 270, 271f, 281
    fringing reef, 280f
lag-to-peak, 127
lahar early warning systems, 381
lahars, 240, 375–377, 380, 381, 385
Lake Agassiz, 458
lake and marine sediment
    and climate, 429–432
Lake Bonneville, 447, 458f
Lake Champlain, 449, 450f
Lake Hitchcock, 312
Lake Missoula 276, 312, 457, 458f
Lake Tanganyika, 400
lakebeds
    and wind erosion, 335
lakes, 239
    deltas of, 273
    drained, *18*
    glacially dammed, 311
    trapped beneath ice, 302
laminar flow, 331
laminar sublayer, 185
land area
    above and below sea level, 9f
land level changes
    and ice sheet volume, 448, 449f
land use
    drainage basin dynamics, 240–244
landform degradation
    dating of, 48–50
landforms
    climate control on, 445
    joint-related, 410
    weathering-dominated, 101–104
Landsat satellite, 45f, 47
landscape behavior
    predicting, 487
landscape change
    and humans, 138–141
    at outcrop and hillslope scales, 61–62

landscape evolution, 461–487
    geographical cycle of, 490
    models of, 467–471
landscape history
    and process and form, 22
    studies of, 44
landscape management, 29
landscape processes
    rates of, 480–487
landscape responses
    to disturbance, 463f
landscape sensitivity, 469
landscape stability, 469
landscape types, 471–479
landscapes, 20–24
    and climate change, 452
    conservation of mass in, 24
    defined, 20
    and soils, 99–101
landslide hazards, 172, 173
landslide prediction, 163
landslide processes, 464
landslide scars, 23
landslides
    Alaska Range, 7
    channel responses, 208
    and channels, *183*
    colluvial valleys, 229f
    and landscape evolution, 466
    periglacial terrain, 319
    and precipitation events, 113, *115*
    rainstorms and, 149
    recurrence intervals, 28
    and volcanoes, 355, 375
landsliding
    and channel initiation, 171
    forested versus clear-cut slopes, 175
    and river incision, 166
    and threshold hillslopes, 168f
land-use planning, 29
Langbein-Schumm curve, 441–442
lapse rate, 297
laser ranging, 44–46
last glacial maximum (LGM), 426, 448
Late Stone Age, 32f
lateral moraines, *48, 310, 291*, 313f
laterites, 98, 103
Laurentide Ice Sheet, 48, 292, 294, 295, 302, 311, 312, 315, 413, 434, 448, 458
lava, 356, 378f
    yield strength of, 360
lava domes, 366, *372*, 373f
lava flow chronosequences, 370
lava flows, 355, 360, 374, 378, 380
    dating, 377
lava sheds, 381
lava tubes, 370, *371*
lava types, 360t
lava viscosity, *365*
lavaka, *150*, 466
leaching, 89

leaves
    and precipitation, 115
leeward slopes, 452
legacy sediments, 241–242, 250
levees, 209
    and deltas, 284
lichen colonization
    and weathering, 78
lichenometry, 51, 52t
lichens, 51
lift and drag forces
    and sediment transport, 194–195
lift forces, 26f
    and wind, 337
Light Detection and Ranging (LiDAR)
    data acquisition, 45f, 46–47, 68, 419
    slope morphology, 172
limestone, 14, *134*
    carbon in, 31
    Grand Canyon, *13*
    weathering of, 80
limestone cave deposits, 56
limonite, 89
linear dunes, *343*, 346
liquid water
    and life, 31
lithogenic sediments, 276
lithology, 13–14
    and landscape evolution, 462
lithosphere
    buoyancy of, 397
    flexural rigidity, 394–395
litter (soil), 95
Little Ice Age, 59, *291*, 428f, 438
littoral currents, 262
loam soil
    infiltration rate, 119t
local relief, 231
lodgment, 310
loess, 307, 336f, 337, 339f, 347–349, 350, 433–434
    global distribution of, 344f
    hillslopes, 147
loess plateau, 336f, 434
log dams
    and channels, 184
log step, *184*
logjams, 193, 198, *204*, 207
    in channels, 184
longshore currents, 262
longshore drift, 262, 274, 275f, 281, 329
longshore transport, 262f
losing streams, 129–130
lowland alluvial channels, 180
luminescence dating, 52t, 56, 57f, *58*
lunar tides, 257
lunettes, 343

maars, 240, 380
Mackenzie River Delta, *322*

macrofossils, 431
macropores, 119, 122, 303
macrotidal coastlines, 258
Madagascar
    gullies, 150
mafic volcanic rocks, 374
magma, 355–356, 363, 369f
    gases in, 366
    viscosity of, 359–361
magma chemistry, 359–460
magma intrusion
    geomorphic effects of, 372–373
magnetic anomaly analysis, 276
magnitude-frequency relationships
    of geomorphic events, 28
main stem basin size, 225
Mammoth Cave, *134*, 136f
manganese
    oxidation, 85
    rock varnish, 50
mangrove vegetation, 270
Manning roughness coefficient,
    185, 186t
Manning's equation, 185
mantle (Earth), 6
    density of, 394
    isostatic uplift, 9
    at plate boundaries, 12f
    thermal convection of, 11
    weathering of, 80
mantle rocks
    density of, 9, 10f
mantle upwelling, 397
Mariana Trench, 279
marine deltas, 273–274
marine gravels, 385
marine landforms and processes,
    276–279
marine organisms, 432
marine oxygen isotope stages (MIS),
    432, 455
marine regression, 255
marine sedimentation, 276
marine sediments
    Earth history, 253
marine settings and drivers, 274–276
marine terraces, 256, 265, 412
    dissected, *429*
marine transform faults, 12f
marine transgression, 255
Mars
    atmosphere of, 31
    floods on, 459
    flows on, 156
    glaciers, 294
    outwash channels on, 478
    polar ice caps, 294
    relict landscape on, *479*
    rock disintegration on, 83
    salt glaciers on, 397
    winds on, 330, 334
    yardangs on, 342

marsh sediment, 274
marshes, 270–271, 281
mass balance
    of glaciers, 322
    of rock-forming elements, 107
mass flows, 156
mass loss
    from rock surfaces, 82
mass movements, 153–159
    channel responses, 208
    and volcanoes, 374–375
mass removal
    hillslope erosion, 164
mass spectrometers, 56
mass spectrometry
    and cosmogenic nuclide analysis, 58
mass wasting, 145
material routing, 24–25
mathematical models
    and landscape evolution, 467,
        469, 471
matrix-supported debris flow deposits,
    157, 239
Mauna Kea, Hawaii
    permafrost, 317
    sorted stripes, *320*
mean local relief, 168f
mean slope, 168f
meander, 193
    geometry of, 199f
meander belt, 199f
meander migration, 192f
meandering channels, 199f, 200, 201, 413
mechanical weathering
    and landscape evolution, 464
medial moraines, *307*, 314
Medieval Warm Period (900–1300 CE),
    438, 441
Mediterranean Sea, 276
megafloods, 458–459
megageomorphology, *see* first-order
    topography
megalandslides, 375, 377
mega-tsunamis, 385
melt-out till, 310
meltwater, 112
    glacially dammed lakes, 311
    and ice sheets and glacier bed, 304
meltwater lakes, 304–305
meltwater streams, 295, 304
mesas, 410
mesotidal coasts, 258
metamorphic rocks, 6, 14
methane, 455
methane gas
    ancient glacial ice, 32f
method of slices, 161, 162f
mica, 87
Mid-Atlantic Ridge, 356, 357f, 400
    volcanoes and, 8f
mid-channel bars, *198*
mid-channel islands, 206f

middens, 431f, 432
mid-ocean ridge volcanism, 361
mid-ocean ridges, 9f, 11, 278–279, 356,
    358f, 392, 399
Milankovitch cycles, 436
military activities
    and aeolian processes, 350
military planning
    and geomorphology, 29–30
mill dam sediments, 244f
mill dams, 243, 246, 248f
mineral cations
    and hydrolysis, 87
mineral dissolution, 86
mineral stability, 84–85
minerals
    nutrients from, 83
mining
    and geomorphic processes, 20
Mississippi River, 206f, 284
    dams and levees, 285, 286f
    deltas, 272, 273, 274
Mississippi River Basin, 127
Mississippi River delta, 285, 286f, 413
Moderate Resolution Imaging
    Spectroradiometer (MODIS), 389
Mojave Desert, *43*, 158, *159*
    lakes, 450
    pavements, 353
    pluvial lakes in, 447
    sediment movement, 65
    sheetwash, 151
Mollisol A horizons, 348
mollisols, 98
monoclines, 409
monogenetic cinder cones, 368
monsoon-driven stream systems, 187
monsoon rainfall
    and landscape evolution, 464
monsoons, 116, 440
montmorillonite, 87
    hydration of, 83
Monument Valley, Arizona, *77*, 410
Moon
    volcanism on, 356
    water ice on, 294
moons (of planets)
    ice on, 294
moraine complexes, 421f
moraines, 428f, 452
    estimating ages of, 314
    of ice sheets, 315
    and lakes, 239
    relict landforms, 446, *447*
    Sierra Nevada, 48
    and threshold lakes, 444
morphology
    quantifying Earth's surface, 43
mottles, 125
mottling, 85
moulins, 304
Mount Mazama, *240*, 356, 367f, 368

Mount Pinatubo eruption (1991), 365, 373f, 380
Mount Rainier, 297, 371, 372, 377, 381
Mount Ruapehu, 375, 377
Mount Shasta, 375, 383
Mount St. Helens, 61, 365, 367f, 368, 372, 374, 375, 376f, 377
Mount St. Helens eruption (1980), 28, 355, 383, 385
    elevation loss, 371–372
    sediment yields, 379f
Mount Vesuvius, 365, 373
Mount Washington, 82, 318
mountain belts, 6
    active and inactive, 11
    and continental collision zones, 11
    erosion of, 11
    physiographic provinces, 22
    see also orogens
mountain drainage bases
    headwater channels, 232
    process domains, 228
mountain landscapes, 212
mountain ranges, 401–404
    and climate, 453f
    crustal roots of, 10f
    decay of, 469
    elevation of, 9f, 10f, 449, 451f
    erosion of, 24, 394
    glacial activity, 309
    and glaciation, 291
    plate boundaries and, 8f
    uplift and erosion temporal-spatial scales, 21f
    see also mid-ocean ridges
mountain stream channel types, 203f
mountain (bedrock) streams, 180
mountain topography
    climatic control of, 449
mountains
    height of, 396
    and precipitation delivery, 16
mudflows, 381
multi-thread stream channels, 199
multispectral remote sensing, 47, 50
Munsell soil color chart, 238
muscovite, 84, 87

Namib Desert, 16, 115
    winds of, 332
Namibia
    precipitation, 115
    sheetwash, 151
Namibian Sand Sea, 329, 341, 343
natural levees, 206
natural resource management, 29
neap tides, 258, 259
Negev Desert, 63, 94, 98, 165, 183
    pavements, 343
neoglacial climate state, 438
neutron capture, 55f
New Madrid Fault Zone, 408

NEXRAD (Next Generation Weather Radar), 113
Niagara Falls, 234, 474, 475f
Nile River delta, 256, 273, 274, 285f
nitrogen atoms, 55f
nitrogen cycling
    human impact on, 33f
nivation, 318
nivation hollows, 318
no-till farming, 108
nor'easters, 114, 268
normal fault scarp, 398
normal faults, 398, 400
normal force, 25, 26f, 34
normal stress, 147–148, 149
    on hillslopes, 160, 161
North America
    coastlines, 281
    physiographic provinces, 22
    western surveys, 44
    winter storms, 114
North American plate, 8f
North Atlantic Deep Water (NADW), 436
North Atlantic Oscillation, 439
Northern Hemisphere
    climate zones, 16
    exposed land in, 9f
    soils, 90
noses, 124
nuclear weapons tests
    measuring cesium-137, 61–62
nuclides, cosmogenic, see cosmogenic nuclides
numerical dating methods, 48, 52–59
numerical models, 66
nunataks, 315
nutrients
    from minerals, 83

O horizon, 92, 93f, 94
obliquity, 436
    Earth's orbit, 437f
obsidian, 363–365, 366
ocean
    and climate dynamics, 426
ocean basins
    abyssal plains, 9f
    elevation of, 6, 8
    isolation of, 276
ocean-bottom topography, 392
ocean circulation
    and terrestrial climates, 16
ocean–continent collision, 402
ocean floor drilling, 276
ocean floor mapping, 383–384
ocean island arc, 356
ocean salinity, 86
ocean thermohaline circulation, 15f
ocean volcanoes, 362
ocean volume, 283
    changes in, 254

ocean water
    saltiness of, 276
oceanic circulation, 15f
oceanic crust, 6, 8–9, 12f, 31, 278
    seafloor spreading centers, 11
    and subduction zones, 11
oceanic hot spots, 358f
oceanic island arcs, 403f
oceanic plates, 12f
    collisions of, 392
offset stream channels, 406
offset streams, 404
offshore bars, 268
offshore sandbanks, 277
offshore zone, 266
Ogallala aquifer, 123, 133, 351
olivine, 84
    oxidation of, 85
Olympic Mountains, 11, 325, 392, 491
100-year flood, 132
open drainage basins, 218
optical surveying, 44, 45f
optically stimulated luminescence (OSL), 56
optimization modeling, 420
orbital forcing
    of climate change, 436, 437f
Oregon Coast Range, 153, 175, 176, 392
    channel-head source area, 171f
organic debris, 204–205
organic fuels
    carbon in, 31
orogenic belts, see orogens
orogenic decay, 21f
orogenic wedges, 472
orogens, 11
orographic effects, 16, 297
    and climate, 453f
orographic precipitation, 452
orthoclase, 87
outburst floods
    and glaciers, 305
outcrops, 108f
outwash plains, 311, 312, 434
outwash streams, 295, 305
    and wind erosion, 334
outwash terraces, 313f, 429
overbank deposits, 205, 206f
    of drainage basins, 223
overgrazing
    and soil erosion, 19
overland flow, 14, 117f, 120
overwash, 269
Owens Valley, 238, 363, 446, 447
Owens Valley Fault Zone, 7, 402
oxbow lakes, 193, 207
oxidation, 83, 84, 85
    and life, 31
oxisols, 98, 99
oxygen isotopic composition
    of seawater, 459f

p-forms, 302
Pacific Decadal Oscillation 439
Pacific Ocean
    seafloor, 278
    subduction zones, 254
Pacific Ocean basin
    volcanoes, 8f
Pacific plate, 8f
pack rats, 431f, 432
pahoehoe lava, 368–370, *371*
paired terraces, 236f, 237
paleoclimate, 53f, 432
paleoflood layers, *429*
paleolake level, 273
paleosols, 100–101, 434
paleostorms, 440
Palouse, 307, 329, 350
    loess deposits, *347*
pan (dry lakebed), *343*
parabolic dunes, 346
paraglacial processes
    geomorphic effects of, 315–316
paraglacial sediment, 222
paraglacial sedimentation, 448
parallel slope retreat, 166, 169f
Paran region (Negev Desert), *94*
parent isotope, 54
parent material, 77
    bioturbation in, 78
    for soils, 90–91
Paricutin cinder cone, 368, 370f
partial melting, 31
particle fall speeds, 333
passive margins, 11, 254, 255f, 277, 400
passive remote sensing, 45f, 47
passive tectonic controls, 392
pater noster lakes, 239
patterned ground, 448
    periglacial terrain, 319–320, 321f
    periglacial zones, 293
pavements (desert), 98, 343, 350, 445
    and military vehicles, 350–351
    names for, 352
    and wind, 352–353
peak discharge, 127
peak flow, 116
peak rainfall intensity, 126f
peak river discharge, 126f
pebble counts, *65*
pediments
    and landscape evolution, 465
pedogenesis, 79, 91
    rate of, 106–107
pedogenic processes, 96f
peds, 95, 96f, 353
pelagic ooze, 278
peneplains, 468, 490
percent modern Carbon (pmC), 54
perennial channels, 180, 199
perennial streams, 130, 183, 217
    effective discharge, 187
periglacial environments, 293–294, 316–322
periglacial landforms, 316–322
periglacial processes, 316–322
    geomorphic effects of, 445
periglacial regions
    rivers and soils in, 322
permafrost, 7, 96, 291, 294, 316f, 317–318, 323
    active layer over, 317
    and plants, 19
permafrost-dominated landscape, 448
permafrost soils, 322
permafrost table, 317
permafrost thickness, 318
permeability, 121–122
Peru-Chile trench, 279
pH values
    and solubility 85
phase transitions (liquid and solid water), 291
phosphorus
    and plant growth, 83
photic zone, 256
photosynthesis, 55f
phreatic eruptions, 380
phreatic zone, 116
phreatomagmatic eruptions, 379–380
phyllosilicate group of minerals, 88f
physical models
    and landscape evolution, 467, 469
physical weathering, 77–80, 80–83
physiographic provinces, 22, 23f
phytoliths, 429
Piedmont, 91, 98, 248f
pigments, 431
pillow basalts, 361
pingos, 321–322
pipe flow (glaciers), 303
pistol-butt trees, *152*
planar landslides, 153
planar slopes, 163, 164f
plane-bed channels, 203f
plane-bed reaches, 204
planetary atmospheres, 31
planetary geomorphology, 30
plant communities
    global distribution of, 18–20
plant roots
    apparent cohesion of, 161
    and rock disintegration, 78
    and soil, 119
plants
    grassland and forest zones, 18
    and landscape evolution, 466
    and soil 79
    and stream morphology, 204–205
    and winds, 330
plastic behavior
    of debris flows, 157
plasticity
    and soil texture, 95
plate boundaries, 6, 8f
    and volcanoes, 357f
    volcanoes and earthquakes at, 8f
plate flexure, 395f
plate tectonic theory, 382
plate tectonics
    development of plates, 11
    and glaciation, 292
    seafloor morphology, 274
    submarine landscapes, 277
    weathering on Earth, *31*
plateaus, 11, 405–406
platforms, 405
playa deposits, *237*
playa lakes, 445
playas, 218, 239, *425*
Playfair, J., 107, 324
Pleistocene Epoch, 33, 55f, 292
    ice sheets, 310
Plinian eruptions, 368, 369f
Pliocene Epoch, 292
plucking, 189, 190–191, 211, 212f
plunge pools, 457f
plunging breakers, 261
plutons, 81f
pluvial lakes, 447
pocket beaches, *265*, 266
point bar/cutback migration, 223
point-bar deposition, 193
point bars, 191, 192f, *193*, 198, 206f
polar convective cell, 15f, 16
polar glaciers, 302
polar regions
    plant communities in, 19
    vegetation zones, 18
polar warming, 323
pollen, 431
pollen analysis, 431
polygenetic stratovolcanoes, 371
polygons
    patterned ground, 319, 320
Pompeii, 365, 373
ponds
    erosion in, 429
pool-riffle channels, 203f
pool-riffle reaches, 204
pore pressures, 149
    on hillslopes, 160, 161
pore spaces, 119f, 120
porosity, 121, 122
potassium feldspar (orthoclase), 87
potassium rocks, 54
potassium/argon (K/Ar) dating, 52t, 54–56
potential energy, 24
    conversion to kinetic energy, 24
    hydrologic cycle, 19f
    and stream power, 189
potential energy loss
    and stream power, 189
potholes, 190
precession
    of Earth's orbit, 437f

precession of the equinoxes, 436
precipitation, 112–116
    and atmospheric circulation, 14–15
    and climate, 4267f
    and climate zones, 16, 17f
    delivery of and mountains, 16
    and elevation, 16
    and erosion and sediment yields, 441f
    and landscape evolution, 464
    monsoons, 440
    potential and kinetic energy of, 24
    and temperature, 440
    and weathering, 79
precipitation changes, 450
precipitation delivery, 114–115
precipitation intensity, 119f
pressure melting point
    of ice, 302
primary minerals, 77–80
primary porosity, 121
primary tributaries, 223
prism pole, 43
process domains
    drainage basins, 228–230
process geomorphology, 22
process models
    and landscape evolution, 470
production-limited slopes,
        see weathering-limited slopes
Providence Canyon State Park, 476
proxy records, 66–67
Puget Sound, 266, 272, 281
pull-apart basins, 399, 404, 407f
pyroclastic debris, see tephra
pyroclastic falls, 365–366
pyroclastic flows, 365, 366, 381
pyroclastic materials, 365–366
pyroxene, 84

quadrangles, 44
quarrying (glacial erosion), 292, 307
quartz, 91
    aeolian sediment, 330
    dating of, 56
    solubility of, 86
    weathering of, 80, 84, 85
quartzite, 14

radar data
    surface roughness 47
radial drainage patterns, 224
radioactive-decay counters, 44
radioactive elements, 31
radioactive isotopes
    carbon-14, 53–54
    see also cosmogenic nuclides
radiocarbon analysis
    sediment deposits, 222
radiocarbon dating, 43, 52t, 53–54,
        55f, 69–71
    sample preparation, 56
radiocarbon years, 54, 70

radiometric dating, 44
radionuclides, 62f
radius of curvature, 201
rain gauges, 112
rain shadows, 16, 474
rainfall
    and climate, 426
    and climate zones, 16
    kinetic energy of, 19f
    overland flow, 14
    pH value, 86t
    potential energy of, 24
    and slope failures, 161
rainsplash, 150–151
rainwater
    pH of, 85
    as weathering agent, 84
rating curve (discharge versus stage), 187
recession constant, 128
recessional moraines, 452
rectangular basins, 225
rectangular drainage patterns, 223, 224
recurrence interval discharge, 139f
recurrence intervals
    of geomorphic events, 28
    of precipitation events, 113–114
redox potential (Eh), 85
reduction, 83, 84, 85
reference condition, 243
refugia, 315
regelation, 306
regolith
    of hillslopes, 164
    and hydrolysis, 87
    and soil, 79
    weathering of, 78
regolith production, 79, 107
regression, marine, 255
regs, 352
relative age information, 43
relative dating methods, 48–52, 52t
    calibrated, 50–52, 52t
    lichenometry, 51
relative humidity, 332
relaxation time, 463
relict calcic horizons, 429
relict ergs, 344
relict landforms, 426, 446–447
relict landscapes, 471, 478–479
relict permafrost landforms, 318
relict pingos, 448
relict soils, 100
relief (drainage basins), 230–231
    and elevation, 9f
repeat photography, 47
residence times
    of groundwater, 133
residual soils, 90, 91
resisting forces, 25, 26f
    and response time, 28
resisting shear strength, 161
resisting stresses, 159–160, 161

response time
    of landscapes and landforms, 27–28
    steady state and thresholds, 27f
restraining bends, 404
resurgent domes, 357, *360*
return flow, 117f, 124f
revegetation
    and coastal dunes, 350
reverse faulting, 402–403, 405f
reverse faults, 398
rhyolite, 54
rhyolitic caldera eruptions, 362
rhyolitic domes, 363
rhyolitic lava, 360, 363
rhyolitic lava flows, *363*
rhythmites, 312
riffles, 191–193
rift-bounding fault, *13*
rift shoulders, 400, *476*,
    at divergent boundaries, 12f, *13*
rift valley systems, 399
rift valleys, 6
    and seafloor spreading centers, 11
rift zones, 393, 397, 399, 400
rifting, 399–400
rifting, ancient, 400
ring width index, 53f
Rio Grande Rift, 399
riparian (near-stream) areas, 180
ripples
    on dunes, 346–347
rising limb, 127, 129
river and floodplain behavior
    model, 246
river channel morphology, 413
river deltas, 283
river-dominated deltas, 274
river flow
    temporal scales, 23
river incisions, 166, 416–417, 454
river longitudinal profiles, 230–231,
    232
river processes
    spatial-temporal scales, 21f
river profile concavity, 228
river stage, 126f
river terraces, 100f, *219*
    and climate change, 427–429
    origin of, *429*
rivers, 483
    and clastic load, 197
    fluvial processes, 179
    gorges, 234
    sediment supply, 415
rivers and streams
    and tectonic forcing, 413
roches moutonnées, 307, *309*
rock avalanches, 158
rock falls, 158
rock flour, 307
rock flows, 158
rock glaciers, 320–321, 448

rock masses
    friction angles, 148
    joint patterns, 80–82
rock structure
    erosion resistance and, 13–14
rock types
    and bedrock erosion, 180
    global distribution of, 6
    weathering of, 14
    weathering rates of, 80
    see also lithology
rock uplift, 10f, 392, 393–394, 481
    coastal landscapes, 412
    and landscape evolution, 463
    patterns, 417
    rates of, 452
rock varnish, 50, 51, 352
rock weathering, 50, 77, 80, 454
    dating method, 52t
    laboratory experiments, 65–66
    soil production, 107
    and subsurface flow, 123
rockfall, 169f
    earthquake-induced, 51
    on glaciers, 309
    from hillslopes, 146f
    Yosemite Valley, 7
rocks
    disaggregation, 80
    erosion-resistant, 14
    melting point of, 393
    shrink-swell behavior, 83
    strength of, 147, 149t
    strength of and weathering, 149–150
    thermal conductivity of, 82
rockslides, 153, 154, 155, 156
    on threshold slopes, 145
rocky coasts, 264–265
Rocky Mountains
    permafrost, 317
    snowpacks, 350
root cohesion, 175f
root decay, 176
root regrowth, 176
root reinforcement, 174
root strength, 149, 161–162, 174, 176
roots
    and channel initiation, 170
    in channels, 184
    and geomorphic processes, 20
    and slope stability, 174–176
Ross Ice Shelf, 300
rotational failures, 153, 155–156, 161, 162f, 183
roughness length, 333
runoff, 119f, 126f
runoff response, 128

sag ponds, 153–154, 404, 405, 407
Sahara Desert, 16, 350, 439
    lakes, 450
    winds of, 332
    yardangs, 342
salinity, 256
salmon
    and geomorphology, 19
Salmon River, 230
salt flats, 218
salt glacier, 397
salt marshes, 271
salt residue, 83
salt weathering, 83
saltating load, 184
saltation, 194f, 195, 197, 337–338, 339
salts
    hydration of, 83
    in sediment, 348
saltwater–freshwater mixing, 272
sample collection, 44
San Andreas Fault, 11, 67f, 393, 398, 404, 417
San Francisco Bay, 271, 272, 404
San Francisco earthquake (1906), 156
sand
    beaches, 266, 267
    entrained, 336f
    transport of, 197
    windblown, 331
sand-bed channels, 198, 232, 233
sand-bedded channel segments, 204
sand dunes, 265, 336f, 346, 426
    and climates, 16
sand ridges, 277
sand seas, see ergs
sand sheets, 351
sand transport
    and dune creation, 345
sandbars, 202
sandstone, 14
    Grand Canyon, 13
sandstone formations
    erosion of, 77
sandurs, 305
sandy loam soil
    infiltration rate, 119t
sandy soil
    infiltration rate, 119t
sanidine, 366
Santa Cruz Mountains, 404
Santorini volcano eruption (1645 BCE), 367, 372–373
saprolite, 147, 150
saturation overland flow, 117f, 123–124
saturation overland flow hydrographs, 128
scaling issues, 65
Scandinavia, 302
    rebound of, 396
scarps
    dating, 420
    and landslides, 154–155, 156
Schmidt rebound hammer, 50
scoria, 368
scoria cones, 369f
scour (ice), 205f, 326f
scouring, 302
scree, 169f
scroll bars, 207
sea caves, 265
sea-cliff erosion, 264–265
sea-cliff retreat, 264
sea cliffs, 253, 264, 265f, 266, 375
sea floor
    volcanic topography, 361
sea level changes, 67, 254–256, 281
    glaciation, 315–316
    and ice sheets, 443f
sea-level heights, 330
sea-level histories, 444f
sea-level rise, 412, 322
    estimates of, 285
sea-level rise and subsidence
    deltas, 285, 286f
sea level rises and falls, 426
    and climate change, 452
sea levels, 442–443
    climate-induced changes, 427
    and glacier flow rates, 301
    and landscape evolution, 463
    predicted rising of, 454
sea stack, 265, 266
seafloor
    abyssal basins, 278
seafloor processes and morphology, 274
seafloor spreading centers, 11, 278
seafloor topography, 12f, 253
seamounts, 356, 358f
seasonal stream, 130
seasonally saturated ground, 123
seasonally saturated zones, 125
seawater
    dissolved load, 276
    salinity of, 256
second-order channels, 225
secondary minerals, 77, 84
    and weathering, 87, 88f
secondary porosity, 121
sector collapse, 375, 381
    geomorphic impacts, 385
sedigraphs, 128f, 129
sediment
    event-driven transport of, 21f
    forces on, 26f
    and mill dams, 246
    and post-glacial sea level rise, 255–256
sediment accumulation rates
    and landscape evolution, 483
sediment budgets, 219–222, 247f
sediment clasts, 212
sediment cores
    cesium-137 concentration in, 62f
sediment deformation
    glaciers, 301

sediment delivery
    and deltas, 284
    river channels, 243
sediment delivery ratio, 245, 246f, 483, 490
sediment deposition
    channels, 272
    coastal environments, 268
    continental shelves, 254
sediment deposits, 66–67
    dating techniques, 222
sediment entrainment, 185
sediment export rates, 492
sediment flux, 27f, 60f, 268
sediment generation
    versus sediment yield, 59–61
sediment grain size
    and flow velocity, 196
sediment loads, 197, 208
    measuring, 60f, 61
    of rivers, 284f, 285
sediment mobility
    and grain size, 196
sediment movement, 65, 66
sediment rating curves, 60f, 61, 220, 221f
sediment residence times
    in drainage basins, 222
sediment routing and storage, 222–223
sediment shedding
    by continents, 449
sediment sinks, 227
sediment sizes
    in fluvial systems, 184
sediment sources
    estuaries, 230
    upland areas, 227
sediment starvation
    deltas, 284
sediment state, 334
sediment storage, 208, 219–221
    bedrock valleys, 229–230
    on floodplains, *222*
sediment supply, 213f
    changes in, 208
    channels, 181
    coastal and marine landscapes, 256
    estuaries, 272
sediment transport, 195f, 245, 481
    alluvial valleys, 230
    beaches, 266
    englacial and subglacial, 301
    grain-size, 222–223
    hillslope, 145, 146
    initiation of, 194–197
    in rivers, 194–199
    slope-dependent, 166
    and slope morphology, 163
    source-to-sink, 25f
    by waves, 262
    wind speeds for, 336f
sediment transport capacity
    of channels, 181, 272
    of fans, 238

sediment transport patterns
    channels, 181
sediment transport rates, 220
sediment-wave asymmetry, 260f, 261
sediment yields, 63, 139, 441–442, 482f, 483, 484, 485f, 486f, 490
    and climate, 449
    and land use, 242f
    short-term, 492, 493
    versus sediment generation, 59–61
sediment yield data, 32–33
sedimentary orogenic wedge, 402
sedimentary rocks, 6, 14
    joint patterns, 82
segregation ice, 318
seismic hazards, 417
Sequoia National Forest, *158*
seric changes
    ocean volume, 254
settling speed, 337
shale, *13*, 14
shallow planar landslides, 153, 154–155
shallow subsurface flow (interflow), 117f, 123
shear plane, 153, 154
shear strain, 153f, 154
shear strength, 147, 148
    of slope-forming materials, 159, 160f, 161
shear stress, 159
    channel beds, 185
    in channels, 182f
    debris flow, 157
    and ice deformation, 299, 300f
    ice flows, 26f
    and rivers, 194
shear zones of transform margins, 404
shearing force, 25–26, 26f, 34
sheet-flood model of pavement formation, 352
sheet flooding, 352
sheetwash, 151
Shenandoah National Park, Virginia, *18*, 130
shield volcanoes, 361, *363*, 369fs, 370
shields, 405
Shields parameter, 196
shoaling
    tsunamis, 264
shoals, 260f, 261
shore-parallel bars, 268
shore-parallel zones, 266
shutter ridges, 404
Shuttle Radar Topography Mission (SRTM), 45f, 46, 47, *75*, 284
Siberian permafrost, 318
side-scan sonar, 276
Sierra Nevada, 22, 48, *125*, 239, 309, 362, 402, *447*
    channel-head source area, 171f
    earthquake-induced rockfall, 51
silcrete, 103, *104*

silica
    in soil, 89
silica dioxide, 276
silica-rich lava flows, 363
silica tetrahedra, 87
silicate
    and global cycle of, *31*
silicate karst, 134
silicate rocks
    weathering of, 454
silicate weathering, 87
silicic volcanism, 454
    landscapes of, 371–372
sills, 272, 380
siltstone, 14
single-thread stream channels, 199
sinkholes, *134*, 135, 136f
sinuosity
    of channels, 199–200
slides, 153, 154–156, 374, 375
slip-face avalanching, 344, *345*
slip faces
    and dune formation, 346f
slope, topographic, 24
slope aspect
    and soils, 90
slope decline, 166
slope failure
    predicting, 160–161
slope failure potential, 163
slope failures, 150
    environmental and time-dependent effects, 161
    predicting, 163
slope-forming materials, 145, 146–150
    shear strength of, 159, 160f
slope morphology, 163–172
slope-profile evolution, 166
slope replacement, 166, 169f
slope stability, 159–163, 161, 162
    and roots, 174–176
slope-stability models, 163
slopes
    windward and leeward, 452
slumps, 153f, 155, *159*, *183*, 374, 384f
smectite, 83
smectite clays, 87, 91
snags, 205, *413*
snow cover
    and plants, 19
snow metamorphism, 296f
snowfall, 115, 452
    potential energy of, 24
snowline, 449, 451f, 452, 464
snowmelt, *123*
soil, 89–98
    and ants and termites, 19, *20*
    definitions of, 79
    pH value, 86t
    properties and depths of, 99f
soil bacteria, 20
soil catenas, 99–100

soil chronosequences, 99, 100f, 106
soil classification, 95–98
soil color, 95
soil conservation measures, 109
soil creep, 151–152
soil depth, 99f
   and soil production, 91, 92f
soil development
   and aeolian material, 352–353
   chronosequence of, 52t
   factors affecting, 90–91
   over time, 99
   rate of, 106–107, 108
   see also pedogenesis
soil development indices, 51, 107
soil erosion, 107, 108f
   and animals, 19
   drainage basins, 246
   Dust Bowl era, *348*
   measuring, *61*, 62
soil falls, 158
soil fertility, 105
soil flux
   hillslopes, 165, 167
soil horizons, 92, 93f
soil infiltration rates, 119–120
soil loss
   from agricultural fields 105, 108f
   and winds, 350
soil-mantled hillslopes, 146, 167, 169
soil orders, 95, 97f, 99
soil organic matter, 95
soil production, 107, 108
   processes and rates of, 91
   from rock weathering, 107–108
soil profiles, 89, 91–95
   hillslope and ridgetop, 246
   rate development, 99
soil residence time, 90, 91
soil strength
   and channel initiation, 170
soil structure, 95, 96f
soil taxonomy, 95
soil texture, 95, 96f
soil thickness
   hillslopes, 167
soil transport rate
   hillslopes, 167
soil variability, *94*
soils
   clays in, 87
   hillslopes, 147
   and landscapes, 99–101
   strength of, 147, 149t
soilwater
   acidity of, 86
   and cation exchange, 89
soilwater flow, 78, 89
solar energy, 452
   and climates, 14
   hydrologic cycle, 19f
   see also insolation

solar forcing, 437f
solar radiation, 292, 436, 464
   and climate, 452
   seasonal distribution, 437f
solar system
   habitable zone, 31
solifluction, 151, 316, 318
solubility, 85
solution (ionization), 83, 84, 85–86, 211
sorted circles, 65
   patterned ground, 319, *320*, 321f
source area
   debris flows, 157
source-to-sink framework, 25
South American plate, 8f
Southern Hemisphere
   wind speeds in, 332
space shuttle
   active remote sensing data, 45f
spalling, 82, 102
spatial and temporal variability
   landscape processes, 483
spatial scales, 22
   geomorphic processes, 21f
spectrometers, 44
spheroidal weathering, 101, *103*
spilling breakers, 261
spillway (gap), 311
spit elongation, 269
spits, 262f, 265f, 268–269, 271f
spodosols, 98
spreading centers, 12f, 356, 358f, 392, 393
spreads, 154–156
spring runoff, 140
spring snowmelt, 116
spring tides, 258–259
springs, 135, 398
springs, volcanic, 377
stadia rods, 44
staff gauges, 60f, 187
stage (water-level elevation), 187
stair-step topography, *77*, *361*
stalagmites, 86
stalactites, 86
star dunes, *329*, *341*, *343*, *345*
stationarity, 114
steady state, 27, 63
steady-state landforms
   and landscape evolution, 468
steady-state landscapes, 468, 471–474, 490–493
steady-state systems, 26, 27f
step-pool channels, 203f
step-pool morphology, *233*
step-pool reaches, 204
stochastic geomorphological processes, 462
Stokes' Law, 333–334, 341
stone stripes, 320
storm events, 492
storm surges, 264, 281, 440
   tidal inlets, 269

storms, *115*
   and beach morphology, 268
   temporal scales, 23–24
straight channels, 199f, 200, 223
strain rate, 299
strath terraces, 235–237, 413, *461*, 481
stratovolcanoes, 358f, 359f, 361, 369f, 371, *372*
   crater lakes, 240
stream capture, 476, 477f
stream channels, 124, 179, 199
   abandoned, *239*
stream discharge, 181
stream junctions, 225
stream load, 194f, 197
stream longitudinal profiles, 234
stream magnitude, 225, 226f
stream orders, 225, 226f
stream power, 189
   derivation of, 189
stream terraces, 405f
streambed mobility, 196
streambeds
   armor layer, 194f
streamborne sediments, 276
streamflow, 181, 189, 194
streams
   braided, 5
   historical land use, 249f
   underfit, 315
strike-slip boundaries, 8f
strike-slip fault zone, *406*, 407
strike-slip faults, 12f, 362, 404
Stromboli volcano, 384, *385*, 368, 369f
structural anticlines (arches), 403
structural domes, 224
structural landforms, 408–411
structural synclines (troughs), 403
structural tectonic controls, 392
subduction
   of water on Earth, 31
subduction zone-driven eruptions, 381
subduction zone volcanism, 361
subduction zones, 11, 392
   continental shelf, 254
   earthquakes and, 8f
   and mountain belts, 11
   and volcanic arcs, 402
   and volcanoes, 356, 357f, 358f, 359f
subglacial carbonate precipitation, *306*
subglacial drainage, 304
subglacial erosion, 306
subglacial eruptions, 380
subglacial floods, 458
subglacial lakes, 303f
subglacial processes, 305–309
subglacial sediment transport, 301
subglacial sediments and landforms, 309–310
sublimation
   glacial, 295

submarine canyons, 27, 278f, *385*
submarine landforms and
  processes, 253
submarine landslides, 263, 277
submerged coastlines, 254, 264
submerged near-shore bars, 268
subpolar glaciers, 302
subsidence, 274
subsurface flow, 14, 123
  in karst, 136f
subsurface flowpaths, 124
subsurface stormflow, 124f, 126f
sub-threshold slopes, 413–414
Suess effect, 69
sugarcane cultivation, 105
sulfur
  oxidation, 85
summer mass balance
  glaciers, 297
summer thunderstorms, 115
supercritical flow, 187, 198
superposed rivers, 225
superposition, 48
superposition mechanism, 225
supraglacial debris, 306, 309
surf, 260
surf zone, 265, 268
surface currents, 275
surface flow
  interactions with groundwater,
    129–130
surface processes
  and climate, 426
surface roughness
  and aeolian sediment, 341
  measuring, 47
surface tension
  and aeolian lift forces, 337
surface uplift, 393–394, 481
  and climate, 426
  and landscape evolution, 462–463
surface water
  through basins, 133
surface-water hydrology, 126–133
surging breakers, 261
surging glaciers, 301–302
suspended load, 194f, 195f, 197, *223*
suspended sediment, 60f, *130*
  floodplains, 206
  sediment-dominated floodplains, 207
  in transport, 129
suspended sediment load, 483, 484f
suspension, 337, 338f
Susquehanna River, *236*, 316, *410*
swales, 124
swash, 261, 262, 266, 267
swash zone, 261
swell (waves), 260
swelling clays, *see* smectite clays
synclinal ridge, 409
Synthetic Aperture Radar (SAR),
  45f, 46

table mountains, *see* tuyas
tafoni, 102
taliks, 317
talus apron, 166
talus, *13*
talus slopes, 146f, 165, 169f
Tambora eruption (1816), 367,
  368, 454
taper, 472, 473f
tarns, 239, 313, 314
tectonic activity
  and zones of weakness, 361
tectonic controls, 392
tectonic deformation
  and landscape evolution, *465*
tectonic estuary, 271f, 272
tectonic extension, 401f
tectonic forcing, 361, 411
  drainage basins, 240
  hillslopes, 413
  rivers and streams, 413
tectonic geomorphology, 391–417
tectonic plates, 6, 391
  and continental margins, 254
  convergence of, 402
  and hot spots, 364f
  movement of, 8f, 11
  and volcanoes, 362
tectonic processes, 391–392, 392–397
tectonic settings, 397–411
tectonics, 6, 11–13
  and climate, 435f
  and erosion, 415–416
  and landscape evolution, 462–463
  landscape response to, 411–417
  pace of, 481
tectonism
  and life, 31
temperate glaciers, 302
temperate regions
  vegetation in, 19
temperature
  and climate zones, 16, 17f
  convective cells of atmospheric
    circulation, 15–16, 15f
  and elevation, 16
  and precipitation, 440
  and weathering, 79
temperature changes, 450
temperature fluctuations
  and weathering, 82
temporal scales (timescales), 23–24
  geomorphic processes, 21f
tensile strength, 147
tephra, 366
tephra dam, *377*
tephrachronology, 366, 433
terminal moraines, 313f, 314
terminations, 435
termite mounds, *20*
termites, mound-building, 19
terra preta soils, 105

terrace formation, 427
terraces, 22
  drainage basins, 235–237
  sediment storage, 219, 222
terracettes, 19
terrain roughness
  and turbulence, 332
terrestrial glaciations, 315
terrestrial sediment
  windblown, 433–434
terrigenous sediments, 276
thalweg, 191
thaw lakes, 319
thaw slumps, 319
thermal buoyancy, 397
  uplift and, 12f
thermal conductivity
  of rocks, 82
thermal contrasts, 397
thermal expansion, 82–83
thermal subsidence
  coral reef, 280f
thermal uplift, 397
thermochronometry, 63, 64f
thermohaline circulation, 436
thermokarst, 319
thermokarst-catalyzed erosion, 323
thermokarst lakes, *322*
thermoluminescence (TL), 56
third-order channels, 225
threshold lakes, 444, *445*
threshold slopes, 166, 168f,
  413–414
thresholds
  and terraces, 235
thresholds, geomorphic, 26–27, 27f
thrust fault scarp, *398*
thrust faults, 402–403, *404*
Tibetan Plateau, *389, 391, 393,* 405,
  408f, 453, 464
  formation of, 11
  permafrost, 317
tidal bores, 259
tidal currents, 270
tidal cycles, 257
tidal flats, 270
tidal flows, 259
tidal range, 258–259
tidal water level, 257
tide-dominated deltas, 274, 275f
tide-generated currents, 259
tides, 257–259
tidewater glaciers, 257f
Tien Shan Mountains, *5, 222,* 319
  Ak-Sai Glacier, *48*
till, 292, *294,* 310, 322, 48
  hillslopes, 147
till plain, 314
tillite, 292, *294*
toe slopes, 124f
toes
  of landslides, 156f, *159*

tombolos, 269
top-down freezing, 318
topographic inversions, 103, 377, 378f, 409
topographic maps, 44
topographic slope, 411
topographic steady state, 490
topography
    and landscape evolution, 462, 464–465
    and soil development, 90
    and weathering, 79–80
    see also first-order topography
topples, 158
topset beds, 273
tors, 102–103
tower karst, 134, 135
toxic dust, 350
trade winds, 15–16, 332
trailing-edge margins, 254, 255f
tranform fault, 404
tranquil flow, 187
transfer functions, 66
transform boundaries, 11
transform faults, 362
transform margins and landforms, 254, 398, 404
transform plate boundaries, 12f
transgression, marine, 255
transgressional margins, 404
transient landscape response, 475
transient landscapes, 471, 474–478
transient response
    of landscapes, 467–468
translational slides, 154–155
translocation
    in soil profiles, 89
transpiration, 111, 116
    hydrologic cycle, 19f
transport capacity
    of mountain streams, 180
transport laws
    and landscape evolution, 470
transport-limited (soil-mantled) slopes, 146f, 165–166
transport zone
    of debris flow, 157
transtensional margins, 404
transverse drainage, 225
transverse dunes, 346
transverse faults, 398
transverse-to-flow ribs, 198–199
travertine, 135
tree blowdown, 28
tree core collecting, 52
tree line, 438
tree ring analysis, 53f
tree rings, 66
tree rooting
    and soil production, 91
tree roots
    and biological weathering, 77, 78
tree-throws, 90, 151

trees
    and apparent cohesion to soils, 161
    along channels, 184
    and climate variability, 438
    as organic debris, 204
    root cohesion, 176
    see also forests
trellis drainage patterns, 223–224
trellis network patterns, 225
trench morphology, 449
trenches, marine, 279
trenching
    of fans, 238
triangular facets, 402
tributaries
    dendritic channel networks, 223
tributary basin size, 225
tributary glaciers, 308
tributary–main stem confluences, 225
trimline
    of glacier ablation zone, 310
tropical beaches
    coral sand, 256
tropical regions
    vegetation in, 19
trough (of wave), 259f
tsunamis, 263, 277, 281, 385, 386
tufa, 135
turbation, 322
turbidites, 277
turbidity currents, 275
    submarine canyons, 277
    submarine currents, 278f
turbulent flow, 187
tuyas, 380
typhoons, 114

U/Th dating, see uranium/thorium (U/Th) dating
ultisols, 98, 99
unconsolidated materials, 120
underfit streams, 315
uniformitarianism, 457
unit stream power, 189
    and gorge incision, 235
United States
    physiographic provinces, 23f
unpaired terraces, 236f, 237
unroofing rates, 63
upland channels, 180
upland land use, 145
uplift
    of geomorphic features, 450f
    types of, 393–397
uplift pulses
    and global climate, 454
uplift rates
    and landscape evolution, 481
uranium/thorium (U/Th) dating, 56
urban land-use change, 139
urbanization
    geomorphic effects of, 139

vadose zone, 120
Valley and Ridge Province, 224, 230, 248f, 410
valley glaciers, 294, 295
valley segments, 22
    drainage basins, 228–230
    shape of, 230
valleys, 230
variable source area concept, 124
varves, 312, 430
vegetation
    and apparent cohesion, 163t
    and channel initiation, 170
    and climate, 440, 445
    and geomorphic responses, 440
    and landscape evolution, 464–465
    and soil, 90
    wind disturbance of, 330
vegetation, riparian, 184
vegetation zones, 18–19
    and climate zones, 17f
ventifacts, 337, 342
vents, 363
vertisols, 97–98
vesicles
    in volcanic rocks, 367
viscosity, 65, 121
visualization tools, 67
volatile elements
    and life, 31
volcanic arcs, 6, 356, 357f, 361, 392, 402
    geomorphology of, 359f
volcanic ash, 356, 365, 368, 433
volcanic bombs, 365, 366
volcanic deformation, 46
volcanic deposits
    weathering of, 374
volcanic domes, 361
volcanic emissions, 381–382
    and atmosphere, 454
volcanic eruptions
    and oceanic crust, 278
    sizes and types, 368
    temporal scales, 23
    undersea, 263
Volcanic Explosivity Index (VEI), 368
volcanic form and process, 369f
volcanic gases, 366–367, 372
volcanic geomorphology, 355–382
volcanic island chains, 393
volcanic islands
    coastal sediment, 256
volcanic landforms, 356, 374
    evolution processes, 372–373
volcanic landscapes, 355, 368–372
volcanic necks, 380, 381
volcanic ocean islands, 375
volcanic processes, 356
volcanic provinces, 361
volcanic terrain, 381
volcanic tremor, 372, 373f

volcanism, 6, 356–363
    basaltic, 12f
    hot spots, 11
    and life, 31
volcano hazard zones, 381
volcano morphology, 359–361
Volcano National Park, 365
volcano–river interaction, 377
volcano sector collapse
    geomorphic effects of, 383–386
volcanoes, 355
    and glaciers, 380
    hydrologic considerations, 377–380
    plate boundaries and, 8f
    radial drainage, 224
    see also island arcs

warm-based glaciers, 302, 303f, 307, 325–326
    sediments and landforms, 309–310
warm-based ice, 304
warm fronts, 115
washload, 197, 198
water and air
    as fluids, 331
water balance
    by latitude, 118f
water budgeting, 133
water budgets, 132–133
water density, 121
water discharge rating curve, 60f
water flow
    potential and kinetic energy of, 24
water movement
    and gravity, 120
water tables, 116–118, 120, 125
water vapor
    potential energy of, 19f
water yields, 118
    and conifer forests, 116
water
    chemical and physical weathering, 79
    hydrologic cycle, 16
    phase diagram for, 292
waterfalls, fossil, 447
watershed flowpaths, 124f
watersheds, *see* drainage basins
wave action
    continental shelves, 254
wave base, 259f, 260f, 261
wave dispersion, 260
wave energy, 263, 281
    and berms, 267f
    sediment flux, 268

wave erosion
    of coastal margin, 254
    sea-cliff retreat, 264–265
wave height, 259, *260*, 261
wave period, 259
wave refraction, 262–263
wave velocity, 259, 263
wave-cut notch, 264, 266
wave-cut platforms, 264–265, *266*
wave-dominated deltas, 274, 275f
wavelength, 259
waves, water, 259–264
weather radars, 113, 132
weather stations, *133*
weathering, 77–80
    on Earth, *31*
    and rock types, 13, 14
    of rocks and atmosphere, 454
    and soil development, 90
    and soils, 104
weathering front, 103, 104f
weathering-limited erosion, 211
weathering-limited slopes, 146f, 163–165
weathering pits, 102
weathering profiles
    continental interior, 404
weathering rates, 80
    and pedogenesis, 91
weathering rind, *50*
wedge sedimentation, 420
weir, 60f, *133*
welded tuff, 366
westerly winds, 332
wetted perimeter, 185
wetting-drying cycles, 83
wind
    climate zones and, 14
    as a geomorphic agent, 329–351
    and waves, 259
    see also entries under aeolian
wind creep, 337, 338–339
wind-deposited sediment, 351
wind direction, 332
    dominant, 452
wind disturbance, 335
wind-driven geomorphic processes, 334–335
wind erosion, 335–337, 350
    dry lakebeds, 335
    glacial outwash stream, 334
wind fences, 350
wind gaps, 476, 477, *478*
wind polish, 337
wind scour, 337
wind speed, 332–333

    and erosion 335–337
    and impact threshold, 337
    and transport, 340f
    vertical distribution of, 333
wind storms, *340*
wind-suspended sediment, 337
wind throw, 335
wind transport processes, 337–341
wind transport profile, *340*
wind-transported sediment
    at glacial margins, 334
windblown terrestrial sediment, 433–434
windward slopes, 452
winter mass balance
    glaciers, 297
winter storms, 114
wood
    as part of river, 197–198
wood removal
    from streams, 241–242
woodland-prairie ecosystem boundary, 438
woody debris
    in channels, 184
worms
    and soil, 19

yardangs, 342
yazoo channels, 206f, 207
Yellowstone caldera, 13, 240, 356, 366
Yellowstone hot spot, 364f
Yellowstone plateau, 13
yield strength
    of flowing debris, 239
Yosemite Valley, 80
    glaciation in, 325
    rockfalls, 7
Younger Dryas, 33, 428f, 436, 458

Zagros Mountains, 397, 403, 465
zircon, 85
zone of accumulation, 92
zone of active meandering, 199f
zone of convexity, 167
zone of eluviation, 92
zone of illuviation, 92
zone of leaching, 91
zones of erosion, 218
zones of sediment deposition, 218
zones of structural uplift, 407f
zones of weakness, 361, 397